WORLD

CHECKLIST

AND BIBLIOGRAPHY OF

Euphorbiaceae 4

(with Pandaceae)

WORLD CHECKLIST AND BIBLIOGRAPHY OF

Euphorbiaceae 4

(with Pandaceae)

Euphorbiaceae:
Pachystemon – Zygospermum
Pandaceae
Doubtful and excluded names and taxa

Rafaël Govaerts, David G. Frodin
and
Alan Radcliffe-Smith

(assisted by Susan Carter, Mike Gilbert and Victor Steinmann for Euphorbiinae, Hans-Jürgen Esser for Hippomaneae, and Petra Hoffman for *Antidesma*)

First published March 2000

World Checklists and Bibliographies, 4.
[The first three in this series were published respectively as Magnoliaceae (October 1996), Fagales (May 1998) and Coniferae (August 1998)].

Address of the principal authors:
Herbarium, Royal Botanic Gardens, Kew, Richmond, Surrey TW9 3AE, United Kingdom

ISBN 1 900347 86 5

Cover photograph: *Phyllanthus epiphyllanthus* L.
        Royal Botanic Gardens, Kew, 1996, *Andrew McRobb*

Cover design by Jeff Eden for Media Resources, Information Services Department, Royal Botanic Gardens, Kew

Page make-up by Media Resources, Information Services Department, Royal Botanic Gardens, Kew, from text generated by David G. Frodin using Microsoft Access 2.0® and Microsoft Word 6.0®

Printed in The European Union by Redwood Books Limited, Trowbridge, Wiltshire, UK.

Published by
The Royal Botanic Gardens, Kew
2000

Summary

Volume 1

Volume 2

Volume 3

Volume 4

Contents, Volume 4

Euphorbiaceae: Pachystemon – Zygospermum

Pandaceae

List of Colour Plates

Plate 1

Synadenium grantii Hook. f. Njombe District, Tanzania, 1956,
    E. Milne-Redhead & P. Taylor (as *S. glaucescens* Pax)   KEW SLIDE COLLECTION
Ricinocarpus pinifolius Desf. Queensland, 1988, *I. K. Ferguson*   KEW SLIDE COLLECTION

Plate 2

Phyllanthus niruri L. India, n.d., Wight Collection, no. 505   KEW ILLUSTRATIONS COLLECTION

Plate 3

Phyllanthus nutans Sw. Jacquin, *Plantarum rariorum horti*
    caesarei schoenbrunnensis descriptiones et icones 2:
        pl. 193 (1797)   KEW ILLUSTRATIONS COLLECTION

Plate 4

Phyllanthus pendulus Roxb. Icones Roxburghianae, no. 265   KEW ILLUSTRATIONS COLLECTION

Plate 5

Phyllanthus tetrandrus Roxb. Icones Roxburghianae, no. 2396

                                           KEW ILLUSTRATIONS COLLECTION

Plate 6

Podadenia sapida Thwaites. Female plant, 1855,
    Thwaites C.P. 3428   KEW ILLUSTRATIONS COLLECTION

Plate 7

Tragia involucrata L. Jacquin, *Icones plantarum rariorum*
    1: pl. 190 (1781-87)   KEW ILLUSTRATIONS COLLECTION

Plate 8

Triadica sebifera (L.) Small. Icones Roxburghianae, no. 989

                                           KEW ILLUSTRATIONS COLLECTION

Pachystemon

Synonyms:
Pachystemon Blume === **Macaranga** Thouars
Pachystemon bancanus Miq. === **Macaranga bancana** (Miq.) Müll.Arg.
Pachystemon depressus Müll.Arg. === **Macaranga depressa** (Müll.Arg.) Müll.Arg.
Pachystemon depressus var. *mollis* Müll.Arg. === **Macaranga indistincta** Whitmore
Pachystemon populifolius Miq. === **Macaranga conifera** (Zoll.) Müll.Arg.

Pachystroma

1 species, middle S. America (E., SE. & S. Brazil and Bolivia); laticiferous forest trees to 20 m or perhaps more with conspicuous sprays of oblong to narrowly oblong, finely-veined, often spiny (and holly-like), spirally arranged leaves. Although Baillon, in placing the genus in the Hippomaneae, was near the mark, from the time of Mueller it became wrongly assigned to Acalyphoideae. Webster has shown that it was after all correctly in the Euphorbioideae, although he retained the separate suprageneric status bestowed by Pax. Esser (1994; see **Euphorbioideae (except Euphorbieae)**), however, has excluded the genus from the Hippomaneae. A relationship with *Ophthalmoblapton*, *Algernonia* and *Tetraplandra* (in the neighbouring tribe Hureae) should be explored. (Euphorbioideae (except Euphorbieae))

Pax, F. (1910). *Pachystroma*. In A. Engler (ed.), Das Pflanzenreich, IV 147 II (Euphorbiaceae-Adrianae): 99-101. Berlin. (Heft 44.) La/Ge. — 1 species, Brazil.

Pachystroma Müll.Arg., Linnaea 34: 177 (1865).
  Bolivia, SE. & S. Brazil. 83 84.
    Acantholoma Gaudich. ex Baill., Adansonia, n.s. 6: 231 (1866).

Pachystroma longifolium (Nees) I.M.Johnst., Contr. Gray Herb. 68: 90 (1923).
  Bolivia, SE. & S. Brazil. 83 BOL 84 BZL BZS. Nanophan. or phan.
    * *Ilex longifolia* Nees, Flora 4: 301 (1821). *Pachystroma ilicifolium* var. *longifolium* (Nees) Müll.Arg., Linnaea 34: 178 (1865).
    Pachystroma ilicifolium Müll.Arg., Linnaea 34: 178 (1865).
    Pachystroma ilicifolium var. *ellipticum* Müll.Arg., Linnaea 34: 178 (1865).
    Pachystroma ilicifolium var. *subintegrum* Müll.Arg., Linnaea 34: 178 (1865).
    Pachystroma ilicifolium var. *heterophyllum* Müll.Arg. in C.F.P.von Martius, Fl. Bras. 11(2): 388 (1874).
    Pachystroma castaneifolium Klotzsch ex Pax in H.G.A.Engler, Pflanzenr., IV, 147, II: 99 (1910), pro syn.

Synonyms:
Pachystroma castaneifolium Klotzsch ex Pax === **Pachystroma longifolium** (Nees) I.M.Johnst.
Pachystroma ilicifolium Müll.Arg. === **Pachystroma longifolium** (Nees) I.M.Johnst.
Pachystroma ilicifolium var. *ellipticum* Müll.Arg. === **Pachystroma longifolium** (Nees) I.M.Johnst.
Pachystroma ilicifolium var. *heterophyllum* Müll.Arg. === **Pachystroma longifolium** (Nees) I.M.Johnst.
Pachystroma ilicifolium var. *longifolium* (Nees) Müll.Arg. === **Pachystroma longifolium** (Nees) I.M.Johnst.
Pachystroma ilicifolium var. *subintegrum* Müll.Arg. === **Pachystroma longifolium** (Nees) I.M.Johnst.

Pachystylidium

1 species, S. and SE. Asia, Jawa, Sulawesi and Philippines; slenderly woody, sometimes sprawling climbers with irritating hairs and soft leaves on open slopes, forest edges or grassy woodland. The distribution suggests a preference for areas with a seasonal climate; some records are from limestone country. The genus is in Asia and Malesia related to *Tragia* (cf. Airy-Shaw, 1969) with *T. novae-hollandiae* providing a possible 'link'. (Acalyphoideae)

> Pax, F. & K. Hoffmann (1919). *Pachystylidium*. In A. Engler (ed.), Das Pflanzenreich, IV 147 IX (Euphorbiaceae-Acalypheae-Plukenetiinae): 108. Berlin. (Heft 68.) La/Ge. — 1 species, Sunda, Philippines.
>
> Airy-Shaw, H. K. (1969). Notes on Malesian and other Asiatic Euphorbiaceae, CXII. Notes on the subtribe Plukenetiinae Pax. Kew Bull. 23: 114-121. En. —*Pachystylidium*, pp. 115-117; full synonymy and extensive discussion, including range extension of *P. hirsutum*.

Pachystylidium Pax & K.Hoffm. in H.G.A.Engler, Pflanzenr., IV, 147, IX: 108 (1919). Trop. Asia. 40 41 42.

Pachystylidium hirsutum (Blume) Pax & K.Hoffm. in H.G.A.Engler, Pflanzenr., IV, 147, IX: 108 (1919).
E. India to Sulawesi. 40 IND 41 CBD LAO THA VIE 42 JAW PHI SUL. (Cl.) cham.
* *Tragia hirsuta* Blume, Bijdr.: 630 (1826).
  Tragia irritans Merr., Philipp. J. Sci., C 9: 491 (1914 publ. 1915).
  Tragia gagei Haines, J. Proc. Asiat. Soc. Bengal 15: 317 (1919 publ. 1920).
  Pachystylidium hirsutum var. *irritans* Pax & K.Hoffm. in H.G.A.Engler, Pflanzenr., IV, 147, IX: 108 (1919).
  Tragia delpyana Gagnep., Bull. Soc. Bot. France 71: 1027 (1924 publ. 1925).

Synonyms:
Pachystylidium hirsutum var. *irritans* Pax & K.Hoffm. === **Pachystylidium hirsutum** (Blume) Pax & K.Hoffm.

Paivaeusa

Synonyms:
Paivaeusa Welw. ex Benth. === **Oldfieldia** Benth. & Hook.f.
Paivaeusa dactylophylla Welw. ex Oliv. === **Oldfieldia dactylophylla** (Welw. ex Oliv.) J.Léonard
Paivaeusa gabonensis A.Chev. === ?
Paivaeusa orientalis Mildbr. === **Oldfieldia somalensis** (Chiov.) Milne-Redh.

Palanostigma

Synonyms:
Palanostigma Mart. ex Klotzsch === **Croton** L.

Palenga

Synonyms:
Palenga Thwaites === **Putranjiva** Wall.
Palenga zeylanica Thwaites === **Putranjiva zeylanica** (Thwaites) Müll.Arg.

Palissya

Synonyms:
Palissya Baill. === **Necepsia** Prain
Palissya castaneifolia Baill. === **Necepsia castaneifolia** (Baill.) Bouchat & J.Léonard

Panopia

Synonyms:
Panopia Noronha ex Thou. === **Macaranga** Thouars

Pantadenia

1 species, SE. Asia (Thailand and Indochina); shrubs with a woody rootstock, the slightly zigzag branchlets bearing leaves with reddish beadlike glands on their undersurface and the inflorescences opposite the most distal. The genus is most closely related to *Blachia* (Airy-Shaw, 1969), although the membraneous leaves in *P. adenanthera* recall *Strophioblachia*. Webster (Synopsis, 1994) also included the Malagasy *Parapantadenia* but this opinion is not here followed. The genus may nevertheless be part of an assemblage representative of an old austral flora (Capuron 1972; see *Parapantadenia*). In the Webster system, all these genera are in Codieae. (Crotonoideae)

Gagnepain, F. (1925). Quelques genres nouveaux d'Euphorbiacées. Bul. Soc. Bot. France 71: 864-879. Fr. — Includes (pp. 873-875) generico-specific protologue of *Pantadenia adenanthera* (described from Laos, Cambodia and southern Vietnam).

Airy-Shaw, H. K. (1969). Notes on Malesian and other Asiatic Euphorbiaceae, CXV. The genus *Pantadenia* Gagnep. in Siam, with a note on affinity. Kew Bull. 23: 122-123. En. — Extension of range and discussion of affinities; conclusion that *Blachia* was most closely related.

Pantadenia Gagnep., Bull. Soc. Bot. France 71: 873 (1924 publ. 1925).
Indo-China. 41.

Pantadenia adenanthera Gagnep., Bull. Soc. Bot. France 71: 873 (1924 publ. 1925).
Thailand, Laos, Cambodia, Vietnam. 41 CBD LAO THA VIE. Cham. or nanophan.

Synonyms:
Pantadenia chauvetiae (Leandri) G.L.Webster === **Parapantadenia chauvetiae** Leandri & Capuron

Paracasearia

Proposed by Boerlage; a synonym of *Drypetes*.

Paracelsea

Synonyms:
Paracelsea Zoll. === **Acalypha** L.

Paracleisthus

Synonyms:
Paracleisthus Gagnep. === **Cleistanthus** Hook.f. ex Planch.
Paracleisthus eberhardtii Gagnep. === **Cleistanthus eberhardtii** (Gagnep.) Croizat

Paracleisthus pierrei Gagnep. === **Cleistanthus pierrei** (Gagnep.) Croizat
Paracleisthus siamensis (Craib) Gagnep. === **Cleistanthus hirsutulus** Hook.f.
Paracleisthus subgracilis Gagnep. === **Cleistanthus sumatranus** (Miq.) Müll.Arg.
Paracleisthus tonkinensis (Jabl.) Gagnep. === **Cleistanthus tonkinensis** Jabl.

Paracroton

4 species, S. & SE. Asia and Malesia to New Guinea; includes *Fahrenheitia* and certain taxa formerly in *Ostodes*. Small to medium forest trees, usually preferring light, with shoots distinguished by long, diffuse, spikes in axils or in terminal panicles. The leaves in the widespread and frequently collected *P. pendulus* are large, coarse, narrowly elliptic, steeply veined and often toothed. *P. sterrhopodus* appears to be very local, on the Cyclops terrane in New Guinea, and remains poorly known. The genus (as *Fahrenheitia*) was somewhat spuriously included by Webster (1994, Synopsis) in his Crotoneae; however, the general impression is one of the Codiaeae. (Crotonoideae)

> Pax, F. (with K. Hoffmann) (1911). *Paracroton*. In A. Engler (ed.), Das Pflanzenreich, IV 147 III (Euphorbiaceae-Cluyticae): 12. Berlin. (Heft 47.) La/Ge. — 1 sp., *Paracroton pendulus*.
>
> Airy-Shaw, H. K. (1963). Notes on Malaysian and other Asiatic Euphorbiaceae, XLI. An overlooked *Ostodes* from South India. Kew Bull. 16: 362-364. En. — Description of *O. integrifolius*. [Now in *Paracroton* following Balakrishnan and Chakrabarty.]
>
> Airy-Shaw, H. K. (1966). Notes on Malaysian and other Asiatic Euphorbiaceae, LXXIX. Realignments in the *Ostodes-Dimorphocalyx*-complex. Kew Bull. 20: 409-413. En. — Key to genera; new combinations with synonymy and indication of distribution; list of *Ostodes* names with their disposition. [Some of the latter were here incorporated into *Fahrenheitia*; more recently, they have been transferred to *Paracroton*.]
>
> Airy-Shaw, H. K. (1972). Notes on Malesian and other Asiatic Euphorbiaceae, CLXVI: Further note on *Fahrenheitia integrifolia*. Kew Bull. 27: 92. En. — Additional collections from southern India reported. [The species was transferred to *Paracroton* by Balakrishnan and Chakrabarty.]
>
> Airy-Shaw, H. K. (1974). Notes on Malesian and other Asiatic Euphorbiaceae, CLXXXIII. A new *Fahrenheitia* (?) from W. New Guinea. Kew Bull. 29: 325-326. En. — Description of *F. sterrhopoda* from near Jayapura. [Now transferred to *Paracroton*.]
>
> • Balakrishnan, N. P. & T. Chakrabarty (1993). The genus *Paracroton* (Euphorbiaceae) in the Indian subcontinent. Kew Bull. 48: 715-726, illus. En. — Treatment of 3 species with key, descriptions, synonymy, types, localities with exsiccatae, indication of phenology, distribution and habitat, illustrations, and commentary. [The fourth recognised species, *P. sterrhopodus*, is endemic to western New Guinea and *P. pendulus* subsp. *pendulus* is outside the range of this paper.]

Paracroton Miq., Fl. Ned. Ind. 1(2): 382 (1859).
> Trop. Asia. 40 41 42.
> > *Desmostemon* Thwaites, Enum. Pl. Zeyl.: 278 (1861).
> > *Fahrenheitia* Rchb. & Zoll. ex Müll.Arg. in A.P.de Candolle, Prodr. 15(2): 1256 (1866).

Paracroton integrifolius (Airy Shaw) N.P.Balakr. & Chakrab., Kew Bull. 48: 718 (1993).
> India (Kerala, Tamil Nadu). 40 IND. Phan.
> > *Ostodes integrifolia* Airy Shaw, Kew Bull. 16: 362 (1963). *Fahrenheitia integrifolia* (Airy Shaw) Airy Shaw, Kew Bull. 20: 410 (1966).

Paracroton pendulus (Hassk.) Miq., Fl. Ned. Ind. 1(2): 382 (1859).
> India to Philippines. 40 IND SRL 41 BMA MLY THA 42 BOR JAW PHI SUM. Phan.
> > *Croton pendulus* Hassk., Pl. Jav. Rar.: 266 (1848). *Ostodes pendula* (Hassk.) A.Meeuse in C.A.Backer, Bekn. Fl. Java 4c(112): 10 (1943). *Fahrenheitia pendula* (Hassk.) Airy Shaw, Kew Bull. 20: 410 (1966).

subsp. **pendulus**

S. Burma, Thailand, Pen. Malaysia, Borneo, Jawa, Sumatera, Philippines. 41 BMA MLY THA 42 BOR JAW PHI SUM. Phan.

Tritaxis macrophylla Müll.Arg., Flora 47: 482 (1864). *Trigonostemon macrophyllus* (Müll.Arg.) Müll.Arg., Linnaea 34: 213 (1865). *Ostodes macrophylla* (Müll.Arg.) Benth. & Hook.f., Gen. Pl. 3: 299 (1880).

Fahrenheitia collina Rchb. & Zoll. ex Müll.Arg. in A.P.de Candolle, Prodr. 15(2): 1256 (1866).

Blumeodendron muelleri Kurz, J. Asiat. Soc. Bengal, Pt. 2, Nat. Hist. 42(2): 245 (1873).

Ostodes serratocrenata Merr., Philipp. J. Sci., C 4: 283 (1909).

Ostodes collina (Rchb.f. & Zoll. ex Müll.Arg.) Pax in H.G.A.Engler, Pflanzenr., IV, 147, III: 21 (1911).

subsp. **zeylanicus** (Thwaites) N.P.Balakr. & Chakrab., Kew Bull. 48: 719 (1993).

SW. India, Sri Lanka. 40 IND SRL. Phan.

* Desmostemon zeylanicus Thwaites, Enum. Pl. Zeyl.: 278 (1861). *Trigonostemon zeylanicus* (Thwaites) Müll.Arg., Linnaea 34: 213 (1865). *Ostodes zeylanica* (Thwaites) Müll.Arg., Linnaea 34: 214 (1865). *Fahrenheitia zeylanica* (Thwaites) Airy Shaw, Kew Bull. 20: 410 (1966).

Paracroton sterrhopodus (Airy Shaw) Radcl.-Sm. & Govaerts, Kew Bull. 52: 189 (1997).

Irian Jaya. 42 NWG. Phan.

* Fahrenheitia sterrhopoda Airy Shaw, Kew Bull. 29: 325 (1974).

Paracroton zeylanicus (Müll.Arg.) N.P.Balakr. & Chakrab., Kew Bull. 48: 723 (1993).

Sri Lanka. 40 SRL. Nanophan. or phan.

Desmostemon zeylanicus var. *minor* Thwaites, Enum. Pl. Zeyl.: 278 (1861). *Ostodes minor* (Thwaites) Müll.Arg., Linnaea 34: 214 (1865). *Ostodes zeylanica* var. *minor* (Thwaites) Bedd., Fl. Sylv. S. India: 213 (1872). *Fahrenheitia minor* (Thwaites) Airy Shaw, Kew Bull. 20: 410 (1966).

* Tritaxis zeylanica Müll.Arg., Flora 47: 482 (1864).

Paradenocline

Synonyms:
Paradenocline Müll.Arg. === **Adenocline** Turcz.

Paradrypetes

2 species, Brazil (one in Atlantic region, the other in SW Amazonia); small to medium trees of riverine or seasonally flooded land with thick, opposite, sometimes prickly leaves. Levin (1992) presents evidence for its inclusion in Oldfieldioideae, closest to *Podocalyx* but meriting a separate subtribe in Podocalyceae; this was followed by Webster. [*P. ilicifolia* of the Atlantic forest was to one writer reminiscent of the Australian *Austrobuxus swainii*; any possible relationship seems as yet unstudied although there are certainly some similarities. In the Webster system, the two genera are in different tribes.] (Oldfieldioideae)

Kuhlmann, J. G. (1935). Novas especies botanicas da Hyléa (Amazonia) e do Rio Doce (Espirito Santo). Arch. Inst. Biol. Veg. Rio de Janeiro 2: 83-89, 7 pls. Pt. — Protologue of *Paradrypetes* and description of one species, *P. ilicifolia* (also new) from the Rio Doce basin in Espirito Santo, Brazil. [Anatomy described by F. R. Milanez in *ibid.*: 133-156.]

Levin, G. A. (1992). Systematics of *Paradrypetes* (Euphorbiaceae). Syst. Bot. 17: 74-83, 7 fig., map. En. — Character review and examination of generic relationships (considered to be nearest *Podocalyx*); descriptive treatment of 2 species (one new) with key, synonymy, references, types, vernacular names, indication of distribution and habitat, localities with exsiccatae, and commentary; list of references at end.

Paradrypetes Kuhlm., Arq. Inst. Biol. Veg. 2: 84 (1935).
Brazil. 84. Phan.

Paradrypetes ilicifolia Kuhlm., Arq. Inst. Biol. Veg. 2: 84 (1935).
Brazil (Espírito Santo, Minas Gerais). 84 BZL. Phan.

Paradrypetes subintegrifolia G.M.Levin, Syst. Bot. 17: 82 (1992).
Brazil (Amazonas). 84 BZN. Phan.

Paragelonium

Synonyms:
Paragelonium Leandri === **Aristogeitonia** Prain
Paragelonium perrieri Leandri === **Aristogeitonia perrieri** (Leandri) Radcl.-Sm.

Paranecepsia

1 species, SE. Africa (including Tanzania and Mozambique); much-branched shrubs or small trees to 12 m of open habitats with *Terminalia*-branching and serrate leaves (reminiscent of some species of *Elaeocarpus*) in tufts. It has been confused locally with *Alchornea* but the two genera are now in different tribes according to the Webster system, *Paranecepsia* in the Bernardieae. (Acalyphoideae)

Radcliffe-Smith, A. (1976). Notes on African Euphorbiaceae, VI. Kew Bull. 30: 675-687. En. — Includes description of genus and of *P. alchorneifolia* from Tanzania and Mozambique (pp. 684-687, with illustration).

Paranecepsia Radcl.-Sm., Kew Bull. 30: 684 (1975 publ. 1976).
E. & S. Trop. Africa. 25 26.

Paranecepsia alchorneifolia Radcl.-Sm., Kew Bull. 30: 684 (1975 publ. 1976).
E. Tanzania, NE. Mozambique (Niassa). 25 TAN 26 MOZ. Phan.

Parapantadenia

1 species, Madagascar; large shrubs or trees of the western dry forest (at least sometimes on sand) with thin, broadly ovate leaves covered with glands on the under surface and inflorescences at ends of branchlets opposite the last leaf. The genus was considered by Capuron (1972, with added notes by Leandri) to be related to the SE. Asian *Pantadenia*; indeed, the two were combined by Webster (Synopsis, 1994). The fruits, however, are quite different (A. Radcliffe-Smith, personal communication). Capuron and Leandri believed the genus to be, with its relatives, 'provenant de l'ancienne flore australe du globe'. In the Webster system both are placed in Codiaeae. (Crotonoideae)

Capuron, R. (1972). Contribution à l'étude de la flore forestière de Madagascar. Adansonia, II, 12: 205-211, illus. Fr. — Comprises two parts, of which the first is a description of *Parapantadenia* along with its only species, *P. chauvetiae* (also new).

Parapantadenia Capuron, Adansonia, n.s., 12: 206 (1972).
Madagascar. 29.

Parapantadenia chauvetiae Leandri & Capuron, Adansonia, n.s., 12: 206 (1972).
Pantadenia chauvetiae (Leandri & Capuron) G.L.Webster, Ann. Missouri Bot. Gard. 81: 106 (1994).
Madagascar. 29 MDG. Nanophan. or phan.

Parodiodendron

1 species, south-central S. America (NW. Argentina and Bolivia); a small tree to 7 m of dry mountain regions, sometimes forming almost pure stands. The small leaves are alternate, closely spaced and finely-veined; superficially they are reminiscent of some species of *Bridelia*. Formerly in *Phyllanthus* but in the Webster system now part of Picrodendreae. (Oldfieldioideae)

Hunziker, A. (1969). *Parodiodendron* gen. nov.: un nuevo genero de Euphorbiaceae (Oldfieldioideae) del Noroeste Argentino. Kurtziana 5: 331-341. Sp. — Protologue of genus; transfer from *Phyllanthus* of a species of Jujuy and Salta (localities with exsiccatae cited). The ecology and affinities are also described.

Parodiodendron Hunz., Kurtziana 5: 331 (1969).
Bolivia, NW. Argentina. 83 85. Phan.

Parodiodendron marginivillosum (Speg.) Hunz., Kurtziana 5: 333 (1969).
Bolivia, Argentina (Jujuy, Salta). 83 BOL 85 AGW. Phan.
* *Phyllanthus marginivillosus* Speg., Physis (Buenos Aires) 3: 167 (1917).

Passaea

Synonyms:
Passaea Baill. === **Bernardia** Houst. ex Mill.
Passaea spartioides Baill. === **Bernardia spartioides** (Baill.) Müll.Arg.

Pausandra

8 species, Americas (southern Central America to Brazil and Bolivia). These small to large (as much as 30 m), possibly quickly-growing trees of primary and secondary forest feature more or less toothed leaves in stiff rosettes at or towards the ends of branches (Lanjouw 1936; Secco 1990). Their closest relationship is, however, with *Dodecastigma*; both are included in the Codiaeae by Webster (1994, Synopsis). The flushes resemble those in *Trigonostemon* and *Paracroton pendulus* though in the Webster system the latter is in a more distant tribe on account of the presence of stellate indumentum. Secco (1990) accepts 6 species with a further 2 doubtful or imperfectly known. (Crotonoideae)

Pax, F. (with K. Hoffmann) (1911). *Pausandra*. In A. Engler (ed.), Das Pflanzenreich, IV 147 III (Euphorbiaceae-Cluytieae): 41-44. Berlin. (Heft 47.) La/Ge. — 4 species, one very likely not properly here; southern C America to S America (Brazil). Further notes in Addenda, p. 110. [Now out of date.]

Lanjouw, J. (1936). The genus *Pausandra* Radlk. Rec. Trav. Bot. Néerl. 33: 758-769, illus. En. — Review of past research; descriptive account of 9 species with key, synonymy, references and citations, localities with exsiccatae, and notes. *P. integrifolia* was excluded (it is now in *Dodecastigma*).

Jablonski, E. (1967). *Pausandra*. Euphorbiaceae, Guayana Highland (Mem. New York Bot. Gard. 17(1)): 153-154. New York. En. — 1 species, *P. martinii*; includes a reduction.

Secco, R. de S. (1987). Uma nova espécie de *Pausandra* Radlk. (Euphorbiaceae-Crotonoideae) da Amazônia. Bol. Mus. Paraense 'Emilio Goeldi', Bot., 3: 59-67, illus. Pt. — Description of *P. fordii* (in honour of the Ford Foundation of New York).

• Secco, R. de S. (1990). Revisão dos gêneros *Anomalocalyx* Ducke, *Dodecastigma* Ducke, *Pausandra* Radlk., *Pogonophora* Miers ex Benth. e *Sagotia* Baill. (Euphorbiaceae-Crotonoideae) para a América do Sul. Belém: Museu Paraense 'Emilio Goeldi'. Pt. — Includes revision of *Pausandra* pp. 50–87; (6 species with a further 2 doubtful or imperfectly known) featuring synonymy, types, descriptions, localities with exsiccatae, commentary, illustrations and maps.

Pausandra Radlk., Flora 53: 92 (1870).
Honduras to Brazil. 80 82 83 84.

Pausandra fordii Secco, Bol. Mus. Paraense Emilio Goeldi, N. S., Bot. 3: 60 (1987).
Brazil (Amapá), Guiana. 82 FRG 84 BZN. Phan.

Pausandra hirsuta Lanj., Recueil Trav. Bot. Néerl. 33: 769 (1936).
Brazil (Acre, Amazonas, Mato Grosso), Peru. 83 PER 84 BZN. Phan.

Pausandra macropetala Ducke, Arch. Jard. Bot. Rio de Janeiro 4: 114 (1925).
Brazil (Amazonas, Pará). 84 BZN. Phan.

Pausandra macrostachya Ducke, Arch. Jard. Bot. Rio de Janeiro 4: 114 (1925).
Brazil (Pará). 84 BZN. Phan. – Imperfectly known.

Pausandra martinii Baill., Adansonia 11: 92 (1873).
Peru, N. Brazil, Venezuela, Guyana, Surinam, Guiana. 82 FRG GUY SUR VEN 83 PER 84 BZN. Phan.
Pausandra flagellorhachis Lanj., Euphorb. Surinam: 30 (1931).

Pausandra megalophylla Müll.Arg. in C.F.P.von Martius, Fl. Bras. 11(2): 504 (1874).
Brazil (Rio de Janeiro). 84 BZL. Phan. – Close to *P. morisiana.*

Pausandra morisiana (Casar.) Radlk., Flora 53: 92 (1870).
Brazil (Rio de Janeiro, Santa Catarina). 84 BZL BZS. Phan.
* *Thouinia morisiana* Casar, Novar. Stirp. Brasil, Decad. 9: 75 (1842).

Pausandra trianae (Müll.Arg.) Baill., Adansonia 11: 92 (1873).
C. America to Peru and Brazil. 80 COS HON NIC PAN 83 CLM PER BOL 84 BZC BZN. Phan.
* *Pogonophora trianae* Müll.Arg., Flora 47: 434 (1864).
Pausandra quadriglandulosa Pax & K.Hoffm. in H.G.A.Engler, Pflanzenr., IV, 147, XIV: 43 (1919).
Pausandra extorris Standl., Trop. Woods 17: 24 (1929).
Pausandra densiflora Lanj., Recueil Trav. Bot. Néerl. 33: 766 (1936).
Pausandra sericea Lanj., Recueil Trav. Bot. Néerl. 33: 767 (1936).
Clavija septentrionalis L. Wms., Fieldiana, Bot. 32(12): 205 (1970).

Synonyms:
Pausandra densiflora Lanj. === **Pausandra trianae** (Müll.Arg.) Baill.
Pausandra extorris Standl. === **Pausandra trianae** (Müll.Arg.) Baill.
Pausandra flagellorhachis Lanj. === **Pausandra martinii** Baill.
Pausandra integrifolia Lanj. === **Dodecastigma integrifolium** (Lanj.) Lanj. & Sandwith
Pausandra quadriglandulosa Pax & K.Hoffm. === **Pausandra trianae** (Müll.Arg.) Baill.
Pausandra sericea Lanj. === **Pausandra trianae** (Müll.Arg.) Baill.

Paxia

Synonyms:
Paxia Herter === **Ditaxis** Vahl ex A.Juss.

Paxiuscula

Synonyms:
Paxiuscula Herter === **Ditaxis** Vahl ex A.Juss.

Peccana

Synonyms:
Peccana Raf. === **Euphorbia** L.

Pedilanthus

17 species, Americas (with a centre of diversity in Mexico); close to part of *Euphorbia* subgen. *Agaloma*. Shrubs or small trees to 5 m with sometimes slightly zig-zag green stems, distichously arranged euphorbioid (and sometimes thick) foliage, and axillary clusters of zygomorphic cyathia. The revision of Dressler (1957) remains standard; there is also a good introduction by Koutnik (1985). The *Euphorbia Journal* volumes moreover contain references to individual species. *P. tithymaloides*, naturally by far the most widely distributed species (although this now is perhaps somewhat obscure), is cultivated almost throughout warmer parts of the world. The name is conserved against *Tithymaloides* Ortega and thus also takes precedence over *Tithymalus* Mill., itself rejected in favour of *Tithymalus* Gaertn. (= Euphorbia L.) (Euphorbioideae (Euphorbieae))

> Poiteau, A., 1812. Observations sur le Pedilanthe (*Pedilanthus* Neck.). *Ann. Mus. Natl. Hist. Nat.* 19: 388–395, 1 plate. Fr. — Revision (3 species); *P. tithymaloides* characteristically near shores and coasts.

> Millspaugh, C. F. (1913). The genera *Pedilanthus* and *Cubanthus*, and other American Euphorbiaceae. Publ. Field Mus. Nat. Hist., Bot. Ser., 2: 353-397. En. —*Pedilanthus*, pp. 353-371 (31 species); key, descriptions (some novelties), synonymy, types with localities, other exsiccatae, and occasional notes. [Superseded by Dressler's revision.]

> • Dressler, R. L. (1957). The genus *Pedilanthus* (Euphorbiaceae). Contr. Gray Herb. 182: 1-188, illus., maps. En. — Monographic treatment of 14 species with key (pp. 98-99), synonymy, references, types, descriptions, indication of distribution and habitat, localities with exsiccatae, distributrion maps, and commentary; summary (pp. 171-176), analyses, list of references, and index at end. The general part includes detailed reviews of characters along with hybridisation patterns and remarks on the origin of the genus (the author considering it to be derived from elements within *Euphorbia* subgen. *Agaloma*). 'Primitive' and 'advanced' features are listed on p. 81 along with an account of evolutionary trends in the genus. [The 'biological species concept' was found useful in drawing taxonomic limits.]

> Jablonski, E. (1967). *Pedilanthus*. Euphorbiaceae, Guayana Highland (Mem. New York Bot. Gard. 17(1)): 189-190. New York. En. — 1 species, *P. tithymaloides*.

> Koutnik, D. (1985). An introduction to the genus *Pedilanthus*. Euphorbia J. 3: 38-42, illus. En. — Illustrated semipopular treatment, with selected references but no key. [A painting by Leah Schwartz on p. 39 shows how highly specialised are the inflorescences.]

Pedilanthus Neck., Elem. Bot. 2: 354 (1790).
  Florida to Brazil. 78 79 80 81 82 83 84.
    Tithymalus Mill., Gard. Dict. Abr. ed. 4: [1231] (1754).
    Tithymaloides Ortega, Tabl. Bot.: 9 (1773).
    Ventenatia Tratt., Gen. Pl.: 86 (1802).
    Crepidaria Haw., Syn. Pl. Succ.: 136 (1812).
    Diadenaria Klotzsch & Garcke, Monatsber. Königl. Preuss. Akad. Wiss. Berlin 1859: 254 (1859).
    Hexadenia Klotzsch & Garcke, Monatsber. Königl. Preuss. Akad. Wiss. Berlin 1859: 253 (1859).
    Tithymalodes Ludw. ex Kuntze, Revis. Gen. Pl. 2: 620 (1891).

Pedilanthus bracteatus (Jacq.) Boiss. in A.P.de Candolle, Prodr. 15(2): 6 (1862).
  Mexico (Sonora to Guerrero). 79 MXE MXN MXS. (Succ.) nanophan.
    *Euphorbia bracteata Jacq., Pl. Hort. Schoenbr. 3: 14 (1798). *Tithymalus bracteatus* (Jacq.) Haw., Syn. Pl. Succ.: 138 (1812).
    Diadenaria articulata Klotzsch & Garcke, Abh. Königl. Akad. Wiss. Berlin 1859: 108 (1860). *Pedilanthus articulatus* (Klotzsch & Garcke) Boiss. in A.P.de Candolle, Prodr. 15(2): 6 (1862).

Diadenaria involucrata Klotzsch & Garcke, Abh. Königl. Akad. Wiss. Berlin 1859: 107 (1860). *Pedilanthus involucratus* (Klotzsch & Garcke) Boiss. in A.P.de Candolle, Prodr. 15(2): 6 (1862).

Diadenaria pavonis Klotzsch & Garcke, Abh. Königl. Akad. Wiss. Berlin 1859: 108 (1860). *Pedilanthus pavonis* (Klotzsch & Garcke) Boiss. in A.P.de Candolle, Prodr. 15(2): 6 (1862).

Pedilanthus rubescens Brandegee, Zoe 5: 209 (1905).

Pedilanthus spectabilis Rob., Proc. Amer. Acad. Arts 43: 23 (1907). *Tithymalus spectabilis* (Rob.) Croizat, Amer. J. Bot. 24: 704 (1937).

Pedilanthus greggii Millsp., Publ. Field Mus. Nat. Hist., Bot. Ser. 2: 363 (1913). *Tithymalus greggii* (Millsp.) Croizat, Amer. J. Bot. 24: 704 (1937).

Pedilanthus olsson-sefferi Millsp., Publ. Field Mus. Nat. Hist., Bot. Ser. 2: 363 (1913). *Tithymalus olsson-sefferi* (Millsp.) Croizat, Amer. J. Bot. 24: 704 (1937).

Tithymalus aztecus Croizat, Amer. J. Bot. 24: 703 (1937).

Tithymalus eochlorus Croizat, Amer. J. Bot. 24: 704 (1937).

Tithymalus subpavonianus Croizat, Amer. J. Bot. 24: 703 (1937).

Pedilanthus calcaratus Schltdl., Linnaea 19: 255 (1847). *Tithymalus calcaratus* (Schltdl.) Croizat, Amer. J. Bot. 24: 704 (1937).
SW. Mexico to Guatemala. 79 MXG MXS MXT 80 GUA.
Pedilanthus ghiesbreghtianus Baill., Adansonia 1: 340 (1861).
Pedilanthus macradenius Donn.Sm., Bot. Gaz. 19: 263 (1894). *Tithymalus macradenius* (Donn.Sm.) Croizat, Amer. J. Bot. 24: 704 (1937).
Pedilanthus purpusii Brandegee, Univ. Calif. Publ. Bot. 4: 377 (1913). *Tithymalus purpusii* (Brandegee) Croizat, Amer. J. Bot. 24: 704 (1937).

Pedilanthus coalcomanensis Croizat, J. Wash. Acad. Sci. 33: 19 (1943).
Mexico (Michoacán). 79 MXS.

Pedilanthus connatus Dressler & Sacamano, Acta Bot. Mex. 18: 21 (1992).
Mexico (Jalisco). 79 MXS.

Pedilanthus cymbiferus Schltdl., Linnaea 19: 253 (1847). *Tithymalus cymbifer* (Schltdl.) Croizat, Amer. J. Bot. 24: 704 (1937).
Mexico (Puebla, Oaxaca). 79 MXC MXS.
Pedilanthus aphyllus Boiss. ex Klotzsch, Abh. Königl. Akad. Wiss. Berlin 1859: 106 (1860).

Pedilanthus diazlunanus J.Lomelí & Sahagun, Acta Bot. Mex. 25: 15 (1993).
Mexico (Jalisco). 79 MXS.

Pedilanthus finkii Boiss. in A.P.de Candolle, Prodr. 15(2): 1261 (1866).
Mexico (Guerrero ?, Veracruz). 79 MXG MXS.

Pedilanthus gracilis Dressler, Contr. Gray Herb. 182: 109 (1957).
Mexico (Guerrero). 79 MXS.

Pedilanthus macrocarpus Benth., Bot. Voy. Sulphur: 49 (1844). *Tithymalus macrocarpus* (Benth.) Croizat, Amer. J. Bot. 24: 704 (1937).
NW. Mexico. 79 MXN. Succ. nanophan.

Pedilanthus millspaughii Pax & K.Hoffm., Repert. Spec. Nov. Regni Veg. 19: 174 (1923). *Tithymalus millspaughii* (Pax & K.Hoffm.) Croizat, Amer. J. Bot. 24: 704 (1937).
Costa Rica. 80 COS.

Pedilanthus nodiflorus Millsp., Publ. Field Columbian Mus., Bot. Ser. 1: 305 (1896). *Tithymalus nodiflorus* (Millsp.) Croizat, Amer. J. Bot. 24: 704 (1937).
Mexico (Yucatán), Honduras. 79 MXT 80 HON.
Pedilanthus personatus Croizat, J. Wash. Acad. Sci. 33: 20 (1943).

Pedilanthus oerstedii Klotzsch, Abh. Königl. Akad. Wiss. Berlin 1859: 106 (1860).
Tithymalus oerstedii (Klotzsch) Croizat, Amer. J. Bot. 24: 704 (1937).
Nicaragua ?, Mexico (Puebla) ? 79 MXC? 80 NIC?

Pedilanthus palmeri Millsp., Publ. Field Mus. Nat. Hist., Bot. Ser. 2: 364 (1913).
SW. Mexico. 79 MXS.
   Pedilanthus peritropoides Millsp., Publ. Field Mus. Nat. Hist., Bot. Ser. 2: 369 (1913).
      Tithymalus peritropoides (Millsp.) Croizat, Amer. J. Bot. 24: 704 (1937).
   Tithymalus koilopremnos Croizat, Amer. J. Bot. 24: 704 (1937).

Pedilanthus pulchellus Dressler, Contr. Gray Herb. 182: 111 (1957).
Mexico (Oaxaca). 79 MXS.

Pedilanthus tehuacanus Brandegee, Univ. Calif. Publ. Bot. 6: 55 (1914). *Tithymalus tehuacanus* (Brandegee) Croizat, Amer. J. Bot. 24: 704 (1937).
C. Mexico. 79 MXC.

Pedilanthus tithymaloides (L.) Poit., Ann. Mus. Natl. Hist. Nat. 19: 390 (1812).
   – FIGURE, p. 1244.
Florida, Mexico, Trop. America. 78 FLA 79 MXE MXG MXS MXT 80 ALL 81 BAH BER cay
   CUB DOM HAI JAM LEE NLA PUE TRT VNA WIN 82 frg GUY SUR VEN 83 CLM ECU
   PER 84 BZL BZN. Nanophan.
   * Euphorbia tithymaloides L., Sp. Pl.: 453 (1753). *Tithymalus tithymaloides* (L.) Croizat,
      Amer. J. Bot. 24: 704 (1937).

subsp. **angustifolius** (Poit.) Dressler, Contr. Gray Herb. 182: 161 (1957).
   Cuba, Hispaniola, Puerto Rico, Leeward Is. 81 CUB DOM HAI LEE PUE. Nanophan.
   * Pedilanthus angustifolius Poit., Ann. Mus. Natl. Hist. Nat. 19: 393 (1812). *Pedilanthus
      tithymaloides* var. *angustifolius* (Poit.) Griseb., Fl. Brit. W. I.: 52 (1859).
   Tithymalus sarissophyllus Croizat, Amer. J. Bot. 24: 703 (1937).

subsp. **bahamensis** (Millsp.) Dressler, Contr. Gray Herb. 182: 165 (1957).
   Bahamas. 81 BAH. Nanophan.
   * Pedilanthus bahamensis Millsp., Publ. Field Mus. Nat. Hist., Bot. Ser. 2: 359 (1913).
      Tithymalus bahamensis (Millsp.) Croizat, Amer. J. Bot. 24: 704 (1937).

subsp. **jamaicensis** (Millsp. & Britton) Dressler, Contr. Gray Herb. 182: 159 (1957).
   S. Jamaica. 81 JAM. Nanophan.
   Pedilanthus grisebachii Millsp. & Britton, Publ. Field Mus. Nat. Hist., Bot. Ser. 2: 361
      (1913). *Tithymalus grisebachii* (Millsp. & Britton) Croizat, Amer. J. Bot. 24: 704 (1937).
   * Pedilanthus jamaicensis Millsp. & Britton, Publ. Field Mus. Nat. Hist., Bot. Ser. 2: 356
      (1913). *Tithymalus jamaicensis* (Millsp. & Britton) Croizat, Amer. J. Bot. 24:
      704 (1937).

subsp. **padifolius** (L.) Dressler, Contr. Gray Herb. 182: 156 (1957).
   Leeward Is., Windward Is. 81 LEE WIN. Nanophan.
   * Euphorbia tithymaloides var. *padifolia* L., Sp. Pl.: 453 (1753). *Pedilanthus padifolius* (L.)
      Poit., Ann. Mus. Natl. Hist. Nat. 19: 393 (1812). *Euphorbia padifolia* (L.) Guss., Cat.
      Pl. Boccadifalco: 25 (1821). *Pedilanthus tithymaloides* var. *padifolius* (L.) Griseb., Fl.
      Brit. W. I.: 52 (1859). *Tithymalus padifolius* (L.) Croizat, Amer. J. Bot. 24: 703 (1937).
   Tithymalus laurocerasifolius Mill., Gard. Dict. ed. 8: 2 (1768). *Pedilanthus laurocerasifolius*
      (Mill.) Wheeler, Contr. Gray Herb. 124: 42 (1939).
   Euphorbia anacampseroides Lam., Encycl. 2: 420 (1788).
   Pedilanthus anacampseroides Klotzsch & Garcke, Monatsber. Königl. Preuss. Akad. Wiss.
      Berlin 1859: 253 (1859).

subsp. **parasiticus** (Klotzsch & Garcke) Dressler, Contr. Gray Herb. 182: 148 (1957).
   Mexico (Veracruz, Yucatán), Belize, Caribbean. 79 MXG MXT 80 BLZ 81 DOM HAI
      JAM LEE PUE. Nanophan.

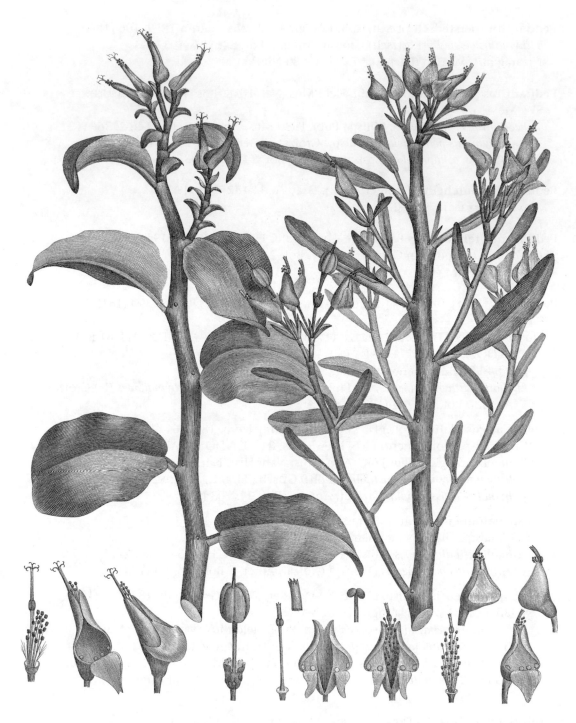

Pedilanthus tithymaloides (L.) Poit. subsp. *tithymaloides* (left) & subsp. *angustifolius* (Poit.) Dressler (right)
(respectively as PP. *tithymaloides* & *angustifolius*)

Artist: A. Poiteau
Ann. Mus. Natl. Hist. Nat. 19: pl. 19 (1812)
KEW ILLUSTRATIONS COLLECTION

* *Pedilanthus parasiticus* Boiss. ex Klotzsch, Abh. Königl. Akad. Wiss. Berlin 1859: 105 (1860). *Tithymalus parasiticus* (Boiss. ex Klotzsch) Croizat, Amer. J. Bot. 24: 704 (1937).
Euphorbia parasitica Pav. ex Boiss. in A.P.de Candolle, Prodr. 15(2): 5 (1862), pro syn.
Pedilanthus ramosissimus Boiss. in A.P.de Candolle, Prodr. 15(2): 5 (1862).
Pedilanthus itzaeus Millsp., Publ. Field Columbian Mus., Bot. Ser. 1: 305 (1896).
Tithymalus itzaeus (Millsp.) Croizat, Amer. J. Bot. 24: 704 (1937).
Pedilanthus latifolius Millsp. & Britton, Ann. Missouri Bot. Gard. 2: 43 (1915).
Tithymalus petanophyllus Croizat, Amer. J. Bot. 24: 704 (1937).

subsp. **retusus** (Benth.) Dressler, Contr. Gray Herb. 182: 154 (1957).
Peru, Brazil (Amazonas, Para, Pernambuco). 83 PER 84 BZE BZN. Nanophan.
* *Pedilanthus retusus* Benth., Hooker's J. Bot. Kew Gard. Misc. 6: 321 (1854).
Tithymalus melanopotamicus Croizat, Amer. J. Bot. 24: 703 (1937).

subsp. **smallii** (Millsp.) Dressler, Contr. Gray Herb. 182: 152 (1957).
S. Florida, N. Cuba. 78 FLA 81 CUB. Nanophan.
* *Pedilanthus smallii* Millsp., Publ. Field Mus. Nat. Hist., Bot. Ser. 2: 358 (1913).
Tithymalus smallii (Millsp.) Small, Man. S.E. Fl.: 804 (1933).

subsp. **tithymaloides**
Mexico, Trop. America. 79 MXE MXG MXS MXT 80 ALL 81 BAH BER cay CUB DOM HAI JAM LEE NLA PUE TRT VNA WIN 82 frg GUY SUR VEN 83 CLM ECU PER 84 BZL BZN. Nanophan.
Euphorbia tithymaloides var. *myrtifolia* L., Sp. Pl.: 453 (1753). *Tithymalus myrtifolius* (L.) Mill., Gard. Dict. ed. 8: 1 (1768). *Euphorbia myrtifolia* (L.) Lam., Encycl. 2: 419 (1788), nom. illeg. *Pedilanthus myrtifolius* (L.) Link, Enum. Hort. Berol. Alt. 2: 18 (1822). *Pedilanthus myrsifolius* (L.) Raf., Fl. Tellur. 4: 117 (1838), sphalm.
Euphorbia carinata Donn, Hortus Cantabrig., ed. 6: 131 (1811). *Pedilanthus carinatus* (Donn) Spreng., Syst. Veg. 3: 802 (1826).
Crepidaria subcarinata Haw., Suppl. Pl. Succ.: 67 (1819). *Pedilanthus subcarinatus* (Haw.) Sweet, Hort. Brit.: 355 (1826).
Euphorbia anacampseroides Descourt., Fl. Méd. Antilles 2: 117 (1822), nom. illeg. *Pedilanthus anacampseroides* Klotzsch & Garcke, Monatsber. Königl. Preuss. Akad. Wiss. Berlin 1859: 253 (1859).
Euphorbia canaliculata Lodd., Bot. Cab.: t .727 (1823), nom. illeg. *Pedilanthus canaliculatus* (Lodd.) Sweet, Hort. Brit.: 355 (1826).
Pedilanthus houlletii Baill., Adansonia 1: 341 (1861).
Pedilanthus fendleri Boiss. in A.P.de Candolle, Prodr. 15(2): 5 (1862).
Pedilanthus pringlei Rob., Proc. Amer. Acad. Arts 29: 322 (1894). *Tithymalus pringlei* (Rob.) Croizat, Amer. J. Bot. 24: 704 (1937).
Pedilanthus gritensis Zahlbr., Ann. K. K. Naturhist. Hofmus. 12: 104 (1897).
Pedilanthus deamii Millsp., Publ. Field Mus. Nat. Hist., Bot. Ser. 2: 356 (1913). *Tithymalus deamii* (Millsp.) Croizat, Amer. J. Bot. 24: 704 (1937).
Pedilanthus campester Brandegee, Univ. Calif. Publ. Bot. 6: 56 (1914).
Pedilanthus petraeus Brandegee, Univ. Calif. Publ. Bot. 10: 411 (1924). *Tithymalus petraeus* (Brandegee) Croizat, Amer. J. Bot. 24: 704 (1937).
Pedilanthus ierensis Britton, Bull. Torrey Bot. Club 53: 468 (1926). *Tithymalus ierensis* (Britton) Croizat, Amer. J. Bot. 24: 704 (1937).
Tithymalus villicus Croizat, Amer. J. Bot. 24: 704 (1937).
Pedilanthus camporum Standl. & Steyerm., Publ. Field Mus. Nat. Hist., Bot. Ser. 23: 124 (1944).

Pedilanthus tomentellus B.L.Rob. & Greenm., Amer. J. Sci., III, 50: 164 (1895). *Tithymalus tomentellus* (B.L.Rob. & Greenm.) Croizat, Amer. J. Bot. 24: 704 (1937).
Mexico (Oaxaca). 79 MXS.

Synonyms:
Pedilanthus anacampseroides Klotzsch & Garcke === **Pedilanthus tithymaloides** subsp. **padifolius** (L.) Dressler
Pedilanthus angustifolius Poit. === **Pedilanthus tithymaloides** subsp. **angustifolius** (Poit.) Dressler
Pedilanthus aphyllus Boiss. ex Klotzsch === **Pedilanthus cymbiferus** Schltdl.
Pedilanthus articulatus (Klotzsch & Garcke) Boiss. === **Pedilanthus bracteatus** (Jacq.) Boiss.
Pedilanthus bahamensis Millsp. === **Pedilanthus tithymaloides** subsp. **bahamensis** (Millsp.) Dressler
Pedilanthus brittonii (Millsp.) Pax & K.Hoffm. === **Euphorbia minutula** Boiss.

Pedilanthus campester Brandegee === **Pedilanthus tithymaloides** (L.) Poit. subsp. **tithymaloides**

Pedilanthus camporum Standl. & Steyerm. === **Pedilanthus tithymaloides** (L.) Poit. subsp. **tithymaloides**

Pedilanthus canaliculatus (Lodd.) Sweet === **Pedilanthus tithymaloides** (L.) Poit. subsp. **tithymaloides**

Pedilanthus carinatus (Donn) Spreng. === **Pedilanthus tithymaloides** (L.) Poit. subsp. **tithymaloides**

Pedilanthus cordatus Spreng. === **Euphorbia cordellata** Haw.

Pedilanthus cordellatus (Haw.) Sweet === **Euphorbia cordellata** Haw.

Pedilanthus deamii Millsp. === **Pedilanthus tithymaloides** (L.) Poit. subsp. **tithymaloides**

Pedilanthus fendleri Boiss. === **Pedilanthus tithymaloides** (L.) Poit. subsp. **tithymaloides**

Pedilanthus ghiesbreghtianus Baill. === **Pedilanthus calcaratus** Schltdl.

Pedilanthus greggii Millsp. === **Pedilanthus bracteatus** (Jacq.) Boiss.

Pedilanthus grisebachii Millsp. & Britton === **Pedilanthus tithymaloides** subsp. **jamaicensis** (Millsp. & Britton) Dressler

Pedilanthus gritensis Zahlbr. === **Pedilanthus tithymaloides** (L.) Poit. subsp. **tithymaloides**

Pedilanthus houlletii Baill. === **Pedilanthus tithymaloides** (L.) Poit. subsp. **tithymaloides**

Pedilanthus ierensis Britton === **Pedilanthus tithymaloides** (L.) Poit. subsp. **tithymaloides**

Pedilanthus involucratus (Klotzsch & Garcke) Boiss. === **Pedilanthus bracteatus** (Jacq.) Boiss.

Pedilanthus itzaeus Millsp. === **Pedilanthus tithymaloides** subsp. **parasiticus** (Klotzsch & Garcke) Dressler

Pedilanthus jamaicensis Millsp. & Britton === **Pedilanthus tithymaloides** subsp. **jamaicensis** (Millsp. & Britton) Dressler

Pedilanthus latifolius Millsp. & Britton === **Pedilanthus tithymaloides** subsp. **parasiticus** (Klotzsch & Garcke) Dressler

Pedilanthus laurocerasifolius (Mill.) Wheeler === **Pedilanthus tithymaloides** subsp. **padifolius** (L.) Dressler

Pedilanthus linearifolius Griseb. === **Cubanthus linearifolius** (Griseb.) Millsp.

Pedilanthus lycioides Baker === **Euphorbia pedilanthoides** Denis

Pedilanthus macradenius Donn.Sm. === **Pedilanthus calcaratus** Schltdl.

Pedilanthus myrsifolius (L.) Raf. === **Pedilanthus tithymaloides** (L.) Poit. subsp. **tithymaloides**

Pedilanthus myrtifolius (L.) Link === **Pedilanthus tithymaloides** (L.) Poit. subsp. **tithymaloides**

Pedilanthus olsson-sefferi Millsp. === **Pedilanthus bracteatus** (Jacq.) Boiss.

Pedilanthus padifolius (L.) Poit. === **Pedilanthus tithymaloides** subsp. **padifolius** (L.) Dressler

Pedilanthus parasiticus Boiss. ex Klotzsch === **Pedilanthus tithymaloides** subsp. **parasiticus** (Klotzsch & Garcke) Dressler

Pedilanthus pavonis (Klotzsch & Garcke) Boiss. === **Pedilanthus bracteatus** (Jacq.) Boiss.

Pedilanthus pectinatus Baker === **Euphorbia sp.**

Pedilanthus peritropoides Millsp. === **Pedilanthus palmeri** Millsp.

Pedilanthus personatus Croizat === **Pedilanthus nodiflorus** Millsp.

Pedilanthus petraeus Brandegee === **Pedilanthus tithymaloides** (L.) Poit. subsp. **tithymaloides**

Pedilanthus pringlei Rob. === **Pedilanthus tithymaloides** (L.) Poit. subsp. **tithymaloides**

Pedilanthus purpusii Brandegee === **Pedilanthus calcaratus** Schltdl.

Pedilanthus ramosissimus Boiss. === **Pedilanthus tithymaloides** subsp. **parasiticus** (Klotzsch & Garcke) Dressler

Pedilanthus retusus Benth. === **Pedilanthus tithymaloides** subsp. **retusus** (Benth.) Dressler

Pedilanthus rubescens Brandegee === **Pedilanthus bracteatus** (Jacq.) Boiss.

Pedilanthus smallii Millsp. === **Pedilanthus tithymaloides** subsp. **smallii** (Millsp.) Dressler

Pedilanthus spectabilis Rob. === **Pedilanthus bracteatus** (Jacq.) Boiss.

Pedilanthus subcarinatus (Haw.) Sweet === **Pedilanthus tithymaloides** (L.) Poit. subsp. **tithymaloides**

Pedilanthus tithymaloides var. *angustifolius* (Poit.) Griseb. === **Pedilanthus tithymaloides** subsp. **angustifolius** (Poit.) Dressler

Pedilanthus tithymaloides var. *padifolius* (L.) Griseb. === **Pedilanthus tithymaloides** subsp. **padifolius** (L.) Dressler

Peltandra

Synonyms:
Peltandra Wight === **Meineckia** Baill.
Peltandra flexuosa Thwaites === **Meineckia parvifolia** (Wight) G.L.Webster
Peltandra longipes Wight === **Meineckia longipes** (Wight) G.L.Webster
Peltandra parvifolia Wight === **Meineckia parvifolia** (Wight) G.L.Webster

Peniculifera

Synonyms:
Peniculifera Ridl. === **Trigonopleura** Hook.f.

Pentabrachion

1 species, WC. tropical Africa (Cameroon to Congo Republic); shrubs or small trees to 10 m of shady undergrowth in high forest on well-drained ground, the leaves very thin but reminiscent of some species of *Bridelia*. Considered to be closely related to *Amanoa*; with that genus it comprises Amanoeae in the Webster system (Webster, Synopsis, 1994). (Phyllanthoideae)

Pax, F. & K. Hoffmann (1922). *Pentabrachium*. In A. Engler (ed.), Das Pflanzenreich, IV 147 XV (Euphorbiaceae-Phyllanthoideae-Phyllantheae): 188-189. Berlin. (Heft 81.) La/Ge. — 1 species, Africa (W and C Africa).

Léonard, J. (1970). *Pentabrachion* Müll. Arg., genre d'Euphorbiaceae nouveau pour la République Démocratique du Congo. Bull. Jard. Bot. Natl. Belg. 40: 349-351. Fr. — Extension of range; includes description.

Pentabrachion Müll.Arg., Flora 47: 532 (1864).
WC. Trop. Africa. 23. Nanophan. or phan.

Pentabrachion reticulatum Müll.Arg., Flora 47: 533 (1864). *Actephila reticulata* (Müll.Arg.) Pax, Bot. Jahrb. Syst. 26: 326 (1899).
WC. Trop. Africa. 23 CMN GAB ZAI. Nanophan. or phan.
Amanoa laurifolia Pax, Bot. Jahrb. Syst. 11: 522 (1893).
Actephila africana Pax in H.G.A.Engler & K.A.E.Prantl, Nat. Pflanzenfam., Nachtr. 1: 210 (1897).

Pentameria

Synonyms:
Pentameria Klotzsch ex Baill. === **Bridelia** Willd.
Pentameria melanthesoides Klotzsch ex Baill. === **Bridelia cathartica** subsp. **melanthesoides** (Klotzsch ex Baill.) J.Léonard

Penteca

Synonyms:
Penteca Raf. === **Croton** L.

Pera

38 species, Americas (Mexico to Bolivia and Brazil), most strongly represented in the Amazon Basin; shrubs or trees, usually relatively small but may reach 25 m or perhaps more (in *P. arborea*). The flower clusters are enclosed in or surrounded by bibracteolate involucres.

The position of this genus in the family has been controversial although all students agree on its distinctness, according it tribal rank. Proposals have also been made for it to have family status (Peraceae), firstly by Klotzsch and, in the twentieth century, by Airy-Shaw and Meeuse. Recent molecular systematic research has shown these views to have some foundation; while family rank appears not to be justified, the genus may form a sister group with all the rest of the 'ACE' Euphorbiaceae (M. Chase, personal communication 1998). No full revision has appeared since 1919. (Acalyphoideae)

Pax, F. & K. Hoffmann (1919). *Pera*. In A. Engler (ed.), Das Pflanzenreich, IV 147 XIII (Euphorbiaceae-Pereae): 2-13. Berlin. (Heft 68.) La/Ge. — 20 species in 6 sections, all American.

Jablonski, E. (1967). *Pera*. Euphorbiaceae, Guayana Highland (Mem. New York Bot. Gard. 17(1)): 147-151. New York. En. — 6 species; 'well represented' in area.

Pera Mutis, Kongl. Vetensk. Acad. Nya Handl. 5: 299 (1784).
Mexico, Trop. America. 79 80 81 82 83 84.
Perula Schreb., Gen.: 703 (1791).
Spixia Leandro, Denkschr. Königl. Akad. Wiss. München 7: 231 (1818-1820 publ. 1821).
Peridium Schott in K.Sprengel, Syst. Veg. 4(2): 410 (1827).
Schismatopera Klotzsch, Arch. Naturgesch. 7: 178 (1841).
Clistranthus Poit. ex Baill., Étude Euphorb.: 434 (1858).

Pera alba Leal, Arch. Jard. Bot. Rio de Janeiro 11: 66 (1951).
Brazil (Brasília D.F.). 84 BZC.

Pera androgyna Rizzini, Revista Brasil. Biol. 32: 321 (1972).
Brazil (?). 84 +.

Pera aperta Croizat, Ann. Missouri Bot. Gard. 29: 353 (1942).
Panama. 80 PAN. – Known only from the type but possibly not distinct from *P. arborea*.

Pera arborea Mutis, Kongl. Vetensk. Acad. Nya Handl. 5: 299 (1784).
Belize to Colombia. 80 BLZ COS PAN 83 CLM. Phan.

Pera bailloniana Müll.Arg. in A.P.de Candolle, Prodr. 15(2): 1030 (1866).
Brazil (Minas Gerais). 84 BZL. Phan.
* *Pera bumeliifolia* Baill., Adansonia 5: 224 (1865).

Pera barbellata Standl., Publ. Field Mus. Nat. Hist., Bot. Ser. 8: 19 (1930).
S. Mexico, Belize, Guatemala. 79 MXT 80 BLZ GUA.

Pera barbinervis (Klotzsch) Pax & K.Hoffm. in H.G.A.Engler, Pflanzenr., IV, 147, XIII: 6 (1919).
Brazil (Bahia, Minas Gerais). 84 BZE BZL. Phan.
* *Spixia barbinervis* Klotzsch, Arch. Naturgesch. 7: 180 (1841).
Pera anisotricha Müll.Arg. in C.F.P.von Martius, Fl. Bras. 11(2): 426 (1874).
Pera bahiana Ule, Bot. Jahrb. Syst. 42: 218 (1908).

Pera benensis Rusby, Descr. S. Amer. Pl.: 49 (1920).
Bolivia. 83 BOL.

Pera bicolor (Klotzsch) Müll.Arg. in A.P.de Candolle, Prodr. 15(2): 1028 (1866).
Guyana, Surinam, Venezuela (Bolívar, Amazonas), Brazil (Amazonas). 82 GUY SUR VEN 84 BZN.
* *Peridium bicolor* Klotzsch, Hooker's J. Bot. Kew Gard. Misc. 2: 44 (1843).
Peridium bicolor var. *nitidum* Benth., Hooker's J. Bot. Gard. Misc. 6: 323 (1845). *Pera nitida* (Benth.) Jabl., Mem. New York Bot. Gard. 17: 148 (1967).

Peridium schomburgkianum Klotzsch in M.R.Schomburgk, Fauna Fl. Brit. Gui.: 1089 (1848). *Pera schomburgkiana* (Klotzsch) Müll.Arg. in A.P.de Candolle, Prodr. 15(2): 1027 (1866).

Pera bumeliifolia Griseb., Nachr. Königl. Ges. Wiss. Georg-Augusts-Univ. 1: 180 (1865).
Bahamas, Cuba, Hispaniola. 81 BAH CUB DOM HAI pue. Phan.
   Pera domingensis Urb., Symb. Antill. 7: 261 (1912).
   Pera depressa Urb. & Ekman, Ark. Bot. 22A(17): 112 (1929).

Pera citriodora Baill., Adansonia 5: 222 (1865).
Venezuela (Amazonas), Brazil (Amazonas). 82 VEN 84 BZN. Nanophan. or phan.

Pera coccinea (Benth.) Müll.Arg. in A.P.de Candolle, Prodr. 15(2): 1028 (1866).
Brazil (Pará). 84 BZN. Phan.
   * *Peridium coccineum* Benth., Hooker's J. Bot. Kew Gard. Misc. 6: 323 (1854).

Pera colombiana Cardiel, Caldasia 16(78): 311 (1991).
Colombia. 83 CLM.

Pera decipiens Müll.Arg., Linnaea 34: 201 (1865).
Guyana, Surinam, S. Venezuela, Brazil (Amazonas). 82 GUY SUR VEN 84 BZN.

Pera distichophylla (Mart.) Baill., Étude Euphorb.: 434 (1858).
Brazil (Amazonas), S. Venezuela. 82 VEN 84 BZN. Phan.
   * *Spixia distichophylla* Mart., Flora 24(2): 30 (1841). *Pera distichophylla* var. *genuina* Müll.Arg. in A.P.de Candolle, Prodr. 15(2): 1026 (1866), nom. inval.
   Schismatopera laurina Benth., Hooker's J. Bot. Kew Gard. Misc. 6: 324 (1854). *Pera distichophylla* var. *laurina* (Benth.) Müll.Arg. in A.P.de Candolle, Prodr. 15(2): 1026 (1866).
   Pera distichophylla var. *martiana* Müll.Arg. in A.P.de Candolle, Prodr. 15(2): 1026 (1866).
   Pera distichophylla var. *lanceolata* Müll.Arg. in C.F.P.von Martius, Fl. Bras. 11(2): 425 (1874).
   Pera heterodoxa Müll.Arg. in C.F.P.von Martius, Fl. Bras. 11(2): 423 (1874).

Pera ekmanii Urb., Symb. Antill. 9: 206 (1924).
NE. Cuba. 81 CUB. Phan.

Pera elliptica Rusby, Mem. New York Bot. Gard. 7: 288 (1927).
Bolivia. 83 BOL.

Pera frutescens Leal, Arch. Jard. Bot. Rio de Janeiro 11: 66 (1951).
Brazil (Pará). 84 BZN.

Pera furfuracea Müll.Arg. in C.F.P.von Martius, Fl. Bras. 11(2): 426 (1874).
Brazil (Bahia). 84 BZE. Phan.

Pera glabrata (Schott) Poepp. ex Baill., Étude Euphorb.: 434 (1858).
Trinidad, Venezuela, Guianas, Brazil (Amazonas, Para). 81 TRT 82 FRG GUY SUR VEN 84 BZE BZL BZN BZS. Nanophan.
   Peridium ferrugineum Schott in K.Sprengel, Syst. Veg. 4(2): 410 (1827). *Pera ferruginea* (Schott) Müll.Arg. in A.P.de Candolle, Prodr. 15(2): 1031 (1866).
   * *Peridium glabratum* Schott in K.Sprengel, Syst. Veg. 4(2): 410 (1827).
   Pera arborea Baill., Adansonia 5: 224 (1865), nom. illeg.
   Pera klotzschiana Baill., Adansonia 5: 225 (1865).
   Pera glabrata var. *parvifolia* Glaz., Bull. Soc. Bot. France 59(3): 626 (1912 publ. 1913).
   Pera glabrata var. *petropolitana* Glaz., Bull. Soc. Bot. France 59(3): 626 (1912 publ. 1913).

Pera glaziovii Taub. ex Pax in H.G.A.Engler, Pflanzenr., IV, 147, XIII: 12 (1919).
Brazil (Rio de Janeiro). 84 BZL.

Pera glomerata Urb., Ark. Bot. 20A(15): 62 (1926).
Haiti (I. Tortuga). 81 HAI. Nanophan.

Pera heteranthera (Schrank) I.M.Johnst., Contr. Gray Herb. 68: 90 (1923).
E. Brazil. 84 BZE BZL. Nanophan. or phan.
* *Spixia heteranthera* Schrank, Denkschr. Königl. Akad. Wiss. München 7: 242 (1818-1920 publ. 1821).
Spixia leandrii Mart., Flora 24(2): 30 (1841). *Pera leandrii* (Mart.) Baill., Étude Euphorb.: 434 (1858). *Pera leandrii* var. *genuina* Müll.Arg. in A.P.de Candolle, Prodr. 15(2): 1027 (1866), nom. inval.
Pera leandrii var. *glabrescens* Müll.Arg. in A.P.de Candolle, Prodr. 15(2): 1027 (1866).

Pera incisa Leal, Arch. Jard. Bot. Rio de Janeiro 11: 65 (1951).
Brazil (Pará). 84 BZN.

Pera longipes Britton & P.Wilson, Mem. Torrey Bot. Club 16: 76 (1920).
Cuba (Sierra de Moa). 81 CUB. Nanophan.

Pera membranacea Leal, Arch. Jard. Bot. Rio de Janeiro 11: 67 (1951).
Brazil (Pará). 84 BZN.

Pera microcarpa Urb., Symb. Antill. 9: 208 (1924).
Cuba (Sierra Maestra). 81 CUB. Phan.

Pera mildbraediana Mansf., Notizbl. Bot. Gart. Berlin-Dahlem 9: 265 (1925).
Peru. 83 PER.

Pera obovata (Klotzsch) Baill., Adansonia 5: 225 (1865).
Brazil (Rio de Janeiro, São Paulo). 84 BZL.
* *Peridium obovatum* Klotzsch, Arch. Naturgesch. 7: 180 (1841).
Pera obtusifolia Müll.Arg. in A.P.de Candolle, Prodr. 15(2): 1030 (1866).

Pera oppositifolia Griseb., Nachr. Königl. Ges. Wiss. Georg-Augusts-Univ. 1: 181 (1865).
W. Cuba (incl. I. de la Juventud). 81 CUB. Phan.

Pera orientensis Borhidi, Acta Bot. Acad. Sci. Hung. 25(12): 43 (1979).
E. Cuba. 81 CUB. Phan.

Pera ovalifolia Urb., Symb. Antill. 9: 207 (1924).
Cuba. 81 CUB. Phan.

Pera pallidifolia Britton & P.Wilson, Mem. Torrey Bot. Club 16: 76 (1920).
NE. Cuba. 81 CUB. Nanophan.

Pera parvifolia (Klotzsch) Müll.Arg. in A.P.de Candolle, Prodr. 15(2): 1031 (1866).
Brazil (Rio de Janeiro). 84 BZL.
* *Peridium parvifolium* Klotzsch, Arch. Naturgesch. 7: 180 (1841).

Pera polylepis Urb., Symb. Antill. 9: 206 (1924).
Cuba. 81 CUB.

subsp. **moaensis** Borhidi, Acta Bot. Acad. Sci. Hung. 25(12): 44 (1979).
E. Cuba. 81 CUB. Nanophan. or phan.

subsp. **polylepis**
    NE. Cuba. 81 CUB. Nanophan. or phan.

Pera pulchrifolia Ducke, Trop. Woods 50: 36 (1937).
    Brazil (Amazonas). 84 BZN.

Pera rubra Leal, Arch. Jard. Bot. Rio de Janeiro 11: 64 (1951).
    Brazil (Rio de Janeiro). 84 BZL.

Pera tomentosa (Benth.) Müll.Arg. in A.P.de Candolle, Prodr. 15(2): 1028 (1866).
    S. Venezuela, Brazil (Amazonas). 82 VEN 84 BZN.
    *Peridium bicolor var. *tomentosum* Benth., Hooker's J. Bot. Kew Gard. Misc. 6:
        323 (1845).
    Pera cinerea Baill., Adansonia 5: 223 (1865), nom. inval.

Synonyms:
Pera anisotricha Müll.Arg. === **Pera barbinervis** (Klotzsch) Pax & K.Hoffm.
Pera arborea Baill. === **Pera glabrata** (Schott) Poepp. ex Baill.
Pera bahiana Ule === **Pera barbinervis** (Klotzsch) Pax & K.Hoffm.
Pera bumeliifolia Baill. === **Pera bailloniana** Müll.Arg.
Pera cinerea Baill. === **Pera tomentosa** (Benth.) Müll.Arg.
Pera corcovadensis Glaz. === ?
Pera depressa Urb. & Ekman === **Pera bumeliifolia** Griseb.
Pera distichophylla var. *genuina* Müll.Arg. === **Pera distichophylla** (Mart.) Baill.
Pera distichophylla var. *lanceolata* Müll.Arg. === **Pera distichophylla** (Mart.) Baill.
Pera distichophylla var. *laurina* (Benth.) Müll.Arg. === **Pera distichophylla** (Mart.) Baill.
Pera distichophylla var. *martiana* Müll.Arg. === **Pera distichophylla** (Mart.) Baill.
Pera domingensis Urb. === **Pera bumeliifolia** Griseb.
Pera echinocarpa Baill. === **Chaetocarpus echinocarpus** (Baill.) Ducke
Pera ferruginea (Schott) Müll.Arg. === **Pera glabrata** (Schott) Poepp. ex Baill.
Pera glabrata var. *parvifolia* Glaz. === **Pera glabrata** (Schott) Poepp. ex Baill.
Pera glabrata var. *petropolitana* Glaz. === **Pera glabrata** (Schott) Poepp. ex Baill.
Pera heterodoxa Müll.Arg. === **Pera distichophylla** (Mart.) Baill.
Pera klotzschiana Baill. === **Pera glabrata** (Schott) Poepp. ex Baill.
Pera leandrii (Mart.) Baill. === **Pera heteranthera** (Schrank) I.M.Johnst.
Pera leandrii var. *genuina* Müll.Arg. === **Pera heteranthera** (Schrank) I.M.Johnst.
Pera leandrii var. *glabrescens* Müll.Arg. === **Pera heteranthera** (Schrank) I.M.Johnst.
Pera nitida (Benth.) Jabl. === **Pera bicolor** (Klotzsch) Müll.Arg.
Pera obtusifolia Müll.Arg. === **Pera obovata** (Klotzsch) Baill.
Pera schomburgkiana (Klotzsch) Müll.Arg. === **Pera bicolor** (Klotzsch) Müll.Arg.

Peridium

Synonyms:
Peridium Schott === **Pera** Mutis
Peridium bicolor Klotzsch === **Pera bicolor** (Klotzsch) Müll.Arg.
Peridium bicolor var. *nitidum* Benth. === **Pera bicolor** (Klotzsch) Müll.Arg.
Peridium bicolor var. *tomentosum* Benth. === **Pera tomentosa** (Benth.) Müll.Arg.
Peridium coccineum Benth. === **Pera coccinea** (Benth.) Müll.Arg.
Peridium ferrugineum Schott === **Pera glabrata** (Schott) Poepp. ex Baill.
Peridium glabratum Schott === **Pera glabrata** (Schott) Poepp. ex Baill.
Peridium obovatum Klotzsch === **Pera obovata** (Klotzsch) Baill.
Peridium parvifolium Klotzsch === **Pera parvifolia** (Klotzsch) Müll.Arg.
Peridium schomburgkianum Klotzsch === **Pera bicolor** (Klotzsch) Müll.Arg.

Periplexis

Synonyms:
Periplexis Wall. === **Drypetes** Vahl

Perula

Synonyms:
Perula Schreb. === **Pera** Mutis

Petalandra

Proposed by F. Mueller and published by Boissier; a synonym of *Euphorbia*.

Petalodiscus

4 species, E. Africa (*P. fadenii*), Madagascar; in the latter the three species currently recognised were referred to *Savia* by Leandri (1958). Shrubs or small trees to 6 m in forest understorey, often on sand; leaves generally small, phyllanthoid and distichously arranged. The exalbuminous seeds differentiate the genus from *Savia* but it remains a member of Wielandieae. [Further revisions to this genus by P. Hoffmann are planned, including further transfers from *Savia*; the list given here is not final. *P. mimosoides* is now in *Blotia* (P. Hoffmann & McPherson, 1998; see there).] (Phyllanthoideae)

> Pax, F. & K. Hoffmann (1922). *Savia*. In A. Engler (ed.), Das Pflanzenreich, IV 147 XV (Euphorbiaceae-Phyllanthoideae-Phyllantheae): 181-188. Berlin. (Heft 81.) La/Ge. — *Petalodiscus* here treated as a section, with 3 species from Madagascar.
>
> Leandri, J. (1958). *Savia*. Fl. Madag. Comores 111 (Euphorbiacées), I: 116-126. Paris. Fr. — Flora treatment (9 species) with key. [Nos. 1, 2, 6, 7, and 9 at least now referred to *Petalodiscus*.]

Petalodiscus (Baill.) Pax in H.G.A.Engler & K.A.E.Prantl, Nat. Pflanzenfam. 3(5): 15 (1890).
E. Trop. Africa, Madagascar. 25 29.

Petalodiscus fadenii (Radcl.-Sm.) Radcl.-Sm., Kew Bull. 47: 679 (1993).
SE. Kenya. 25 KEN. Nanophan. or phan. – Provisionally accepted.
  * *Savia fadenii* Radcl.-Sm., Kew Bull. 27: 508 (1972).

Petalodiscus laureola (Baill.) Pax in H.G.A.Engler & K.A.E.Prantl, Nat. Pflanzenfam. 3(5): 15 (1890).
Madagascar. 29 MDG. Nanophan. or phan.
  * *Savia laureola* Baill., Étude Euphorb.: 572 (1858).

Petalodiscus platyrachis (Baill.) Pax in H.G.A.Engler & K.A.E.Prantl, Nat. Pflanzenfam. 3(5): 15 (1890).
Madagascar. 29 MDG. Nanophan. or phan.
  * *Savia platyrachis* Baill., Étude Euphorb.: 572 (1858).
    Savia platyrachis var. *microphylla* Leandri, in Fl. Madag. 111: 120 (1958).

Petalodiscus pulchella (Baill.) Pax in H.G.A.Engler & K.A.E.Prantl, Nat. Pflanzenfam. 3(5): 15 (1890).
E. Madagascar. 29 MDG. Nanophan.
  * *Savia pulchella* Baill., Étude Euphorb.: 573 (1858).

Synonyms:
Petalodiscus oblongifolius (Baill.) Pax === **Blotia oblongifolia** (Baill.) Leandri
Petalodiscus mimosoides (Baill.) Pax === **Blotia mimosoides** (Baill.) Petra Hoffm. & McPherson

Petaloma

A Rafinesque segregate of *Euphorbia*, established for *E. marginata* (subgen. *Agaloma*) but only published by Boissier (1862) in the synonymy of his sect. *Petaloma*. It did not actually appear in the author's *Atlantic Journal and Friend of Knowledge* 6: 177 (1833) as Boissier suggested (Wheeler 1943; see **Euphorbia**). It is a later homonym of *Petaloma* Sw. (= *Mouriri*, Melastomataceae) and *Petaloma* Roxb. (= *Lumnitzera*, Combretaceae).

Synonyms:
Petaloma Raf. ex Boiss. === **Euphorbia** L.

Petalostigma

6 species, Australia (essentially in the north) with one, *P. pubescens*, also in New Guinea. Shrubs or small trees with distichously arranged foliage, the leaves sometimes whitish-pubescent on the under surface. With a distinctive pollen morphology the genus was placed by Webster (Synopsis, 1994) into its own subtribe, Petalostigmatinae, in Caletieae. The most recent revision is by Airy-Shaw (1980). Further studies have since been made by Paul Forster. (Oldfieldioideae)

Pax, F. & K. Hoffmann (1922). *Petalostigma*. In A. Engler (ed.), Das Pflanzenreich, IV 147 XV (Euphorbiaceae-Phyllanthoideae-Phyllantheae): 281-283. Berlin. (Heft 81.) La/Ge. — 3 species; Australia.

Airy-Shaw, H. K. (1974). Notes on Malesian and other Asiatic Euphorbiaceae, CLXXIV: The genus *Petalostigma* in New Guinea. Kew Bull. 29: 303. En. — First record of genus in New Guinea; extension of *P. pubescens* from northern Australia into the southern 'bulge'.

Airy-Shaw, H. K. (1980). *Petalostigma*. Kew Bull. 35: 661-665. (Euphorbiaceae-Platylobeae of Australia.) En. — Treatment of 6 species.

Forster, P. I. & P. C. van Welzen (1999). The Malesian species of *Choriceras*, *Fontainea* and *Petalostigma* (Euphorbiaceae). Blumea 44: 99-107, illus. En. — *Petalostigma*, pp. 104-106; treatment of 1 species (in southern New Guinea and Australia) with genus and species descriptions, synonymy, references, types, indication of distribution and ecology, and general notes; general references and list of specimens seen at end of paper.

Petalostigma F.Muell., Hooker's J. Bot. Kew Gard. Misc. 9: 16 (1857).
Papua New Guinea, Australia. 42 50. Nanophan. or phan.
Hylococcus R.Br. ex Benth., Fl. Austral. 6: 92 (1873).

Petalostigma banksii Britten & S.Moore, J. Bot. 41: 225 (1903).
Northern Territory, Queensland. 50 NTA QLD. Phan.

Petalostigma nummularium Airy Shaw, Kew Bull. 31: 373 (1976).
Western Australia, Northern Territory, Queensland. 50 NTA QLD WAU. Nanophan.
Petalostigma quadriloculare var. *nigrum* Ewart & O.B.Davies, Fl. N. Territory: 166 (1917).

Petalostigma pachyphyllum Airy Shaw, Kew Bull. 31: 372 (1976).
Queensland (Leichhardt). 50 QLD. Nanophan.

Petalostigma pubescens Domin, Biblioth. Bot. 89: 317 (1927).
W. Papua New Guinea, Western Australia, Northern Territory, Queensland, NE. New South Wales. 42 NWG 50 NSW NTA QLD WAU. Phan.
Petalostigma quadriloculare var. *pubescens* Müll.Arg., Flora 47: 481 (1864).

Petalostigma quadriloculare F.Muell., Hooker's J. Bot. Kew Gard. Misc. 9: 17 (1857).
  Petalostigma quadriloculare var. *genuina* Müll.Arg. in A.P.de Candolle, Prodr. 15(2): 273 (1866), nom. inval.
  Western Australia, Northern Territory, Queensland. 50 NTA QLD WAU. Nanophan.
    Petalostigma quadriloculare var. *sericeum* Müll.Arg., Flora 47: 481 (1864).
    Petalostigma humilis W.Fitzg., J. Roy. Soc. W. Australia 3: 163 (1918).
    Petalostigma haplocladum Pax & K.Hoffm. in H.G.A.Engler, Pflanzenr., IV, 147, XV: 283 (1922).
    Petalostigma micrandrum Domin, Biblioth. Bot. 89: 317 (1927).

Petalostigma triloculare Müll.Arg., Flora 47: 471 (1864).
  Queensland. 50 QLD. Phan.
    Petalostigma australianum Baill., Adansonia 7: 356 (1867), nom. illeg.
    Petalostigma quadriloculare var. *glabrescens* Benth., Fl. Austral. 6: 92 (1873). *Petalostigma glabrescens* (Benth.) Domin, Biblioth. Bot. 89: 317 (1927).

Synonyms:
Petalostigma australianum Baill. === **Petalostigma triloculare** Müll.Arg.
Petalostigma glabrescens (Benth.) Domin === **Petalostigma triloculare** Müll.Arg.
Petalostigma haplocladum Pax & K.Hoffm. === **Petalostigma quadriloculare** F.Muell.
Petalostigma humilis W.Fitzg. === **Petalostigma quadriloculare** F.Muell.
Petalostigma micrandrum Domin === **Petalostigma quadriloculare** F.Muell.
Petalostigma quadriloculare var. *genuina* Müll.Arg. === **Petalostigma quadriloculare** F.Muell.
Petalostigma quadriloculare var. *glabrescens* Benth. === **Petalostigma triloculare** Müll.Arg.
Petalostigma quadriloculare var. *nigrum* Ewart & O.B.Davies === **Petalostigma nummularium** Airy Shaw
Petalostigma quadriloculare var. *pubescens* Müll.Arg. === **Petalostigma pubescens** Domin
Petalostigma quadriloculare var. *sericeum* Müll.Arg. === **Petalostigma quadriloculare** F.Muell.

Phaedra

Synonyms:
Phaedra Klotzsch === **Bernardia** Houst. ex Mill.

Philyra

1 species, S. America (SE. Brazil to Paraguay); sometimes arborescent, often slenderly spiny shrubs of dry forest and scrub to 5 m with small, tufted leaves recalling those in *Paranecepsia*. The spines represent modified stipules. The genus is in the Webster system grouped with *Ditaxis* and *Argythamnia* (and the now-separate *Chiropetalum*) but the woody habit is distinctive. (Acalyphoideae)

  Pax, F. (with K. Hoffmann) (1912). *Philyra*. In A. Engler (ed.), Das Pflanzenreich, IV 147 VI (Euphorbiaceae-Acalypheae-Chrozophorinae): 49-51. Berlin. (Heft 57.) La/Ge. — 1 species, SE Brazil.

Philyra Klotzsch, Arch. Naturgesch. 1: 199 (1841).
  Paraguay, SE. Brazil. 84 85.

Philyra brasiliensis Klotzsch, Arch. Naturgesch. 1: 199 (1841). *Ditaxis brasiliensis* (Klotzsch) Baill., Adansonia 4: 269 (1864). *Argythamnia brasiliensis* (Klotzsch) Müll.Arg., Linnaea 34: 144 (1865).
  Paraguay, SE. Brazil. 84 BZL 85 PAR. Nanophan. or phan.

PLATE 1

A. *Synadenium grantii*   **B.** *Ricinocarpus pinifolius*

PLATE 2

Phyllanthus niruri

PLATE 3

Phyllanthus nutans

PLATE 4

Phyllanthus pendulus

PLATE 5

Phyllanthus tetrandrus

PLATE 6

Podadenis sapida

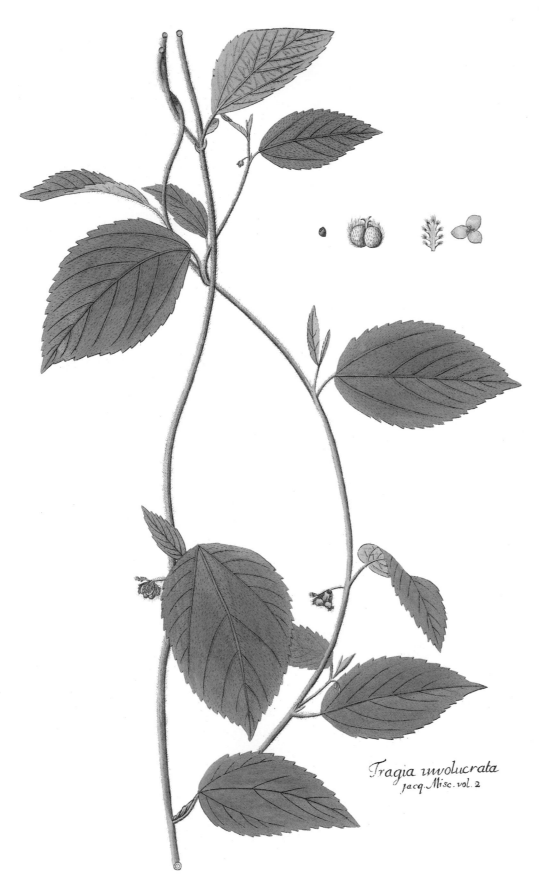

Tragia involucrata
Jacq. Misc. vol. 2

PLATE 7

Tragia involucrata

PLATE 8

Triadica sebifera

Phocea

Synonyms:
Phocea Seem. === **Macaranga** Thouars

Phyllanoa

1 species, S America (Colombia); known only from the type collection. Small trees of montane forest to 8 m with phyllanthoid branching, distichously arranged, finely serrate or dentate leaves, and persistent perianth-lobes. Webster (Synopsis, 1994) proposed an alliance with *Thecacoris* in Antidesminae. (Phyllanthoideae)

> Croizat, L. (1943). Euphorbiaceae Cactaceaeque novae vel criticae colombianae, I. Caldasia 2(7): 123-139, illus. En. — Includes (pp. 123-126) protologue of *Phyllanoa* along with description of 1 species, *P. colombiana*, from the Cordillera Occidental in Depto. Valle de Cauca.

Phyllanoa Croizat, Caldasia 2: 123, 124 (1943).
Colombian Andes. 83. Phan.

Phyllanoa colombiana Croizat, Caldasia 2: 124 (1943).
Colombia. 83 CLM. Phan.

Phyllanthidea

Synonyms:
Phyllanthidea Didr. === **Andrachne** L.
Phyllanthidea microphylla (Lam.) Didr. === **Andrachne microphylla** (Lam.) Baill.

Phyllanthodendron

Now usually combined with *Phyllanthus* (a practice followed here) but remains accepted in some quarters.

Synonyms:
Phyllanthodendron Hemsl. === **Phyllanthus** L.
Phyllanthodendron album Craib & Hutch. === **Phyllanthus roseus** (Craib & Hutch.) Beille
Phyllanthodendron anthopotamicum (Hand.-Mazz.) Croizat === **Phyllanthus anthopotamicus** Hand.-Mazz.
Phyllanthodendron breynioides P.T.Li === **Phyllanthus breynioides** (P.T.Li) Govaerts & Radcl.-Sm.
Phyllanthodendron carinatum (Beille) Croizat === **Phyllanthus carinatus** Beille
Phyllanthodendron caudatifolium P.T.Li === **Phyllanthus lii** Govaerts & Radcl.-Sm.
Phyllanthodendron cavaleriei H.Lév. === **Phyllanthus dunnianus** (H.Lév.) Hand.-Mazz. ex Rehder
Phyllanthodendron chevalieri (Gagnep.) Airy Shaw === **Phyllanthus arachnodes** Govaerts & Radcl.-Sm.
Phyllanthodendron coriaceum Gage === **Phyllanthus ridleyanus** Airy Shaw
Phyllanthodendron dubium (Ridl.) Gage === **Phyllanthus roseus** (Craib & Hutch.) Beille
Phyllanthodendron dunnianum H.Lév. === **Phyllanthus dunnianus** (H.Lév.) Hand.-Mazz. ex Rehder
Phyllanthodendron lativenium Croizat === **Phyllanthus lativenius** (Croizat) Govaerts & Radcl.-Sm.
Phyllanthodendron lingulatum (Beille) Croizat === **Phyllanthus lingulatus** Beille
Phyllanthodendron minutiflorum (Ridl.) Airy Shaw === **Phyllanthus ridleyanus** Airy Shaw
Phyllanthodendron mirabile (Müll.Arg.) Hemsl. === **Phyllanthus mirabilis** Müll.Arg.

Phyllanthodendron moi (P.T.Li) P.T.Li === **Phyllanthus moi** P.T.Li

Phyllanthodendron orbicularifolium P.T.Li === **Phyllanthus orbicularifolius** (P.T.Li) Govaerts & Radcl.-Sm.

Phyllanthodendron petraeum P.T.Li === **Phyllanthus guanxiensis** Govaerts & Radcl.-Sm.

Phyllanthodendron poilanei (Beille) Croizat === **Phyllanthus poilanei** Beille

Phyllanthodendron roseum Craib & Hutch. === **Phyllanthus roseus** (Craib & Hutch.) Beille

Phyllanthodendron roseum var. *glabrum* Craib ex Hosseus === **Phyllanthus roseus** (Craib & Hutch.) Beille

Phyllanthodendron roseum var. *siamense* (Pax & K.Hoffm.) Craib === **Phyllanthus roseus** (Craib & Hutch.) Beille

Phyllanthodendron siamense (Pax & K.Hoffm.) Hosseus === **Phyllanthus roseus** (Craib & Hutch.) Beille

Phyllanthodendron yunnanense Croizat === **Phyllanthus yunnanensis** (Croizat) Govaerts & Radcl.-Sm.

Phyllanthus

833 species, widespread in tropical and subtropical regions; includes *Phyllanthodendron* among numerous other one-time segregates. Annual or perennial herbs or shrubs or small trees, most with specialised 'phyllanthoid' branching but this is more or less absent from subgen. *Isocladus* (usually considered the least specialised infrageneric taxon). There are several centres of greater or lesser secondary radiation (e.g. New Guinea with 35 species, Australia with 50 or so (though a recent revision is wanting), the Guayana Highland in northern South America with 58, Madagascar with 58, the West Indies with 80-odd (a large number of them being in Cuba), and New Caledonia with 111 native). The genus was not revised for *Das Pflanzenreich*, and the only current overall treatment remains that by Pax and Hoffmann in the 2nd edn. of Pflanzenfamilien (1931, cited below). This in turn was only to some extent modified from that of Mueller who had recognised 44 sections (including what is now *Glochidion*). Their limits were similar to those currently accepted save for *Margaritaria*, now recognised as a distinct genus. Infrageneric classification has since been reviewed by Webster (1956-58), Brunel (1975, 1987) and Holm-Nielsen (1979). 10 subgenera were recognised by Holm-Nielsen, the most recent formally published account. Brunel (1987: 461-464) has furnished a scheme calling for 13 subgenera and 72 sections, keyed out on pollen characters. In this scheme sect. *Isocladus* is reduced to 3 sections (2 unnamed) in place of the 4 recognised by Webster. There is also a general but rather miscellaneous review by Nozeran, Rossignol-Bancilhon and Mangenot (1984). The last-named have presented a diagram of presumed evolutionary trends («movements») with particular reference to gross organisation. In Holm-Nielsen (1979) there is an isochor-type map of sections with the greatest diversity shown to occur in West Malesia and the West Indies. As in some other widespread and large tropical/subtropical genera, Africa is by contrast less rich. Detailed revisions and other treatments since the nineteenth century have been regional, either as parts of floras or otherwise. Recently, Stuppy (1995; see **Phyllanthoideae**) has suggested that as presently circumscribed the genus is unnatural; seed structure was too diverse to be usefully characteristic. In the opinion of the writer of these lines (D. Frodin), reassessment should in particular begin with subgen. *Isocladus*: is it really in fact a series of derived lines reflecting greater or lesser degrees of paedomorphosis or indeed one or more segregate genera? (Phyllanthoideae)

Baillon, H. (1860). Monographie des *Phyllanthus*. Adansonia 1: 24-43. Fr. — An exposition of 16 species, precursory to a projected monograph; includes synonymy, references, descriptions, localities with exsiccatae, and commentary. [Mainly of historical interest.]

Dingler, H. (1885). Die Flachsprosse der Phanerogamen: vergleichend morphologisch-anatomische Studien. Erstes Heft: *Phyllanthus* sect. *Xylophylla*. iv, 153 pp., 63 figs. in 3 pls. Munich: Ackermann. Ge. — Includes a comprehensive and accurate account of growth form and its modifications in *Phyllanthus* as part of a study on phylloclades (the latter clearly being derived from more generalized shoot structures). [Reviewed by Webster (1956-58) but not cited by Nozeran et al. (1985).]

Froembling, W. (1896). Anatomisch-systematische Untersuchung von Blatt und Axe der Crotoneen und Euphyllantheen. Bot. Centralbl. 16(65): 129-139, 177-192, 241-249, 289-297, 321-329, 369-378, 403-411, 433-441; 2 pls. (Diss., Univ. München; reprinted, 76 pp., 2 pls., Cassel.) Ge. — An anatomical study of *Croton* and its immediate allies as well as of *Phyllanthus* and its immediate allies. Most of the paper is on *Croton* where a great diversity of structure was observed; such did not obtain in *Phyllanthus* for which coverage is limited to the last installment (with some attention given to the phenomenon of phylloclady in part of the genus). All species studied are listed. Captions to the two plates appear in the last installment.

Robinson, C. B. (1909). Philippine Phyllanthinae. Philip. J. Sci. 4, Bot.: 71-105. En. — Regional revision covering 7 genera (*Phyllanthus* with 19 species, pp. 75-86; *Cicca* with 1 species, p. 87); features synonymy, descriptions of novelties, localities with exsiccatae, and commentary along with a key for *Phyllanthus*. Sectional names are attached to each species of *Phyllanthus*.

Pax, F. (with K. Hoffmann) (1911). *Uranthera*. In A. Engler (ed.), Das Pflanzenreich, IV 147 III (Euphorbiaceae-Cluytieae): 95. Berlin. (Heft 47.) La/Ge. — 1 species, *U. siamensis* (now *Phyllanthus roseus* (Craib and Hutch.) Beille); referred by some to *Phyllanthodendron*.

Pax, F. & K. Hoffmann (1922). *Aporosella*. In A. Engler (ed.), Das Pflanzenreich, IV 147 XV (Euphorbiaceae-Phyllanthoideae-Phyllantheae): 105-107. Berlin. (Heft 81.) La/Ge. — 1 species, S America (Paraguay, Brazil). [Now included in *Phyllanthus*.]

Diels, L. (1931). Aufklärung der Gattung *Leichhardtia* F.v.M. Notizbl. Bot. Gart. Mus. Berlin-Dahlem 11: 308-310. Ge. — With 1 species, *L. clamboides*, the genus was originally thought of as menispermaceous. Additional collections (mainly from New Guinea) showed its membership of *Phyllanthus* (with *Leichhardtia* proposed as a sectional name in place of *Nymania* [a step no longer allowed]). 9 species (3 new) accounted for, their centre in Papuasia; key. [The group now also includes *Hexaspermum*.]

• Pax, F. & K. Hoffmann (1931). *Phyllanthus*. In A. Engler (ed.), Die natürlichen Pflanzenfamilien, 2. Aufl., 19c: 60-66, illus. Leipzig. Ge. — Synopsis with description of genus; 30 sections but no subsections recognised, with in each sectional account a listing of representative species. [Never revised for Pflanzenreich. This account remains the most recent *overall* survey of the genus, though in arrangement it is little changed from that of Muller in 1866 and in recent decades has been much criticised, with alternatives proposed by Webster, Brunel and others. The authors estimated a total for the genus of c. 480 species.]

Croizat, L. (1942). On certain Euphorbiaceae from the tropical Far East. J. Arnold Arbor. 23: 29-54. En. — Extensive treatment, including key to sections of the segregate genus *Phyllanthodendron* (pp. 32-38); some novelties.

Webster, G. (1956-58). A monographic study of the West Indian species of *Phyllanthus*. J. Arnold Arbor. 37: 91-122, 217-268, 340-359; 38: 51-80, 170-198, 295-373; 39: 49-100, 111-212; 32 pls. En. — Complete regional revision; 80-odd species, with keys (main key in 3rd installment), descriptions, citations of exsiccatae, and commentary. General part in the first two of the eight installments; author comments that the logistical problems of a large genus are compounded in *Phyllanthus* by great biological diversity. [A new system of the genus was also proposed which, with modifications, is the best available though not comprehensive.]

Leandri, J. (1958). *Phyllanthus*. Fl. Madag. Comores 111 (Euphorbiacées), I: 30-105. Paris. Fr. — Flora treatment (58 species) with key.

Airy-Shaw, H. K. (1960). Notes on Malaysian Euphorbiaceae, XVI. On the identity of the genus *Arachnodes* Gagnep. Kew Bull. 14: 469-471. En. —*Arachnodes* a section of *Phyllanthodendron*, with sect. *Phyllanthodendron* restricted to *P. mirabile*; key to 4 species in SE Asia and Peninsular Malaysia. [*Phyllanthodendron* since reduced to *Phyllanthus*.]

Webster, G. (1960). Supplement to a monographic study of the West Indian species of *Phyllanthus*. J. Arnold Arbor. 41: 279-286. En. — Additions to the author's 1956-58 revision.

Airy-Shaw, H. K. (1963). Notes on Malaysian and other Asiatic Euphorbiaceae, XXII. Limestone association of *Phyllanthodendron minutiflorum*. Kew Bull. 16: 343. En. — Habitat note. [Genus now part of *Phyllanthus*.]

Johnston, M. C. & B. H. Warnock (1963). *Phyllanthus* and *Reverchonia* (Euphorbiaceae) in far western Texas. Southwestern Nat. 8: 15-22. En. — Incoudes descriptive coverage of 3 species of *Phyllanthus*, with key; no exsiccatae cited.

Steenis, C. G. G. J. van (1963). *Phyllanthodendron mirabile* (Euphorbiaceae). Dansk Bot. Ark. 23: 93-100.-105. En. — Natural history of plant, a caudiciform shrub of limestone hills and cliffs. [Now included in *Phyllanthus*.]

Airy-Shaw, H. K. (1966). Notes on Malaysian and other Asiatic Euphorbiaceae, LXVIII. New species of *Phyllanthus* L. Kew Bull. 20: 383-386. En. — 2 species, New Guinea.

Jablonski, E. (1967). *Phyllanthus*. Euphorbiaceae, Guayana Highland (Mem. New York Bot. Gard. 17(1)): 85-118. New York. En. — 58 species; key and extensive discussion with statistics. Only 6 species (mostly weedy) are common with the West Indies; the relationship of the two centres is otherwise not close. 19 species are high-altitude endemics, 13 described here for the first time.

Airy-Shaw, H. K. (1969). Notes on Malesian and other Asiatic Euphorbiaceae, XCIX. New or noteworthy species of *Phyllanthus* L. Kew Bull. 23: 26-40. En. — Treatment of 18 species (some new), with indication of sections; no keys.

Webster, G. (1970). A revision of *Phyllanthus* in the continental United States. Brittonia 22: 44-76, illus., maps. En. — Critical treatment of 12 species (8 native) with key, descriptions, synonymy, localities with exsiccatae, figures, maps and commentary. The general part includes in particular a biogeographical analysis. [Affinities are scattered with the native species being in sects. *Paraphyllanthus* and *Loxopodium* (subgen. *Isocladus*) and *Phyllanthus* (subgen. *Phyllanthus*); the last-named has two subsections at least partly native. Only *P. carolinensis* extends well northwards, reaching the Middle West as well as Pennsylvania.]

Airy-Shaw, H. K. (1971). Notes on Malesian and other Asiatic Euphorbiaceae, CXXV. New or noteworthy species of *Phyllanthus* L. Kew Bull. 25: 493-495. En. — 2 species, one new; Thailand and W Malaysia.

• Bancilhon, L. (1971). Contribution à l'étude taxonomique du genre *Phyllanthus*. 81 pp. (Boissiera 18). Geneva. (Originally part of a D.É. dissertation, Univ. de Paris-Orsay.) Fr. — A morphogenetic, organographic and architectural study and character analysis; a short historical review is also included. Among other areas of enquiry the progressions in plant architecture are examined using a range of species (but not as fully as might be desired). [See also Nozeran et al., 1984. The author here noted (p. 11) that Webster in his work of the 1950s and 1960s believed that 'malgre tout, .. la classification à laquelle il a abouti reflète encore très insuffisamment les vraies relations entre les différents sous-genres qu'il reconnus'.]

Webster, G. & H. K. Airy-Shaw (1971). A provisional synopsis of the New Guinea taxa of *Phyllanthus* (Euphorbiaceae). Kew Bull. 26: 85-109. En. — 35 species in 5 subgenera; key, synonymy, distribution and exsiccatae.

Airy-Shaw, H. K. (1972). Notes on Malesian and other Asiatic Euphorbiaceae, CLIV: A new *Phyllanthus* from New Guinea. Kew Bull. 27: 74-75. En. — From central New Guinea; named after one of the authors of *World Checklist and Bibliography of Euphorbiaceae*.

Airy-Shaw, H. K. (1974). Notes on Malesian and other Asiatic Euphorbiaceae, CLXXII. New or noteworthy species of *Phyllanthus* L. Kew Bull. 29: 294-296. En. — Novelties and notes, with for the novelties indication of sections.

Brunel, J. F. (1975). Contribution à l'étude de quelques *Phyllanthus* africaines et à la taxonomie du genre *Phyllanthus* L. (Euphorbiaceae). 2 fascicles (text, illustrations). Strasbourg. (Thèse du doctorat de spécialité (troisième cycle), Univ. Louis Pasteur, Strasbourg.) Fr. — An earlier version of the first part of Brunel (1987). The general part (pp. 7-39) encompasses an overall review of *Phyllanthus* with coverage of vegetative and reproductive features along with data from karyology, palynology, ecology and biology. The 'special' part comprises a critique of the infrageneric division of *Phyllanthus* with special reference to Africa (and particularly West Africa) as well as to the scheme of Webster, covering in all six subgenera; there are, however, no keys (save to eight sections in subgen. *Kirganelia*) nor any phenetic or phylogenetic diagrams. Infrageneric taxa from elsewhere in the range of the genus are only casually referred to. Bibliography, pp. 138-144. [The author has used this study mainly to bring into play the value of 'new' morphological and other features, but due to its geographical limitation it represents only a partial attempt towards a better understanding of the genus as a whole. Readers are better referred to his 1987 survey.]

Airy-Shaw, H. K. (1978). Notes on Malesian and other Asiatic Euphorbiaceae, CCIX. *Phyllanthus* L. Kew Bull. 33: 35-37. En. — Descriptions of three species from New Guinea, two in sect. *Nymania*.

Airy-Shaw, H. K. (1978). Notes on Malesian and other Asiatic Euphorbiaceae, CLXXXIX. New or noteworthy species of *Phyllanthus* L. Kew Bull. 32: 367-370. En. — Treatment of 3 species (one new); sections indicated.

Nozeran, R., L. Rossignol-Bancilhon & R. Haïcour (1978). Une espèce rudérale, pantropicale en cours de diversification, *Phyllanthus urinaria* L. Rev. Gén. Bot. 75: 201-210. Fr. — A biological study, with special reference to Vietnam populations exhibiting unusual 'polymorphy'. For a possible resolution, see Rossignol-Bancilhon et al. (1987).

Holm-Nielsen, L. B. (1979). Comments on the distribution and evolution of the genus *Phyllanthus* (Euphorbiaceae). In K. Larsen and L. B. Holm-Nielsen (eds), *Tropical Botany*: 277-290, maps. London: Academic Press. En. — Review of classification and biogeography of genus, as well as of special studies in embryology, cytology, and palynology (the latter in somewhat cursory fashion); isochor map of 'evolutionary centres' (p. 288); list of literature at end. [The greatest diversity at sectional level is in West Malesia, where 13 are represented. The Americas have two centres with 12 sections, while in Africa the maximum is in the south-central zone where 7 sections are present. The author indicates that 'the picture of possible subgeneric relationships .. may be the result of a reticulate evolutionary system'. This is to all intents and purposes a confession of ignorance or is indicative of underlying problems with available systems of the genus. (A modified Webster scheme with 10 subgenera, each mapped with all sections as accepted by the author, was adopted; see pp. 278-284. However, some of these sections have not been published!)]

Airy-Shaw, H. K. (1980). New or noteworthy Australian Euphorbiaceae, II. Muelleria 4: 207-241. En. —*Phyllanthus*, pp. 214-216; notes with on p. 214 the reduction of *Hexaspermum paniculatum* to *Phyllanthus clamboides*. [The genus was listed as 'incertae sedis' in Pax and Hoffmann, 1931.]

Airy-Shaw, H. K. (1980). Notes on Euphorbiaceae from Indomalesia, Australia and the Pacific, CCXXXII. *Phyllanthus* L. Kew Bull. 35: 386-389. En. — Descriptions of two new Australian species, one a segregate from a 'Bentham' species.

Brunel, J. F. & J. Roux (1981). *Phyllanthus* de Madagascar, I. À propos de deux *Phyllanthus* de la sous-section *Swartziani*. Adansonia, II, 20: 393-403, illus. Fr. —*P. amarus* in Madagascar in place of *P. niruri*; description of *P. andranovatensis* (related to *P. fraternus* Webster), with key distinguishing it from *P. amarus*.

Brunel, J. F. & J. Roux (1981). *Phyllanthus* de Madagascar, II. À propos du «complexe» *Phyllanthus nummulariaefolius* Poir.— *P. tenellus* Roxb. Bull. Mus. Natl. Hist. Nat., IV, B (Adansonia, III), 3: 185-199, illus. Fr. — A biosystematic study, including character analyses but lacking a key; the two species are distinct, the latter with an additional subspecies. References, p. 199.

Radcliffe-Smith, A. (1981). Notes on African Euphorbiaceae, IX. Kew Bull. 35: 763-777. En. — Includes many species of *Phyllanthus*. [Precursory to *Flora of Tropical East Africa*.]

Meewis, B. & W. Punt (1983). Pollen morphology and taxonomy of the subgenus *Kirganelia* (Juss.) Webster (genus *Phyllanthus*, Euphorbiaceae) from Africa. Rev. Palaeobot. Palynol. 39: 131-160, illus. En. — Treatment of 19 species (of 35) in 2 sections (*Floribundi* and *Anisonema*) with key to pollen types, list of vouchers, descriptions of pollen and commentary; scheme of relationships (p. 141). Taxonomic correlations follow on pp. 144-146 with particular reference to the 8 sections proposed (some not formally described) by Brunel (1975); the results from the present study were found to be only in partial agreement. The author indicated that 'a taxonomic treatment of the species at the sectional level is urgently needed' (p. 146).

Brunel, J. F. (1984). Southeast Asian Phyllantheae, II: Some *Phyllanthus* of sous-section *Swartziani*. Nordic J. Bot. 4: 469-473. En. — Detailed study of 3 species, with description of *P. airy-shawii* as new; figures but no map or key.

Nozeran, R., L. Rossignol-Bancilhon & G. Mangenot (1984). Les recherches sur le genre *Phyllanthus* (Euphorbiaceae): acquis et perspectives. Bot. Helvet. 94: 199-233. Fr. — Includes bibliography.

Rossignol-Bancilhon, L. & M. Rossignol (1985). Architecture et tendances évolutives dans le genre *Phyllanthus* (Euphorbiaceae). Bull. Mus. Natl. Hist. Nat., IV, B (Adansonia, III), 7: 67-80. Fr. — Evolution from the perspective of gross morphology and organization; emphasis on plant and shoot architecture, with comparisons with the architectural classification (in 4 types) of Webster (1956-58) and an attempt to explore, partly through experimentation, possible mechanisms relating to observed phyletic trends (among them chemical substances with a bacteriostatic action and trophic movements). On the other hand, no relationship was found between pollen, floral and vegetative trends, suggesting strong dyschrony. No formal taxonomic treatment is presented.

Brunel, J. F. (1987). *Plagiocladus*. In *idem*, Sur le genre *Phyllanthus* L. et quelques genres voisins de la tribu des Phyllantheae Dumort. Strasbourg. Fr. — Description of this (new) genus, a segregate of *Phyllanthus*; 1 species. [Not validly published and not currently supported as distinct from the larger genus.]

• Brunel, J. F. (1987). Sur le genre *Phyllanthus* L. et quelques genres voisins de la tribu des Phyllantheae Dumort. (Euphorbiaceae, Phyllantheae) en Afrique intertropicale et à Madagascar. 472, 196 pp., illus., plates, maps. Strasbourg. (Thèse du doctorat, Sciences Naturelles, Univ. Louis Pasteur, Strasbourg.) Fr. — The main part of this work includes, firstly, a historical review covering the various approaches of Linnaeus, de Jussieu, Baillon, Mueller, Pax and Webster (for a synoptic key to the last-named's system, see pp. 36-37). The historical review is followed by a survey of the whole genus (pp. 41-97) on the basis of the Webster system (42 sections in 10 subgenera). Chapters 3-7 encompass studies and evaulations of different morphological and other characters (with Chapter 3 covering vegetative morphology including plant architecture — 6 groups with 16 types recognised — and Chapter 5 on karyological and palynological evidence). Chapter 8 comprises a revised survey of the genus and its allies (listed on p. 241) for Africa and Madagascar, while chapter 9 covers phytogeography (with many maps). Conclusions and 'perspectives' may be found on pp. 451-472, with an essay into taximetry and (pp. 470-471) presentation of a dendrogram. A full illustrated descriptive treatment (only of the tropical African species), without keys but with citations of exsiccatae, appears in the Annex (for synoptic lists, see pp. 1-2); a bibliography of 360 references and an index conclude the work. [In the review of Webster's system this work has the most useful illustrated introduction to the genus available. There is, however, little or no evolutionary or phylogenetic analysis and the phytogeographical treatment is largely limited to Africa.]

Punt, W. (1987). A survey of pollen morphology in Euphorbiaceae with special reference to *Phyllanthus*. Bot. J. Linn. Soc. 94: 127-142, illus. En. — Introduction with review of scope and quality of sources; survey of pollen morphology in relation to the family system of Webster (1975); systematic survey of pollen in *Phyllanthus* in relation to the subgeneric scheme of Webster (9 covered). [A great variety of pollen types was described, but no conclusions drawn.]

Rossignol-Bancilhon, L., M. Rossignol & R. Haïcour (1987). A systematic revision of *Phyllanthus* subsection *Urinaria* (Euphorbiaceae). Amer. J. Bot. 74: 1853-1862. En. — A biosystematic study in extension of Nozeran et al. (1978). Recognition of 4 species and 1 additional subspecies centered around *P. urinaria*, with deswcriptions of 2 novelties; keys to subsections of sect. *Urinaria* and to species in the subsection. [Covers Africa, Madagascar and Asia.]

Moreira Santiago, L. J. (1988). Estudos preliminares da seção *Choretropsis* Müll. Arg., gênero *Phyllanthus* L. (Euphorbiaceae). Bradea 5(2): 44-49, 2 illus. Pt. — Transfer of S American species of sect. *Xylophylla* to sect. *Choretropsis* as a new subsection *Applanata*; 2 new species also described. [One of the new species, *P. goianensis*, is an erect shrub with slender photosynthetic branches and flowers in extremely congested cyathium-like axillary inflorescences. It is related to *P. spartioides*. The other species, *P. edmundoi*, has flattened cladode-branches like *P. angustifolius*; it is related to *P. klotzschianus.* — Sect. *Choretropsis* is one of those treated as *incertae sedis* by Brunel (1987) as it was not then covered by Webster; his view was that it was probably in subgen. *Phyllanthus* and thus *convergent* with respect to sect. *Xylophylla* in a stricter sense. In the Mueller system as used in *Flora brasiliensis* the two sections were adjacent.]

Schmid, M. (1991). *Phyllanthus.* Fl. Nouvelle-Calédonie, 17 (Euphorbiacées, II): 31-320. Paris. Fr. — Flora treatment (116 species: 1 insufficiently known, 111 indigenous, 3 certainly synanthropic); keys.

Radcliffe-Smith, A. (1992). Notes on African Euphorbiaceae, XXVIII. Kew Bull. 47: 677-683, illus. En. — Pp. 679-681 on *Phyllanthus in E. and SC. Africa; includes two new species and one new nothospecies.*

Amaral, L. de G. & M. Ulysséa (1993(1994)). Considerações sobre a identificação de espécies de *Phyllanthus* (quebra-pedra). Insula 22: 21-38, illus. Pt. — Notes on identification, in aid of studies of medicinal and other properties of *PP. tenellus, niruri* and *urinaria.* Results are presented in the form of a comparative table (p. 29) enabling multi-access, but no analytical dichotomous keys appear. [These species, all present on the campus of the Unversidade Federal de Santa Catarina, are widespread weeds and this study may well be found useful elsewhere.]

Haïcour, R. et al. (1994). Patterns of diversification and evolution in *Phyllanthus odontadenius* (Euphorbiaceae). Ann. Missouri Bot. Gard. 81: 289-301, illus. En. — A biosystematic study emphasizing comparative morphology and genetics, with an examination of levels of reproductive incompatability in relation to morphological distinction. A case is made for viewing *P. odontadenius* sens. lat. and *P. urinaria* sens. lat. as complexes of closely related species (p. 298); dendrogram of possible evolutionary directions between and among the complexes, p. 300. No keys or formal taxonomic treatment are presented. [Subsect. Odontadenii.]

Hunter, J. T. & J. J. Bruhl (1996). Three new species of *Phyllanthus* (Euphorbiaceae: Phyllantheae) in South Australia. J. Adelaide Bot. Gard. 17: 127-136, illus. En. — General remarks; descriptions of novelties (with very full documentation); key to *Phyllanthus* in South Australia. [Precursory to a collectively authored treatment of the family for *Flora of Australia.*]

Radcliffe-Smith, A. (1996). Notes on African Euphorbiaceae, XXX. *Phyllanthus* (v) & c. Kew Bull. 51: 301-331, illus. En. — Notes and novelties preparatory to *Flora Zambesiaca*; includes (pp. 305-330) 25 new taxa of *Phyllanthus* of which 18 are new species. [Many of the new species were initially proposed by Brunel (1987) but not effectively published by him.]

Radcliffe-Smith, A. & R. Govaerts (1997). New names and combinations in Euphorbiaceae-Phyllanthoïdeae. Kew Bull. 51: 175-178. En. — Miscellaneous names and combinations, many in *Phyllanthus* including transfer of taxa formerly in *Phyllanthodendron.*

Phyllanthus L., Sp. Pl.: 981 (1753).
Trop. & Subtrop. 20 21 22 23 24 25 26 27 28 29 34 35 36 38 40 41 42 50 60 61 62 63 77 78 79 80 81 82 83 84 85.
Niruri Adans., Fam. Pl. 2: 356 (1763).
Cicca L., Mant. Pl. 1: 124 (1767).
Xylophylla L., Mant. Pl. 2: 147 (1771).
Conami Aubl., Hist. Pl. Guiane 2: 926 (1775).
Urinaria Medik., Malvenfam.: 80 (1787).
Genesiphylla L'Hér., Sert. Angl.: 29 (1788).

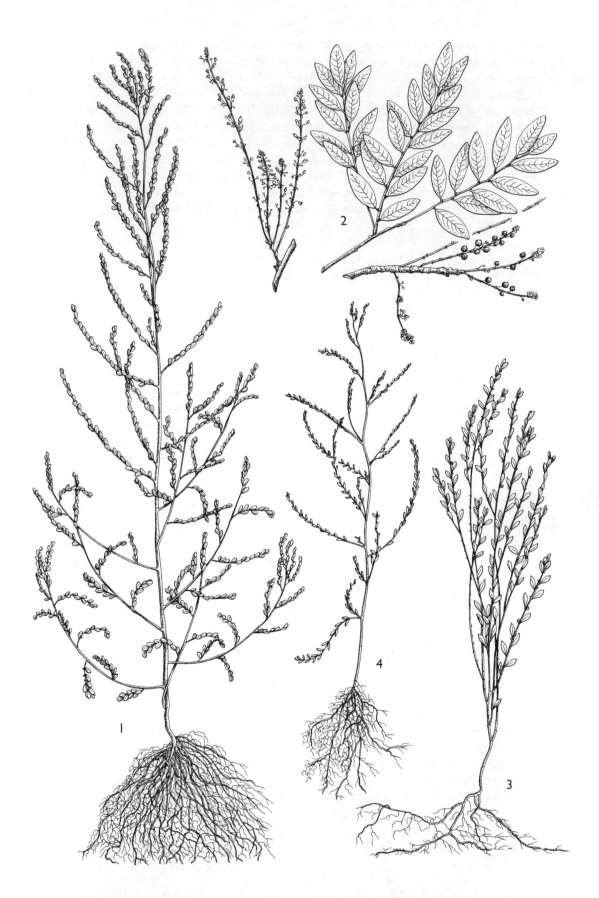

Phyllanthus amarus Schumach. & Thonn. in C.F. Schumacher (1), *Phyllanthus casticum* P. Willemet (2), *Phyllanthus maderaspatensis* L. var. *frazieri* Fosberg (3), *Phyllanthus mackenzei* Fosberg (4)

Artist: Ann Davies
Fosberg & Renvoize, *Fl. Aldabra*: 241, fig. 38 (1980)
KEW ILLUSTRATIONS COLLECTION

Kirganelia Juss., Gen. Pl.: 387 (1789).

Cathetus Lour., Fl. Cochinch.: 607 (1790).

Emblica Gaertn., Fruct. Sem. Pl. 2: 122 (1790).

Nymphanthus Lour., Fl. Cochinch.: 543 (1790).

Tephranthus Neck., Elem. Bot. 2: 235 (1790).

Tricarium Lour., Fl. Cochinch.: 557 (1790).

Rhopium Schreb., Gen. Pl.: 608 (1791).

Epistylium Sw., Fl. Ind. Occid. 2: 1095 (1800).

Cycca Batsch, Tab. Affin. Regni Veg.: 168 (1802), orth. var.

Anisonema A.Juss., Euphorb. Gen.: 19 (1824).

Ardinghalia Comm. ex A.Juss., Euphorb. Gen.: 21 (1824).

Geminaria Raf., First Cat. Gard. Transylv. Univ.: 14 (1824).

Menarda Comm. ex A.Juss., Euphorb. Gen.: 23 (1824).

Synexemia Raf., Neogenyton: 2 (1825).

Scepasma Blume, Bijdr.: 582 (1826).

Genesiphyla Raf., Sylva Tellur.: 92 (1838).

Hexadena Raf., Sylva Tellur.: 92 (1838).

Lomanthes Raf., Sylva Tellur.: 92 (1838).

Moeroris Raf., Sylva Tellur.: 91 (1838).

Nellica Raf., Sylva Tellur.: 92 (1838).

Niruris Raf., Sylva Tellur.: 91 (1838).

Frankia Steud., Nomencl. Bot., ed. 2, 1: 646 (1840).

Asterandra Klotzsch, Arch. Naturgesch. 7: 200 (1841).

Eriococcus Hassk., Tijdschr. Natuurl. Gesch. Physiol. 10: 143 (1843).

Ceramanthus Hassk., Cat. Hort. Bot. Bogor.: 240 (1844).

Macraea Wight, Icon. Pl. Ind. Orient. 5(2): 27 (1852).

Reidia Wight, Icon. Pl. Ind. Orient. 5(2): 27 (1852).

Chorisandra Wight, Icon. Pl. Ind. Orient. 6: 13 (1853).

Dichelactina Hance in W.G.Walpers, Ann. Bot. Syst. 3: 375 (1853).

Staurothyrax Griff., Not. Pl. Asiat. 4: 476 (1854).

Hemicicca Baill., Étude Euphorb.: 645 (1858).

Orbicularia Baill., Étude Euphorb.: 616 (1858).

Oxalistylis Baill., Étude Euphorb.: 628 (1858).

Williamia Baill., Étude Euphorb.: 559 (1858).

Dichrophyllum Klotzsch & Garcke, Monatsber. Königl. Preuss. Akad. Wiss. Berlin 1859: 249 (1859).

Clambus Miers, Ann. Mag. Hist. Nat., III, 13: 123 (1864).

Leichhardtia F.Muell., Fragm. 10: 67 (1876).

Diasperus L. ex Kuntze, Rev. Gen. Pl. 2: 596 (1891).

Phyllanthodendron Hemsl., Hooker's Icon. Pl. 26: t. 2563 (1898).

Aporosella Chodat, Bull. Herb. Boissier, II, 5: 488 (1905).

Flueggeopsis K.Schum. in K.M.Schumann & C.A.G.Lauterbach, Fl. Schutzgeb. Südsee, Nachtr.: 289 (1905).

Hemiglochidion K.Schum. in K.M.Schumann & C.A.G.Lauterbach, Fl. Schutzgeb. Südsee, Nachtr.: 289 (1905).

Nymania K.Schum. in K.M.Schumann & C.A.G.Lauterbach, Fl. Schutzgeb. Südsee, Nachtr.: 291 (1905).

Uranthera Pax & K.Hoffm. in H.G.A.Engler, Pflanzenr., IV, 147, III: 95 (1911).

Dimorphocladium Britton, Mem. Torrey Bot. Club 16: 74 (1920).

Ramsdenia Britton, Mem. Torrey Bot. Club 16: 72 (1920).

Roigia Britton, Mem. Torrey Bot. Club 16: 73 (1920).

Dendrophyllanthus S.Moore, J. Linn. Soc., Bot. 45: 395 (1921).

Pseudoglochidion Gamble, Bull. Misc. Inform. Kew 1925: 329 (1925).

Hexaspermum Domin, Biblioth. Bot. 89: 315 (1927).

Arachnodes Gagnep., Notul. Syst. (Paris) 14: 32 (1950).

Phyllanthus abditus G.L.Webster, Contr. Gray Herb. 176: 50 (1955).
SW. Haiti (Massif de la Hotte). 81 HAI. Cham.

Phyllanthus abnormis Baill., Adansonia 1: 42 (1860).
Texas, Florida, N. Mexico. 77 TEX 78 FLA 79 MXE. Ther.

> var. **abnormis**
> Texas, Florida, N. Mexico. 77 TEX 78 FLA 79 MXE. Ther.
> *Phyllanthus drummondii* Small, Fl. S.E. U.S.: 692 (1903).
> *Phyllanthus garberi* Small, Fl. S.E. U.S.: 692 (1903).

> var. **riograndensis** G.L.Webster, Ann. Missouri Bot. Gard. 54: 198 (1967).
> SE. Texas. 77 TEX. Ther.

Phyllanthus acacioides Urb., Symb. Antill. 3: 287 (1902).
Tobago. 81 TRT. Nanophan. or phan.

Phyllanthus acidus (L.) Skeels, U.S.D.A. Bur. Pl. Industr. Bull. 148: 17 (1909).
Brazil (Pará), naturalised elswere. (81) bah jam lee pue win 84 BZN (25) tan uga (40) pak ind (41) tha (42) bor phi. Nanophan. or phan. – Fruit edible.
> * *Averrhoa acida* L., Sp. Pl.: 428 (1753). *Cicca acida* (L.) Merr., Interpr. Herb. Amboin.: 314 (1917).
> *Cicca disticha* L., Mant. Pl. 1: 124 (1767). *Phyllanthus distichus* (L.) Müll.Arg. in A.P.de Candolle, Prodr. 15(2): 413 (1866).
> *Cicca nodiflora* Lam., Encycl. 2: 1 (1786). *Phyllanthus distichus* f. *nodiflorus* (Lam.) Müll.Arg. in A.P.de Candolle, Prodr. 15(2): 414 (1866).
> *Cicca racemosa* Lour., Fl. Cochinch.: 556 (1790).
> *Tricarium cochinchinense* Lour., Fl. Cochinch.: 557 (1790). *Phyllanthus cochinchinensis* (Lour.) Müll.Arg. in A.P.de Candolle, Prodr. 15(2): 417 (1866), nom. illeg.
> *Phyllanthus longifolius* Jacq., Pl. Hort. Schoenbr. 2: 36 (1797).
> *Cicca acidissima* Blanco, Fl. Filip.: 700 (1837). *Phyllanthus acidissimus* (Blanco) Müll.Arg., Linnaea 32: 50 (1863), nom. illeg.
> *Phyllanthus cicca* Müll.Arg., Linnaea 32: 50 (1863).
> *Phyllanthus cicca* var. *bracteosa* Müll.Arg., Linnaea 32: 50 (1863).

Phyllanthus acinacifolius Airy Shaw & G.L.Webster, Kew Bull. 26: 95 (1971).
NE. Papua New Guinea. 42 NWG. Nanophan.

Phyllanthus actephilifolius J.J.Sm., Nova Guinea 12: 543 (1917).
Irian Jaya. 42 NWG. Phan.

Phyllanthus acuminatus Vahl, Symb. Bot. 2: 95 (1791).
Caribbean, Mexico to Argentina. 79 MXE MXN MXS MXT 80 BLZ COS ELS GUA HON PAN 81 CUB JAM LEE TRT WIN 82 GUY SUR VEN 83 CLM PER 84 BZL BZN 85 AGE AGW PAR. Nanophan. or phan. – Used as fish poison.
> *Phyllanthus lycioides* Kunth in F.W.H.von Humboldt, A.J.A.Bonpland & C.S.Kunth, Nov. Gen. Sp. 2: 112 (1817).
> *Phyllanthus mucronatus* Kunth in F.W.H.von Humboldt, A.J.A.Bonpland & C.S.Kunth, Nov. Gen. Sp. 2: 112 (1817).
> *Phyllanthus ruscoides* Kunth in F.W.H.von Humboldt, A.J.A.Bonpland & C.S.Kunth, Nov. Gen. Sp. 2: 113 (1817).
> *Phyllanthus averrhoifolius* Steud., Nomencl. Bot., ed. 2, 2: 326 (1841).
> *Phyllanthus cumanensis* Willd. ex Steud., Nomencl. Bot., ed. 2, 2: 327 (1841).
> *Phyllanthus foetidus* Pav. ex Baill., Adansonia 1: 34 (1860), pro syn.

Phyllanthus acutifolius Poir. ex Spreng., Syst. Veg. 3: 21 (1826).
SE. Brazil. 84 BZL. Cham.
> *Phyllanthus oxyphyllus* Müll.Arg., Linnaea 32: 40 (1863).

Phyllanthus acutissimus Miq., Fl. Ned. Ind. 1(2): 369 (1859).
S. Burma, Thailand, Sumatera, Jawa. 41 BMA THA 42 JAW SUM. Nanophan.
Scepasma longifolium Hassk., Tijdschr. Natuurl. Gesch. Physiol. 10: 143 (1843).
Phyllanthus curtipes Airy Shaw, Kew Bull. 23: 37 (1969).

Phyllanthus adenodiscus Müll.Arg., Linnaea 32: 23 (1863).
Mexico (San Luis Potosí, Veracruz). 79 MXE MXG. Nanophan.
Phyllanthus glaucescens Schltdl. & Cham., Linnaea 6: 364 (1831), nom. illeg.

Phyllanthus adenophyllus Müll.Arg., Linnaea 32: 24 (1863).
Brazil (Amazonas). 84 BZN.

Phyllanthus adianthoides Klotzsch, London J. Bot. 2: 51 (1843).
Guyana, Surinam. 82 GUY SUR. Nanophan.

Phyllanthus aeneus Baill., Adansonia 2: 231 (1862). *Glochidion aeneum* (Baill.) Müll.Arg., Linnaea 32: 59 (1863).
New Caledonia. 60 NWC. Nanophan. or phan.

var. **aeneus**
New Caledonia. 60 NWC. Nanophan. or phan.
Glochidion aeneum (Baill.) Müll.Arg., Linnaea 32: 59 (1863).
Phyllanthus durus S.Moore, J. Linn. Soc., Bot. 45: 396 (1921).
Phyllanthus maytenifolius S.Moore, J. Linn. Soc., Bot. 45: 398 (1921).
Phyllanthus salacioides S.Moore, J. Linn. Soc., Bot. 45: 398 (1921).
Phyllanthus olivaceus Müll.Arg. ex Guillaumin, Arch. Bot. Mém. 2(3): 3 (1929).

var. **cordifolius** M.Schmid, in Fl. N. Caled. & Depend. 17: 208 (1991).
SE. New Caledonia (Yaté-Ounia Reg.). 60 NWC. Nanophan.

var. **longistylis** M.Schmid, in Fl. N. Caled. & Depend. 17: 206 (1991).
SE. New Caledonia. 60 NWC. Nanophan.

var. **nepouiensis** M.Schmid, in Fl. N. Caled. & Depend. 17: 208 (1991).
New Caledonia (Népoui Pen.). 60 NWC. Nanophan.

var. **papillosus** M.Schmid, in Fl. N. Caled. & Depend. 17: 209 (1991).
New Caledonia (Mt. Koniambo). 60 NWC. Nanophan.

Phyllanthus affinis Müll.Arg., Linnaea 32: 48 (1863). *Reidia affinis* (Müll.Arg.) Alston in H.Trimen, Handb. Fl. Ceylon 6(Suppl.): 258 (1931).
Sri Lanka. 40 SRL.

Phyllanthus airy-shawii Brunel & J.P.Roux, Nordic J. Bot. 4: 470 (1984).
N. Thailand. 41 THA. Ther. – Related to *P. debilis*.

Phyllanthus albidiscus (Ridl.) Airy Shaw, Kew Bull. 23: 26 (1969).
Thailand (incl. Lankawi I.) to Jawa. 41 THA 42 JAW MLY. Nanophan.
Ceramanthus gracilis Hassk., Cat. Hort. Bot. Bogor.: 240 (1844). *Phyllanthus gracilis* (Hassk.) Baill., Étude Euphorb.: 630 (1858), nom. illeg. *Phyllanthus ceramanthus* G.L.Webster, J. Arnold Arbor. 37: 233 (1956).
Cleistanthus albidiscus Ridl., Bull. Misc. Inform. Kew 1923: 360 (1923).

Phyllanthus albizzioides (Kurz) Hook.f., Fl. Brit. India 5: 289 (1887).
Burma. 41 BMA.
Cicca albizzioides Kurz, J. Asiat. Soc. Bengal, Pt. 2, Nat. Hist. 42(2): 239 (1873).

Phyllanthus albolapidosi Leandri, Mém. Inst. Sci. Madagascar, Sér. B, Biol. Vég. 8: 224 (1957).
W. Madagascar. 29 MDG. Nanophan.

Phyllanthus oxycoccifolius Hutch. (1), *P. retinervis* Hutch. (2), *P. crassinervius* Radcl.-Sm. (3), *P. harrisii* Radcl.-Sm. (4), *P. irriguus* Radcl.-Sm. (5), *P. micromeris* Radcl.-Sm. (6), *P. rhizomatosus* Radcl.-Sm. (7), *P. ukagurensis* Radcl.-Sm. (8) & *P. wingfieldii* Radcl.-Sm. (9)

Artist: Christine Grey-Wilson
Fl. Trop. East Africa, Euphorbiaceae 1: 22, fig. 3 (1987)

Phyllanthus almadensis Müll.Arg. in C.F.P.von Martius, Fl. Bras. 11(2): 38 (1873).
Brazil (Bahia). 84 BZE.

Phyllanthus alpestris Beille, Bull. Soc. Bot. France 55(8): 56 (1908).
Guinea, Sierra Leone, Liberia, Ivory Coast. 22 GUI IVO LBR SIE. Cham.
Phyllanthus leonensis Hutch., Bull. Misc. Inform. Kew 1917: 232 (1917).
Phyllanthus monticola Hutch. & Dalziel, Fl. W. Trop. Afr. 1: 291 (1928).

Phyllanthus amarus Schumach. & Thonn. in C.F.Schumacher, Beskr. Guin. Pl.: 421 (1827). –
FIGURE, p. 1262. *Phyllanthus niruri* var. *amarus* (Schumach. & Thonn.) Leandri, in Fl. Madag.
111: 73 (1958).
Florida, Mexico, Trop. America, introduced elsewere. 78 FLA 79 MXE 80 ALL 81 ALL 82
ALL 82 BOL CLM ECU PER 84 ALL 85 AGE CLN PAR URU (2) all (29) mdg (36) chc chh
chs (38) tai (41) tha (42) bor jaw mly nwg phi (50) nsw qld wau (60) nwc (61) sci. Ther.
Phyllanthus swartzii Kostel., Allg. Med.-Pharm. Fl. 5: 1771 (1836).
Phyllanthus scabrellus Webb in W.J.Hooker, Niger Fl.: 175 (1849). *Phyllanthus niruri* var.
scabrellus (Webb) Müll.Arg., Linnaea 32: 43 (1863).
Phyllanthus nanus Hook.f., Fl. Brit. India 5: 298 (1887).
Phyllanthus amarus var. *baronianus* Leandri, Notul. Syst. (Paris) 7: 183 (1939).
Phyllanthus niruri var. *baronianus* (Leandri) Leandri, in Fl. Madag. 111: 74 (1958).
Phyllanthus niruroides var. *madagascariensis* Leandri, Notul. Syst. (Paris) 7: 184 (1939).

Phyllanthus ambatovolanus Leandri, Notul. Syst. (Paris) 6: 191 (1938).
E. Madagascar. 29 MDG. Nanophan.

Phyllanthus americanus Sessé & Moç., Pl. Nov. Hisp.: 159 (1890).
Mexico (Sinaloa). 79 MXN.

Phyllanthus amicorum G.L.Webster, Pacific Sci. 40: 100 (1986 publ. 1988).
Tonga. 60 TON.

Phyllanthus amieuensis Guillaumin, Mém. Mus. Natl. Hist. Nat., B, Bot. 8: 242 (1962).
New Caledonia (Col d'Amieu). 60 NWC. Nanophan.

Phyllanthus amnicola G.L.Webster, Contr. Gray Herb. 176: 54 (1955).
Dominican Rep. (Monte Cristi). 81 DOM. Ther. or cham.

Phyllanthus amoenus Müll.Arg. in C.F.P. de Martius, Fl. Bras. 11(2): 66 (1873).
Brazil (Minas Gerais). 84 BZL.

Phyllanthus ampandrandavae Leandri, Mém. Inst. Sci. Madagascar, Sér. B, Biol. Vég. 8:
224 (1957).
W. Madagascar. 29 MDG. Ther.

Phyllanthus anabaptizatus Müll.Arg. in A.P.de Candolle, Prodr. 15(2): 421 (1866).
Sri Lanka. 40 SRL.
* *Reidia polyphylla* Wight, Icon. Pl. Ind. Orient. 5(2): 28, t. 1904(4) (1852). *Epistylium*
polyphyllum (Wight) Thwaites, Enum. Pl. Zeyl.: 283 (1861).
Epistylium zeylanicum Baill., Étude Euphorb.: 648 (1858), nom. nud. *Phyllanthus*
zeylanicus Müll.Arg., Linnaea 32: 49 (1863).

Phyllanthus anadenus Jabl., Mem. New York Bot. Gard. 17: 93 (1967).
Venezuela (Amazonas: Cerro Sipapo). 82 VEN. Nanophan.

Phyllanthus analamerae Leandri, Mém. Inst. Sci. Madagascar, Sér. B, Biol. Vég. 8: 225 (1957).
W. Madagascar (Diego-Suarez). 29 MDG. Nanophan. or phan.

Phyllanthus anamalayanus (Gamble) G.L.Webster, Ann. Missouri Bot. Gard. 81: 45 (1994).
India (Anamalai Hills). 40 IND.
   * *Pseudoglochidion anamalayanum* Gamble, Bull. Misc. Inform. Kew 1925: 330 (1925).

Phyllanthus andalangiensis Leandri, Mém. Inst. Sci. Madagascar, Sér. B, Biol. Vég. 8: 226 (1957).
C. & E. Madagascar. 29 MDG. Nanophan.

Phyllanthus andamanicus N.P.Balakr. & N.G.Nair, Bull. Bot. Surv. India 24: 35 (1982 publ. 1983). *Phyllanthus balakrishnairii* Govaerts & Radcl.-Sm., Kew Bull. 51: 176 (1996), nom. illeg.
Andaman Is. 41 AND.

Phyllanthus anderssonii Müll.Arg. in A.P.de Candolle, Prodr. 15(2): 395 (1866).
Windward Is. (Barbados). 81 WIN. Nanophan.
   Phyllanthus barbadensis Urb., Symb. Antill. 3: 287 (1902).

Phyllanthus andranovatensis Brunel & J.P.Roux, Adansonia, n.s., 20: 400 (1981).
SE. Madagascar. 29 MDG.

Phyllanthus angavensis (Leandri) Leandri, Mém. Inst. Sci. Madagascar, Sér. B, Biol. Vég. 8: 222 (1957).
SW. Madagascar. 29 MDG. Nanophan.
   * *Phyllanthus casticum* subsp. *angavensis* Leandri, Notul. Syst. (Paris) 7: 174, 178 (1939).

Phyllanthus angkorensis Beille in H.Lecomte, Fl. Indo-Chine 5: 583 (1927).
Cambodia. 41 CBD.

Phyllanthus angolensis Müll.Arg., J. Bot. 2: 329 (1864).
Malawi, Zambia, Angola. 26 ANG MLW ZAM. Ther. or cham.

Phyllanthus angustatus Hutch. in R.E.Fries, Wiss. Erg. Schwed. Rhod.-Kongo Exped. 1: 121 (1911-1912 publ. 1914).
Zambia. 26 ZAM. Hemicr.

Phyllanthus angustifolius (Sw.) Sw., Fl. Ind. Occid. 2: 1111 (1800).
Caribbean. 81 CAY CUB DOM JAM SWC. Nanophan.
   * *Xylophylla angustifolia* Sw., Prodr.: 28 (1788). *Phyllanthus angustifolius* var. *genuinus* Müll.Arg. in A.P.de Candolle, Prodr. 15(2): 430 (1866), nom. inval.
   Phyllanthus cognatus Spreng., Syst. Veg. 3: 23 (1826). *Phyllanthus linearis* var. *cognatus* (Spreng.) Müll.Arg. in A.P.de Candolle, Prodr. 15(2): 430 (1866).
   Xylophylla contorta Britton, Bull. Torrey Bot. Club 37: 353 (1910).

Phyllanthus angustissimus Müll.Arg., Linnaea 32: 55 (1863).
Brazil (Minas Gerais). 84 BZL. Cham.

Phyllanthus anisophyllioides Merr., Philipp. J. Sci. 26: 465 (1925).
Philippines. 42 PHI.

Phyllanthus ankarana Leandri, Bull. Soc. Bot. France 81: 543 (1934).
W. Madagascar. 29 MDG. Cham. or nanophan.

Phyllanthus annamensis Beille in H.Lecomte, Fl. Indo-Chine 5: 585 (1927).
Hainan, Vietnam. 36 CHH 41 VIE. Nanophan. or phan.

Phyllanthus anthopotamicus Hand.-Mazz., Symb. Sin. 7: 223 (1931). *Phyllanthodendron anthopotamicum* (Hand.-Mazz.) Croizat, J. Arnold Arbor. 23: 37 (1942).
S. China. 36 CHC CHS. Nanophan.

Phyllanthus aoraiensis Nadeaud, Enum. Pl. Tahiti: 73 (1873).
Society Is. (Tahiti). 61 SCI. Nanophan.

Phyllanthus aoupinieensis M.Schmid, in Fl. N. Caled. & Depend. 17: 275 (1991).
New Caledonia (Massif de l'Aoupinié). 60 NWC. Nanophan.

Phyllanthus aphanostylus Airy Shaw & G.L.Webster, Kew Bull. 26: 106 (1971).
New Guinea (incl. D'Entrecasteaux Is.). 42 NWG. Phan.

   var. **aphanostylus**
      Papua New Guinea (incl. D'Entrecasteaux Is.). 42 NWG. Phan.

   var. **tristis** Airy Shaw, Kew Bull., Addit. Ser. 8: 183 (1980).
      Irian Jaya, NW. Papua New Guinea. 42 NWG. Phan.

Phyllanthus apiculatus Merr., Philipp. J. Sci. 16: 540 (1920).
Philippines. 42 PHI.

Phyllanthus arachnodes Govaerts & Radcl.-Sm., Kew Bull. 51: 175 (1996).
Cambodia. 41 CBD.
    * *Arachnodes chevalieri* Gagnep., Notul. Syst. (Paris) 14: 32 (1950). *Phyllanthodendron chevalieri* (Gagnep.) Airy Shaw, Kew Bull. 14: 471 (1960). *Phyllanthus chevalieri* (Gagnep.) G.L.Webster, Ann. Missouri Bot. Gard. 81: 45 (1994), nom. illeg.

Phyllanthus arbuscula (Sw.) J.F.Gmel., Syst. Nat. 2: 204 (1791).
Jamaica. 81 JAM. Nanophan. or phan.
    Xylophylla angustifolia var. *linearis* Sw., Prodr.: 28 (1788). *Phyllanthus linearis* (Sw.) Sw., Fl. Ind. Occid. 2: 1113 (1800). *Xylophylla linearis* (Sw.) Steud., Nomencl. Bot., ed. 2, 2: 793 (1841).
    * *Xylophylla arbuscula* Sw., Prodr.: 28 (1788).
    Phyllanthus speciosus Jacq., Collectanea 2: 360 (1789). *Genesiphylla speciosa* (Jacq.) Raf., Sylva Tellur.: 92 (1838).
    Phyllanthus coxianus Fawc., J. Bot. 57: 66 (1919).
    Phyllanthus inaequaliflorus Fawc. & Rendle, J. Bot. 57: 66 (1919).
    Phyllanthus swartzii Fawc. & Rendle, J. Bot. 57: 66 (1919), nom. illeg.
    Phyllanthus dingleri G.L.Webster, J. Arnold Arbor. 37: 413 (1956).

Phyllanthus archboldianus Airy Shaw & G.L.Webster, Kew Bull. 26: 88 (1971).
New Guinea. 42 NWG. Nanophan. or phan.

Phyllanthus ardisianthus Airy Shaw & G.L.Webster, Kew Bull. 26: 94 (1971).
Irian Jaya. 42 NWG. Nanophan. or phan.

Phyllanthus arenarius Beille in H.Lecomte, Fl. Indo-Chine 5: 587 (1927).
Vietnam, S. China. 36 CHC CHH CHS 41 VIE. Hemicr.

   var. **arenarius**
      Vietnam, China (Guangdong, Hainan). 36 CHH CHS 41 VIE. Hemicr.

   var. **yunnanensis** T.L.Chin, Acta Phytotax. Sin. 19: 350 (1981).
      China (SW. Yunnan). 36 CHC. Hemicr.

Phyllanthus arenicola Casar., Nov. Stirp. Bras.: 88 (1845).
SE. Brazil. 84 BZL. Cham.

Phyllanthus argyi H.Lév., Mem. Real Acad. Ci. Barcelona 12(22): 10 (1916).
S. China. 36 CHS.

Phyllanthus aridus Benth., Fl. Austral. 6: 110 (1873).
N. Western Australia, Northern Territory. 50 NTA WAU.
  Phyllanthus polycladus W.Fitzg., J. Roy. Soc. W. Australia 3: 164 (1918), nom. illeg.
  Phyllanthus hesperonotos Govaerts & Radcl.-Sm., Kew Bull. 51: 177 (1996).

Phyllanthus armstrongii Benth., Fl. Austral. 6: 112 (1873).
Northern Territory. 50 NTA. Nanophan. or phan.

Phyllanthus arnhemicus S.Moore, J. Linn. Soc., Bot. 45: 215 (1920).
Northern Territory. 50 NTA.

Phyllanthus artensis M.Schmid, in Fl. N. Caled. & Depend. 17: 242 (1991).
New Caledonia (N. Ile Art). 60 NWC. Nanophan.

Phyllanthus arvensis Müll.Arg., J. Bot. 2: 332 (1864).
Zaire to S. Africa. 23 ZAI 25 TAN 26 ANG MLW ZAM ZIM 27 NAT. Cham. or nanophan.

Phyllanthus aspersus Brunel & J.P.Roux, Willdenowia 14: 386 (1984 publ. 1985).
Cameroon. 23 CMN.

Phyllanthus asperulatus Hutch., Bull. Misc. Inform. Kew 1920: 27 (1920).
Northern Prov., Botswana, Mozamboque, Zimbabwe. 26 MOZ ZIM 27 BOT TVL.

Phyllanthus atabapoensis Jabl., Mem. New York Bot. Gard. 17: 110 (1967).
Brazil (Amazonas), Venezuela (Amazonas). 82 VEN 84 BZN. Nanophan. or phan.

Phyllanthus attenuatus Miq., Linnaea 21: 479 (1848).
Guiana, Surinam, Guyana, Venezuela (Amazonas), Brazil (Roraima, Para, Maranhao). 82
  FRG GUY SUR VEN 84 BZE BZN. Nanophan.
  Phyllanthus guianensis auct.; Müll.Arg. in A.P.de Candolle, Prodr. 15(2): 376 (1866),
  misappl.

Phyllanthus augustinii Baill., Adansonia 5: 354 (1865).
SE. Brazil. 84 BZL. Cham.

Phyllanthus australis Hook.f., London J. Bot. 6: 284 (1847).
S. & E. Australia, Tasmania. 50 QLD SOA TAS VIC.

Phyllanthus austroparensis Radcl.-Sm., Kew Bull. 47: 679 (1992).
Tanzania. 25 TAN.

Phyllanthus avanguiensis M.Schmid, in Fl. N. Caled. & Depend. 17: 182 (1991).
New Caledonia (Mt. Boulinda). 60 NWC. Nanophan.

Phyllanthus avicularis Müll.Arg., Linnaea 32: 32 (1863).
Brazil (Minas Gerais). 84 BZL. Cham.

Phyllanthus axillaris (Sw.) Müll.Arg. in A.P.de Candolle, Prodr. 15(2): 412 (1866).
W. Jamaica. 81 JAM. Nanophan.
  Omphalea axillaris Sw., Prodr.: 95 (1788). *Epistylium axillare* (Sw.) Sw., Fl. Ind. Occid. 2:
  1095 (1800).
  Omphalea epistylium Poir. in J.B.A.M.de Lamarck, Encycl., Suppl. 4: 140 (1816).
  Phyllanthus epistylium (Poir.) Griseb., Fl. Brit. W. I.: 33 (1859).

Phyllanthus baeckeoides J.T.Hunter & J.J.Bruhl, Nuytsia 11: 149 (1997).
Western Australia (near Laverton). 50 WAU. Nanophan.

Phyllanthus baeobotryoides Wall. ex Müll.Arg., Linnaea 32: 15 (1863).
Bangladesh, Burma. 40 BAN 41 BMA. Nanophan.

Phyllanthus bahiensis Müll.Arg., Linnaea 32: 20 (1863).
NE. Brazil. 84 BZE. Nanophan.

Phyllanthus baillonianus Müll.Arg., Linnaea 32: 47 (1863). *Reidia bailloniana* (Müll.Arg.)
Gamble, Fl. Madras: 1293 (1925).
Sri Lanka. 40 SRL.
   Epistylium cordifolium Baill., Étude Euphorb.: 648 (1858), nom. nud. *Reidia cordifolia*
      Alston in H.Trimen, Handb. Fl. Ceylon 6(Suppl.): 258 (1931).
   Epistylium latifolium Thwaites, Enum. Pl. Zeyl.: 283 (1861).

Phyllanthus baladensis Baill., Adansonia 2: 233 (1862). *Glochidion baladense* (Baill.)
Müll.Arg., Linnaea 32: 59 (1863).
NW. to C. New Caledonia. 60 NWC. Nanophan.
   Phyllanthus adenandrus Müll.Arg. ex Guillaumin, Arch. Bot. Mém. 2(3): 3 (1929).

Phyllanthus balansae Beille in H.Lecomte, Fl. Indo-Chine 5: 602 (1927).
Vietnam. 41 VIE.

Phyllanthus balansaeanus Guillaumin, Arch. Bot. Mém. 2(3): 4 (1929).
SW. New Caledonia. 60 NWC. Nanophan.
   Phyllanthus balansaeanus var. *glaber* M.Schmid, in Fl. N. Caled. & Depend. 17: 148 (1991).

Phyllanthus bancilhonae Brunel & J.P.Roux, Boissiera 32: 175 (1980).
W. Trop. Africa. 22 GUI IVO LBR SIE TOG.

Phyllanthus baraouaensis M.Schmid, in Fl. N. Caled. & Depend. 17: 272 (1991).
New Caledonia (W. Mé Maoya). 60 NWC. Nanophan.

Phyllanthus barbarae M.C.Johnst., Syst. Bot. 11: 35 (1986).
Mexico (Tamaulipas). 79 MXE. – Close to *P. galeottianus*.

Phyllanthus bathianus Leandri, Bull. Soc. Bot. France 80: 371 (1933).
C. Madagascar. 29 MDG. Cham.

Phyllanthus beddomei (Gamble) M.Mohanan, J. Econ. Taxon. Bot. 6: 480 (1985).
India. 40 IND.
   * *Reidia beddomei* Gamble, Bull. Misc. Inform. Kew 1925: 331 (1925).

Phyllanthus beillei Hutch. in D.Oliver, Fl. Trop. Afr. 6(1): 733 (1912). *Phyllanthus*
welwitschianus var. *beillei* (Hutch.) Radcl.-Sm., Kew Bull. 35: 775 (1981).
Trop. Africa, Indo-China. 22 GUI NGA SIE TOG 23 CAF CMN ZAI 25 KEN TAN 26 ANG
MLW MOZ ZAM ZIM 41 CBD THA. Nanophan.
   Phyllanthus stolzianus Pax & K.Hoffm. in H.G.A.Engler & C.G.O.Drude, Veg. Erde 9,
      III(2): 29 (1921).
   Phyllanthus nyassae Pax & K.Hoffm., Notizbl. Bot. Gart. Berlin-Dahlem 10: 383 (1928).
   Phyllanthus grahamii Hutch. & M.B.Moss, Bull. Misc. Inform. Kew 1937: 413 (1937).

Phyllanthus benguelensis Müll.Arg., J. Bot. 2: 331 (1864).
Angola. 26 ANG. Ther.

Phyllanthus benguetensis C.B.Rob., Philipp. J. Sci., C 4: 78 (1909).
　　Philippines. 42 PHI.

Phyllanthus bequaertii Robyns & Lawalrée, Bull. Jard. Bot. État 18: 265 (1947).
　　WC. & E. Trop. Africa. 23 BUR ZAI 25 KEN TAN UGA.

Phyllanthus bernardii Jabl., Mem. New York Bot. Gard. 17: 112 (1967).
　　Venezuela. 82 VEN. Nanophan.

Phyllanthus bernerianus Baill. ex Müll.Arg. in A.P.de Candolle, Prodr. 15(2): 361 (1866).
　　Mozambique, Madagascar. 26 MOZ 29 MDG. Nanophan.

　　var. **bernerianus**
　　　　Madagascar. 29 MDG. Nanophan.

　　var. **glaber** Radcl.-Sm., Kew Bull. 51: 305 (1996).
　　　　Mozambique. 26 MOZ. Nanophan.

Phyllanthus berteroanus Müll.Arg., Linnaea 32: 44 (1863).
　　N. Hispaniola. 81 DOM HAI. Cham.
　　　　Phyllanthus anisophyllus Urb., Repert. Spec. Nov. Regni Veg. 18: 364 (1922).

Phyllanthus betsileanus Leandri, Bull. Soc. Bot. France 80: 372 (1933).
　　C. Madagascar. 29 MDG. Nanophan.

Phyllanthus biantherifer Croizat, Trop. Woods 78: 7 (1944).
　　Brazil (Amazonas: Humayta). 84 BZN. Nanophan.

Phyllanthus bicolor Vis., Atti Reale Ist. Veneto Sc. Lett. Arti, III, 4: 139 (1858).
　　Venezuela. 82 VEN.
　　　　Phyllanthus spectabilis Hügel ex Vis., Atti Reale Ist. Veneto Sc. Lett. Arti, III, 4: 139
　　　　　　(1858), pro syn.

Phyllanthus biflorus Rusby, Mem. New York Bot. Gard. 7: 282 (1927).
　　Bolivia. 83 BOL.

Phyllanthus binhii Thin, Euphorb. Vietnam: 48 (1995).
　　Vietnam. 41 VIE. Nanophan.

Phyllanthus birmanicus Müll.Arg., Linnaea 32: 47 (1863).
　　Burma. 41 BMA.

Phyllanthus blanchetianus Müll.Arg., Linnaea 32: 38 (1863).
　　Brazil (Bahia). 84 BZE. Cham.

Phyllanthus blancoanus Müll.Arg., Linnaea 32: 49 (1863).
　　Philippines. 42 PHI.
　　　　* *Phyllanthus tetrander* Blanco, Fl. Filip., ed. 2: 480 (1845), nom. illeg.

Phyllanthus bodinieri (H.Lév.) Rehder, J. Arnold Arbor. 18: 212 (1937).
　　China (SE. Guangxi, SE. Guizhou). 36 CHC CHS. Nanophan.
　　　　* *Sterculia bodinieri* H.Lév., Fl. Kouy-Tchéou: 406 (1915).

Phyllanthus boehmii Pax, Bot. Jahrb. Syst. 15: 525 (1893).
　　Zaire, Ethiopia, Uganda, Kenya, Tanzania, Malawi, Mozambique. 23 ZAI 24 ETH 25 KEN
　　　　TAN UGA 26 MLW MOZ. Ther.

var. **boehmii**
Zaire, Ethiopia, Uganda, Kenya, Tanzania, Malawi, Mozambique. 23 ZAI 24 ETH 25 KEN TAN UGA 26 MLW MOZ. Ther.

var. **humilis** Radcl.-Sm., Kew Bull. 35: 766 (1981).
Kenya, Tanzania. 25 KEN TAN. Cham.

Phyllanthus boguenensis M.Schmid, in Fl. N. Caled. & Depend. 17: 286 (1991).
C. New Caledonia. 60 NWC. Cham. or nanophan.

Phyllanthus bojerianus (Baill.) Müll.Arg. in A.P.de Candolle, Prodr. 15(2): 343 (1866).
Madagascar. 29 MDG. Nanophan.
* *Kirganelia bojeriana* Baill., Adansonia 2: 47 (1861).

var. **bojerianus**
C. & W. Madagascar. 29 MDG. Nanophan.
Phyllanthus atroviridis Müll.Arg., Linnaea 32: 8 (1863).

var. **meridionalis** Leandri, Mém. Inst. Sci. Madagascar, Sér. B, Biol. Vég. 8: 226 (1957).
S. Madagascar. 29 MDG. Nanophan.

Phyllanthus bolivarensis Steyerm., Fieldiana, Bot. 28: 317 (1952).
Venezuela (Bolívar). 82 VEN. Nanophan.

Phyllanthus bolivianus Pax & K.Hoffm., Meded. Rijks-Herb. 40: 18 (1921).
Bolivia. 83 BOL.

Phyllanthus boninsimae Nakai, Bot. Mag. (Tokyo) 26: 96 (1912).
Ogasawara-shoto. 62 OGA.

Phyllanthus borenensis M.G.Gilbert, Kew Bull. 42: 354 (1987).
Ethiopia (Sidamo). 24 ETH. Nanophan. – Close to *P. sepialis*.

Phyllanthus borjaensis Jabl., Mem. New York Bot. Gard. 17: 108 (1967).
Venezuela (Apure, Bolívar: Cerro San Borja). 82 VEN. Nanophan.

Phyllanthus botryanthus Müll.Arg. in A.P.de Candolle, Prodr. 15(2): 323 (1866).
Glochidion botryanthum (Müll.Arg.) Pax & K.Hoffm. in H.G.A.Engler, Nat. Pflanzenfam. ed. 2, 19c: 58 (1931).
Colombia, Venezuela, Aruba, Curaçao, Bonaire. 81 ARU NLA 82 VEN 83 CLM. Nanophan. or phan.
Phyllanthus euwensii Bold., Fl. Dutch W. Ind. Is. 2: 50 (1914).

Phyllanthus bourgeoisii Baill., Adansonia 2: 235 (1862). *Glochidion bourgeoisii* (Baill.) Müll.Arg., Linnaea 32: 70 (1863).
Irian Jaya, Bismarck Archip., NW. to SC. New Caledonia. 42 BIS NWG 60 NWC. Nanophan.
Phyllanthus neocaledonicus Müll.Arg. ex Guillaumin, Arch. Bot. Mém. 2(3): 5 (1929).

Phyllanthus brachyphyllus Urb., Repert. Spec. Nov. Regni Veg. 13: 452 (1914).
Haiti. 81 HAI. Cham.

Phyllanthus brandegei Millsp., Proc. Calif. Acad. Sci., II, 2: 218 (1889).
Mexico (Baja California). 79 MXN.

Phyllanthus brasiliensis (Aubl.) Poir. in J.B.A.M.de Lamarck, Encycl. 5: 296 (1804).
Guianas, Venezuela, Brazil, Peru. 82 FRG GUY SUR VEN 83 PER 84 BZN. Nanophan.

* *Conami brasiliensis* Aubl., Hist. Pl. Guiane 2: 926 (1775). *Phyllanthus conami* Sw., Prodr.: 28 (1788). *Conami conami* (Sw.) Britton, Bot. Porto Rico 5: 475 (1924). *Cicca brasiliensis* (Aubl.) Baill., Étude Euphorb.: 618 (1858), nom. nud. *Phyllanthus fruticosus* Baill., Adansonia 5: 356 (1865).

Phyllanthus brassii C.T.White, Proc. Roy. Soc. Queensland 47: 81 (1935 publ. 1936).
Queensland (Thornton Peak). 50 QLD.

Phyllanthus brevipes Hook.f., Fl. Brit. India 5: 297 (1887).
Arunachal Pradesh (Mishmi Hills). 40 EHM. Nanophan.

Phyllanthus breynioides (P.T.Li) Govaerts & Radcl.-Sm., Kew Bull. 51: 176 (1996).
China (Guangxi). 36 CHS. Nanophan.
* *Phyllanthodendron breynioides* P.T.Li, Bull. Bot. Res., Harbin 7(3): 6 (1987).

Phyllanthus brunelii J.P.Roux, Nordic J. Bot. 4: 50 (1984).
Thailand. 41 THA. – Close to *P. reticulatus*.

Phyllanthus buchii Urb., Symb. Antill. 3: 288 (1902).
Hispaniola. 81 DOM HAI. Nanophan.

Phyllanthus bupleuroides Baill., Adansonia 2: 233 (1862). *Glochidion bupleuroides* (Baill.) Müll.Arg., Linnaea 32: 59 (1863).
New Caledonia. 60 NWC. Cham. or nanophan.

   var. **bupleuroides**
     New Caledonia (Canala Reg.). 60 NWC. Cham. or nanophan.

   var. **latiaxialis** M.Schmid, in Fl. N. Caled. & Depend. 17: 224 (1991).
     New Caledonia (Thio to Poindimié). 60 NWC. Cham. or nanophan.

   var. **meoriensis** M.Schmid, in Fl. N. Caled. & Depend. 17: 224 (1991).
     New Caledonia (Mt. Ori). 60 NWC. Nanophan.

   var. **ngoyensis** (Schltr.) M.Schmid, in Fl. N. Caled. & Depend. 17: 222 (1991).
     EC. & SE. New Caledonia. 60 NWC. Cham. or nanophan.
     * *Phyllanthus ngoyensis* Schltr., Bot. Jahrb. Syst. 39: 146 (1906).

   var. **poroensis** M.Schmid, in Fl. N. Caled. & Depend. 17: 224 (1991).
     New Caledonia (Canala to Houaïlou). 60 NWC. Cham. or nanophan.

Phyllanthus buxifolius (Blume) Müll.Arg., Linnaea 32: 50 (1863).
Jawa, Borneo (Sabah, SE. Kalimantan), Philippines. 42 BOR JAW PHI. Phan.
   * *Scepasma buxifolia* Blume, Bijdr.: 583 (1826).

Phyllanthus buxoides Guillaumin, Arch. Bot. Mém. 2(3): 6 (1929).
NW. New Caledonia. 60 NWC. Nanophan.

Phyllanthus cacuminum Müll.Arg., Flora 48: 371 (1865).
Jawa. 42 JAW.

Phyllanthus caesius Airy Shaw & G.L.Webster, Kew Bull. 26: 90 (1971).
New Guinea. 42 NWG. Nanophan. or phan.

Phyllanthus caespitosus Brenan, Kew Bull. 21: 258 (1967).
Malawi, Zambia. 26 MLW ZAM. Cham.

Phyllanthus calcicola M.Schmid, in Fl. N. Caled. & Depend. 17: 118 (1991).
NW. New Caledonia. 60 NWC. Nanophan. or phan.

Phyllanthus caligatus Brunel & J.P.Roux, Willdenowia 14: 384 (1984 publ. 1985).
Cameroon. 23 CMN.

Phyllanthus calycinus Labill., Nov. Holl. Pl. 2: 75 (1806).
Western Australia, South Australia. 50 SOA WAU.
Phyllanthus pimeleoides A.DC., Rapp. [Not.] Pl. Rar. Genève 9: 15 (1826).
Phyllanthus cygnorum Endl. in S.L.Endlicher & al., Enum. Pl.: 19 (1837).
Phyllanthus pulchellus Endl. in S.L.Endlicher & al., Enum. Pl.: 19 (1837).
Phyllanthus preissianus Klotzsch in J.G.C.Lehmann, Pl. Preiss. 1: 179 (1845).
Clutia berberifolia Pax in H.G.A.Engler, Pflanzenr., IV, 147, III: 83 (1911).

Phyllanthus caribaeus Urb., Symb. Antill. 5: 382 (1908).
Leeward Is., Trinidad, Tobago. 81 LEE TRT. Ther.

Phyllanthus carinatus Beille, Bull. Soc. Bot. France 72: 160 (1925). *Phyllanthodendron carinatum* (Beille) Croizat, J. Arnold Arbor. 23: 36 (1942).
Vietnam. 41 VIE.

Phyllanthus carlottae M.Schmid, in Fl. N. Caled. & Depend. 17: 258 (1991).
SE. New Caledonia. 60 NWC. Nanophan. or phan.

Phyllanthus carnosulus Müll.Arg., Linnaea 32: 30 (1863).
SE. Cuba. 81 CUB. Hemicr.
Phyllanthus callitrichoides Griseb., Nachr. Königl. Ges. Wiss. Georg-Augusts-Univ. 1: 166 (1865).
Phyllanthus haplocladus Urb., Repert. Spec. Nov. Regni Veg. 28: 214 (1930).

Phyllanthus caroliniensis Walter, Fl. Carol.: 228 (1788).
C. & E. U.S.A. to Argentina. 74 ILL KAN MSO OKL 75 INI OHI PEN 77 TEX 78 ALL 80 COS PAN 81 BAH CUB DOM HAI JAM SWC TRT WIN 82 FRG GUY SUR VEN 83 BOL CLM ECU GAL PER 84 BZL BZN 85 AGE PAR URU. Ther.

subsp. **caroliniensis**
C. & E. U.S.A. to Argentina. 74 ILL KAN MSO OKL 75 INI OHI PEN 77 TEX 78 ALL 81 JAM SWC TRT WIN 82 FRG GUY SUR VEN 83 BOL CLM ECU GAL PER 84 BZL 85 AGE PAR URU. Ther.
Phyllanthus obovatus Muhl. ex Willd., Sp. Pl. 4: 574 (1805).
Synexemia pumila Raf., Neogenyton: 2 (1825). *Andrachne pumila* (Raf.) Rydb., Brittonia 1: 96 (1931).
Phyllanthus graminicola Britton, Bull. Torrey Bot. Club 48: 333 (1921 publ. 1922), nom. illeg.
Andrachne schaffneriana Pax & K.Hoffm. in H.G.A.Engler, Pflanzenr., IV, 147, XV: 179 (1922).

subsp. **guianensis** (Klotzsch) G.L.Webster, Contr. Gray Herb. 167: 46 (1955).
N. South America, N. Brazil. 82 FRG GUY SUR 84 BZN. Ther.
* *Phyllanthus guianensis* Klotzsch, London J. Bot. 2: 51 (1843). *Phyllanthus schomburgkianus* var. *guianensis* (Klotzsch) Müll.Arg. in A.P.de Candolle, Prodr. 15(2): 387 (1866). *Cicca guianensis* (Klotzsch) Splitg. ex Lanj., Euphorb. Surinam: 109 (1931).
Phyllanthus caroliniensis var. *antillanus* Müll.Arg., Linnaea 32: 36 (1863). *Phyllanthus schomburgkianus* var. *antillanus* (Müll.Arg.) Müll.Arg. in A.P.de Candolle, Prodr. 15(2): 387 (1866).
Phyllanthus guianensis var. *acuminatus* Müll.Arg., Linnaea 32: 29 (1863). *Phyllanthus schomburgkianus* var. *acuminatus* (Müll.Arg.) Müll.Arg. in A.P.de Candolle, Prodr. 15(2): 387 (1866).
Phyllanthus guianensis var. *anceps* Müll.Arg., Linnaea 32: 29 (1863). *Phyllanthus schomburgkianus* var. *anceps* (Müll.Arg.) Müll.Arg. in A.P.de Candolle, Prodr. 15(2): 387 (1866).

subsp. **saxicola** (Small) G.L.Webster, Contr. Gray Herb. 176: 46 (1955).
S. Florida, Bahamas, Cuba, Hispaniola, Jamaica. 78 FLA 81 BAH CUB DOM HAI JAM. Ther.
  * *Phyllanthus saxicola* Small, Bull. New York Bot. Gard. 3: 428 (1905).

subsp. **stenopterus** (Müll.Arg.) G.L.Webster, J. Arnold Arbor. 37: 348 (1956).
Costa Rica, Panama, Colombia, Guiana, Venezuela. 80 COS PAN 82 FRG VEN 83 CLM. Ther.
  * *Phyllanthus stenopterus* Müll.Arg., Linnaea 34: 74 (1865).

Phyllanthus carpentariae Müll.Arg., Linnaea 34: 72 (1865).
Northern Territory, Queensland, New South Wales. 50 NSW NTA QLD.
  Phyllanthus grandisepalus F.Muell. ex Müll.Arg., Linnaea 34: 72 (1865).
  Phyllanthus hebecarpus Benth., Fl. Austral. 6: 108 (1873).
  Phyllanthus fuernrohrii var. *suffruticosus* Domin, Biblioth. Bot. 89(22): 321 (1927).

Phyllanthus carrenoi Steyerm., Bol. Soc. Venez. Ci. Nat. 32: 343 (1976).
Venezuela (Bolívar: Jaua Mts.). 82 VEN.

Phyllanthus casearioides S.Moore, J. Linn. Soc., Bot. 45: 397 (1921).
New Caledonia (Touho Reg.). 60 NWC. Nanophan.

Phyllanthus cassioides Rusby, Bull. New York Bot. Gard. 8: 100 (1912).
Bolivia. 83 BOL.

Phyllanthus casticum P.Willemet, Ann. Bot. (Usteri) 18: 55 (1796). *Phyllanthus casticum* var. *genuinus* Müll.Arg. in A.P.de Candolle, Prodr. 15(2): 348 (1866), nom. inval.
Madagascar, Mascarenes. 29 MAU MDG REU SEY. Nanophan.

var. **casticum** – FIGURE, p. 1262.
Madagascar, Mascarenes. 29 MAU MDG REU SEY. Nanophan.
Kirganelia virginea J.F.Gmel., Syst. Nat.: 1008 (1792). *Phyllanthus virgineus* (J.F.Gmel.) Pers., Syn. Pl. 2: 591 (1807).
Kirganelia phyllanthoides A.Juss., Euphorb. Gen.: 108 (1824).
Kirganelia elegans Juss. ex Spreng., Syst. Veg. 3: 48 (1826).
Phyllanthus casticum f. *parvifolius* Müll.Arg. in A.P.de Candolle, Prodr. 15(2): 348 (1866).
  Phyllanthus casticum var. *madagascariensis* Leandri, in Fl. Madag. 111: 60 (1958).
Phyllanthus schimperianus Hemsl., J. Bot. 55: 287 (1917).
Phyllanthus casticum f. *fiherenensis* Leandri, Notul. Syst. (Paris) 7: 175, 179 (1939).

var. **kirganelia** (Willd.) Govaerts in R.Govaerts, D.G.Frodin & A.Radcliffe-Smith, World Checklist Bibliogr. Euphorbiaceae: 1276 (2000).
C. & E. Madagascar. 29 MDG. Nanophan.
Phyllanthus fasciculatus Poir. in J.B.A.M.de Lamarck, Encycl. 5: 304 (1804). *Phyllanthus kirganelia* var. *fasciculatus* (Poir.) Müll.Arg., Linnaea 32: 12 (1863). *Phyllanthus casticum* var. *fasciculatus* (Poir.) Müll.Arg. in A.P.de Candolle, Prodr. 15(2): 348 (1866).
  * *Phyllanthus kirganelia* Willd., Sp. Pl. 4: 587 (1805).
Phyllanthus kirganelia var. *sieberi* Müll.Arg., Linnaea 32: 13 (1863).
Phyllanthus timoriensis Müll.Arg., Linnaea 32: 13 (1863).

Phyllanthus castus S.Moore, J. Linn. Soc., Bot. 45: 401 (1921). *Phyllanthus pancherianus* var. *castus* (S.Moore) Guillaumin, Arch. Bot. Mém. 2(3): 13 (1929).
SE. New Caledonia. 60 NWC. Cham. or nanophan.

Phyllanthus caudatifolius Merr., Philipp. J. Sci. 30: 403 (1926).
Philippines. 42 PHI.

Phyllanthus caudatus Müll.Arg. in A.P.de Candolle, Prodr. 15(2): 321 (1866).
SE. New Caledonia. 60 NWC. Nanophan. or phan.

var. **caudatus**
SE. New Caledonia. 60 NWC. Nanophan. or phan.

var. **pubescens** M.Schmid, in Fl. N. Caled. & Depend. 17: 74 (1991).
SE. New Caledonia. 60 NWC. Nanophan. or phan.

Phyllanthus cauliflorus (Sw.) Griseb., Fl. Brit. W. I.: 33 (1859).
W. Jamaica. 81 JAM. Phan.
* *Omphalea cauliflora* Sw., Prodr.: 95 (1788). *Epistylium cauliflorum* (Sw.) Sw., Fl. Ind.
Occid. 2: 1095 (1800).

Phyllanthus cauticola J.T.Hunter & J.J.Bruhl, Nuytsia 11: 151 (1997).
Northern Territory. 50 NTA. Cham. or nanophan.

Phyllanthus caymanensis G.L.Webster & Proctor, Rhodora 86(846): 121 (28 June 1984).
Cayman Is. 81 CAY.

Phyllanthus cedrelifolius Verdc., Bull. Misc. Inform. Kew 1924: 259 (1924).
KwaZulu-Natal, Cape Prov. 27 CPP NAT.

Phyllanthus celebicus Koord., Meded. Lands Plantentuin 19: 588 627 (1898).
Sulawesi. 42 SUL.

Phyllanthus ceratostemon Brenan, Kew Bull. 21: 259 (1967).
Central African Rep., Chad, Zaire, Zambia. 23 CAF ZAI 24 CHA 26 ZAM.

Phyllanthus chacoensis Morong, Ann. New York Acad. Sci. 7: 218 (1892).
S. Brazil, Paraguay, Bolivia. 83 BOL 84 BZS 85 PAR. Phan.
Aporosella hassleriana Chodat, Bull. Herb. Boissier, II, 5: 489 (1905).
Aporosella chacoensis Speg., Cat. Descr. Maderas: 349, 373 (1910).
Aporosella chacorensis Pax & K.Hoffm., Meded. Rijks-Herb. 40: 19 (1921), sphalm.

Phyllanthus chamaecerasus Baill., Adansonia 2: 235 (1862). *Glochidion chamaecerasus*
(Baill.) Müll.Arg., Linnaea 32: 70 (1863).
New Caledonia. 60 NWC. Nanophan.

var. **aoupinieensis** M.Schmid, in Fl. N. Caled. & Depend. 17: 128 (1991).
New Caledonia (Houaïlou Reg.). 60 NWC. Nanophan.

var. **chamaecerasus**
NW. New Caledonia (Balade). 60 NWC. Nanophan.

var. **intermedius** M.Schmid, in Fl. N. Caled. & Depend. 17: 126 (1991).
C. & S. New Caledonia. 60 NWC. Nanophan.

var. **longipedicellatus** M.Schmid, in Fl. N. Caled. & Depend. 17: 127 (1991).
New Caledonia (Bourail Reg.). 60 NWC. Nanophan.

var. **meoriensis** M.Schmid, in Fl. N. Caled. & Depend. 17: 128 (1991).
SC. New Caledonia. 60 NWC. Nanophan.

f. **ripicola** (Guillaumin) M.Schmid, in Fl. N. Caled. & Depend. 17: 130 (1991).
S. New Caledonia. 60 NWC. Nanophan.
* *Dendrophyllanthus ripicola* Guillaumin, Mém. Mus. Natl. Hist. Nat., B, Bot. 8: 178 (1959).

var. **vieillardii** (Baill.) M.Schmid, in Fl. N. Caled. & Depend. 17: 129 (1991).
SC. New Caledonia. 60 NWC. Nanophan.

* *Phyllanthus vieillardii* Baill., Adansonia 2: 236 (1862).
  Glochidion cataractarum Müll.Arg., Linnaea 32: 70 (1863). *Phyllanthus cataractarum*
  (Müll.Arg.) Müll.Arg. in A.P.de Candolle, Prodr. 15(2): 320 (1866).

Phyllanthus chamaecristoides Urb., Symb. Antill. 9: 185 (1924).
  Cuba. 81 CUB. Nanophan.

  subsp. **baracoensis** (Urb.) G.L.Webster, J. Arnold Arbor. 39: 134 (1958).
    E. Cuba (near coasts). 81 CUB. Nanophan.
    * *Phyllanthus baracoensis* Urb., Symb. Antill. 9: 186 (1924).
    Phyllanthus coelophyllus Urb., Symb. Antill. 9: 187 (1924).

  subsp. **chamaecristoides**
    Cuba (Sierra de Nipe). 81 CUB. Nanophan.
    Phyllanthus apiculatus Urb., Symb. Antill. 9: 184 (1924), nom. illeg.

Phyllanthus chandrabosei Govaerts & Radcl.-Sm., Kew Bull. 51: 176 (1996).
  India. 40 IND.
    * *Reidia stipulacea* Gamble, Bull. Misc. Inform. Kew 1925: 332 (1925). *Phyllanthus
    stipulaceus* (Gamble) Kumari & Chandrab., in Fl. Tamil Nadu, India 2: 239 (1987),
    nom. illeg.

Phyllanthus chantrieri André, Rev. Hort. 1883: 537 (1883).
  Vietnam. 41 VIE.

Phyllanthus chekiangensis Croizat & Metcalf, Lingnan Sci. J. 20: 194 (1942).
  SE. China. 36 CHS. Nanophan.
    Phyllanthus kiangsiensis Croizat & Metcalf, Lingnan Sci. J. 20: 195 (1942).
    Phyllanthus leptoclados var. *pubescens* P.T.Li & D.Y.Liu, Bull. Bot. Res., Harbin 6: 181 (1986).

Phyllanthus cherrieri M.Schmid, in Fl. N. Caled. & Depend. 17: 278 (1991).
  New Caledonia (Mt. Arago). 60 NWC. Nanophan.

Phyllanthus chevalieri Beille, Bull. Soc. Bot. France 55(8): 57 (1908).
  Chad, Sudan, Kenya, Tanzania. 24 CHA SUD 25 KEN TAN. Ther.

Phyllanthus chiapensis Sprague, Bull. Misc. Inform. Kew 1909: 264 (1909).
  Mexico (Chiapas: Cacote). 79 MXT.

Phyllanthus chimantae Jabl., Mem. New York Bot. Gard. 17: 100 (1967).
  Venezuela (Bolívar: Chimanta Mts.). 82 VEN. Nanophan.

Phyllanthus choretroides Müll.Arg., Linnaea 32: 52 (1863).
  Brazil (Minas Gerais: Serra da Lapa). 84 BZL.

Phyllanthus chrysanthus Baill., Adansonia 2: 238 (1862).
  New Caledonia. 60 NWC. Cham.

  var. **chrysanthus**
    New Caledonia. 60 NWC. Cham.
    Phyllanthus persimilis Müll.Arg., Linnaea 32: 34 (1863).

  var. **deverdensis** M.Schmid, in Fl. N. Caled. & Depend. 17: 53 (1991).
    New Caledonia (Cap Deverd). 60 NWC. Cham.

  var. **micrantheoides** (Baill.) M.Schmid, in Fl. N. Caled. & Depend. 17: 52 (1991).
    New Caledonia. 60 NWC. Cham.
    * *Phyllanthus micrantheoides* Baill., Adansonia 2: 238 (1862).
    Phyllanthus rufidulus Müll.Arg., Linnaea 32: 29 (1863).

Phyllanthus rufidulus var. *kafeateensis* Guillaumin, Mém. Mus. Natl. Hist. Nat., B, Bot. 8(3): 247 (1962), holotype not designated.

Phyllanthus chryseus Howard, J. Arnold Arbor. 28: 121 (1947).
E. Cuba (Sierra de Moa). 81 CUB. Nanophan.

Phyllanthus ciccoides Müll.Arg., Linnaea 32: 13 (1863).
Papua New Guinea to Vanuatu. 42 BIS NWG 50 NTA QLD WAU 60 NWC SOL VAN. Nanophan. or phan.

var. **ciccoides**
Papua New Guinea, Solomon Is., N. Australia, NW. New Caledonia, Vanuatu. 42 NWG 50 NTA WAU 60 NWC SOL VAN. Nanophan. or phan.
Phyllanthus novaehollandiae Baill., Adansonia 6: 343 (1866), nom. illeg.
Phyllanthus baccatus F.Muell. ex Benth., Fl. Austral. 6: 102 (1873).
Flueggeopsis microspermus K.Schum. in K.M.Schumann & C.A.G.Lauterbach, Fl. Schutzgeb. Südsee, Nachtr.: 289 (1905).
Flueggeopsis pelas K.Schum. in K.M.Schumann & C.A.G.Lauterbach, Fl. Schutzgeb. Südsee, Nachtr.: 290 (1905). *Phyllanthus pelas* (K.Schum.) Pax & K.Hoffm. in H.G.A.Engler, Nat. Pflanzenfam. ed. 2, 19c: 62 (1931).

var. **puberulus** Airy Shaw, Muelleria 4: 215 (1980).
Papua New Guinea, Bismarck Archip., Queensland, Solomon Is., Vanuatu. 42 BIS NWG 50 QLD 60 SOL VAN. Nanophan. or phan.

Phyllanthus ciliaris Baill., Adansonia 11: 373 (1876).
New Caledonia (near Balade). 60 NWC. – Provisionally accepted.

Phyllanthus cinctus Urb., Symb. Antill. 9: 191 (1924).
E. Cuba. 81 CUB. Cham. or nanophan.
Conami ovalifolia Britton, Mem. Torrey Bot. Club 16: 73 (1920).
Phyllanthus brittonii Alain, Contr. Ocas. Mus. Hist. Nat. Colegio "De La Salle" 11: 1 (1952)

Phyllanthus cinereus Müll.Arg., Linnaea 32: 48 (1863). *Reidia cinerea* (Müll.Arg.) Alston in H.Trimen, Handb. Fl. Ceylon 6(Suppl.): 258 (1931).
Sri Lanka. 40 SRL.
Epistylium floribundum Thwaites, Enum. Pl. Zeyl.: 283 (1861).

Phyllanthus cladanthus Müll.Arg., Linnaea 32: 46 (1863).
C. & E. Jamaica. 81 JAM. Phan.

Phyllanthus clamboides (F.Muell.) Diels, Notizbl. Bot. Gart. Berlin-Dahlem 11: 309 (1931).
New Guinea, Queensland, Solomon Is. 42 NWG 50 QLD 60 SOL. Phan.
* *Leichhardtia clamboides* F.Muell., Fragm. 10: 68 (1876).
Phyllanthus paniculatus Oliv., Hooker's Icon. Pl. 24: t. 2372 (1895). *Hexaspermum paniculatum* (Oliv.) Domin, Biblioth. Bot. 89: 316 (1927).
Nymania insignis K.Schum. in K.M.Schumann & C.A.G.Lauterbach, Fl. Schutzgeb. Südsee, Nachtr.: 292 (1905). *Phyllanthus insignis* (K.Schum.) J.J.Sm., Nova Guinea 8: 781 (1912), nom. illeg. *Phyllanthus schumannianus* L.S.Sm., N. Queensland Naturalist 15: 2 (1947).
Phyllanthus choristylus Diels, Notizbl. Bot. Gart. Berlin-Dahlem 11: 309 (1931).

Phyllanthus clarkei Hook.f., Fl. Brit. India 5: 297 (1887).
Bhutan to S. China, N. Thailand, Vietnam. 36 CHC CHS CHT 40 ASS EHM 41 BMA THA VIE. Cham.
Phyllanthus simplex var. *tonkinensis* Beille in H.Lecomte, Fl. Indo-Chine 5: 578 (1927).

Phyllanthus claussenii Müll.Arg., Linnaea 32: 40 (1863).
SE. Brazil. 84 BZL.
Phyllanthus claussenii var. *oblongifolius* Müll.Arg. in C.F.P.von Martius, Fl. Bras. 11(2): 61 (1873).

Phyllanthus coalcomanensis Croizat, J. Wash. Acad. Sci. 33: 13 (1943).
Mexico (Michoacán). 79 MXS. Phan.

Phyllanthus coccineus Noronha, Verh. Batav. Genootsch. Kunsten 5(4): 22 (1790).
Jawa. 42 JAW.

Phyllanthus cochinchinensis Spreng., Syst. Veg. 3: 21 (1826).
Assam to S. China. 36 CHC CHH CHS CHT 40 ASS BAN 41 CBD LAO VIE. Nanophan.
Cathetus fasciculata Lour., Fl. Cochinch.: 608 (1790). *Phyllanthus fasciculatus* (Lour.) Müll.Arg. in A.P.de Candolle, Prodr. 15(2): 350 (1866), nom. illeg.
Phyllanthus cinerascens Hook. & Arn., Bot. Beechey Voy.: 211 (1837).
Phyllanthus roeperianus Wall. ex Müll.Arg., Linnaea 32: 28 (1863).
Phyllanthus serissifolius Wall. ex Müll.Arg. in A.P.de Candolle, Prodr. 15(2): 350 (1866), pro syn.
Phyllanthus embergeri Haicour & Rossignol, Amer. J. Bot. 74: 1860 (1987 publ. 1988).

Phyllanthus collinsiae Craib, Bull. Misc. Inform. Kew 1913: 72 (1913).
Thailand, Vietnam, Lesser Sunda Is. 41 THA VIE 42 LSI. Nanophan. or phan.

var. **collinsiae**
Thailand, Vietnam. 41 THA VIE. Nanophan. or phan.
Phyllanthus bienhoensis Beille in H.Lecomte, Fl. Indo-Chine 5: 588 (1927).

var. **leiocarpus** Airy Shaw, Kew Bull. 29: 295 (1974).
Lesser Sunda Is. (Sula I., Sumbawa, W. Flores). 42 LSI. Nanophan. or phan.

Phyllanthus × collinum-misuku Radcl.-Sm., Kew Bull. 47: 680 (1992). P. ovalifolius × P. reticulatus.
Malawi. 26 MLW.

Phyllanthus collinus Domin, Biblioth. Bot. 89: 320 (1927).
Queensland. 50 QLD.

Phyllanthus columnaris Müll.Arg., Linnaea 32: 15 (1863).
Burma, Thailand, N. Pen. Malaysia. 41 BMA THA 42 MLY. Phan.

Phyllanthus coluteoides Baill. ex Müll.Arg. in A.P.de Candolle, Prodr. 15(2): 335 (1866).
W. Madagascar, Mozambique Channel Is. (Juan de Nova I.). 29 MCI MDG. Nanophan.

Phyllanthus comorensis Leandri, Notul. Syst. (Paris) 6: 194 (1938).
Comoros. 29 COM. Cham. or nanophan.

Phyllanthus comosus Urb., Repert. Spec. Nov. Regni Veg. 13: 451 (1914).
NE. Cuba. 81 CUB. Nanophan.

Phyllanthus compressus Kunth in F.W.H.von Humboldt, A.J.A.Bonpland & C.S.Kunth, Nov. Gen. Sp. 2: 109 (1817).
Mexico to Peru. 79 MXG 80 COS 83 PER 84 BZN. Cham.

Phyllanthus comptonii S.Moore, J. Linn. Soc., Bot. 45: 398 (1921).
New Caledonia (Ngoye Valley). 60 NWC-. Nanophan. or phan.

Phyllanthus comptus G.L.Webster, Contr. Gray Herb. 176: 61 (1955).
Cuba (Cajalbana Reg.). 81 CUB. Nanophan.

Phyllanthus confusus Brenan, Mem. New York Bot. Gard. 9: 68 (1954).
Malawi (Mt. Mulanje). 26 MLW. Cham.

Phyllanthus conjugatus M.Schmid, in Fl. N. Caled. & Depend. 17: 279 (1991).
New Caledonia. 60 NWC. Cham.

var. **conjugatus**
New Caledonia (Vallée de la Kalouehola). 60 NWC. Cham.

var. **ducosensis** M.Schmid, in Fl. N. Caled. & Depend. 17: 281 (1991).
New Caledonia (Ducos Pen.). 60 NWC. Cham.

var. **maaensis** M.Schmid, in Fl. N. Caled. & Depend. 17: 280 (1991).
New Caledonia (Mt. Maa). 60 NWC. Cham.

Phyllanthus consanguineus Müll.Arg. in A.P.de Candolle, Prodr. 15(2): 378 (1866).
Réunion. 29 REU. Nanophan.

Phyllanthus cordatulus C.B.Rob., Philipp. J. Sci., C 4: 76 (1909).
Philippines. 42 PHI.

Phyllanthus cordobensis (Kuntze) K.Schum., Just's Bot. Jahresber. 26(1): 350 (1898).
NC. Argentina. 85 AGE.
  * *Diasperus cordobensis* Kuntze, Revis. Gen. Pl. 3(2): 285 (1898).

Phyllanthus cornutus Baill., Adansonia 2: 236 (1862). *Glochidion cornutum* (Baill.)
Müll.Arg., Linnaea 32: 71 (1863).
New Caledonia (incl. I. des Pins). 60 NWC. Nanophan.

Phyllanthus coursii Leandri, Mém. Inst. Sci. Madagascar, Sér. B, Biol. Vég. 8: 226 (1957).
C. Madagascar. 29 MDG. Nanophan.

Phyllanthus crassinervius Radcl.-Sm., Kew Bull. 35: 766 (1981). – FIGURE, p. 1266.
Zaire, Tanzania (Mbeya), Malawi, Zambia. 23 ZAI 25 TAN 26 MLW ZAM. Hemicr. – Close
to *P. caespitosus*.

Phyllanthus cristalensis Urb., Repert. Spec. Nov. Regni Veg. 28: 212 (1930).
E. Cuba (Sierra del Cristal). 81 CUB. Nanophan.

Phyllanthus croizatii Steyerm., Fieldiana, Bot. 28: 317 (1952).
Venezuela. 82 VEN.

Phyllanthus cryptophilus (Comm. ex A.Juss.) Müll.Arg., Linnaea 32: 8 (1863).
E. Madagascar. 29 MDG. Cham. or nanophan.
  * *Menarda cryptophila* Comm. ex A.Juss., Euphorb. Gen.: 109 (1824).

Phyllanthus curranii C.B.Rob., Philipp. J. Sci., C 4: 77 (1909).
Philippines. 42 PHI.

Phyllanthus cuscutiflorus S.Moore, J. Bot. 43: 148 (1905).
Papua New Guinea, Queensland. 42 NWG 50 QLD. Phan.

Phyllanthus cuspidatus Müll.Arg., Flora 48: 377 (1865).
Samoa. 60 SAM.

Phyllanthus cyrtophylloides Müll.Arg. in A.P.de Candolle, Prodr. 15(2): 1270 (1866).
Jawa. 42 JAW.

Phyllanthus daclacensis Thin, J. Biol. (Vietnam) 14(2): 23 (1992).
Vietnam. 41 VIE.

Phyllanthus dallachyanus Benth., Fl. Austral. 6: 104 (1873).
N. Australia. 50 NTA QLD.

Phyllanthus dawsonii Steyerm., Los Angeles County Mus. Contr. Sci. 21: 13 (1958).
Brazil (Goiás). 84 BZC.

Phyllanthus dealbatus Alston in H.Trimen, Handb. Fl. Ceylon 6(Suppl.): 257 (1931).
Sri Lanka. 40 SRL.

Phyllanthus debilis Klein ex Willd., Sp. Pl. 4: 582 (1805). *Phyllanthus niruri* var. *debilis*
(Klein ex Willd.) Müll.Arg. in A.P.de Candolle, Prodr. 15(2): 407 (1866).
S. India to Jawa, introduced elsewere. 40 IND SRL 41 THA 42 JAW MLY nwg SUM (50) nta
qld (62) crl mrn (63) haw (81) lee. Ther.
   Phyllanthus niruri var. *javanicus* Müll.Arg., Linnaea 32: 43 (1863).
   Phyllanthus leai S.Moore, J. Linn. Soc., Bot. 45: 217 (1920).

Phyllanthus deciduiramus Däniker, Vierteljahrschr. Naturf. Ges. Zürich 76: 167 (1931).
New Caledonia (Mt. Kaala). 60 NWC. Cham. or nanophan.

Phyllanthus decipiens (Baill.) Müll.Arg. in A.P.de Candolle, Prodr. 15(2): 347 (1866).
Comoros, Madagascar (incl. Nosi Bé I.). 29 COM MDG. Nanophan. or phan.
  * *Kirganelia decipiens* Baill., Adansonia 2: 49 (1861). *Phyllanthus decipiens* var. *genuinus*
   Müll.Arg. in A.P.de Candolle, Prodr. 15(2): 347 (1866), nom. inval. *Phyllanthus*
   casticum f. *decipiens* (Baill.) Leandri, Notul. Syst. (Paris) 7: 175, 180 (1939).

  var. **antisihanakensis** Leandri, in Fl. Madag. 111: 54 (1958).
   Madagascar. 29 MDG. Nanophan. or phan.

  var. **boivinianus** (Baill.) Leandri, in Fl. Madag. 111: 54 (1958).
   Madagascar (incl. Nosi Bé I.). 29 MDG. Nanophan. or phan.
   * *Kirganelia boiviniana* Baill., Adansonia 2: 49 (1861). *Phyllanthus casticum* f. *boivinianus*
    (Baill.) Müll.Arg. in A.P.de Candolle, Prodr. 15(2): 348 (1866). *Phyllanthus boivinianus*
    (Baill.) Voeltzk., Fl. Fauna Comoren: 448 (1916).

  var. **decipiens**
   Madagascar (incl. Nosi Bé I.). 29 COM MDG. Nanophan. or phan.
   Phyllanthus casticum f. *occidentalis* Leandri, Notul. Syst. (Paris) 7: 175, 179 (1939).

  f. **trilocularis** (Baill.) Leandri, in Fl. Madag. 111: 54 (1958).
   Comoros (Mayotte), Madagascar (Nosi Bé I.). 29 COM MDG.
   * *Kirganelia trilocularis* Baill., Étude Euphorb.: 614 (1858). *Phyllanthus decipiens* var.
    trilocularis (Baill.) Müll.Arg. in A.P.de Candolle, Prodr. 15(2): 347 (1866).

Phyllanthus delagoensis Hutch., Bull. Misc. Inform. Kew 1920: 28 (1920).
SC. Trop. & S. Africa. 26 MOZ ZIM 27 NAT SWZ TVL.

Phyllanthus delicatissimus Jabl., Mem. New York Bot. Gard. 17: 107 (1967).
Venezuela (Amazonas). 82 VEN. Nanophan.

Phyllanthus delpyanus Hutch. in D.Oliver, Fl. Trop. Afr. 6(1): 1047 (1913).
Gabon, E. Zaire, Kenya, Tanzania, Zambia. 23 GAB ZAI 25 KEN TAN 26 ZAM. Nanophan.
or phan.

Phyllanthus deplanchei Müll.Arg., Linnaea 32: 13 (1863).
New Caledonia (incl. I. des Pins). 60 NWC. Nanophan. or phan.
 * *Kirganelia vieillardii* Baill., Adansonia 2: 231 (1862).

Phyllanthus dewildeorum M.G.Gilbert, Kew Bull. 42: 356 (1987).
W. Ethiopia. 24 ETH. Ther. or cham.

Phyllanthus diandrus Pax in H.G.A.Engler & C.G.O.Drude, Veg. Erde 9, III(2): 29 (1921).
Cameroon. 23 CMN.

Phyllanthus dictyophlebsis Radcl.-Sm., Kew Bull. 47: 680 (1992).
Tanzania. 25 TAN.

Phyllanthus dictyospermus Müll.Arg. in A.P.de Candolle, Prodr. 15(2): 394 (1866).
Brazil (Minas Gerais). 84 BZL.

Phyllanthus dimorphus Britton & P.Wilson, Mem. Torrey Bot. Club 16: 75 (1920).
Cuba (Sierra de Trinidad). 81 CUB. Cham. or nanophan.

Phyllanthus dinizii Huber, Bull. Soc. Bot. Genève 6: 182 (1914 publ. 1915).
Brazil (Pará). 84 BZN.

Phyllanthus dinklagei Pax, Bot. Jahrb. Syst. 19: 77 (1894).
WC. Trop. Africa. 23 CMN CON GAB GGI ZAI.

Phyllanthus dinteri Pax, Bot. Jahrb. Syst. 43: 75 (1909).
Namibia. 27 NAM.

Phyllanthus discolor Poepp. ex Spreng., Syst. Veg. 3: 21 (1826).
C. & W. Cuba. 81 CUB. Nanophan.
 Phyllanthus pruinosus Poepp. ex A.Rich. in R.de la Sagra, Hist. Fis. Cuba, Bot. 2: 216 (1850).
 Phyllanthus decander Sessé & Moç., Fl. Mexic., ed. 2: 212 (1894).

Phyllanthus distichus Hook. & Arn., Bot. Beechey Voy.: 95 (1832).
Hawaiian Is. 63 HAW. Nanophan.

var. **degeneri** (Sherff) Govaerts & Radcl.-Sm., Kew Bull. 51: 176 (1996).
Hawaiian Is. 63 HAW. Nanophan.
 * *Phyllanthus sandwicensis* var. *degeneri* Sherff, Publ. Field Mus. Nat. Hist., Bot. Ser. 17: 567 (1939).

var. **distichus**
Hawaiian Is. (Oahu, Maui, Kauai). 63 HAW. Nanophan.
 Phyllanthus argentatus Noronha, Verh. Batav. Genootsch. Kunsten 5(4): 22 (1790), nom. nud.
 Phyllanthus cheremela Roxb., Hort. Bengal.: 104 (1814), nom. nud.
 Phyllanthus sandwicensis Müll.Arg., Linnaea 32: 31 (1863).
 Phyllanthus sandwicensis var. *oblongifolius* Müll.Arg., Linnaea 32: 31 (1863).
 Phyllanthus sandwicensis f. *grandifolia* Wawra, Flora 58: 149 (1875).

var. **ellipticus** (Müll.Arg.) Govaerts & Radcl.-Sm., Kew Bull. 51: 176 (1996).
Hawaiian Is. 63 HAW.
 * *Phyllanthus sandwicensis* var. *ellipticus* Müll.Arg., Linnaea 32: 31 (1863).
 Phyllanthus sandwicensis var. *parvifolius* Müll.Arg., Linnaea 32: 32 (1863). *Phyllanthus sandwicensis* f. *parvifolius* (Müll.Arg.) Wawra, Flora 58: 149 (1875).
 Phyllanthus sandwicensis var. *radicans* Müll.Arg., Linnaea 32: 32 (1863).
 Phyllanthus sandwicensis f. *rufidus* Fosberg, Occas. Pap. Bernice Pauahi Bishop Mus. 12(15): 6 (1936).

Phyllanthus dongmoensis Thin, J. Biol. (Vietnam) 14(2): 16 (1992).
Vietnam. 41 VIE.

Phyllanthus dorotheae M.Schmid, in Fl. N. Caled. & Depend. 17: 300 (1991).
C. & EC. New Caledonia. 60 NWC. Cham. or nanophan.

Phyllanthus dracunculoides Baill., Adansonia 2: 239 (1862). *Glochidion salicifolium* var.
dracunculoides (Baill.) Müll.Arg., Linnaea 32: 59 (1863). *Phyllanthus salicifolius* var.
dracunculoides (Baill.) Müll.Arg. in A.P.de Candolle, Prodr. 15(2): 317 (1866).
New Caledonia. 60 NWC. Nanophan.

   var. **amieuensis** M.Schmid, in Fl. N. Caled. & Depend. 17: 297 (1991).
   C. New Caledonia. 60 NWC. Nanophan.

   var. **dracunculoides**
   NW. New Caledonia. 60 NWC. Nanophan.
   Phyllanthus triquetrus S.Moore, J. Linn. Soc., Bot. 45: 396 (1921).

   var. **tiwakaensis** M.Schmid, in Fl. N. Caled. & Depend. 17: 297 (1991).
   NW. New Caledonia. 60 NWC. Nanophan.

Phyllanthus duidae Gleason, Bull. Torrey Bot. Club 58: 382 (1931).
Venezuela (Amazonas: Cerro Duida). 82 VEN. Nanophan.

Phyllanthus dumbeaensis M.Schmid, in Fl. N. Caled. & Depend. 17: 200 (1991).
New Caledonia (Dumbéa-Couvelée). 60 NWC. Nanophan.

Phyllanthus dumetosus Poir. in J.B.A.M.de Lamarck, Encycl. 5: 303 (1804). *Kirganelia*
dumetosa (Poir.) Spreng., Syst. Veg. 3: 48 (1826).
Rodrigues. 29 ROD. Nanophan.
   Phyllanthus thesioides Müll.Arg., Linnaea 32: 36 (1863).

Phyllanthus dumosus C.B.Rob., Philipp. J. Sci., C 4: 79 (1909).
Philippines. 42 PHI.

Phyllanthus dunnianus (H.Lév.) Hand.-Mazz. ex Rehder, J. Arnold Arbor. 14: 230 (1933).
China (Guizhou, Yunnan, Guangxi), Vietnam. 36 CHC CHS 41 VIE. Nanophan. or phan.
   Phyllanthodendron cavaleriei H.Lév., Repert. Spec. Nov. Regni Veg. 9: 454 (1911).
   *Phyllanthodendron dunnianum H.Lév., Repert. Spec. Nov. Regni Veg. 9: 324 (1911).

Phyllanthus dusenii Hutch., Bull. Misc. Inform. Kew 1911: 314 (1911).
Nigeria, Cameroon. 22 NGA 23 CMN. Nanophan.

Phyllanthus dzumacensis M.Schmid, in Fl. N. Caled. & Depend. 17: 105 (1991).
New Caledonia (Dzumac Mts.). 60 NWC. Nanophan. or phan.

Phyllanthus echinospermus C.Wright, Anales Acad. Ci. Méd. Habana 7: 108 (1870).
W. Cuba. 81 CUB. Cham.
   Phyllanthus minimus C.Wright, Anales Acad. Ci. Méd. Habana 7: 108 (1870).

Phyllanthus edmundoi L.J.M.Santiago, Bradea 5: 46 (1988).
Brazil (Bahia). 84 BZE. Nanophan.

Phyllanthus effusus S.Moore, J. Bot. 61(Suppl.): 45 (1923).
Papua New Guinea. 42 NWG. Phan.
   Phyllanthus oreadum S.Moore, J. Bot. 61(Suppl.): 45 (1923).
   Phyllanthus carolii Diels, Notizbl. Bot. Gart. Berlin-Dahlem 11: 310 (1931).

Phyllanthus ekmanii G.L.Webster, Contr. Gray Herb. 176: 60 (1955).
Cuba (Sierra de Nipe). 81 CUB. Cham. or nanophan.

Phyllanthus elegans Wall. ex Müll.Arg., Linnaea 32: 46 (1863).
Burma, Thailand, N. Pen. Malaysia, Vietnam. 41 BMA THA VIE 42 MLY. Nanophan.

Phyllanthus eliae (Brunel & J.P.Roux) Brunel ex Govaerts & Radcl.-Sm., Kew Bull. 51:
176 (1996).
Togo. 22 TOG. Ther.
  * *Phyllanthus sublanatus* subsp. *eliae* Brunel & J.P.Roux, Bull. Soc. Bot. France 123:
     375 (1976).
     Phyllanthus sublanatus subsp. *eyademae* Brunel, Ann. Univ. Benin, Sci 6: 113 (1979).

Phyllanthus elsiae Urb., Repert. Spec. Nov. Regni Veg. 15: 405 (1919).
Trinidad, Tobago, SW. Mexico, C. America, Venezuela, Guyana, Surinam, Colombia. 79
MXS MXT 80 BLZ ELS PAN 81 TRT 82 GUY SUR VEN 83 CLM. Phan.
•

Phyllanthus emblica L., Sp. Pl.: 982 (1753). *Cicca emblica* (L.) Kurz, Forest Fl. Burma 2:
352 (1877).
Trop. & Subtrop. Asia. (29) mau 36 CHC CHH CHS 38 TAI 40 IND PAK SRL 41 CBD LAO
THA VIE 42 BOR JAW LSI MLY SUM. Phan. – The edible fruits are probably the richest
known natural source of Vitamin C.
Emblica officinalis Gaertn., Fruct. Sem. Pl. 2: 122 (1790).
Phyllanthus mimosifolius Salisb., Prodr. Stirp. Chap. Allerton: 391 (1796).
Phyllanthus taxifolius D.Don, Prodr. Fl. Nepal.: 63 (1825).
Emblica arborea Raf., Sylva Tellur.: 91 (1838).
Phyllanthus glomeratus Roxb. ex Wall., Numer. List: 7903 (1847), nom. inval.
Phyllanthus mairei H.Lév., Bull. Acad. Int. Géogr. Bot. 25: 23 (1915).

Phyllanthus engleri Pax in H.G.A.Engler, Pflanzenw. Ost-Afrikas C: 236 (1895).
Tanzania, Mozambique, Zambia, Zimbabwe. 25 TAN 26 MOZ ZAM ZIM. Nanophan.
or phan.

Phyllanthus epiphyllanthus L., Sp. Pl.: 981 (1753). *Xylophylla epiphyllanthus* (L.) Hornem.,
Hort. Bot. Hafn.: 961 (1815). *Phyllanthus epiphyllanthus* var. *genuinus* Müll.Arg. in A.P.de
Candolle, Prodr. 15(2): 428 (1866), nom. inval.
Caribbean. 81 BAH CAY CUB DOM HAI LEE PUE TRT WIN. Nanophan.

subsp. **dilatatus** (Müll.Arg.) G.L.Webster, J. Arnold Arbor. 39: 203 (1958).
E. Cuba. 81 CUB. Nanophan.
  * *Phyllanthus epiphyllanthus* var. *dilatatus* Müll.Arg. in A.P.de Candolle, Prodr. 15(2):
     428 (1866).

subsp. **domingensis** G.L.Webster, J. Arnold Arbor. 39: 202 (1958).
N. Hispaniola. 81 DOM HAI. Nanophan.

subsp. **epiphyllanthus**
Caribbean. 81 BAH CAY CUB DOM HAI LEE PUE TRT WIN. Nanophan.
Xylophylla falcata Sw., Prodr.: 28 (1788). *Phyllanthus falcatus* (Sw.) J.F.Gmel., Syst. Nat.:
204 (1791).
Xylophylla epiphyllanthus Britton in J.K.Small, Fl. Florida Keys: 76 (1913).

Phyllanthus ericoides Torr. in W.H.Emory, Rep. U.S. Mex. Bound. 2(1): 193 (1858).
Mexico (Chihuahua), Texas (Terrell Co.). 77 TEX 79 MXE. Cham.

Phyllanthus erwinii J.T.Hunter & J.J.Bruhl, J. Adelaide Bot. Gard. 17: 130 (1996).
W. & C. Australia. 50 NTA SOA WAU. Hemicr. or cham.
Phyllanthus lacunarius var. *deuterocalyx* Gauba, Victorian Naturalist 65: 184 (1948).

Phyllanthus erythrotrichus C.B.Rob., Philipp. J. Sci., C 6: 333 (1911).
Philippines. 42 PHI.

Phyllanthus eurisladro Mart. ex Colla, Herb. Pedem. 5: 106 (1836).
Brazil. 84 +. – Provisionally accepted.

Phyllanthus eutaxioides S.Moore, J. Linn. Soc., Bot. 45: 216 (1920).
Queensland (Darwin), Northern Territory (Gove Pen.). 50 NTA QLD.

Phyllanthus evanescens Brandegee, Zoe 5: 207 (1905).
Mexico to Nicaragua. 79 MXN 80 NIC. Ther.

Phyllanthus everettii C.B.Rob., Philipp. J. Sci., C 4: 80 (1909).
Philippines. 42 PHI.

Phyllanthus evrardii Beille in H.Lecomte, Fl. Indo-Chine 5: 599 (1927).
Vietnam. 41 VIE.

Phyllanthus excisus Urb., Repert. Spec. Nov. Regni Veg. 13: 449 (1914).
Cuba (Sierra Sagua Baracoa). 81 CUB. Nanophan.

Phyllanthus exilis S.Moore, J. Bot. 64: 97 (1926).
N. Australia. 50 NTA QLD WAU.

Phyllanthus eximius G.L.Webster & Proctor, J. Arnold Arbor. 41: 283 (1960).
Jamaica. 81 JAM. Nanophan. or phan.

Phyllanthus fadyenii Urb., Symb. Antill. 6: 13 (1909).
Jamaica. 81 JAM. Cham.

Phyllanthus faguetii Baill., Adansonia 2: 237 (1862). *Glochidion faguetii* (Baill.) Müll.Arg.,
Linnaea 32: 59 (1863).
New Caledonia. 60 NWC. Nanophan.

   var. **brevipedicellatus** M.Schmid, in Fl. N. Caled. & Depend. 17: 168 (1991).
   NC. New Caledonia (Touho). 60 NWC. Nanophan.

   var. **faguetii**
   W. & S. New Caledonia. 60 NWC. Nanophan.
   Micrantheum inversum Pancher ex Baill., Adansonia 2: 237 (1862).

   var. **gracilipedicellatus** M.Schmid, in Fl. N. Caled. & Depend. 17: 170 (1991).
   NC. New Caledonia (Touho). 60 NWC. Nanophan.

   var. **lifuensis** (Guillaumin) M.Schmid, in Fl. N. Caled. & Depend. 17: 167 (1991).
   NW. & NC. New Caledonia. 60 NWC. Nanophan.
   *Phyllanthus lifuensis Guillaumin, Arch. Bot. Mém. 2(3): 11 (1929).

   var. **rhombifolius** M.Schmid, in Fl. N. Caled. & Depend. 17: 170 (1991).
   New Caledonia (Poindimié-Tuoho Reg.). 60 NWC. Nanophan.

Phyllanthus fallax Müll.Arg., Linnaea 34: 73 (1865).
Brazil (Minas Gerais). 84 BZL. Cham.

Phyllanthus fangchengensis P.T.Li, Acta Phytotax. Sin. 25: 377 (1987).
China (SE. Guangxi). 36 CHS. Nanophan.

Phyllanthus fastigiatus Mart. ex Müll.Arg., Linnaea 32: 45 (1863).
Brazil (Minas Gerais). 84 BZL.

Phyllanthus favieri M.Schmid, in Fl. N. Caled. & Depend. 17: 262 (1991).
New Caledonia. 60 NWC. Nanophan.

var. **favieri**
New Caledonia (Koniambo). 60 NWC. Nanophan.

var. **kaalaensis** M.Schmid, in Fl. N. Caled. & Depend. 17: 264 (1991).
New Caledonia (Mt. Kaala). 60 NWC. Nanophan.

Phyllanthus ferax Standl., Publ. Field Mus. Nat. Hist., Bot. Ser. 11: 134 (1932).
Belize, Guatemala. 80 BLZ GUA.

Phyllanthus filicifolius Gage, Bull. Misc. Inform. Kew 1914: 241 (1914).
Pen. Malaysia (Kedah). 42 MLY.

Phyllanthus fimbriatitepalus Guillaumin, Bull. Mus. Natl. Hist. Nat., II, 9: 300 (1937).
Vanuatu. 60 VAN.

Phyllanthus fimbriatus (Wight) Müll.Arg., Linnaea 32: 47 (1863).
India. 40 IND.
  * *Reidia fimbriata* Wight, Icon. Pl. Ind. Orient. 5(2): 28, t. 1904(1) (1852).

Phyllanthus fimbricalyx P.T.Li, Acta Phytotax. Sin. 25: 380 (1987).
China (SW. Yunnan). 36 CHC. Nanophan.

Phyllanthus finschii K.Schum., Bot. Jahrb. Syst. 9: 205 (1887). *Hemiglochidion finschii*
(K.Schum.) K.Schum. in K.M.Schumann & C.A.G.Lauterbach, Fl. Schutzgeb. Südsee,
Nachtr.: 289 (1905).
New Guinea, Bismarck Archip., Tenimbar I. 42 BIS NWG. Nanophan. or phan.
    Phyllanthus cupuliformis Warb., Bot. Jahrb. Syst. 13: 356 (1891).
    Hemiglochidion cupuliforme K.Schum. in K.M.Schumann & C.A.G.Lauterbach, Fl.
      Schutzgeb. Südsee, Nachtr.: 289 (1905).
    Hemiglochidion hylodendron K.Schum. in K.M.Schumann & C.A.G.Lauterbach, Fl.
      Schutzgeb. Südsee, Nachtr.: 289 (1905). *Phyllanthus hylodendron* (K.Schum.) Airy
      Shaw & G.L.Webster, Kew Bull. 26: 99 (1971).

Phyllanthus fischeri Pax, Bot. Jahrb. Syst. 19: 77 (1894).
Ethiopia to Tanzania. 24 ETH 25 KEN TAN UGA. Ther., cham. or nanophan.

Phyllanthus flagellaris Benth., Fl. Austral. 6: 106 (1873).
Northern Territory (Darwin). 50 NTA.

Phyllanthus flagelliformis Müll.Arg., Linnaea 32: 54 (1863).
Brazil (Bahia). 84 BZE.
    Phyllanthus linearis Mart., Nova Acta Acad. Caes. Leop-Carol. German. Nat. Cur. 11:
      38 (1823).
    Phyllanthus flagelliformis var. *demonstrans* Müll.Arg., Linnaea 32: 55 (1863).

Phyllanthus flaviflorus (Lauterb. & K.Schum.) Airy Shaw, Kew Bull. 23: 39 (1969).
Papua New Guinea. 42 NWG. Nanophan. or phan.
  * *Actephila flaviflora* K.Schum. & Lauterb., Fl. Schutzgeb. Südsee: 388 (1900).
    Phyllanthus nervosus Airy Shaw, Kew Bull. 20: 383 (1966).

Phyllanthus flexuosus (Siebold & Zucc.) Müll.Arg. in A.P.de Candolle, Prodr. 15(2):
324 (1866).
S. China, S. Japan. 36 CHC CHS 38 JAP. Nanophan.

* *Cicca flexuosa* Siebold & Zucc., Fl. Jap. Fam. Nat. 1: 35 (1845). *Glochidion flexuosum* (Siebold & Zucc.) F.Muell. ex Miq., Ann. Mus. Bot. Lugduno-Batavi 3: 128 (1867). *Hemicicca flexuosa* (Siebold & Zucc.) Hurus., Bot. Mag. (Tokyo) 60: 71 (1947). *Hemicicca japonica* Baill., Étude Euphorb.: 646 (1858). *Phyllanthus japonicus* (Baill.) Müll.Arg., Linnaea 32: 52 (1863).

Phyllanthus fluitans Benth. ex Müll.Arg., Linnaea 32: 36 (1863).
Brazil (Amazonas), Peru (Loreto), Ecuador, Venezuela, Paraguay. 82 VEN 83 ECU PER 84 BZN 85 PAR. Hel.

Phyllanthus fluminis-athi Radcl.-Sm., Kew Bull. 29: 439 (1974).
SC. Kenya. 25 KEN. Nanophan.

Phyllanthus × fluminis-sabi Radcl.-Sm., Kew Bull. 51: 306 (1996). P. muellerianum × P. reticulatum.
Mozambique. 26 MOZ.

Phyllanthus fluminis-zambesi Radcl.-Sm., Kew Bull. 51: 306 (1996).
Zambia. 26 ZAM.

Phyllanthus formosus Urb., Repert. Spec. Nov. Regni Veg. 13: 450 (1914).
E. Cuba. 81 CUB. Cham.

Phyllanthus forrestii W.W.Sm., Notes Roy. Bot. Gard. Edinburgh 8: 195 (1914).
SC. China. 36 CHC. Cham.
Phyllanthus echinocarpus T.L.Chin, Acta Phytotax. Sin. 19: 347 (1981).

Phyllanthus fractiflexus M.Schmid, in Fl. N. Caled. & Depend. 17: 238 (1991).
New Caledonia (Mt. Koniambo). 60 NWC. Nanophan.

Phyllanthus fraguensis M.C.Johnst., Syst. Bot. 10: 300 (1985).
Mexico. 79 MXE.

Phyllanthus francavillanus Müll.Arg., Linnaea 32: 20 (1863).
Brazil (Pará: Cano). 84 BZN.

Phyllanthus franchetianus H.Lév., Bull. Acad. Int. Géogr. Bot. 25: 23 (1915).
China (Guizhou, Sichuan, Yunnan). 36 CHC. Nanophan.
Phyllanthus leiboensis T.L.Chin, Acta Phytotax. Sin. 19: 348 (1981).

Phyllanthus francii Guillaumin, Bull. Mus. Natl. Hist. Nat. 33: 273 (1927).
S. New Caledonia. 60 NWC. Cham. or nanophan.

Phyllanthus fraternus G.L.Webster, Contr. Gray Herb. 176: 53 (1955).
Pakistan to NW. India, naturalised elsewere. (21) cvi (22) gam gnb gui mli nga ngr sen tog (23) rwa zai (24) cha dji eth som sud (25) ken tan (26) all (27) bot nam (29) com (35) oma yem 40 IND PAK WHM (78) fla (81) ber trt win. Ther.
Phyllanthus lonphali Wall., Numer. List: 7895 (1847), nom. inval.
Phyllanthus fraternus subsp. *togoensis* Brunel & J.P.Roux, Bull. Soc. Bot. France 122: 161 (1975).

Phyllanthus frazieri Radcl.-Sm., Kew Bull. 37: 425 (1982).
Tanzania. 25 TAN. Ther.

Phyllanthus friesii Hutch. in R.E.Fries, Wiss. Erg. Schwed. Rhod.-Kongo Exped. 1: 121 (1911-1912 publ. 1914).
Tanzania (Iringa), Zambia. 25 TAN 26 ZAM. Hemicr.

Phyllanthus frodinii Airy Shaw, Kew Bull. 27: 74 (1972).
NE. Papua New Guinea. 42 NWG. Phan.

Phyllanthus fuernrohrii F.Muell., Trans. Philos. Soc. Victoria 1: 15 (1855).
Australia. 50 NSW NTA QLD SOA WAU.

Phyllanthus fuertesii Urb., Repert. Spec. Nov. Regni Veg. 13: 451 (1914). *Phyllanthus micranthus* subsp. *fuertesii* (Urb.) G.L.Webster, Contr. Gray Herb. 176: 55 (1955).
Hispaniola. 81 DOM HAI. Ther. or cham.

Phyllanthus fuscoluridus Müll.Arg. in A.P.de Candolle, Prodr. 15(2): 346 (1866).
Madagascar. 29 MDG. Nanophan. or phan.

   var. **borealis** Leandri, in Fl. Madag. 111: 49 (1958).
      C. Madagascar. 29 MDG. Nanophan. or phan.

   var. **decaryanus** Leandri, in Fl. Madag. 111: 47 (1958).
      E. Madagascar. 29 MDG. Nanophan. or phan.

   subsp. **fuscoluridus**
      C. Madagascar. 29 MDG. Nanophan. or phan.
      Phyllanthus bojerianus Baill., Adansonia 2: 53 (1861), nom. nud.
      Phyllanthus emirnensis Müll.Arg., Linnaea 32: 14 (1863).

   var. **parviflorus** Leandri, Notul. Syst. (Paris) 7: 180 (1939).
      E. Madagascar. 29 MDG. Nanophan. or phan.

   subsp. **villosus** Leandri, Notul. Syst. (Paris) 6: 196 (1938).
      C. Madagascar. 29 MDG. Nanophan. or phan.

Phyllanthus gageanus (Gamble) M.Mohanan, J. Econ. Taxon. Bot. 6: 480 (1985).
S. India. 40 IND.
   * *Reidia gageana* Gamble, Bull. Misc. Inform. Kew 1925: 331 (1925).

Phyllanthus gagnioevae Brunel & J.P.Roux, Boissiera 32: 175 (1980).
W. Trop. Africa to Rwanda. 22 GUI IVO LBR NGA SIE TOG 23 CAF CMN RWA.

Phyllanthus galeottianus Baill., Adansonia 1: 32 (1860).
Mexico (Baja California Sur, Sonora to Michoacan). 79 MXE MXG MXN MXS.

Phyllanthus gallinetae Jabl., Mem. New York Bot. Gard. 17: 111 (1967).
Guyana, Venezuela (Amazonas), Brazil (Amazonas). 82 GUY VEN 84 BZN. Nanophan. or phan.

Phyllanthus geayi Leandri ex Humbert, Notul. Syst. (Paris) 6: 193 (1938).
E. Madagascar. 29 MDG. Cham. or nanophan.

Phyllanthus gentryi G.L.Webster, Ann. Missouri Bot. Gard. 75: 1096 (1988).
Panama. 80 PAN. Nanophan. or phan.

Phyllanthus geoffrayi Beille in H.Lecomte, Fl. Indo-Chine 5: 584 (1927).
E. Thailand, Laos. 41 LAO THA. Nanophan.

Phyllanthus gillettianus Brunel ex Radcl.-Sm., Kew Bull. 51: 306 (1996).
Zaire, Kenya, Zambia, Zimbabwe, Botswana. 23 ZAI 25 KEN 26 ZAM ZIM 27 BOT. Ther.

Phyllanthus gjellerupi J.J.Sm., Nova Guinea 8: 780 (1912).
Irian Jaya. 42 NWG. Nanophan.

Phyllanthus glabrescens (Miq.) Müll.Arg., Linnaea 32: 48 (1863).
Lesser Sunda Is. (Sumbawa). 42 LSI.
* Reidia glabrescens Miq., Fl. Ned. Ind. 1(2): 374 (1859).

Phyllanthus gladiatus Müll.Arg., Linnaea 32: 52 (1863).
Brazil (Bahia). 84 BZE.

Phyllanthus glaucinus (Miq.) Müll.Arg., Linnaea 32: 13 (1863).
Sumatera. 42 SUM.
* Anisonema glaucinum Miq., Fl. Ned. Ind., Eerste Bijv.: 449 (1861).

Phyllanthus glaucophyllus Sond., Linnaea 23: 133 (1850).
Trop. & S. Africa. 22 GNB GUI IVO LBR SEN SIE 23 CAF ZAI 25 KEN TAN UGA 25 ANG MLW MOZ ZAM ZIM 27 NAT OFS SWZ TVL. Cham.
Phyllanthus glaucophyllus var. *major* Müll.Arg., Flora 47: 514 (1864).

Phyllanthus glaucoviridis Jabl., Mem. New York Bot. Gard. 17: 101 (1967).
Venezuela (Amazonas: Rio Atabapo). 82 VEN. Nanophan. or phan.

Phyllanthus glaucus Wall. ex Müll.Arg., Linnaea 32: 14 (1863). *Flueggeopsis glauca* (Wall. ex Müll.Arg.) D.Das in U.N.Kanjilal & al., Fl. Assam 4: 158 (1940). *Hemicicca glauca* (Wall. ex Müll.Arg.) Hurus. & Yu.Tanaka, Fl. E. Himal.: 179 (1966).
Nepal to S. China. 36 CHC CHH CHS CHT 40 ASS EHM NEP. Nanophan. or phan.
Phyllanthus flueggeiformis Müll.Arg. in A.P.de Candolle, Prodr. 15(2): 349 (1866).

Phyllanthus glaziovii Müll.Arg. in C.F.P.von Martius, Fl. Bras. 11(2): 41 (1873).
SE. Brazil. 84 BZL.

Phyllanthus glochidioides Elmer, Leafl. Philipp. Bot. 4: 1302 (1911).
Philippines. 42 PHI.

Phyllanthus gneissicus S.Moore, J. Linn. Soc., Bot. 45: 399 (1921).
New Caledonia. 60 NWC. Nanophan.

var. **broumoiriensis** M.Schmid, in Fl. N. Caled. & Depend. 17: 256 (1991).
New Caledonia (Mé Broumoiri). 60 NWC. Nanophan.

var. **gneissicus**
NW. New Caledonia. 60 NWC. Nanophan.

var. **ramosus** M.Schmid, in Fl. N. Caled. & Depend. 17: 256 (1991).
New Caledonia (Ponérihouen to Touho). 60 NWC. Nanophan.

var. **toninensis** (S.Moore) M.Schmid, in Fl. N. Caled. & Depend. 17: 255 (1991).
NW. New Caledonia. 60 NWC. Nanophan.
* Phyllanthus toninensis S.Moore, J. Linn. Soc., Bot. 45: 400 (1921).

Phyllanthus goianensis L.J.M.Santiago, Bradea 5(2): 45 (1988).
Brazil (Goiás). 84 BZC. Cham.

Phyllanthus golonensis M.Schmid, in Fl. N. Caled. & Depend. 17: 304 (1991).
New Caledonia (Golone). 60 NWC. Cham. or nanophan.

Phyllanthus gomphocarpus Hook.f., Fl. Brit. India 5: 301 (1887). *Reidia gomphocarpa* (Hook.f.) C.E.C.Fisch., Bull. Misc. Inform. Kew 1927: 314 (1927).
Pen. Malaysia, Jawa. 42 JAW MLY.
Phyllanthus accrescens J.J.Sm. in S.H.Koorders & T.Valeton, Bijdr. Boomsoort. Java 12: 101 (1910).

Phyllanthus goniostemon Radcl.-Sm., Kew Bull. 51: 307 (1996).
Uganda. 25 UGA. Ther.

Phyllanthus gossweileri Hutch., Bull. Misc. Inform. Kew 1911: 315 (1911).
Angola, Zaire, Zambia. 23 ZAI 26 ANG ZAM.

Phyllanthus goudotianus (Baill.) Müll.Arg., Linnaea 32: 8 (1863).
E. Madagascar. 29 MDG. Nanophan.
* *Menarda goudotiana* Baill., Adansonia 2: 62 (1861).

Phyllanthus gracilipes (Miq.) Müll.Arg., Linnaea 32: 47 (1863).
China (W. Guangxi), Thailand (incl. Lankawi I.), Vietnam, W. Malesia. 36 CHS 41 THA
VIE 42 BOR JAW MLY SUM. Nanophan.
Eriococcus gracilis Hassk., Tijdschr. Natuurl. Gesch. Physiol. 10: 143 (1843). *Reidia
gracilis* (Hassk.) Miq., Fl. Ned. Ind. 1(2): 373 (1859).
* *Reidia gracilipes* Miq., Fl. Ned. Ind. 1(2): 374 (1859).
Phyllanthus concinnus Ridl., J. Straits Branch Roy. Asiat. Soc. 59: 171 (1911).
Phyllanthus hullettii Ridl., Bull. Misc. Inform. Kew 1923: 363 (1923).
Phyllanthus discofractus Croizat, J. Arnold Arbor. 23: 31 (1942).

Phyllanthus graminicola Hutch., J. Linn. Soc., Bot. 40: 191 (1911).
Zimbabwe, Mozambique, Northern Prov. 26 MOZ ZIM 27 TVL.
Phyllanthus rogersii Hutch. ex S.Moore, J. Bot. 57: 160 (1919).

Phyllanthus grandifolius L., Sp. Pl.: 981 (1753). *Phyllanthus grandifolius* var. *genuinus*
Müll.Arg. in A.P.de Candolle, Prodr. 15(2): 329 (1866), nom. inval. *Asterandra grandifolia*
(L.) Britton, Brooklyn Bot. Gard. Mem. 1: 61 (1918).
SE. Mexico. 79 MXT 84 BZE?
Phyllanthus grandiflorus Crantz, Inst. Rei Herb. 1: 144 (1766).
Andrachne arborea Mill., Gard. Dict. ed. 8: 3 (1768).
Phyllanthus glaucescens Kunth in F.W.H.von Humboldt, A.J.A.Bonpland & C.S.Kunth,
Nov. Gen. Sp. 2: 115 (1817).
Agyneia berteri Spreng., Syst. Veg. 3: 19 (1826).
Phyllanthus averrhoifolius Müll.Arg. in A.P.de Candolle, Prodr. 15(2): 329 (1866), pro syn.

Phyllanthus graveolens Kunth in F.W.H.von Humboldt, A.J.A.Bonpland & C.S.Kunth, Nov.
Gen. Sp. 2: 112 (1817).
Peru, Brazil (Amazonas). 83 PER 84 BZN. Nanophan.
Phyllanthus tenellus Benth., Bot. Voy. Sulphur: 165 (1846), nom. illeg. *Phyllanthus
graveolens* var. *tenellus* (Benth.) Müll.Arg. in A.P.de Candolle, Prodr. 15(2): 383 (1866).
Phyllanthus benthamianus Müll.Arg., Linnaea 32: 29 (1863).

Phyllanthus greenei Elmer, Leafl. Philipp. Bot. 3: 929 (1910).
Philippines. 42 PHI.

Phyllanthus griffithii Müll.Arg., Linnaea 32: 27 (1863).
Assam, Bhutan. 40 ASS EHM. Nanophan.
Phyllanthus stylosus Griff., Itin. Pl. Khasyah Mts.: 33 (1848), nom. nud.

Phyllanthus guangdongensis P.T.Li, Acta Phytotax. Sin. 25: 376 (1987).
China (WSW. Guangdong). 36 CHS. Nanophan.

Phyllanthus guanxiensis Govaerts & Radcl.-Sm., Kew Bull. 51: 176 (1996).
China (SW. Guangxi). 36 CHS. Nanophan.
* *Phyllanthodendron petraeum* P.T.Li, Bull. Bot. Res., Harbin 7(3): 4 (1987).

Phyllanthus guillauminii Däniker, Vierteljahrschr. Naturf. Ges. Zürich 76: 167 (1931).
New Caledonia (Tiébaghi Mts.). 60 NWC. Nanophan.

Phyllanthus gunnii Hook.f., London J. Bot. 6: 284 (1847).
E. & SE. Australia. 50 NSW QLD TAS VIC.
Phyllanthus gasstroemii Müll.Arg. in A.P.de Candolle, Prodr. 15(2): 358 (1866).
Phyllanthus indigoferoides A.Cunn. ex Benth., Fl. Austral. 6: 104 (1873), pro syn.

Phyllanthus gypsicola McVaugh, Brittonia 13: 194 (1961).
Mexico (Colima). 79 MXE. Nanophan.

Phyllanthus hainanensis Merr., Lingnan Sci. J. 14: 20 (1935).
Hainan. 36 CHH. Nanophan.

Phyllanthus hakgalensis Thwaites ex Trimen, J. Bot. 23: 242 (1885, as 'uakgalensis'); Syst.
Cat. Fl. Pl. Ceylon: 80 (1885).
Sri Lanka (Hakgala Hill). 40 SRL. Nanophan. – Known only from the type.

Phyllanthus harmandii Beille in H.Lecomte, Fl. Indo-Chine 5: 586 (1927).
Cambodia. 41 CBD.

Phyllanthus harrimanii G.L.Webster, Rhodora 80: 570 (1978).
Mexico (Tamaulipas). 79 MXE.

Phyllanthus harrisii Radcl.-Sm., Kew Bull. 35: 768 (1981). – FIGURE, p. 1266.
Kenya, Tanzania (incl. Zanzibar). 25 KEN TAN. Ther.

Phyllanthus hasskarlianus Müll.Arg., Linnaea 32: 16 (1863).
Jawa. 42 JAW.
 * *Agyneia multiflora* Hassk., Cat. Hort. Bot. Bogor.: 240 (1844).

Phyllanthus helenae M.Schmid, in Fl. N. Caled. & Depend. 17: 77 (1991).
C. New Caledonia. 60 NWC. Nanophan.

Phyllanthus heliotropus C.Wright ex Griseb., Nachr. Königl. Ges. Wiss. Georg-Augusts-
Univ. 1: 167 (1865).
W. Cuba (incl. I. de la Juventud). 81 CUB. Hemicr.

Phyllanthus heteradenius Müll.Arg. in C.F.P.von Martius, Fl. Bras. 11(2): 63 (1873).
Brazil (?). 84 +. Biennial.

Phyllanthus heterophyllus E.Mey. ex Müll.Arg., Linnaea 32: 43 (1863).
KwaZulu-Natal, Cape Prov. 27 CPP NAT.
Phyllanthus incurvus Sond., Linnaea 23: 135 (1850), nom. illeg.
Phyllanthus andrachniformis Pax, Bull. Herb. Boissier, II, 8: 634 (1908).

Phyllanthus heterotrichus Lundell, Contr. Univ. Michigan Herb. 7: 24 (1942).
Mexico (San Luis Potosí). 79 MXE. Nanophan.

Phyllanthus hexadactylus McVaugh, Brittonia 13: 195 (1961).
W. Mexico. 79 MXS. Ther.

Phyllanthus heyneanus Müll.Arg., Linnaea 32: 49 (1863).
S. India. 40 IND. Nanophan.
Reidia ovalifolia Wight, Icon. Pl. Ind. Orient. 5(2): 28, t. 1904(3) (1852). *Eriococcus*
ovalifolia (Wight) K.K.N.Nair, J. Bombay Nat. Hist. Soc. 79: 454 (1982 publ. 1983).
Phyllanthus longiflorus B.Heyne ex Müll.Arg., Linnaea 32: 49 (1863), pro syn.
Phyllanthus nephradenius Müll.Arg. in A.P.de Candolle, Prodr. 15(2): 423 (1866).
Reidia longiflora Gamble, Fl. Madras: 1293 (1925).

Phyllanthus hildebrandtii Pax, Bot. Jahrb. Syst. 15: 526 (1893).
Ethiopia, N. Somalia. 24 ETH SOM. Nanophan. – Perhaps not distinct from *P. sepialis*.

Phyllanthus hirtellus F.Muell. ex Müll.Arg., Linnaea 32: 22 (1863). *Phyllanthus thymoides*
var. *hirtellus* (F.Muell. ex Müll.Arg.) Müll.Arg. in A.P.de Candolle, Prodr. 15(2): 372 (1866).
SE. Australia. 50 NSW VIC.
Phyllanthus hirtellus var. *ledifolius* Müll.Arg., Linnaea 32: 22 (1863). *Phyllanthus thymoides*
var. *ledifolius* (Müll.Arg.) Müll.Arg. in A.P.de Candolle, Prodr. 15(2): 372 (1866).
Phyllanthus hirtellus var. *thymoides* Müll.Arg., Linnaea 32: 22 (1863). *Phyllanthus thymoides*
(Müll.Arg.) Sieber ex Müll.Arg. in A.P.de Candolle, Prodr. 15(2): 372 (1866).
Phyllanthus thymoides var. *glabratus* Müll.Arg. in A.P.de Candolle, Prodr. 15(2): 372 (1866).
Phyllanthus ledifolius A.Cunn. ex Benth., Fl. Austral. 6: 109 (1873).
Phyllanthus thymoides var. *parviflorus* J.M.Black, Fl. S. Austral. 2: 352 (1924).
Phyllanthus hirtellus var. *parviflorus* (J.M.Black) Eichler, Suppl. J.M. Blacki Fl. Austral.
ed. 2: 209 (1965).

Phyllanthus hodjelensis Schweinf., Bull. Herb. Boissier 7: 304 (1899).
Yemen. 35 YEM.

Phyllanthus holostylus Milne-Redh., Bull. Misc. Inform. Kew 1937: 414 (1937).
Zaire, Angola, Zambia. 23 ZAI 26 ANG ZAM. Cham.

Phyllanthus hortensis Govaerts & Radcl.-Sm., Kew Bull. 51: 177 (1996).
Origin unknown. – Described from a specimen cultivated in the Jardin des Plantes, Paris.
* *Phyllanthus ovatus* Desf. ex Poir. in J.B.A.M.de Lamarck, Encycl., Suppl. 4: 404 (1816),
nom. illeg.

Phyllanthus houailouensis M.Schmid, in Fl. N. Caled. & Depend. 17: 93 (1991).
C. New Caledonia (Houaïlou). 60 NWC. Nanophan.

Phyllanthus huallagensis Standl. ex Croizat, J. Wash. Acad. Sci. 33: 13 (1943).
Peru (San Martin). 83 PER. Phan.

Phyllanthus humbertianus Leandri, Notul. Syst. (Paris) 6: 194 (1938).
SW. Madagascar. 29 MDG. Nanophan.

Phyllanthus hutchinsonianus S.Moore, J. Linn. Soc., Bot. 40: 192 (1911).
Tanzania, Mozambique, Zimbabwe. 25 TAN 26 MOZ ZIM. Cham.

Phyllanthus hypoleucus Müll.Arg., Linnaea 32: 40 (1863). *Phyllanthus lacteus* Müll.Arg. in
A.P.de Candolle, Prodr. 15(2): 402 (1866), nom. superfl.
Brazil (Bahia). 84 BZE. Cham.

Phyllanthus hypospodius F.Muell., Victorian Naturalist 8: 177 (Mar. 1892).
Queensland (South Kennedy). 50 QLD.

Phyllanthus hyssopifolioides Kunth in F.W.H.von Humboldt, A.J.A.Bonpland &
C.S.Kunth, Nov. Gen. Sp. 2: 108 (1817).
Dominican Rep., Trinidad, Costa Rica to Paraguay. 80 COS PAN 81 DOM TRT 82 GUI
VEN 84 BZN 85 PAR. Ther.
Phyllanthus hyssopifolius Müll.Arg. in A.P.de Candolle, Prodr. 15(2): 390 (1866), sphalm.
Phyllanthus monocladus Urb., Repert. Spec. Nov. Regni Veg. 15: 404 (1919).

Phyllanthus ibonensis Rusby, Mem. New York Bot. Gard. 7: 281 (1927).
Bolivia. 83 BOL.

Phyllanthus ichthyomethius Rusby, Mem. New York Bot. Gard. 7: 282 (1927).
  Bolivia. 83 BOL.

Phyllanthus imbricatus G.L.Webster, Contr. Gray Herb. 176: 56 (1955).
  Cuba (SW. I. de la Juventud). 81 CUB. Hemicr.
    Phyllanthus nanus Millsp. ex Britton, Bull. Torrey Bot. Club 43: 464 (1916), nom. illeg.

Phyllanthus incrustatus Urb., Repert. Spec. Nov. Regni Veg. 13: 449 (1914).
  NE. Cuba. 81 CUB. Nanophan.

Phyllanthus incurvus Thunb., Prodr. Pl. Cap.: 24 (1794).
  SC. Trop. & S. Africa. 26 MOZ ZIM 27 BOT CPP NAM NAT OFS SWZ TVL.
    Phyllanthus capensis Spreng. ex Sond., Linnaea 23: 135 (1850).
    Phyllanthus genistoides Sond., Linnaea 23: 134 (1850).
    Phyllanthus multicaulis Müll.Arg., Linnaea 32: 18 (1863).

Phyllanthus indigoferoides Benth., Fl. Austral. 6: 110 (1873).
  N. Western Australia, Northern Territory. 50 NTA WAU.

Phyllanthus indofischeri Bennet, Indian Forester 109: 221 (1983).
  India. 40 IND.
    * *Emblica fischeri* Gamble, Bull. Misc. Inform. Kew 1925: 330 (1925). *Phyllanthus fischeri*
      (Gamble) J.L.Ellis, Bull. Bot. Surv. India 22: 193 (1980 publ. 1982), no basionym vol.
      or page. *Phyllanthus cecilfischeri* J.L.Ellis, Bull. Bot. Surv. India 24: 209 (1982 publ.
      1983), nom. superfl.

Phyllanthus inflatus Hutch., Bull. Misc. Inform. Kew 1920: 334 (1920).
  S. Sudan to Mozambique. 24 SUD 25 KEN TAN UGA 26 MLW MOZ ZAM ZIM. Nanophan.
    or phan.

Phyllanthus insulae-japen Airy Shaw, Kew Bull. 33: 37 (1978).
  Irian Jaya. 42 NWG. Nanophan.

Phyllanthus insulensis Beille in H.Lecomte, Fl. Indo-Chine 5: 604 (1927).
  Vietnam. 41 VIE.

Phyllanthus involutus J.T.Hunter & J.J.Bruhl, Telopea 7: 155 (1997).
  S. Queensland, New South Wales. 50 NSW QLD. Cham. or nanophan.

Phyllanthus iratsiensis Leandri, Notul. Syst. (Paris) 6: 193 (1938).
  EC. Madagascar. 29 MDG. Nanophan.

Phyllanthus irriguus Radcl.-Sm., Kew Bull. 29: 440 (1974). – FIGURE, p. 1266.
  Tanzania (Songea). 25 TAN. Ther. or hel.

Phyllanthus isalensis (Leandri) Leandri, Bull. Soc. Bot. France 81: 450 (1934).
  WC. Madagascar. 29 MDG. Cham. or nanophan.
    * *Phyllanthus betsileanus* var. *isalensis* Leandri, Notul. Syst. (Paris) 6: 188 (1938).

Phyllanthus isomonensis Leandri, Mém. Inst. Sci. Madagascar, Sér. B, Biol. Vég. 8: 229 (1957).
  C. Madagascar. 29 MDG. Cham. or nanophan.

Phyllanthus itatiaiensis Brade, Arch. Jard. Bot. Rio de Janeiro 15: 9 (1957).
  SE. Brazil. 84 BZL.

Phyllanthus ivohibeus Leandri, Notul. Syst. (Paris) 6: 197 (1938).
　　C. Madagascar. 29 MDG. Nanophan.

Phyllanthus jablonskianus Steyerm. & Luteyn, Ann. Missouri Bot. Gard. 71: 317 (1984).
　　Venezuela (Amazonas: Cerro de la Neblina). 82 VEN.

Phyllanthus jaegeri Brunel & J.P.Roux, Boissiera 32: 176 (1980).
　　Sierra Leone. 22 SIE.

Phyllanthus jaffrei M.Schmid, in Fl. N. Caled. & Depend. 17: 113 (1991).
　　C. New Caledonia. 60 NWC. Nanophan.

Phyllanthus janeirensis Müll.Arg. in C.F.P.von Martius, Fl. Bras. 11(2): 45 (1873).
　　Brazil (Rio de Janeiro). 84 BZL.

Phyllanthus jauaensis Jabl., Mem. New York Bot. Gard. 23: 865 (1972).
　　Venezuela (Bolívar: Jaua Mts.). 82 VEN.

Phyllanthus jaubertii Vieill. ex Guillaumin, Arch. Bot. Mém. 2(3): 9 (1929).
　　New Caledonia. 60 NWC. Nanophan.

　var. **brachypoda** Guillaumin, Arch. Bot. Mém. 2(3): 10 (1929).
　　New Caledonia (Canala). 60 NWC. Nanophan.

　var. **jaubertii**
　　New Caledonia (Poindimié-Touho Reg.). 60 NWC. Nanophan.

Phyllanthus javanicus (Miq.) Müll.Arg., Linnaea 32: 50 (1863).
　　Jawa. 42 JAW.
　　　* *Hedycarpus javanicus* Miq., Fl. Ned. Ind. 1(2): 359 (1859).

Phyllanthus juglandifolius Willd., Enum. Pl., Suppl.: 64 (1814).
　　Trop. America. 81 CUB DOM HAI PUE LEE TRT 82 FRG GUY VEN 83 CLM ECU PER 84
　　BZE BZN. Nanophan. or phan.

　subsp. **cornifolius** (Kunth) G.L.Webster, J. Arnold Arbor. 39: 151 (1958).
　　Trinidad, Venezuela, N. & E. Brazil, Ecuador, Peru. 81 TRT 82 VEN 83 ECU PER 84 BZE
　　BZN. Nanophan. or phan.
　　* *Phyllanthus cornifolius* Kunth in F.W.H.von Humboldt, A.J.A.Bonpland & C.S.Kunth,
　　　Nov. Gen. Sp. 2: 115 (1817). *Asterandra cornifolia* (Kunth) Klotzsch, Arch.
　　　Naturgesch. 7: 200 (1841). *Phyllanthus grandifolius* var. *cornifolius* (Kunth) Müll.Arg.
　　　in A.P.de Candolle, Prodr. 15(2): 329 (1866).
　　Phyllanthus grandifolius var. *salzmannii* Müll.Arg. in A.P.de Candolle, Prodr. 15(2):
　　　329 (1866).

　subsp. **juglandifolius**
　　Trop. America. 81 CUB DOM HAI PUE LEE 82 FRG GUY VEN 83 CLM ECU PER 84 BZE
　　BZN. Nanophan. or phan.
　　Phyllanthus quinquefidus Sessé & Moç. ex Baill., Adansonia 1: 39 (1860).

Phyllanthus junceus Müll.Arg. in A.P.de Candolle, Prodr. 15(2): 411 (1866).
　　W. Cuba (incl. I. de la Juventud). 81 CUB. Hemicr. or cham.
　　　Phyllanthus pruinosus var. *subnudus* C.Wright ex Griseb., Cat. Pl. Cub.: 16 (1866).
　　　Phyllanthus squamatus C.Wright ex Sauvalle, Anales Acad. Ci. Méd. Habana 7: 109 (1870).

Phyllanthus kaessneri Hutch., Bull. Misc. Inform. Kew 1911: 315 (1911).
　　Zaire, Kenya, Tanzania, Zambia. 23 ZAI 25 KEN TAN 26 ZAM. Nanophan. or phan.

var. **kaessneri**
Kenya, Tanzania, Zambia. 25 KEN TAN 26 ZAM. Nanophan. or phan.

var. **polycytotrichus** Radcl.-Sm., Kew Bull. 35: 769 (1981).
Zaire, Kenya, Tanzania, Zambia. 23 ZAI 25 KEN TAN 26 ZAM. Nanophan. or phan.

Phyllanthus kampotensis Beille in H.Lecomte, Fl. Indo-Chine 5: 606 (1927).
Cambodia, Vietnam. 41 CBD VIE.

Phyllanthus kanalensis Baill., Adansonia 2: 234 (1862). *Glochidion lenormandii* Müll.Arg.,
Linnaea 32: 59 (1863).
New Caledonia. 60 NWC. Nanophan.
Phyllanthus eugenioides Guillaumin, Mém. Mus. Natl. Hist. Nat., B, Bot. 8: 244 (1962).

Phyllanthus kerrii Airy Shaw, Kew Bull. 23: 32 (1969).
NE. Thailand. 41 THA. Nanophan.

Phyllanthus kerstingii Brunel, Willdenowia 15: 251 (1985).
W. Trop. Africa. 22 GHA IVO MLI NGA SEN SIE TOG.

Phyllanthus keyensis Warb., Bot. Jahrb. Syst. 13: 355 (1891). *Flueggeopsis keyensis* (Warb.)
K.Schum. in K.M.Schumann & C.A.G.Lauterbach, Fl. Schutzgeb. Südsee, Nachtr.: 290 (1905).
New Guinea (Kei I.). 42 NWG.

Phyllanthus kinabaluicus Airy Shaw, Kew Bull. 29: 294 (1974).
Borneo (Sabah). 42 BOR. Nanophan.

Phyllanthus klotzschianus Müll.Arg., Linnaea 32: 53 (1863).
Guyana, Brazil (Minas Gerais, Bahia). 82 GUY 84 BZE BZL. Nanophan.
Phyllanthus klotzschianus var. *gardneri* Müll.Arg., Linnaea 32: 53 (1863).
Phyllanthus klotzschianus var. *linearis* Müll.Arg., Linnaea 32: 53 (1863).
Phyllanthus klotzschianus var. *major* Müll.Arg., Linnaea 32: 53 (1863).
Phyllanthus klotzschianus var. *minor* Müll.Arg., Linnaea 32: 53 (1863).
Phyllanthus klotzschianus var. *pallidiflorus* Müll.Arg., Linnaea 32: 53 (1863).
Phyllanthus klotzschianus var. *racemulosus* Müll.Arg., Linnaea 32: 53 (1863).
Phyllanthus klotzschianus var. *robustus* Müll.Arg., Linnaea 32: 53 (1863).

Phyllanthus koghiensis Guillaumin, Arch. Bot. Mém. 2(3): 10 (1929).
SE. New Caledonia (Massif des Koghis). 60 NWC. Nanophan. or phan.

Phyllanthus koniamboensis M.Schmid, in Fl. N. Caled. & Depend. 17: 136 (1991).
New Caledonia. 60 NWC. Nanophan. or phan.

var. **koniamboensis**
NW. New Caledonia. 60 NWC. Nanophan. or phan.

var. **taomensis** M.Schmid, in Fl. N. Caled. & Depend. 17: 139 (1991).
New Caledonia (Massif du Taom). 60 NWC. Nanophan.

Phyllanthus kostermansii Airy Shaw, Kew Bull. 29: 296 (1974).
Irian Jaya. 42 NWG. Nanophan. or phan.

Phyllanthus kouaouaensis M.Schmid, in Fl. N. Caled. & Depend. 17: 114 (1991).
C. New Caledonia. 60 NWC. Nanophan.

Phyllanthus koumacensis Guillaumin, State Univ. Iowa Stud. Nat. Hist. 20: 35 (1965).
NW. New Caledonia. 60 NWC. Nanophan.

var. **brevitepalus** M.Schmid, in Fl. N. Caled. & Depend. 17: 86 (1991).
    NW. New Caledonia. 60 NWC. Nanophan.

var. **koumacensis**
    NW. New Caledonia (Tiébaghi). 60 NWC. Nanophan.

Phyllanthus lacerosus Airy Shaw, Kew Bull. 35: 386 (1980).
    N. & E. Australia. 50 NSW NTA QLD WAU.

Phyllanthus laciniatus C.B.Rob., Philipp. J. Sci., C 4: 84 (1909).
    Philippines. 42 PHI.

Phyllanthus lacunarius F.Muell., Trans. Philos. Soc. Victoria 1: 14 (1855).
    Australia. 50 NSW NTA QLD SOA VIC.

Phyllanthus lacunellus Airy Shaw, Kew Bull. 35: 387 (1980).
    Australia. 50 NSW NTA QLD SOA WAU.

Phyllanthus lagoensis Müll.Arg. in C.F.P.von Martius, Fl. Bras. 11(2): 65 (1873).
    Brazil (Minas Gerais). 84 BZL. Cham.

Phyllanthus lamprophyllus Müll.Arg. in A.P.de Candolle, Prodr. 15(2): 324 (1866).
    Jawa (Madura I.), Philippines (Palawan), Lesser Sunda Is., New Guinea, Queensland. 42
        JAW LSI NWG PHI 50 QLD. Nanophan.
        Phyllanthus hellwigii Warb., Bot. Jahrb. Syst. 18: 198 (1894).

Phyllanthus lanceifolius Merr., Philipp. J. Sci., C 9: 489 (1914 publ. 1915).
    Philippines. 42 PHI.

Phyllanthus lanceolatus Poir. in J.B.A.M.de Lamarck, Encycl. 5: 299 (1804).
    Mauritius. 29 MAU. Cham. or nanophan.

Phyllanthus larensis Steyerm., Fieldiana, Bot. 28: 318 (1952).
    Venezuela. 82 VEN.

Phyllanthus lasiogyuus Müll.Arg. in A.P.de Candolle, Prodr. 15(2): 357 (1866).
    Paraguay (Ypanema). 85 PAR. Phan.

Phyllanthus latifolius (L.) Sw., Fl. Ind. Occid. 2: 1109 (1800).
    S. Jamaica. 81 JAM. Nanophan. or phan.
        *Xylophylla latifolia L., Mant. Pl. 2: 221 (1771).
        Genesiphylla asplenifolia L'Hér., Sert. Angl.: 29 (1788). *Xylophylla asplenifolia* (L'Hér.)
            Salisb., Prodr. Stirp. Chap. Allerton: 390 (1796).
        Phyllanthus isolepis Urb., Symb. Antill. 3: 290 (1902).

Phyllanthus lativenius (Croizat) Govaerts & Radcl.-Sm., Kew Bull. 51: 177 (1996).
    China (Guizhou). 36 CHS. Nanophan.
        *Phyllanthodendron lativenium Croizat, J. Arnold Arbor. 23: 36 (1942).

Phyllanthus lawii J.Graham, Cat. Pl. Bombay: 181 (1839).
    India. 40 IND.
        Phyllanthus spinulosus B.Heyne ex Wall., Numer. List: 7897A (1847), nom. inval.
        Phyllanthus polyphyllus Dalzell & Gibson, Bombay Fl.: 234 (1861).
        Phyllanthus juniperinoides Müll.Arg., Linnaea 32: 18 (1863).

Phyllanthus laxiflorus Benth., Pl. Hartw.: 90 (1842).
    Guatemala. 80 GUA.

Phyllanthus lebrunii Robyns & Lawalrée, Bull. Jard. Bot. État 18: 264 (1947).
Rwanda, Zaire. 23 RWA ZAI.

Phyllanthus lediformis Jabl., Mem. New York Bot. Gard. 17: 103 (1967).
Venezuela (Amazonas: Serrania Yutaje). 82 VEN. Nanophan.

Phyllanthus leonardianus Lisowski, Malaisse & Symoens, Bull. Soc. Roy. Bot. Belgique 107: 200 (1974).
Zaire. 23 ZAI.

Phyllanthus leptobotryosus Donn.Sm., Bot. Gaz. 54: 241 (1912).
Costa Rica. 80 COS.

Phyllanthus leptocaulos Müll.Arg. in C.F.P.von Martius, Fl. Bras. 11(2): 47 (1873).
Brazil (Minas Gerais). 84 BZL. Ther.

Phyllanthus leptoclados Benth., Fl. Hongk.: 312 (1861). *Epistylium leptocladon* (Benth.) Hance, Ann. Sci. Nat., Bot., IV, 18: 229 (1862).
S. China, Nansei-shoto. 36 CHC CHS 38 NNS. Nanophan.
　　Phyllanthus liukiuensis Hayata, J. Coll. Sci. Imp. Univ. Tokyo 20(3): 11 (1904).
　　Phyllanthus glabricapsulus F.P.Metcalf, Lingnan Sci. J. 10: 483 (1931).

Phyllanthus leptoneurus Urb., Symb. Antill. 7: 246 (1912).
Hispaniola. 81 DOM HAI. Cham. or nanophan.
　　Phyllanthus trigonus Urb. & Ekman, Ark. Bot. 20A(15): 46 (1926).

Phyllanthus leptophyllus Müll.Arg. in A.P.de Candolle, Prodr. 15(2): 411 (1866).
Brazil (Minas Gerais). 84 BZL.

Phyllanthus leschenaultii Müll.Arg., Linnaea 32: 37 (1863).
India. 40 IND.

Phyllanthus leucanthus Pax, Bot. Jahrb. Syst. 15: 524 (1893).
E. Africa, Zaire (Shaba), Rwanda. 23 RWA ZAI 24 DJI ETH SOM 25 KEN TAN UGA 26 MLW MOZ ZAM ZIM. Ther. or nanophan. – Close to *P. odontadensis*.
　　Phyllanthus eylesii S.Moore, J. Bot. 58: 79 (1920).

Phyllanthus leucocalyx Hutch., Bull. Misc. Inform. Kew 1911: 316 (1911).
Somalia ?, Kenya, Uganda, Tanzania, Mozambique. 24 SOM? 25 KEN TAN UGA 26 MOZ. Ther. or cham.

Phyllanthus leytensis Elmer, Leafl. Philipp. Bot. 1: 307 (1908).
Philippines (Leyte). 42 PHI.

Phyllanthus liebmannianus Müll.Arg. in A.P.de Candolle, Prodr. 15(2): 366 (1866).
Florida, Mexico, C. America. 78 FLA 79 MXE MXG MXT 80 BLZ GUA. Hemicr.

subsp. **liebmannianus**
　　E. Mexico. 79 MXE MXG MXT 80 BLZ GUA. Hemicr.

subsp. **platylepis** (Small) G.L.Webster, Brittonia 22: 57 (1970).
　　NW. Florida. 78 FLA. Hemicr.
　　Phyllanthus platylepis Small, Fl. S.E. U.S., ed. 2: 1347 (1913).

Phyllanthus ligustrifolius S.Moore, J. Linn. Soc., Bot. 45: 402 (1921).
New Caledonia. 60 NWC. Nanophan.

var. **boulindaensis** M.Schmid, in Fl. N. Caled. & Depend. 17: 237 (1991).
WC. New Caledonia. 60 NWC. Nanophan.

var. **colnettensis** M.Schmid, in Fl. N. Caled. & Depend. 17: 238 (1991).
C. New Caledonia. 60 NWC. Nanophan.

var. **ligustrifolius**
NW. New Caledonia. 60 NWC. Nanophan.

Phyllanthus lii Govaerts & Radcl.-Sm., Kew Bull. 51: 177 (1996).
China (Guangxi). 36 CHS. Nanophan.
  * *Phyllanthodendron caudatifolium* P.T.Li, Bull. Bot. Res., Harbin 7(3): 7 (1987).

Phyllanthus limmuensis Cufod., Österr. Akad. Wiss., Math.-Naturwiss. Kl., Sitzungsber.,
Abt. 1, Biol. 156: 484 (1947).
SW. Ethiopia. 24 ETH.

Phyllanthus lindbergii Müll.Arg. in C.F.P.von Martius, Fl. Bras. 11(2): 35 (1873).
Brazil (Minas Gerais). 84 BZL.

Phyllanthus lindenianus Baill., Adansonia 2: 13 (1861). *Phyllanthus cyclanthera* var.
lindenianus (Baill.) Müll.Arg. in A.P.de Candolle, Prodr. 15(2): 408 (1866).
Cuba, Haiti, W. Dominican Rep. 81 CUB DOM HAI. Ther. or cham.

var. **inaequifolius** (G.L.Webster) G.L.Webster, J. Arnold Arbor. 38: 187 (1957).
S. Haiti, Dominican Rep. 81 DOM HAI. Ther.
  * *Phyllanthus inaequifolius* G.L.Webster, Contr. Gray Herb. 176: 48 (1955). *Phyllanthus*
    lindenianus subsp. *inaequifolius* (G.L.Webster) Borhidi, Bot. Közlem. 58: 176 (1971).

var. **jimenezii** G.L.Webster, J. Arnold Arbor. 38: 185 (1957). *Phyllanthus lindenianus* subsp.
jimenezii (G.L.Webster) Borhidi, Bot. Közlem. 58: 176 (1971).
Dominican Rep. (Constanza). 81 DOM. Ther.

var. **leonardorum** (G.L.Webster) G.L.Webster, J. Arnold Arbor. 38: 186 (1957).
C. & N. Haiti. 81 HAI. Ther.
  * *Phyllanthus leonardorum* G.L.Webster, Contr. Gray Herb. 176: 50 (1955). *Phyllanthus*
    lindenianus subsp. *leonardorum* (G.L.Webster) Borhidi, Bot. Közlem. 58: 176 (1971).

var. **lindenianus**
Cuba, Haiti, W. Dominican Rep. 81 CUB DOM HAI. Ther. or cham.
Phyllanthus gracillimus Baill., Adansonia 2: 14 (1861). *Phyllanthus cyclanthera* var.
  gracillimus (Baill.) Müll.Arg., Linnaea 32: 44 (1863).
Phyllanthus cyclanthera var. *scabrellus* Müll.Arg., Linnaea 32: 44 (1863).

Phyllanthus lingulatus Beille, Bull. Soc. Bot. France 72: 161 (1925). *Phyllanthodendron*
lingulatum (Beille) Croizat, J. Arnold Arbor. 23: 34 (1942).
S. Thailand, Vietnam. 41 THA VIE. Cl. nanophan.

Phyllanthus loandensis Welw. ex Müll.Arg., J. Bot. 2: 329 (1864).
C. & E. Trop. & S. Africa. 23 BUR ZAI 25 TAN 26 ANG MLW MOZ ZAM ZIM 27 TVL. Cham.

Phyllanthus lokohensis Leandri, Mém. Inst. Sci. Madagascar, Sér. B, Biol. Vég. 8: 229 (1957).
E. Madagascar. 29 MDG. Cham. or nanophan.

Phyllanthus longeramosus Guillaumin ex M.Schmid, in Fl. N. Caled. & Depend. 17:
292 (1991).
New Caledonia (Dumbéa-Païta Reg.). 60 NWC. Nanophan.
Phyllanthus longeramosus Guillaumin, Mém. Mus. Natl. Hist. Nat., B, Bot. 8: 245 (1962),
  nom. inval.

Phyllanthus longistylus Jabl., Mem. New York Bot. Gard. 17: 100 (1967).
Venezuela (Amazonas, Bolívar). 82 VEN. Nanophan.

Phyllanthus loranthoides Baill., Adansonia 2: 238 (1862).
New Caledonia. 60 NWC. Nanophan.

var. **longifolius** M.Schmid, in Fl. N. Caled. & Depend. 17: 284 (1991).
NW. New Caledonia. 60 NWC. Nanophan.

var. **loranthoides**
NW. New Caledonia. 60 NWC. Nanophan.

var. **ripicola** M.Schmid, in Fl. N. Caled. & Depend. 17: 286 (1991).
NW. & C. New Caledonia. 60 NWC. Nanophan.

Phyllanthus luciliae M.Schmid, in Fl. N. Caled. & Depend. 17: 269 (1991).
New Caledonia (Cap Bocage). 60 NWC. Nanophan.

Phyllanthus lunifolius Gilbert & Thulin, Nordic J. Bot. 13: 171 (1993).
C. Somalia. 24 SOM. Cham.

Phyllanthus macgregorii C.B.Rob., Philipp. J. Sci., C 6: 334 (1911).
Philippines. 42 PHI.

Phyllanthus mackenziei Fosberg, Kew Bull. 33: 189 (1978). – FIGURE, p. 1262
Aldabra, Seychelles. 29 ALD SEY.

Phyllanthus macphersonii M.Schmid, in Fl. N. Caled. & Depend. 17: 142 (1991).
New Caledonia (Ouaco). 60 NWC. Nanophan.

Phyllanthus macraei Müll.Arg., Linnaea 32: 29 (1863).
India (S. Kerala). 40 IND.
  * *Macraea rheedii* Wight, Icon. Pl. Ind. Orient. 5(2): 27 (1852).

Phyllanthus macranthus Pax, Bot. Jahrb. Syst. 19: 77 (1894).
Tanzania to Angola. 23 ZAI 25 TAN 26 ANG MOZ ZAM ZIM 27 TVL. Nanophan.

var. **gilletii** (De Wild.) Brunel ex Radcl.-Sm., Kew Bull. 51: 308 (1996).
Zaire, Angola, Zambia ?, Mozambique. 23 ZAI 26 ANG MOZ ZAM.
  * *Phyllanthus gilletii* De Wild., Ann. Mus. Congo Belge, Bot., V, 2: 266 (1908).

var. **macranthus**
Tanzania, Mozambique, Zimbabwe, Angola, Northern Prov. 25 TAN 26 ANG MOZ ZIM
27 TVL. Nanophan.

Phyllanthus macrocalyx Müll.Arg., Linnaea 32: 48 (1863). *Reidia macrocalyx* (Müll.Arg.)
Gamble, Fl. Madras: 1292 (1925).
S. India. 40 IND.
  Reidia latifolia Wight, Icon. Pl. Ind. Orient. 5(2): 28, t. 1904(2) (1852).

Phyllanthus macrochorion Baill., Adansonia 2: 232 (1862). *Glochidion macrochorion* (Baill.)
Müll.Arg., Linnaea 32: 59 (1863).
NW. New Caledonia. 60 NWC. Nanophan.

Phyllanthus madagascariensis Müll.Arg., Linnaea 32: 35 (1863).
C. & E. Madagascar. 29 MDG. Cham. or nanophan.

var. **kalambatitrensis** Leandri, Mém. Inst. Sci. Madagascar, Sér. B, Biol. Vég. 8: 230 (1957).
C. Madagascar. 29 MDG. Cham. or nanophan.

var. **madagascariensis**
> C. & E. Madagascar. 29 MDG. Cham. or nanophan.
> *Phyllanthus chapelieri* Baill., Adansonia 2: 52 (1861), nom. nud.

Phyllanthus madeirensis Croizat, Trop. Woods 78: 7 (1944).
> Brazil (Amazonas: Humayta). 84 BZN. Nanophan. – Close to *P. racemigerus*.

Phyllanthus maderaspatensis L., Sp. Pl.: 982 (1753).
> Africa, Madagascar, Mascarenes, Arabian Pen., Pakistan to N. Australia. 22 GHA IVO
> NGA SEN TOG 23 CMN RWA ZAI 24 ETH SOC 25 KEN TAN UGA 27 ALL 29 ALD MAU
> MDG REU 35 YEM 36 CHS 40 IND PAK SRL 42 JAW NWG 50 NSW NTA QLD WAU.
> Ther. or cham.

var. **exasperata** Radcl.-Sm., Kew Bull. 47: 681 (1992).
> Kenya. 25 KEN.

var. **frazieri** Fosberg, Kew Bull. 33: 188 (1978). – FIGURE, p. 1262
> Aldabra. 29 ALD.

var. **maderaspatensis**
> Africa, Madagascar, Mascarenes, Arabian Pen., Pakistan, India, Sri Lanka, China,
> Malesia, N. & E. Australia. 22 GHA IVO NGA SEN TOG 23 CMN RWA ZAI 24 ETH
> SOC 25 KEN TAN UGA 27 ALL 29 MAU MDG REU 35 YEM 36 CHS 40 IND PAK SRL
> 42 JAW NWG 50 NSW NTA QLD WAU. Ther. or cham.
> *Phyllanthus andrachnoides* Willd., Sp. Pl. 4: 575 (1805).
> *Phyllanthus cuneatus* Willd., Enum. Pl., Suppl.: 65 (1814).
> *Phyllanthus obcordatus* Willd., Enum. Pl., Suppl.: 65 (1814).
> *Phyllanthus javanicus* Poir. ex Spreng., Syst. Veg. 3: 21 (1826), pro syn.
> *Phyllanthus thonningii* Schumach. & Thonn. in C.F.Schumacher, Beskr. Guin. Pl.: 418
> (1827). *Phyllanthus maderaspatensis* var. *thonningii* (Schumach. & Thonn.) Müll.Arg.
> in A.P.de Candolle, Prodr. 15(2): 362 (1866).
> *Phyllanthus gracilis* Roxb., Fl. Ind. ed. 1832, 3: 654 (1832), nom. illeg.
> *Phyllanthus stipulaceus* Bojer, Hortus Maurit.: 280 (1837).
> *Phyllanthus arabicus* Hochst. ex Steud., Nomencl. Bot., ed. 2, 2: 326 (1841).
> *Phyllanthus longifolius* Sond., Linnaea 23: 135 (1850).
> *Phyllanthus venosus* Hochst. ex A.Rich., Tent. Fl. Abyss. 2: 254 (1850).
> *Phyllanthus vaccinioides* Klotzsch in W.C.H.Peters, Naturw. Reise Mossambique: 105
> (1861), nom. illeg.
> *Phyllanthus gueinzii* Müll.Arg., Linnaea 32: 18 (1863).
> *Phyllanthus brachypodus* F.Muell. ex Benth., Fl. Austral. 6: 103 (1873), pro syn.
> *Phyllanthus maderaspatensis* f. *fastigiatus* Leandri, in Fl. Madag. 111: 77 (1958).
> *Phyllanthus maderaspatensis* f. *thermarum* Leandri, in Fl. Madag. 111: 77 (1958).

Phyllanthus maestrensis Urb., Symb. Antill. 9: 193 (1924).
> Cuba (Sierra Maestra). 81 CUB. Cham. or nanophan.

Phyllanthus mafingensis Radcl.-Sm., Kew Bull. 51: 308 (1996).
> Malawi (Mafinga Mts.). 26 MLW. Cham.

Phyllanthus magnificens Brunel & J.P.Roux, Willdenowia 11: 82 (1981).
> Ghana, Togo. 22 GHA TOG.

Phyllanthus maguirei Jabl., Mem. New York Bot. Gard. 17: 105 (1967).
> Venezuela (Amazonas: Cerro de la Neblina). 82 VEN. Nanophan.

Phyllanthus majus Steyerm., Fieldiana, Bot. 28: 318 (1952).
> S. Venezuela, Guyana. 82 GUY VEN. Nanophan.

Phyllanthus maleolens Urb. & Ekman, Ark. Bot. 22A(8): 60 (1928).
Hispaniola. 81 DOM HAI. Nanophan. or phan.

Phyllanthus mananarensis Leandri, Mém. Inst. Sci. Madagascar, Sér. B, Biol. Vég. 8: 230 (1957).
SW. Madagascar. 29 MDG. Nanophan.

Phyllanthus manausensis W.A.Rodrigues, Acta Amazon. 1: 17 (1971).
Brazil (Amazonas). 84 BZN.

Phyllanthus mandjeliaensis M.Schmid, in Fl. N. Caled. & Depend. 17: 315 (1991).
NW. New Caledonia. 60 NWC. Nanophan.

Phyllanthus mangenotii M.Schmid, in Fl. N. Caled. & Depend. 17: 156 (1991).
C. New Caledonia (Monéo reg.). 60 NWC. Nanophan. or phan.

Phyllanthus manicaensis Brunel ex Radcl.-Sm., Kew Bull. 51: 309 (1996).
Mozambique. 26 MOZ. Cham.

Phyllanthus mannianus Müll.Arg., Flora 47: 514 (1864).
Nigeria, Cameroon. 22 NGA 23 CMN. Nanophan.
Phyllanthus pseudoreticulatus Pax & K.Hoffm., Bot. Jahrb. Syst. 45: 235 (1910).

Phyllanthus mantsakariva Leandri, Mém. Inst. Sci. Madagascar, Sér. B, Biol. Vég. 8: 232 (1957).
N. Madagascar. 29 MDG. Nanophan.

Phyllanthus margaretae M.Schmid, in Fl. N. Caled. & Depend. 17: 316 (1991).
C. New Caledonia (Mt. Aoupinié). 60 NWC. Cham. or nanophan.

Phyllanthus marianus Müll.Arg., Linnaea 32: 17 (1863).
Marianas. 62 MRN.

Phyllanthus maritimus J.J.Sm., Nova Guinea 8: 779 (1912).
Irian Jaya. 42 NWG. Nanophan.

Phyllanthus martii Müll.Arg. in C.F.P.von Martius, Fl. Bras. 11(2): 27 (1873).
Brazil (Amazonas). 84 BZN.

Phyllanthus martinii Radcl.-Sm., Kew Bull. 51: 311 (1996).
Zambia. 26 ZAM. Ther.

Phyllanthus matitanensis Leandri, Notul. Syst. (Paris) 6: 196 (1938).
C. & E. Madagascar. 29 MDG. Nanophan.

Phyllanthus megacarpus (Gamble) Kumari & Chandrab., in Fl. Tamil Nadu, India 2: 238 (1987).
India. 40 IND.
* *Reidia megacarpa* Gamble, Bull. Misc. Inform. Kew 1925: 332 (1925).

Phyllanthus megalanthus C.B.Rob., Philipp. J. Sci., C 6: 334 (1911).
Philippines. 42 PHI.

Phyllanthus megapodus G.L.Webster, Contr. Gray Herb. 176: 62 (1955).
Windward Is. (Dominica, Martinique). 81 WIN. Nanophan. or phan.
Phyllanthus mimosoides lusus *macrophyllus* Müll.Arg. in A.P.de Candolle, Prodr. 15(2): 381 (1866).

Phyllanthus melleri Müll.Arg., Flora 47: 514 (1864).
  Madagascar. 29 MDG. Cham. or nanophan.
    Phyllanthus melleri var. *campenonii* Leandri, Notul. Syst. (Paris) 7: 185 (1939).

Phyllanthus memaoyaensis M.Schmid, in Fl. N. Caled. & Depend. 17: 274 (1991).
  C. New Caledonia. 60 NWC. Nanophan.

Phyllanthus mendesii Brunel ex Radcl.-Sm., Kew Bull. 51: 312 (1996).
  Angola, Namibia, Botswana, Zambia, Zimbabwe. 26 ANG ZAM ZIM 27 BOT NAM. Ther.

Phyllanthus mendoncae Brunel ex Radcl.-Sm., Kew Bull. 51: 313 (1996).
  Ethiopia to Mozambique. 23 BUR 24 ETH 25 KEN TAN 26 MOZ. Cham.

Phyllanthus merinthopodus Diels, Notizbl. Bot. Gart. Berlin-Dahlem 11: 310 (1931).
  New Guinea. 42 NWG. Phan.

Phyllanthus meuieensis M.Schmid, in Fl. N. Caled. & Depend. 17: 249 (1991).
  New Caledonia (Mé Ouié). 60 NWC. Nanophan.

Phyllanthus mexiae Croizat, J. Wash. Acad. Sci. 33: 14 (1943).
  Ecuador. 83 ECU.

Phyllanthus meyerianus Müll.Arg., Linnaea 32: 42 (1863).
  S. Africa. 27 CPP NAT SWZ.
    Phyllanthus woodii Hutch., Bull. Misc. Inform. Kew 1914: 336 (1914).

Phyllanthus mickelii McVaugh, Brittonia 13: 196 (1961).
  Mexico (Colima). 79 MXE. Nanophan.

Phyllanthus micrandrus Müll.Arg., Linnaea 32: 27 (1863).
  Mexico, Guatemala, Venezuela. 79 MXN MXS MXT 80 GUA 82 VEN.
    Phyllanthus pringlei S.Watson, Proc. Amer. Acad. Arts 26: 147 (1891).

Phyllanthus micranthus A.Rich. in R.de la Sagra, Hist. Fis. Cuba, Bot. 2: 216 (1850).
  SE. Cuba. 81 CUB. Cham.

Phyllanthus microcladus Müll.Arg., Linnaea 34: 71 (1865). *Sauropus albiflorus* subsp.
  microcladus (Müll.Arg.) Airy Shaw, Kew Bull. 35: 672 (1980), no basionym ref.
  SE. Queensland, E. New South Wales. 50 NSW QLD.
    Phyllanthus microcladus var. *microphyllus* Müll.Arg., Linnaea 34: 72 (1865).
    Phyllanthus microcladus var. *puberulus* Müll.Arg., Linnaea 34: 71 (1865).
    Phyllanthus pusillifolius S.Moore, J. Linn. Soc., Bot. 45: 216 (1920).

Phyllanthus microdendron Müll.Arg., J. Bot. 2: 830 (1864).
  Angola, Zambia. 26 ANG ZAM. Phan.

  var. **asper** Radcl.-Sm., Kew Bull. 51: 315 (1996).
    Zambia. 26 ZAM.

  var. **microdendron**
    Angola. 26 ANG. Phan.
    Phyllanthus antunesii Pax, Bot. Jahrb. Syst. 23: 519 (1897).

Phyllanthus microdictyus Urb., Symb. Antill. 9: 183 (1924).
  NE. Cuba. 81 CUB. Nanophan.

Phyllanthus micromalus McVaugh, Brittonia 13: 198 (1961).
Mexico (Nayarit). 79 MXS. Cham.

Phyllanthus micromeris Radcl.-Sm., Kew Bull. 35: 769 (1981). – FIGURE, p. 1266.
Tanzania, Malawi. 25 TAN 26 MLW. Cham.

    var. **micromeris**
        Tanzania, Malawi. 25 TAN 26 MLW. Cham.

    var. **mughessensis** Radcl.-Sm., Kew Bull. 51: 316 (1996).
        Malawi. 26 MLW.

    var. **sesbanioides** Radcl.-Sm., Kew Bull. 51: 316 (1996).
        Malawi. 26 MLW.

Phyllanthus microphyllinus Müll.Arg., J. Bot. 2: 332 (1864).
Angola. 26 ANG. Hemicr.

Phyllanthus microphyllus Kunth in F.W.H.von Humboldt, A.J.A.Bonpland & C.S.Kunth,
Nov. Gen. Sp. 2: 87 (1817).
Venezuela ?, Brazil (Para). 82 VEN? 84 BZN. Ther.

Phyllanthus mieschii Brunel & J.P.Roux, Willdenowia 11: 87 (1981).
Congo, Zaire. 23 CON ZAI.

Phyllanthus millei Standl., Publ. Field Mus. Nat. Hist., Bot. Ser. 22: 87 (1940).
Ecuador. 83 ECU.

Phyllanthus mimicus G.L.Webster, Contr. Gray Herb. 176: 52 (1955).
Tobago, Trinidad. 81 TRT. Ther.

Phyllanthus mimosoides Sw., Prodr.: 27 (1788).
Leeward Is., Windward Is., Trinidad. 81 LEE TRT WIN. Nanophan. or phan.

Phyllanthus minahassae Koord., Meded. Lands Plantentuin 19: 588, 627 (1898).
Sulawesi. 42 SUL.

Phyllanthus minarum Standl. & Steyerm., Publ. Field Mus. Nat. Hist., Bot. Ser. 23: 125 (1944).
Guatemala. 80 GUA.

Phyllanthus mindorensis C.B.Rob., Philipp. J. Sci., C 4: 82 (1909).
Philippines (Mindoro). 42 PHI.

Phyllanthus minutifolius Jabl., Mem. New York Bot. Gard. 17: 115 (1967).
Venezuela (Amazonas: Cerro Sipapo). 82 VEN. Cham.

Phyllanthus minutulus Müll.Arg. in C.F.P.von Martius, Fl. Bras. 11(2): 54 (1873).
Brazil (Goiás, Minas Gerais). 84 BZC BZL.

Phyllanthus mirabilis Müll.Arg., Flora 47: 513 (1864). *Phyllanthodendron mirabile*
(Müll.Arg.) Hemsl., Hooker's Icon. Pl. 26: t. 2563-2564 (1898).
Thailand. 41 THA. Phan.

Phyllanthus mirificus G.L.Webster, Contr. Gray Herb. 176: 58 (1955).
NE. Cuba. 81 CUB. Nanophan.

Phyllanthus missionis Hook.f., Fl. Brit. India 5: 297 (1887).
S. India. 40 IND.

Phyllanthus mitchellii Benth., Fl. Austral. 6: 103 (1873).
Queensland (Leichhardt). 50 QLD.
Micrantheum triandrum Hook. in T.L.Mitchell, J. Exped. Trop. Australia: 342 (1848).
Phyllanthus triandrus (Hook.) Druce, Bot. Soc. Exch. Club Brit. Isles 1916: 639 (1917),
nom. illeg.
Micrantheum nervosum Pancher ex Guillaumin, Ann. Inst. Bot.-Géol. Colon. Marseille,
II, 9: 222 (1911).

Phyllanthus mittenianus Hutch. in D.Oliver, Fl. Trop. Afr. 6(1): 725 (1912).
Kenya (Teita), Tanzania (Uluguru Mts.). 25 KEN TAN. Nanophan. – Close to *P. sacleuxii*.
Phyllanthus taitensis Hutch. in D.Oliver, Fl. Trop. Afr. 6(1): 1049 (1913).
Phyllanthus sacleuxii Radcl.-Sm., Kew Bull. 29: 440 (1974).

Phyllanthus mocinianus Baill., Adansonia 1: 35 (1860).
Mexico to Peru. 79 MXE MXN MXT 80 COS PAN 83 CLM PER. Nanophan. or phan.
Phyllanthus arboreus Sessé & Moç. ex Baill., Adansonia 1: 35 (1860), pro syn.
Phyllanthus anisolobus Müll.Arg. in A.P.de Candolle, Prodr. 15(2): 382 (1866).
Phyllanthus pittieri Pax, Anales Inst. Fis.-Geog. Nac. Costa Rica 9: 195 (1898).

Phyllanthus mocquerysianus A.DC., Bull. Herb. Boissier, II, 1: 564 (1901).
E. Madagascar, Comoros ? 29 COM? MDG. Nanophan.
Kirganelia glaucescens Baill., Adansonia 2: 48 (1861). *Phyllanthus casticum* f. *glaucescens*
(Baill.) Müll.Arg. in A.P.de Candolle, Prodr. 15(2): 348 (1866).
Phyllanthus kirganelia var. *glaber* Müll.Arg., Linnaea 32: 13 (1863).

Phyllanthus moeroensis De Wild., Ann. Mus. Congo Belge, Bot., V, 1: 273 (1906).
Zaire. 23 ZAI.
Phyllanthus ringoetii De Wild., Contr. Fl. Katanga, Suppl. 4: 43 (1932).

Phyllanthus moi P.T.Li, Guihaia 3: 167 (1983). *Phyllanthodendron moi* (P.T.Li) P.T.Li, Bull.
Bot. Res., Harbin 7(3): 9 (1987).
China (W. Guangxi). 36 CHS. Nanophan.

Phyllanthus montanus (Sw.) Sw., Fl. Ind. Occid. 2: 1117 (1800).
Jamaica. 81 JAM. Nanophan. or phan.
* *Xylophylla montana* Sw., Prodr.: 28 (1788).

Phyllanthus montevidensis Müll.Arg., Linnaea 32: 37 (1863).
Uruguay. 85 URU.

Phyllanthus montis-fontium M.Schmid, in Fl. N. Caled. & Depend. 17: 78 (1991).
SE. New Caledonia. 60 NWC. Nanophan. or phan.

Phyllanthus montrouzieri Guillaumin, Ann. Soc. Bot. Lyon 38: 109 (1913 publ. 1914).
NW. to WC. New Caledonia. 60 NWC. Nanophan. or phan.

var. **montrouzieri**
NW. to WC. New Caledonia. 60 NWC. Nanophan. or phan.

var. **pandopensis** M.Schmid, in Fl. N. Caled. & Depend. 17: 215 (1991).
NW. New Caledonia (S. of Koumac). 60 NWC. Nanophan. or phan.

var. **poyaensis** M.Schmid, in Fl. N. Caled. & Depend. 17: 215 (1991).
WC. New Caledonia (Massif du Boulinda). 60 NWC. Nanophan. or phan.

Phyllanthus mooneyi M.G.Gilbert, Kew Bull. 42: 357 (1987).
Ethiopia. 24 ETH. Hemicr., cham. or nanophan.

Phyllanthus moorei M.Schmid, in Fl. N. Caled. & Depend. 17: 159 (1991).
New Caledonia. 60 NWC. Nanophan. or phan.
 Dendrophyllanthus comptonii S.Moore, J. Linn. Soc., Bot. 45: 395 (1921).

var. **acutitepalus** M.Schmid, in Fl. N. Caled. & Depend. 17: 162 (1991).
New Caledonia. 60 NWC. Nanophan. or phan.

var. **moorei**
New Caledonia. 60 NWC. Nanophan. or phan.

Phyllanthus moramangicus (Leandri) Leandri, in Fl. Madag. 111: 84 (1958).
EC. Madagascar. 29 MDG. Nanophan.
 Phyllanthus melleri subsp. *moramangicus* Leandri, Notul. Syst. (Paris) 6: 197 (1938).

Phyllanthus moratii M.Schmid, in Fl. N. Caled. & Depend. 17: 313 (1991).
NW. New Caledonia (Tiwaka-Amoa Reg.). 60 NWC. Nanophan.

Phyllanthus mouensis M.Schmid, in Fl. N. Caled. & Depend. 17: 104 (1991).
SE. New Caledonia. 60 NWC. Nanophan.

Phyllanthus mozambicensis Gand., Bull. Soc. Bot. France 66: 287 (1919 publ. 1920).
Mozambique (Delagoa Bay). 26 MOZ.

Phyllanthus muellerianus (Kuntze) Exell, Cat. Vasc. Pl. S. Tome: 290 (1944).
Trop. Africa. 22 BEN GHA GNB GUI IVO LBR MLI NGA SIE 23 CMN GAB GGI ZAI 24
SUD 25 KEN TAN UGA 26 ANG MLW MOZ ZAM. (Cl.) nanophan. or phan.
 Kirganelia floribunda Baill., Adansonia 1: 83 (1860). *Phyllanthus floribundus* (Baill.)
Müll.Arg., Linnaea 32: 14 (1863), nom. illeg.
 Diasperus muellerianus Kuntze, Revis. Gen. Pl. 2: 597 (1891).

Phyllanthus mukerjeeanus D.Mitra & Bennet, Bull. Bot. Soc. Bengal 19: 145 (1967).
India. 40 IND.

Phyllanthus multiflorus Poir. in J.B.A.M.de Lamarck, Encycl. 5: 299 (1804). *Anisonema
multiflorum* (Poir.) Wight, Icon. Pl. Ind. Orient. 5: t. 1899 (1852).
N. & W. Madagascar. 29 MDG. Nanophan.
 Phyllanthus grandiflorus Spreng., Syst. Veg. 3: 22 (1826).
 Menarda pulchella Baill., Étude Euphorb.: 609 (1858).

Phyllanthus muriculatus J.J.Sm. in S.H.Koorders & T.Valeton, Bijdr. Boomsoort. Java 12:
93 (1910).
Jawa. 42 JAW.

Phyllanthus muscosus Ridl., J. Fed. Malay States Mus. 4: 61 (1909).
Pen. Malaysia. 42 MLY.

Phyllanthus myriophyllus Urb., Ark. Bot. 17(7): 36 (1922).
S. Haiti. 81 HAI. Nanophan. or phan.

Phyllanthus myrsinites Kunth in F.W.H.von Humboldt, A.J.A.Bonpland & C.S.Kunth, Nov.
Gen. Sp. 2: 111 (1817).
Venezuela (Amazonas). 82 VEN. Nanophan. – Close to *P. glaucoviridis*.

Phyllanthus myrtaceus Sond., Linnaea 23: 134 (1850).
Zimbabwe, Mozambique, S. Africa. 26 MOZ ZIM 27 CPP NAT.
 Phyllanthus revolutus E.Mey. ex Sond., Linnaea 23: 135 (1850).
 Phyllanthus bachmannii Pax, Bot. Jahrb. Syst. 23: 520 (1897).

Phyllanthus myrtifolius (Wight) Müll.Arg. in A.P.de Candolle, Prodr. 15(2): 396 (1866).
Sri Lanka. (36) chh chs (38) tai 40 SRL. Cham.
  * *Macraea myrtifolia* Wight, Icon. Pl. Ind. Orient. 5(2): 27 (1852).

Phyllanthus myrtilloides Griseb., Mem. Amer. Acad. Arts, n.s., 8: 158 (1860).
E. Cuba. 81 CUB. Nanophan.

subsp. **alainii** G.L.Webster, J. Arnold Arbor. 39: 129 (1958).
  E. Cuba (Sierra del Cristal). 81 CUB. Nanophan.

subsp. **erythrinus** (Müll.Arg.) G.L.Webster, J. Arnold Arbor. 39: 125 (1958).
  NE. Cuba. 81 CUB. Nanophan.
  Phyllanthus purpureus C.Wright ex Griseb., Nachr. Königl. Ges. Wiss. Georg-Augusts-
    Univ. 1: 168 (1865), nom. illeg.
  * *Phyllanthus erythrinus* Müll.Arg. in A.P.de Candolle, Prodr. 15(2): 332 (1866).
  Orbicularia foveolata Britton, Mem. Torrey Bot. Club 16: 73 (1920). *Phyllanthus foveolatus*
    (Britton) Alain, Contr. Ocas. Mus. Hist. Nat. Colegio "De La Salle" 11: 2 (1952).
  Phyllanthus cardiophyllus Urb., Symb. Antill. 9: 190 (1924).
  Phyllanthus melanodiscus Urb., Symb. Antill. 9: 190 (1924).

subsp. **myrtilloides**
  E. Cuba. 81 CUB. Nanophan.

subsp. **shaferi** (Urb.) G.L.Webster, J. Arnold Arbor. 39: 130 (1958).
  E. Cuba (Sierra de Nipe). 81 CUB. Nanophan.
  * *Phyllanthus shaferi* Urb., Repert. Spec. Nov. Regni Veg. 13: 448 (1914).

subsp. **spathulifolius** (Griseb.) G.L.Webster, J. Arnold Arbor. 39: 131 (1958).
  NE. Cuba. 81 CUB. Nanophan.
  * *Phyllanthus spathulifolius* Griseb., Nachr. Königl. Ges. Wiss. Georg-Augusts-Univ. 1: 169
    (1865). *Phyllanthus myrtilloides* var. *spathulifolius* (Griseb.) M.Gómez, Anales Soc. Esp.
    Hist. Nat. 23: 53 (1894).

Phyllanthus nanellus P.T.Li, Acta Phytotax. Sin. 25: 376 (1987).
Hainan. 36 CHH. Nanophan.

Phyllanthus narayanswamii Gamble, Bull. Misc. Inform. Kew 1925: 329 (1925).
India (Tamil Nadu). 40 IND.

Phyllanthus natoensis M.Schmid, in Fl. N. Caled. & Depend. 17: 171 (1991).
New Caledonia (Ponérihouen Reg.). 60 NWC. Nanophan.

Phyllanthus neblinae Jabl., Mem. New York Bot. Gard. 17: 107 (1967).
Venezuela (Amazonas: Cerro de la Neblina). 82 VEN. Nanophan.

Phyllanthus nemorum Russell ex Müll.Arg., Linnaea 34: 70 (1865).
S. India. 40 IND. Nanophan.

Phyllanthus neoleonensis Croizat, J. Wash. Acad. Sci. 33: 14 (1943).
Mexico (Nuevo León). 79 MXE.

Phyllanthus nhatrangensis Beille in H.Lecomte, Fl. Indo-Chine 5: 601 (1927).
S. Vietnam. 41 VIE.

Phyllanthus nigericus Brenan, Kew Bull. 5: 215 (1950).
Nigeria, Bioko. 22 NGA 23 GGI. Ther.

Phyllanthus nigrescens (Blanco) Müll.Arg. in A.P.de Candolle, Prodr. 15(2): 348 (1866).
New Caledonia (Mt. Ninga). 60 NWC. Nanophan.
Philippines. 42 PHI. Nanophan. – Provisionally accepted.
   * *Kirganelia nigrescens* Blanco, Fl. Filip., ed. 2: 712 (1845).

Phyllanthus niinamii Hayata, J. Coll. Sci. Imp. Univ. Tokyo 20(3): 14 (1904).
Taiwan. 38 TAI.

Phyllanthus ningaensis M.Schmid, in Fl. N. Caled. & Depend. 17: 144 (1991).
New Caledonia (Mt. Ninga). 60 NWC. Nanophan.

Phyllanthus niruri L., Sp. Pl.: 981 (1753). *Nymphanthus niruri* (L.) Lour., Fl. Cochinch.: 545
(1790). *Phyllanthus niruri* var. *genuinus* Müll.Arg. in A.P.de Candolle, Prodr. 15(2): 406
(1866), nom. inval.
Caribbean, S. Texas to Argentina. 77 TEX 79 MXE 80 COS 81 DOM HAI LEE PUE WIN 82
   VEN 84 BZL 85 AGE. Ther.

   subsp. **lathyroides** (Kunth) G.L.Webster, Contr. Gray Herb. 176: 52 (1955).
      S. Texas, N. Mexico, Costa Rica, Venezuela, Brazil, Argentina. 77 TEX 79 MXE 80 COS
         82 VEN 84 BZL 85 AGE. Ther. or cham.
      * *Phyllanthus lathyroides* Kunth in F.W.H.von Humboldt, A.J.A.Bonpland & C.S.Kunth,
         Nov. Gen. Sp. 2: 110 (1817).
      Phyllanthus purpurascens Kunth in F.W.H.von Humboldt, A.J.A.Bonpland & C.S.Kunth,
         Nov. Gen. Sp. 2: 110 (1817).
      Phyllanthus microphyllus Mart. in C.F.P.von Martius & J.B.von Spix, Reise Bras. 1:
         285 (1823).
      Phyllanthus mimosoides Lodd., Bot. Cab.: 721 (1823).
      Phyllanthus parvifolius Steud., Nomencl. Bot., ed. 2, 2: 327 (1841).
      Phyllanthus chlorophaeus Baill., Adansonia 1: 27 (1861).
      Phyllanthus lathyroides f. *microphyllus* Müll.Arg., Linnaea 32: 42 (1863).
      Phyllanthus lathyroides f. *minor* Müll.Arg., Linnaea 32: 42 (1863).
      Phyllanthus lathyroides f. *oblongifolius* Müll.Arg., Linnaea 32: 42 (1863).
      Phyllanthus lathyroides f. *purpurascens* Müll.Arg., Linnaea 32: 42 (1863).
      Phyllanthus lathyroides f. *rosellus* Müll.Arg., Linnaea 32: 42 (1863).
      Phyllanthus rosellus Müll.Arg. in C.F.P.von Martius, Fl. Bras. 11(2): 53 (1873).
      Phyllanthus williamsii Standl., Publ. Field Mus. Nat. Hist., Bot. Ser. 17: 266 (1937).

   subsp. **niruri**
      Caribbean. 81 DOM HAI LEE PUE WIN. Ther.
      Phyllanthus humilis Salisb., Prodr. Stirp. Chap. Allerton: 391 (1796).
      Phyllanthus carolinianus Blanco, Fl. Filip.: 691 (1837).
      Phyllanthus moeroris Oken, Allg. Naturgesch. 3(3): 1601 (1841).
      Phyllanthus kirganelia Blanco, Fl. Filip., ed. 2: 480 (1845).
      Phyllanthus filiformis Pav. ex Baill., Adansonia 1: 29 (1860).
      Phyllanthus ellipticus Buckley, Proc. Acad. Nat. Sci. Philadelphia 1862: 7 (1863), nom. illeg.
      Phyllanthus lathyroides var. *commutatus* Müll.Arg., Linnaea 32: 41 (1863).

Phyllanthus niruroides Müll.Arg., J. Bot. 2: 331 (1864).
W. Trop. Africa, São Tomé. 22 GNB GUI IVO LBR NGA SIE TOG 23 GGI. Ther.

Phyllanthus nitens M.Schmid, in Fl. N. Caled. & Depend. 17: 239 (1991).
New Caledonia (Massif du Boulinda). 60 NWC. Nanophan.

Phyllanthus nitidulus Müll.Arg. in A.P.de Candolle, Prodr. 15(2): 1272 (1866).
Jawa. 42 JAW.

Phyllanthus nothisii M.Schmid, in Fl. N. Caled. & Depend. 17: 179 (1991).
WC. New Caledonia. 60 NWC. Nanophan.

var. **alticola** M.Schmid, in Fl. N. Caled. & Depend. 17: 182 (1991).
WC. New Caledonia (Mt. Boulinda, Mt. Paéoua: above 1000 m). 60 NWC. Nanophan.

var. **nothisii**
WC. New Caledonia (Mt. Boulinda). 60 NWC. Nanophan.

Phyllanthus novaehollandiae Müll.Arg. in A.P.de Candolle, Prodr. 15(2): 346 (1866).
Papua New Guinea, Queensland. 42 NWG 50 QLD. Nanophan.
Phyllanthus uberiflorus F.Muell. ex Baill., Adansonia 6: 343 (Sept. 1866).

Phyllanthus nozeranianus Brunel & J.P.Roux, Willdenowia 14: 382 (1984 publ. 1985).
Ivory Coast. 22 IVO.

Phyllanthus nummulariifolius Poir. in J.B.A.M.de Lamarck, Encycl. 5: 302 (1804). *Menarda nummulariifolia* (Poir.) Baill., Étude Euphorb.: 609 (1858).
Trop. & S. Africa, Madagascar, Mascarenes, Seychelles. 22 GAM GHA GUI LBR NGA SIE TOG 23 BUR CMN GAB GGI RWA ZAI 24 SUD 25 KEN TAN UGA 26 ANG MLW MOZ ZAM ZIM 27 BOT NAT TVL 29 MAU MDG REU SEY. Cham. or nanophan.

var. **capillaris** (Schumach. & Thonn.) Radcl.-Sm., Kew Bull. 51: 316 (1996).
Trop. Africa, Ascencion. 22 GAM GHA GUI LBR NGA SIE TOG 23 CMN GAB GGI ZAI 25 KEN TAN UGA 26 ANG MLW MOZ ZAM ZIM 28 ASC. Cham. or nanophan.
* *Phyllanthus capillaris* Schumach. & Thonn. in C.F.Schumacher, Beskr. Guin. Pl.: 417 (1827).
Phyllanthus stuhlmanii Pax in H.G.A.Engler, Pflanzenw. Ost-Afrikas C: 236 (1895).
Phyllanthus capillaris var. *stuhlmannii* (Pax) Hutch. in D.Oliver, Fl. Trop. Afr. 6(1): 709 (1912).

subsp. **nummulariifolius**
Trop. & S. Africa, Madagascar, Mascarenes, Seychelles. 23 BUR CMN GAB GGI RWA ZAI 24 SUD 25 KEN TAN UGA 26 MLW MOZ ZAM ZIM 27 BOT NAT TVL 29 MAU MDG REU SEY. Cham. or nanophan.
Menarda capillaris Baill., Adansonia 1: 85 (1860).
Phyllanthus maderaspatensis Thouars ex Baill., Adansonia 2: 46 (1861), nom. illeg.
Phyllanthus peduncularis Boivin ex Baill., Adansonia 2: 45 (1861).
Phyllanthus tenellus var. *nossibeensis* Müll.Arg., Linnaea 32: 7 (1863).

subsp. **vinanibeae** Brunel & J.P.Roux, Bull. Mus. Natl. Hist. Nat., B, Adansonia 3: 198 (1981).
Madagascar. 29 MDG. Cham. or nanophan.

Phyllanthus nummularioides Müll.Arg., Linnaea 32: 5 (1863).
W. Dominican Rep. 81 DOM. Nanophan.

Phyllanthus nutans Sw., Prodr.: 27 (1788).
Cayman Is., Jamaica, E. Cuba. 81 CAY CUB JAM. Nanophan. or phan.

subsp. **grisebachianus** (Müll.Arg.) G.L.Webster, J. Arnold Arbor. 39: 61 (1958).
Cayman Is., E. Cuba. 81 CAY CUB. Nanophan. or phan.
* *Phyllanthus grisebachianus* Müll.Arg., Linnaea 32: 26 (1863).

subsp. **nutans**
Cayman Is., Jamaica. 81 CAY JAM. Nanophan. or phan.
Phyllanthus nutans var. *purdiaeana* Baill., Adansonia 2: 15 (1861).
Phyllanthus nutans var. *trojanus* G.L.Webster, Contr. Gray Herb. 176: 47 (1955).

Phyllanthus nyikae Radcl.-Sm., Kew Bull. 51: 317 (1996).
Malawi. 26 MLW. Cham.

Phyllanthus oaxacanus Brandegee, Univ. Calif. Publ. Bot. 6: 185 (1915).
Mexico (Oaxaca: San Geronimo). 79 MXS.

Phyllanthus obdeltophylla Leandri, Mém. Inst. Sci. Madagascar, Sér. B, Biol. Vég. 8: 232 (1957).
C. Madagascar. 29 MDG. Nanophan.

Phyllanthus obfalcatus Lasser & Maguire, Brittonia 7: 79 (1950).
Venezuela (Amazonas). 82 VEN. Nanophan.

Phyllanthus oblanceolatus J.T.Hunter & J.J.Bruhl, J. Adelaide Bot. Gard. 17: 128 (1996).
C. & EC. Australia. 50 NSW NYA SOA. Cl. cham.

Phyllanthus oblongiglans M.G.Gilbert, Kew Bull. 42: 359 (1987).
Ethiopia. 24 ETH. Hemicr. or hel.

Phyllanthus obtusatus (Billb.) Müll.Arg. in A.P.de Candolle, Prodr. 15(2): 433 (1866).
Brazil (?). 84 +.
   * *Xylophylla obtusata* Billb., Pl. Bras. 1: 12 (1817).

Phyllanthus occidentalis J.T.Hunter & J.J.Bruhl, Telopea 7: 157 (1997).
SE. Queensland, E. New South Wales. 50 NSW QLD. Cham. or nanophan.

Phyllanthus octomerus Müll.Arg. in C.F.P.von Martius, Fl. Bras. 11(2): 30 (1873).
Brazil (Bahia). 84 BZE. Nanophan.

Phyllanthus odontadenius Müll.Arg., J. Bot. 2: 331 (1864).
Trop. Africa. 22 BEN GHA GNB IVO LBR NGA SIE TOG 23 BUR CAF CMN CON GAB GGI
   ZAI RWA 24 SUD 25 KEN TAN UGA 26 ANG MLW ZAM ZIM. Ther. or nanophan.
   Phyllanthus braunii Pax, Bot. Jahrb. Syst. 15: 525 (1893). *Phyllanthus odontadenius* var.
      braunii (Pax) Hutch. in D.Oliver, Fl. Trop. Afr. 6(1): 728 (1912).
   Phyllanthus santhomensis Beille, Bull. Soc. Bot. France 55(8): 56 (1908).

Phyllanthus oligospermus Hayata, Icon. Pl. Formosan. 9: 93 (1920).
Taiwan. 38 TAI. Nanophan.

Phyllanthus omahakensis Dinter & Pax, Bot. Jahrb. Syst. 45: 234 (1910).
Zimbabwe, Zambia, Botswana, Angola, Namibia. 26 ANG ZAM ZIM 27 BOT NAM.

Phyllanthus oppositifolius Baill. ex Müll.Arg., Linnaea 32: 24 (1863).
Mauritius. 29 MAU. Nanophan.

Phyllanthus orbicularifolius (P.T.Li) Govaerts & Radcl.-Sm., Kew Bull. 51: 177 (1996).
China (Guangxi). 36 CHS. Nanophan.
   * *Phyllanthodendron orbicularifolium* P.T.Li, Bull. Bot. Res., Harbin 7(3): 5 (1987).

Phyllanthus orbicularis Kunth in F.W.H.von Humboldt, A.J.A.Bonpland & C.S.Kunth, Nov.
Gen. Sp. 2: 111 (1817). *Orbicularia orbicularis* (Kunth) Moldenke, Revista Sudamer. Bot. 6:
178 (1940).
Cuba to Puerto Rico. 81 CUB DOM HAI PUE. Nanophan.
   Orbicularia phyllanthoides Baill., Étude Euphorb.: 617 (1858).
   Phyllanthus rotundifolius Sessé & Moç., Fl. Mexic., ed. 2: 212 (1894), nom. illeg.
   Andrachne cuneifolia Britton, Mem. Torrey Bot. Club 16: 72 (1920).
   Phyllanthus cuneifolius (Britton) Croizat, J. Wash. Acad. Sci. 33: 12 (1943).

Phyllanthus orbiculatus Rich., Actes Soc. Hist. Nat. Paris 1: 113 (1792). *Phyllanthus orbiculatus* var. *genuinus* Müll.Arg. in A.P.de Candolle, Prodr. 15(2): 401 (1866), nom. inval.
Trinidad to Paraguay. 81 TRT 82 FRG SUR VEN 83 BOL CLM PER 84 BZN 85 PAR. Ther.
Phyllanthus poiretianus Müll.Arg., Linnaea 32: 39 (1863).
Phyllanthus orbiculatus var. *acutifolius* Müll.Arg. in A.P.de Candolle, Prodr. 15(2): 401 (1866).
Phyllanthus orbicularis var. *lignescens* Müll.Arg. in C.F.P.von Martius, Fl. Bras. 11(2): 62 (1873).
Phyllanthus orbiculatus var. *intermedius* Müll.Arg. in C.F.P.von Martius, Fl. Bras. 11(2): 62 (1873).

Phyllanthus oreophilus Müll.Arg., Linnaea 32: 49 (1863).
Sri Lanka. 40 SRL.
Epistylium montanum Thwaites, Enum. Pl. Zeyl.: 283 (1861). *Reidia montana* (Thwaites) Alston in H.Trimen, Handb. Fl. Ceylon 6(Suppl.): 258 (1931).

Phyllanthus orientalis (Craib) Airy Shaw, Kew Bull. 25: 495 (1971).
NW. Thailand. 41 THA. Nanophan.
* *Chorisandra orientalis* Craib, Bull. Misc. Inform. Kew 1914: 285 (1914).

Phyllanthus orinocensis Steyerm., Fieldiana, Bot. 28: 321 (1952).
Venezuela (Amazonas: Cerro Duida). 82 VEN. Nanophan.

Phyllanthus ouveanus Däniker, Vierteljahrschr. Naturf. Ges. Zürich 76: 168 (1931).
New Caledonia (Loyalty Is.). 60 NWC. Nanophan.
Phyllanthus mareensis Guillaumin, Mém. Mus. Natl. Hist. Nat., B, Bot. 8: 246 (1962), no type designated.

Phyllanthus ovalifolius Forssk., Fl. Aegypt.-Arab.: 159 (1775).
Yemen, Trop. Africa. 22 NGA 23 CMN RWA ZAI 24 ETH 25 KEN TAN UGA 26 ANG MLW MOZ ZAM ZIM 35 YEM. (Cl.) nanophan. or phan.
Phyllanthus guineensis Pax, Bull. Herb. Boissier 6: 732 (1898).
Phyllanthus lalambensis Schweinf., Bull. Herb. Boissier 7(App. 2): 302 (1899).
Phyllanthus ugandensis Rendle, J. Linn. Soc., Bot. 37: 210 (1905).

Phyllanthus ovatifolius J.J.Sm., Bull. Jard. Bot. Buitenzorg, III, 1: 390 (1920).
Maluku (Key I.). 42 MOL.

Phyllanthus ovatus Poir. in J.B.A.M.de Lamarck, Encycl. 5: 297 (1804). *Glochidion ovatum* (Poir.) Müll.Arg., Linnaea 32: 71 (1863).
Martinique. 81 WIN. Nanophan. or phan.
Phyllanthus grandifolius Spreng., Syst. Veg. 3: 22 (1826), nom. illeg.

Phyllanthus oxycarpus Müll.Arg. in A.P.de Candolle, Prodr. 15(2): 1270 (1866).
Sumatera. 42 SUM.

Phyllanthus oxycoccifolius Hutch. in D.Oliver, Fl. Trop. Afr. 6(1): 735 (1912).
– FIGURE, p. 1266.
Tanzania, Angola. 25 TAN 26 ANG. Cham.

Phyllanthus oxyphyllus Miq., Fl. Ned. Ind., Eerste Bijv.: 448 (1861).
S. Thailand, Pen. Malaysia, Sumatera. 41 THA 42 MLY SUM. Nanophan.
Phyllanthus frondosus Wall. ex Müll.Arg., Linnaea 32: 17 (1863).
Phyllanthus kunstleri Hook.f., Fl. Brit. India 5: 292 (1887).

Phyllanthus pachyphyllus Müll.Arg. in A.P.de Candolle, Prodr. 15(2): 353 (1866).
   S. Thailand, Vietnam, Hainan, Burma, Pen. Malaysia(incl. Singapore), Borneo (E.
   Kalimantan). 36 CHH 41 BMA? THA VIE 42 BOR MLY. Nanophan.
   Phyllanthus klossii Ridl., J. Fed. Malay States Mus. 10: 114 (1920).
   Phyllanthus campanulatus Ridl., Bull. Misc. Inform. Kew 1923: 362 (1923).
   Phyllanthus frondosus var. *rigidus* Ridl., Fl. Malay Penins. 3: 203 (1924).
   hyllanthus sciadiostylus Airy Shaw, Kew Bull. 23: 30 (1969).

Phyllanthus pachystylus Urb., Symb. Antill. 3: 286 (1902).
   E. Cuba (Sierra Sagua Baracoa). 81 CUB. Nanophan.

Phyllanthus pacificus Müll.Arg., Linnaea 32: 31 (1863).
   Marquesas. 61 MRQ. Cham. or nanophan.
   Phyllanthus pacificus var. *quaylei* F.Br., Bernice P. Bishop Mus. Bull. 130: 138 (1935).
   Phyllanthus pacificus var. *uahukensis* F.Br., Bernice P. Bishop Mus. Bull. 130: 139 (1935).
   Phyllanthus pacificus var. *uapensis* F.Br., Bernice P. Bishop Mus. Bull. 130: 137 (1935).

Phyllanthus pacoensis Thin, J. Biol. (Vietnam) 14(2): 19 (1992).
   Vietnam. 41 VIE.

Phyllanthus paezensis Jabl., Mem. New York Bot. Gard. 17: 113 (1967).
   Venezuela (Bolívar: Is. in Rio Orinoco). 82 VEN. Nanophan.

Phyllanthus palauensis Hosok., Trans. Nat. Hist. Soc. Taiwan 25: 19 (1935).
   Caroline Is. (Palau). 62 CRL.

Phyllanthus × pallidus C.Wright ex Griseb., Nachr. Königl. Ges. Wiss. Georg-Augusts-Univ.
   1: 168 (1865). P. discolor × P. ? *Phyllanthus discolor* var. *pallidus* (C.Wright ex Griseb.)
   G.L.Webster, Contr. Gray Herb. 176: 57 (1955).
   W. Cuba. 81 CUB. Nanophan.
   Phyllanthus × sagraeanus Urb., Symb. Antill. 9: 182 (1924).

Phyllanthus panayensis Merr., Philipp. J. Sci. 16: 539 (1920).
   Philippines. 42 PHI.

Phyllanthus pancherianus Baill., Adansonia 2: 235 (1862). *Glochidion pancherianum* (Baill.)
   Müll.Arg., Linnaea 32: 71 (1863).
   New Caledonia. 60 NWC. Nanophan.

   var. **kopetoensis** M.Schmid, in Fl. N. Caled. & Depend. 17: 110 (1991).
      NC. New Caledonia (Mt. Kopéto). 60 NWC. Nanophan.

   var. **memaoyaensis** M.Schmid, in Fl. N. Caled. & Depend. 17: 111 (1991).
      C. New Caledonia (S. of Mé Maoya). 60 NWC. Nanophan.

   var. **nakadaensis** M.Schmid, in Fl. N. Caled. & Depend. 17: 111 (1991).
      SE. Central New Caledonia (Mt. Nakada). 60 NWC. Nanophan.

   var. **pancherianus**
      C. New Caledonia. 60 NWC. Nanophan.

Phyllanthus papuanus Gage, Nova Guinea 12: 479 (1917).
   Irian Jaya. 42 NWG. Nanophan.

Phyllanthus paraguayensis Parodi, Anales Soc. Ci. Argent. 11: 50 (1881).
   Paraguay. 85 PAR.

Phyllanthus parainduratus M.Schmid, in Fl. N. Caled. & Depend. 17: 228 (1991).
EC. & WC. New Caledonia. 60 NWC. Cham. or nanophan.

Phyllanthus parangoyensis M.Schmid, in Fl. N. Caled. & Depend. 17: 250 (1991).
New Caledonia (Canala). 60 NWC. Nanophan.

Phyllanthus paraqueensis Jabl., Mem. New York Bot. Gard. 17: 104 (1967).
Venezuela (Amazonas: Cerro Sipapo). 82 VEN. Nanophan.

Phyllanthus parvifolius Buch.-Ham. ex D.Don, Prodr. Fl. Nepal.: 63 (1825).
N. Pakistan, Kashmir, N. India, Nepal, Bhutan, Assam, China. 36 CHC CHH CHS 40 ASS
EHM IND NEP PAK WHM. Nanophan.
Phyllanthus juniperinus Wall., Numer. List: 7901 (1847), nom. inval.
Phyllanthus praetervisus Müll.Arg., Linnaea 34: 73 (1865).

Phyllanthus parvulus Sond., Linnaea 23: 132 (1850).
Zaire to S. Africa. 23 ZAI 26 MOZ ZAM ZIM 27 BOT CPP NAM NAT OFS SWZ TVL.

var. **garipensis** (Müll.Arg.) Radcl.-Sm., Kew Bull. 51: 317 (1996).
SC. Trop. & S. Africa. 26 MOZ ZAM 27 CPP NAT OFS SWZ TVL.
Phyllanthus garipensis E.Mey. in J.F.Drège, Zwei Pflanzengeogr. Dokum.: 93 (1843),
nom. nud.
Phyllanthus burchellii Müll.Arg., Linnaea 32: 7 (1863).
* *Phyllanthus tenellus* var. *garipensis* Müll.Arg., Linnaea 32: 7 (1863).

var. **parvulus**
Zaire, South Africa, Namibia, Botswana, Zimbabwe, Mozambique. 23 ZAI 26 MOZ ZIM
27 BOT CPP NAM OFS TVL.

Phyllanthus parvus Hutch., Bull. Misc. Inform. Kew 1911: 316 (1911).
Zaire (Shaba), W. & S. Tanzania, Malawi, Zambia, Angola. 23 ZAI 25 TAN 26 ABG MLW
ZAM. Ther.

Phyllanthus paucitepalus M.Schmid, in Fl. N. Caled. & Depend. 17: 140 (1991).
New Caledonia (Col de Mô). 60 NWC. Nanophan.

Phyllanthus pavonianus Baill., Adansonia 1: 30 (1860).
Peru. 83 PER.
Phyllanthus oxycladus Müll.Arg., Linnaea 32: 26 (1863).

Phyllanthus paxianus Dinter, Repert. Spec. Nov. Regni Veg. 22: 379 (1926).
Zaire, Angolia, Namibia. 23 ZAI 26 ANG 27 NAM.

Phyllanthus paxii Hutch., Bull. Misc. Inform. Kew 1911: 316 (1911).
Zaire, Burundi, Tanzania, Mozambique, Malawi, Zambia, Angola. 23 BUR ZAI 25 TAN 26
ANG MLW MOZ ZAM. Hemicr. or cham.

Phyllanthus pectinatus Hook.f., Fl. Brit. India 5: 290 (1887). *Emblica pectinata* (Hook.f.)
Ridl., Fl. Malay Penins. 3: 217 (1924).
Pen. Malaysia. 42 MLY. – Close to *P. emblica*.

Phyllanthus peltatus Guillaumin, Arch. Bot. Mém. 2(3): 13 (1929).
NW. New Caledonia. 60 NWC. Nanophan.

Phyllanthus pendulus Roxb., Fl. Ind. ed. 1832, 3: 663 (1832).
Bangladesh. 40 BAN. Biennial.

Phyllanthus pentandrus Schumach. & Thonn. in C.F.Schumacher, Beskr. Guin. Pl.: 419 (1827).

Trop. & S. Africa. 22 BEN GAM GHA GNB GUI IVO MLI NGA SEN 23 CMN ZAI 24 ETH 25 TAN 26 ANG MLW MOZ ZAM ZIM 27 BOT NAM NAT SWZ TVL. Ther.
Phyllanthus piluliferus Fenzl, Flora 27: 312 (1844), nom. nud.
Phyllanthus scoparius Welw., Apont.: 591 (1859).
Menarda linifolia Baill., Adansonia 1: 84 (1860), nom. nud.
Phyllanthus linifolius Vahl ex Baill., Adansonia 1: 84 (1860), pro syn.
Phyllanthus linoides Hochst. ex Baill., Adansonia 1: 84 (1860), pro syn.
Phyllanthus deflexus Klotzsch in W.C.H.Peters, Naturw. Reise Mossambique: 104 (1861).
Phyllanthus dilatatus Klotzsch in W.C.H.Peters, Naturw. Reise Mossambique: 106 (1861).

Phyllanthus pentaphyllus C.Wright ex Griseb., Nachr. Königl. Ges. Wiss. Georg-Augusts-Univ. 1: 167 (1865).

S. Florida, Caribbean, Venezuela. 78 FLA 81 BAH CUB DOM HAI LEE NLA PUE 82 VEN. Cham.

var. **floridanus** G.L.Webster, Contr. Gray Herb. 176: 56 (1955).
S. Florida. 78 FLA. Hemicr.

subsp. **pentaphyllus**
Bahamas, Cuba, Hispaniola, Curaçao, Venezuela. 81 BAH CUB DOM HAI NLA 82 VEN. Cham.
Phyllanthus cyclanthera Baill., Adansonia 1: 31 (1860), nom rejic.
Phyllanthus niruri var. *radicans* Müll.Arg., Linnaea 32: 44 (1863). *Phyllanthus radicans* (Müll.Arg.) Small, Fl. S.E. U.S.: 692 (1903).
Phyllanthus polycladus var. *curassavicus* Urb., Symb. Antill. 5: 384 (1908).

subsp. **polycladus** (Urb.) G.L.Webster, Contr. Gray Herb. 176: 56 (1955).
Puerto Rico, Leeward Is. 81 LEE PUE. Hemicr.
* *Phyllanthus polycladus* Urb., Symb. Antill. 1: 333 (1899).
Phyllanthus polycladus var. *guadeloupensis* Urb., Symb. Antill. 1: 333 (1899).

Phyllanthus pergracilis Gillespie, Bernice P. Bishop Mus. Bull. 91: 18 (1932).
Fiji. 60 FIJ.

Phyllanthus perpusillus Baill., Adansonia 5: 358 (1865).
Brazil (Minas Gerais). 84 BZL.

Phyllanthus pervilleanus (Baill.) Müll.Arg., Linnaea 32: 13 (1863).
Comoros, Madagascar (Nosi Bé). 29 COM MDG. Nanophan.
* *Kirganelia pervilleana* Baill., Adansonia 2: 50 (1861).
Phyllanthus pervilleanus var. *glaber* Müll.Arg. in A.P.de Candolle, Prodr. 15(2): 347 (1866).
Phyllanthus humblotianus Baill. in A.Grandidier, Hist. Phys. Madagascar, Atlas: 221 (1891).

Phyllanthus petaloideus Paul G.Wilson, Hooker's Icon. Pl. 36: t. 3589 (1962).
Mexico (México State). 79 MXC. Nanophan.

Phyllanthus petchikaraensis M.Schmid, in Fl. N. Caled. & Depend. 17: 90 (1991).
New Caledonia (Col de Petchikara). 60 NWC. Nanophan.

Phyllanthus petelotii Croizat, J. Arnold Arbor. 23: 30 (1942).
Vietnam. 41 VIE. Nanophan.

Phyllanthus petenensis Lundell, Phytologia 57: 367 (1985).
Guatemala. 80 GUA.

Phyllanthus petiolaris Roxb., Fl. Ind. ed. 1832, 3: 664 (1832).
India. 40 IND. – Provisionally accepted.

Phyllanthus petraeus A.Chev. & Beille ex Beille, Bull. Soc. Bot. France 55(8): 58 (1908).
Sierra Leone, Liberia, Guinea. 22 GUI LBR SIE. Nanophan.

Phyllanthus philippioides Leandri, Bull. Soc. Bot. France 80: 373 (1933).
C. Madagascar. 29 MDG. Cham.

Phyllanthus phillyreifolius Poir. in J.B.A.M.de Lamarck, Encycl. 5: 299 (1804). *Phyllanthus polymorphus* Müll.Arg., Linnaea 32: 24 (1863), nom. illeg. *Phyllanthus phillyreifolius* var. *genuinus* Müll.Arg. in A.P.de Candolle, Prodr. 15(2): 378 (1866), nom. inval.
Mascarenes. 29 MAU REU. Nanophan.

   var. **commersonii** Müll.Arg. in A.P.de Candolle, Prodr. 15(2): 377 (1866).
     Mauritius. 29 MAU. Nanophan.
     Phyllanthus commersonianus Baill., Adansonia 2: 53 (1861), nom. nud.
     Phyllanthus phillyreifolius var. *longifolius* Müll.Arg. in A.P.de Candolle, Prodr. 15(2): 377 (1866).

   var. **crassistigma** Coode, Kew Bull. 33: 116 (1978).
     Réunion. 29 REU. Nanophan.

   var. **gracilipes** Coode, Kew Bull. 33: 118 (1978).
     Mauritius. 29 MAU. Nanophan.

   var. **phillyreifolius**
     Mascarenes. 29 MAU REU. Nanophan.
     Phyllanthus phillyreifolius var. *angustifolius* Müll.Arg. in A.P.de Candolle, Prodr. 15(2): 377 (1866).
     Phyllanthus phillyreifolius var. *ellipticus* Müll.Arg. in A.P.de Candolle, Prodr. 15(2): 377 (1866).
     Phyllanthus phillyreifolius var. *lanceolatus* Müll.Arg. in A.P.de Candolle, Prodr. 15(2): 377 (1866).
     Phyllanthus phillyreifolius var. *parvifolius* Müll.Arg. in A.P.de Candolle, Prodr. 15(2): 377 (1866).

   var. **stylifer** Coode, Kew Bull. 33: 118 (1978).
     Mauritius. 29 MAU. Nanophan.

   var. **telfairianus** Wall. ex Müll.Arg. in A.P.de Candolle, Prodr. 15(2): 377 (1866).
     Mauritius. 29 MAU. Nanophan.
     Phyllanthus subcordatus Baill., Adansonia 2: 53 (1861), nom. nud.
     Phyllanthus phillyreifolius var. *oblongifolius* Müll.Arg. in A.P.de Candolle, Prodr. 15(2): 377 (1866).
     Phyllanthus phillyreifolius var. *rotundifolius* Müll.Arg. in A.P.de Candolle, Prodr. 15(2): 377 (1866).

   var. **triangularis** Müll.Arg. in A.P.de Candolle, Prodr. 15(2): 377 (1866).
     Mauritius. 29 MAU. Nanophan.

Phyllanthus phlebocarpus Urb., Symb. Antill. 9: 189 (1924).
NE. Cuba. 81 CUB. Nanophan.
     Phyllanthus breviramis Urb., Symb. Antill. 9: 192 (1924).
     Phyllanthus estrellensis Urb., Symb. Antill. 9: 188 (1924).
     Phyllanthus norlindii Urb., Symb. Antill. 9: 187 (1924).

Phyllanthus phuquocensis Beille in H.Lecomte, Fl. Indo-Chine 5: 581 (1927).
Vietnam. 41 VIE.

Phyllanthus physocarpus Müll.Arg., Flora 47: 515 (1864).
Principe. 23 GGI. Nanophan. or phan.

Phyllanthus pileostigma Coode, Kew Bull. 33: 119 (1978).
Mauritius. 29 MAU. Nanophan.

Phyllanthus pilifer M.Schmid, in Fl. N. Caled. & Depend. 17: 120 (1991).
New Caledonia. 60 NWC. Nanophan.

var. **grandieensis** M.Schmid, in Fl. N. Caled. & Depend. 17: 123 (1991).
New Caledonia (Mt. Grandié). 60 NWC. Nanophan.

var. **pilifer**
C. New Caledonia. 60 NWC. Nanophan.
Phyllanthus cataractarum var. *villosus* Däniker, Vierteljahrschr. Naturf. Ges. Zürich
77(19): 211 (1932).

Phyllanthus pimichinianus Jabl., Mem. New York Bot. Gard. 17: 111 (1967).
Venezuela (Amazonas: Rio Guainia reg.). 82 VEN. Nanophan. or phan.

Phyllanthus pindaiensis M.Schmid, in Fl. N. Caled. & Depend. 17: 183 (1991).
WC. New Caledonia (Népoui Pen.). 60 NWC. Nanophan.

Phyllanthus pinifolius Baill., Adansonia 5: 353 (1865).
S. Brazil. 84 BZL. Nanophan.

Phyllanthus pinjenensis M.Schmid, in Fl. N. Caled. & Depend. 17: 92 (1991).
NW. New Caledonia (near Koné). 60 NWC. Cham. or nanophan.

Phyllanthus pinnatus (Wight) G.L.Webster, J. Arnold Arbor. 38: 52 (1957).
Kenya, Tanzania, C. Mozambique, S. Malawi, Zimbabwe, Northern Prov. (Zoutpansberg),
E. India, Sri Lanka. 25 KEN TAN 26 MLW MOZ ZIM 27 TVL 40 IND SRL. Nanophan.
* *Chorisandra pinnata* Wight, Icon. Pl. Ind. Orient. 6: 13 (1853).
Phyllanthus wightianus Müll.Arg., Linnaea 32: 6 (1863). *Phyllanthus obliquus* Wall. ex
Hook.f., Fl. Brit. India 5: 303 (1887), pro syn.
Phyllanthus kirkianus Müll.Arg., Flora 47: 486 (1864).
Cluytiandra schinzii Pax, Bull. Herb. Boissier, II, 8: 635 (1908).
Phyllanthus senensis Müll.Arg. in A.P.de Candolle, Prodr. 15(2): 335 (1966).

Phyllanthus pireyi Beille in H.Lecomte, Fl. Indo-Chine 5: 605 (1927).
Vietnam. 41 VIE.

Phyllanthus piscatorum Kunth in F.W.H.von Humboldt, A.J.A.Bonpland & C.S.Kunth,
Nov. Gen. Sp. 2: 90, 113 (1817).
Venezuela, N. Brazil, Peru. 82 VEN 83 PER 84 BZN. Nanophan.

Phyllanthus platycalyx Müll.Arg. in A.P.de Candolle, Prodr. 15(2): 318 (1866).
Glochidion platycalyx (Müll.Arg.) Pax & K.Hoffm. in H.G.A.Engler, Nat. Pflanzenfam.
ed. 2, 19c: 58 (1931).
New Caledonia. 60 NWC. Nanophan.

var. **angustifolius** Guillaumin, Arch. Bot. Mém. 2(3): 14 (1929).
New Caledonia (Tiwaka). 60 NWC. Nanophan.

var. **platycalyx**
NW. & NC. New Caledonia. 60 NWC. Nanophan.

Phyllanthus poeppigianus (Müll.Arg.) Müll.Arg. in A.P.de Candolle, Prodr. 15(2): 323 (1866).
  Brazil (Amazonas), Peru (Loreto). 83 PER 84 BZN. Nanophan.
    * *Glochidion poeppigianum* Müll.Arg., Linnaea 32: 71 (1863).

Phyllanthus pohlianus Müll.Arg. in C.F.P.von Martius, Fl. Bras. 11(2): 49 (1873).
  Brazil (Minas Gerais). 84 BZL. Cham.

Phyllanthus poilanei Beille, Bull. Soc. Bot. France 72: 162 (1925). *Phyllanthodendron poilanei*
  (Beille) Croizat, J. Arnold Arbor. 23: 34 (1942).
  S. Vietnam. 41 VIE. Cl. nanophan.
    Phyllanthus lignulatus var. *tonkinensis* Baill. in H.Lecomte, Fl. Indo-Chine 5: 593 (1927).

Phyllanthus poliborealis Airy Shaw, Kew Bull. 33: 36 (1978).
  Irian Jaya. 42 NWG. Nanophan. or phan.

Phyllanthus polyanthus Pax, Bot. Jahrb. Syst. 28: 19 (1899).
  C. & E. Trop. & S. Africa. 23 CAF CMN CON GAB GGI ZAI 25 KEN TAN 26 ANG MLW
    MOZ ZAM 27 NAT. Nanophan. or phan.
    Phyllanthus pynaertii De Wild., Ann. Mus. Congo Belge, Bot., V, 2: 267 (1908).
    Phyllanthus klainei Hutch. in D.Oliver, Fl. Trop. Afr. 6(1): 1048 (1913).

Phyllanthus polygonoides Nutt. ex Spreng., Syst. Veg. 3: 23 (1826).
  SW. Missouri to SE. New Mexico & N. Mexico. 74 MSO OKL 77 NWM TEX 78 LOU 79
    MXE MXN. Hemicr. or cham.

Phyllanthus polygynus M.Schmid, in Fl. N. Caled. & Depend. 17: 97 (1991).
  NW. New Caledonia. 60 NWC. Nanophan. or phan.

Phyllanthus polyphyllus Willd., Sp. Pl. 4: 586 (1805).
  S. India, Sri Lanka, Thailand. 40 IND SRL 41 THA. Nanophan. or phan.

  var. **polyphyllus**
    S. India, Sri Lanka. 40 IND SRL. Nanophan. or phan.
    Phyllanthus emblicoides Müll.Arg., Linnaea 32: 15 (1863).

  var. **siamensis** Airy Shaw, Kew Bull. 23: 33 (1969).
    NE. Thailand. 41 THA. Nanophan. or phan.

Phyllanthus pomiferus Hook.f., Fl. Brit. India 5: 289 (1887).
  S. Burma. 41 BMA.

Phyllanthus popayanensis Pax, Bot. Jahrb. Syst. 26: 503 (1899).
  Colombia. 83 CLM.

Phyllanthus poueboensis M.Schmid, in Fl. N. Caled. & Depend. 17: 154 (1991).
  NW. New Caledonia (Pouébo Reg.). 60 NWC. Nanophan.

Phyllanthus poumensis Guillaumin, Arch. Bot. Mém. 2(3): 15 (1929).
  NW. New Caledonia. 60 NWC. Nanophan.

  var. **longistylis** M.Schmid, in Fl. N. Caled. & Depend. 17: 135 (1991).
    NW. New Caledonia. 60 NWC. Nanophan.

  var. **longitepalus** M.Schmid, in Fl. N. Caled. & Depend. 17: 135 (1991).
    NW. New Caledonia. 60 NWC. Nanophan.

  var. **poumensis**
    NW. New Caledonia. 60 NWC. Nanophan.

Phyllanthus praelongipes Airy Shaw & G.L.Webster, Kew Bull. 26: 100 (1971).
Papua New Guinea, Queensland. 42 NWG 50 QLD. Phan.

Phyllanthus prainianus Collett & Hemsl., J. Linn. Soc., Bot. 28: 123 (1890).
Burma. 41 BMA.

Phyllanthus procerus C.Wright, Anales Acad. Ci. Méd. Habana 7: 149 (1870).
Cuba (incl. I. de la Juventud). 81 CUB. Ther.

Phyllanthus proctoris G.L.Webster, J. Arnold Arbor. 39: 195 (1958).
W. Jamaica. 81 JAM. Nanophan.

Phyllanthus profusus N.E.Br., J. Linn. Soc., Bot. 37: 113 (1905).
Guinea to Congo. 22 GHA GUI IVO LBR 23 CMN CON. Nanophan.
  Phyllanthus wildemannii Beille, Bull. Soc. Bot. France 61(8): 293 (1914 publ. 1917).

Phyllanthus prominulatus J.T.Hunter & J.J.Bruhl, Nuytsia 11: 153 (1997).
Northern Territory. 50 NTA. Hemicr.

Phyllanthus pronyensis Guillaumin, Bull. Mus. Natl. Hist. Nat. 33: 273 (1927).
SE. New Caledonia. 60 NWC. Cham. or nanophan.

Phyllanthus prostratus Müll.Arg., J. Bot. 2: 330 (1864).
Angola, Zimbabwe. 26 ANG ZIM. Hemicr. or cham.

Phyllanthus prunifolius Rusby, Mem. New York Bot. Gard. 7: 283 (1927).
Bolivia. 83 BOL.

Phyllanthus pseudocarunculatus Radcl.-Sm., Kew Bull. 51: 318 (1996).
S. Zaire, Zambia. 23 ZAI 26 ZAM. Ther.

Phyllanthus pseudocicca Griseb., Nachr. Königl. Ges. Wiss. Georg-Augusts-Univ. 1: 166 (1865).
E. Cuba. 81 CUB. Nanophan.
  Phyllanthus cicca Griseb., Nachr. Königl. Ges. Wiss. Georg-Augusts-Univ. 1: 166 (1865), pro syn.
  Phyllanthus brevistipulus Urb., Symb. Antill. 9: 183 (1924).
  Phyllanthus punctulatus Urb., Symb. Antill. 9: 184 (1924).

Phyllanthus pseudoconami Müll.Arg. in C.F.P.von Martius, Fl. Bras. 11(2): 43 (1873).
Peru, Brazil (Amazonas). 83 PER 84 BZN. Nanophan.
  Phyllanthus pseudoconami var. *pubescens* Müll.Arg. in C.F.P.von Martius, Fl. Bras. 11(2): 43 (1873).

Phyllanthus pseudoguyanensis Herter & Mansf., Revista Sudamer. Bot. 5: 33 (1936 publ. 1937).
Uruguay. 85 URU.

Phyllanthus pseudoniruri Müll.Arg., Flora 47: 539 (1864).
Cameroon & Ethiopia to Botswana. 23 BUR CMN RWA ZAI 24 ETH SOM SUD 25 KEN TAN UGA 26 MLW ZAM ZIM 27 BOT. Ther. or nanophan. – Close to *P. leucanthus*.

Phyllanthus pseudonobilis Rusby, Mem. New York Bot. Gard. 7: 281 (1927).
Bolivia. 83 BOL.

Phyllanthus pseudotrichopodus M.Schmid, in Fl. N. Caled. & Depend. 17: 245 (1991).
EC. New Caledonia (Houaïlou to Touho). 60 NWC. Nanophan.

Phyllanthus pterocladus S.Moore, J. Linn. Soc., Bot. 45: 400 (1921).
NW. New Caledonia. 60 NWC. Nanophan. or phan.
  Phyllanthus kaalaensis Guillaumin, State Univ. Iowa Stud. Nat. Hist. 20: 34 (1965).

Phyllanthus pudens Wheeler, Contr. Gray Herb. 127: 50 (1939).
SE. Texas, SC. Louisiana. 77 TEX 78 LOU. Ther.
  * *Phyllanthus avicularis* Small, Bull. Torrey Bot. Club 27: 278 (1900), nom. illeg.

Phyllanthus pulcher Wall. ex Müll.Arg., Linnaea 32: 49 (1863).
China (Yunnan, Guangxi), Indo-China, W. Malesia. 36 CHC CHS 41 BMA CBD LAO THA
    VIE 42 BOR JAW MLY SUM. Nanophan.
  Epistylium glaucescens Baill., Étude Euphorb.: 648 (1858), nom. nud.
  Epistylium phyllanthoides Baill., Étude Euphorb.: 648 (1858), nom. nud.
  Epistylium pulchrum Baill., Étude Euphorb.: 648 (1858), nom. nud.
  Anisonema zollingeri Miq., Fl. Ned. Ind. 1(2): 375 (1859). *Phyllanthus zollingeri* (Miq.)
    Müll.Arg., Linnaea 32: 47 (1863).
  Reidia glaucescens Miq., Fl. Ned. Ind. 1(2): 374 (1859).
  Phyllanthus pallidus Müll.Arg. in A.P.de Candolle, Prodr. 15(2): 283 (1866).
  Phyllanthus asteranthos Croizat, J. Jap. Bot. 16: 655 (1940).
  Phyllanthus lacerilobus Croizat, Caldasia 3: 21 (1944).

Phyllanthus pulcherrimus Herter ex Arechav., Anales Mus. Nac. Montevideo, II, 1:
72 (1910).
Uruguay. 85 URU.

Phyllanthus pulchroides Beille in H.Lecomte, Fl. Indo-Chine 5: 597 (1927).
Vietnam. 41 VIE.

Phyllanthus pullenii Airy Shaw & G.L.Webster, Kew Bull. 26: 105 (1971).
New Guinea. 42 NWG. Nanophan. or phan.

Phyllanthus pulverulentus Urb., Symb. Antill. 9: 192 (1924).
SE. Cuba. 81 CUB. Cham.

Phyllanthus pumilus (Blanco) Müll.Arg. in A.P.de Candolle, Prodr. 15(2): 349 (1866).
Philippines. 42 PHI.
  * *Kirganelia pumila* Blanco, Fl. Filip.: 712 (1837).

Phyllanthus purpureus Müll.Arg., J. Bot. 2: 329 (1864).
Angola, Namibia. 26 ANG 27 NAM. Nanophan.

Phyllanthus purpusii Brandegee, Univ. Calif. Publ. Bot. 6: 55 (1914).
Mexico (Chiapas: Cerro del Boqueron). 79 MXT.

Phyllanthus pycnophyllus Müll.Arg. in A.P.de Candolle, Prodr. 15(2): 322 (1866).
Guyana, Venezuela (Bolívar). 82 GUY VEN. Nanophan.
  Glochidion microphyllum Müll.Arg., Linnaea 32: 69 (1863).

Phyllanthus quintuplinervis M.Schmid, in Fl. N. Caled. & Depend. 17: 148 (1991).
SC. New Caledonia. 60 NWC. Nanophan.

  var. **meoriensis** M.Schmid, in Fl. N. Caled. & Depend. 17: 153 (1991).
    SC. New Caledonia (Mé Ori). 60 NWC. Nanophan.

var. **quintuplinervis**
SC. New Caledonia. 60 NWC. Nanophan.

Phyllanthus racemigerus Müll.Arg., Linnaea 32: 23 (1863).
Venezuela. 82 VEN. Nanophan.

Phyllanthus ramillosus Müll.Arg., Linnaea 32: 36 (1863).
S. Brazil. 84 BZS. Nanophan.
Phyllanthus empetrifolius Baill., Adansonia 5: 352 (1865).

Phyllanthus ramosii Quisumb. & Merr., Philipp. J. Sci. 37: 160 (1928).
Philippines. 42 PHI.

Phyllanthus ramosus Vell., Fl. Flumin. 10: t. 17 (1831).
SE. Brazil. 84 BZL. Ther.
Phyllanthus cladotrichus Müll.Arg., Linnaea 32: 25 (1863).

Phyllanthus rangoloakensis Leandri, Notul. Syst. (Paris) 6: 198 (1938).
C. Madagascar. 29 MDG. Nanophan.

Phyllanthus raynalii Brunel & J.P.Roux, Willdenowia 14: 387 (1984 publ. 1985).
Cameroon, Gabon. 23 CMN GAB.

Phyllanthus regnellianus Müll.Arg. in C.F.P.von Martius, Fl. Bras. 11(2): 58 (1873).
Brazil (Minas Gerais). 84 BZL. Nanophan.

Phyllanthus reticulatus Poir. in J.B.A.M.de Lamarck, Encycl. 5: 298 (1804). *Anisonema reticulatum* (Poir.) A.Juss., Euphorb. Gen.: 4 (1824). *Kirganelia reticulata* (Poir.) Baill., Étude Euphorb.: 613 (1858). *Cicca reticulata* (Poir.) Kurz, Forest Fl. Burma 2: 354 (1877).
Trop. & S. Africa, Trop. & Subtrop. Asia, N. Australia. 22 BEN GHA GUI MLI NGA SEN SIE 23 CMN ZAI 24 SOM 25 KEN TAN UGA 26 MLW MOZ ZAM ZIM 27 BOT NAM 36 CHC CHH CHS 38 NNS 40 IND PAK SRL 41 CBD LAO THA VIE 42 BOR JAW MOL NWG PHI SUL SUM (81) jam win 50 NTA QLD WAU. (Cl.) nanophan. or phan.

var. **glaber** (Thwaites) Müll.Arg., Linnaea 32: 12 (1863).
Trop. & S. Africa, Trop. & Subtrop. Asia. 22 BEN GHA GUI MLI NGA SEN SIE 23 CMN ZAI 24 SOM 25 KEN TAN UGA 26 MLW MOZ ZAM ZIM 27 BOT NAM NAT SWZ TVL 36 CHC CHH CHS 38 TAI 40 IND PAK SRL 41 THA VIE 42 BOR JAW MOL NWG PHI SUL SUM. Nanophan.
Phyllanthus polyspermus Schumach. & Thonn. in C.F.Schumacher, Beskr. Guin. Pl.: 416 (1827).
Kirganelia prieuriana var. *glabra* Baill., Adansonia 1: 83 (1860). *Phyllanthus prieurianus* var. *glaber* (Baill.) Müll.Arg., Linnaea 32: 12 (1863).
*Kirganelia multiflora var. *glabra* Thwaites, Enum. Pl. Zeyl.: 282 (1861).
Kirganelia zanzibariensis Baill., Adansonia 2: 48 (1861).

var. **orae-solis** Radcl.-Sm., Kew Bull. 51: 319 (1996).
Mozambique. 26 MOZ.

var. **reticulatus**
Trop. & S. Africa, Trop. & Subtrop. Asia, N. Australia. 22 BEN GHA GUI MLI NGA SEN SIE 23 CMN ZAI 24 SOM 25 KEN TAN UGA 26 MLW MOZ ZAM ZIM 27 BOT NAM 36 CHC CHH CHS 38 NNS TAI 40 IND PAK SRL 41 CBD LAO THA VIE 42 BOR JAW MOL NWG PHI SUL SUM (81) jam win 50 NTA QLD WAU. (Cl.) nanophan. or phan.
Phyllanthus multiflorus Willd., Sp. Pl. 4: 581 (1805). *Kirganelia multiflora* (Willd.) Baill., Étude Euphorb.: 614 (1858).
Anisonema dubium Blume, Bijdr.: 589 (1826). *Kirganelia dubia* (Blume) Baill., Étude Euphorb.: 614 (1858).

Anisonema intermedium Decne., Nouv. Ann. Mus. Hist. Nat. 4: 482 (1831). *Kirganelia intermedia* (Decne.) Baill., Étude Euphorb.: 614 (1858).

Cicca decandra Blanco, Fl. Filip.: 487, 701 (1837).

Phyllanthus scandens Roxb. ex Dillwyn, Rev. Hortus Malab.: 7 (1839).

Melanthesa oblongifolia Oken, Allg. Naturgesch. 3(3): 1602 (1841).

Phyllanthus chamissonis Klotzsch, Nova Acta Acad. Caes. Leop-Carol. German. Nat. Cur. 19(Suppl. 1): 420 (1843).

Phyllanthus griseus Wall., Numer. List: 7918A (1847), nom. inval.

Phyllanthus spinescens Wall., Numer. List: 7934 (1847), nom. inval.

Anisonema puberulum Baill., Étude Euphorb.: 614 (1858).

Anisonema wrightianum Baill., Étude Euphorb.: 614 (1858), nom. nud.

Kirganelia eglandulosa Baill., Étude Euphorb.: 614 (1858), nom. nud.

Kirganelia puberula Baill., Étude Euphorb.: 614 (1858), nom. inval.

Kirganelia sinensis Baill., Étude Euphorb.: 614 (1858), nom. nud. *Phyllanthus sinensis* Müll.Arg., Linnaea 32: 12 (1863).

Phyllanthus puberulus Miq. ex Baill., Étude Euphorb.: 614 (1858).

Phyllanthus jamaicensis Griseb., Fl. Brit. W. I.: 34 (1859). *Anisonema jamaicense* (Griseb.) Griseb., Fl. Brit. W. I.: 716 (1864).

Kirganelia prieuriana Baill., Adansonia 1: 82 (1860). *Phyllanthus prieurianus* (Baill.) Müll.Arg., Linnaea 32: 12 (1863).

Phyllanthus alaternoides Rchb. ex Baill., Adansonia 1: 83 (1860).

Cicca microcarpa Benth., Fl. Hongk.: 312 (1861). *Phyllanthus microcarpus* (Benth.) Müll.Arg., Linnaea 32: 51 (1863). *Kirganelia microcarpa* (Benth.) Hurus. & Yu.Tanaka in H.Hara, Fl. E. Himal.: 179 (1966).

Phyllanthus pentandrus Roxb. ex Thwaites, Enum. Pl. Zeyl.: 282 (1861).

Phyllanthus oblongifolius Pax in H.G.A.Engler & K.A.E.Prantl, Nat. Pflanzenfam. 3(5): 19 (1890).

Phyllanthus dalbergioides Wall. ex J.J.Sm. in S.H.Koorders & T.Valeton, Bijdr. Boomsoort. Java 12: 67, 69 (1910).

Glochidion microphyllum Ridl., J. Straits Branch Roy. Asiat. Soc. 59: 173 (1911), nom. illeg.

Phyllanthus takaoensis Hayata, Icon. Pl. Formosan. 9: 94 (1920).

Phyllanthus erythrocarpus Ridl., Bull. Misc. Inform. Kew 1923: 362 (1923).

Kirganelia lineata Alston in H.Trimen, Handb. Fl. Ceylon 6(Suppl.): 259 (1931).

Phyllanthus retinervis Hutch. in D.Oliver, Fl. Trop. Afr. 6(1): 735 (1912). – FIGURE, p. 1266.
Zaire, Tanzania, Malawi, Zambia. 23 ZAI 25 TAN 26 MLW ZAM. Cham. or nanophan.

Phyllanthus retroflexus Brade, Arch. Jard. Bot. Rio de Janeiro 15: 8 (1957).
Brazil (Espírito Santo). 84 BZL.

Phyllanthus revaughanii Coode, Kew Bull. 33: 120 (1978).
Mozambique Channel Is. (Europa I., Juan de Nova I.), Mascarenes. 29 MAU MCI REU. Cham.
  *Phyllanthus longifolius Lam. ex Poir. in J.B.A.M.de Lamarck, Encycl. 5: 303 (1804), nom. illeg.

Phyllanthus rhabdocarpus Müll.Arg. in A.P.de Candolle, Prodr. 15(2): 1271 (1866).
Jawa. 42 JAW.

Phyllanthus rheedii Wight, Icon. Pl. Ind. Orient. 5: t. 1895 (1852).
S. India, Sri Lanka. 40 IND SRL. Ther.
  Phyllanthus flaccidus Thwaites, Enum. Pl. Zeyl.: 283 (1861).

Phyllanthus rheophilus Airy Shaw, Kew Bull. 20: 385 (1966).
Papua New Guinea. 42 NWG. Nanophan.

Phyllanthus rhizomatosus Radcl.-Sm., Kew Bull. 37: 427 (1982).– FIGURE, p. 1266
Tanzania (Morogoro). 25 TAN. Cham. – Probably a sport of *P. suffrutescens.*

Phyllanthus rhodocladus S.Moore, J. Linn. Soc., Bot. 45: 397 (1921).
New Caledonia (E. Ignambi). 60 NWC. Nanophan.

Phyllanthus ridleyanus Airy Shaw, Kew Bull. 26: 323 (1972).
S. Thailand, Pen. Malaysia. 41 THA 42 MLY. Cl. nanophan. or phan.
* *Cleistanthus minutiflorus* Ridl., J. Straits Branch Roy. Asiat. Soc. 59: 169 (1911).
   Phyllanthodendron minutiflorum (Ridl.) Airy Shaw, Kew Bull. 14: 471 (1960).
   Phyllanthodendron coriaceum Gage, Rec. Bot. Surv. India 9: 219 (1922).

Phyllanthus riedelianus Müll.Arg., Linnaea 32: 16 (1863). *Glochidion riedelianum*
(Müll.Arg.) Pax & K.Hoffm. in H.G.A.Engler, Nat. Pflanzenfam. ed. 2, 19c: 58 (1931).
Brazil (Rio de Janeiro). 84 BZL.

Phyllanthus rivae Pax, Annuario Reale Ist. Bot. Roma 6: 182 (1897).
Zaire, Ethiopia. 23 ZAI 24 ETH. – Close to *P. macranthus.*

Phyllanthus robinsonii Merr., Philipp. J. Sci., C 7: 405 (1912 publ. 1913).
Philippines. 42 PHI.

Phyllanthus robustus Mart. ex Colla, Herb. Pedem. 5: 106 (1836).
Brazil (Rio de Janeiro).84 BZL. – Provisionally accepted.

Phyllanthus roseus (Craib & Hutch.) Beille in H.Lecomte, Fl. Indo-Chine 5: 590 (1927).
China (Yunnan), Indo-China, Pen. Malaysia. 36 CHC 41 LAO THA VIE 42 MLY. Nanophan.
   or phan.
   Phyllanthodendron album Craib & Hutch., Bull. Misc. Inform. Kew 1910: 279 (1910).
* *Phyllanthodendron roseum* Craib & Hutch., Bull. Misc. Inform. Kew 1910: 23 (1910).
   Cleistanthus dubius Ridl., J. Straits Branch Roy. Asiat. Soc. 59: 168 (1911).
      Phyllanthodendron dubium (Ridl.) Gage, Rec. Bot. Surv. India 9: 219 (1922).
   Glochidion flavum Ridl., J. Straits Branch Roy. Asiat. Soc. 59: 173 (1911).
   Phyllanthodendron roseum var. *glabrum* Craib ex Hosseus, Beih. Bot. Centralbl. 28(2): 406
      (1911). *Phyllanthus roseus* var. *glabrum* (Craib ex Hosseus) Craib ex Beille in
      H.Lecomte, Fl. Indo-Chine 5: 590 (1927). *Phyllanthus albus* (Craib ex Hosseus) Beille
      in H.Lecomte, Fl. Indo-Chine 5: 590 (1927), nom. illeg.
   Uranthera siamensis Pax & K.Hoffm. in H.G.A.Engler, Pflanzenr., IV, 147, III: 95 (1911).
      Phyllanthodendron siamense (Pax & K.Hoffm.) Hosseus, Repert. Spec. Nov. Regni Veg.
      10: 116 (1911). *Phyllanthodendron roseum* var. *siamense* (Pax & K.Hoffm.) Craib, Bull.
      Misc. Inform. Kew 1911: 460 (1911).

Phyllanthus rosmarinifolius Müll.Arg. in C.F.P.von Martius, Fl. Bras. 11(2): 60 (1873).
S. Brazil. 84 BZL. Nanophan.

Phyllanthus rosselensis Airy Shaw & G.L.Webster, Kew Bull. 26: 103 (1971).
New Guinea (Louisiade Archip.). 42 NWG. Nanophan.

Phyllanthus rotundatus Poir. in J.B.A.M.de Lamarck, Encycl. 5: 297 (1804).
India. 40 IND. – Provisionally accepted.

Phyllanthus rotundifolius Klein ex Willd., Sp. Pl. 4: 584 (1805).
Cape Verde Is., Trop. Africa, Arabian Pen., Pakistan, India, Sri Lanka. 20 EGY 21 CVI 22
   MLI MTN SEN 23 RWA 24 ETH SOM SUD 25 KEN TAN UGA 35 SAU YEM 40 IND PAK
   SRL. Ther. or cham.
   Phyllanthus scabrifolius Hook.f., Fl. Brit. India 5: 299 (1887).
   Phyllanthus aspericaulis Pax, Bot. Jahrb. Syst. 43: 218 (1909).

Phyllanthus rouxii Brunel, Bull. Soc. Bot. France 127: 489 (1980 publ. 1981).
Ghana, Togo. 22 GHA TOG.

Phyllanthus rozennae M.Schmid, in Fl. N. Caled. & Depend. 17: 303 (1991).
New Caledonia (S. I. Art). 60 NWC. Nanophan.

Phyllanthus ruber (Lour.) Spreng., Syst. Veg. 3: 22 (1826).
Hainan, Vietnam. 36 CHH 41 VIE.
   * *Nymphanthus ruber* Lour., Fl. Cochinch.: 544 (1790). *Phyllanthus tsiangii* P.T.Li, Acta
      Phytotax. Sin. 25: 375 (1987).

Phyllanthus rubescens Beille in H.Lecomte, Fl. Indo-Chine 5: 602 (1927).
Vietnam. 41 VIE.

Phyllanthus rubicundus Beille, Bull. Soc. Bot. France 72: 162 (1925).
Vietnam. 41 VIE.

Phyllanthus rubriflorus J.J.Sm., Nova Guinea 8: 781 (1912).
New Guinea. 42 NWG. Nanophan. or phan.

Phyllanthus rubristipulus Govaerts & Radcl.-Sm., Kew Bull. 51: 177 (1996).
Papua New Guinea. 42 NWG.
   * *Phyllanthus rubriflorus* Beille in H.Lecomte, Fl. Indo-Chine 5: 600 (1927), nom. illeg.

Phyllanthus rupestris Kunth in F.W.H.von Humboldt, A.J.A.Bonpland & C.S.Kunth, Nov.
Gen. Sp. 2: 110 (1817).
Colombia, Venezuela (Apure, Amazonas), Brazil (Amazonas). 82 VEN 83 CLM 84 BZN.
   Nanophan. or phan.
   Phyllanthus brachycladus Müll.Arg., Linnaea 32: 35 (1863).
   Phyllanthus brachycladus var. *oblongifolius* Müll.Arg., Linnaea 32: 35 (1863).
   Phyllanthus microcarpoides Croizat, Trop. Woods 78: 6 (1944).

Phyllanthus rupicola Elmer, Leafl. Philipp. Bot. 3: 927 (1910).
Philippines. 42 PHI.

Phyllanthus rupiinsularis Hosok., Trans. Nat. Hist. Soc. Taiwan 25: 19 (1935).
Caroline Is. (Palau). 62 CRL.

Phyllanthus ruscifolius Müll.Arg. in A.P.de Candolle, Prodr. 15(2): 358 (1866).
Colombia. 83 CLM.

Phyllanthus saffordii Merr., Philipp. J. Sci., C 9: 104 (1914).
Marianas. 62 MRN.

Phyllanthus salicifolius Baill., Adansonia 2: 239 (1862). *Glochidion salicifolium* (Baill.)
Müll.Arg., Linnaea 32: 59 (1863). *Phyllanthus salicifolius* var. *genuinus* Müll.Arg. in A.P.de
Candolle, Prodr. 15(2): 316 (1866), nom. inval.
NW. New Caledonia. 60 NWC. Nanophan.

Phyllanthus salomonis Airy Shaw, Kew Bull. 32: 368 (1978).
New Guinea (Louisiade Archip.), Solomon Is. 42 NWG 60 SOL. Phan.

Phyllanthus salviifolius Kunth in F.W.H.von Humboldt, A.J.A.Bonpland & C.S.Kunth, Nov.
Gen. Sp. 2: 116 (1817). *Kirganelia salviifolia* (Kunth) Spreng., Syst. Veg. 3: 48 (1826).
Phyllanthus salviifolius var. *genuinus* Müll.Arg. in A.P.de Candolle, Prodr. 15(2): 330 (1866),
nom. inval.

Costa Rica, Peru, Ecuador, Colombia, Venezuela. 80 COS 82 VEN 83 CLM ECU PER. Phan.
Phyllanthus floribundus Kunth in F.W.H.von Humboldt, A.J.A.Bonpland & C.S.Kunth, Nov. Gen. Sp. 2: 116 (1817). *Kirganelia floribunda* (Kunth) Spreng., Syst. Veg. 3: 48 (1826). *Phyllanthus salviifolius* var. *floribundus* (Kunth) Müll.Arg. in A.P.de Candolle, Prodr. 15(2): 331 (1866).
Phyllanthus kunthiana Baill., Étude Euphorb.: 629 (1858).
Phyllanthus salviifolius var. *glabrescens* Müll.Arg. in A.P.de Candolle, Prodr. 15(2): 330 (1866).

Phyllanthus samarensis Müll.Arg., Linnaea 34: 73 (1865).
Philippines (Samar). 42 PHI.

Phyllanthus sanjappae Chakrab. & M.G.Gangop., J. Bombay Nat. Hist. Soc. 90: 69 (1993).
Andaman Is. 41 AND.

Phyllanthus sarasinii Guillaumin, Arch. Bot. Mém. 2(3): 16 (1929).
C. New Caledonia. 60 NWC. Nanophan.

Phyllanthus sarothamnoides Govaerts & Radcl.-Sm., Kew Bull. 51: 177 (1996).
Brazil (Bahia: Serra da Lapa). 84 BZE.

Phyllanthus sauropodoides Airy Shaw, Muelleria 4: 216 (1980).
Queensland. 50 QLD.

Phyllanthus savannicola Domin, Biblioth. Bot. 89: 321 (1927).
Queensland. 50 QLD. – Provisionally accepted.

Phyllanthus saxosus F.Muell., Linnaea 25: 441 (1852). *Phyllanthus gunnii* var. *saxosus* (F.Muell.) Müll.Arg., Linnaea 32: 21 (1863).
Western Australia, South Australia. 50 SOA WAU.

Phyllanthus scaber Klotzsch in J.G.C.Lehmann, Pl. Preiss. 1: 179 (1845).
W. & S. Western Australia. 50 WAU.
Phyllanthus scaber var. *angustifolius* Müll.Arg. in A.P.de Candolle, Prodr. 15(2): 372 (1866).
Phyllanthus scaber var. *pallidiflorus* Müll.Arg. in A.P.de Candolle, Prodr. 15(2): 372 (1866).
Phyllanthus maitlandianus Diels, Bot. Jahrb. Syst. 35: 338 (1904).

Phyllanthus schliebenii Mansf. ex Radcl.-Sm., Kew Bull. 35: 772 (1981).
Tanzania (Lindi). 25 TAN. Nanophan.

Phyllanthus schomburgkianus Müll.Arg. in A.P.de Candolle, Prodr. 15(2): 387 (1866).
Guyana, Surinam, Brazil. 82 GUY SUR 84 BZC BZL BZN.

Phyllanthus scopulorum (Britton) Urb., Symb. Antill. 9: 187 (1924).
NE. Cuba. 81 CUB. Nanophan.
* *Orbicularia scopulorum* Britton, Mem. Torrey Bot. Club 16: 73 (1920).

Phyllanthus securinegoides Merr., Philipp. J. Sci., C 9: 490 (1914 publ. 1915).
Philippines. 42 PHI.
Phyllanthus urdanetensis Elmer, Leafl. Philipp. Bot. 7: 2650 (1915).

Phyllanthus selbyi Britton & P.Wilson, Mem. Torrey Bot. Club 16: 74 (1920).
Cuba (incl. I. de la Juventud). 81 CUB. Hemicr.
Phyllanthus pinosius Urb., Repert. Spec. Nov. Regni Veg. 28: 213 (1930).

Phyllanthus sellowianus (Klotzsch) Müll.Arg., Linnaea 32: 37 (1863).
S. Brazil, Paraguay, Uruguay, N. Argentina. 84 BZS 85 AGE PAR URU. Nanophan.
   Phyllanthus ziziphoides Baill. ex Gibert, Enum. Pl. Montev.: 54 , nom. nud.
   * *Asterandra sellowiana* Klotzsch, Arch. Naturgesch. 7: 200 (1841).

Phyllanthus sepialis Müll.Arg., Naturwiss. Verein Abh. Bremen 7: 25 (1880).
S. Sudan, S. Ethiopia, Kenya, Uganda, Tanzania. 24 ETH SUD 25 KEN TAN UGA. Nanophan.
   Phyllanthus meruensis Pax, Bot. Jahrb. Syst. 15: 526 (1893).
   Phyllanthus conradii Pax, Bot. Jahrb. Syst. 43: 75 (1909).

Phyllanthus sequoiifolius Jabl., Mem. New York Bot. Gard. 17: 93 (1967).
Venezuela (Bolívar: Chimanta Mts.). 82 VEN. Nanophan.

Phyllanthus serpentinicola Radcl.-Sm., Kew Bull. 51: 320 (1996).
Zimbabwe. 26 ZIM. Cham.

Phyllanthus serpentinus S.Moore, J. Linn. Soc., Bot. 45: 399 (1921).
NW. New Caledonia. 60 NWC. Nanophan.

Phyllanthus sessilis Warb., Bot. Jahrb. Syst. 13: 357 (1891).
New Guinea. 42 NWG. – Provisionally accepted.

Phyllanthus seyrigii Leandri, Mém. Inst. Sci. Madagascar, Sér. B, Biol. Vég. 8: 236 (1957).
Madagascar. 29 MDG. Phan.
   Phyllanthus casticum f. *trilocularis* Leandri, Notul. Syst. (Paris) 7: 175, 180 (1939).

Phyllanthus sibuyanensis Elmer, Leafl. Philipp. Bot. 3: 928 (1910).
Philippines. 42 PHI.

Phyllanthus sikkimensis Müll.Arg., Linnaea 32: 48 (1863).
Sikkim, Bhutan, Assam, Burma, Thailand, N. Pen. Malaysia. 40 ASS EHM 41 BMA THA 42
   MLY. Nanophan.
   * *Agyneia tetrandra* Buch.-Ham., Trans. Linn. Soc. London 15: 125 (1826).
   Phyllanthus hamiltonianus Müll.Arg., Linnaea 34: 75 (1865). *Reidia hamiltoniana*
      (Mull.Arg.) A.M.Cowan & Cowan, Trees N. Bengal: 117 (1929). *Eriococcus*
      hamiltonianus (Müll.Arg.) Hurus. & Yu.Tanaka, Fl. E. Himal.: 177 (1966).
   Phyllanthus perlisensis Ridl., J. Straits Branch Roy. Asiat. Soc. 59: 171 (1911).
   Phyllanthus secundiflora Ridl., J. Straits Branch Roy. Asiat. Soc. 59: 170 (1911).

Phyllanthus similis Müll.Arg., Linnaea 34: 71 (1865).
Queensland, New South Wales. 50 NSW QLD.

Phyllanthus simplicicaulis Müll.Arg., Linnaea 32: 38 (1863).
Brazil (Minas Gerais). 84 BZL.

Phyllanthus singalensis (Miq.) Müll.Arg., Linnaea 32: 48 (1863).
Sumatera. 42 SUM.
   * *Reidia singalensis* Miq., Fl. Ned. Ind., Eerste Bijv.: 449 (1861).

Phyllanthus singampattianus (Sebast. & A.N.Henry) Kumari & Chandrab., in Fl. Tamil
Nadu, India 2: 238 (1987).
India. 40 IND.
   * *Reidia singampattiana* Sebast. & A.N.Henry, Bull. Bot. Surv. India 2: 437 (1960).

Phyllanthus skutchii Standl., Publ. Field Mus. Nat. Hist., Bot. Ser. 22: 346 (1940).
Costa Rica. 80 COS. Phan.

Phyllanthus smithianus G.L.Webster, Pacific Sci. 40: 99 (1986 publ. 1988).
Fiji. 60 FIJ.

Phyllanthus societatis Müll.Arg. in A.P.de Candolle, Prodr. 15(2): 364 (1866).
S. Pacific. 60 NRU 61 COO TUA. Cham. or nanophan.

Phyllanthus somalensis Hutch. in D.Oliver, Fl. Trop. Afr. 6(1): 710 (1912).
S. Somalia, NE. Kenya. 24 SOM 25 KEN. Nanophan.

Phyllanthus songboiensis Thin, J. Biol. (Vietnam) 14(2): 18 (1992).
Vietnam. 41 VIE.

Phyllanthus sootepensis Craib, Bull. Misc. Inform. Kew 1911: 459 (1911).
China (S. Yunnan), N. Thailand. 36 CHC 41 THA. Nanophan.
   Phyllanthus subpulchellus Croizat, J. Jap. Bot. 16: 652 (1940).

Phyllanthus spartioides Pax & K.Hoffm., Repert. Spec. Nov. Regni Veg. 19: 174 (1923).
Brazil (Bahia). 84 BZE.

Phyllanthus spinosus Chiov., Fl. Somala 1: 305 (1929).
Somalia. 24 SOM. Nanophan.

Phyllanthus spirei Beille in H.Lecomte, Fl. Indo-Chine 5: 606 (1927).
Laos. 41 LAO.

Phyllanthus sponiifolius Müll.Arg., Linnaea 32: 25 (1863).
Ecuador. 83 ECU.

Phyllanthus spruceanus Müll.Arg., Linnaea 32: 40 (1863).
Brazil (Amazonas). 84 BZN.

Phyllanthus squamifolius (Lour.) Stokes, Bot. Mat. Med. 4: 364 (1812).
Vietnam. 41 VIE.
   * Nymphanthus squamifolia Lour., Fl. Cochinch.: 544 (1790).

Phyllanthus standleyi McVaugh, Brittonia 13: 199 (1961).
SW. Mexico. 79 MXS.
   * Phyllanthus perpusillus Standl., Amer. Midl. Naturalist 36: 178 (1946), nom. illeg.

Phyllanthus stenophyllus Guillaumin, Arch. Bot. Mém. 2(3): 17 (1929).
New Caledonia (Poindimié-Touho). 60 NWC. Nanophan.

Phyllanthus stipitatus M.Schmid, in Fl. N. Caled. & Depend. 17: 178 (1991).
NC. New Caledonia (Massif du Boulinda). 60 NWC. Nanophan.

Phyllanthus stipularis Merr., Philipp. J. Sci. 1(Suppl.): 75 (1906).
Philippines. 42 PHI.

Phyllanthus stipulatus (Raf.) G.L.Webster, Contr. Gray Herb. 176: 53 (1955).
Trop. America. 80 COS 81 CUB DOM HAI JAM LEE PUE TRT WIN 82 FRG GUY SUR VEN
   83 PER 84 BZN. Cham.
   * Moeroris stipulata Raf., Sylva Tellur.: 91 (1838).
   Phyllanthus microphyllus Klotzsch, London J. Bot. 2: 51 (1843).
   Phyllanthus diffusus Klotzsch in B.Seemann, Bot. Voy. Herald: 105 (1853). *Phyllanthus diffusus* var. *genuinus* Müll.Arg. in A.P.de Candolle, Prodr. 15(2): 409 (1866), nom. inval.
   Phyllanthus hoffmeisteri Klotzsch, Bot. Ergebn. Reise Waldemar: 117 (1862).

Phyllanthus hoffmannseggii Müll.Arg., Linnaea 32: 45 (1863).
Phyllanthus hoffmannseggii var. *oblongifolius* Müll.Arg., Linnaea 32: 45 (1863).
   Phyllanthus diffusus var. *oblongifolius* (Müll.Arg.) Müll.Arg. in A.P.de Candolle, Prodr.
   15(2): 410 (1866).
Phyllanthus aquaticus C.Wright ex Sauvalle, Anales Acad. Ci. Méd. Habana 7: 110 (1870).

Phyllanthus striaticaulis J.T.Hunter & J.J.Bruhl, J. Adelaide Bot. Gard. 17: 133 (1996).
   South Australia. 50 SOA. Hemicr. or cham.

Phyllanthus strobilaceus Jabl., Mem. New York Bot. Gard. 17: 96 (1967).
   S. Venezuela. 82 VEN. Nanophan. or phan.

Phyllanthus stultitiae Airy Shaw, Kew Bull. 32: 368 (1978).
   Irian Jaya. 42 NWG. Nanophan.
   * *Glochidion pulchellum* Airy Shaw, Kew Bull. 23: 22 (1969). *Phyllanthus concinnus* Airy
      Shaw, Hooker's Icon. Pl. 38: t. 3706 (June 1974), nom. illeg.

Phyllanthus subapicalis Jabl., Mem. New York Bot. Gard. 17: 101 (1967).
   Venezuela (Amazonas). 82 VEN. Nanophan.

Phyllanthus subcarnosus C.Wright ex Griseb., Nachr. Königl. Ges. Wiss. Georg-Augusts-
   Univ. 1: 168 (1865).
   Cuba, Haiti. 81 CUB HAI. Nanophan. or phan.
      Phyllanthus leonis Alain, Contr. Ocas. Mus. Hist. Nat. Colegio "De La Salle" 12: 1 (1953).

Phyllanthus subcrenulatus F.Muell., Fragm. 1: 108 (1859).
   Queensland, New South Wales. 50 NSW QLD.

Phyllanthus subcuneatus Greenm., Proc. Amer. Acad. Arts 33: 478 (1898).
   Mexico (Puebla). 79 MXC.

Phyllanthus subemarginatus Müll.Arg., Linnaea 32: 39 (1863).
   SE. Brazil. 84 BZL.

Phyllanthus subglomeratus Poir. in J.B.A.M.de Lamarck, Encycl. 5: 304 (1804).
   Lesser Antilles. 81 LEE TRT WIN.
      Phyllanthus brasiliensis var. *oblongifolius* Müll.Arg., Linnaea 32: 27 (1863).

Phyllanthus sublanatus Schumach. & Thonn. in C.F.Schumacher, Beskr. Guin. Pl.: 420 (1827).
   W. Trop. Africa to Chad. 22 BEN GHA MLI NGA SIE TOG 23 CON ZAI 24 CHA. Ther.

Phyllanthus submarginalis Airy Shaw, Kew Bull. 37: 33 (1982).
   Lesser Sunda Is. (W. Flores). 42 LSI.

Phyllanthus subobscurus Müll.Arg. in A.P.de Candolle, Prodr. 15(2): 1271 (1866).
   Jawa. 42 JAW.

Phyllanthus suffrutescens Pax, Bot. Jahrb. Syst. 15: 523 (1893).
   S. Ethiopia, Somalia, Kenya, Uganda, Tanzania. 24 ETH SOM 25 KEN TAN UGA. Hemicr.
      Phyllanthus myrtilloides Chiov., Ann. Bot. (Rome) 10: 405 (1912), nom. illeg.

Phyllanthus sulcatus J.T.Hunter & J.J.Bruhl, Nuytsia 11: 15 (1997).
   Northern Territory, Queensland. 50 NTA QLD. Ther.

Phyllanthus sylvincola S.Moore, J. Linn. Soc., Bot. 45: 401 (1921).
   SE. New Caledonia. 60 NWC. Cham. or nanophan.

Phyllanthus symphoricarpoides Kunth in F.W.H.von Humboldt, A.J.A.Bonpland & C.S.Kunth, Nov. Gen. Sp. 2: 114 (1817).
Colombia. 83 CLM.

Phyllanthus tabularis Airy Shaw, Kew Bull. 34: 598 (1980).
Papua New Guinea. 42 NWG.

Phyllanthus tagulae Airy Shaw & G.L.Webster, Kew Bull. 26: 102 (1971).
New Guinea (Louisiade Archip.). 42 NWG. Phan.

Phyllanthus talbotii Sedgw., J. Indian Bot. 2: 124 (1921).
W. India. 40 IND.

Phyllanthus tampinensis Leandri, Mém. Inst. Sci. Madagascar, Sér. B, Biol. Vég. 8: 233 (1957).
E. Madagascar. 29 MDG. Nanophan.

Phyllanthus tangoensis M.Schmid, in Fl. N. Caled. & Depend. 17: 162 (1991).
New Caledonia (Plateau de Tango). 60 NWC. Nanophan.

Phyllanthus tatei F.Muell., S. Sci. Rec. 2: 55 (1882).
S. Australia. 50 SOA.
   Phyllanthus rigidus Tate, Trans. Roy. Soc. South Australia 6: 101 (1883).
   Micrantheum tatei J.M.Black, Fl. S. Austral. 2: 356 (1924).

Phyllanthus taxodiifolius Beille in H.Lecomte, Fl. Indo-Chine 5: 605 (1927).
China (S. Yunnan, SW. Guangxi), Thailand, Cambodia, Vietnam. 36 CHC CHS 41 CBD THA VIE. Nanophan.

Phyllanthus taylorianus Brunel ex Radcl.-Sm., Kew Bull. 51: 322 (1996).
Cameroon to Ethiopia and Zimbabwe. 23 BUR CAF CMN ZAI 24 ETH 24 TAN UGA 26 MLW ZAM ZIM. Ther.

Phyllanthus tenellus Roxb., Fl. Ind. ed. 1832, 3: 668 (1832).
Yemen, Tanzania, Mozambique, Mascarenes, introduced elsewhere. 25 TAN 26 MOZ 29 MAU? REU 35 YEM (40) ind (42) nwc (50) nsw nta qld wau (60) van (63) haw (78) fla (81) jam lee win. Ther.

  var. **arabicus** Müll.Arg. in A.P.de Candolle, Prodr. 15(2): 339 (1866).
  Yemen. 35 YEM.
  Phyllanthus maderaspatensis Forssk., Fl. Aegypt.-Arab.: 159 (1775), nom. illeg.

  var. **tenellus**
  Tanzania, Mozambique, Mascarenes, introduced elsewhere. 25 TAN 26 MOZ 29 MAU? REU (40) ind (42) nwc (50) nsw qld (60) van (63) haw (78) fla (81) jam lee win. Ther.
  Phyllanthus tenellus var. *roxburghii* Müll.Arg., Linnaea 32: 7 (1863).
  Phyllanthus corcovadensis Müll.Arg. in C.F.P.von Martius, Fl. Bras. 11(2): 30 (1873).
  Phyllanthus brisbanicus F.M.Bailey, Queensl. Fl. 5: 1418 (1902).
  Phyllanthus minor Fawc. & Rendle, J. Bot. 57: 65 (1919).

Phyllanthus tener Radcl.-Sm., Kew Bull. 51: 323 (1996).
Zambia. 26 ZAM. Hemicr.

Phyllanthus tenuicaulis Müll.Arg., Linnaea 32: 44 (1863).
E. Cuba, Haiti. 81 CUB HAI.

  var. **haitiensis** G.L.Webster, Contr. Gray Herb. 176: 48 (1955). *Phyllanthus tenuicaulis* subsp. *haitiensis* (G.L.Webster) Borhidi, Bot. Közlem. 58: 176 (1971).
  N. Haiti. 81 HAI. Cham.

var. **tenuicaulis**
E. Cuba. 81 CUB.

Phyllanthus tenuipedicellatus M.Schmid, in Fl. N. Caled. & Depend. 17: 98 (1991).
New Caledonia. 60 NWC. Cham. or nanophan.

var. **kaloueholaensis** M.Schmid, in Fl. N. Caled. & Depend. 17: 103 (1991).
New Caledonia (S. of Tontouta). 60 NWC. Cham. or nanophan.

var. **tenuipedicellatus**
New Caledonia. 60 NWC. Cham. or nanophan.

var. **tontoutaensis** M.Schmid, in Fl. N. Caled. & Depend. 17: 103 (1991).
New Caledonia (Tontouta). 60 NWC. Cham. or nanophan.

Phyllanthus tenuipes C.B.Rob., Philipp. J. Sci., C 4: 78 (1909).
Philippines. 42 PHI.

Phyllanthus tenuirhachis J.J.Sm., Icon. Bogor.: t. 263 (1908).
Maluku, Sulawesi, Irian Jaya. 42 MOL NWG SUL. Nanophan.

Phyllanthus tenuis Radcl.-Sm., Kew Bull. 51: 323 (1996).
Zambia. 26 ZAM. Ther.

Phyllanthus tepuicola Steyerm., Acta Bot. Venez. 10: 236 (1975 publ. 1976).
Venezuela. 82 VEN.

Phyllanthus tequilensis B.L.Rob. & Greenm., Proc. Amer. Acad. Arts 29: 392 (1894).
Mexico (Jalisco). 79 MXS.

Phyllanthus tessmannii Hutch. in D.Oliver, Fl. Trop. Afr. 6(1): 704 (1912).
São Tomé, Equatorial Guinea. 23 EQG GGI.
Phyllanthus gracilipes Pax, Bot. Jahrb. Syst. 45: 234 (1910), nom. illeg.

Phyllanthus tetrandrus Roxb., Fl. Ind. ed. 1832, 3: 674 (1832). *Epistylium roxburghii* Baill.,
Étude Euphorb.: 648 (1858), nom. illeg. *Phyllanthus roxburghii* Müll.Arg., Linnaea 32: 47
(1863), nom. illeg. *Reidia tetrandra* (Roxb.) V.Naray. in U.N.Kanjilal & al., Fl. Assam 4:
erratum (1940).
Assam, Bangladesh. 40 ASS BAN.

Phyllanthus thaii Thin, J. Biol. (Vietnam) 14(2): 22 (1992).
Vietnam. 41 VIE.

Phyllanthus thulinii Radcl.-Sm., Kew Bull. 35: 774 (1981).
Tanzania (Morogoro). 25 TAN. Cham. or nanophan.

Phyllanthus tiebaghiensis M.Schmid, in Fl. N. Caled. & Depend. 17: 94 (1991).
NW. New Caledonia (Massif de la Tiébaghi). 60 NWC. Cham. or nanophan.

Phyllanthus tireliae M.Schmid, in Fl. N. Caled. & Depend. 17: 246 (1991).
New Caledonia (Massif du Boulinda). 60 NWC. Nanophan. or phan.

Phyllanthus tixieri M.Schmid, in Fl. N. Caled. & Depend. 17: 241 (1991).
New Caledonia (Kouaoua Reg.). 60 NWC. Cham. or nanophan.

Phyllanthus torrentium Müll.Arg. in A.P.de Candolle, Prodr. 15(2): 316 (1866).
New Caledonia. 60 NWC. Cham. or nanophan.

var. **induratus** (S.Moore) M.Schmid, in Fl. N. Caled. & Depend. 17: 234 (1991).
　　C. New Caledonia. 60 NWC. Cham. or nanophan.
　　Phyllanthus induratus S.Moore, J. Linn. Soc., Bot. 45: 397 (1921).

var. **torrentium**
　　WC. New Caledonia. 60 NWC. Cham. or nanophan.

Phyllanthus touranensis Beille in H.Lecomte, Fl. Indo-Chine 5: 608 (1927).
　　Vietnam. 41 VIE.

Phyllanthus trichopodus Leandri, Bull. Soc. Bot. France 81: 452 (1934).
　　W. Madagascar. 29 MDG. Nanophan.

Phyllanthus trichosporus Adelb., Blumea 5: 507 (1945).
　　Jawa, Sulawesi. 42 JAW SUL.

Phyllanthus trichotepalus Brenan, Kew Bull. 8: 91 (1953).
　　Uganda, Rwanda, Burundi, Zaire. 23 BUR RWA ZAI 25 UGA. Ther.

Phyllanthus triphlebius C.B.Rob., Philipp. J. Sci., C 4: 82 (1909).
　　Philippines. 42 PHI.

Phyllanthus tritepalus M.Schmid, in Fl. N. Caled. & Depend. 17: 153 (1991).
　　EC. New Caledonia (Canala). 60 NWC. Nanophan.
　　　Glochidion vieillardii Müll.Arg., Linnaea 32: 70 (1863).

Phyllanthus trungii Thin, J. Biol. (Vietnam) 14(2): 21 (1992).
　　Vietnam. 41 VIE.

Phyllanthus tsarongensis W.W.Sm., Notes Roy. Bot. Gard. Edinburgh 13: 177 (1921).
　　Tibet, China (Sichuan, Yunnan). 36 CHC CHT. Nanophan.

Phyllanthus tsetserrae Brunel ex Radcl.-Sm., Kew Bull. 51: 325 (1996).
　　Mozambique. 26 MOZ. Hemicr.

Phyllanthus tuerckheimii G.L.Webster, Ann. Missouri Bot. Gard. 54: 195 (1967).
　　Guatemala. 80 GUA.

Phyllanthus tulearicus Leandri, Notul. Syst. (Paris) 6: 199 (1938).
　　SW. Madagascar. 29 MDG. Nanophan.

Phyllanthus udoricola Radcl.-Sm., Kew Bull. 51: 326 (1996).
　　Malawi, Zambia, Zimbabwe. 26 MLW ZAM ZIM. Ther.

Phyllanthus ukagurensis Radcl.-Sm., Kew Bull. 35: 774 (1981). – FIGURE, p. 1266.
　　NE. Tanzania. 25 TAN. Nanophan.

Phyllanthus umbratus Müll.Arg. in A.P.de Candolle, Prodr. 15(2): 356 (1866).
　　Brazil (Rio de Janeiro). 84 BZL. Nanophan.

Phyllanthus umbricola Guillaumin, Arch. Bot. Mém. 2(3): 18 (1929).
　　SE. New Caledonia. 60 NWC. Cham. or nanophan.

Phyllanthus unifoliatus M.Schmid, in Fl. N. Caled. & Depend. 17: 198 (1991).
　　WC. New Caledonia (Pindai Pen.). 60 NWC. Nanophan.

Phyllanthus unioensis M.Schmid, in Fl. N. Caled. & Depend. 17: 139 (1991).
  SC. New Caledonia (Table Unio). 60 NWC. Nanophan.

Phyllanthus urbanianus Mansf., Repert. Spec. Nov. Regni Veg. 32: 86 (1933).
  Haiti (Massif de la Hotte). 81 HAI. Nanophan.

Phyllanthus urceolatus Baill., Adansonia 2: 239 (1862).
  Society Is. (Moorea, Raiatea, Tahiti). 61 SCI. Nanophan.

Phyllanthus urinaria L., Sp. Pl.: 982 (1753).
  Trop. America, W. & WC. Trop. Africa, Mascarenes, Trop. & Subtrop. Asia, N. Australia. 22
    GHA IVO NGA SIE 23 CON 29 MAU mdg REU 36 CHC CHH CHN CHS 38 JAP NNS
    TAI 40 IND PAK SRL 41 BMA CBD LAO THA VIE 42 BOR JAW MLY NWG PHI SUM 50
    NTA WAU (60) nwc 80 COS 81 DOM HAI JAM LEE TRT WIN 82 ALL 83 PER 84 BZE
    BZN. Ther. or cham. – Perhaps only native to Asia.

  var. **hookeri** (Müll.Arg.) Hook.f., Fl. Brit. India 5: 294 (1887).
    Trop. & Subtrop. Asia. 36 CHS 38 JAP TAI 40 IND SRL 41 THA VIE 42 MLY PHI. Ther.
      or cham.
    * *Phyllanthus hookeri* Müll.Arg., Linnaea 32: 19 (1863).

  subsp. **nudicarpus** Rossignol & Haicour, Amer. J. Bot. 74: 1861 (1987 publ. 1988).
    Vietnam, S. China, Hainan, Philippines. 36 CHH CHS 41 VIE 42 PHI. Ther. or cham.

  subsp. **urinaria**
    Trop. America, Trop. Africa, Mascarenes, Trop. & Subtrop. Asia. 22 GHA IVO NGA SIE
      23 CON 29 MAU mdg REU 36 CHC CHH CHN CHS 38 NNS TAI 40 IND PAK SRL 41
      CBD LAO THA VIE 42 BOR JAW NWG PHI SUM (60) nwc 80 COS 81 JAM LEE TRT
      WIN 82 ALL 83 PER 84 BZE BZN. Ther. or cham.
    Phyllanthus cantoniensis Hornem., Enum. Pl. Hort. Hafn.: 29 (1807).
    Phyllanthus alatus Blume, Bijdr.: 594 (1826).
    Phyllanthus lepidocarpus Siebold & Zucc., Fl. Jap. 1: 35 (1835).
    Phyllanthus echinatus Buch.-Ham. ex Wall., Numer. List: 7893B (1847), nom. inval.
    Phyllanthus muricatus Wall., Numer. List: 7898D (1847), nom. inval.
    Phyllanthus leprocarpus Wight, Icon. Pl. Ind. Orient. 5: t. 1895 (1852).
    Phyllanthus rubens Bojer ex Baker, Fl. Mauritius: 309 (1877).
    Phyllanthus chamaepeuce Ridl., Trans. Linn. Soc. London, Bot. 3: 345 (1893).
    Phyllanthus lauterbachianus Pax, Repert. Spec. Nov. Regni Veg. 8: 325 (1910).
    Phyllanthus verrucosus Elmer, Leafl. Philipp. Bot. 7: 2649 (1915).
    Phyllanthus quangtriensis Beille in H.Lecomte, Fl. Indo-Chine 5: 584 (1927).
    Phyllanthus nozeranii Rossignol & Haicour, Amer. J. Bot. 74: 1858 (1987 publ. 1988).

Phyllanthus ussuriensis Rupr. & Maxim., Bull. Cl. Phys.-Math. Acad. Imp. Sci. Saint-
  Pétersbourg 15: 222 (1857). *Phyllanthus simplex* var. *ussuriensis* (Rupr. & Maxim.)
  Müll.Arg., Linnaea 32: 33 (1863).
  Mongolia, Russian Far East, Japan, Korea, Nansei-shoto, Taiwan, China. 31 KHA 36 CHC
    CHM CHN CHS 37 MON 38 JAP KOR NNS TAI. Ther.
    Phyllanthus anceps Benth., Fl. Hongk.: 311 (1861), nom. illeg. *Phyllanthus wilfordii*
      Croizat & Metcalf, Lingnan Sci. J. 20: 194 (1942).
    Phyllanthus simplex var. *chinensis* Müll.Arg., Linnaea 32: 33 (1863). *Phyllanthus virgatus*
      var. *chinensis* (Müll.Arg.) G.L.Webster, J. Jap. Bot. 46: 68 (1971).
    Phyllanthus matsumurae Hayata ex Fabe, Bot. Mag. (Tokyo) 18: 12 (1904).

Phyllanthus utricularis Airy Shaw & G.L.Webster, Kew Bull. 26: 101 (1971).
  Irian Jaya. 42 NWG. Phan.

Phyllanthus vacciniifolius (Müll.Arg.) Müll.Arg. in A.P.de Candolle, Prodr. 15(2): 322 (1866).
S. Venezuela, Guyana. 82 GUY VEN. Nanophan.
Glochidion vacciniifolium Müll.Arg., Linnaea 32: 69 (1863).

subsp. **vacciniifolius**
Venezuela (Bolívar), Guyana. 82 GUY VEN. Nanophan.

subsp. **vinillaensis** Steyerm., Ann. Missouri Bot. Gard. 71: 317 (1984).
Venezuela (Amazonas: Serrania Vinilla). 82 VEN. Nanophan.

Phyllanthus vakinankaratrae Leandri, Mém. Inst. Sci. Madagascar, Sér. B, Biol. Vég. 8: 233 (1957).
C. Madagascar. 29 MDG. Cham. or nanophan.
Phyllanthus vakinankaratrae var. *concolor* Leandri, in Fl. Madag. 111: 91 (1958).

Phyllanthus valerii Standl., Publ. Field Mus. Nat. Hist., Bot. Ser. 18: 619 (1937).
Costa Rica. 80 COS. Nanophan. or phan.

Phyllanthus valleanus Croizat, Ciencia (Mexico) 6: 354 (1946).
Colombia. 83 CLM.

Phyllanthus vanderystii Hutch. & De Wild., Pl. Bequaert. 5(4): 470 (1932).
Zaire. 23 ZAI.

Phyllanthus vatovaviensis Leandri, Mém. Inst. Sci. Madagascar, Sér. B, Biol. Vég. 8: 233 (1957).
E. Madagascar (Mt. Vatovavy). 29 MDG. Nanophan.

Phyllanthus veillonii M.Schmid, in Fl. N. Caled. & Depend. 17: 130 (1991).
New Caledonia (S. I. Art). 60 NWC. Nanophan.

Phyllanthus ventricosus G.L.Webster, Ann. Missouri Bot. Gard. 54: 198 (1967).
Peru (Loreto). 83 PER. Nanophan.

Phyllanthus ventuarii Jabl., Mem. New York Bot. Gard. 17: 104 (1967).
Venezuela (Amazonas: Serrania Paru). 82 VEN. Cham. or nanophan.

Phyllanthus venustulus Leandri, Notul. Syst. (Paris) 6: 198 (1938).
C. Madagascar. 29 MDG. Cham.

Phyllanthus verdickii De Wild., Ann. Mus. Congo Belge, Bot., V, 1: 274 (1906).
WC. Trop. Africa to Angola. 23 CAF CMN CON GAB ZAI 26 ANG.
Phyllanthus dekindtii Hutch. in D.Oliver, Fl. Trop. Afr. 6(1): 719 (1912).

Phyllanthus vergens Baill. in A.Grandidier, Hist. Phys. Madagascar, Atlas: 225 (1892).
N. Madagascar. 29 MDG. Nanophan.

Phyllanthus verrucicaulis Airy Shaw, Kew Bull. 33: 35 (1978).
New Guinea, Bismarck Archip. 42 BIS NWG. Nanophan. or phan.

Phyllanthus vespertilio Baill., Adansonia 2: 233 (1862). *Glochidion vespertilio* (Baill.) Müll.Arg., Linnaea 32: 59 (1863).
New Caledonia (Canala, Col d'Amos). 60 NWC. Cham. or nanophan.

Phyllanthus vichadensis Croizat, J. Arnold Arbor. 26: 181 (1945).
Colombia. 83 CLM.

Phyllanthus villosus Poir. in J.B.A.M.de Lamarck, Encycl. 5: 297 (1804).
S. China. 36 CHS.
Nymphanthus chinensis Lour., Fl. Cochinch.: 544 (1790).

Phyllanthus vincentae J.F.Macbr., Publ. Field Mus. Nat. Hist., Bot. Ser. 13(3A1): 47 (1951).
Peru (Loreto). 83 PER. Nanophan.
 * *Xylosma minutiflora* J.F.Macbr., Candollea 5: 392 (1934).

Phyllanthus virgatus G.Forst., Fl. Ins. Austr.: 65 (1786). *Phyllanthus simplex* var. *virgatus* (G.Forst.) Müll.Arg., Linnaea 32: 32 (1863), nom. illeg.
China, Trop. Asia, Oceania. 36 CHC CHH CHN CHS 38 NNS 40 IND PAK SRL WHM 41 THA VIE 42 BOR JAW NWG PHI SUL SUM 50 NSW NTA QLD WAU 60 NWC. Ther. or cham.

var. **gardnerianus** (Wight) Govaerts & Radcl.-Sm., Kew Bull. 51: 177 (1996).
Sri Lanka. 40 SRL.
 * *Macraea gardneriana* Wight, Icon. Pl. Ind. Orient. 5(2): 27 (1852). *Phyllanthus gardneri* Thwaites, Enum. Pl. Zeyl.: 282 (1861), nom illeg. *Phyllanthus simplex* var. *gardnerianus* (Wight) Müll.Arg., Linnaea 32: 33 (1863).
Phyllanthus virgatus var. *oblongifolius* Müll.Arg., Linnaea 32: 32 (1863).

var. **virgatus**
China, Trop. Asia, Oceania. 36 CHC CHH CHN CHS 38 NNS 40 IND PAK WHM 41 THA VIE 42 BOR JAW NWG PHI SUL SUM 50 QLD 60 NWC. Ther. or cham.
Phyllanthus simplex Retz., Observ. Bot. 5: 29 (1789). *Phyllanthus simplex* var. *genuinus* Müll.Arg. in A.P.de Candolle, Prodr. 15(2): 391 (1866), nom. inval.
Phyllanthus anceps Vahl, Symb. Bot. 2: 95 (1791). *Melanthesa anceps* (Vahl) Miq., Fl. Ned. Ind. 1(2): 371 (1859).
Phyllanthus pedunculatus Kostel., Allg. Med.-Pharm. Fl. 5: 1769 (1836).
Phyllanthus depressus Buch.-Ham. ex Dillwyn, Rev. Hortus Malab. (1839), nom. illeg.
Phyllanthus fruticosus B.Heyne ex Wall., Numer. List: 7899A (1847), nom. inval.
Phyllanthus marginatus B.Heyne ex Wall., Numer. List: 7899A (1847), nom. inval.
Macraea oblongifolia Wight, Icon. Pl. Ind. Orient. 5(2): 27 (1852).
Macraea ovalifolia Wight, Icon. Pl. Ind. Orient. 5(2): 27 (1852).
Melanthesa rupestris Miq., Fl. Ned. Ind. 1(2): 371 (1859).
Phyllanthus gardneri Thwaites, Enum. Pl. Zeyl.: 282 (1861), nom illeg.
Phyllanthus pratensis Pancher ex Baill., Adansonia 2: 237 (1862).
Phyllanthus conterminus Müll.Arg., Linnaea 32: 32 (1863).
Phyllanthus gardnerianus Baill. ex Müll.Arg., Linnaea 32: 33 (1863), pro syn.
Phyllanthus miquelianus Müll.Arg., Linnaea 32: 33 (1863).
Phyllanthus patens Miq. ex Müll.Arg., Linnaea 32: 34 (1863).
Phyllanthus simplex var. *myriocladus* Müll.Arg., Linnaea 32: 33 (1863).
Phyllanthus simplex var. *pratensis* Müll.Arg., Linnaea 32: 33 (1863).
Phyllanthus beckleri Müll.Arg., Linnaea 34: 74 (1865).
Phyllanthus minutiflorus F.Muell. ex Müll.Arg., Linnaea 34: 75 (1865). *Phyllanthus simplex* var. *minutiflorus* (F.Muell. ex Müll.Arg.) Domin, Biblioth. Bot. 22: 877 (1927). *Phyllanthus virgatus* var. *minutiflorus* (F.Muell. ex Müll.Arg.) Airy Shaw, Kew Bull., Addit. Ser. 8: 190 (1980), pro syn.
Phyllanthus simplex var. *brevipes* Müll.Arg. in A.P.de Candolle, Prodr. 15(2): 392 (1866).
Phyllanthus filicaulis Benth., Fl. Austral. 6: 111 (1873). *Phyllanthus simplex* var. *filicaulis* (Benth.) Domin, Biblioth. Bot. 22: 876 (1927).
Phyllanthus gracillimus F.Muell. ex Benth., Fl. Austral. 6: 112 (1873), pro syn. *Phyllanthus minutiflorus* var. *gracillimus* (F.Muell. ex Benth.) Benth., Fl. Austral. 6: 112 (1873). *Phyllanthus simplex* var. *gracillimus* Domin, Biblioth. Bot. 22: 877 (1927).
Phyllanthus simplex var. *leiospermus* Benth., Fl. Austral. 6: 111 (1873).
Phyllanthus trachygyne Benth., Fl. Austral. 6: 103 (1873).
Phyllanthus weinlandii K.Schum. in K.M.Schumann & C.A.G.Lauterbach, Fl. Schutzgeb. Südsee, Nachtr.: 287 (1905).

Phyllanthus eboracensis S.Moore, J. Linn. Soc., Bot. 45: 216 (1920).
Phyllanthus simplex var. *myrtifolius* Domin, Biblioth. Bot. 22: 876 (1927).
Phyllanthus simplex var. *pinifolius* Domin, Biblioth. Bot. 22: 877 (1927).
Phyllanthus virgatus var. *hirtellus* Airy Shaw, Kew Bull., Addit. Ser. 8: 195 (1980).

Phyllanthus virgulatus Müll.Arg., J. Bot. 2: 330 (1864).
Zaire, Angola, Zambia, Malawi. 23 ZAI 26 ANG MLW ZAM. Cham.

Phyllanthus virgultiramus Däniker, Vierteljahrschr. Naturf. Ges. Zürich 76: 169 (1931).
NW. New Caledonia (Masssif du Koniambo). 60 NWC. Nanophan.

Phyllanthus viridis M.E.Jones, Contr. W. Bot. 18: 47 (1933).
Mexico (Baja California). 79 MXN. – Provisionally accepted.

Phyllanthus volkensii Engl., Pflanzenw. Ost-Afrikas C: 236 (1895).
Kenya, Tanzania. 24 KEN TAN. Nanophan.

Phyllanthus vulcani Guillaumin, Mém. Mus. Natl. Hist. Nat., B, Bot. 8(3): 248 (1962).
EC. to SE. New Caledonia. 60 NWC. Nanophan.

> var. **baumannii** Guillaumin ex M.Schmid, in Fl. N. Caled. & Depend. 17: 268 (1991).
> SE. New Caledonia. 60 NWC. Nanophan.
> *Phyllanthus baumannii* Guillaumin, Mém. Mus. Natl. Hist. Nat., B, Bot. 8: 242 (1962), no type designated.
> *Phyllanthus hurlimannii* Guillaumin, Mém. Mus. Natl. Hist. Nat., B, Bot. 8: 244 (1962), no type designated.

> var. **vulcani**
> EC. to SE. New Caledonia. 60 NWC. Nanophan.

Phyllanthus warburgii K.Schum. in K.M.Schumann & C.A.G.Lauterbach, Fl. Schutzgeb. Südsee, Nachtr.: 286 (1905). *Hemiglochidion warburgii* (K.Schum.) K.Schum. in K.M.Schumann & C.A.G.Lauterbach, Fl. Schutzgeb. Südsee, Nachtr.: 289 (1905).
New Guinea. 42 NWG. Nanophan.
  * *Phyllanthus columnaris* Warb., Bot. Jahrb. Syst. 13: 356 (1891), nom. illeg.

Phyllanthus watsonii Airy Shaw, Kew Bull. 25: 493 (1971).
Pen. Malaysia. 42 MLY.

Phyllanthus websterianus Steyerm., Los Angeles County Mus. Contr. Sci. 21: 17 (1958).
Brazil (Goiás). 84 BZC.

Phyllanthus welwitschianus Müll.Arg., J. Bot. 2: 330 (1864).
Tanzania, Malawi, Mozambique, Zambia, Zaire (Shaba), Angola. 23 ZAI 25 TAN 26 ANG MLW MOZ ZAM. Nanophan.

Phyllanthus wheeleri G.L.Webster, Kew Bull. 50: 266 (1995).
Sri Lanka. 40 SRL. Nanophan.
  Phyllanthus gardnerianus var. *pubescens* Thwaites, Enum. Pl. Zeyl.: 282 (1861).
    Phyllanthus simplex f. *pubescens* (Thwaites) Müll.Arg., Linnaea 32: 33 (1863).
  Phyllanthus gardneri var. *pubescens* Thwaites, Enum. Pl. Zeyl.: 282 (1861).

Phyllanthus wightianus Müll.Arg., Linnaea 32: 47 (1863). *Phyllanthus obliquus* Wall. ex Hook.f., Fl. Brit. India 5: 303 (1887), pro syn.
S. India. 40 IND.
  * *Reidia floribunda* Wight, Icon. Pl. Ind. Orient. 5(2): 28, t. 1903 (1852).

Phyllanthus wilkesianus Müll.Arg. in A.P.de Candolle, Prodr. 15(2): 396 (1866).
Fiji. 60 FIJ.

Phyllanthus williamioides Griseb., Nachr. Königl. Ges. Wiss. Georg-Augusts-Univ. 1: 169 (1865).
E. Cuba (Sierra Sagua Baracoa). 81 CUB. Nanophan.

Phyllanthus wingfieldii Radcl.-Sm., Kew Bull. 35: 776 (1981). – FIGURE, p. 1266.
E. Tanzania. 25 TAN. Nanophan.

Phyllanthus winitii Airy Shaw, Kew Bull. 23: 36 (1969).
N. Thailand. 41 THA. Nanophan.

Phyllanthus wittei Robyns & Lawalrée, Bull. Jard. Bot. État 18: 266 (1947).
Zaire. 23 ZAI.

Phyllanthus womersleyi Airy Shaw & G.L.Webster, Kew Bull. 26: 86 (1971).
Papua New Guinea. 42 NWG. Nanophan.

Phyllanthus xiphephorus Radcl.-Sm., Kew Bull. 51: 327 (1996).
Zambia. 26 ZAM. Ther.

Phyllanthus yaouhensis Schltr., Bot. Jahrb. Syst. 39: 146 (1906).
New Caledonia (Nouméa). 60 NWC. Cham. or nanophan.
Phyllanthus erythranthus Guillaumin, Mém. Mus. Natl. Hist. Nat., B, Bot. 8: 244 (1962).

Phyllanthus yunnanensis (Croizat) Govaerts & Radcl.-Sm., Kew Bull. 51: 178 (1996).
China (Guizhou, Yunnan). 36 CHC. Nanophan. or phan.
* *Phyllanthodendron yunnanense* Croizat, J. Arnold Arbor. 23: 36 (1942).

Phyllanthus yvettae M.Schmid, in Fl. N. Caled. & Depend. 17: 309 (1991).
EC. & SE. New Caledonia. 60 NWC. Nanophan.

Phyllanthus zambicus Radcl.-Sm., Kew Bull. 51: 328 (1996).
Zambia. 26 ZAM. Ther.

Phyllanthus zanthoxyloides Steyerm., Fieldiana, Bot. 28: 321 (1952).
Venezuela. 82 VEN.

Phyllanthus zippelianus Müll.Arg. in A.P.de Candolle, Prodr. 15(2): 433 (1866).
Lesser Sunda Is. (Timor). 42 LSI.
* *Phyllanthus cantoniensis* Zipp. ex Span., Linnaea 15: 347 (1841), nom. illeg.

Phyllanthus zornioides Radcl.-Sm., Kew Bull. 51: 328 (1996).
Malawi, Zambia, Zimbabwe. 26 MLW ZAM ZIM. Ther.

Synonyms:
Phyllanthus accrescens J.J.Sm. === **Phyllanthus gomphocarpus** Hook.f.
Phyllanthus acidissimus Noronha === **Sauropus androgynus** (L.) Merr.
Phyllanthus acidissimus (Blanco) Müll.Arg. === **Phyllanthus acidus** (L.) Skeels
Phyllanthus acuminatissimus C.B.Rob. === **Flueggea flexuosa** Müll.Arg.
Phyllanthus adamii Müll.Arg. === **Sauropus glaucus** (F.Muell.) Airy Shaw
Phyllanthus adenandrus Müll.Arg. ex Guillaumin === **Phyllanthus baladensis** Baill.
Phyllanthus agynus Hunter ex Ridl. === **Breynia coronata** Hook.f.
Phyllanthus alaternoides Rchb. ex Baill. === **Phyllanthus reticulatus** Poir. var. **reticulatus**
Phyllanthus alatus Blume === **Phyllanthus urinaria** L. subsp. **urinaria**

Phyllanthus albicans Wall. === **Flueggea leucopyrus** Willd.

Phyllanthus albiflorus F.Muell. ex Müll.Arg. === **Sauropus albiflorus** (F.Muell. ex Müll.Arg.) Airy Shaw

Phyllanthus albus (Craib ex Hosseus) Beille === **Phyllanthus roseus** (Craib & Hutch.) Beille

Phyllanthus albus Noronha === ?

Phyllanthus albus (Blanco) Müll.Arg. === **Glochidion album** (Blanco) Boerl.

Phyllanthus alegrensis Glaz. === ?

Phyllanthus amapondensis Sim === **Margaritaria discoidea** var. **nitida** (Pax) Radcl.-Sm.

Phyllanthus amarus var. *baronianus* Leandri === **Phyllanthus amarus** Schumach. & Thonn.

Phyllanthus amentuliger Müll.Arg. === **Glochidion amentuligerum** (Müll.Arg.) Croizat

Phyllanthus anceps Benth. === **Phyllanthus ussuriensis** Rupr. & Maxim.

Phyllanthus anceps Vahl === **Phyllanthus virgatus** G.Forst. var. **virgatus**

Phyllanthus andamanicus Kurz === **Glochidion helferi** (Müll.Arg.) Hook.f.

Phyllanthus andersonii Müll.Arg. === **Glochidion ellipticum** Wight

Phyllanthus andrachniformis Pax === **Phyllanthus heterophyllus** E.Mey. ex Müll.Arg.

Phyllanthus andrachnoides Willd. === **Phyllanthus maderaspatensis** L. var. **maderaspatensis**

Phyllanthus angulatus Schumach. & Thonn. === **Flueggea virosa** (Roxb. ex Willd.) Voigt subsp. **virosa**

Phyllanthus angustifolius var. *genuinus* Müll.Arg. === **Phyllanthus angustifolius** (Sw.) Sw.

Phyllanthus anisolobus Müll.Arg. === **Phyllanthus mocinianus** Baill.

Phyllanthus anisophyllus Urb. === **Phyllanthus berteroanus** Müll.Arg.

Phyllanthus anomalus (Baill.) Müll.Arg. === **Margaritaria anomala** (Baill.) Fosberg

Phyllanthus anomalus subsp. *erythroxyloides* (Müll.Arg.) Leandri === **Margaritaria anomala** (Baill.) Fosberg

Phyllanthus antillanus (A.Juss.) Müll.Arg. === **Margaritaria nobilis** L.f.

Phyllanthus antillanus var. *concolor* Müll.Arg. === **Margaritaria nobilis** L.f.

Phyllanthus antillanus var. *glaucescens* (Griseb.) Müll.Arg. === **Margaritaria scandens** (Wright ex Griseb.) G.L.Webster

Phyllanthus antillanus var. *hypomalacus* (Standl.) Lundell === **Margaritaria nobilis** L.f.

Phyllanthus antillanus var. *pedicellaris* Müll.Arg. === **Margaritaria nobilis** L.f.

Phyllanthus antillanus var. *virens* (Griseb.) Müll.Arg. === **Margaritaria tetracocca** (Baill.) G.L.Webster

Phyllanthus antunesii Pax === **Phyllanthus microdendron** Müll.Arg. var. **microdendron**

Phyllanthus apiculatus Urb. === **Phyllanthus chamaecristoides** Urb. subsp. **chamaecristoides**

Phyllanthus aquaticus C.Wright ex Sauvalle === **Phyllanthus stipulatus** (Raf.) G.L.Webster

Phyllanthus arabicus Hochst. ex Steud. === **Phyllanthus maderaspatensis** L. var. **maderaspatensis**

Phyllanthus arborescens (Blume) Müll.Arg. === **Glochidion zeylanicum** var. **arborescens** (Blume) Chakrab. & M.G.Gangop.

Phyllanthus arboreus Müll.Arg. === **Glochidion candolleanum** (Wight & Arn.) Chakrab. & M.G.Gangop.

Phyllanthus arboreus Sessé & Moç. ex Baill. === **Phyllanthus mocinianus** Baill.

Phyllanthus argentatus Noronha === **Phyllanthus distichus** Hook. & Arn. var. **distichus**

Phyllanthus arnottianus Müll.Arg. === **Glochidion zeylanicum** var. **talbotii** (Hook.f.) Haines

Phyllanthus aspericaulis Pax === **Phyllanthus rotundifolius** Klein ex Willd.

Phyllanthus asperus Müll.Arg. === **Glochidion heyneanum** (Wight & Arn.) Wight

Phyllanthus assamicus Müll.Arg. === **Glochidion ellipticum** Wight

Phyllanthus asteranthos Croizat === **Phyllanthus pulcher** Wall. ex Müll.Arg.

Phyllanthus atropurpureus Bojer === ?

Phyllanthus atroviridis Müll.Arg. === **Phyllanthus bojerianus** (Baill.) Müll.Arg. var. **bojerianus**

Phyllanthus austinii Standl. === **Astrocasia austinii** (Standl.) G.L.Webster

Phyllanthus averrhoifolius Müll.Arg. === **Phyllanthus grandifolius** L.

Phyllanthus averrhoifolius Steud. === **Phyllanthus acuminatus** Vahl

Phyllanthus avicularis Small === **Phyllanthus pudens** Wheeler

Phyllanthus baccatus F.Muell. ex Benth. === **Phyllanthus ciccoides** Müll.Arg. var. **ciccoides**

Phyllanthus bacciformis L. === **Sauropus bacciformis** (L.) Airy Shaw

Phyllanthus bachmannii Pax === **Phyllanthus myrtaceus** Sond.

Phyllanthus bahamensis Urb. === **Margaritaria scandens** (Wright ex Griseb.) G.L.Webster

Phyllanthus balakrishnairii Govaerts & Radcl.-Sm. === **Phyllanthus andamanicus** N.P.Balakr. & N.G.Nair

Phyllanthus balansaeanus var. *glaber* M.Schmid === **Phyllanthus balansaeanus** Guillaumin

Phyllanthus banksii A.Cunn. ex Benth. === **Neoroepera banksii** Benth.

Phyllanthus baracoensis Urb. === **Phyllanthus chamaecristoides** subsp. **baracoensis** (Urb.) G.L.Webster

Phyllanthus barbadensis Urb. === **Phyllanthus anderssonii** Müll.Arg.

Phyllanthus bartlettii Standl. === **Meineckia bartlettii** (Standl.) G.L.Webster

Phyllanthus baumannii Guillaumin === **Phyllanthus vulcani** var. **baumannii** Guillaumin ex M.Schmid

Phyllanthus beckleri Müll.Arg. === **Phyllanthus virgatus** G.Forst. var. **virgatus**

Phyllanthus benthamianus Müll.Arg. === **Glochidion lanceolarium** (Roxb.) Voigt

Phyllanthus benthamianus Müll.Arg. === **Phyllanthus graveolens** Kunth

Phyllanthus betsileanus var. *isalensis* Leandri === **Phyllanthus isalensis** (Leandri) Leandri

Phyllanthus bicolor Müll.Arg. === **Glochidion triandrum** (Blanco) C.B.Rob.

Phyllanthus bienhoensis Beille === **Phyllanthus collinsiae** Craib var. **collinsiae**

Phyllanthus billardieri (Baill.) Müll.Arg. === **Glochidion billardieri** Baill.

Phyllanthus blumei Steud. === **Breynia cernua** (Poir.) Müll.Arg.

Phyllanthus boivinianus (Baill.) Voeltzk. === **Phyllanthus decipiens** var. **boivinianus** (Baill.) Leandri

Phyllanthus bojerianus Baill. === **Phyllanthus fuscoluridus** Müll.Arg. subsp. **fuscoluridus**

Phyllanthus borneensis Müll.Arg. === **Glochidion borneense** (Müll.Arg.) Boerl.

Phyllanthus boroniacus F.Muell. ex Grüning === **Micrantheum hexandrum** Hook.f.

Phyllanthus bossiaeoides A.Cunn. ex Benth. === **Sauropus glaucus** (F.Muell.) Airy Shaw

Phyllanthus brachycladus Müll.Arg. === **Phyllanthus rupestris** Kunth

Phyllanthus brachycladus var. *oblongifolius* Müll.Arg. === **Phyllanthus rupestris** Kunth

Phyllanthus brachylobus (Müll.Arg.) Müll.Arg. === **Glochidion coriaceum** Thwaites

Phyllanthus brachypodus F.Muell. ex Benth. === **Phyllanthus maderaspatensis** L. var. **maderaspatensis**

Phyllanthus brasiliensis var. *oblongifolius* Müll.Arg. === **Phyllanthus subglomeratus** Poir.

Phyllanthus braunii Pax === **Phyllanthus odontadenius** Müll.Arg.

Phyllanthus breviramis Urb. === **Phyllanthus phlebocarpus** Urb.

Phyllanthus brevistipulus Urb. === **Phyllanthus pseudocicca** Griseb.

Phyllanthus brisbanicus F.M.Bailey === **Phyllanthus tenellus** Roxb. var. **tenellus**

Phyllanthus brittonii Alain === **Phyllanthus cinctus** Urb.

Phyllanthus brunonis S.Moore === **Sauropus brunonis** (S.Moore) Airy Shaw

Phyllanthus burchellii Müll.Arg. === **Phyllanthus parvulus** var. **garipensis** (Müll.Arg.) Radcl.-Sm.

Phyllanthus caledonicus (Müll.Arg.) Müll.Arg. === **Glochidion caledonicum** Müll.Arg.

Phyllanthus callitrichoides Griseb. === **Phyllanthus carnosulus** Müll.Arg.

Phyllanthus calycinus Wall. === **Breynia vitis-idaea** (Burm.f.) C.E.C.Fisch.

Phyllanthus campanulatus Ridl. === **Phyllanthus pachyphyllus** Müll.Arg.

Phyllanthus canaranus Müll.Arg. === **Glochidion zeylanicum** (Gaertn.) A.Juss. var. **zeylanicum**

Phyllanthus cantoniensis Zipp. ex Span. === **Phyllanthus zippelianus** Müll.Arg.

Phyllanthus cantoniensis Hornem. === **Phyllanthus urinaria** L. subsp. **urinaria**

Phyllanthus capensis Spreng. ex Sond. === **Phyllanthus incurvus** Thunb.

Phyllanthus capillariformis Vatke & Pax ex Pax === **Meineckia phyllanthoides** subsp. **capillariformis** (Vatke & Pax ex Pax) G.L.Webster

Phyllanthus capillaris Schumach. & Thonn. === **Phyllanthus nummulariifolius** var. **capillaris** (Schumach. & Thonn.) Radcl.-Sm.

Phyllanthus capillaris var. *stuhlmannii* (Pax) Hutch. === **Phyllanthus nummulariifolius** var. **capillaris** (Schumach. & Thonn.) Radcl.-Sm.

Phyllanthus capilliformis Pax & Vatke ex Engl. === **Meineckia phyllanthoides** subsp. **capillariformis** (Vatke & Pax ex Pax) G.L.Webster

Phyllanthus capillipes S.F.Blake === **Meineckia capillipes** (Blake) G.L.Webster

Phyllanthus cardiophyllus Urb. === **Phyllanthus myrtilloides** subsp. **erythrinus** (Müll.Arg.) G.L.Webster

Phyllanthus carolii Diels === **Phyllanthus effusus** S.Moore

Phyllanthus carolinianus Blanco === **Phyllanthus niruri** L. subsp. **niruri**

Phyllanthus caroliniensis var. *antillanus* Müll.Arg. === **Phyllanthus caroliniensis** subsp. **guianensis** (Klotzsch) G.L.Webster

Phyllanthus casticum subsp. *angavensis* Leandri === **Phyllanthus angavensis** (Leandri) Leandri

Phyllanthus casticum f. *boivinianus* (Baill.) Müll.Arg. === **Phyllanthus decipiens** var. **boivinianus** (Baill.) Leandri

Phyllanthus casticum f. *decipiens* (Baill.) Leandri === **Phyllanthus decipiens** (Baill.) Müll.Arg.

Phyllanthus casticum var. *fasciculatus* (Poir.) Müll.Arg. === **Phyllanthus casticum** var. **kirganelia** (Willd.) Govaerts

Phyllanthus casticum f. *fiherenensis* Leandri === **Phyllanthus casticum** P.Willemet. var. **casticum**

Phyllanthus casticum var. *genuinus* Müll.Arg. === **Phyllanthus casticum** P.Willemet

Phyllanthus casticum f. *glaucescens* (Baill.) Müll.Arg. === **Phyllanthus mocquerysianus** A.DC.

Phyllanthus casticum var. *madagascariensis* Leandri === **Phyllanthus casticum** P.Willemet. var. **casticum**

Phyllanthus casticum f. *occidentalis* Leandri === **Phyllanthus decipiens** (Baill.) Müll.Arg. var. **decipiens**

Phyllanthus casticum f. *parvifolius* Müll.Arg. === **Phyllanthus casticum** P.Willemet. var. **casticum**

Phyllanthus casticum var. *roxburghianus* Müll.Arg. === **Phyllanthus sp.**

Phyllanthus casticum f. *trilocularis* Leandri === **Phyllanthus seyrigii** Leandri

Phyllanthus casuarinoides Glaz. === **?**

Phyllanthus cataractarum (Müll.Arg.) Müll.Arg. === **Phyllanthus chamaecerasus** var. **vieillardii** (Baill.) M.Schmid

Phyllanthus cataractarum var. *villosus* Däniker === **Phyllanthus pilifer** M.Schmid var. **pilifer**

Phyllanthus cecilfischeri J.L.Ellis === **Phyllanthus indofischeri** Bennet

Phyllanthus celastroides Müll.Arg. === **Glochidion celastroides** (Müll.Arg.) Pax

Phyllanthus ceramanthus G.L.Webster === **Phyllanthus albidiscus** (Ridl.) Airy Shaw

Phyllanthus ceramica Pers. === **Exocarpos ceramicus** (L.) Endl. (Santalaceae)

Phyllanthus cernuus Poir. === **Breynia cernua** (Poir.) Müll.Arg.

Phyllanthus chamaepeuce Ridl. === **Phyllanthus urinaria** L. subsp. **urinaria**

Phyllanthus chamissonis Klotzsch === **Phyllanthus reticulatus** Poir. var. **reticulatus**

Phyllanthus chapelieri Baill. === **Phyllanthus madagascariensis** Müll.Arg. var. **madagascariensis**

Phyllanthus cheloniphorbe Hutch. === **Margaritaria anomala** (Baill.) Fosberg

Phyllanthus cheremela Roxb. === **Phyllanthus distichus** Hook. & Arn. var. **distichus**

Phyllanthus chevalieri (Gagnep.) G.L.Webster === **Phyllanthus arachnodes** Govaerts & Radcl.-Sm.

Phyllanthus chlorophaeus Baill. === **Phyllanthus niruri** subsp. **lathyroides** (Kunth) G.L.Webster

Phyllanthus choristylus Diels === **Phyllanthus clamboides** (F.Muell.) Diels

Phyllanthus cicca Müll.Arg. === **Phyllanthus acidus** (L.) Skeels

Phyllanthus cicca Griseb. === **Phyllanthus pseudocicca** Griseb.

Phyllanthus cicca var. *bracteosa* Müll.Arg. === **Phyllanthus acidus** (L.) Skeels

Phyllanthus ciliatoglandulosus Millsp. === **Andrachne microphylla** (Lam.) Baill.

Phyllanthus cinerascens (Miq.) Müll.Arg. === **Alphitonia cinerascens** (Miq.) Hoogl. (Rhamnaceae)

Phyllanthus cinerascens Hook. & Arn. === **Phyllanthus cochinchinensis** Spreng.

Phyllanthus cinereoviridis Pax === **Cissampelos capensis** L.f. (Menispermaceae)

Phyllanthus cladotrichus Müll.Arg. === **Phyllanthus ramosus** Vell.

Phyllanthus claussenii var. *oblongifolius* Müll.Arg. === **Phyllanthus claussenii** Müll.Arg.

Phyllanthus coccineus (Buch.-Ham.) Müll.Arg. === **Glochidion coccineum** (Buch.-Ham.) Müll.Arg.

Phyllanthus cochinchinensis (Lour.) Müll.Arg. === **Phyllanthus acidus** (L.) Skeels

Phyllanthus coelophyllus Urb. === **Phyllanthus chamaecristoides** subsp. **baracoensis** (Urb.) G.L.Webster

Phyllanthus cognatus Spreng. === **Phyllanthus angustifolius** (Sw.) Sw.

Phyllanthus columnaris Warb. === **Phyllanthus warburgii** K.Schum.

Phyllanthus commersonianus Baill. === **Phyllanthus phillyreifolius** var. **commersonii** Müll.Arg.

Phyllanthus compressicaulis (Kurz ex Teijsm. & Binn.) Müll.Arg. === **Glochidion philippicum** (Cav.) C.B.Rob.

Phyllanthus conami Sw. === **Phyllanthus brasiliensis** (Aubl.) Poir.

Phyllanthus concinnus Airy Shaw === **Phyllanthus stultitiae** Airy Shaw

Phyllanthus concinnus Ridl. === **Phyllanthus gracilipes** (Miq.) Müll.Arg.

Phyllanthus concolor (Müll.Arg.) Müll.Arg. === **Glochidion concolor** Müll.Arg.

Phyllanthus concolor var. *ellipticus* Müll.Arg. === **Glochidion concolor** Müll.Arg.

Phyllanthus concolor var. *obovatus* (Müll.Arg.) Müll.Arg. === **Glochidion concolor** Müll.Arg.

Phyllanthus congestus Benth. ex Müll.Arg. === **Jablonskia congesta** (Benth. ex Müll.Arg.) G.L.Webster

Phyllanthus conradii Pax === **Phyllanthus sepialis** Müll.Arg.

Phyllanthus conterminus Müll.Arg. === **Phyllanthus virgatus** G.Forst. var. **virgatus**

Phyllanthus corcovadensis Müll.Arg. === **Phyllanthus tenellus** Roxb. var. **tenellus**

Phyllanthus cordatus (Seem. ex Müll.Arg.) Müll.Arg. === **Glochidion cordatum** Seem. ex Müll.Arg.

Phyllanthus cordifolius Wall. ex Decne. === **Andrachne cordifolia** (Wall. ex Decne.) Müll.Arg.

Phyllanthus coriaceus (Thwaites) Müll.Arg. === **Glochidion coriaceum** Thwaites

Phyllanthus cornifolius Kunth === **Phyllanthus juglandifolius** subsp. **cornifolius** (Kunth) G.L.Webster

Phyllanthus coxianus Fawc. === **Phyllanthus arbuscula** (Sw.) J.F.Gmel.

Phyllanthus crassifolius Müll.Arg. === **Sauropus crassifolius** (Müll.Arg.) Airy Shaw

Phyllanthus crotalaroides Zipp. ex Span. === **?**

Phyllanthus cumanensis Willd. ex Steud. === **Phyllanthus acuminatus** Vahl

Phyllanthus cumingii (Müll.Arg.) Müll.Arg. === **Glochidion album** (Blanco) Boerl.

Phyllanthus cuneatus Willd. === **Phyllanthus maderaspatensis** L. var. **maderaspatensis**

Phyllanthus cuneifolius (Britton) Croizat === **Phyllanthus orbicularis** Kunth

Phyllanthus cupuliformis Warb. === **Phyllanthus finschii** K.Schum.

Phyllanthus curtipes Airy Shaw === **Phyllanthus acutissimus** Miq.

Phyllanthus cyanospermus (Gaertn.) Müll.Arg. === **Margaritaria cyanosperma** (Gaertn.) Airy Shaw

Phyllanthus cyclanthera Baill. === **Phyllanthus pentaphyllus** C.Wright ex Griseb. subsp. **pentaphyllus**

Phyllanthus cyclanthera var. *gracillimus* (Baill.) Müll.Arg. === **Phyllanthus lindenianus** Baill. var. **lindenianus**

Phyllanthus cyclanthera var. *lindenianus* (Baill.) Müll.Arg. === **Phyllanthus lindenianus** Baill.

Phyllanthus cyclanthera var. *scabrellus* Müll.Arg. === **Phyllanthus lindenianus** Baill. var. **lindenianus**

Phyllanthus cygnorum Endl. === **Phyllanthus calycinus** Labill.

Phyllanthus cyrtophyllus (Miq.) Müll.Arg. === **Glochidion cyrtostylum** Miq.

Phyllanthus cyrtostylus (Miq.) Müll.Arg. === **Glochidion cyrtostylum** Miq.

Phyllanthus dalbergioides Wall. ex J.J.Sm. === **Phyllanthus reticulatus** Poir. var. **reticulatus**

Phyllanthus daltonii Müll.Arg. === **Glochidion daltonii** (Müll.Arg.) Kurz

Phyllanthus dasyanthus (Kurz ex Teijsm. & Binn.) Müll.Arg. === **Glochidion dasyanthum** Kurz ex Teijsm. & Binn.

Phyllanthus decander Sessé & Moç. === **Phyllanthus discolor** Poepp. ex Spreng.

Phyllanthus decaryanus Leandri === **Margaritaria decaryana** (Leandri) G.L.Webster

Phyllanthus decaryanus var. *manambia* Leandri === **Margaritaria decaryana** (Leandri) G.L.Webster

Phyllanthus decipiens var. *genuinus* Müll.Arg. === **Phyllanthus decipiens** (Baill.) Müll.Arg.

Phyllanthus decipiens var. *trilocularis* (Baill.) Müll.Arg. === **Phyllanthus decipiens** f. **trilocularis** (Baill.) Leandri

Phyllanthus deflexus Klotzsch === **Phyllanthus pentandrus** Schumach. & Thonn.

Phyllanthus dekindtii Hutch. === **Phyllanthus verdickii** De Wild.

Phyllanthus depressus Buch.-Ham. ex Dillwyn === **Phyllanthus virgatus** G.Forst. var. **virgatus**

Phyllanthus diffusus Klotzsch === **Phyllanthus stipulatus** (Raf.) G.L.Webster

Phyllanthus diffusus var. *genuinus* Müll.Arg. === **Phyllanthus stipulatus** (Raf.) G.L.Webster

Phyllanthus diffusus var. *oblongifolius* (Müll.Arg.) Müll.Arg. === **Phyllanthus stipulatus** (Raf.) G.L.Webster

Phyllanthus dilatatus Klotzsch === **Phyllanthus pentandrus** Schumach. & Thonn.

Phyllanthus dingleri G.L.Webster === **Phyllanthus arbuscula** (Sw.) J.F.Gmel.

Phyllanthus dioicus Schumach. & Thonn. === **Flueggea virosa** (Roxb. ex Willd.) Voigt subsp. **virosa**

Phyllanthus discofractus Croizat === **Phyllanthus gracilipes** (Miq.) Müll.Arg.

Phyllanthus discoideus (Baill.) Müll.Arg. === **Margaritaria discoidea** (Baill.) G.L.Webster

Phyllanthus discolor var. *pallidus* (C.Wright ex Griseb.) G.L.Webster === **Phyllanthus** × **pallidus** C.Wright ex Griseb.

Phyllanthus distichus (L.) Müll.Arg. === **Phyllanthus acidus** (L.) Skeels

Phyllanthus distichus f. *nodiflorus* (Lam.) Müll.Arg. === **Phyllanthus acidus** (L.) Skeels

Phyllanthus ditassoides Müll.Arg. === **Sauropus ditassoides** (Müll.Arg.) Airy Shaw

Phyllanthus diversifolius Miq. === **Glochidion rubrum** Blume var. **rubrum**

Phyllanthus diversifolius var. *longifolius* Müll.Arg. === **Glochidion ellipticum** Wight

Phyllanthus diversifolius var. *wightiana* Müll.Arg. === **Glochidion ellipticum** Wight

Phyllanthus dongfangensis P.T.Li === **Cleistanthus concinnus** Croizat

Phyllanthus dregeanus Scheele === **Andrachne ovalis** (E.Mey. ex Sond.) Müll.Arg.

Phyllanthus drummondii Small === **Phyllanthus abnormis** Baill. var. **abnormis**

Phyllanthus durus S.Moore === **Phyllanthus aeneus** Baill. var. **aeneus**

Phyllanthus eboracensis S.Moore === **Phyllanthus virgatus** G.Forst. var. **virgatus**

Phyllanthus echinatus Buch.-Ham. ex Wall. === **Phyllanthus urinaria** L. subsp. **urinaria**

Phyllanthus echinocarpus T.L.Chin === **Phyllanthus forrestii** W.W.Sm.

Phyllanthus eglandulosus (Baill.) Leandri === **Margaritaria anomala** (Baill.) Fosberg

Phyllanthus elachophyllus F.Muell. ex Benth. === **Sauropus elachophyllus** (F.Muell. ex Benth.) Airy Shaw

Phyllanthus ellipticus Desf. ex Poir. === ?

Phyllanthus ellipticus Buckley === **Phyllanthus niruri** L. subsp. **niruri**

Phyllanthus ellipticus Kunth === **Flueggea elliptica** (Spreng.) Baill.

Phyllanthus elongatus Mart. ex Colla === ?

Phyllanthus × *elongatus* (Jacq.) Steud. === **Phyllanthus 'Elongatus'**

Phyllanthus embergeri Haicour & Rossignol === **Phyllanthus cochinchinensis** Spreng.

Phyllanthus emblicoides Müll.Arg. === **Phyllanthus polyphyllus** Willd. var. **polyphyllus**

Phyllanthus emirnensis Müll.Arg. === **Phyllanthus fuscoluridus** Müll.Arg. subsp. **fuscoluridus**

Phyllanthus empetrifolius Baill. === **Phyllanthus ramillosus** Müll.Arg.

Phyllanthus epiphyllanthus var. *dilatatus* Müll.Arg. === **Phyllanthus epiphyllanthus** subsp. **dilatatus** (Müll.Arg.) G.L.Webster

Phyllanthus epiphyllanthus var. *genuinus* Müll.Arg. === **Phyllanthus epiphyllanthus** L.

Phyllanthus epistylium (Poir.) Griseb. === **Phyllanthus axillaris** (Sw.) Müll.Arg.

Phyllanthus eriocarpus (Champ. ex Benth.) Müll.Arg. === **Glochidion eriocarpum** Champ. ex Benth.

Phyllanthus erythranthus Guillaumin === **Phyllanthus yaouhensis** Schltr.

Phyllanthus erythrinus Müll.Arg. === **Phyllanthus myrtilloides** subsp. **erythrinus** (Müll.Arg.) G.L.Webster

Phyllanthus erythrocarpus Ridl. === **Phyllanthus reticulatus** Poir. var. **reticulatus**

Phyllanthus erythroxyloides Müll.Arg. === **Margaritaria anomala** (Baill.) Fosberg

Phyllanthus essequiboensis Klotzsch === ?

Phyllanthus estrellensis Urb. === **Phyllanthus phlebocarpus** Urb.

Phyllanthus eugenioides Guillaumin === **Phyllanthus kanalensis** Baill.

Phyllanthus euwensii Bold. === **Phyllanthus botryanthus** Müll.Arg.

Phyllanthus eylesii S.Moore === **Phyllanthus leucanthus** Pax

Phyllanthus fagifolius Müll.Arg. === **Glochidion hohenackeri** (Müll.Arg.) Bedd.
   var. **hohenackeri**

Phyllanthus falcatus (Sw.) J.F.Gmel. === **Phyllanthus epiphyllanthus** L. subsp. **epiphyllanthus**

Phyllanthus fasciculatus Poir. === **Phyllanthus casticum** var. **kirganelia** (Willd.) Govaerts

Phyllanthus fasciculatus (Lour.) Müll.Arg. === **Phyllanthus cochinchinensis** Spreng.

Phyllanthus ferdinandii Müll.Arg. === **Glochidion ferdinandii** (Müll.Arg.) F.M.Bailey

Phyllanthus ferdinandii var. *mollis* Benth. === **Glochidion philippicum** (Cav.) C.B.Rob.

Phyllanthus ferdinandii var. *supra-axillaris* Benth. === **Glochidion zeylanicum** (Gaertn.)
   A.Juss. var. **zeylanicum**

Phyllanthus filicaulis Benth. === **Phyllanthus virgatus** G.Forst. var. **virgatus**

Phyllanthus filiformis Pav. ex Baill. === **Phyllanthus niruri** L. subsp. **niruri**

Phyllanthus filipes Balf.f. === **Meineckia filipes** (Balf.f.) G.L.Webster

Phyllanthus finlaysonianus Wall. === ?

Phyllanthus fischeri (Gamble) J.L.Ellis === **Phyllanthus indofischeri** Bennet

Phyllanthus flaccidus Thwaites === **Phyllanthus rheedii** Wight

Phyllanthus flacourtioides Hutch. === **Margaritaria discoidea** var. **nitida** (Pax) Radcl.-Sm.

Phyllanthus flagelliformis var. *demonstrans* Müll.Arg. === **Phyllanthus flagelliformis**
   Müll.Arg.

Phyllanthus flavidus (Kurz ex Teijsm. & Binn.) Müll.Arg. === **Glochidion flavidum** Kurz ex
   Teijsm. & Binn.

Phyllanthus floribundus Kunth === **Phyllanthus salviifolius** Kunth

Phyllanthus floribundus (Baill.) Müll.Arg. === **Phyllanthus muellerianus** (Kuntze) Exell

Phyllanthus flueggeiformis Müll.Arg. === **Phyllanthus glaucus** Wall. ex Müll.Arg.

Phyllanthus fluggeoides Müll.Arg. === **Flueggea suffruticosa** (Pall.) Baill.

Phyllanthus foetidus Pav. ex Baill. === **Phyllanthus acuminatus** Vahl

Phyllanthus forskahlii Lepr. ex Baill. === **Euphorbia forskalii** J.Gay

Phyllanthus foveolatus (Britton) Alain === **Phyllanthus myrtilloides** subsp. **erythrinus**
   (Müll.Arg.) G.L.Webster

Phyllanthus fraternus subsp. *togoensis* Brunel & J.P.Roux === **Phyllanthus fraternus**
   G.L.Webster

Phyllanthus fraxinifolius Lodd. === **Glochidion lanceolarium** (Roxb.) Voigt

Phyllanthus frondosus Wall. ex Müll.Arg. === **Phyllanthus oxyphyllus** Miq.

Phyllanthus frondosus var. *rigidus* Ridl. === **Phyllanthus pachyphyllus** Müll.Arg.

Phyllanthus fruticosus Baill. === **Phyllanthus brasiliensis** (Aubl.) Poir.

Phyllanthus fruticosus B.Heyne ex Wall. === **Phyllanthus virgatus** G.Forst. var. **virgatus**

Phyllanthus fuernrohrii var. *suffruticosus* Domin === **Phyllanthus carpentariae** Müll.Arg.

Phyllanthus fulvirameus (Miq.) Müll.Arg. === **Glochidion fulvirameum** Miq.

Phyllanthus fuscus Müll.Arg. === **Glochidion varians** Miq.

Phyllanthus garberi Small === **Phyllanthus abnormis** Baill. var. **abnormis**

Phyllanthus gardneri Thwaites === **Phyllanthus virgatus** G.Forst. var. **virgatus**

Phyllanthus gardneri var. *pubescens* Thwaites === **Phyllanthus wheeleri** G.L.Webster

Phyllanthus gardnerianus Baill. ex Müll.Arg. === **Phyllanthus virgatus** G.Forst. var. **virgatus**

Phyllanthus gardnerianus var. *pubescens* Thwaites === **Phyllanthus wheeleri** G.L.Webster

Phyllanthus garipensis E.Mey. === **Phyllanthus parvulus** var. **garipensis** (Müll.Arg.) Radcl.-Sm.

Phyllanthus gasstroemii Müll.Arg. === **Phyllanthus gunnii** Hook.f.

Phyllanthus gaudichaudii Müll.Arg. === **Glochidion gaudichaudii** (Müll.Arg.) Boerl.

Phyllanthus genistoides Sond. === **Phyllanthus incurvus** Thunb.

Phyllanthus gigantifolius Vidal === **Glochidion gigantifolium** (Vidal) J.J.Sm.

Phyllanthus gilletii De Wild. === **Phyllanthus macranthus** var. **gilletii** (De Wild.) Brunel ex
   Radcl.-Sm.

Phyllanthus glabellus (L.) Fawc. & Rendle === **Croton glabellus** L.

Phyllanthus glabricapsulus F.P.Metcalf === **Phyllanthus leptoclados** Benth.

Phyllanthus glaucescens Schltdl. & Cham. === **Phyllanthus adenodiscus** Müll.Arg.

Phyllanthus glaucescens Kunth === **Phyllanthus grandifolius** L.

Phyllanthus glaucifolius (Müll.Arg.) Ridl. === **Glochidion lutescens** Blume

Phyllanthus glaucogynus Müll.Arg. === **Glochidion moonii** Thwaites var. **moonii**

Phyllanthus glaucophyllus var. *major* Müll.Arg. === **Phyllanthus glaucophyllus** Sond.

Phyllanthus glomeratus Roxb. ex Wall. === **Phyllanthus emblica** L.

Phyllanthus glomerulatus (Miq.) Müll.Arg. === **Glochidion glomerulatum** (Miq.) Boerl.

Phyllanthus goniocladus Merr. & Chun === **Sauropus bacciformis** (L.) Airy Shaw

Phyllanthus gracilentus Müll.Arg. === **Glochidion gracilentum** (Müll.Arg.) Boerl.

Phyllanthus gracilipes Pax === **Phyllanthus tessmannii** Hutch.

Phyllanthus gracilis Roxb. === **Phyllanthus maderaspatensis** L. var. **maderaspatensis**

Phyllanthus gracilis (Hassk.) Baill. === **Phyllanthus albidiscus** (Ridl.) Airy Shaw

Phyllanthus gracillimus F.Muell. ex Benth. === **Phyllanthus virgatus** G.Forst. var. **virgatus**

Phyllanthus gracillimus Baill. === **Phyllanthus lindenianus** Baill. var. **lindenianus**

Phyllanthus grahamii Hutch. & M.B.Moss === **Phyllanthus beillei** Hutch.

Phyllanthus graminicola Britton === **Phyllanthus caroliniensis** Walter subsp. **caroliniensis**

Phyllanthus grandiflorus Spreng. === **Phyllanthus multiflorus** Poir.

Phyllanthus grandiflorus Crantz === **Phyllanthus grandifolius** L.

Phyllanthus grandifolius Spreng. === **Phyllanthus ovatus** Poir.

Phyllanthus grandifolius var. *cornifolius* (Kunth) Müll.Arg. === **Phyllanthus juglandifolius** subsp. **cornifolius** (Kunth) G.L.Webster

Phyllanthus grandifolius var. *genuinus* Müll.Arg. === **Phyllanthus grandifolius** L.

Phyllanthus grandifolius var. *salzmannii* Müll.Arg. === **Phyllanthus juglandifolius** subsp. **cornifolius** (Kunth) G.L.Webster

Phyllanthus grandisepalus F.Muell. ex Müll.Arg. === **Phyllanthus carpentariae** Müll.Arg.

Phyllanthus graveolens var. *tenellus* (Benth.) Müll.Arg. === **Phyllanthus graveolens** Kunth

Phyllanthus grayanus Müll.Arg. === **Glochidion grayanum** (Müll.Arg.) Florence

Phyllanthus grisebachianus Müll.Arg. === **Phyllanthus nutans** subsp. **grisebachianus** (Müll.Arg.) G.L.Webster

Phyllanthus griseus Wall. === **Phyllanthus reticulatus** Poir. var. **reticulatus**

Phyllanthus gueinzii Müll.Arg. === **Phyllanthus maderaspatensis** L. var. **maderaspatensis**

Phyllanthus guianensis (Aubl.) Müll.Arg. === **Meborea guianensis** Aubl.

Phyllanthus guianensis Klotzsch === **Phyllanthus caroliniensis** subsp. **guianensis** (Klotzsch) G.L.Webster

Phyllanthus guianensis auct.; Müll.Arg. === **Phyllanthus attenuatus** Miq.

Phyllanthus guianensis var. *acuminatus* Müll.Arg. === **Phyllanthus caroliniensis** subsp. **guianensis** (Klotzsch) G.L.Webster

Phyllanthus guianensis var. *anceps* Müll.Arg. === **Phyllanthus caroliniensis** subsp. **guianensis** (Klotzsch) G.L.Webster

Phyllanthus guineensis Pax === **Phyllanthus ovalifolius** Forssk.

Phyllanthus gunnii var. *saxosus* (F.Muell.) Müll.Arg. === **Phyllanthus saxosus** F.Muell.

Phyllanthus gymnanthus Baill. === **Bischofia javanica** Blume

Phyllanthus hamiltonianus Müll.Arg. === **Phyllanthus sikkimensis** Müll.Arg.

Phyllanthus hamrur Forssk. === **Flueggea virosa** (Roxb. ex Willd.) Voigt subsp. **virosa**

Phyllanthus haplocladus Urb. === **Phyllanthus carnosulus** Müll.Arg.

Phyllanthus hebecarpus Benth. === **Phyllanthus carpentariae** Müll.Arg.

Phyllanthus helferi Müll.Arg. === **Glochidion helferi** (Müll.Arg.) Hook.f.

Phyllanthus hellwigii Warb. === **Phyllanthus lamprophyllus** Müll.Arg.

Phyllanthus hesperonotos Govaerts & Radcl.-Sm. === **Phyllanthus aridus** Benth.

Phyllanthus heterodoxus Müll.Arg. === **Glochidion heterodoxum** (Müll.Arg.) Pax & K.Hoffm.

Phyllanthus heteromorphus Rusby === **Margaritaria nobilis** L.f.

Phyllanthus heyneanus (Wight) Müll.Arg. === **Glochidion heyneanum** (Wight & Arn.) Wight

Phyllanthus hirsutus (Roxb.) Müll.Arg. === **Glochidion zeylanicum** var. **talbotii** (Hook.f.) Haines

Phyllanthus hirtellus (F.Muell.) Müll.Arg. === **Sauropus hirtellus** (F.Muell.) Airy Shaw

Phyllanthus hirtellus var. *ledifolius* Müll.Arg. === **Phyllanthus hirtellus** F.Muell. ex Müll.Arg.

Phyllanthus hirtellus var. *parviflorus* (J.M.Black) Eichler === **Phyllanthus hirtellus** F.Muell. ex Müll.Arg.

Phyllanthus hirtellus var. *thymoides* Müll.Arg. === **Phyllanthus hirtellus** F.Muell. ex Müll.Arg.

Phyllanthus hoffmannseggii Müll.Arg. === **Phyllanthus stipulatus** (Raf.) G.L.Webster

Phyllanthus hoffmannseggii var. *oblongifolius* Müll.Arg. === **Phyllanthus stipulatus** (Raf.) G.L.Webster

Phyllanthus hoffmeisteri Klotzsch === **Phyllanthus stipulatus** (Raf.) G.L.Webster

Phyllanthus hohenackeri Müll.Arg. === **Glochidion hohenackeri** (Müll.Arg.) Bedd.

Phyllanthus hongkongensis (Müll.Arg.) Müll.Arg. === **Glochidion zeylanicum** (Gaertn.) A.Juss. var. **zeylanicum**

Phyllanthus hookeri Müll.Arg. === **Phyllanthus urinaria** var. **hookeri** (Müll.Arg.) Hook.f.

Phyllanthus hotteanus Urb. & Ekman === **Margaritaria hotteana** (Urb. & Ekman) G.L.Webster

Phyllanthus hullettii Ridl. === **Phyllanthus gracilipes** (Miq.) Müll.Arg.

Phyllanthus humblotianus Baill. === **Phyllanthus pervilleanus** (Baill.) Müll.Arg.

Phyllanthus humilis Pax === **?**

Phyllanthus humilis Salisb. === **Phyllanthus niruri** L. subsp. **niruri**

Phyllanthus huntii Ewart & O.B.Davies === **Sauropus huntii** (Ewart & Davies) Airy Shaw

Phyllanthus hurlimannii Guillaumin === **Phyllanthus vulcani** var. **baumannii** Guillaumin ex M.Schmid

Phyllanthus hylodendron (K.Schum.) Airy Shaw & G.L.Webster === **Phyllanthus finschii** K.Schum.

Phyllanthus hypoleucus (Miq.) Müll.Arg. === **Glochidion lutescens** Blume

Phyllanthus hyssopifolius Müll.Arg. === **Phyllanthus hyssopifolioides** Kunth

Phyllanthus hysteranthus Müll.Arg. === **Margaritaria anomala** (Baill.) Fosberg

Phyllanthus inaequaliflorus Fawc. & Rendle === **Phyllanthus arbuscula** (Sw.) J.F.Gmel.

Phyllanthus inaequalis Rusby === **Astrocasia jacobinensis** (Müll.Arg.) G.L.Webster

Phyllanthus inaequifolius G.L.Webster === **Phyllanthus lindenianus** var. **inaequifolius** (G.L.Webster) G.L.Webster

Phyllanthus incurvus Sond. === **Phyllanthus heterophyllus** E.Mey. ex Müll.Arg.

Phyllanthus indicus (Dalzell) Müll.Arg. === **Margaritaria indica** (Dalzell) Airy Shaw

Phyllanthus indicus f. *vestita* J.J.Sm. === **Margaritaria indica** (Dalzell) Airy Shaw

Phyllanthus indigoferoides A.Cunn. ex Benth. === **Phyllanthus gunnii** Hook.f.

Phyllanthus induratus S.Moore === **Phyllanthus torrentium** var. **induratus** (S.Moore) M.Schmid

Phyllanthus insignis Müll.Arg. === **Glochidion insigne** (Müll.Arg.) J.J.Sm.

Phyllanthus insignis (K.Schum.) J.J.Sm. === **Phyllanthus clamboides** (F.Muell.) Diels

Phyllanthus insulanus (Müll.Arg.) Müll.Arg. === **Glochidion insulanum** Müll.Arg.

Phyllanthus introductus Steud. === **Breynia fruticosa** (L.) Hook.f.

Phyllanthus isolepis Urb. === **Phyllanthus latifolius** (L.) Sw.

Phyllanthus jacobinensis Müll.Arg. === **Astrocasia jacobinensis** (Müll.Arg.) G.L.Webster

Phyllanthus jamaicensis Griseb. === **Phyllanthus reticulatus** Poir. var. **reticulatus**

Phyllanthus japonicus (Baill.) Müll.Arg. === **Phyllanthus flexuosus** (Siebold & Zucc.) Müll.Arg.

Phyllanthus jardinii Müll.Arg. === **?**

Phyllanthus javanicus Poir. ex Spreng. === **Phyllanthus maderaspatensis** L. var. **maderaspatensis**

Phyllanthus jullienii Beille === **Flueggea jullienii** (Beille) G.L.Webster

Phyllanthus juniperinoides Müll.Arg. === **Phyllanthus lawii** J.Graham

Phyllanthus juniperinus Wall. === **Phyllanthus parvifolius** Buch.-Ham. ex D.Don

Phyllanthus jussieuianus (Wight) Müll.Arg. === **Glochidion stellatum** (Retz.) Bedd.

Phyllanthus kaalaensis Guillaumin === **Phyllanthus pterocladus** S.Moore

Phyllanthus kanalophilus Müll.Arg. === **Glochidion billardieri** Baill.

Phyllanthus khasicus Müll.Arg. === **Glochidion khasicum** (Müll.Arg.) Hook.f.

Phyllanthus kiangsiensis Croizat & Metcalf === **Phyllanthus chekiangensis** Croizat & Metcalf

Phyllanthus kipareh Müll.Arg. === **Glochidion obscurum** (Roxb. ex Willd.) Blume

Phyllanthus kirganelia Blanco === **Phyllanthus niruri** L. subsp. **niruri**

Phyllanthus kirganelia Willd. === **Phyllanthus casticum** var. **kirganelia** (Willd.) Govaerts

Phyllanthus kirganelia var. *fasciculatus* (Poir.) Müll.Arg. === **Phyllanthus casticum** var. **kirganelia** (Willd.) Govaerts

Phyllanthus kirganelia var. *glaber* Müll.Arg. === **Phyllanthus mocquerysianus** A.DC.

Phyllanthus kirganelia var. *sieberi* Müll.Arg. === **Phyllanthus casticum** var. **kirganelia** (Willd.) Govaerts

Phyllanthus kirkianus Müll.Arg. === **Phyllanthus pinnatus** (Wight) G.L.Webster

Phyllanthus klainei Hutch. === **Phyllanthus polyanthus** Pax

Phyllanthus klossii Ridl. === **Phyllanthus pachyphyllus** Müll.Arg.

Phyllanthus klotzschianus var. *gardneri* Müll.Arg. === **Phyllanthus klotzschianus** Müll.Arg.

Phyllanthus klotzschianus var. *linearis* Müll.Arg. === **Phyllanthus klotzschianus** Müll.Arg.

Phyllanthus klotzschianus var. *major* Müll.Arg. === **Phyllanthus klotzschianus** Müll.Arg.

Phyllanthus klotzschianus var. *minor* Müll.Arg. === **Phyllanthus klotzschianus** Müll.Arg.

Phyllanthus klotzschianus var. *pallidiflorus* Müll.Arg. === **Phyllanthus klotzschianus** Müll.Arg.

Phyllanthus klotzschianus var. *racemulosus* Müll.Arg. === **Phyllanthus klotzschianus** Müll.Arg.

Phyllanthus klotzschianus var. *robustus* Müll.Arg. === **Phyllanthus klotzschianus** Müll.Arg.

Phyllanthus kollmannianus Müll.Arg. === **Glochidion lutescens** Blume

Phyllanthus korthalsii Müll.Arg. === **Glochidion korthalsii** (Müll.Arg.) Boerl.

Phyllanthus kunstleri Hook.f. === **Phyllanthus oxyphyllus** Miq.

Phyllanthus kunthiana Baill. === **Phyllanthus salviifolius** Kunth

Phyllanthus kurzianus Müll.Arg. === **Glochidion philippicum** (Cav.) C.B.Rob.

Phyllanthus lacerilobus Croizat === **Phyllanthus pulcher** Wall. ex Müll.Arg.

Phyllanthus lacteus Müll.Arg. === **Phyllanthus hypoleucus** Müll.Arg.

Phyllanthus lacunarius var. *deuterocalyx* Gauba === **Phyllanthus erwinii** J.T.Hunter & J.J.Bruhl

Phyllanthus laevigatus Müll.Arg. === **Glochidion lutescens** Blume

Phyllanthus lalambensis Schweinf. === **Phyllanthus ovalifolius** Forssk.

Phyllanthus lambertianus (Müll.Arg.) Müll.Arg. === **Glochidion impuber** (Roxb.) Govaerts

Phyllanthus lanceilimbus (Merr.) Merr. === **Glochidion lanceilimbum** Merr.

Phyllanthus lanceolarius (Roxb.) Müll.Arg. === **Glochidion lanceolarium** (Roxb.) Voigt

Phyllanthus lathyroides Kunth === **Phyllanthus niruri** subsp. **lathyroides** (Kunth) G.L.Webster

Phyllanthus lathyroides var. *commutatus* Müll.Arg. === **Phyllanthus niruri** L. subsp. **niruri**

Phyllanthus lathyroides f. *microphyllus* Müll.Arg. === **Phyllanthus niruri** subsp. **lathyroides** (Kunth) G.L.Webster

Phyllanthus lathyroides f. *minor* Müll.Arg. === **Phyllanthus niruri** subsp. **lathyroides** (Kunth) G.L.Webster

Phyllanthus lathyroides f. *oblongifolius* Müll.Arg. === **Phyllanthus niruri** subsp. **lathyroides** (Kunth) G.L.Webster

Phyllanthus lathyroides f. *purpurascens* Müll.Arg. === **Phyllanthus niruri** subsp. **lathyroides** (Kunth) G.L.Webster

Phyllanthus lathyroides f. *rosellus* Müll.Arg. === **Phyllanthus niruri** subsp. **lathyroides** (Kunth) G.L.Webster

Phyllanthus laurifolius A.Rich. === **Savia sessiliflora** (Sw.) Willd.

Phyllanthus lauterbachianus Pax === **Phyllanthus urinaria** L. subsp. **urinaria**

Phyllanthus leai S.Moore === **Phyllanthus debilis** Klein ex Willd.

Phyllanthus ledifolius A.Cunn. ex Benth. === **Phyllanthus hirtellus** F.Muell. ex Müll.Arg.

Phyllanthus leiboensis T.L.Chin === **Phyllanthus franchetianus** H.Lév.

Phyllanthus leonardorum G.L.Webster === **Phyllanthus lindenianus** var. **leonardorum** (G.L.Webster) G.L.Webster

Phyllanthus leonensis Hutch. === **Phyllanthus alpestris** Beille

Phyllanthus leonis Alain === **Phyllanthus subcarnosus** C.Wright ex Griseb.

Phyllanthus lepidocarpus Siebold & Zucc. === **Phyllanthus urinaria** L. subsp. **urinaria**

Phyllanthus leprocarpus Wight === **Phyllanthus urinaria** L. subsp. **urinaria**

Phyllanthus leptoclados var. *pubescens* P.T.Li & D.Y.Liu === **Phyllanthus chekiangensis** Croizat & Metcalf

Phyllanthus leptogynus Müll.Arg. === **Glochidion nemorale** Thwaites

Phyllanthus leschenaultii var. *tenellus* Müll.Arg. === **Sauropus quadrangularis** (Willd.) Müll.Arg.

Phyllanthus leucogynus Müll.Arg. === **Glochidion leucogynum** Miq.

Phyllanthus leucophyllus Strachey & Winterb. ex Baill. === **Flueggea virosa** (Roxb. ex Willd.) Voigt subsp. **virosa**

Phyllanthus leucopyrus (Willd.) D.Koenig ex Roxb. === **Flueggea leucopyrus** Willd.

Phyllanthus lhotzkyanus Hochst. ex Steud. === **Micrantheum ericoides** Desf.

Phyllanthus lifuensis Guillaumin === **Phyllanthus faguetii** var. **lifuensis** (Guillaumin) M.Schmid

Phyllanthus lignulatus var. *tonkinensis* Baill. === **Phyllanthus poilanei** Beille

Phyllanthus lindenianus subsp. *inaequifolius* (G.L.Webster) Borhidi === **Phyllanthus lindenianus** var. **inaequifolius** (G.L.Webster) G.L.Webster

Phyllanthus lindenianus subsp. *jimenezii* (G.L.Webster) Borhidi === **Phyllanthus lindenianus** var. **jimenezii** G.L.Webster

Phyllanthus lindenianus subsp. *leonardorum* (G.L.Webster) Borhidi === **Phyllanthus lindenianus** var. **leonardorum** (G.L.Webster) G.L.Webster

Phyllanthus linearis Mart. === **Phyllanthus flagelliformis** Müll.Arg.

Phyllanthus linearis (Sw.) Sw. === **Phyllanthus arbuscula** (Sw.) J.F.Gmel.

Phyllanthus linearis var. *cognatus* (Spreng.) Müll.Arg. === **Phyllanthus angustifolius** (Sw.) Sw.

Phyllanthus linifolius Vahl ex Baill. === **Phyllanthus pentandrus** Schumach. & Thonn.

Phyllanthus linoides Hochst. ex Baill. === **Phyllanthus pentandrus** Schumach. & Thonn.

Phyllanthus lissocarpus S.Moore === **Sauropus stenocladus** (Müll.Arg.) J.T.Hunter & J.J.Bruhl subsp. **stenocladus**

Phyllanthus littoralis (Blume) Müll.Arg. === **Glochidion littorale** Blume

Phyllanthus liukiuensis Hayata === **Phyllanthus leptoclados** Benth.

Phyllanthus llanosii (Müll.Arg.) Müll.Arg. === **Sauropus villosus** (Blanco) Merr.

Phyllanthus lobocarpus Benth. === **Glochidion lobocarpum** (Benth.) F.M.Bailey

Phyllanthus longeramosus Guillaumin === **Phyllanthus longeramosus** Guillaumin ex M.Schmid

Phyllanthus longfieldiae Ridl. === **Glochidion longfieldiae** (Ridl.) F.Br.

Phyllanthus longiflorus B.Heyne ex Müll.Arg. === **Phyllanthus heyneanus** Müll.Arg.

Phyllanthus longifolius Lam. ex Poir. === **Phyllanthus revaughanii** Coode

Phyllanthus longifolius Jacq. === **Phyllanthus acidus** (L.) Skeels

Phyllanthus longifolius Sond. === **Phyllanthus maderaspatensis** L. var. **maderaspatensis**

Phyllanthus longipes Steyerm. === **Gymnanthes belizensis** G.L.Webster

Phyllanthus longipes (Wight) Müll.Arg. === **Meineckia longipes** (Wight) G.L.Webster

Phyllanthus lonphali Wall. === **Phyllanthus fraternus** G.L.Webster

Phyllanthus loureirii Müll.Arg. === **Bridelia tomentosa** Blume

Phyllanthus lucena B.Heyne ex Roth === **Flueggea leucopyrus** Willd.

Phyllanthus lucens Poir. === **Breynia fruticosa** (L.) Hook.f.

Phyllanthus lucidus Steud. === **Flueggea virosa** (Roxb. ex Willd.) Voigt subsp. **virosa**

Phyllanthus lucidus (Blume) Müll.Arg. === **Glochidion lucidum** Blume

Phyllanthus lutescens (Blume) Müll.Arg. === **Glochidion lutescens** Blume

Phyllanthus luzoniensis Merr. === **Margaritaria luzoniensis** (Merr.) Airy Shaw

Phyllanthus lycioides Kunth === **Phyllanthus acuminatus** Vahl

Phyllanthus macahensis Glaz. === **?**

Phyllanthus macrocarpus (Blume) Müll.Arg. === **Glochidion macrocarpum** Blume

Phyllanthus macrophyllus (Labill.) Müll.Arg. === **Glochidion macphersonii** Govaerts & Radcl.-Sm.

Phyllanthus macropus Hook.f. === **Meineckia macropus** (Hook.f.) G.L.Webster

Phyllanthus macrostachyus Müll.Arg. === **Picramnia antidesma** Sw. (Picramniaceae)

Phyllanthus maderaspatensis Thouars ex Baill. === **Phyllanthus nummulariifolius** Poir. subsp. **nummulariifolius**

Phyllanthus maderaspatensis Forssk. === **Phyllanthus tenellus** var. **arabicus** Müll.Arg.

Phyllanthus maderaspatensis f. *fastigiatus* Leandri === **Phyllanthus maderaspatensis** L. var. **maderaspatensis**

Phyllanthus maderaspatensis f. *thermarum* Leandri === **Phyllanthus maderaspatensis** L. var. **maderaspatensis**

Phyllanthus maderaspatensis var. *thonningii* (Schumach. & Thonn.) Müll.Arg. === **Phyllanthus maderaspatensis** L. var. **maderaspatensis**

Phyllanthus mairei H.Lév. === **Phyllanthus emblica** L.

Phyllanthus maitlandianus Diels === **Phyllanthus scaber** Klotzsch

Phyllanthus malabaricus Müll.Arg. === **Glochidion ellipticum** Wight

Phyllanthus manono (Baill. ex Müll.Arg.) Müll.Arg. === **Glochidion manono** Baill. ex Müll.Arg.

Phyllanthus mareensis Guillaumin === **Phyllanthus ouveanus** Däniker

Phyllanthus marginatus B.Heyne ex Wall. === **Phyllanthus virgatus** G.Forst. var. **virgatus**

Phyllanthus marginivillosus Speg. === **Parodiodendron marginivillosum** (Speg.) Hunz.

Phyllanthus matsumurae Hayata ex Fabe === **Phyllanthus ussuriensis** Rupr. & Maxim.

Phyllanthus maytenifolius S.Moore === **Phyllanthus aeneus** Baill. var. **aeneus**

Phyllanthus melanodiscus Urb. === **Phyllanthus myrtilloides** subsp. **erythrinus** (Müll.Arg.) G.L.Webster

Phyllanthus melleri var. *campenonii* Leandri === **Phyllanthus melleri** Müll.Arg.

Phyllanthus melleri subsp. *moramangicus* Leandri === **Phyllanthus moramangicus** (Leandri) Leandri

Phyllanthus meruensis Pax === **Phyllanthus sepialis** Müll.Arg.

Phyllanthus micrantheoides Baill. === **Phyllanthus chrysanthus** var. **micrantheoides** (Baill.) M.Schmid

Phyllanthus micranthus subsp. *fuertesii* (Urb.) G.L.Webster === **Phyllanthus fuertesii** Urb.

Phyllanthus microcarpoides Croizat === **Phyllanthus rupestris** Kunth

Phyllanthus microcarpus (Benth.) Müll.Arg. === **Phyllanthus reticulatus** Poir. var. **reticulatus**

Phyllanthus microcladus var. *microphyllus* Müll.Arg. === **Phyllanthus microcladus** Müll.Arg.

Phyllanthus microcladus var. *puberulus* Müll.Arg. === **Phyllanthus microcladus** Müll.Arg.

Phyllanthus microphyllus Klotzsch === **Phyllanthus stipulatus** (Raf.) G.L.Webster

Phyllanthus microphyllus Mart. === **Phyllanthus niruri** subsp. **lathyroides** (Kunth) G.L.Webster

Phyllanthus mimosifolius Salisb. === **Phyllanthus emblica** L.

Phyllanthus mimosoides Lodd. === **Phyllanthus niruri** subsp. **lathyroides** (Kunth) G.L.Webster

Phyllanthus mimosoides lusus *macrophyllus* Müll.Arg. === **Phyllanthus megapodus** G.L.Webster

Phyllanthus minensis Glaz. === ?

Phyllanthus minimus C.Wright === **Phyllanthus echinospermus** C.Wright

Phyllanthus minor Fawc. & Rendle === **Phyllanthus tenellus** Roxb. var. **tenellus**

Phyllanthus minusculus Glaz. === ?

Phyllanthus minutiflorus F.Muell. & Tate === **Sauropus trachyspermus** (F.Muell.) Airy Shaw

Phyllanthus minutiflorus F.Muell. ex Müll.Arg. === **Phyllanthus virgatus** G.Forst. var. **virgatus**

Phyllanthus minutiflorus var. *gracillimus* (F.Muell. ex Benth.) Benth. === **Phyllanthus virgatus** G.Forst. var. **virgatus**

Phyllanthus miquelianus Müll.Arg. === **Phyllanthus virgatus** G.Forst. var. **virgatus**

Phyllanthus moeroris Oken === **Phyllanthus niruri** L. subsp. **niruri**

Phyllanthus mollis (Blume) Müll.Arg. === **Glochidion molle** Blume

Phyllanthus moluccanus (Blume) Müll.Arg. === **Glochidion moluccanum** Blume

Phyllanthus monocladus Urb. === **Phyllanthus hyssopifolioides** Kunth

Phyllanthus monticola Leandri === **Glochidion oreichtitum** (Leandri) Leandri

Phyllanthus monticola Hutch. & Dalziel === **Phyllanthus alpestris** Beille

Phyllanthus moonii (Thwaites) Müll.Arg. === **Glochidion moonii** Thwaites

Phyllanthus mucronatus Kunth === **Phyllanthus acuminatus** Vahl

Phyllanthus multicaulis Müll.Arg. === **Phyllanthus incurvus** Thunb.

Phyllanthus multiflorus Willd. === **Phyllanthus reticulatus** Poir. var. **reticulatus**

Phyllanthus multilocularis (Rottler ex Willd.) Müll.Arg. === **Glochidion multiloculare** (Rottler ex Willd.) Voigt

Phyllanthus muricatus Wall. === **Phyllanthus urinaria** L. subsp. **urinaria**

Phyllanthus myrianthus Müll.Arg. === **Glochidion myrianthum** (Müll.Arg.) Pax & K.Hoffm.

Phyllanthus myrtilloides Chiov. === **Phyllanthus suffrutescens** Pax

Phyllanthus myrtilloides var. *spathulifolius* (Griseb.) M.Gómez === **Phyllanthus myrtilloides** subsp. **spathulifolius** (Griseb.) G.L.Webster

Phyllanthus myrtillus Wall. === **Sauropus quadrangularis** (Willd.) Müll.Arg.

Phyllanthus nanogynus Müll.Arg. === **Glochidion glomerulatum** (Miq.) Boerl.

Phyllanthus nanus Hook.f. === **Phyllanthus amarus** Schumach. & Thonn.

Phyllanthus nanus Millsp. ex Britton === **Phyllanthus imbricatus** G.L.Webster

Phyllanthus naviluri Miq. ex Müll.Arg. === **Breynia retusa** (Dennst.) Alston

Phyllanthus neilgherrensis (Wight) Müll.Arg. === **Glochidion candolleanum** (Wight & Arn.) Chakrab. & M.G.Gangop.

Phyllanthus nemoralis (Thwaites) Müll.Arg. === **Glochidion nemorale** Thwaites

Phyllanthus neocaledonicus Müll.Arg. ex Guillaumin === **Phyllanthus bourgeoisii** Baill.

Phyllanthus neogranatensis Müll.Arg. === **Meineckia neogranatensis** (Müll.Arg.) G.L.Webster

Phyllanthus neopeltandrus Griseb. === **Chascotheca neopeltandra** (Griseb.) Urb.

Phyllanthus nepalensis Müll.Arg. === **Glochidion heyneanum** (Wight & Arn.) Wight

Phyllanthus nephradenius Müll.Arg. === **Phyllanthus heyneanus** Müll.Arg.

Phyllanthus nervosus Airy Shaw === **Phyllanthus flaviflorus** (Lauterb. & K.Schum.) Airy Shaw

Phyllanthus neurocarpus Müll.Arg. === **Astrocasia neurocarpa** (Müll.Arg.) I.M.Johnst. ex Standl.

Phyllanthus ngoyensis Schltr. === **Phyllanthus bupleuroides** var. **ngoyensis** (Schltr.) M.Schmid

Phyllanthus niruri var. *amarus* (Schumach. & Thonn.) Leandri === **Phyllanthus amarus** Schumach. & Thonn.

Phyllanthus niruri var. *baronianus* (Leandri) Leandri === **Phyllanthus amarus** Schumach. & Thonn.

Phyllanthus niruri var. *debilis* (Klein ex Willd.) Müll.Arg. === **Phyllanthus debilis** Klein ex Willd.

Phyllanthus niruri var. *genuinus* Müll.Arg. === **Phyllanthus niruri** L.

Phyllanthus niruri var. *javanicus* Müll.Arg. === **Phyllanthus debilis** Klein ex Willd.

Phyllanthus niruri var. *radicans* Müll.Arg. === **Phyllanthus pentaphyllus** C.Wright ex Griseb. subsp. **pentaphyllus**

Phyllanthus niruri var. *scabrellus* (Webb) Müll.Arg. === **Phyllanthus amarus** Schumach. & Thonn.

Phyllanthus niruroides var. *madagascariensis* Leandri === **Phyllanthus amarus** Schumach. & Thonn.

Phyllanthus nitidus Müll.Arg. === **Eurya japonica** Thunb. (Theaceae)

Phyllanthus nitidus (Roxb.) Reinw. ex Blume === **Glochidion zeylanicum** (Gaertn.) A.Juss. var. **zeylanicum**

Phyllanthus nivosus W.Bull === **Breynia disticha** J.R.Forst. & G.Forst.

Phyllanthus nobilis (L.f.) Müll.Arg. === **Margaritaria nobilis** L.f.

Phyllanthus nobilis var. *antillanus* (A.Juss.) Müll.Arg. === **Margaritaria nobilis** L.f.

Phyllanthus nobilis var. *brasiliensis* Müll.Arg. === **Margaritaria nobilis** L.f.

Phyllanthus nobilis var. *guyanensis* Müll.Arg. === **Margaritaria nobilis** L.f.

Phyllanthus nobilis var. *hypomalacus* Standl. === **Margaritaria nobilis** L.f.

Phyllanthus nobilis var. *martii* Müll.Arg. === **Margaritaria nobilis** L.f.

Phyllanthus nobilis var. *panamensis* Müll.Arg. === **Margaritaria nobilis** L.f.

Phyllanthus nobilis var. *pavonianus* (Baill.) Müll.Arg. === **Margaritaria nobilis** L.f.

Phyllanthus nobilis var. *peruvianus* Müll.Arg. === **Margaritaria nobilis** L.f.

Phyllanthus nobilis var. *riedelianus* Müll.Arg. === **Margaritaria nobilis** L.f.

Phyllanthus norlindii Urb. === **Phyllanthus phlebocarpus** Urb.

Phyllanthus novaehollandiae Baill. === **Phyllanthus ciccoides** Müll.Arg. var. **ciccoides**

Phyllanthus nozeranii Rossignol & Haicour === **Phyllanthus urinaria** L. subsp. **urinaria**

Phyllanthus nutans var. *purdiaeana* Baill. === **Phyllanthus nutans** Sw. subsp. **nutans**

Phyllanthus nutans var. *trojanus* G.L.Webster === **Phyllanthus nutans** Sw. subsp. **nutans**

Phyllanthus nyassae Pax & K.Hoffm. === **Phyllanthus beillei** Hutch.

Phyllanthus obcordatus Willd. === **Phyllanthus maderaspatensis** L. var. **maderaspatensis**

Phyllanthus obliquus Wall. ex Hook.f. === **Phyllanthus wightianus** Müll.Arg.

Phyllanthus obliquus (Willd.) Müll.Arg. === **Glochidion zeylanicum** (Gaertn.) A.Juss. var. **zeylanicum**

Phyllanthus oblongifolius Pax === **Phyllanthus reticulatus** Poir. var. **reticulatus**

Phyllanthus obovatus (Siebold & Zucc.) Müll.Arg. === **Glochidion obovatum** Siebold & Zucc.

Phyllanthus obovatus Muhl. ex Willd. === **Phyllanthus caroliniensis** Walter subsp. **caroliniensis**

Phyllanthus obscurus Roxb. ex Willd. === **Glochidion obscurum** (Roxb. ex Willd.) Blume

Phyllanthus obtusus Schrank === **Flueggea virosa** (Roxb. ex Willd.) Voigt subsp. **virosa**

Phyllanthus ochrophyllus Benth. === **Sauropus ochrophyllus** (Benth.) Airy Shaw

Phyllanthus odontadenius var. *braunii* (Pax) Hutch. === **Phyllanthus odontadenius** Müll.Arg.

Phyllanthus oligotrichus Müll.Arg. === **Glochidion oligotrichum** (Müll.Arg.) Boerl.

Phyllanthus olivaceus Müll.Arg. ex Guillaumin === **Phyllanthus aeneus** Baill. var. **aeneus**

Phyllanthus orbicularis var. *lignescens* Müll.Arg. === **Phyllanthus orbiculatus** Rich.

Phyllanthus orbiculatus var. *acutifolius* Müll.Arg. === **Phyllanthus orbiculatus** Rich.

Phyllanthus orbiculatus var. *genuinus* Müll.Arg. === **Phyllanthus orbiculatus** Rich.

Phyllanthus orbiculatus var. *intermedius* Müll.Arg. === **Phyllanthus orbiculatus** Rich.

Phyllanthus oreadum S.Moore === **Phyllanthus effusus** S.Moore

Phyllanthus oreichtitus Leandri === **Glochidion oreichtitum** (Leandri) Leandri

Phyllanthus ornatus (Kurz ex Teijsm. & Binn.) Müll.Arg. === **Glochidion ornatum** Kurz ex Teijsm. & Binn.

Phyllanthus ovalis E.Mey. ex Sond. === **Andrachne ovalis** (E.Mey. ex Sond.) Müll.Arg.

Phyllanthus ovatus Desf. ex Poir. === **Phyllanthus hortensis** Govaerts & Radcl.-Sm.

Phyllanthus oxycladus Müll.Arg. === **Phyllanthus pavonianus** Baill.

Phyllanthus oxyphyllus Müll.Arg. === **Phyllanthus acutifolius** Poir. ex Spreng.

Phyllanthus pacificus var. *quaylei* F.Br. === **Phyllanthus pacificus** Müll.Arg.

Phyllanthus pacificus var. *uahukensis* F.Br. === **Phyllanthus pacificus** Müll.Arg.

Phyllanthus pacificus var. *uapensis* F.Br. === **Phyllanthus pacificus** Müll.Arg.

Phyllanthus pallidus Müll.Arg. === **Phyllanthus pulcher** Wall. ex Müll.Arg.

Phyllanthus pancherianus var. *castus* (S.Moore) Guillaumin === **Phyllanthus castus** S.Moore

Phyllanthus paniculatus Oliv. === **Phyllanthus clamboides** (F.Muell.) Diels

Phyllanthus parahybensis Glaz. === ?

Phyllanthus parvifolius Steud. === **Phyllanthus niruri** subsp. **lathyroides** (Kunth) G.L.Webster

Phyllanthus patens Miq. ex Müll.Arg. === **Phyllanthus virgatus** G.Forst. var. **virgatus**

Phyllanthus patens Roxb. === **Breynia retusa** (Dennst.) Alston

Phyllanthus peduncularis Boivin ex Baill. === **Phyllanthus nummulariifolius** Poir. subsp. **nummulariifolius**

Phyllanthus pedunculatus Kostel. === **Phyllanthus virgatus** G.Forst. var. **virgatus**

Phyllanthus pedunculatus Warb. === **Glochidion** sp.

Phyllanthus pelas (K.Schum.) Pax & K.Hoffm. === **Phyllanthus ciccoides** Müll.Arg. var. **ciccoides**

Phyllanthus peltandrus Müll.Arg. === **Meineckia parvifolia** (Wight) G.L.Webster

Phyllanthus penangensis Müll.Arg. === **Glochidion rubrum** Blume var. **rubrum**

Phyllanthus pentandrus Roxb. ex Thwaites === **Phyllanthus reticulatus** Poir. var. **reticulatus**

Phyllanthus perlisensis Ridl. === **Phyllanthus sikkimensis** Müll.Arg.

Phyllanthus perpusillus Standl. === **Phyllanthus standleyi** McVaugh

Phyllanthus perrottetianus Müll.Arg. === **Glochidion candolleanum** (Wight & Arn.) Chakrab. & M.G.Gangop.

Phyllanthus persimilis Müll.Arg. === **Phyllanthus chrysanthus** Baill. var. **chrysanthus**

Phyllanthus pervilleanus var. *glaber* Müll.Arg. === **Phyllanthus pervilleanus** (Baill.) Müll.Arg.

Phyllanthus philippinensis Müll.Arg. === **Glochidion philippicum** (Cav.) C.B.Rob.

Phyllanthus phillyreifolius var. *angustifolius* Müll.Arg. === **Phyllanthus phillyreifolius** Poir. var. **phillyreifolius**

Phyllanthus phillyreifolius var. *ellipticus* Müll.Arg. === **Phyllanthus phillyreifolius** Poir. var. **phillyreifolius**

Phyllanthus phillyreifolius var. *genuinus* Müll.Arg. === **Phyllanthus phillyreifolius** Poir.

Phyllanthus phillyreifolius var. *lanceolatus* Müll.Arg. === **Phyllanthus phillyreifolius** Poir. var. **phillyreifolius**

Phyllanthus phillyreifolius var. *longifolius* Müll.Arg. === **Phyllanthus phillyreifolius** var. **commersonii** Müll.Arg.

Phyllanthus phillyreifolius var. *oblongifolius* Müll.Arg. === **Phyllanthus phillyreifolius** var. **telfairianus** Wall. ex Müll.Arg.

Phyllanthus phillyreifolius var. *parvifolius* Müll.Arg. === **Phyllanthus phillyreifolius** Poir. var. **phillyreifolius**

Phyllanthus phillyreifolius var. *rotundifolius* Müll.Arg. === **Phyllanthus phillyreifolius** var. **telfairianus** Wall. ex Müll.Arg.

Phyllanthus pilosus (Lour.) Müll.Arg. === **Glochidion pilosum** (Lour.) Merr.

Phyllanthus piluliferus Fenzl === **Phyllanthus pentandrus** Schumach. & Thonn.

Phyllanthus pimeleoides A.DC. === **Phyllanthus calycinus** Labill.

Phyllanthus pinosius Urb. === **Phyllanthus selbyi** Britton & P.Wilson

Phyllanthus pittieri Pax === **Phyllanthus mocinianus** Baill.

Phyllanthus placenta Hassk. === **Sauropus rhamnoides** Blume

Phyllanthus placentatus Noronha === **Sauropus rhamnoides** Blume

Phyllanthus platylepis Small === **Phyllanthus liebmannianus** subsp. **platylepis** (Small) G.L.Webster

Phyllanthus podenzanae S.Moore === **Sauropus podenzanae** (S.Moore) Airy Shaw

Phyllanthus podocarpus Müll.Arg. === **Glochidion podocarpum** (Müll.Arg.) C.B.Rob.

Phyllanthus poilanei Beille === **?**

Phyllanthus poiretianus Müll.Arg. === **Phyllanthus orbiculatus** Rich.

Phyllanthus polycarpus Müll.Arg. === **Glochidion borneense** (Müll.Arg.) Boerl.

Phyllanthus polycladus W.Fitzg. === **Phyllanthus aridus** Benth.

Phyllanthus polycladus Urb. === **Phyllanthus pentaphyllus** subsp. **polycladus** (Urb.) G.L.Webster

Phyllanthus polycladus var. *curassavicus* Urb. === **Phyllanthus pentaphyllus** C.Wright ex Griseb. subsp. **pentaphyllus**

Phyllanthus polycladus var. *guadeloupensis* Urb. === **Phyllanthus pentaphyllus** subsp. **polycladus** (Urb.) G.L.Webster

Phyllanthus polygamus Hochst. ex A.Rich. === **Flueggea virosa** (Roxb. ex Willd.) Voigt subsp. **virosa**

Phyllanthus polymorphus Müll.Arg. === **Phyllanthus phillyreifolius** Poir.

Phyllanthus polyphyllus Dalzell & Gibson === **Phyllanthus lawii** J.Graham

Phyllanthus polyspermus Schumach. & Thonn. === **Phyllanthus reticulatus** var. **glaber** (Thwaites) Müll.Arg.

Phyllanthus pomaceus Moon === **Breynia retusa** (Dennst.) Alston

Phyllanthus portoricensis (Kuntze) Urb. === **Flueggea virosa** (Roxb. ex Willd.) Voigt subsp. **virosa**

Phyllanthus praetervisus Müll.Arg. === **Phyllanthus parvifolius** Buch.-Ham. ex D.Don

Phyllanthus pratensis Pancher ex Baill. === **Phyllanthus virgatus** G.Forst. var. **virgatus**

Phyllanthus preissianus Klotzsch === **Phyllanthus calycinus** Labill.

Phyllanthus prieurianus (Baill.) Müll.Arg. === **Phyllanthus reticulatus** Poir. var. **reticulatus**

Phyllanthus prieurianus var. *glaber* (Baill.) Müll.Arg. === **Phyllanthus reticulatus** var. **glaber** (Thwaites) Müll.Arg.

Phyllanthus pringlei S.Watson === **Phyllanthus micrandrus** Müll.Arg.

Phyllanthus protoguayanensis Herter === **Meborea guianensis** Aubl.

Phyllanthus pruinosus Poepp. ex A.Rich. === **Phyllanthus discolor** Poepp. ex Spreng.

Phyllanthus pruinosus var. *subnudus* C.Wright ex Griseb. === **Phyllanthus junceus** Müll.Arg.

Phyllanthus pseudoconami var. *pubescens* Müll.Arg. === **Phyllanthus pseudoconami** Müll.Arg.

Phyllanthus pseudoreticulatus Pax & K.Hoffm. === **Phyllanthus mannianus** Müll.Arg.

Phyllanthus puberulus Miq. ex Baill. === **Phyllanthus reticulatus** Poir. var. **reticulatus**

Phyllanthus puberus (L.) Müll.Arg. === **Glochidion puberum** (L.) Hutch.

Phyllanthus pubescens Klotzsch === **Sauropus villosus** (Blanco) Merr.

Phyllanthus pubescens Moon === **Glochidion moonii** Thwaites var. **moonii**

Phyllanthus pubigerus A.Rich. === **Savia sessiliflora** (Sw.) Willd.

Phyllanthus pulchellus Endl. === **Phyllanthus calycinus** Labill.

Phyllanthus punctulatus Urb. === **Phyllanthus pseudocicca** Griseb.

Phyllanthus purpurascens Kunth === **Phyllanthus niruri** subsp. **lathyroides** (Kunth) G.L.Webster

Phyllanthus purpureus C.Wright ex Griseb. === **Phyllanthus myrtilloides** subsp. **erythrinus** (Müll.Arg.) G.L.Webster

Phyllanthus pusillifolius S.Moore === **Phyllanthus microcladus** Müll.Arg.

Phyllanthus pycnocarpus Müll.Arg. === **Glochidion candolleanum** (Wight & Arn.) Chakrab. & M.G.Gangop.

Phyllanthus pynaertii De Wild. === **Phyllanthus polyanthus** Pax

Phyllanthus quadrangularis Willd. === **Sauropus quadrangularis** (Willd.) Müll.Arg.

Phyllanthus quangtriensis Beille === **Phyllanthus urinaria** L. subsp. **urinaria**

Phyllanthus quercinus Müll.Arg. === **Glochidion philippicum** (Cav.) C.B.Rob.

Phyllanthus quinquefidus Sessé & Moç. ex Baill. === **Phyllanthus juglandifolius** Willd. subsp. **juglandifolius**

Phyllanthus racemifer Steud. === **Breynia racemosa** (Blume) Müll.Arg. var. **racemosa**

Phyllanthus racemosus L.f. === **Sauropus bacciformis** (L.) Airy Shaw

Phyllanthus radicans (Müll.Arg.) Small === **Phyllanthus pentaphyllus** C.Wright ex Griseb. subsp. **pentaphyllus**

Phyllanthus ramiflorus (J.R.Forst. & G.Forst.) Müll.Arg. === **Glochidion ramiflorum** J.R.Forst. & G.Forst.

Phyllanthus ramiflorus (Aiton) Pers. === **Flueggea suffruticosa** (Pall.) Baill.

Phyllanthus ramiflorus var. *lanceolatus* Müll.Arg. === **Glochidion concolor** Müll.Arg.

Phyllanthus ramosissimus (F.Muell.) Müll.Arg. === **Sauropus ramosissimus** (F.Muell.) Airy Shaw

Phyllanthus reclinatus Roxb. === **Breynia reclinata** (Roxb.) Hook.f.

Phyllanthus reichenbachianus Sieber ex Baill. === **Flueggea virosa** (Roxb. ex Willd.) Voigt subsp. **virosa**

Phyllanthus reinwardtii Müll.Arg. === **Glochidion reinwardtii** (Müll.Arg.) Boerl.

Phyllanthus retusus Dennst. === **Breynia retusa** (Dennst.) Alston

Phyllanthus revolutus E.Mey. ex Sond. === **Phyllanthus myrtaceus** Sond.

Phyllanthus rhamnoides Bojer ex Müll.Arg. === **Breynia racemosa** (Blume) Müll.Arg. var. **racemosa**

Phyllanthus rhamnoides Retz. === **Breynia vitis-idaea** (Burm.f.) C.E.C.Fisch.

Phyllanthus rhamnoides Roxb. === **Sauropus quadrangularis** (Willd.) Müll.Arg.

Phyllanthus rhomboidalis (Baill.) Müll.Arg. === **Margaritaria rhomboidalis** (Baill.) G.L.Webster

Phyllanthus rhytidospermus F.Muell. ex Müll.Arg. === **Sauropus trachyspermus** (F.Muell.) Airy Shaw

Phyllanthus rigens (F.Muell.) Müll.Arg. === **Sauropus rigens** (F.Muell.) Airy Shaw

Phyllanthus rigidulus F.Muell. ex Müll.Arg. === **Sauropus rigidulus** (F.Muell. ex Müll.Arg.) Airy Shaw

Phyllanthus rigidus Tate === **Phyllanthus tatei** F.Muell.

Phyllanthus ringoetii De Wild. === **Phyllanthus moeroensis** De Wild.

Phyllanthus roemerianus Scheele === **Andrachne phyllanthoides** (Nutt.) Müll.Arg.

Phyllanthus roeperianus Wall. ex Müll.Arg. === **Phyllanthus cochinchinensis** Spreng.

Phyllanthus rogersii Hutch. ex S.Moore === **Phyllanthus graminicola** Hutch.

Phyllanthus rosellus Müll.Arg. === **Phyllanthus niruri** subsp. **lathyroides** (Kunth) G.L.Webster

Phyllanthus roseus var. *glabrum* (Craib ex Hosseus) Craib ex Beille === **Phyllanthus roseus** (Craib & Hutch.) Beille

Phyllanthus rotundifolius Sessé & Moç. === **Phyllanthus orbicularis** Kunth

Phyllanthus roxburghii Müll.Arg. === **Phyllanthus tetrandrus** Roxb.

Phyllanthus rubellus Müll.Arg. === **Flueggea elliptica** (Spreng.) Baill.

Phyllanthus rubens Bojer ex Baker === **Phyllanthus urinaria** L. subsp. **urinaria**

Phyllanthus ruber Noronha === **Breynia cernua** (Poir.) Müll.Arg.

Phyllanthus rubriflorus Beille === **Phyllanthus rubristipulus** Govaerts & Radcl.-Sm.

Phyllanthus rufidulus Müll.Arg. === **Phyllanthus chrysanthus** var. **micrantheoides** (Baill.) M.Schmid

Phyllanthus rufidulus var. *kafeateensis* Guillaumin === **Phyllanthus chrysanthus** var. **micrantheoides** (Baill.) M.Schmid

Phyllanthus rufoglaucus Müll.Arg. === **Glochidion rufoglaucum** (Müll.Arg.) Boerl.

Phyllanthus rupestris var. *oblongifolius* Müll.Arg. === **Phyllanthus ruprestris**

Phyllanthus ruscoides Kunth === **Phyllanthus acuminatus** Vahl

Phyllanthus sacleuxii Radcl.-Sm. === **Phyllanthus mittenianus** Hutch.

Phyllanthus × *sagraeanus* Urb. === **Phyllanthus** × **pallidus** C.Wright ex Griseb.

Phyllanthus salacioides S.Moore === **Phyllanthus aeneus** Baill. var. **aeneus**

Phyllanthus salicifolius var. *dracunculoides* (Baill.) Müll.Arg. === **Phyllanthus dracunculoides** Baill.

Phyllanthus salicifolius var. *genuinus* Müll.Arg. === **Phyllanthus salicifolius** Baill.

Phyllanthus salviifolius var. *floribundus* (Kunth) Müll.Arg. === **Phyllanthus salviifolius** Kunth

Phyllanthus salviifolius var. *genuinus* Müll.Arg. === **Phyllanthus salviifolius** Kunth

Phyllanthus salviifolius var. *glabrescens* Müll.Arg. === **Phyllanthus salviifolius** Kunth

Phyllanthus sambiranensis Leandri === **Glochidion sambiranense** (Leandri) Leandri

Phyllanthus sandwicensis Müll.Arg. === **Phyllanthus distichus** Hook. & Arn. var. **distichus**

Phyllanthus sandwicensis var. *degeneri* Sherff === **Phyllanthus distichus** var. **degeneri** (Sherff) Govaerts & Radcl.-Sm.

Phyllanthus sandwicensis var. *ellipticus* Müll.Arg. === **Phyllanthus distichus** var. **ellipticus** (Müll.Arg.) Govaerts & Radcl.-Sm.

Phyllanthus sandwicensis f. *grandifolia* Wawra === **Phyllanthus distichus** Hook. & Arn. var. **distichus**

Phyllanthus sandwicensis var. *hypoglaucus* H.Lév. === **Breynia disticha** J.R.Forst. & G.Forst.

Phyllanthus sandwicensis var. *oblongifolius* Müll.Arg. === **Phyllanthus distichus** Hook. & Arn. var. **distichus**

Phyllanthus sandwicensis f. *parvifolius* (Müll.Arg.) Wawra === **Phyllanthus distichus** var. **ellipticus** (Müll.Arg.) Govaerts & Radcl.-Sm.

Phyllanthus sandwicensis var. *parvifolius* Müll.Arg. === **Phyllanthus distichus** var. **ellipticus** (Müll.Arg.) Govaerts & Radcl.-Sm.

Phyllanthus sandwicensis var. *radicans* Müll.Arg. === **Phyllanthus distichus** var. **ellipticus** (Müll.Arg.) Govaerts & Radcl.-Sm.

Phyllanthus sandwicensis f. *rufidus* Fosberg === **Phyllanthus distichus** var. **ellipticus** (Müll.Arg.) Govaerts & Radcl.-Sm.

Phyllanthus santhomensis Beille === **Phyllanthus odontadenius** Müll.Arg.

Phyllanthus saxicola Small === **Phyllanthus caroliniensis** subsp. **saxicola** (Small) G.L.Webster

Phyllanthus scaber var. *angustifolius* Müll.Arg. === **Phyllanthus scaber** Klotzsch

Phyllanthus scaber var. *pallidiflorus* Müll.Arg. === **Phyllanthus scaber** Klotzsch

Phyllanthus scabrellus Webb === **Phyllanthus amarus** Schumach. & Thonn.

Phyllanthus scabrifolius Hook.f. === **Phyllanthus rotundifolius** Klein ex Willd.

Phyllanthus scandens (C.Wright ex Griseb.) Müll.Arg. === **Margaritaria scandens** (Wright ex Griseb.) G.L.Webster

Phyllanthus scandens Roxb. ex Dillwyn === **Phyllanthus reticulatus** Poir. var. **reticulatus**

Phyllanthus schimperianus Hemsl. === **Phyllanthus casticum** P.Willemet. var. **casticum**

Phyllanthus schomburgkianus var. *acuminatus* (Müll.Arg.) Müll.Arg. === **Phyllanthus caroliniensis** subsp. **guianensis** (Klotzsch) G.L.Webster

Phyllanthus schomburgkianus var. *anceps* (Müll.Arg.) Müll.Arg. === **Phyllanthus caroliniensis** subsp. **guianensis** (Klotzsch) G.L.Webster

Phyllanthus schomburgkianus var. *antillanus* (Müll.Arg.) Müll.Arg. === **Phyllanthus caroliniensis** subsp. **guianensis** (Klotzsch) G.L.Webster

Phyllanthus schomburgkianus var. *guianensis* (Klotzsch) Müll.Arg. === **Phyllanthus caroliniensis** subsp. **guianensis** (Klotzsch) G.L.Webster

Phyllanthus schumannianus L.S.Sm. === **Phyllanthus clamboides** (F.Muell.) Diels

Phyllanthus sciadiostylus Airy Shaw === **Phyllanthus pachyphyllus** Müll.Arg.

Phyllanthus scoparius Welw. === **Phyllanthus pentandrus** Schumach. & Thonn.

Phyllanthus scoparius Müll.Arg. === **Phyllanthus sp.**

Phyllanthus secundiflora Ridl. === **Phyllanthus sikkimensis** Müll.Arg.

Phyllanthus seemannianus (Müll.Arg.) Müll.Arg. === **Glochidion seemannii** Müll.Arg.

Phyllanthus semicordatus Müll.Arg. === **Glochidion semicordatum** (Müll.Arg.) Boerl.

Phyllanthus senaei Glaz. === ?

Phyllanthus senensis Müll.Arg. === **Phyllanthus pinnatus** (Wight) G.L.Webster

Phyllanthus sepiarius Roxb. ex Wall. === **Breynia vitis-idaea** (Burm.f.) C.E.C.Fisch.

Phyllanthus sericeus (Blume) Müll.Arg. === **Glochidion sericeum** (Blume) Zoll. & Moritzi

Phyllanthus serissifolius Wall. ex Müll.Arg. === **Phyllanthus cochinchinensis** Spreng.

Phyllanthus shaferi Urb. === **Phyllanthus myrtilloides** subsp. **shaferi** (Urb.) G.L.Webster

Phyllanthus silheticus Müll.Arg. === **Glochidion zeylanicum** var. **arborescens** (Blume) Chakrab. & M.G.Gangop.

Phyllanthus simplex Retz. === **Phyllanthus virgatus** G.Forst. var. **virgatus**

Phyllanthus simplex var. *brevipes* Müll.Arg. === **Phyllanthus virgatus** G.Forst. var. **virgatus**

Phyllanthus simplex var. *chinensis* Müll.Arg. === **Phyllanthus ussuriensis** Rupr. & Maxim.

Phyllanthus simplex var. *filicaulis* (Benth.) Domin === **Phyllanthus virgatus** G.Forst. var. **virgatus**

Phyllanthus simplex var. *gardnerianus* (Wight) Müll.Arg. === **Phyllanthus virgatus** var. **gardnerianus** (Wight) Govaerts & Radcl.-Sm.

Phyllanthus simplex var. *genuinus* Müll.Arg. === **Phyllanthus virgatus** G.Forst. var. **virgatus**

Phyllanthus simplex var. *gracillimus* Domin === **Phyllanthus virgatus** G.Forst. var. **virgatus**

Phyllanthus simplex var. *leiospermus* Benth. === **Phyllanthus virgatus** G.Forst. var. **virgatus**

Phyllanthus simplex var. *minutiflorus* (F.Muell. ex Müll.Arg.) Domin === **Phyllanthus virgatus** G.Forst. var. **virgatus**

Phyllanthus simplex var. *myriocladus* Müll.Arg. === **Phyllanthus virgatus** G.Forst. var. **virgatus**

Phyllanthus simplex var. *myrtifolius* Domin === **Phyllanthus virgatus** G.Forst. var. **virgatus**

Phyllanthus simplex var. *pinifolius* Domin === **Phyllanthus virgatus** G.Forst. var. **virgatus**

Phyllanthus simplex var. *pratensis* Müll.Arg. === **Phyllanthus virgatus** G.Forst. var. **virgatus**

Phyllanthus simplex f. *pubescens* (Thwaites) Müll.Arg. === **Phyllanthus wheeleri** G.L.Webster

Phyllanthus simplex var. *tonkinensis* Beille === **Phyllanthus clarkei** Hook.f.

Phyllanthus simplex var. *ussuriensis* (Rupr. & Maxim.) Müll.Arg. === **Phyllanthus ussuriensis** Rupr. & Maxim.

Phyllanthus simplex var. *virgatus* (G.Forst.) Müll.Arg. === **Phyllanthus virgatus** G.Forst.

Phyllanthus simsianus Wall. === **Breynia fruticosa** (L.) Hook.f.

Phyllanthus sinensis Müll.Arg. === **Phyllanthus reticulatus** Poir. var. **reticulatus**

Phyllanthus sinicus Müll.Arg. === **Margaritaria nobilis** L.f.

Phyllanthus smilacifolius Griseb. === ?

Phyllanthus spathulifolius Griseb. === **Phyllanthus myrtilloides** subsp. **spathulifolius** (Griseb.) G.L.Webster

Phyllanthus speciosus Jacq. === **Phyllanthus arbuscula** (Sw.) J.F.Gmel.

Phyllanthus speciosus Noronha === **Sauropus androgynus** (L.) Merr.

Phyllanthus spectabilis Hügel ex Vis. === **Phyllanthus bicolor** Vis.

Phyllanthus sphaerogynus Müll.Arg. === **Glochidion sphaerogynum** (Müll.Arg.) Kurz

Phyllanthus spinescens Wall. === **Phyllanthus reticulatus** Poir. var. **reticulatus**

Phyllanthus spinulosus B.Heyne ex Wall. === **Phyllanthus lawii** J.Graham

Phyllanthus squamatus C.Wright ex Sauvalle === **Phyllanthus junceus** Müll.Arg.

Phyllanthus stellatus Retz. === **Glochidion stellatum** (Retz.) Bedd.

Phyllanthus stenocladus Müll.Arg. === **Sauropus stenocladus** (Müll.Arg.) J.T.Hunter & J.J.Bruhl

Phyllanthus stenopterus Müll.Arg. === **Phyllanthus caroliniensis** subsp. **stenopterus** (Müll.Arg.) G.L.Webster

Phyllanthus stipulaceus (Gamble) Kumari & Chandrab. === **Phyllanthus chandrabosei** Govaerts & Radcl.-Sm.

Phyllanthus stipulaceus Bojer === **Phyllanthus maderaspatensis** L. var. **maderaspatensis**

Phyllanthus stocksii Müll.Arg. === **Margaritaria indica** (Dalzell) Airy Shaw

Phyllanthus stolzianus Pax & K.Hoffm. === **Phyllanthus beillei** Hutch.

Phyllanthus strictus Roxb. === **Sauropus androgynus** (L.) Merr.

Phyllanthus stuhlmanii Pax === **Phyllanthus nummulariifolius** var. **capillaris** (Schumach. & Thonn.) Radcl.-Sm.

Phyllanthus stylosus Griff. === **Phyllanthus griffithii** Müll.Arg.

Phyllanthus subcordatus Baill. === **Phyllanthus phillyreifolius** var. **telfairianus** Wall. ex Müll.Arg.

Phyllanthus suberosus Wight ex Müll.Arg. === **Meineckia parvifolia** (Wight) G.L.Webster

Phyllanthus sublanatus subsp. *eliae* Brunel & J.P.Roux === **Phyllanthus eliae** (Brunel & J.P.Roux) Brunel ex Govaerts & Radcl.-Sm.

Phyllanthus sublanatus subsp. *eyademae* Brunel === **Phyllanthus eliae** (Brunel & J.P.Roux) Brunel ex Govaerts & Radcl.-Sm.

Phyllanthus submollis K.Schum. & Lauterb. === **Glochidion submolle** (K.Schum. & Lauterb.) Airy Shaw

Phyllanthus submollis var. *glabra* Lauterb. & K.Schum. === **Glochidion novoguineense** K.Schum.

Phyllanthus subpulchellus Croizat === **Phyllanthus sootepensis** Craib

Phyllanthus subscandens (Zoll. & Moritzi) Müll.Arg. === **Glochidion zeylanicum** (Gaertn.) A.Juss. var. **zeylanicum**

Phyllanthus suffultus Wall. === **Breynia retusa** (Dennst.) Alston

Phyllanthus sundaicus (Kurz ex Teijsm. & Binn.) Müll.Arg. === **Margaritaria indica** (Dalzell) Airy Shaw

Phyllanthus superbus (Baill. ex Mull.Arg.) Müll.Arg. === **Glochidion superbum** Baill. ex Müll.Arg.

Phyllanthus swartzii Kostel. === **Phyllanthus amarus** Schumach. & Thonn.

Phyllanthus swartzii Fawc. & Rendle === **Phyllanthus arbuscula** (Sw.) J.F.Gmel.

Phyllanthus sylvaticus Steud. === **Breynia virgata** (Blume) Müll.Arg.

Phyllanthus symplocoides Müll.Arg. === **Glochidion moonii** Thwaites var. **moonii**

Phyllanthus taitensis Hutch. === **Phyllanthus mittenianus** Hutch.

Phyllanthus taitensis (Baill. ex Müll.Arg.) Müll.Arg. === **Glochidion taitense** Baill. ex Müll.Arg.

Phyllanthus takaoensis Hayata === **Phyllanthus reticulatus** Poir. var. **reticulatus**

Phyllanthus taxifolius D.Don === **Phyllanthus emblica** L.

Phyllanthus tenellus Benth. === **Phyllanthus graveolens** Kunth

Phyllanthus tenellus var. *garipensis* Müll.Arg. === **Phyllanthus parvulus** var. **garipensis** (Müll.Arg.) Radcl.-Sm.

Phyllanthus tenellus var. *nossibeensis* Müll.Arg. === **Phyllanthus nummulariifolius** Poir. subsp. **nummulariifolius**

Phyllanthus tenellus var. *roxburghii* Müll.Arg. === **Phyllanthus tenellus** Roxb. var. **tenellus**

Phyllanthus tenuicaulis subsp. *haitiensis* (G.L.Webster) Borhidi === **Phyllanthus tenuicaulis** var. **haitiensis** G.L.Webster

Phyllanthus ternauxii Glaz. === ?

Phyllanthus tetrander Blanco === **Phyllanthus blancoanus** Müll.Arg.

Phyllanthus teysmannii Müll.Arg. === **Glochidion zeylanicum** var. **arborescens** (Blume) Chakrab. & M.G.Gangop.

Phyllanthus thesioides Benth. === **Sauropus thesioides** (Benth.) Airy Shaw

Phyllanthus thesioides Müll.Arg. === **Phyllanthus dumetosus** Poir.

Phyllanthus thomsonii Müll.Arg. === **Glochidion thomsonii** (Müll.Arg.) Hook.f.

Phyllanthus thonningii Schumach. & Thonn. === **Phyllanthus maderaspatensis** L. var. **maderaspatensis**

Phyllanthus thwaitesianus Müll.Arg. === **Meineckia parvifolia** (Wight) G.L.Webster

Phyllanthus thymoides (Müll.Arg.) Sieber ex Müll.Arg. === **Phyllanthus hirtellus** F.Muell. ex Müll.Arg.

Phyllanthus thymoides var. *glabratus* Müll.Arg. === **Phyllanthus hirtellus** F.Muell. ex Müll.Arg.

Phyllanthus thymoides var. *hirtellus* (F.Muell. ex Müll.Arg.) Müll.Arg. === **Phyllanthus hirtellus** F.Muell. ex Müll.Arg.

Phyllanthus thymoides var. *hirtellus* (F.Muell.) Müll.Arg. === **Sauropus hirtellus** (F.Muell.) Airy Shaw

Phyllanthus thymoides var. *ledifolius* (Müll.Arg.) Müll.Arg. === **Phyllanthus hirtellus var. thymoides**

Phyllanthus thymoides var. *parviflorus* J.M.Black === **Phyllanthus hirtellus** F.Muell. ex Müll.Arg.

Phyllanthus timoriensis Müll.Arg. === **Phyllanthus casticum** var. **kirganelia** (Willd.) Govaerts

Phyllanthus tinctorius Vahl ex Baill. === **Breynia vitis-idaea** (Burm.f.) C.E.C.Fisch.

Phyllanthus tomentosus (Dalzell) Müll.Arg. === **Glochidion zeylanicum** var. **talbotii** (Hook.f.) Haines

Phyllanthus tomentosus Noronha === **Glochidion molle** Blume

Phyllanthus toninensis S.Moore === **Phyllanthus gneissicus** var. **toninensis** (S.Moore) M.Schmid

Phyllanthus tonkinensis Gentil === ?

Phyllanthus trachygyne Benth. === **Phyllanthus virgatus** G.Forst. var. **virgatus**

Phyllanthus trachyspermus F.Muell. === **Sauropus trachyspermus** (F.Muell.) Airy Shaw

Phyllanthus tremulus Griseb. === **Astrocasia tremula** (Griseb.) G.L.Webster

Phyllanthus triandrus (Hook.) Druce === **Phyllanthus mitchellii** Benth.

Phyllanthus triandrus (Blanco) Müll.Arg. === **Glochidion triandrum** (Blanco) C.B.Rob.

Phyllanthus trichogynus (Müll.Arg.) Müll.Arg. === **Glochidion trichogynum** Müll.Arg.

Phyllanthus trigonocladus Ohwi === **Flueggea suffruticosa** (Pall.) Baill.

Phyllanthus trigonus Urb. & Ekman === **Phyllanthus leptoneurus** Urb.

Phyllanthus trinervius Wall. === **Sauropus trinervius** Hook.f. & Thomson ex Müll.Arg.

Phyllanthus triquetrus S.Moore === **Phyllanthus dracunculoides** Baill. var. **dracunculoides**

Phyllanthus tristis A.Juss. === **Breynia vitis-idaea** (Burm.f.) C.E.C.Fisch.

Phyllanthus tsiangii P.T.Li === **Phyllanthus ruber** (Lour.) Spreng.

Phyllanthus turbinatus Noronha === **Breynia virgata** (Blume) Müll.Arg.

Phyllanthus turbinatus K.D.Koenig ex Roxb. === **Breynia retusa** (Dennst.) Alston

Phyllanthus turbinatus Sims === **Breynia fruticosa** (L.) Hook.f.

Phyllanthus uberiflorus F.Muell. ex Baill. === **Phyllanthus novaehollandiae** Müll.Arg.

Phyllanthus udicola Mart. ex Colla === ?

Phyllanthus ugandensis Rendle === **Phyllanthus ovalifolius** Forssk.

Phyllanthus urdanetensis Elmer === **Phyllanthus securinegoides** Merr.

Phyllanthus vaccinioides Scheele === **Clutia heterophylla** Thunb.

Phyllanthus vaccinioides Klotzsch === **Phyllanthus maderaspatensis** L. var. **maderaspatensis**

Phyllanthus vakinankaratrae var. *concolor* Leandri === **Phyllanthus vakinankaratrae** Leandri

Phyllanthus valeriae M.Schmid === ?

Phyllanthus varians Müll.Arg. === **Glochidion varians** Miq.

Phyllanthus velutinus (Wight) Müll.Arg. === **Glochidion heyneanum** (Wight & Arn.) Wight

Phyllanthus venosus Hochst. ex A.Rich. === **Phyllanthus maderaspatensis** L. var. **maderaspatensis**

Phyllanthus venulosus Müll.Arg. === **Glochidion venulosum** (Müll.Arg.) P.T.Li

Phyllanthus verrucosus Elmer === **Phyllanthus urinaria** L. subsp. **urinaria**

Phyllanthus verrucosus Thunb. === **Flueggea verrucosa** (Thunb.) G.L.Webster

Phyllanthus vieillardii Baill. === **Phyllanthus chamaecerasus** var. **vieillardii** (Baill.) M.Schmid

Phyllanthus virens (Griseb.) Müll.Arg. === **Margaritaria tetracocca** (Baill.) G.L.Webster

Phyllanthus virgatus var. *chinensis* (Müll.Arg.) G.L.Webster === **Phyllanthus ussuriensis** Rupr. & Maxim.

Phyllanthus virgatus var. *hirtellus* Airy Shaw === **Phyllanthus virgatus** G.Forst. var. **virgatus**

Phyllanthus virgatus var. *minutiflorus* (F.Muell. ex Müll.Arg.) Airy Shaw === **Phyllanthus virgatus** G.Forst. var. **virgatus**

Phyllanthus virgatus var. *oblongifolius* Müll.Arg. === **Phyllanthus virgatus** var. **gardnerianus** (Wight) Govaerts & Radcl.-Sm.

Phyllanthus virgineus (J.F.Gmel.) Pers. === **Phyllanthus casticum** P.Willemet. var. **casticum**

Phyllanthus virosus Roxb. ex Willd. === **Flueggea virosa** (Roxb. ex Willd.) Voigt

Phyllanthus vitiensis Müll.Arg. === **Glochidion vitiense** (Müll.Arg.) Gillespie

Phyllanthus vitis-idea (Burm.f.) D.Koenig ex Roxb. === **Breynia vitis-idaea** (Burm.f.) C.E.C.Fisch.

Phyllanthus wagapensis Müll.Arg. === **Glochidion billardieri** Baill.

Phyllanthus wallichianus (Müll.Arg.) Müll.Arg. === **Glochidion glomerulatum** (Miq.) Boerl.

Phyllanthus weinlandii K.Schum. === **Phyllanthus virgatus** G.Forst. var. **virgatus**

Phyllanthus welwitschianus var. *beillei* (Hutch.) Radcl.-Sm. === **Phyllanthus beillei** Hutch.

Phyllanthus wightianus Müll.Arg. === **Phyllanthus pinnatus** (Wight) G.L.Webster

Phyllanthus wildemannii Beille === **Phyllanthus profusus** N.E.Br.

Phyllanthus wilfordii Croizat & Metcalf === **Phyllanthus ussuriensis** Rupr. & Maxim.

Phyllanthus williamsii Standl. === **Phyllanthus niruri** subsp. **lathyroides** (Kunth) G.L.Webster

Phyllanthus woodii Hutch. === **Phyllanthus meyerianus** Müll.Arg.

Phyllanthus wrightii (Benth.) Müll.Arg. === **Glochidion wrightii** Benth.

Phyllanthus xerocarpus O.Schwarz === **Glochidion xerocarpum** (O.Schwartz) Airy Shaw

Phyllanthus zeylanicus Müll.Arg. === **Phyllanthus anabaptizatus** Müll.Arg.

Phyllanthus zeylanicus (Gaertn.) Müll.Arg. === **Glochidion zeylanicum** (Gaertn.) A.Juss.

Phyllanthus ziziphoides Baill. ex Gibert === **Phyllanthus sellowianus** (Klotzsch) Müll.Arg.

Phyllanthus zollingeri (Miq.) Müll.Arg. === **Phyllanthus pulcher** Wall. ex Müll.Arg.

Phyllanthus zollingeri (Miq.) Müll.Arg. === **Glochidion zollingeri** Miq.

Phyllanthus zygophylloides Müll.Arg. === ?

Phyllaurea

Synonyms:
Phyllaurea Lour. === **Codiaeum** Rumph. ex A.Juss.

Phyllera

An orthographic variant of *Philyra* proposed by Endlicher.

Picrodendron

1 species, West Indies (Greater Antilles (except Puerto Rico), Bahamas and Swan Is.). Many different families have in the past been proposed as the proper place for this genus of deciduous trees with trifoliolate leaves. Fawcett and Rendle in 1917 were the first to suggest Euphorbiaceae; however, Pax and Hoffmann (1931: 232) again excluded it from the family. It was never covered by them for *Das Pflanzenreich*. Hayden (1977), Hayden et al. (1984) and Hakki (1985) in detailed studies support its inclusion, with Hayden further placing it within Oldfieldioideae. Stuppy (1995; see **Phyllanthoideae**) has called for separate family status. An older systematic treatment of the genus is in J. K. Small's account of Simaroubaceae for *North American Flora* (25: 227-239. 1917); his three species have, however, now been united. (Oldfieldioideae)

Hayden, W. J. (1977). Comparative anatomy and systematics of *Picrodendron*, genus incertae sedis. J. Arnold Arbor. 58: 257-279, illus. En. — Anatomical studies related to establishment of proper affinities; placement in Euphorbiaceae subfamily Oldfieldioideae supported. [Numerous other connections, along with separate family status, had been proposed.]

• Hayden, W. J. et al. (1984). Systematics and palynology of *Picrodendron*: further evidence for relationship with the Oldfieldioideae (Euphorbiaceae). J. Arnold Arbor. 65: 105-127, illus., map En. — Includes a full systematic treatment with descriptions, synonymy, references, types, localities with exsiccatae, indication of distribution and habitat, illustrations, map, and commentary. [The tree was first recorded by Sloane in 1696 and the earliest illustration dates from 1725. It was thought by Linnaeus to be a walnut (*Juglans*) and this link persisted in literature for more then two centuries.]

Hakki, M. I. (1985). Studies on West Indian plants 3. On floral morphology, anatomy and relationship of *Picrodendron baccatum* (L.) Krug et Urban. Bot. Jahrb. Syst. 107: 379-394. En. — Detailed morphological, anatomical and embryological study; provides support for Hayden's (1977) placement of genus in Euphorbiaceae.

Picrodendron Planch., London J. Bot. 5: 579 (1846).
 Caribbean. 81. Phan.

Picrodendron baccatum (L.) Krug & Urb. ex Urb., Bot. Jahrb. Syst. 15: 308 (1892).
 Caribbean. 81 BAH CAY CUB DOM HAI JAM SWC. Phan.
  *Juglans baccata L., Syst. Nat. ed. 10, 2: 1272 (1759). *Picrodendron juglans* Griseb., Fl. Brit. W. I.: 177 (1861), nom. illeg.
  Schmidelia macrocarpa A.Rich. in R.de la Sagra, Hist. Fis. Cuba, Bot. 11: 283 (1850). *Picrodendron macrocarpum* (A.Rich.) Britton, Bull. New York Bot. Gard. 4: 139 (1906).
  Picrodendron medium Small, J. New York Bot. Gard. 18: 185 (1917).

Synonyms:

Picrodendron arboreum (Mill.) Planch. === **Allophylus cobbe** (L.) Raeusch. (Sapindaceae)
Picrodendron calunga Mart. ex Engl. === **Simaba ferruginea** (Simaroubaceae)
Picrodendron juglans Griseb. === **Picrodendron baccatum** (L.) Krug & Urb. ex Urb.
Picrodendron macrocarpum (A.Rich.) Britton === **Picrodendron baccatum** (L.) Krug & Urb. ex Urb.
Picrodendron medium Small === **Picrodendron baccatum** (L.) Krug & Urb. ex Urb.

Pierardia

Synonyms:

Pierardia Roxb. === **Baccaurea** Lour.
Pierardia barteri Baill. === **Maesobotrya barteri** (Baill.) Hutch.
Pierardia costulata (Miq.) Müll.Arg. === **Baccaurea costulata** (Miq.) Müll.Arg.
Pierardia courtallensis Wight === **Baccaurea courtallensis** (Wight) Müll.Arg.
Pierardia dasystachya Miq. === **Baccaurea dasystachya** (Miq.) Müll.Arg.
Pierardia dulcis Jack === **Baccaurea dulcis** (Jack) Müll.Arg.
Pierardia flaccida Wall. === **Baccaurea flaccida** Müll.Arg.
Pierardia griffoniana Baill. === **Maesobotrya griffoniana** (Baill.) Pierre ex Hutch.
Pierardia macrocarpa Miq. === **Baccaurea macrocarpa** (Miq.) Müll.Arg.
Pierardia macrophylla Müll.Arg. === **Baccaurea macrophylla** (Müll.Arg.) Müll.Arg.
Pierardia macrostachya Wight & Arn. === **Baccaurea ramiflora** Lour.
Pierardia motleyana Müll.Arg. === **Baccaurea motleyana** (Müll.Arg.) Müll.Arg.
Pierardia parviflora Müll.Arg. === **Baccaurea parviflora** (Müll.Arg.) Müll.Arg.
Pierardia pubera Miq. === **Baccaurea pubera** (Miq.) Müll.Arg.
Pierardia pyrrhodasya Miq. === **Terminalia** sp. (Combretaceae)

Pierardia racemosa (Reinw. ex Blume) Blume === **Baccaurea racemosa** (Reinw. ex Blume) Müll.Arg.

Pierardia rhakodiskus (Hassk.) Hassk. === **Drypetes rhakodiskos** (Hassk.) Bakh.f.

Pierardia sapida Roxb. === **Baccaurea ramiflora** Lour.

Pierardia seemannii Müll.Arg. === **Baccaurea seemannii** (Müll.Arg.) Müll.Arg.

Pilinophytum

A Klotzsch segregate from *Croton*.

Pimelodendron

5 species, Malesia to NE. Australia and the Solomon Islands; includes *Stomatocalyx*. Glabrous laticiferous forest trees to 20 m or more (with *P. griffithianum* up to 39 m by 2.4 m gbh). *P. amboinicum*, commonly collected in Papuasia and often a leading component of its lowland forests, features rhythmic shoot growth and coarsely crenate leaves held at right angles to the stem with a distinct 'twist' or pulvinus at the top of the petiole. The genus has traditionally been placed in Euphorbioideae; Webster assigns it to Stomatocalyceae. On the other hand, recent studies suggest that its exclusion from the subfamily is called for (H.-J. Esser, personal communication). (Euphorbioideae (except Euphorbieae))

Pax, F. (with K. Hoffmann) (1912). *Pimeleodendron*. In A. Engler (ed.), Das Pflanzenreich, IV 147 V (Euphorbiaceae-Hippomaneae): 54-56. Berlin. (Heft 52.) La/Ge. — 4 species in 2 groups.

Airy-Shaw, H. K. (1971). Notes on Malesian and other Asiatic Euphorbiaceae, CXLVII. *Pimeleodendron macrocarpum* in Malaya and Borneo. Kew Bull. 25: 551-552. En. — Range extension; previously known only from Sumatra.

Pimelodendron Hassk., Verslagen Meded. Afd. Natuurk. Kon. Akad. Wetensch. 4: 140 (1855). Trop. Asia, N. Australia. 41 42 50 60.

Stomatocalyx Müll.Arg. in A.P.de Candolle, Prodr. 15(2): 1142 (1866).

Pimelodendron amboinicum Hassk., Verslagen Meded. Afd. Natuurk. Kon. Akad. Wetensch. 4: 140 (1855).
Sulawesi to Solomon Is., Queensland (Cook). 42 BIS MOL NWG SUL 50 QLD 60 SOL. Phan. – Latex used for gluing or cementing wooden articles together.

Pimelodendron papuanum Warb., Bot. Jahrb. Syst. 18: 198 (1894).

Pimelodendron griffithianum (Müll.Arg.) Benth. in G.Bentham & J.D.Hooker, Gen. Pl. 3: 332 (1880).
Pen. Malaysia, Sumatera (incl. Bangka), Borneo. 42 BOR MLY SUM. Phan.

* *Stomatocalyx griffithianus* Müll.Arg. in A.P.de Candolle, Prodr. 15(2): 1142 (1866).

Pimelodendron borneense Warb., Bot. Jahrb. Syst. 18: 199 (1894).

Pimelodendron acuminatum Merr., Philipp. J. Sci., C 11: 74 (1916).

Pimelodendron papaveroides J.J.Sm., Bull. Jard. Bot. Buitenzorg, III, 6: 104 (1924).

Pimelodendron macrocarpum J.J.Sm., Bull. Jard. Bot. Buitenzorg, III, 6: 103 (1924).
Pen. Malaysia, S. Sumatera, Borneo (E. Kalimantan). 42 BOR MLY SUM. Phan.

Pimelodendron naumannianum Pax & K.Hoffm. in H.G.A.Engler, Pflanzenr., IV, 147, XVII: 200 (1924).
Lesser Sunda Is. (Timor). 42 LSI. Phan.

Pimelodendron zoanthogyne J.J.Sm., Bull. Jard. Bot. Buitenzorg, III, 6: 105 (1924).
Borneo (W. Kalimantan). 42 BOR. Phan.

Synonyms:
Pimelodendron acuminatum Merr. === **Pimelodendron griffithianum** (Müll.Arg.) Benth.
Pimelodendron borneense Warb. === **Pimelodendron griffithianum** (Müll.Arg.) Benth.
Pimelodendron dispersum Elmer === **Actephila excelsa** var. **javanica** (Miq.) Pax & K.Hoffm.
Pimelodendron papaveroides J.J.Sm. === **Pimelodendron griffithianum** (Müll.Arg.) Benth.
Pimelodendron papuanum Warb. === **Pimelodendron amboinicum** Hassk.

Piranhea

4 species, scattered through C. & S. America; now includes *Celaenodendron* (Radcliffe-Smith & Ratter 1996). Trees to 50 m tall, the leaves palmately 3-5-foliolate. It is related to *Picrodendron* and the somewhat diffrerent *Parodiodendron* (Webster, Synopsis, 1994); all these form his subtribe Picrodendrinae in tribe Picrodendreae. (Oldfieldioideae)

Pax, F. & K. Hoffmann (1922). *Piranhea*. In A. Engler (ed.), Das Pflanzenreich, IV 147 XV (Euphorbiaceae-Phyllanthoideae-Phyllantheae): 295. Berlin. (Heft 81.) La/Ge. — 1 species; S America (Brazil, Guianas).

Standley, P. C. (1927). *Celaenodendron*. In R. S. Ferris, Preliminary report on the flora of the Tres Marías Islands. Contr. Dudley Herb. 1: 57-86 (incl. 3 pls), 1 pl. En. — Includes (pp. 76-77) protologue of *Celaenodendron* (from Tres Marías Is., Mexico). [The genus is now united with *Piranhea*.]

Jablonski, E. (1967). *Piranhea*. Euphorbiaceae, Guayana Highland (Mem. New York Bot. Gard. 17(1)): 121-122. New York. En. — 2 species, N. South America; key.

• Radcliffe-Smith, A. & J. A. Ratter (1996). A new *Piranhea* from Brazil, and the subsumption of the genus *Celaenodendron* (Euphorbiaceae: Oldfieldioideae). Kew Bull. 51: 543-548, illus. En. — Novelties (including description of *P. securinega*) and a reduction; includes key to 4 species (p. 547).

Piranhea Baill., Adansonia 6: 235 (1866).
Mexico to N. Brazil. 79 82 84. Phan.
Celaenodendron Standl., Contr. Dudley Herb. 1: 76 (1927).

Piranhea longepedunculata Jabl., Mem. New York Bot. Gard. 17: 122 (1967).
Venezuela (E. Bolívar). 82 VEN. Phan.

Piranhea mexicana (Standl.) Radcl.-Sm., Kew Bull. 51: 546 (1996).
Mexico (Sinaloa). 79 MXN.
* *Celaenodendron mexicanum* Standl., Contr. Dudley Herb. 1: 76 (1927).

Piranhea securinega Radcl.-Sm. & Ratter, Kew Bull. 51: 543 (1996).
Brazil (Minas Gerais). 84 BZL. Phan.

Piranhea trifoliata Baill., Adansonia 6: 236 (1866).
Brazil (Amazonas, Pará), Venezuela (Amazonas, Bolívar), Guyana. 82 GUY VEN 84 BZN. Phan.

var. **pubescens** Radcl.-Sm., Kew Bull. 51: 547 (1996).
S. Venezuela. 82 VEN. Phan.

var. **trifoliata**
Brazil (Amazonas, Pará), Venezuela (Amazonas, Bolívar), Guyana. 82 GUY VEN 84 BZN. Phan.

Piscaria

Synonyms:
Piscaria Piper === **Croton** L.

Plagianthera

Synonyms:
Plagianthera Rchb.f. & Zoll. === **Mallotus** Lour.

Plagiocladus

An unpublished Brunel segregate of *Phyllanthus* credited to Africa and Madagascar.

Plagiostyles

1 species, W. and C. Africa (Nigeria to Zaïre); small to medium glabrous laticiferous trees to 25 m frequently encountered in forest. The genus is closely related to *Pimelodendron*, differing in free sepals and a unilocular ovary; revision may show, however, that these distinctions are insigificant in relation to the marked similarities. The leaves are distinctly brochidodromous. Like its Malesian relative, those who once included it with the former Crotoneae are more nearly right than wrong. (Euphorbioideae (except Euphorbieae))

> Pax, F. (with K. Hoffmann) (1914). *Plagiostyles.* In A. Engler (ed.), Das Pflanzenreich, IV 147 VII [Euphorbiaceae-Additamentum V]: 420-421. Berlin. (Heft 63.) La/Ge. — 1 species, W and C Africa.
>
> Stapf, O. (1915). *Plagiostyles africana* Prain. Ic. Pl. 31: pl. 3010. En. — Plant portrait with descriptive text and commentary. [The author assigned it to the Benthamian tribe Crotoneae. He also noted that there may be a second, abortive ovule in each carpel.]

Plagiostyles Pierre, Bull. Mens. Soc. Linn. Paris 2: 1327 (1897).
  W. & WC. Trop. Africa. 22 23.

Plagiostyles africana (Müll.Arg.) Prain, Bull. Misc. Inform. Kew 1912: 107 (1912).
  Nigeria, Cameroon, Equatorial Guinea, Gabon, Congo, Zaire. 22 NGA 23 CMN CON EQG GAB ZAI. Phan.
  * *Daphniphyllum africanum* Müll.Arg., Flora 47: 536 (1864).
  Plagiostyles klaineana Pierre, Bull. Mens. Soc. Linn. Paris 2: 1327 (1897).

Synonyms:
Plagiostyles klaineana Pierre === **Plagiostyles africana** (Müll.Arg.) Prain

Platygyna

7 species, Cuba; twining woody climbers with alternate oblong toothed leaves and stinging hairs, found in thickets and stony places. Alain (1971) reduced the genus to *Tragia*. According to the last revision by Borhidi (1972) only one of the species, *P. hexandra*, is within Cuba relatively widely distributed; some of the remainder are quite local. (Acalyphoideae)

> Pax, F. & K. Hoffmann (1919). *Platygyne.* In A. Engler (ed.), Das Pflanzenreich, IV 147 IX (Euphorbiaceae-Acalypheae-Plukenetiinae): 26-27. Berlin. (Heft 68.) La/Ge. — 1 species, Cuba.
>
> Alain, Hno. (Alain H. Liogier) (1971). Novitates antillanae, IV. Mem. New York Bot. Gard. 21(2): 107-157. En. — Pp. 132-133 comprise a discussion of differences between *Platygyna* and *Tragia* with reduction of 4 Cuban species of *Platygyna* to the latter genus.
>
> Borhidi, A. (1972). La taxonomía del género *Platygyne* Merc. Ann. Hist.-Nat. Mus. Natl. Hung. 64: 89-94, map. Sp. — Revision (7 species, two new) with key, brief descriptions, synonymy, localities with exsiccatae for the novelties, and indication of life-form, distribution, habitats and vegetation types; distribution map, p. 91.

Platygyna Mercier, Ser. Bull. Bot. 1: 168 (1830).
　　Cuba. 81.
　　　Acanthocaulon Klotzsch in S.L.Endlicher, Gen. Pl., Suppl. 4(3): 88 (1850).

Platygyna dentata Alain, Contr. Ocas. Mus. Hist. Nat. Colegio "De La Salle" 11: 6 (1952).
　　Tragia dentata (Alain) Alain, Mem. New York Bot. Gard. 21 :133 (1971), nom. illeg.
　　SE. Cuba. 81 CUB. Cham.

Platygyna hexandra (Jacq.) Müll.Arg. in A.P.de Candolle, Prodr. 15(2): 914 (1866).
　　Cuba. 81 CUB. Cham.
　　　* *Tragia hexandra* Jacq., Enum. Syst. Pl.: 81 (1763).
　　　Platygyna urens Mercier, Ser. Bull. Bot. 1: 168 (1830).
　　　Tragia pruricus Willd. ex Endl., Gen. Pl., Suppl. 4(3): 88 (1850). *Platygyna pruriens*
　　　　(Willd. ex Endl.) Baill., Hist. Pl. 5: 215 (1874).

Platygyna leonis Alain, Contr. Ocas. Mus. Hist. Nat. Colegio "De La Salle" 11: 6 (1952).
　　Tragia leonis (Alain) Alain, Mem. New York Bot. Gard. 21 :133 (1971).
　　E. Cuba. 81 CUB. Cham.

Platygyna obovata Borhidi, Ann. Hist.-Nat. Mus. Natl. Hung. 64: 93 (1972). *Tragia obovata*
　　(Borhidi) Borhidi, Acta Bot. Acad. Sci. Hung. 19: 44 (1973).
　　E. Cuba. 81 CUB.

Platygyna parvifolia Alain, Contr. Ocas. Mus. Hist. Nat. Colegio "De La Salle" 11: 8 (1952).
　　Tragia parvifolia (Alain) Alain, Mem. New York Bot. Gard. 21: 133 (1971), nom. illeg.
　　EC. & E. Cuba. 81 CUB.

Platygyna triandra Borhidi, Ann. Hist.-Nat. Mus. Natl. Hung. 64: 92 (1972). *Tragia triandra*
　　(Borhidi) Borhidi, Acta Bot. Acad. Sci. Hung. 19: 44 (1973), nom. illeg.
　　E. Cuba. 81 CUB.

Platygyna volubilis Howard, J. Arnold Arbor. 28: 120 (1947). *Tragia howardii* Alain, Mem.
　　New York Bot. Gard. 21: 133 (1971).
　　E. Cuba (Sierra de Moa). 81 CUB.

Synonyms:
Platygyna pruriens (Willd. ex Endl.) Baill. === **Platygyna hexandra** (Jacq.) Müll.Arg.
Platygyna urens Mercier === **Platygyna hexandra** (Jacq.) Müll.Arg.

Pleiostemon

Now reduced to *Flueggea* (Webster 1984; see that genus) but long maintained as distinct.

Synonyms:
Pleiostemon Sond. === **Flueggea** Willd.
Pleiostemon verrucosum (Thunb.) Sond. === **Flueggea verrucosa** (Thunb.) G.L.Webster

Pleopadium

Synonyms:
Pleopadium Raf. === **Croton** L.

Plesiatropha

A Pierre ms. name published by Hutchinson in *Flora of Tropical Africa*, 6(1): 799 (1912) as a synonym of *Mildbraedia*.

Pleuradena

Proposed by Rafinesque with a single species, *P. coccinea*, a synonym of *Euphorbia pulcherrima* Willd. ex Klotzsch. The name is nomenclaturally prior to *Poinsettia* Graham but is a later homonym of *Pleuradenia* Raf. (= *Collinsonia* L., Lamiaceae).

Synonyms:
Pleuradena Raf. === **Euphorbia** L.

Plukenetia

13 species, Americas (Mexico to Brazil, 12) and Madagascar (1); laticiferous woody climbers of moist to dry forest or in more open habitats, distinguished by 4-angled fruits and pinnately or palmately veined leaves. In *P. polyadenia* (including *Elaeophora abutaefolia*), a high climber, these fruits are indehiscent and reach 10 by 11 cm (illustrated by Ducke, 1925). *P. madagascariensis* is a vine of dry scrub and spiny forest. Circumscription of the genus has varied, but in the Pax & Hoffmann system it was relatively narrow. Gillespie (1993) has taken about the opposite point of view. She has so far reduced *Apodandra*, *Elaeophora*, *Eleutherostigma* and *Vigia*. In support of the reduction of the last-named, she showed that *Fragariopsis paxii* Pittier was based on a plant from a Venezuelan population intermediate between *Plukenetia volubilis* and *P. stipellata*, referring it to the former. Her studies of the genus are continuing and it is likely that *Pterococcus* (Africa, Asia and Malesia) and *Tetracarpidium* (Africa) will also be reduced. (Acalyphoideae)

Pax, F. & K. Hoffmann (1919). *Apodandra*. In A. Engler (ed.), Das Pflanzenreich, IV 147 IX (Euphorbiaceae-Acalypheae-Plukenetiinae): 20-21. Berlin. (Heft 68.) La/Ge. — 2 species, S America (Amazon Basin).

Pax, F. & K. Hoffmann (1919). *Eleutherostigma*. In A. Engler (ed.), Das Pflanzenreich, IV 147 IX (Euphorbiaceae-Acalypheae-Plukenetiinae): 11-12. Berlin. (Heft 68.) La/Ge. — 1 species. [Now included in *Plukenetia*.]

Pax, F. & K. Hoffmann (1919). *Fragariopsis*. In A. Engler (ed.), Das Pflanzenreich, IV 147 IX (Euphorbiaceae-Acalypheae-Plukenetiinae): 19-20. Berlin. (Heft 68.) La/Ge. — Monotypic, Brazil.

Pax, F. & K. Hoffmann (1919). *Plukenetia*. In A. Engler (ed.), Das Pflanzenreich, IV 147 IX (Euphorbiaceae-Acalypheae-Plukenetiinae): 12-17. Berlin. (Heft 68.) La/Ge. — 6 species in 2 sections, all American.

Ducke, A. (1925). *Elaeophora*. Arq. Jard. Bot. Rio de Janeiro 4 (in Plantes nouvelles .. III): 112-113, illus. Fr/La. — Protologue of genus with transfer from *Plukenetia* of one species, *P. abutaefolia* (illustrated in plate 9). [An additional species is described in *ibid.*, 5: 145-146 (1930). The genus was by Webster incorporated into *Plukenetia*.]

Jablonski, E. (1967). *Plukenetia*. Euphorbiaceae, Guayana Highland (Mem. New York Bot. Gard. 17(1)): 142-143. New York. En. — 4 species, none new; 'poorly collected' in area.

• Gillespie, L. J. (1993). A synopsis of neotropical *Plukenetia* (Euphorbiaceae) including two new species. Syst. Bot. 18(4): 575-592. En. — Introduction (genus now inclusive of *Eleutherostigma* and *Fragariopsis* (*Vigia*)); primary morphological features; synoptic treatment of 11 species with synonymy, descriptions of novelties, indication of distribution, citation of exsiccatae, and commentary; references at end. [Since publication of this revision, an additional species, *P. carabasiae*, has been described.]

Plukenetia L., Sp. Pl.: 1192 (1753).
Mexico, Trop. America, Madagascar. 29 79 80 81 82 83 84.

Vigia Vell., Fl. Flumin. 9: 128 (1831).
Accia A.St.-Hil., Leçons Bot.: 499 (1840).
Fragariopsis A.St.-Hil., Leçons Bot.: 426 (1840).
Botryanthe Klotzsch, Arch. Naturgesch. 2: 190 (1841).
Apodandra Pax & K.Hoffm. in H.G.A.Engler, Pflanzenr., IV, 147, IX: 20 (1919).
Eleutherostigma Pax & K.Hoffm. in H.G.A.Engler, Pflanzenr., IV, 147, IX: 11 (1919).
Elaeophora Ducke, Arch. Jard. Bot. Rio de Janeiro 4: 112 (1925).

Plukenetia brachybotrya Müll.Arg., Linnaea 34: 158 (1865). *Apodandra brachybotrya*
(Müll.Arg.) J.F.Macbr., Publ. Field Mus. Nat. Hist., Bot. Ser. 13(3A1): 117 (1951).
Bolivia, Brazil (Acre, Amazonas ?, Mato Grosso ?, Pará ?), Ecuador, Peru,. 83 BOL ECU PER
84 BZC BZN. Cl. nanophan.
Plukenetia buchtienii Pax, Repert. Spec. Nov. Regni Veg. 7: 110 (1909). *Apodandra*
buchtienii (Pax) Pax in H.G.A.Engler, Pflanzenr., IV, 147, IX: 21 (1919).

Plukenetia carabiasiae J.Jiménez Ram., Anales Inst. Biol. Univ. Nac. Auton. Mexico, Bot.
64: 55 (1993).
Mexico (Oaxaca). 79 MXS.

Plukenetia lehmanniana (Pax & K.Hoffm.) Huft & Gillespie, Syst. Bot. 18: 584 (1993).
Colombia, Ecuador. 83 CLM ECU. Cl. nanophan.
  * *Eleutherostigma lehmannianum* Pax & K.Hoffm. in H.G.A.Engler, Pflanzenr., IV, 147, IX:
    11 (1919).
  Plukenetia chaponensis Croizat, Caldasia 2: 431 (1944).

Plukenetia loretensis Ule, Verh. Bot. Vereins Prov. Brandenburg 50: 81 (1908 publ. 1909).
Apodandra loretensis (Ule) Pax & K.Hoffm. in H.G.A.Engler, Pflanzenr., IV, 147, IX: 21 (1919).
SE. Guyana, S. Venezuela, Peru, Bolivia, Brazil (Amazonas, Mato Grosso, Rondônia). 82
GUY VEN 83 BOL PER 84 BZC BZN. Cl. nanophan.

Plukenetia madagascariensis Leandri, Bull. Soc. Bot. France 85: 527 (1938 publ. 1939).
Madagascar. 29 MDG.

Plukenetia multiglandulosa Jabl., Mem. New York Bot. Gard. 17: 143 (1967).
Venezuela (Amazonas: Mt. Cerro Paru). 82 VEN. Cl. nanophan.

Plukenetia penninervia Müll.Arg., Linnaea 34: 158 (1865).
SW. & S. Mexico, C. America, Venezuela, Colombia, Peru. 79 MXS MXT 80 BLZ COS GUA
NIC PAN 82 VEN 83 CLM PER. Cl. nanophan.
Plukenetia angustifolia Standl., Publ. Field Mus. Nat. Hist., Bot. Ser. 4: 314 (1929).

Plukenetia polyadenia Müll.Arg. in C.F.P.von Martius, Fl. Bras. 11(2): 335 (1874).
S. Trop. America. 82 FRG GUY SUR VEN 83 ECU PER 84 BZN. Cl. nanophan.
Elaeophora abutifolia Ducke, Arch. Jard. Bot. Rio de Janeiro 4: 112 (1925). *Plukenetia*
abutifolia (Ducke) Pax & K.Hoffm. in H.G.A.Engler, Nat. Pflanzenfam. ed. 2, 19c:
141 (1931).

Plukenetia serrata (Vell.) L.J.Gillespie, Syst. Bot. 18: 587 (1993).
E. Brazil. 84 BZE BZL. Cl. nanophan.
  * *Vigia serrata* Vell., Fl. Flumin. 9: 128 (1831).
  Fragariopsis scandens A.St.-Hil., Leçons Bot.: 426 (1840). *Plukenetia scandens* (A.St.-Hil.)
    Pax in H.G.A.Engler & K.A.E.Prantl, Nat. Pflanzenfam. 3(5): 67 (1890).
  Botryanthe concolor Klotzsch, Arch. Naturgesch. 2: 191 (1841).
  Botryanthe discolor Klotzsch, Arch. Naturgesch. 2: 191 (1841).
  Fragariopsis warmingii Müll.Arg. in C.F.P.von Martius, Fl. Bras. 11(2): 338 (1874).
    Plukenetia warmingii (Müll.Arg.) Pax in H.G.A.Engler & K.A.E.Prantl, Nat.
    Pflanzenfam. 3(5): 67 (1890).

Plukenetia stipellata L.J.Gillespie, Syst. Bot. 18: 588 (1993).
S. Mexico, C. America. 79 MXT 80 COS GUA NIC PAN 83 CLM? Cl. nanophan.

Plukenetia supraglandulosa L.J.Gillespie, Syst. Bot. 18: 589 (1993).
Guiana, Brazil (Amapá). 82 FRG 84 BZN. Cl. nanophan.

Plukenetia verrucosa Sm., Nova Acta Regiae Soc. Sci. Upsal. 6: 4 (1799).
Guianas, Trinidad, Brazil (Amapa, N. Amazonas, N. Para). 81 TRT 82 FRG GUY SUR 84
BZN. Cl. nanophan.
Plukenetia volubilis L.f., Suppl. Pl.: 421 (1782), nom. illeg.
Plukenetia integrifolia Vahl, Eclog. Amer. 3: 43 (1807).

Plukenetia volubilis L., Sp. Pl.: 1192 (1753).
Windward Is., S. Trop. America. 81 WIN 82 SUR VEN 83 BOL CLM ECU PER 84 BZN.
Cl. nanophan.
Plukenetia peruviana Müll.Arg., Linnaea 34: 157 (1865).
Plukenetia macrostyla Ule, Verh. Bot. Vereins Prov. Brandenburg 50: 80 (1908 publ. 1909).
Fragariopsis paxii Pittier, J. Wash. Acad. Sci. 19: 351 (1929).

Synonyms:
Plukenetia abutifolia (Ducke) Pax & K.Hoffm. === **Plukenetia polyadenia** Müll.Arg.
Plukenetia africana Sond. === **Pterococcus africanus** (Sond.) Pax & K.Hoffm.
Plukenetia angustifolia Standl. === **Plukenetia penninervia** Müll.Arg.
Plukenetia buchtienii Pax === **Plukenetia brachybotrya** Müll.Arg.
Plukenetia chaponensis Croizat === **Plukenetia lehmanniana** (Pax & K.Hoffm.) Huft & Gillespie
Plukenetia conophora Müll.Arg. === **Tetracarpidium conophorum** (Müll.Arg.) Hutch. & Dalziel
Plukenetia corniculata Sm. === **Pterococcus corniculatus** (Sm.) Pax & K.Hoffm.
Plukenetia hastata Müll.Arg. === **Pterococcus africanus** (Sond.) Pax & K.Hoffm.
Plukenetia integrifolia Vahl === **Plukenetia verrucosa** Sm.
Plukenetia macrostyla Ule === **Plukenetia volubilis** L.
Plukenetia occidentalis Leandro ex Baill. === **Romanoa tamnoides** (A.Juss.) Radcl.-Sm.
var. **tamnoides**
Plukenetia peruviana Müll.Arg. === **Plukenetia volubilis** L.
Plukenetia procumbens Prain === **Pterococcus procumbens** (Prain) Pax & K.Hoffm.
Plukenetia scandens (A.St.-Hil.) Pax === **Plukenetia serrata** (Vell.) L.J.Gillespie
Plukenetia sinuata Ule === **Romanoa tamnoides** var. **sinuata** (Ule) Radcl.-Sm.
Plukenetia tamnoides (A.Juss.) Müll.Arg. === **Romanoa tamnoides** (A.Juss.) Radcl.-Sm.
Plukenetia volubilis L.f. === **Plukenetia verrucosa** Sm.
Plukenetia warmingii (Müll.Arg.) Pax === **Plukenetia serrata** (Vell.) L.J.Gillespie
Plukenetia zenkeri Pax === **Hamilcoa zenkeri** (Pax) Prain

Podadenia

1 species, Sri Lanka; sometimes not separated from *Ptychopyxis* (e.g. in *Revised Handbook to the Flora of Ceylon* 11; see **Asia**). Small to large trees to 30 m in forests, the leaves pinnately veined and brownish-pubescent below. Further study is required (Webster, Synopsis, 1994); in that system it is also close to *Blumeodendron*. (Acalyphoideae)

Pax, F. (with K. Hoffmann) (1914). *Podadenia*. In A. Engler (ed.), Das Pflanzenreich, IV 147 VII (Euphorbiaceae-Acalypheae-Mercurialinae): 19-21. Berlin. (Heft 63.) La/Ge. — 2 species, Sri Lanka, Java. [The Javan *P. javanica* is now in *Ptychopyxis*.]

Croizat, L. (1942). On certain Euphorbiaceae from the tropical Far East. J. Arnold Arbor. 23: 29-54. En. —*Podadenia* herein reduced to *Ptychopyxis*; new combination, *P. thwaitesii*.

Airy-Shaw, H. K. (1960). Notes on Malaysian Euphorbiaceae, VIII. A synopsis of the genus *Ptychopyxis* Miq. Kew Bull. 14: 363-374. En. —*Podadenia* treated, following Croizat, as a subgenus of *Ptychopyxis* but author hints that generic status is probably to be preferred; one species in Sri Lanka.

Podadenia Thwaites, Enum. Pl. Zeyl.: 273 (1861).
  Sri Lanka. 40.

Podadenia sapida Thwaites, Enum. Pl. Zeyl.: 274 (1861).
  Sri Lanka. 40 SRL. Phan.
    Stylanthus thwaitesii Baill., Étude Euphorb.: 426 (1858), nom. nud.
    Podadenia thwaitesii Müll.Arg. in A.P.de Candolle, Prodr. 15(2): 791 (1866). *Ptychopyxis thwaitesii* (Müll.Arg.) Croizat, J. Arnold Arbor. 23: 48 (1942).

Synonyms:
Podadenia javanica J.J.Sm. === **Ptychopyxis javanica** (J.J.Sm.) Croizat
Podadenia thwaitesii Müll.Arg. === **Podadenia sapida** Thwaites

Podocalyx

1 species, northern S. America (Venezuela and Brazil); shrubs or small trees with alternate entire leaves. Not now considered to be in the flora of the Guianas although listed for Guyana by Pax & Hoffmann (1922). In the past treated as a section of *Richeria* (Phyllanthoideae) but pollen characters point to Oldfieldioideae (Webster, Synopsis, 1994). (Oldfieldioideae)

Pax, F. & K. Hoffmann (1922). *Richeria*. In A. Engler (ed.), Das Pflanzenreich, IV 147 XV (Euphorbiaceae-Phyllanthoideae-Phyllantheae): 26-30. Berlin. (Heft 81.) La/Ge. — 6 species in 2 sections, the latter, *Podocalyx* (pp. 29-30), comprising *R. loranthoides* of Amazonian S America. [Now at generic rank; the remaining 5 are in *Richeria* s.s.]

Jablonski, E. (1967). *Richeria*. Euphorbiaceae, Guayana Highland (Mem. New York Bot. Gard. 17(1)): 124-127. New York. En. — Genus still part of *Richeria* following Pax and Hoffmann; of the 4 species treated here only *R. loranthoides* is in *Podocalyx*.

Podocalyx Klotzsch, Arch. Naturgesch. 7: 202 (1841).
  N. South America (Venezuela, N. Brazil). 82 84.

Podocalyx loranthoides Klotzsch, Arch. Naturgesch. 7: 202 (1841). *Richeria loranthoides* (Klotzsch) Müll.Arg. in A.P.de Candolle, Prodr. 15(2): 469 (1866).
  Venezuela (Amazonas, Bolívar), Brazil (Amazonas, Pará). 82 VEN 84 BZN.

Podostachys

Synonyms:
Podostachys Klotzsch === **Croton** L.
Podostachys lundianus Didr. === **Croton lundianus** (Didr.) Müll.Arg.
Podostachys sclerocalyx Didr. === **Croton sclerocalyx** (Didr.) Müll.Arg.
Podostachys subfloccosa Didr. === **Croton lundianus** (Didr.) Müll.Arg.

Poggeophyton

Synonyms:
Poggeophyton Pax === **Erythrococca** Benth.

Pogonophora

3 species, Amazonian and eastern S. America (1) and in WC. Africa (Gabon, 2); small to large trees (but to no more than 8 m in Africa). *P. schomburgkiana* occupies a wide variety of habitats but usually along rivers or in low-lying land. The genus has been assigned to its own tribe or subtribe since the time of Müller, though its pollen is similar to that of

Clutia (Webster, Synopsis, 1994). The two African species do not really resemble their S. American congener and whether or not they belong in the genus is a matter for further investigation. (Acalyphoideae)

Pax, F. (with K. Hoffmann) (1911). *Pogonophora*. In A. Engler (ed.), Das Pflanzenreich, IV 147 III (Euphorbiaceae-Cluytieae): 108-110. Berlin. (Heft 47.) La/Ge. — 2 species, S America. [*P. trianae* is, however, homotypic with *Pausandra trianae* and is properly in that genus.]

Jablonski, E. (1967). *Pogonophora*. Euphorbiaceae, Guayana Highland (Mem. New York Bot. Gard. 17(1)): 153. New York. En. — 1 species, *P. schomburgkiana*; reductions.

Letouzey, R. (1969). Présence au Gabon du genre *Pogonophora* Miers ex Bentham, Euphorbiacée d'Amérique du Sud tropicale. Adansonia, II, 9: 273-276. Fr. — Range extension of genus with description of *P. africana*.

Secco, R. de S. (1990). Revisão dos gêneros *Anomalocalyx* Ducke, *Dodecastigma* Ducke, *Pausandra* Radlk., *Pogonophora* Miers ex Benth. e *Sagotia* Baill. (Euphorbiaceae-Crotonoideae) para a América do Sul. 133 pp., illus., maps. Belém: Museu Paraense 'Emilio Goeldi'. Pt. — Includes (pp. 87-95) revision of American *Pogonophora*; one species accepted, with full documentation.]

Pogonophora Miers ex Benth., Hooker's J. Bot. Kew Gard. Misc. 6: 372 (1854).
S. Trop. America, Gabon. 23 82 84.
Poraresia Gleason, Bull. Torrey Bot. Club 58: 385 (1931).

Pogonophora africana Letouzey, Adansonia, n.s. 9: 275 (1969).
Gabon. 23 GAB. Nanophan. or phan.

Pogonophora letouzeyi Feuillet, Novon 3: 23 (1993).
Gabon. 23 GAB.

Pogonophora schomburgkiana Miers ex Benth., Hooker's J. Bot. Kew Gard. Misc. 6: 372 (1854).
Colombia, S. Venezuela, Guyana, Surinam, Guiana, Brazil. 82 FRG GUY SUR VEN 83 CLM 84 BZE BZL BZN. Nanophan. or phan.
Pogonophora schomburgkiana var. *longifolia* Miers ex Benth., Hooker's J. Bot. Kew Gard. Misc. 6: 372 (1854).
Pogonophora schomburgkiana f. *elliptica* Pax in H.G.A.Engler, Pflanzenr., IV, 147, III: 109 (1911).
Pogonophora glaziovii Taub. ex Glaziou, Bull. Soc. Bot. France 59(3): 626 (1912 publ. 1913), nom. nud.

Synonyms:
Pogonophora cunurii Baill. === **Micrandra spruceana** (Baill.) R.E.Schult.
Pogonophora glaziovii Taub. ex Glaziou === **Pogonophora schomburgkiana** Miers ex Benth.
Pogonophora schomburgkiana f. *elliptica* Pax === **Pogonophora schomburgkiana** Miers ex Benth.
Pogonophora schomburgkiana var. *longifolia* Miers ex Benth. === **Pogonophora schomburgkiana** Miers ex Benth.
Pogonophora trianae Müll.Arg. === **Pausandra trianae** (Müll.Arg.) Baill.

Pogonophyllum

Synonyms:
Pogonophyllum Didr. === **Micrandra** Benth.
Pogonophyllum elatum Didr. === **Micrandra elata** (Didr.) Müll.Arg.

Poilaniella

Synonyms:
Poilaniella Gagnep. === **Trigonostemon** Blume
Poilaniella fragilis Gagnep. === **Trigonostemon fragilis** (Gagnep.) Airy Shaw

Poinsettia

Usually included with *Euphorbia* at subgeneric rank but as a fairly distinctive group, and with *E. pulcherrima* ubiquitous in cultivation, considered by some authors as separate. Notable exponents of the latter view have been Klotzsch, the Britton 'school' including Millspaugh, and Dressler (for the latter, see Dressler 1962 under **Euphorbia**).

Synonyms:
Poinsettia Graham === **Euphorbia** L.
Poinsettia barbellata (Engelm.) Small === **Euphorbia cyathophora** Murr.
Poinsettia coccinea Dressler === **Euphorbia hormorrhiza** Radcl.-Sm.
Poinsettia colorata (Engelm.) Dressler === **Euphorbia colorata** Engelm.
Poinsettia cornastra Dressler === **Euphorbia cornastra** (Dressler) Radcl.-Sm.
Poinsettia cuphosperma (Engelm.) Small === **Euphorbia dentata** Michx.
Poinsettia cyathophora (Murray) Klotzsch & Garcke === **Euphorbia cyathophora** Murr.
Poinsettia dentata (Michx.) Klotzsch & Garcke === **Euphorbia dentata** Michx.
Poinsettia edwardsii Klotzsch & Garcke === **Euphorbia cyathophora** Murr.
Poinsettia eriantha (Benth.) Rose & Standl. === **Euphorbia eriantha** Benth.
Poinsettia frangulifolia (Kunth) Klotzsch & Garcke === **Euphorbia heterophylla** L.
Poinsettia geniculata (Ortega) Klotzsch & Garcke === **Euphorbia heterophylla** L.
Poinsettia graminifolia (Michx.) Millsp. === **Euphorbia cyathophora** Murr.
Poinsettia havanensis Small === **Euphorbia heterophylla** L.
Poinsettia heterophylla (L.) Klotzsch & Garcke === **Euphorbia heterophylla** L.
Poinsettia inornata Dressler === **Euphorbia limaensis** Oudejans
Poinsettia insulana (Vell.) Klotzsch & Garcke === **Euphorbia insulana** Vell.
Poinsettia lancifolia (Schltdl.) Klotzsch & Garcke === **Euphorbia lancifolia** Schltdl.
Poinsettia morisoniana (Klotzsch) Klotzsch & Garcke === **Euphorbia heterophylla** L.
Poinsettia oerstediana Klotzsch & Garcke === **Euphorbia oerstediana** (Klotzsch & Garcke) Boiss.
Poinsettia pedunculata Klotzsch & Garcke === **Euphorbia strigosa** Hook. & Arn.
Poinsettia pentadactyla (Griseb.) Dressler === **Euphorbia pentadactyla** Griseb.
Poinsettia pinetorum Small === **Euphorbia cyathophora** Murr.
Poinsettia prunifolia (Jacq.) Klotzsch & Garcke === **Euphorbia heterophylla** L.
Poinsettia pulcherrima (Willd. ex Klotzsch) Graham === **Euphorbia pulcherrima** Willd. ex Klotzsch
Poinsettia punicea (Sw.) Klotzsch & Garcke === **Euphorbia punicea** Sw.
Poinsettia radians (Benth.) Klotzsch & Garcke === **Euphorbia radians** Benth.
Poinsettia restiacea (Benth.) Dressler === **Euphorbia restiacea** Benth.
Poinsettia ruiziana Klotzsch & Garcke === **Euphorbia heterophylla** L.
Poinsettia schiedeana Klotzsch & Garcke === **Euphorbia dentata** Michx.
Poinsettia strigosa (Hook. & Arn.) Arthur === **Euphorbia strigosa** Hook. & Arn.
Poinsettia xalapensis (Kunth) Klotzsch & Garcke === **Euphorbia xalapensis** Kunth

Polyandra

1 species, northern S. America (western Amazon Basin in Brazil); reputedly imperfectly known (only staminate material being so far recorded). Described as an 'árvore grande'; occurs in lowland forest on well-drained ground. An affinity with *Mallotus* has been suggested for this and other Alchorneae (Webster, Synopsis, 1994). (Acalyphoideae)

Leal, C. G. (1951). Contribuição ao estudo da família Euphorbiaceae. Arq. Jard. Bot. Rio de Janeiro 11: 63-70. Pt. — Protologue of *Polyandra* and description of a new species, *P. bracteosa*, a large tree from Amazonas (p. 64).

Polyandra Leal, Arch. Jard. Bot. Rio de Janeiro 11: 63 (1951).
   N. Brazil. 84.

Polyandra bracteosa Leal, Arch. Jard. Bot. Rio de Janeiro 11: 64 (1951).
   Brazil (Amazonas). 84 BZN. Phan.

Polyboea

Synonyms:
Polyboea Klotzsch === **Bernardia** Houst. ex Mill.

Polydragma

Synonyms:
Polydragma Hook.f. === **Spathiostemon** Blume

Poranthera

8 species, Australia (with *P. microphylla* extending to the South Island of New Zealand); little-branched annual or perennial herbs or subshrubs to 0.60 m of grassland, woodland and scrub, often on sand. The presence of narrow cotyledons resulted it being placed by Mueller in his 'Stenolobae'; this was followed by Pax and Hoffmann. As with other Australasian genera of that group, however, it appears instead to be a specialised offshoot from the *Antidesma* line (Webster, Synopsis, 1994). No revision has appeared since 1913. (Phyllanthoideae)

   Gruening, G. (1913). *Poranthera*. In A. Engler (ed.), Das Pflanzenreich, IV 147 [Stenolobieae] (Euphorbiaceae-Porantheroideae et Ricinocarpoideae): 13-21. Berlin. (Heft 58.) La/Ge. — 7 species in 2 subgenera; Australia, New Zealand. [Species 7, *P. alpina*, now separated as *Oreoporanthera*, leaving the genus exclusively Australian.]

Poranthera Rudge, Trans. Linn. Soc. London 10: 302 (1811).
   Australia, New Zealand. 50 51. Ther. or cham.

Poranthera coerulea O.Schwarz, Repert. Spec. Nov. Regni Veg. 24: 87 (1927).
   Northern Territory. 50 NTA. Ther.

Poranthera corymbosa Al.Brongn. in L.I.Duperrey, Voy. Monde: 220 (1829). *Poranthera corymbosa* var. *genuina* Müll.Arg. in A.P.de Candolle, Prodr. 15(2): 192 (1866), nom. inval. – FIGURE, p. 1368.
   Queensland (Sandy I.), New South Wales. 50 NSW QLD. Ther. or cham.

   var. **arbuscula** Müll.Arg. in A.P.de Candolle, Prodr. 15(2): 192 (1866).
      New South Wales. 50 NSW. Ther. or cham.
    * *Poranthera arbuscula* Sieber ex Benth., Fl. Austral. 6: 56 (1873), pro syn.

   var. **corymbosa**
      Queensland (Sandy I.), New South Wales. 50 NSW QLD. Ther. or cham.
      Poranthera laxa Siebold ex Gruning in H.G.A.Engler, Pflanzenr., IV, 147: 19 (1913).

   var. **linarioides** (Siebold ex Benth.) Grüning in H.G.A.Engler, Pflanzenr., IV, 147: 19 (1913).
      New South Wales. 50 NSW. Ther. or cham.
    * *Poranthera linarioides* Sieber ex Baill., Étude Euphorb.: 574 (1858).

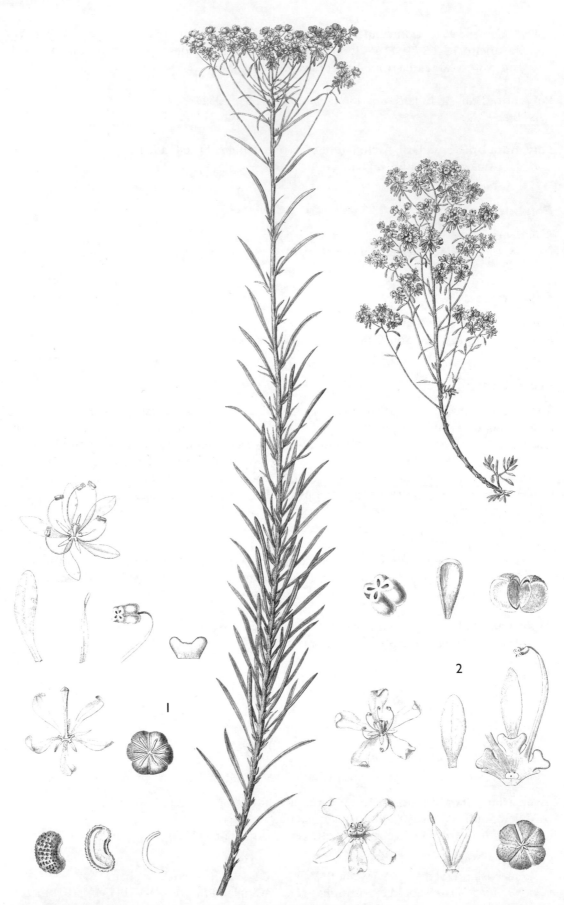

Poranthera corymbosa Brongn. var. *corymbosa* (as *P. corymbosa*) (left) (1)
Poranthera microphylla Brongn. var. *microphylla* (as *P. microphylla*) (right) (2)

Artist: Bessa
Duperrey, Voy. Monde, pl. 50 [1834]
KEW ILLUSTRATIONS COLLECTION

var. **sparsifolia** Grüning in H.G.A.Engler, Pflanzenr., IV, 147: 20 (1913).
New South Wales. 50 NSW. Ther. or cham.

Poranthera drummondii Klotzsch in J.G.C.Lehmann, Pl. Preiss. 2: 231 (1848). *Poranthera microphylla* var. *drummondii* (Klotzsch) Müll.Arg. in A.P.de Candolle, Prodr. 15(2): 193 (1866).
SW. Australia. 50 WAU. Ther.

Poranthera ericifolia Rudge, Trans. Linn. Soc. London 10: 302 (1811).
E. Australia. 50 NSW QLD. Cham.
Poranthera ericifolia f. *estriolata* Müll.Arg. in A.P.de Candolle, Prodr. 15(2): 191 (1866).

Poranthera ericoides Klotzsch in J.G.C.Lehmann, Pl. Preiss. 2: 232 (1848).
S. & SW. Australia. 50 SOA WAU. Cham.
Poranthera glauca Klotzsch in J.G.C.Lehmann, Pl. Preiss. 2: 231 (1848).
Poranthera piceoides Klotzsch in J.G.C.Lehmann, Pl. Preiss. 2: 232 (1848).
Poranthera arbuscula Sond., Linnaea 28: 567 (1856).
Poranthera cicatricosa F.Muell. ex Benth., Fl. Austral. 6: 55 (1873), pro syn.

Poranthera huegelii Klotzsch in J.G.C.Lehmann, Pl. Preiss. 2: 231 (1848).
SW. Australia. 50 WAU. Ther.

Poranthera microphylla Brongn. in L.I.Duperrey, Voy. Monde: 221 (1829). *Poranthera microphylla* var. *genuina* Müll.Arg. in A.P.de Candolle, Prodr. 15(2): 193 (1866), nom. inval. – FIGURE, p. 1368.
E. & S. Australia, Tasmania, New Zealand South I. 50 NSW NTA? QLD SOA TAS VIC WAU VIC 51 NZS. Ther. or cham.

var. **diffusa** Müll.Arg. in A.P.de Candolle, Prodr. 15(2): 193 (1866).
Victoria, Tasmania. 50 TAS VIC. Ther. or cham.

var. **glauca** Müll.Arg. in A.P.de Candolle, Prodr. 15(2): 193 (1866).
Australia. 50 NSW QLD WAU VIC. Ther. or cham.

var. **intermedia** Müll.Arg. in A.P.de Candolle, Prodr. 15(2): 193 (1866).
Western Australia. 50 WAU. Ther. or cham.

var. **microphylla**
E. & S. Australia, Tasmania, New Zealand South I. 50 NSW NTA? QLD SOA TAS VIC WAU 51 NZS. Ther. or cham.

var. **procera** Grüning in H.G.A.Engler, Pflanzenr., IV, 147: 18 (1913).
Western Australia. 50 WAU. Ther. or cham.

Poranthera triandra J.M.Black, Trans. Roy. Soc. South Australia 40: 66 (1916).
S. Australia. 50 SOA.

Synonyms:
Poranthera alpina Cheeseman ex Hook.f. === **Oreoporanthera alpina** (Cheeseman ex Hook.f.) Hutch.
Poranthera arbuscula Sieber ex Benth. === **Poranthera corymbosa** var. **arbuscula** Müll.Arg.
Poranthera arbuscula Sond. === **Poranthera ericoides** Klotzsch
Poranthera cicatricosa F.Muell. ex Benth. === **Poranthera ericoides** Klotzsch
Poranthera corymbosa var. *genuina* Müll.Arg. === **Poranthera corymbosa** Al.Brongn.
Poranthera ericifolia f. *estriolata* Müll.Arg. === **Poranthera ericifolia** Rudge
Poranthera glauca Klotzsch === **Poranthera ericoides** Klotzsch
Poranthera laxa Siebold ex Gruning === **Poranthera corymbosa** Al.Brongn. var. **corymbosa**
Poranthera linarioides Sieber ex Baill. === **Poranthera corymbosa** var. **linarioides** (Siebold ex Benth.) Grüning

Poranthera microphylla var. *drummondii* (Klotzsch) Müll.Arg. === **Poranthera drummondii** Klotzsch

Poranthera microphylla var. *genuina* Müll.Arg. === **Poranthera microphylla** Brongn.

Poranthera piceoides Klotzsch === **Poranthera ericoides** Klotzsch

Poraresia

Synonyms:
Poraresia Gleason === **Pogonophora** Miers ex Benth.

Prosartema

Synonyms:
Prosartema Gagnep. === **Trigonostemon** Blume
Prosartema gaudichaudii Gagnep. === **Trigonostemon gagnepainianus** Airy Shaw
Prosartema stellare Gagnep. === **Trigonostemon stellaris** (Gagnep.) Airy Shaw

Prosorus

Synonyms:
Prosorus Dalzell === **Margaritaria** L.f.
Prosorus cyanospermus (Gaertn.) Thwaites === **Margaritaria cyanosperma** (Gaertn.) Airy Shaw
Prosorus gaertneri Thwaites === **Margaritaria cyanosperma** (Gaertn.) Airy Shaw
Prosorus indicus Dalzell === **Margaritaria indica** (Dalzell) Airy Shaw
Prosorus luzoniensis (Merr.) Airy Shaw === **Margaritaria luzoniensis** (Merr.) Airy Shaw

Protomegabaria

3 species, W. & WC. tropical Africa (Ghana to Zaïre); small to medium forest trees to 25 m allied to *Baccaurea* and placed by Webster (Synopsis, 1994) in Antidesmeae subtribe Scepinae. *P. stapfiana* often occurs in swampy situations. (Phyllanthoideae)

Hutchinson, J. (1911). *Protomegabaria stapfiana* (Beille) Hutch. Ic. Pl. 30: pl. 2929. En. — Plant portrait with description and commentary; includes protologue of genus and transfer of *Maesobotrya stapfiana*.

Pax, F. & K. Hoffmann (1922). *Protomegabaria*. In A. Engler (ed.), Das Pflanzenreich, IV 147 XV (Euphorbiaceae-Phyllanthoideae-Phyllantheae): 43-44. Berlin. (Heft 81.) La/Ge. — 2 species, Africa (W Africa).

Léonard, J. (1995). *Protomegabaria meiocarpa* J. Léonard, espèce nouvelle d'un genre nouveau pour le Zaïre (Euphorbiaceae). Bull. Jard. Bot. Natl. Belg. 64: 53-63, illus. Fr. — Novelty with an extension of the generic range; extensive discussion of the genus. Only *P. meiocarpa* is considered in detail. [For additional treatment, see pp. 79-82 in *idem, Flore d'Afrique Centrale: Euphorbiaceae* 2: 79-82, illus. (1995).]

Protomegabaria Hutch., Hooker's Icon. Pl. 30: t. 2929 (1911).
W. & WC. Trop. Africa. 22 23. Phan.

Protomegabaria macrophylla (Pax) Hutch., Hooker's Icon. Pl. 30: t. 2929 (1911).
Ghana, S. Nigeria, W. Cameroon. 22 GHA NGA 23 CAM. Phan. Probably not distinct from *P. stapfiana*.
 * *Baccaurea macrophylla* Pax, Bot. Jahrb. Syst. 28: 21 (1899). *Megabaria macrophylla* (Pax) Pierre ex A.Chev., Veg. Ut. Afr. Trop. Franc. 9: 292 (1917).
 Megabaria klaineana Pierre ex Pax in H.G.A.Engler, Pflanzenr., IV, 147, XV: 44 (1922), nom. illeg.

Protomegabaria meiocarpa J.Léonard, Bull. Jard. Bot. Belg. 64: 60 (1995).
  E. Zaire. 23 ZAI. Phan.

Protomegabaria stapfiana (Beille) Hutch., Hooker's Icon. Pl. 30: t. 2929 (1911).
  – FIGURE, p. 1372.
  W. & WC. Trop. Africa. 22 GHA IVO LBR NGA SIE 23 CMN GAB. Phan.
    * *Maesobotrya stapfiana* Beille, Bull. Soc. Bot. France 57(8): 121 (1910).
    Megabaria obovata Pierre ex Hutch., Hooker's Icon. Pl. 30: t. 2929 (1911).
    Spondianthus obovatus Engl., Notizbl. Bot. Gart. Berlin-Dahlem 5: 242 (1911).

Pseudagrostistachys

2 species, W., C. and E. tropical Africa (Ghana and Nigeria to Uganda and Zambia; also in Bioko and São Tomé). Shrubs or small to medium trees to 20 m with many-veined large leaves, generally found at higher elevations with both species often found in swamp forest understorey or along rivers. The genus is by Webster (Synopsis, 1994) grouped with the Asian *Agrostistachys* as well as two others in Agrostistachydeae. (Acalyphoideae)

> Pax, F. (with K. Hoffmann) (1912). *Pseudagrostistachys.* In A. Engler (ed.), Das Pflanzenreich, IV 147 VI (Euphorbiaceae-Acalypheae-Chrozophorinae): 96-98. Berlin. (Heft 57.) La/Ge. — 1 species, Africa.
> Lebrun, J. (1934). Note sur le genre *Pseudagrostistachys* Pax et K. Hoffm. (Euphorbiacées). Bull. Soc. Roy. Bot. Belg. 67: 97-100. Fr. — General discussion, with key; synoptic treatment of 3 species (one new, including description) with indication of synonymy and distribution.

Pseudagrostistachys Pax & K.Hoffm. in H.G.A.Engler, Pflanzenr., IV, 147, VI: 96 (1912).
  WC. & E. Trop. Africa. 23 25.

Pseudagrostistachys africana (Müll.Arg.) Pax & K.Hoffm. in H.G.A.Engler, Pflanzenr., IV, 147, VI: 97 (1912).
  Ghana, Nigeria, Bioko, São Tomé, Zaire. 22 GHA NGA 23 GGI ZAI. Phan.
    * *Agrostistachys africana* Müll.Arg., Flora 47: 534 (1864).

  subsp. **africana**
    Ghana, Nigeria, Bioko, São Tomé. 22 GHA NGA 23 GGI. Phan.

  subsp. **humbertii** (Lebrun) J.Léonard, in Fl. Congo Rwanda-Burundi 8(1): 184 (1962).
    Zaire. 23 ZAI. Phan.
    * *Pseudagrostistachys humbertii* Lebrun, Bull. Soc. Roy. Bot. Belgique 67: 99 (1934).

Pseudagrostistachys ugandensis (Hutch.) Pax & K.Hoffm. in H.G.A.Engler, Pflanzenr., IV, 147, XVII: 180 (1924).
  Uganda, Zaire (Orientale, Kivu), Zambia. 23 ZAI 25 UGA 26 ZAM. Phan.
    * *Agrostistachys ugandensis* Hutch., Bull. Misc. Inform. Kew 1917: 233 (1917).

Synonyms:
Pseudagrostistachys humbertii Lebrun === **Pseudagrostistachys africana** subsp. **humbertii** (Lebrun) J.Léonard

Pseudanthus

11 species, Australia; shrubs or subshrubs of scrub or woodland with minute or small narrow leaves, sometimes also in rocky or sandy places. The presence of narrow cotyledons resulted it being placed by Mueller in his 'Stenolobae'; this was followed by Pax and Hoffmann. Like other Australian genera of that group, however, it appears

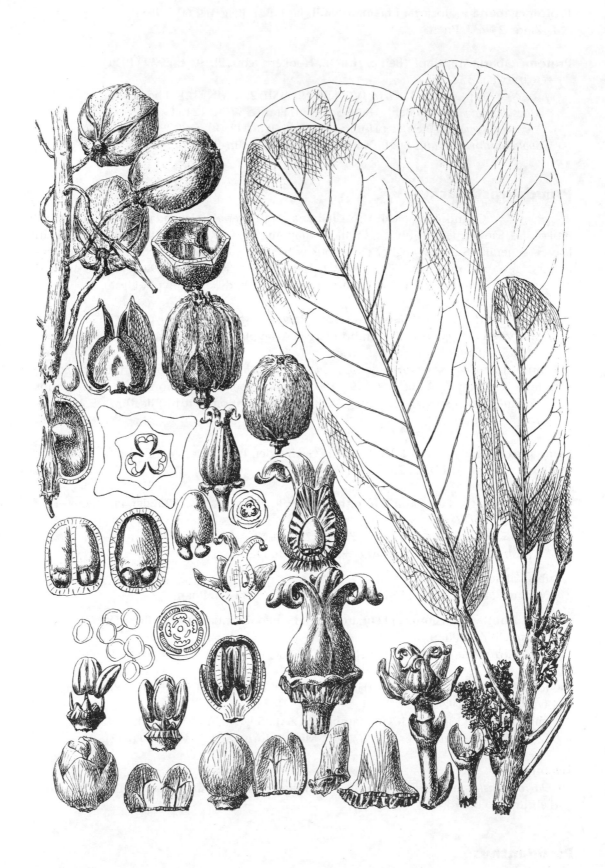

Protomegabaria stapfiana (Beille) Hutch. (as *Megabaria obovatum* Pierre in sched.)
Artist: E. Delpy
Unpublished

instead to be a specialised derivative, in this case within Oldfieldioideae. It now includes *Stachystemon*, still kept separate by Webster (Synopsis, 1994). The genus is close to *Micrantheum* but differs in the absence of foliose stipules and 1-seeded fruits. Both genera are in the Webster system part of subtribe Pseudanthinae in the Caletieae. No recent revision is available but Radcliffe-Smith (1994) has excluded *P. tasmanicus* from the genus and from Euphorbiaceae. (Oldfieldioideae)

Gruening, G. (1913). *Pseudanthus*. In A. Engler (ed.), Das Pflanzenreich, IV 147 [Stenolobieae] (Euphorbiaceae-Porantheroideae et Ricinocarpoideae): 26-32. Berlin. (Heft 58.) La/Ge. — 7 species in 3 sections, all in Australia.

Gruening, G. (1913). *Stachystemon*. In A. Engler (ed.), Das Pflanzenreich, IV 147 [Stenolobieae] (Euphorbiaceae-Porantheroideae et Ricinocarpoideae): 32-35. Berlin. (Heft 58.) La/Ge. — 3 species; W. Australia. [Genus now reduced to *Pseudanthus*.]

Weber, J. Z.; del. J. Morley (1985). Plant portraits, 16: *Pseudanthus micranthus* Benth. (Euphorbiaceae). J. Adelaide Bot. Gard. 7: 212, 214-215, col. illus. En. — Colour portrait (p. 212) with description and commentary.

Radcliffe-Smith, A. (1993). Notes on Australian Euphorbiaceae, 2: *Pseudanthus* and *Stachystemon*. Kew Bull. 48: 165-168. En. — Comparison of genera with table of androecial features for each species; on the basis of the recently described and 'intermediate' *Stachystemon axillaris* that genus was reduced.

Radcliffe-Smith, A. (1994). A further note on *Pseudanthus tasmanicus* Rodway. Kew Bull. 49: 260. En. —*P. tasmanicus* referable to *Muehlenbeckia axillaris* (Hook. f.) Endl. (Polygonaceae).

Pseudanthus Sieber ex Spreng., Syst. Veg. 4(2): 25 (1827).
Australia. 50. Nanophan.
  Stachystemon Planch., London J. Bot. 4: 471 (1845).
  Chrysostemon Klotzsch in J.G.C.Lehmann, Pl. Preiss. 2: 232 (1848).
  Chorizotheca Müll.Arg., Linnaea 32: 76 (1863).

Pseudanthus axillaris (A.S.George) Radcl.-Sm., Kew Bull. 48: 167 (1993).
Western Australia. 50 WAU. Nanophan.
  * *Stachystemon axillaris* George, J. Roy. Soc. W. Australia 50(4): 97 (1967).

Pseudanthus brachyphyllus (Müll.Arg.) F.Muell., Trans. Roy. Soc. New South Wales: 11 (1881).
Western Australia. 50 WAU. Nanophan.
  * *Stachystemon brachyphyllus* Müll.Arg., Linnaea 32: 76 (1863).
  Stachystemon brevifolius Planch. ex Grüning in H.G.A.Engler, Pflanzenr., IV, 147: 34 (1913), pro syn.

Pseudanthus divaricatissimus (Müll.Arg.) Benth., Fl. Austral. 6: 60 (1873).
New South Wales, Victoria. 50 NSW VIC. Nanophan.
  * *Caletia divaricatissima* Müll.Arg., Linnaea 32: 79 (1863). *Caletia divaricatissima* var. *genuinus* Müll.Arg., Flora 22: 486 (1864), nom. inval.
  Caletia divaricatissima var. *orbicularis* Müll.Arg., Flora 22: 486 (1864). *Pseudanthus divaricatissimus* var. *orbicularis* (Müll.Arg.) Benth., Fl. Austral. 6: 60 (1873).
  Caletia orientalis var. *orbicularis* Baill., Adansonia 6: 327 (1866).
  Pseudanthus divaricatissimus var. *genuinus* (Müll.Arg.) Grüning in H.G.A.Engler, Pflanzenr., IV, 147: 30 (1913), nom. inval.

Pseudanthus micranthus Benth., Fl. Austral. 6: 59 (1873).
South Australia. 50 SOA. Nanophan.

Pseudanthus nematophorus F.Muell., Fragm. 2: 14 (1861).
Western Australia. 50 WAU. Nanophan.

Pseudanthus orientalis F.Muell., Fragm. 2: 14 (1861). *Caletia orientalis* (F.Muell.) Baill., Adansonia 6: 327 (1866).

New South Wales, Queensland. 50 NSW QLD. Nanophan.

    Caletia linearis Müll.Arg., Linnaea 32: 79 (1863).

    Pseudanthus brunonis Endl. ex Grüning in H.G.A.Engler, Pflanzenr., IV, 147: 28 (1913).

Pseudanthus ovalifolius F.Muell., Trans. & Proc. Philos. Inst. Victoria 2: 66 (1858). *Caletia ovalifolia* (F.Muell.) Müll.Arg., Linnaea 34: 55 (1865).

New South Wales, Victoria, Tasmania. 50 NSW TAS VIC. Nanophan.

    Caletia wilhelmii F.Muell. ex Müll.Arg. in A.P.de Candolle, Prodr. 15(2): 194 (1866).

Pseudanthus pimeleoides Sieber ex Spreng., Syst. Veg. 4(2): 25 (1827).

New South Wales, Victoria. 50 NSW VIC. Nanophan.

Pseudanthus polyandrus F.Muell., Fragm. 2: 153 (1861). *Stachystemon polyandrus* (F.Muell.) Benth., Fl. Austral. 6: 62 (1873).

Western Australia. 50 WAU. Cham. or nanophan.

    Pseudanthus chryseus Müll.Arg., Flora 47: 486 (1864).

Pseudanthus vermicularis (Planch.) F.Muell., Trans. Roy. Soc. New South Wales: 11 (1881).
– FIGURE, p. 1375.

Western Australia. 50 WAU. Nanophan.

    * *Stachystemon vermicularis* Planch., London J. Bot. 4: 472 (1845).

Pseudanthus virgatus (Klotzsch) Müll.Arg., Linnaea 34: 56 (1865).

Western Australia. 50 WAU. Nanophan.

    * *Chrysostemon virgatus* Klotzsch in J.G.C.Lehmann, Pl. Preiss. 2: 232 (1848).

    Pseudanthus occidentalis F.Muell., Fragm. 1: 107 (1859).

    Chorizotheca micrantheoides Müll.Arg., Linnaea 32: 76 (1863).

    Pseudanthus nitidus Müll.Arg. in A.P.de Candolle, Prodr. 15(2): 197 (1866).

Synonyms:

Pseudanthus brunonis Endl. ex Grüning === **Pseudanthus orientalis** F.Muell.

Pseudanthus chryseus Müll.Arg. === **Pseudanthus polyandrus** F.Muell.

Pseudanthus divaricatissimus var. *genuinus* (Müll.Arg.) Grüning === **Pseudanthus divaricatissimus** (Müll.Arg.) Benth.

Pseudanthus divaricatissimus var. *orbicularis* (Müll.Arg.) Benth. === **Pseudanthus divaricatissimus** (Müll.Arg.) Benth.

Pseudanthus nitidus Müll.Arg. === **Pseudanthus virgatus** (Klotzsch) Müll.Arg.

Pseudanthus occidentalis F.Muell. === **Pseudanthus virgatus** (Klotzsch) Müll.Arg.

Pseudanthus tasmanicus Rodway === **Muhlenbeckia axillaris** (Hook.f.) Endl.
(Polygonaceae)

Pseudoglochidion

Synonyms:

Pseudoglochidion Gamble === **Phyllanthus** L.

Pseudoglochidion anamalayanum Gamble === **Phyllanthus anamalayanus** (Gamble) G.L.Webster

Pseudolachnostylis

1 species, Africa (Zaïre and Tanzania southwards); much-branched trees to 18 m of bush- and woodland, sometimes in low-lying places, the leaves distichously arranged and more or less pubescent on the under surface. Opinions have varied on the number of species worthy of recognition and, more recently, whether or not any infraspecific taxa should be accepted; here the genus is treated as monospecific with four varieties following Radcliffe-Smith

Pseudanthus vermicularis (Planch.) F. Muell. (as *Stachystemon vermiculare* Planch.)

Artist: W.F. & M. Planchon
Lond. J. Bot. 4: pl. 15 (1845)
KEW ILLUSTRATIONS COLLECTION

(1978). It was assigned to Pseudolachnostylidinae by Pax & Hoffmann but the tribe may be unnatural (Webster, Synopsis, 1994). Pollen similarity has been given as evidence for an affinity with *Amanoa* but this was not accepted by Webster who places it next to *Keayodendron*. (Phyllanthoideae)

Pax, F. & K. Hoffmann (1922). *Pseudolachnostylis*. In A. Engler (ed.), Das Pflanzenreich, IV 147 XV (Euphorbiaceae-Phyllanthoideae-Phyllantheae): 206-209. Berlin. (Heft 81.) La/Ge. — 4-5 species, Africa. [In FTEA considerably reduced.]

Radcliffe-Smith, A. (1978). Notes on African Euphorbiaceae, VIII. Kew Bull. 33: 233-242, illus. En. —*Pseudolachnostylis*, p. 242; contains a discussion of the progress of reduction of former segregate species with a suggestion that some formal infraspecific taxa should be recognised.

Pseudolachnostylis Pax, Bot. Jahrb. Syst. 28: 19 (1899).
 Trop. Africa. 23 25 26. Nanophan. or phan.

Pseudolachnostylis maprouneifolia Pax, Bot. Jahrb. Syst. 28: 20 (1899).
 – FIGURE, p. 1377.
 Tanzania to S. Africa. 23 BUR ZAI 25 TAN 26 ANG MLW MOZ ZAM ZIM 27 BOT NAM TVL. Nanophan. or phan.

var. **dekindtii** (Pax) Radcl.-Sm., Kew Bull. 33: 242 (1978).
 Zaire (Shaba), Tanzania, Malawi, Zambia, Zimbabwe, Mozambique, Angola, Namibia, Botswana, Transvaal. 23 ZAI 25 TAN 26 ANG MLW MOZ ZAM ZIM 27 BOT NAM TVL. Nanophan. or phan.
 * *Pseudolachnostylis dekindtii* Pax, Bot. Jahrb. Syst. 28: 20 (1899).

var. **glabra** (Pax) Brenan, Mem. New York Bot. Gard. 9: 67 (1954).
 Zaire (Shaba), Burundi, Tanzania to S. Africa. 23 BUR ZAI 25 TAN 26 ANG MLW MOZ ZAM ZIM 27 BOT NAM TVL. Nanophan. or phan.
 Cleistanthus glaucus Hiern, Cat. Afr. Pl. 1: 955 (1900). *Pseudolachnostylis glauca* (Hiern) Hutch. in D.Oliver, Fl. Trop. Afr. 6(1): 671 (1912).
 Pseudolachnostylis verdickii De Wild., Ann. Mus. Congo Belge, Bot., IV, 1: 205 (1903).
 * *Pseudolachnostylis dekindtii* var. *glabra* Pax, Bot. Jahrb. Syst. 43: 75 (1909).
 Pseudolachnostylis bussei Pax ex Hutch. in D.Oliver, Fl. Trop. Afr. 6(1): 672 (1912).

var. **maprouneifolia**
 Zaire (Shaba), Tanzania, Burundi, Malawi, Mozambique, Zambia, Zimbabwe, Botswana. 23 BUR ZAI 25 TAN 26 MLW MOZ ZAM ZIM 27 BOT. Nanophan. or phan.

var. **polygyna** (Pax & K.Hoffm.) Radcl.-Sm., Kew Bull. 33: 242 (1978).
 Tanzania, Malawi, Zambia. 25 TAN 26 MLW ZAM. Nanophan. or phan.
 * *Pseudolachnostylis polygyna* Pax & K.Hoffm. in H.G.A.Engler & C.G.O.Drude, Veg. Erde 9, III(2): 31 (1921).

Synonyms:
Pseudolachnostylis bussei Pax ex Hutch. === **Pseudolachnostylis maprouneifolia** var. **glabra** (Pax) Brenan
Pseudolachnostylis dekindtii Pax === **Pseudolachnostylis maprouneifolia** var. **dekindtii** (Pax) Radcl.-Sm.
Pseudolachnostylis dekindtii var. *glabra* Pax === **Pseudolachnostylis maprouneifolia** var. **glabra** (Pax) Brenan
Pseudolachnostylis glauca (Hiern) Hutch. === **Pseudolachnostylis maprouneifolia** var. **glabra** (Pax) Brenan
Pseudolachnostylis polygyna Pax & K.Hoffm. === **Pseudolachnostylis maprouneifolia** var. **polygyna** (Pax & K.Hoffm.) Radcl.-Sm.
Pseudolachnostylis verdickii De Wild. === **Pseudolachnostylis maprouneifolia** var. **glabra** (Pax) Brenan

Pseudolachnostylis maprouneifolia Pax
Artist: Matilda Smith
Ic. Pl. 31: pl. 3011 (1915)

Pseudosagotia

Now reduced to *Croizatia* (q.v.)

Synonyms:
Pseudosagotia Secco === **Croizatia** Steyerm.
Pseudosagotia brevipetiolata Secco === **Croizatia brevipetiolata** (Secco) Govaerts

Pseudotragia

Synonyms:
Pseudotragia Pax === **Pterococcus** Hassk.

Psilostachys

Synonyms:
Psilostachys Turcz. === **Cleidion** Blume

Pterococcus

3 species, Africa (2,) Asia and Malesia (1); slender woody climbers. Very close to *Plukenetia* and likely to be united with it (Gillespie 1993; see that genus); the floral differences are minor against a basic likeness. Webster (Synopsis, 1994) has also made the reduction. (Acalyphoideae)

Pax, F. & K. Hoffmann (1919). *Pterococcus*. In A. Engler (ed.), Das Pflanzenreich, IV 147 IX (Euphorbiaceae-Acalypheae-Plukenetiinae): 21-24. Berlin. (Heft 68.) La/Ge. — 3 species, Africa, Asia, W Malesia.

Pterococcus Hassk., Flora 25(2): 41 (1842).
Trop. Asia. 40 41 42.
    Ceratococcus Meisn., Pl. Vasc. Gen.: 369 (1843).
    Hedraiostylus Hassk., Tijdschr. Ned.-Indië 10: 141 (1843).
    Sajorium Endl., Gen. Pl., Suppl. 3: 98 (1843).
    Pseudotragia Pax, Bull. Herb. Boissier, II, 8: 635 (1908).

Pterococcus africanus (Sond.) Pax & K.Hoffm. in H.G.A.Engler, Pflanzenr., IV, 147, IX: 22 (1919).
Namibia to Mozambique. 26 MOZ 27 BOT NAM TVL. Cl. hemicr. or cham.
  * *Plukenetia africana* Sond., Linnaea 23: 110 (1850).
    Plukenetia hastata Müll.Arg., Flora 47: 469 (1864).
    Tragia schultzeana Dinter ex Pax & K.Hoffm. in H.G.A.Engler, Pflanzenr., IV, 147, IX: 23 (1919), pro syn.

Pterococcus corniculatus (Sm.) Pax & K.Hoffm. in H.G.A.Engler, Pflanzenr., IV, 147, IX: 22 (1919).
Bhutan, Sikkim, Assam, Burma, Thailand, W. Sumatera, Borneo (Sarawak), Philippines, Flores, Timor, Moluccas. 40 ASS EHM 41 BMA THA 42 BOR LSI MOL PHI SUM. Cl. nanophan.
  * *Plukenetia corniculata* Sm., Nova Acta Regiae Soc. Sci. Upsal. 6: 4 (1799).
    Pterococcus glaberrimus Hassk., Flora 25(2): 41 (1842).

Pterococcus procumbens (Prain) Pax & K.Hoffm. in H.G.A.Engler, Pflanzenr., IV, 147, IX: 23 (1919).
Angola. 26 ANG. Cl. hemicr.
  * *Plukenetia procumbens* Prain, Bull. Misc. Inform. Kew 1912: 240 (1912).

Synonyms:
Pterococcus glaberrimus Hassk. === **Pterococcus corniculatus** (Sm.) Pax & K.Hoffm.

Ptychopyxis

13 species, SE. Asia and Malesia (W. Malesia and New Guinea). Small to medium forest trees, occasionally to 27 m (*P. caput-medusae*); their capsules are large, usually furry, and often rough to spiny. The section *Podadenia* is now treated as a separate genus following Webster (Synopsis, 1994); both remain in his subtribe Blumeodendrinae of the Pycnocomeae. Our current understanding of this genus is relatively recent. A synoptic treatment was produced by Airy-Shaw (1960). Of individual species, only that in the former *Clarorivinia* (along with some others then in *Mallotus*) was covered by Pax in *Pflanzenreich*. *P. costata* was treated along with the genus by Pax & Hoffmann (1931: 230) as 'unsicher'. A further revision has been prepared by Cornelius Sri Murdo Yuwono. (Acalyphoideae)

> Hooker, J. D. (1887). *Ptychopyxis costata*. Ic. Pl. 18: pl. 1703. En. — Plant portrait with descriptive text and discussion.
>
> Pax, F. (with K. Hoffmann) (1914). *Clarorivinia*. In A. Engler (ed.), Das Pflanzenreich, IV 147 VII (Euphorbiaceae-Acalypheae-Mercurialinae): 17. Berlin. (Heft 63.) La/Ge. — 1 species, New Guinea. [For an additional species, see ibid., XIV (Additamentum VI): 13 (1919). Now part of *Ptychopyxis*.]
>
> Croizat, L. (1942). On certain Euphorbiaceae from the tropical Far East. J. Arnold Arbor. 23: 29-54. En. —*Podadenia* and *Calpigyne* in synonymy; 9 spp. treated, incl. 3 new (pp. 48-50).
>
> Airy-Shaw, H. K. (1960). Notes on Malaysian Euphorbiaceae, VIII. A synopsis of the genus *Ptychopyxis* Miq. Kew Bull. 14: 363-374. En. — 12 species in sect. *Ptychopyxis*; key, descriptions of novelties and synonymy. [A second section, *Podadenia*, is now treated as a separate genus as the author hints here.]
>
> Airy-Shaw, H. K. (1963). Notes on Malaysian and other Asiatic Euphorbiaceae, XXIX. Further notes on *Ptychopyxis* Miq. Kew Bull. 16: 347. En. — Additional records for three species.
>
> Airy-Shaw, H. K. (1966). Notes on Malaysian and other Asiatic Euphorbiaceae, LX. New *Ptychopyxis* from Sumatra and Borneo. Kew Bull. 20: 27-28. En. — Description of *P. nervosa*.
>
> Airy-Shaw, H. K. (1972). Notes on Malesian and other Asiatic Euphorbiaceae, CLX: An overlooked synonym of *Ptychopyxis kingii*. Kew Bull. 27: 86. En. — New synonym.

Ptychopyxis Miq., Fl. Ned. Ind., Eerste Bijv.: 402 (1861).
  Trop. Asia. 40 41 42.
    Clarorivinia Pax & K.Hoffm. in H.G.A.Engler, Pflanzenr., IV, 147, VII: 17 (1914).

Ptychopyxis arborea (Merr.) Airy Shaw, Kew Bull. 14: 369 (1960).
  Borneo (Sarawak, Sabah, E. Kalimantan). 42 BOR. Phan. – Fruit edible.
    * Mallotus arboreus Merr., Univ. Calif. Publ. Bot. 15: 159 (1929).

Ptychopyxis bacciformis Croizat, J. Arnold Arbor. 23: 49 (1942).
  Vietnam, Borneo (Sarawak, Sabah, E. Kalimantan). 41 VIE 42 BOR. Phan.
    Ptychopyxis poilanei Croizat, J. Arnold Arbor. 23: 50 (1942).

Ptychopyxis caput-medusae (Hook.f.) Ridl., Fl. Malay Penins. 3: 295 (1924).
  Pen. Malaysia. 42 MLY.
    * Mallotus caput-medusae Hook.f., Fl. Brit. India 5: 449 (1887).

Ptychopyxis chrysantha (K.Schum.) Airy Shaw, Kew Bull. 14: 370 (1960).
  Papua New Guinea. 42 NWG. Phan.

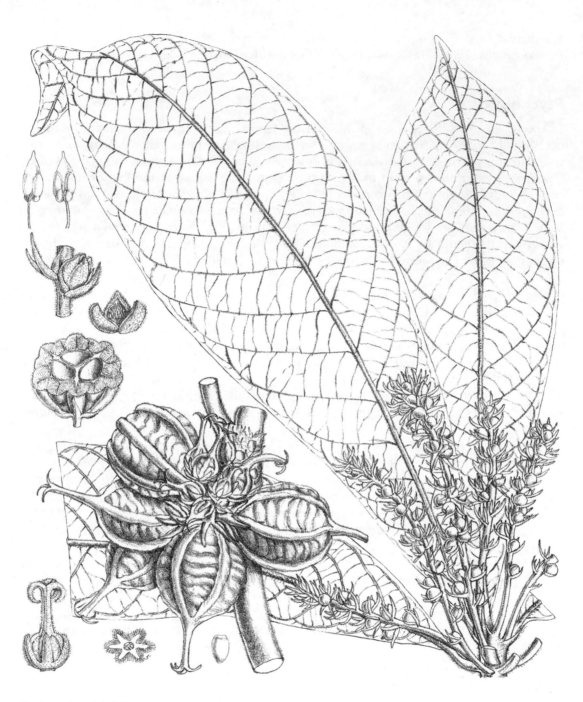

Ptychopyxis costata Miq.
Artist: Matilda Smith
Ic. Pl. 18: pl. 1703 (1887)
KEW ILLUSTRATIONS COLLECTION

* *Mallotus chrysanthus* K.Schum. in K.M.Schumann & U.M.Hollrung, Fl. Kais. Wilh.
    Land: 78 (1889). *Clarorivinia chrysantha* (K.Schum.) Pax & K.Hoffm. in H.G.A.Engler,
    Pflanzenr., IV, 147, VII: 17 (1914).
    Clarorivinia grandifolia Pax & K.Hoffm. in H.G.A.Engler, Pflanzenr., IV, 147, XIV: 13 (1919).

Ptychopyxis costata Miq., Fl. Ned. Ind., Eerste Bijv.: 402 (1861). – FIGURE, p. 1380.
    W. Malesia. 42 BOR MLY SUM. Phan.

var. **costata**
    W. Sumatera. 42 SUM. Phan.

var. **oblanceolata** Airy Shaw, Kew Bull. 14: 366 (1960).
    Pen. Malaysia, Borneo (C. Sarawak). 42 BOR MLY. Phan.

Ptychopyxis glochidiifolia Airy Shaw, Kew Bull. 14: 373 (1960).
Borneo (Sarawak, Brunei, E. Kalimantan), Papua New Guinea ? 42 BOR NWG? Phan.

Ptychopyxis grandis Airy Shaw, Kew Bull. 14: 367 (1960).
Borneo (Sarawak). 42 BOR. Phan.

Ptychopyxis javanica (J.J.Sm.) Croizat, J. Arnold Arbor. 23: 49 (1942).
S. Thailand, Pen. Malaysia, Sumatera ?, Borneo (Sabah, E. Kalimantan?), Jawa. 41 THA 42 BOR JAW MLY SUM? Phan.
* *Podadenia javanica* J.J.Sm., Meded. Dept. Landb. Ned.-Indië 10: 388 (1910).
Ptychopyxis angustifolia Gage, Rec. Bot. Surv. India 9: 248 (1922).
Ptychopyxis arborea var. *cacuminum* Airy Shaw, Kew Bull. 20: 28 (1966).

Ptychopyxis kingii Ridl., Fl. Malay Penins. 3: 296 (1924).
Pen. Malaysia, E. Sumatera, Borneo (Sarawak, Sabah). 42 BOR MLY SUM. Phan.
– Fruit edible.
Mallotus arboreus var. *platyphyllus* Merr., Pap. Michigan Acad. Sci. 19: 161 (1933).

Ptychopyxis nervosa Airy Shaw, Kew Bull. 20: 27 (1966).
Sumatera. 42 SUM. Phan.

Ptychopyxis philippina Croizat, J. Arnold Arbor. 23: 49 (1942).
Philippines. 42 PHI.

Ptychopyxis plagiocarpa Airy Shaw, Kew Bull. 14: 372 (1960).
SW. Thailand. 41 THA. Phan.

Ptychopyxis triradiata Airy Shaw, Kew Bull. 14: 366 (1960).
Pen. Malaysia. 42 MLY.

Synonyms:
Ptychopyxis angustifolia Gage === **Ptychopyxis javanica** (J.J.Sm.) Croizat
Ptychopyxis arborea var. *cacuminum* Airy Shaw === **Ptychopyxis javanica** (J.J.Sm.) Croizat
Ptychopyxis frutescens (Blume) Croizat === **Koilodepas frutescens** (Blume) Airy Shaw
Ptychopyxis poilanei Croizat === **Ptychopyxis bacciformis** Croizat
Ptychopyxis thwaitesii (Müll.Arg.) Croizat === **Podadenia sapida** Thwaites

Putranjiva

4 species, S. Asia, Sri Lanka, and parts of Malesia (including New Guinea); questionably distinct from *Drypetes* (the main distinguishing characters being absence of a disk, 2-4 instead of 3-25 or more stamens, extrorse anthers and dilated styles). Medium to large forest trees; *P. roxburghii*, whose distribution suggests a preference for seasonal conditions, is often cultivated. Never revised for *Das Pflanzenreich* (save for *P. roxburghii*, included in *Drypetes* by Pax & Hoffmann (1922) as no. 137, *D. timorensis*) and only briefly surveyed in Pax & Hoffmann (1931; 4 species in S., SE. and E. Asia). All known species were reviewed by Hurusawa (1954; see **Asia**); with little discussion, he opted for their inclusion in two sections in a subgeneric *Putrajiva* within *Drypetes*. Airy-Shaw also followed Hurusawa's lead, choosing reduction of the whole genus (*Euphorbiaceae of Siam*; see Airy-Shaw, 1972, under **Asia**). Webster and Radcliffe-Smith, however, have continued to follow Pax & Hoffmann. They were in turn followed by Philcox for *Revised Handbook to the Flora of Ceylon* (vol. 11, 1997) which includes the members of Hurusawa's sect. *Roxburghianae* (= sect. *Putranjiva*), *PP. roxburghii* and *zeylanica*. Both are listed below, along with two of those of sect. *Matsumuranae* (inclusive of *Liodendron*). The latter section is, however, closer to *Drypetes*, particularly *D. integerrima*. [Along with *Drypetes*, the genus is part of Drypeteae for which separate family status has been proposed (as Putranjivaceae); for further discussion, see *Drypetes*.] (Phyllanthoideae)

Pax, F. & K. Hoffmann (1931). *Putranjiva*. In A. Engler (ed.), Die natürlichen Pflanzenfamilien, 2. Aufl., 19c: 59. Leipzig. Ge. — Synopsis with description of genus; 4 species, S Asia, Sri Lanka, Borneo. The only survey available.

Keng, H. (1951). New or critical Euphorbiaceae from eastern Asia. J. Washington Acad. Sci. 41: 200-205. En. — *Liodendron*, pp. 201-203; protologue of genus, three species transfers (including *L. integerrimum*), and discussion. [Of these three species, *LL. integerrimum* and *matsumurae* were accounted for under *Putranjiva* by Pax and Hoffman (1931).

Hurusawa, I. (1954). Eine nochmalige Durchsicht des herkömmlichen Systems der Euphorbiaceen in weiterem Sinne. J. Fac. Sci. Univ. Tokyo III, 6: 209-342, illus., 4 pls. Ge. — Contains (pp. 335-338; as sect. *Matsumuranae*) a revision of four E. Asian species formerly included in *Putranjiva* and *Liodendron* but here assigned to *Drypetes*.

Keng, H. (1955). The Euphorbiaceae of Taiwan. Taiwania, 6: 27-66. En. — *Liodendron*, pp. 54-56; 2 species. *L. integerrimum* was, after reinvestigation referred to *Drypetes*, thus following Hurusawa. There is also a discussion of Hurusawa's broad concept of *Drypetes* (which Keng thought too wide, crossing three Pax & Hoffmann subtribes).

Tokuoka, T. & Peng Ching-i (1997(1998)). See **Drypetes**.

Putranjiva Wall., Tent. Fl. Napal.: 61 (1826).
Trop. Asia. 40 41 42. Phan.
   Nageia Roxb., Fl. Ind. ed. 1832, 3: 766 (1832).
   Palenga Thwaites, Hooker's J. Bot. Kew Gard. Misc. 8: 270 (1856).
   Liodendron H.Keng, J. Wash. Acad. Sci. 41: 201 (1951).

Putranjiva formosana Kaneh. & Sasaki ex Shimada, Cat. Governm. Herb. (Form.): 312 (1930). *Drypetes formosana* (Kaneh. & Sasaki ex Shimada) Kaneh., Formos. Trees, ed. rev.: 336 (1936). *Liodendron formosanum* (Kaneh. & Sasaki ex Shimada) H.Keng, J. Wash. Acad. Sci. 41: 202 (1951).
China (Guangdong), Taiwan. 36 CHS 38 TAI. Phan.

Putranjiva matsumurae Koidz., Bot. Mag. (Tokyo) 33: 116 (1919). *Drypetes matsumurae* (Koidz.) Kaneh., Formos. Trees, ed. rev.: 337 (1936). *Liodendron matsumurae* (Koidz.) H.Keng, J. Wash. Acad. Sci. 41: 202 (1951).
Nansei-shoto, Japan (Kyushu). 38 JAP NNS. Phan.
   Drypetes liukiuensis Hayata ex Hurus., J. Fac. Sci. Univ. Tokyo, Sect. 3, Bot. 6: 337 (1954).

Putranjiva roxburghii Wall., Tent. Fl. Napal.: 61 (1826). *Nageia putranjiva* Roxb., Fl. Ind. ed. 1832, 3: 766 (1832), nom. illeg. *Drypetes roxburghii* (Wall.) Hurus., J. Fac. Sci. Univ. Tokyo, Sect. 3, Bot. 6: 335 (1954). – FIGURE, p. 1383.
Trop. Asia. (36) chs 40 ASS EHM IND PAK SRL WHM 41 CMN LAO THA 42 BOR JAW LSI MLY MOL NWG. Phan.
   Cyclostemon racemosus Zipp. ex Span., Linnaea 15: 348 (1841).
   Pycnosandra timorensis Blume, Mus. Bot. 2: 192 (1856). *Drypetes timorensis* (Blume) Pax & K.Hoffm. in H.G.A.Engler, Pflanzenr., IV, 147, XV: 278 (1922). *Drypetes roxburghii* var. *timorensis* (Blume) Airy Shaw, Kew Bull., Addit. Ser. 4: 107 (1975).
   Putranjiva amblyocarpa Müll.Arg. in A.P.de Candolle, Prodr. 15(2): 444 (1866).
   Putranjiva sphaerocarpa Müll.Arg. in A.P.de Candolle, Prodr. 15(2): 443 (1866).

Putranjiva zeylanica (Thwaites) Müll.Arg. in A.P.de Candolle, Prodr. 15(2): 444 (1866).
Sri Lanka. 40 SRL. Phan.
   * *Palenga zeylanica* Thwaites, Hooker's J. Bot. Kew Gard. Misc. 8: 271 (1856). *Drypetes zeylanica* (Thwaites) Hurus., J. Fac. Sci. Univ. Tokyo, Sect. 3, Bot. 6: 335 (1954).

Synonyms:
Putranjiva amblyocarpa Müll.Arg. === **Putranjiva roxburghii** Wall.

Putranjiva roxburghii Wall.
Artist: W.H. Fitch
Stewart & Brandis, Forest Fl. N.W. India, Atlas: pl. 53 (1874)
KEW ILLUSTRATIONS COLLECTION

Putranjiva integerrima Koidz. === **Drypetes integerrima** (Koidz.) Hosok.
Putranjiva sphaerocarpa Müll.Arg. === **Putranjiva roxburghii** Wall.

Pycnocoma

18 species, W. to S. Tropical Africa (most) and Madagascar (1); shrubs or small trees in understorey of primary or secondary forest. *P. angustifolia* (West Africa) is monocaulous, with axillary inflorescences. The genus is related to *Argomuellera* and by some the two have

been united. The two are the main constituents of Webster's subtribe Pycnocominae. Léonard in his list of the genus (1959) accounted for 15 species; his revision for Zaïre (1996) covered 10 of which 4 were new. In the latter country they are mostly relatively local vicariants related to habitat and altitude. (Acalyphoideae)

Pax, F. (with K. Hoffmann) (1914). *Pycnocoma*. In A. Engler (ed.), Das Pflanzenreich, IV 147 VII (Euphorbiaceae-Acalypheae-Mercurialinae): 52-59. Berlin. (Heft 63.) La/Ge. — 12 species, Africa.

Léonard, J. (1959). Observations sur les genres *Pycnocoma* et *Argomuellera* (Euphorbiacées africaines). Bull. Soc. Roy. Bot. Belg. 91: 267-281. Fr. — General discussion of the two genera with evaluation of characters and key; synopsis of *Pycnocoma* (pp. 273-274; 15 species) with synonymy but no key or indication of distribution.

• Léonard, J. (1996). Révision des espèces zaïroises du genre *Pycnocoma* Benth. (Euphorbiaceae). Bull. Jard. Bot. Natl. Belg. 65: 37-72, illus., maps. Fr. — Treatment of 10 species (4 new) with key, synonymy, references and citations, types, localities with exsiccatae, indication of habitat and chorology, and notes on uses and biology; chorological summary (pp. 66-71) and index at end. A key for distinguishing the genus from *Argomuellera* appers in the general part. [The species are mostly relatively local vicariants based on habitat and altitude.]

Pycnocoma Benth. in W.J.Hooker, Niger Fl.: 508 (1849).
  Trop. Africa, Madagascar. 22 23 25 26 29.
    Comopyena Kuntze in T.E.von Post & C.E.O.Kuntze, Lex. Gen. Phan.: 138 (1903).

Pycnocoma angustifolia Prain, Bull. Misc. Inform. Kew 1908: 439 (1908).
  Sierra Leone, Liberia, Ivory Cost. 22 IVO LBR SIE. Nanophan.
    Pycnocoma beillei A.Chev. ex Hutch. & Dalziel, Fl. W. Trop. Afr. 1: 307 (1928), nom. illeg.

Pycnocoma bampsiana J.Léonard, Bull. Jard. Bot. Belg. 65: 62 (1996).
  C. Zaire. 23 ZAI. Nanophan.

Pycnocoma chevalieri Beille, Bull. Soc. Bot. France 55(8): 81 (1908).
  Congo to N. Uganda. 23 CAF CON ZAI 24 SUD 25 UGA. Nanophan. or phan.
    Pycnocoma lucida Pax & K.Hoffm. in H.G.A.Engler, Pflanzenr., IV, 147, VII: 56 (1914).

Pycnocoma cornuta Müll.Arg., Flora 47: 483 (1864).
  Ghana, Nigeria. 22 GHA NGA. Nanophan.

Pycnocoma dentata Hiern, Cat. Afr. Pl. 1: 983 (1900).
  Angola. 26 ANG. Nanophan. or phan.

Pycnocoma devredii J.Léonard, Bull. Jard. Bot. Belg. 65: 50 (1996).
  Zaire (Kasai). 23 ZAI. Nanophan.

Pycnocoma elua J.Léonard, Bull. Jard. Bot. État 25: 295 (1955).
  C. Zaire. 23 ZAI. Nanophan.

Pycnocoma insularum J.Léonard, Bull. Jard. Bot. État 25: 297 (1955).
  C. Zaire. 23 ZAI. Nanophan.

Pycnocoma littoralis Pax, Bot. Jahrb. Syst. 19: 100 (1894).
  SE. Kenya, E. Tanzania. 25 KEN TAN. Nanophan. or phan.

Pycnocoma louisii J.Léonard, Bull. Jard. Bot. Belg. 65: 52 (1996).
  Zaire. 23 ZAI. Nanophan.

Pycnocoma macrantha Pax ex Engl., Abh. Königl. Akad. Wiss. Berlin 1894: 44 (1894).
   – FIGURE, p. 1386.
   Tanzania (Lushoto). 25 TAN. Nanophan. or phan.

Pycnocoma macrophylla Benth. in W.J.Hooker, Niger Fl.: 508 (1849). *Pycnocoma*
   macrophylla var. *genuina* Pax & K.Hoffm. in H.G.A.Engler, Pflanzenr., IV, 147, VII: 55
   (1914), nom. inval.
   W. & WC. Trop. Africa. 22 GHA IVO NGA 23 CMN GGI ZAI. Nanophan.
      Pycnocoma zenkeri Pax, Bot. Jahrb. Syst. 26: 329 (1899). *Pycnocoma macrophylla* var.
         zenkeri (Pax) Pax in H.G.A.Engler, Pflanzenr., IV, 147, VII: 55 (1914).
      Pycnocoma brachystachya Pax, Bot. Jahrb. Syst. 43: 82 (1909).
      Pycnocoma macrophylla var. *microsperma* Pax & K.Hoffm. in H.G.A.Engler, Pflanzenr., IV,
         147, VII: 55 (1914).
      Pycnocoma macrophylla var. *longicornuta* J.Léonard, Bull. Jard. Bot. État 25: 299 (1955).

Pycnocoma minor Müll.Arg., Flora 47: 483 (1864).
   Equatorial Guinea, Gabon. 23 EQG GAB. Nanophan.
      Pycnocoma petiolaris Pierre ex Prain in D.Oliver, Fl. Trop. Afr. 6(1): 961 (1913).

Pycnocoma reticulata Baill., Adansonia 1: 259 (1861). *Argomuellera reticulata* (Baill.) Pax &
   K.Hoffm. in H.G.A.Engler, Nat. Pflanzenfam. ed. 2, 19c: 108 (1931).
   Madagascar. 29 MDG.

Pycnocoma reygaertii De Wild., Repert. Spec. Nov. Regni Veg. 13: 382 (1914).
   C. Zaire. 23 ZAI. Nanophan. or phan.
      Pycnocoma mortehanii De Wild., Pl. Bequaert. 3: 497 (1926).

Pycnocoma subflava J.Léonard, Bull. Jard. Bot. Belg. 65: 57 (1996).
   C. Zaire. 23 ZAI. Nanophan. or phan.

Pycnocoma thollonii Prain, Bull. Misc. Inform. Kew 1912: 193 (1912).
   Congo. 23 CON. Nanophan.

Pycnocoma thonneri Pax ex De Wild. & T.Durand, Ann. Mus. Congo Belge, Bot., II, 1:
   51 (1899).
   Zaire, Congo. 23 CON ZAI. Nanophan. or phan.
      Pycnocoma trilobata De Wild., Miss. Ém. Laurent 1: 132 (1905).
      Pycnocoma longipes Pax, Bot. Jahrb. Syst. 43: 324 (1909).

Synonyms:
Pycnocoma beillei A.Chev. ex Hutch. & Dalziel === **Pycnocoma angustifolia** Prain
Pycnocoma brachystachya Pax === **Pycnocoma macrophylla** Benth.
Pycnocoma danguyana Leandri === **Argomuellera danguyana** (Leandri) J.Léonard
Pycnocoma decaryana Leandri === **Argomuellera decaryana** (Leandri) J.Léonard
Pycnocoma gigantea Baill. === **Argomuellera gigantea** (Baill.) Pax & K.Hoffm.
Pycnocoma gigantea var. *calcicola* Leandri === **Argomuellera calcicola** (Leandri) J.Léonard
Pycnocoma hirsuta Prain === **Argomuellera macrophylla** Pax
Pycnocoma hutchinsonii Beille === **Argomuellera macrophylla** Pax
Pycnocoma laurentii De Wild. === **Argomuellera macrophylla** Pax
Pycnocoma longipes Pax === **Pycnocoma thonneri** Pax ex De Wild. & T.Durand
Pycnocoma lucida Pax & K.Hoffm. === **Pycnocoma chevalieri** Beille
Pycnocoma macrophylla var. *genuina* Pax & K.Hoffm. === **Pycnocoma macrophylla** Benth.
Pycnocoma macrophylla var. *longicornuta* J.Léonard === **Pycnocoma macrophylla** Benth.
Pycnocoma macrophylla var. *microsperma* Pax & K.Hoffm. === **Pycnocoma macrophylla**
   Benth.
Pycnocoma macrophylla var. *zenkeri* (Pax) Pax === **Pycnocoma macrophylla** Benth.

Pycnocoma macrantha Pax
Artist: Judy C. Dunkley
Fl. Trop. East Africa, Euphorbiaceae 1: 230, fig. 46 (1987)
KEW ILLUSTRATIONS COLLECTION

Pycnocoma mortehanii De Wild. === **Pycnocoma reygaertii** De Wild.
Pycnocoma parviflora Pax === **Argomuellera macrophylla** Pax
Pycnocoma perrieri Leandri === **Argomuellera perrieri** (Leandri) J.Léonard
Pycnocoma petiolaris Pierre ex Prain === **Pycnocoma minor** Müll.Arg.
Pycnocoma rigidifolia Baill. === **Droceloncia rigidifolia** (Baill.) J.Léonard
Pycnocoma sapinii De Wild. === **Argomuellera macrophylla** Pax
Pycnocoma sassandrae Beille === **Argomuellera macrophylla** Pax
Pycnocoma trewioides Baill. === **Argomuellera trewioides** (Baill.) Pax & K.Hoffm.
Pycnocoma trilobata De Wild. === **Pycnocoma thonneri** Pax ex De Wild. & T.Durand
Pycnocoma zenkeri Pax === **Pycnocoma macrophylla** Benth.

Pycnosandra

Typified by *Pycnosandra serrata* and thus referable to *Drypetes*.

Synonyms:
Pycnosandra Blume === **Drypetes** Vahl
Pycnosandra serrata (Blume) Blume === **Drypetes serrata** (Blume) Pax & K.Hoffm.
Pycnosandra timorensis Blume === **Putranjiva roxburghii** Wall.

Pythius

Synonyms:
Pythius Raf. === **Euphorbia** L.

Quadrasia

Synonyms:
Quadrasia Elmer === **Claoxylon** A.Juss.
Quadrasia euphorbioides Elmer === **Claoxylon euphorbioides** (Elmer) Merr.

Ramelia

Synonyms:
Ramelia Baill. === **Bocquillonia** Baill.
Ramelia codonostylis Baill. === **Bocquillonia codonostylis** (Baill.) Airy Shaw

Ramsdenia

Synonyms:
Ramsdenia Britton === **Phyllanthus** L.

Redia

Synonyms:
Redia Casar. === **Cleidion** Blume

Regnaldia

Synonyms:
Regnaldia Baill. === **Chaetocarpus** Thwaites

Reidia

A former segregate of *Phyllanthus*.

Synonyms:
Reidia Wight === **Phyllanthus** L.
Reidia affinis (Müll.Arg.) Alston === **Phyllanthus affinis** Müll.Arg.
Reidia bailloniana (Müll.Arg.) Gamble === **Phyllanthus baillonianus** Müll.Arg.
Reidia beddomei Gamble === **Phyllanthus beddomei** (Gamble) M.Mohanan
Reidia cinerea (Müll.Arg.) Alston === **Phyllanthus cinereus** Müll.Arg.
Reidia cordifolia Alston === **Phyllanthus baillonianus** Müll.Arg.
Reidia fimbriata (Müll.Arg.) Wight === **Phyllanthus fimbriatus** Müll.Arg.
Reidia floribunda Wight === **Phyllanthus wightianus** Müll.Arg.
Reidia gageana Gamble === **Phyllanthus gageanus** (Gamble) M.Mohanan
Reidia glabrescens Miq. === **Phyllanthus glabrescens** (Miq.) Müll.Arg.
Reidia glaucescens Miq. === **Phyllanthus pulcher** Wall. ex Müll.Arg.
Reidia gomphocarpa (Hook.f.) C.E.C.Fisch. === **Phyllanthus gomphocarpus** Hook.f.
Reidia gracilipes Miq. === **Phyllanthus gracilipes** (Miq.) Müll.Arg.
Reidia gracilis (Hassk.) Miq. === **Phyllanthus gracilipes** (Miq.) Müll.Arg.
Reidia hakgalensis (Thwaites ex Trimen) Alston === **Phyllanthus hakgalensis**
Reidia hamiltoniana (Müll.Arg.) A.M.Cowan & Cowan === **Phyllanthus sikkimensis** Müll.Arg.
Reidia latifolia Wight === **Phyllanthus macrocalyx** Müll.Arg.
Reidia longiflora Gamble === **Phyllanthus heyneanus** Müll.Arg.
Reidia macrocalyx (Müll.Arg.) Gamble === **Phyllanthus macrocalyx** Müll.Arg.
Reidia megacarpa Gamble === **Phyllanthus megacarpus** (Gamble) Kumari & Chandrab.
Reidia montana (Thwaites) Alston === **Phyllanthus oreophilus** Müll.Arg.
Reidia ovalifolia Wight === **Phyllanthus heyneanus** Müll.Arg.
Reidia polyphylla Wight === **Phyllanthus anabaptizatus** Müll.Arg.
Reidia singalensis Miq. === **Phyllanthus singalensis** (Miq.) Müll.Arg.
Reidia singampattiana Sebast. & A.N.Henry === **Phyllanthus singampattianus** (Sebast. & A.N.Henry) Kumari & Chandrab.
Reidia stipulacea Gamble === **Phyllanthus chandrabosei** Govaerts & Radcl.-Sm.
Reidia tetrandra (Roxb.) V.Naray. === **Phyllanthus tetrandrus** Roxb.
Reidia tricocca Casar. === **Cleidion tricoccum** (Casar.) Baill.

Reissipa

Synonyms:
Reissipa Steud. ex Klotzsch === **Monotaxis** Brongn.

Reutealis

1 species, Philippines; trees to 20 m resembling *Aleurites*. *R. trisperma* is reportedly naturalised in parts of Java and also is cultivated there and elsewhere as well as in the Philippines. Formerly treated (e.g. by Pax, 1910) in *Aleurites* as sect. *Reutealis*. There is a good descriptive account by J.J. Smith in Meded. Dept. Landb. Ned. Ind. 10: 556-560 (1910, as *Aleurites trisperma*). It also appears in colour plate 296 in the *Gran Edición* of Blanco's *Flora de Filipinas*. A second species, from southern China, is soon to be described. Recent studies of the genus and its relatives by Wolfgang Stuppy and two collaborators suggest that in phyletic terms it is basal to *Aleurites* and *Vernicia*. (Crotonoideae)

Langeron, M. (1902). See *Aleurites*.
Pax, F. (1910). *Aleurites*. In A. Engler (ed.), Das Pflanzenreich, IV 147 [I] (Euphorbiaceae-Jatropheae): 128-133. Berlin. (Heft 42.) La/Ge. — 4 species, of which 1 in sect. *Reutiales* (=*Reutealis*; p. 131).

Airy-Shaw, H. K. (1966). Notes on Malaysian and other Asiatic Euphorbiaceae, LXXII. Generic segregation in the affinity of *Aleurites* J. R. et G. Forst. Kew Bull. 20: 393-395. En. — *Reutealis* separated from *Aleurites*; one new combination.

Airy-Shaw, H. K. (1972). Notes on Malesian and other Asiatic Euphorbiaceae, CLIX: Additional literature reference to *Reutealis trisperma*. Kew Bull. 27: 85-86. En. — A reference to W. H. Brown's *Minor products of Philippine forests* (1921) where there is a description and illustration of this tree.

Stuppy, W. et al. (1999). Revision of genera *Aleurites*, *Reutealis* and *Vernicia* (Euphorbiaceae). Blumea 44: 73-98, illus. En. — *Reutealis*, pp. 85-88; treatment of 1 species with description, synonymy, references, types, indication of distribution and habitat, and commentary; all general references, identification list and index to botanical names at end of paper.

Reutealis Airy Shaw, Kew Bull. 20: 394 (1966).
  Philippines. 42.

Reutealis trisperma (Blanco) Airy Shaw, Kew Bull. 20: 395 (1966).
  Philippines. 42 jaw PHI. Phan.
  * *Aleurites trisperma* Blanco, Fl. Filip.: 755 (1837).
    Aleurites saponaria Blanco, Fl. Filip., ed. 2: 519 (1845).

Reverchonia

1 species, SW. North America (Oklahoma to Utah, south just into Chihuahua); related to *Phyllanthus*. The single species, *R. arenaria*, is a specialised annual subshrub of sandhills with alternate, linear-oblanceolate leaves. The male disk is about all that separates the genus from *Phyllanthus* sect. *Isocladus* (Webster & Miller 1963). At the same time, these authors raised the possibility of a relationship with *Securinega* (i.e. *Flueggea*). Like *Tetracoccus*, the genus was thought likely to be relictual. The first full study is that of Webster & Miller (1963); there never was a *Pflanzenreich* revision. (Phyllanthoideae)

Pax, F. & K. Hoffmann (1931). *Reverchonia*. In A. Engler (ed.), Die natürlichen Pflanzenfamilien, 2. Aufl., 19c: 66. Leipzig. Ge. — Synopsis with description of genus; 1 species, SW. N America.

• Webster, G. & K. I. Miller (1963). The genus *Reverchonia* (Euphorbiaceae). Rhodora 65: 193-207, illus., map. En. — Full treatment with extensive preliminary discussion, description, type, localities with exsiccatae, and discussion of distribution, habitat, phenology and variability. The general part includes a consideration of the importance of ovular morphology in Euphorbiaceae (pp. 198-199).

Reverchonia A.Gray, Proc. Amer. Acad. Arts 16: 107 (1880).
  C. & SW. U.S.A., N. Mexico. 74 76 77 79. Ther.

Reverchonia arenaria A.Gray, Proc. Amer. Acad. Arts 16: 107 (1880).
  C. & SW. U.S.A., N. Mexico. 74 OKL 76 ARI UTA 77 TEX 79 MXE. Ther.

Rhopalostylis

Synonyms:
Rhopalostylis Klotzsch ex Baill. === **Dalechampia** Plum. ex L.

Rhopium

Synonyms:
Rhopium Schreb. === **Phyllanthus** L.

Rhytis

Synonyms:

Rhytis Lour. === **Antidesma** L.

Rhytis fruticosa Lour. === **Antidesma fruticosum** (Lour.) Müll.Arg.

Richeria

5 species, Americas (Costa Rica and Windward Islands to Peru and SE Brazil; in S. America wholly east of the Andes); shrubs or trees 2-25 m with leathery foliage and flowers in axillary spikes, the sprays reminiscent of *Baccaurea* in the Old World. Since the treatments of Pax and Hoffmann (1922) and Jablonski (1967) part of the genus has been transferred to *Podocalyx*. The remainder was partially revised by Secco & Webster (1990) without, however, *R. tomentosa* Huft. As well as *Baccaurea*, the genus is closest to *Maesobotrya* (Africa) and *Aporusa* (eastern tropics); all are in Webster's subtribe Scepinae (tribe Antidesmeae). (Phyllanthoideae)

Pax, F. & K. Hoffmann (1922). *Richeria*. In A. Engler (ed.), Das Pflanzenreich, IV 147 XV (Euphorbiaceae-Phyllanthoideae-Phyllantheae): 26-30. Berlin. (Heft 81.) La/Ge. — 6 species, Americas. [Part of genus since removed to *Podocalyx*.]

Jablonski, E. (1967). *Richeria*. Euphorbiaceae, Guayana Highland (Mem. New York Bot. Gard. 17(1)): 124-127. New York. En. — 4 species, none new. [*R. loranthoides* now referred to *Podocalyx* in the Oldfieldioideae.]

• Secco, R. de S. & G. L. Webster (1990). Materiaes para a flora amazônica, IX. Ensaio sobre a sistemática do gênero *Richeria* Vahl (Euphorbiaceae). Bol. Mus. Paraense 'Emilio Goeldi', Bot. 6: 141-158, illus., map. Pt. — Partial revision (2 species and 1 additional variety) with keys, descriptions, types, synonymy with references, vernacular names, indication of distribution, localities with exsiccatae, and extensive concluding commentary; map, figures and photographs at end. [*R. tomentosa* Huft is not included.]

Richeria Vahl, Eclog. Amer. 1: 30 (1797).
C. & S. America, Caribbean. 80 81 82 83 84.
Guarania Wedd. ex Baill., Étude Euphorb.: 598 (1858).

Richeria australis Müll.Arg. in C.F.P.von Martius, Fl. Bras. 11(2): 17 (1873).
Brazil (São Paulo). 84 BZL. Phan.

Richeria dressleri G.L.Webster, Ann. Missouri Bot. Gard. 75: 1094 (1988).
Costa Rica, Panama. 80 COS PAN. Phan.

Richeria grandis Vahl, Eclog. Amer. 1: 30 (1797). *Richeria grandis* var. *genuina* Müll.Arg. in A.P.de Candolle, Prodr. 15(2): 468 (1866), nom. inval.
Trop. America. 80 PAN 81 LEE WIN TRT 82 FRG GUY VEN 83 BOL CLM PER 84 BZC BZE BZL BZN BZS. Phan.

var. **gardneriana** (Baill.) Müll.Arg. in A.P.de Candolle, Prodr. 15(2): 468 (1866).
Brazil. 84 BZC BZE BZL BZN. Nanophan. or phan.
* Guarania gardneriana Baill., Étude Euphorb.: 598 (1858). *Richeria gardneriana* (Baill.) Baill. ex Müll.Arg. in A.P.de Candolle, Prodr. 15(2): 468 (1866), pro syn.
Guarania longifolia Baill., Adansonia 5: 349 (1865). *Richeria longifolia* (Baill.) Baill. ex Müll.Arg. in A.P.de Candolle, Prodr. 15(2): 467 (1866). *Richeria grandis* var. *longifolia* (Baill.) Müll.Arg. in A.P.de Candolle, Prodr. 15(2): 467 (1866).
Richeria grandis var. *pohliana* Müll.Arg. in C.F.P.von Martius, Fl. Bras. 11(2): 17 (1873).
Richeria grandis var. *latifolia* Pax & K.Hoffm. in H.G.A.Engler, Pflanzenr., IV, 147, XV: 28 (1922).

var. **grandis**
Trop. America. 80 PAN 81 LEE WIN TRT 82 FRG GUY VEN 83 BOL CLM PER 84 BZE BZN BZS. Phan.

Amanoa divaricata Poepp. & Endl., Nov. Gen. Sp. Pl. 3: 22 (1841). *Richeria grandis* var. *divaricata* (Poepp. & Endl.) Müll.Arg. in A.P.de Candolle, Prodr. 15(2): 467 (1866).

Amanoa racemosa Poepp. & Endl., Nov. Gen. Sp. Pl. 3: 23 (1841). *Richeria grandis* var. *racemosa* (Poepp. & Endl.) Müll.Arg. in A.P.de Candolle, Prodr. 15(2): 467 (1866). *Richeria racemosa* (Poepp. & Endl.) Pax & K.Hoffm. in H.G.A.Engler, Pflanzenr., IV, 147, XV: 28 (1922).

Amanoa ramiflora Poepp. & Endl., Nov. Gen. Sp. Pl. 3: 23 (1841).

Amanoa purpurascens Poepp. ex Baill., Étude Euphorb.: 598 (1858).

Guarania ramiflora Wedd. ex Baill., Étude Euphorb.: 598 (1858).

Antidesma longifolium Decne. ex Baill., Adansonia 5: 349 (1865), nom. illeg.

Guarania laurifolia Baill., Adansonia 5: 348 (1865). *Richeria laurifolia* (Baill.) Baill. ex Müll.Arg. in A.P.de Candolle, Prodr. 15(2): 468 (1866). *Richeria grandis* var. *laurifolia* (Baill.) Müll.Arg. in A.P.de Candolle, Prodr. 15(2): 468 (1866).

Guarania purpurascens Wedd. ex Baill., Adansonia 5: 347 (1865). *Richeria purpurascens* (Wedd. ex Baill.) Baill. ex Müll.Arg. in A.P.de Candolle, Prodr. 15(2): 467 (1866), pro syn.

Guarania spruceana Baill., Adansonia 5: 348 (1865). *Richeria spruceana* (Baill.) Baill. ex Müll.Arg. in A.P.de Candolle, Prodr. 15(2): 468 (1866), pro syn.

Richeria submembranacea Steyerm., Publ. Field Mus. Nat. Hist., Bot. Ser. 17: 419 (1938).

Richeria olivieri Philcox, Kew Bull. 32: 224 (1977).

Richeria obovata (Müll.Arg.) Pax & K.Hoffm. in H.G.A.Engler, Pflanzenr., IV, 147, XV: 29 (1922).
Costa Rica, Panama, Venezuela (Amazonas, Bolivar). 80 COS PAN 82 VEN. Phan.
* *Richeria grandis* var. *obovata* Müll.Arg. in A.P.de Candolle, Prodr. 15(2): 468 (1866).

Richeria tomentosa Huft, Ann. Missouri Bot. Gard. 76: 1078 (1989).
Colombia. 83 CLM. Phan.

Synonyms:
Richeria gardneriana (Baill.) Baill. ex Müll.Arg. === **Richeria grandis** var. **gardneriana** (Baill.) Müll.Arg.

Richeria grandis var. *divaricata* (Poepp. & Endl.) Müll.Arg. === **Richeria grandis** Vahl var. **grandis**

Richeria grandis var. *genuina* Müll.Arg. === **Richeria grandis** Vahl

Richeria grandis var. *latifolia* Pax & K.Hoffm. === **Richeria grandis** var. **gardneriana** (Baill.) Müll.Arg.

Richeria grandis var. *laurifolia* (Baill.) Müll.Arg. === **Richeria grandis** Vahl var. **grandis**

Richeria grandis var. *longifolia* (Baill.) Müll.Arg. === **Richeria grandis** var. **gardneriana** (Baill.) Müll.Arg.

Richeria grandis var. *obovata* Müll.Arg. === **Richeria obovata** (Müll.Arg.) Pax & K.Hoffm.

Richeria grandis var. *pohliana* Müll.Arg. === **Richeria grandis** var. **gardneriana** (Baill.) Müll.Arg.

Richeria grandis var. *racemosa* (Poepp. & Endl.) Müll.Arg. === **Richeria grandis** Vahl var. **grandis**

Richeria laurifolia (Baill.) Baill. ex Müll.Arg. === **Richeria grandis** Vahl var. **grandis**

Richeria longifolia (Baill.) Baill. ex Müll.Arg. === **Richeria grandis** var. **gardneriana** (Baill.) Müll.Arg.

Richeria loranthoides (Klotzsch) Müll.Arg. === **Podocalyx loranthoides** Klotzsch

Richeria olivieri Philcox === **Richeria grandis** Vahl var. **grandis**

Richeria purpurascens (Wedd. ex Baill.) Baill. ex Müll.Arg. === **Richeria grandis** Vahl var. **grandis**

Richeria racemosa (Poepp. & Endl.) Pax & K.Hoffm. === **Richeria grandis** Vahl var. **grandis**

Richeria spruceana (Baill.) Baill. ex Müll.Arg. === **Richeria grandis** Vahl var. **grandis**

Richeria submembranacea Steyerm. === **Richeria grandis** Vahl var. **grandis**

Richeriella malayana Hend.
Artist: Mary Grierson
Ic. Pl. 38: pl. 3703 (1974)

Richeriella

2 species, SE. China (Hainan), S. Thailand and Malesia; closely related to *Flueggea* and *Margaritaria*. Shrubs or small trees to 12 m. with indefinite shoot growth, distichous foliage and slender branching axillary panicles. *R. malayana* is recorded from limestone hills in both Peninsular Malaysia and Borneo. The very similar *R. gracilis* was first described from Palawan. (Phyllanthoideae)

Pax, F. & K. Hoffmann (1922). *Richeriella*. In A. Engler (ed.), Das Pflanzenreich, IV 147 XV (Euphorbiaceae-Phyllanthoideae-Phyllantheae): 30. Berlin. (Heft 81.) La/Ge. — 1 species, Philippines. [Subsequent additions have been made.]

Airy-Shaw, H. K. (1971). Notes on Malesian and other Asiatic Euphorbiaceae, CXXIII. New records for *Richeriella* Pax and Hoffm., with a note on generic affinity. Kew Bull. 25: 489-491. En. — 2 species, *R. gracilis* and *R. malayana*.

Richeriella Pax & K.Hoffm. in H.G.A.Engler, Pflanzenr., IV, 147, XV: 30 (1922).
Hainan, Indo-China, Malesia. 36 41 42. Phan.

Richeriella gracilis (Merr.) Pax & K.Hoffm. in H.G.A.Engler, Pflanzenr., IV, 147, XV: 30 (1922).
Thailand, Hainan, Philippines (Luzon, Palawan). 36 CHH 41 THA 42 PHI. Phan.
* *Baccaurea gracilis* Merr., Philipp. J. Sci. 1(Suppl.): 203 (1906).

Richeriella malayana Hend., Gard. Bull. Straits Settlem. 7: 122 (1933). – FIGURE, p. 1392.
Pen. Malaysia, Borneo. 42 BOR MLY. Phan.
Richeriella malayana var. *macrocarpa* Airy Shaw, Kew Bull. 25: 490 (1971).

Synonyms:
Richeriella malayana var. *macrocarpa* Airy Shaw === **Richeriella malayana** Hend.

Ricinella

Synonyms:
Ricinella Müll.Arg. === **Adelia** L.
Ricinella membranifolia Müll.Arg. === **Adelia membranifolia** (Müll.Arg.) Chodat & Hassl.
Ricinella oaxacana Müll.Arg. === **Adelia oaxacana** (Müll.Arg.) Hemsl.
Ricinella peduncularis Kuntze === **Adelia peduncularis** (Kuntze) Pax & K.Hoffm.
Ricinella pedunculosa (A.Rich.) Müll.Arg. === **Adelia ricinella** L.
Ricinella triloba Müll.Arg. === **Adelia triloba** (Müll.Arg.) Hemsl.

Ricinocarpos

16 species, Australia; shrubs or undershrubs of scrub and woodland featuring inrolled, often needle-like leaves with contrasting surfaces and terminal inflorescences. The presence of narrow cotyledons caused Mueller to group it with superficially similar genera in his 'Stenolobae'; this was followed by Pax and Hoffmann. Like other Australian genera of that group, however, it appears instead to be specialised, in this case allied to *Alphandia* and closest to Crotoneae (Webster 1994). No revision has appeared since 1913. (Crotonoideae)

Gruening, G. (1913). *Ricinocarpus*. In A. Engler (ed.), Das Pflanzenreich, IV 147 [Stenolobieae] (Euphorbiaceae-Porantheroideae et Ricinocarpoideae): 37-49. Berlin. (Heft 58.) La/Ge. — 15 species in 4 sections.

Ricinocarpos Desf., Mém. Mus. Hist. Nat. 3: 459 (1817).
Australia. 50.
Ricinocarpus A.Juss., Euphorb. Gen.: 36 (1824).
Roeperia Spreng., Syst. Veg. 3: 13, 147 (1826).
Echinosphaera Sieber ex Steud., Nomencl. Bot., ed. 2, 1: 538 (1840).

Ricinocarpos bowmanii F.Muell., Fragm. 1: 181 (1859). *Ricinocarpos bowmanii* var. *genuina*
Grüning in H.G.A.Engler, Pflanzenr., IV, 147: 43 (1913), nom. inval.
SE. Queensland, New South Wales. 50 NSW QLD. Cham. or nanophan.
    Ricinocarpos puberulus Baill., Étude Euphorb.: 344 (1858), nom. nud.
    Ricinocarpos bowmanii var. *plana* Grüning in H.G.A.Engler, Pflanzenr., IV, 147:
      43 (1913).

Ricinocarpos cyanescens Müll.Arg., Linnaea 34: 60 (1865).
Western Australia. 50 WAU. Nanophan.
    Ricinocarpos taxifolia Klotzsch ex Grüning in H.G.A.Engler, Pflanzenr., IV, 147:
      47 (1913).

Ricinocarpos glaucus Endl. in S.L.Endlicher & al., Enum. Pl.: 18 (1837). *Ricinocarpos glaucus*
var. *genuinus* Müll.Arg. in A.P.de Candolle, Prodr. 15(2): 205 (1866), nom. inval.
SW. Australia. 50 WAU. Cham. or nanophan.
    Ricinocarpos undulatus Lehm., Pl. Preiss. 2: 370 (1848). *Ricinocarpos glaucus* var.
      undulatus (Lehm.) Müll.Arg. in A.P.de Candolle, Prodr. 15(2): 205 (1866).
    Ricinocarpos glaucus var. *jasminoides* Baill., Adansonia 6: 295 (1866).

Ricinocarpos gloria-medii J.H.Willis, Muelleria 3(2): 95 (1975).
S. Northern Territory. 50 NTA. Nanophan.

Ricinocarpos ledifolius F.Muell., Fragm. 1: 76 (1858).
N. New South Wales, Queensland. 50 NSW QLD. Nanophan.

Ricinocarpos major Müll.Arg., Linnaea 34: 59 (1865).
Tasmania. 50 TAS. Nanophan.

Ricinocarpos marginatus Benth., Fl. Austral. 6: 73 (1873).
Western Australia (Gardner). 50 WAU. Nanophan.
    Croton marginatus A.Cunn. ex Benth., Fl. Austral. 6: 73 (1873), pro syn.

Ricinocarpos muricatus Müll.Arg., Linnaea 34: 61 (1865).
Western Australia. 50 WAU. Nanophan.

Ricinocarpos pinifolius Desf., Mém. Mus. Hist. Nat. 3: 459 (1817).
Queensland (incl. Moreton Is.), New South Wales, Victoria, Tasmania. 50 NSW QLD TAS
VIC. Nanophan.
    Croton corollatus Sol. ex Baill., Étude Euphorb.: 351 (1858).
    Ricinocarpos sidifolius Baill., Étude Euphorb.: 344 (1858).
    Ricinocarpos sidiformis F.Muell. ex Benth., Fl. Austral. 6: 70 (1873).
    Ricinocarpos megalanthus Gand., Bull. Soc. Bot. France 66: 287 (1919 publ. 1920).
    Ricinocarpos proximus Gand., Bull. Soc. Bot. France 66: 287 (1919 publ. 1920).

Ricinocarpos psilocladus (Müll.Arg.) Benth., Fl. Austral. 6: 71 (1873).
Western Australia (Irwin). 50 WAU. Nanophan.
    *Bertya gummifera var. *psiloclada* Müll.Arg., Flora 47: 471 (1864).
    Bertya psiloclada Müll.Arg. ex B.D.Jacks., Index Kew. 1: 296 (1893), nom. illeg.

Ricinocarpos rosmarinifolius Benth., Fl. Austral. 6: 72 (1873).
Western Australia (Gardner). 50 WAU. Nanophan.
    Croton rosmarinifolius A.Cunn. ex Benth., Fl. Austral. 6: 73 (1873), pro syn.

Ricinocarpos speciosus Müll.Arg., Flora 47: 470 (1864).
Queensland, NE. New South Wales. 50 NSW QLD. Nanophan.

Ricinocarpos stylosus Diels, Bot. Jahrb. Syst. 35: 335 (1904).
  Western Australia (S. Coolgardie, Eyre). 50 WAU. Nanophan.
    Bertya andrewsii Fitzg., J. Western Austral. Nat. Hist. Soc. 2: 31 (1905).

Ricinocarpos trichophorus Müll.Arg., Linnaea 34: 60 (1865).
  Western Australia (Eyre, Roe). 50 WAU. Nanophan.

Ricinocarpos tuberculatus Müll.Arg., Linnaea 34: 60 (1865).
  SW. Western Australia. 50 WAU. Nanophan.

Ricinocarpos velutinus F.Muell., Fragm. 9: 2 (1875).
  SW. Western Australia. 50 WAU. Nanophan.

Synonyms:
Ricinocarpos bowmanii var. *genuina* Grüning === **Ricinocarpos bowmanii** F.Muell.
Ricinocarpos bowmanii var. *plana* Grüning === **Ricinocarpos bowmanii** F.Muell.
Ricinocarpos glaucus var. *genuinus* Müll.Arg. === **Ricinocarpos glaucus** Endl.
Ricinocarpos glaucus var. *jasminoides* Baill. === **Ricinocarpos glaucus** Endl.
Ricinocarpos glaucus var. *undulatus* (Lehm.) Müll.Arg. === **Ricinocarpos glaucus** Endl.
Ricinocarpos megalanthus Gand. === **Ricinocarpos pinifolius** Desf.
Ricinocarpos mitchellii Sond. === **Bertya mitchellii** (Sond.) Müll.Arg.
Ricinocarpos neocaledonicus S.Moore === **Baloghia neocaledonica** (S.Moore) McPherson
Ricinocarpos proximus Gand. === **Ricinocarpos pinifolius** Desf.
Ricinocarpos puberulus Baill. === **Ricinocarpos bowmanii** F.Muell.
Ricinocarpos sidifolius Baill. === **Ricinocarpos pinifolius** Desf.
Ricinocarpos sidiformis F.Muell. ex Benth. === **Ricinocarpos pinifolius** Desf.
Ricinocarpos tasmanicus Sond. & F.Muell. === **Bertya tasmanica** (Sond.) Müll.Arg.
Ricinocarpos taxifolia Klotzsch ex Grüning === **Ricinocarpos cyanescens** Müll.Arg.
Ricinocarpos undulatus Lehm. === **Ricinocarpos glaucus** Endl.

Ricinocarpus

Ricinocarpus A. Juss. is an orthographic variant of *Ricinocarpos*, while *Ricinocarpus* Burm. ex Kuntze covers a misguided series of transfers fro *Acalypha* by Otto Kuntze.

Synonyms:
Ricinocarpus A.Juss. === **Ricinocarpos** Desf.
Ricinocarpus Burm. ex Kuntze === **Acalypha** L.
Ricinocarpus accedens (Müll.Arg.) Kuntze === **Acalypha accedens** Müll.Arg.
Ricinocarpus acuminatus (Benth.) Kuntze === **Acalypha acuminata** Benth.
Ricinocarpus adenostachyus (Müll.Arg.) Kuntze === **Acalypha adenostachya** Müll.Arg.
Ricinocarpus adenotrichus (A.Rich.) Kuntze === **Acalypha ornata** Hochst. ex A.Rich.
Ricinocarpus alnifolius (Klein ex Willd.) Kuntze === **Acalypha alnifolia** Klein ex Willd.
Ricinocarpus alopecuroides (Jacq.) Kuntze === **Acalypha alopecuroides** Jacq.
Ricinocarpus alternifolius (Hochst. ex Baill.) Kuntze === **Acalypha hochstetteriana**
    Müll.Arg.
Ricinocarpus amblyodontus (Müll.Arg.) Kuntze === **Acalypha amblyodonta** (Müll.Arg.)
    Müll.Arg.
Ricinocarpus anemioides (Kunth) Kuntze === **Acalypha anemioides** Kunth
Ricinocarpus angatensis (Blanco) Kuntze === **Acalypha amentacea** var. **velutina**
    (Müll.Arg.) Fosberg
Ricinocarpus angolensis (Müll.Arg.) Kuntze === **Acalypha welwitschiana** Müll.Arg.
Ricinocarpus angustifolius (Sw.) Kuntze === **Acalypha angustifolia** Sw.
Ricinocarpus anisodontus (Müll.Arg.) Kuntze === **Acalypha insulana** var. **anisodonta**
    (Müll.Arg.) Govaerts

Ricinocarpus arcianus (Baill.) Kuntze === **Acalypha arciana** (Baill.) Müll.Arg.

Ricinocarpus aristatus (Kunth) Kuntze === **Acalypha aristata** Kunth

Ricinocarpus arvensis (Poepp. & Endl.) Kuntze === **Acalypha aristata** Kunth

Ricinocarpus australis (L.) Kuntze === **Acalypha australis** L.

Ricinocarpus baillonianus (Müll.Arg.) Kuntze === **Acalypha indica** L.

Ricinocarpus benguelensis (Müll.Arg.) Kuntze === **Acalypha benguelensis** Müll.Arg.

Ricinocarpus berteroanus (Müll.Arg.) Kuntze === **Acalypha berteroana** Müll.Arg.

Ricinocarpus bipartitus (Müll.Arg.) Kuntze === **Acalypha bipartita** Müll.Arg.

Ricinocarpus bisetosus (Bertol. ex Spreng.) Kuntze === **Acalypha bisetosa** Bertol. ex Spreng.

Ricinocarpus blancoanus Kuntze === **Acalypha amentacea** Roxb. subsp. **amentacea**

Ricinocarpus boivinianus (Baill.) Kuntze === **Acalypha boiviniana** Baill.

Ricinocarpus boliviensis (Müll.Arg.) Kuntze === **Acalypha boliviensis** Müll.Arg.

Ricinocarpus botterianus (Müll.Arg.) Kuntze === **Acalypha botteriana** Müll.Arg.

Ricinocarpus brachyandrus (Baill.) Kuntze === **Acalypha accedens** Müll.Arg.

Ricinocarpus brachycladus (Müll.Arg.) Kuntze === **Acalypha brachyclada** Müll.Arg.

Ricinocarpus brachystachyus (Hornem.) Kuntze === **Acalypha supera** Forssk.

Ricinocarpus bracteatus (Miq.) Kuntze === **Acalypha** sp.

Ricinocarpus brasiliensis (Müll.Arg.) Kuntze === **Acalypha brasiliensis** Müll.Arg.

Ricinocarpus brevibracteatus (Müll.Arg.) Kuntze === **Acalypha brevibracteata** Müll.Arg.

Ricinocarpus brevicaulis (Müll.Arg.) Kuntze === **Acalypha brevicaulis** Müll.Arg.

Ricinocarpus brevipes (Müll.Arg.) Kuntze === **Acalypha vellamea** Baill.

Ricinocarpus bullatus (Müll.Arg.) Kuntze === **Acalypha bullata** Müll.Arg.

Ricinocarpus californicus (Benth.) Kuntze === **Acalypha californica** Benth.

Ricinocarpus callosus (Benth.) Kuntze === **Acalypha macrostachya** Jacq.

Ricinocarpus cancanus (Müll.Arg.) Kuntze === **Acalypha macrostachya** Jacq.

Ricinocarpus capillipes (Müll.Arg.) Kuntze === **Acalypha capillipes** Müll.Arg.

Ricinocarpus capitatus (Willd.) Kuntze === **Acalypha alnifolia** Klein ex Willd.

Ricinocarpus carolinianus Kuntze === **Acalypha ostryifolia** Riddell

Ricinocarpus carpinifolius (Poir.) Kuntze === **Acalypha angustifolia** Sw.

Ricinocarpus carthagenensis (Jacq.) Kuntze === **Acalypha carthagenensis** Jacq.

Ricinocarpus caturus (Blume) Kuntze === **Acalypha caturus** Blume

Ricinocarpus chamaedrifolius (Lam.) Kuntze === **Acalypha chamaedrifolia** (Lam.) Müll.Arg.

Ricinocarpus chorisandrus (Baill.) Kuntze === **Acalypha chorisandra** Baill.

Ricinocarpus ciliatus (Forssk.) Kuntze === **Acalypha ciliata** Forssk.

Ricinocarpus cinctus (Müll.Arg.) Kuntze === **Acalypha cincta** Müll.Arg.

Ricinocarpus claussenii (Turcz.) Kuntze === **Acalypha claussenii** (Turcz.) Müll.Arg.

Ricinocarpus codonocalyx (Baill.) Kuntze === **Acalypha codonocalyx** Baill.

Ricinocarpus communis (Müll.Arg.) Kuntze === **Acalypha communis** Müll.Arg.

Ricinocarpus consimilis (Müll.Arg.) Kuntze === **Acalypha amentacea** var. **grandis** (Benth.) Fosberg

Ricinocarpus conspicuus (Müll.Arg.) Kuntze === **Acalypha conspicua** Müll.Arg.

Ricinocarpus conterminus (Müll.Arg.) Kuntze === **Acalypha contermina** Müll.Arg.

Ricinocarpus controversus Kuntze === **Acalypha controversa** (Kuntze) K.Schum.

Ricinocarpus cordobensis (Müll.Arg.) Kuntze === **Acalypha communis** Müll.Arg.

Ricinocarpus costaricensis Kuntze === **Acalypha costaricensis** (Kuntze) Knobl.

Ricinocarpus crenatus (Hochst. ex A.Rich.) Kuntze === **Acalypha crenata** Hochst. ex A.Rich.

Ricinocarpus cuneatus (Poepp. & Endl.) Kuntze --- **Acalypha cuneata** Poepp. & Endl.

Ricinocarpus cunninghamii (Müll.Arg.) Kuntze === **Acalypha nemorum** F.Muell. ex Müll.Arg.

Ricinocarpus cuspidatus (Jacq.) Kuntze === **Acalypha cuspidata** Jacq.

Ricinocarpus deciduus (Forssk.) Kuntze === **Acalypha indica** L.

Ricinocarpus decumbens (Thunb.) Kuntze === **Acalypha capensis** (L.f.) Prain

Ricinocarpus dentatus (Schumach. & Thonn.) Kuntze === **Mallotus oppositifolius** (Geiseler) Müll.Arg. var. **oppositifolius**

Ricinocarpus depauperatus (Müll.Arg.) Kuntze === **Acalypha depauperata** Müll.Arg.

Ricinocarpus depressinervius Kuntze === **Acalypha depressinervia** (Kuntze) K.Schum.

Ricinocarpus dictyoneurus (Müll.Arg.) Kuntze === **Acalypha dictyoneura** Müll.Arg.

Ricinocarpus digyneius (Raf.) Kuntze === **?**

Ricinocarpus digynostachyus (Baill.) Kuntze === **Acalypha digynostachya** Baill.

Ricinocarpus dimorphus (Müll.Arg.) Kuntze === **Acalypha dimorpha** Müll.Arg.

Ricinocarpus distans (Müll.Arg.) Kuntze === **Acalypha distans** Müll.Arg.

Ricinocarpus divaricatus (Müll.Arg.) Kuntze === **Acalypha divaricata** Müll.Arg.

Ricinocarpus diversifolius (Jacq.) Kuntze === **Acalypha diversifolia** Jacq.

Ricinocarpus dumetorum (Müll.Arg.) Kuntze === **Acalypha dumetorum** Müll.Arg.

Ricinocarpus ecklonii (Baill.) Kuntze === **Acalypha ecklonii** Baill.

Ricinocarpus ellipticus (Sw.) Kuntze === **Acalypha elliptica** Sw.

Ricinocarpus emirnensis (Baill.) Kuntze === **Acalypha emirnensis** Baill.

Ricinocarpus eremorum (Müll.Arg.) Kuntze === **Acalypha eremorum** Müll.Arg.

Ricinocarpus erythrostachyus (Müll.Arg.) Kuntze === **Acalypha padifolia** Kunth

Ricinocarpus exaltatus (Baill.) Kuntze === **Acalypha amentacea** var. **grandis** (Benth.) Fosberg

Ricinocarpus fallax (Müll.Arg.) Kuntze === **Acalypha lanceolata** Willd. var. **lanceolata**

Ricinocarpus fasciculatus (Müll.Arg.) Kuntze === **Acalypha fasciculata** Müll.Arg.

Ricinocarpus firmulus (Müll.Arg.) Kuntze === **Acalypha firmula** Müll.Arg.

Ricinocarpus forsterianus (Müll.Arg.) Kuntze === **Acalypha forsteriana** Müll.Arg.

Ricinocarpus fournieri (Müll.Arg.) Kuntze === **Acalypha fournieri** Müll.Arg.

Ricinocarpus fruticosus (Forssk.) Kuntze === **Acalypha fruticosa** Forssk.

Ricinocarpus fruticulosus (Raf.) Kuntze === **?**

Ricinocarpus fuscescens (Müll.Arg.) Kuntze === **Acalypha fuscescens** Müll.Arg.

Ricinocarpus glabratus (Thunb.) Kuntze === **Acalypha glabrata** Thunb.

Ricinocarpus glabratus var. *latifolius* (Sond.) Kuntze === **Acalypha glabrata** Thunb. f. **glabrata**

Ricinocarpus glabratus f. *pilosior* Kuntze === **Acalypha glabrata** f. **pilosior** (Kuntze) Prain & Hutch.

Ricinocarpus glandulosus Kuntze === **Acalypha hassleriana** Chodat

Ricinocarpus gracilis (Spreng.) Kuntze === **Acalypha gracilis** Spreng.

Ricinocarpus grandidentatus (Müll.Arg.) Kuntze === **Acalypha capensis** (L.f.) Prain

Ricinocarpus grandis (Benth.) Kuntze === **Acalypha amentacea** var. **grandis** (Benth.) Fosberg

Ricinocarpus grisebachianus Kuntze === **Acalypha grisebachiana** (Kuntze) Pax & K.Hoffm.

Ricinocarpus havanensis (Müll.Arg.) Kuntze === **Acalypha havanensis** Müll.Arg.

Ricinocarpus hederaceus (Torr.) Kuntze === **Acalypha monostachya** Cav.

Ricinocarpus hernandiifolius (Sw.) Kuntze === **Acalypha hernandiifolia** Sw.

Ricinocarpus heterodontus (Müll.Arg.) Kuntze === **Acalypha macrostachya** Jacq.

Ricinocarpus hispidus (Burm.f.) Kuntze === **Acalypha hispida** Burm.f.

Ricinocarpus indicus (L.) Kuntze === **Acalypha indica** L.

Ricinocarpus infestus (Poepp. & Endl.) Kuntze === **Acalypha infesta** Poepp. & Endl.

Ricinocarpus insulanus (Müll.Arg.) Kuntze === **Acalypha insulana** Müll.Arg.

Ricinocarpus integrifolius (Willd.) Kuntze === **Acalypha integrifolia** Willd.

Ricinocarpus irazuensis Kuntze === **Acalypha septemloba** Müll.Arg.

Ricinocarpus jardinii (Müll.Arg.) Kuntze === **Acalypha jardinii**

Ricinocarpus laevifolius (Müll.Arg.) Kuntze === **Acalypha repanda** var. **denudata** (Müll.Arg.) A.C.Sm.

Ricinocarpus laevigatus (Sw.) Kuntze === **Acalypha laevigata** Sw.

Ricinocarpus lagoensis (Müll.Arg.) Kuntze === **Acalypha lagoensis** Müll.Arg.

Ricinocarpus lanceolatus (Willd.) Kuntze === **Acalypha lanceolata** Willd.

Ricinocarpus langianus (Müll.Arg.) Kuntze === **Acalypha langiana** Müll.Arg.

Ricinocarpus languidus Kuntze === **Acalypha brachiata** Krauss

Ricinocarpus latifolius (Müll.Arg.) Kuntze === **Acalypha insulana** Müll.Arg. var. **insulana**

Ricinocarpus laxiflorus (Müll.Arg.) Kuntze === **Acalypha laxiflora** Müll.Arg.

Ricinocarpus lepinei (Müll.Arg.) Kuntze === **Acalypha lepinei** Müll.Arg.

Ricinocarpus leptocladus (Benth.) Kuntze === **Acalypha leptoclada** Benth.

Ricinocarpus leptopodus (Müll.Arg.) Kuntze === **Acalypha leptopoda** Müll.Arg.

Ricinocarpus leptorhachis (Müll.Arg.) Kuntze === **Acalypha leptorhachis** Müll.Arg.

Ricinocarpus liebmannii (Müll.Arg.) Kuntze === **Acalypha liebmanniana** Müll.Arg.

Ricinocarpus lindenianus (Müll.Arg.) Kuntze === **Acalypha lindeniana** Müll.Arg.

Ricinocarpus lindheimeri (Müll.Arg.) Kuntze === **Acalypha phleoides** Cav.

Ricinocarpus livingstonianus (Müll.Arg.) Kuntze === **Acalypha ornata** Hochst. ex A.Rich.

Ricinocarpus longispicatus (Müll.Arg.) Kuntze === **Acalypha longispicata** Müll.Arg.

Ricinocarpus longistipularis (Müll.Arg.) Kuntze === **Acalypha longistipularis** Müll.Arg.

Ricinocarpus macrodontus (Müll.Arg.) Kuntze === **Acalypha macrodonta** Müll.Arg.

Ricinocarpus macrostachyoides (Müll.Arg.) Kuntze === **Acalypha macrostachyoides** Müll.Arg.

Ricinocarpus macrostachyus (Jacq.) Kuntze === **Acalypha macrostachya** Jacq.

Ricinocarpus malabaricus (Müll.Arg.) Kuntze === **Acalypha malabarica** Müll.Arg.

Ricinocarpus mandonii (Müll.Arg.) Kuntze === **Acalypha mandonii** Müll.Arg.

Ricinocarpus mannianus (Müll.Arg.) Kuntze === **Acalypha manniana** Müll.Arg.

Ricinocarpus marginatus (Poir.) Kuntze === **Acalypha integrifolia** subsp. **marginata** (Poir.) Coode

Ricinocarpus martianus (Müll.Arg.) Kuntze === **Acalypha martiana** Müll.Arg.

Ricinocarpus melochiifolius (Müll.Arg.) Kuntze === **Acalypha melochiifolia** Müll.Arg.

Ricinocarpus membranaceus (A.Rich.) Kuntze === **Acalypha membranacea** A.Rich.

Ricinocarpus mexicanus (Müll.Arg.) Kuntze === **Acalypha mexicana** Müll.Arg.

Ricinocarpus microcephalus (Müll.Arg.) Kuntze === **Acalypha microcephala** Müll.Arg.

Ricinocarpus mollis (Kunth) Kuntze === **Acalypha mollis** Kunth

Ricinocarpus monostachyus (Cav.) Kuntze === **Acalypha monostachya** Cav.

Ricinocarpus multicaulis (Müll.Arg.) Kuntze === **Acalypha multicaulis** Müll.Arg.

Ricinocarpus nemorum (F.Muell. ex Müll.Arg.) Kuntze === **Acalypha nemorum** F.Muell. ex Müll.Arg.

Ricinocarpus neocaledonicus (Müll.Arg.) Kuntze === **Acalypha pancheriana** Baill.

Ricinocarpus neogranatensis (Müll.Arg.) Kuntze === **Acalypha macrostachya** Jacq.

Ricinocarpus neomexicanus (Müll.Arg.) Kuntze === **Acalypha neomexicana** Müll.Arg.

Ricinocarpus nigritianus (Müll.Arg.) Kuntze === **Acalypha ornata** Hochst. ex A.Rich.

Ricinocarpus obscurus (Müll.Arg.) Kuntze === **Acalypha obscura** Müll.Arg.

Ricinocarpus ocymoides (Kunth) Kuntze === **Acalypha ocymoides** Kunth

Ricinocarpus oliganthus (Müll.Arg.) Kuntze === **Acalypha oligantha** Müll.Arg.

Ricinocarpus oligodontus (Müll.Arg.) Kuntze === **Acalypha oligodonta** Müll.Arg.

Ricinocarpus ornatus (Hochst. ex A.Rich.) Kuntze === **Acalypha ornata** Hochst. ex A.Rich.

Ricinocarpus oxyodontus (Müll.Arg.) Kuntze === **Acalypha oxyodonta** (Müll.Arg.) Müll.Arg.

Ricinocarpus padifolius (Kunth) Kuntze === **Acalypha padifolia** Kunth

Ricinocarpus pancherianus (Baill.) Kuntze === **Acalypha pancheriana** Baill.

Ricinocarpus parvifolius (Müll.Arg.) Kuntze === **Acalypha microphylla** Klotzsch var. **microphylla**

Ricinocarpus parvulus (Hook.f.) Kuntze === **Acalypha parvula** Hook.f.

Ricinocarpus patens (Müll.Arg.) Kuntze === **Acalypha patens** Müll.Arg.

Ricinocarpus peckoltii (Müll.Arg.) Kuntze === **Acalypha peckoltii** Müll.Arg.

Ricinocarpus peduncularis (Meisn. ex C.Krauss) Kuntze === **Acalypha peduncularis** Meisn. ex C.Krauss

Ricinocarpus persimilis (Müll.Arg.) Kuntze === **Acalypha ostryifolia** Riddell

Ricinocarpus peruvianus (Müll.Arg.) Kuntze === **Acalypha peruviana** Müll.Arg.

Ricinocarpus petiolaris (Hochst. ex C.Krauss) Kuntze --- **Acalypha brachiata** Krauss

Ricinocarpus philippinensis (Müll.Arg.) Kuntze === **Acalypha amentacea** Roxb. subsp. **amentacea**

Ricinocarpus phleoides (Torr.) Kuntze === **Acalypha phleoides** Cav.

Ricinocarpus pilosus (Cav.) Kuntze === **Acalypha pilosa** Cav.

Ricinocarpus platyphyllus (Müll.Arg.) Kuntze === **Acalypha platyphylla** Müll.Arg.

Ricinocarpus plicatus (Müll.Arg.) Kuntze === **Acalypha plicata** Müll.Arg.

Ricinocarpus pohlianus (Müll.Arg.) Kuntze === **Acalypha pohliana** Müll.Arg.

Ricinocarpus poiretii (Spreng.) Kuntze === **Acalypha poiretii** Spreng.

Ricinocarpus polymorphus (Müll.Arg.) Kuntze === **Acalypha polymorpha** Müll.Arg.

Ricinocarpus polystachyus (Jacq.) Kuntze === **Acalypha polystachya** Jacq.

Ricinocarpus portoricensis (Müll.Arg.) Kuntze === **Acalypha portoricensis** Müll.Arg.

Ricinocarpus prunifolius (Nees & Mart.) Kuntze === **Acalypha prunifolia** Nees & Mart.

Ricinocarpus pruriens (Nees & Mart.) Kuntze === **Acalypha pruriens** Nees & Mart.

Ricinocarpus psilostachyus (Hochst. ex A.Rich.) Kuntze === **Acalypha psilostachya** Hochst. ex A.Rich.

Ricinocarpus pubiflorus (Klotzsch) Kuntze === **Acalypha pubiflora** (Klotzsch) Baill.

Ricinocarpus purpurascens (Kunth) Kuntze === **Acalypha purpurascens** Kunth

Ricinocarpus radians (Torr.) Kuntze === **Acalypha radians** Torr.

Ricinocarpus radicans (Müll.Arg.) Kuntze === **Acalypha radicans** Müll.Arg.

Ricinocarpus reflexus (Müll.Arg.) Kuntze === **Acalypha reflexa** Müll.Arg.

Ricinocarpus repandus (Müll.Arg.) Kuntze === **Acalypha repanda** Müll.Arg.

Ricinocarpus reticulatus (Poir.) Kuntze === **Acalypha filiformis** Poir. subsp. **filiformis**

Ricinocarpus rhombifolius (Schltdl.) Kuntze === **Acalypha rhombifolia** Schltdl.

Ricinocarpus richardianus (Baill.) Kuntze === **Acalypha richardiana** Baill.

Ricinocarpus riedelianus (Baill.) Kuntze === **Acalypha riedeliana** Baill.

Ricinocarpus rivularis (Seem.) Kuntze === **Acalypha rivularis** Seem.

Ricinocarpus rottleroides (Baill.) Kuntze === **Acalypha rottleroides** Baill.

Ricinocarpus ruizianus (Müll.Arg.) Kuntze === **Acalypha ruiziana** Müll.Arg.

Ricinocarpus salicifolius (Müll.Arg.) Kuntze === **Acalypha salicifolia** Müll.Arg.

Ricinocarpus samydifolius (Poepp. & Endl.) Kuntze === **Acalypha diversifolia** Jacq.

Ricinocarpus scabrosus (Sw.) Kuntze === **Acalypha scabrosa** Sw.

Ricinocarpus scandens (Benth.) Kuntze === **Acalypha scandens** Benth.

Ricinocarpus schiedeanus (Schltdl.) Kuntze === **Acalypha schiedeana** Schltdl.

Ricinocarpus schlechtendalianus (Müll.Arg.) Kuntze === **Acalypha schlechtendaliana** Müll.Arg.

Ricinocarpus schlumbergeri (Müll.Arg.) Kuntze === **Acalypha schlumbergeri** Müll.Arg.

Ricinocarpus segetalis (Müll.Arg.) Kuntze === **Acalypha segetalis** Müll.Arg.

Ricinocarpus seminudus (Müll.Arg.) Kuntze === **Acalypha seminuda** Müll.Arg.

Ricinocarpus senensis (Klotzsch) Kuntze === **Acalypha brachiata** Krauss

Ricinocarpus senilis (Baill.) Kuntze === **Acalypha senilis** Baill.

Ricinocarpus setosus (A.Rich.) Kuntze === **Acalypha setosa** A.Rich.

Ricinocarpus sidifolius (A.Rich.) Kuntze === **Acalypha brachiata** Krauss

Ricinocarpus sonderianus (Müll.Arg.) Kuntze === **Acalypha sonderiana** Müll.Arg.

Ricinocarpus spachianus (Baill.) Kuntze === **Acalypha spachiana** Baill.

Ricinocarpus spiciflorus (L.) Kuntze === **Acalypha caturus** Blume

Ricinocarpus stipulaceus (Klotzsch) Kuntze === **Acalypha amentacea** Roxb. subsp. **amentacea**

Ricinocarpus strictus (Poepp. & Endl.) Kuntze === **Acalypha stricta** Poepp. & Endl.

Ricinocarpus subtomentosus (Lag.) Kuntze === **Acalypha subtomentosa** Lag.

Ricinocarpus subvillosus (Müll.Arg.) Kuntze === **Acalypha subvillosa** Müll.Arg.

Ricinocarpus tarapotensis (Müll.Arg.) Kuntze === **Acalypha macrostachya** Jacq.

Ricinocarpus tenuifolius (Müll.Arg.) Kuntze === **Acalypha tenuifolia** Müll.Arg.

Ricinocarpus tenuirameus (Müll.Arg.) Kuntze === **Acalypha tenuiramea** Müll.Arg.

Ricinocarpus tenuis (Müll.Arg.) Kuntze === **Acalypha brachiata** Krauss

Ricinocarpus tomentosus (Sw.) Kuntze === **Acalypha tomentosa** Sw.

Ricinocarpus tricholobus (Müll.Arg.) Kuntze === **Acalypha tricholoba** Müll.Arg.

Ricinocarpus trilobus (Müll.Arg.) Kuntze === **Acalypha triloba** Müll.Arg.

Ricinocarpus unibracteatus (Müll.Arg.) Kuntze === **Acalypha unibracteata** Müll.Arg.

Ricinocarpus urostachyus (Baill.) Kuntze === **Acalypha urostachya** Baill.

Ricinocarpus urticifolius (Raf.) Kuntze === **Acalypha virginica** L. var. **virginica**

Ricinocarpus vagans (Cav.) Kuntze === **Acalypha vagans** Cav.

Ricinocarpus vahlianus (Oliv.) Kuntze === **Acalypha crenata** Hochst. ex A.Rich.

Ricinocarpus vellameus (Baill.) Kuntze === **Acalypha vellamea** Baill.

Ricinocarpus villicaulis (Hochst. ex A.Rich.) Kuntze === **Acalypha brachiata** Krauss
Ricinocarpus villosus (Jacq.) Kuntze === **Acalypha villosa** Jacq.
Ricinocarpus virgatus (L.) Kuntze === **Acalypha virgata** L.
Ricinocarpus virginicus (L.) Kuntze === **Acalypha virginica** L.
Ricinocarpus weddellianus (Baill.) Kuntze === **Acalypha weddelliana** Baill.
Ricinocarpus welwitschianus (Müll.Arg.) Kuntze === **Acalypha welwitschiana** Müll.Arg.
Ricinocarpus wilkesianus (Müll.Arg.) Kuntze === **Acalypha amentacea** subsp. **wilkesiana** (Müll.Arg.) Fosberg
Ricinocarpus zambesicus (Müll.Arg.) Kuntze === **Acalypha brachiata** Krauss
Ricinocarpus zeyheri (Baill.) Kuntze === **Acalypha zeyheri** Baill.
Ricinocarpus zollingeri (Müll.Arg.) Kuntze === **Acalypha zollingeri** Müll.Arg.

Ricinodendron

1 species, Africa; light-demanding deciduous trees to 46 m with digitate foliage, terminal inflorescences and soft wood, the bole to 25 m or more. Fast-growing and found in a wide range of habitats but prefers drier forest and regrowth. Another species formerly included in the genus, *R. rautanenii*, is here referred to *Schinziophyton*. Along with *Givotia*, the two genera comprise tribe Ricinodendreae in the Webster system. (Crotonoideae)

Pax, F. (with K. Hoffmann) (1911). *Ricinodendron*. In A. Engler (ed.), Das Pflanzenreich, IV 147 III (Euphorbiaceae-Cluytieae): 45-49. Berlin. (Heft 47.) La/Ge. — 3 species in 2 subgenera. [The authors thought subgen. *Heteroricinodendron* questionably in the genus; its sole species, *R. staudtii*, is now known to have been based on a mixture of *Cola pachycarpa* and *Lannea welwitschii*, neither of them in the family. *R. rautanenii*, the 'Manketti' in southern Africa, is now in *Schinziophyton*.]

Léonard, J. (1961). Notulae systematicae XXXII. Observations sur des espèces africaines de *Clutia*, *Ricinodendron* et *Sapium* (Euphorbiacées). Bull. Jard. Bot. État 31: 391-406. Fr. —*Ricinodendron*, pp. 396-401; treatment of *R. heudelotii* with reduction of *R. africanum* to subspecific rank.

Ricinodendron Müll.Arg., Flora 47: 533 (1864).
　Trop. Africa. 22 23 24 25 26.
　　Barrettia Sim, Forest Fl. Port. E. Afr.: 103 (1909).

Ricinodendron heudelotii (Baill.) Heckel, Ann. Inst. Bot.-Géol. Colon. Marseille 5(2): 40 (1898).
　Trop. Africa. 22 GHA GNB GUI IVO LBR NGA SIE 23 CAB CAF CMN CON EQG GAB GGI ZAI 24 SUD 25 KEN TAN UGA 26 ANG MOZ. Phan.
　　*Jatropha heudelotii Baill., Adansonia 1: 64 (1860).

subsp. **africanum** (Müll.Arg.) J.Léonard, Bull. Jard. Bot. État 31: 398 (1961). – FIGURE, p. 1401.
　　Trop. Africa. 22 NGA 23 CAB CAF CMN CON EQG GAB GGI ZAI 24 SUD 25 TAN UGA 26 ANG MOZ. Phan.
　　* Ricinodendron africanum Müll.Arg., Flora 47: 533 (1864).
　　Ricinodendron gracilius Mildbr., Notizbl. Bot. Gart. Berlin-Dahlem 12: 516 (1935).

subsp. **heudelotii**
　　W. Trop. Africa. 22 GHA GNB GUI IVO LBR SIE. Phan.
　　Barrettia umbrosa Sim, Forest Fl. Port. E. Afr.: 103 (1909).

var. **tomentellum** (Hutch. & E.A.Bruce) Radcl.-Sm., Kew Bull. 27: 507 (1972).
　　Kenya, Tanzania. 25 KEN TAN. Phan.
　　* Ricinodendron tomentellum Hutch. & E.A.Bruce, Bull. Misc. Inform. Kew 1931: 270 (1931).
　　Ricinodendron schliebenii Mildbr., Notizbl. Bot. Gart. Berlin-Dahlem 12: 516 (1935).

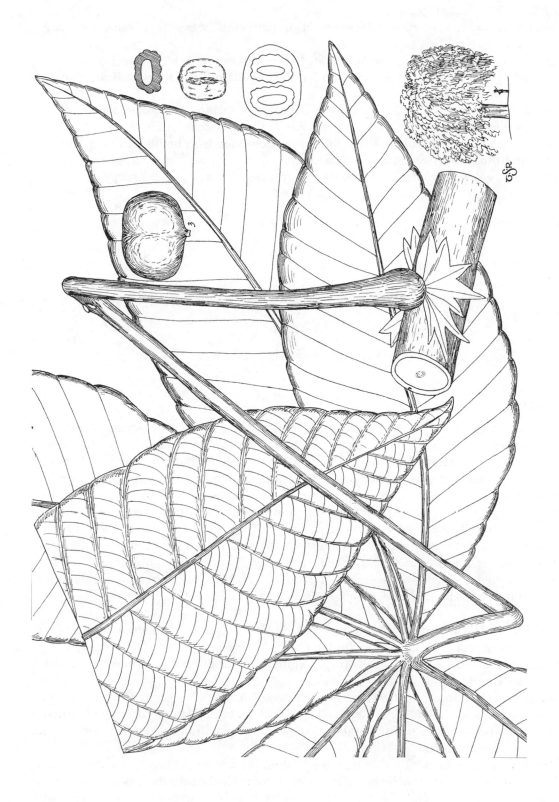

Ricinodendron heudelotii (Baill.) Heckel subsp. *africanum* (Müll. Arg.) J. Léonard

Ricinodendron heudelotii (Baill.) Heckel subsp. *africanum* (Müll. Arg.) J. Léonard (as *Barrettia umbrosa* Sim)
Artist: T.R. Sim
Sim, Forest Fl. Port. E. Afr.: pl. 71 (1909)

Synonyms:

Ricinodendron africanum Müll.Arg. === **Ricinodendron heudelotii** subsp. **africanum** (Müll.Arg.) J.Léonard

Ricinodendron gracilius Mildbr. === **Ricinodendron heudelotii** subsp. **africanum** (Müll.Arg.) J.Léonard

Ricinodendron rautanenii Schinz === **Schinziophyton rautanenii** (Schinz) Radcl.-Sm.

Ricinodendron schliebenii Mildbr. === **Ricinodendron heudelotii** var. **tomentellum** (Hutch. & E.A.Bruce) Radcl.-Sm.

Ricinodendron staudtii Pax === leaves, **Cola pachycarpa** K. Schum. (Sterculiaceae); inflorescence, **Lannea welwitschii** (Hiern) Engl. (Anacardiaceae)

Ricinodendron tomentellum Hutch. & E.A.Bruce === **Ricinodendron heudelotii** var. **tomentellum** (Hutch. & E.A.Bruce) Radcl.-Sm.

Ricinodendron viticoides Mildbr. === **Schinziophyton rautanenii** (Schinz) Radcl.-Sm.

Ricinoides

Synonyms:
Ricinoides Mill. === **Jatropha** L.

Ricinoides

Synonyms:
Ricinoides Tourn. ex Moench === **Chrozophora** Neck. ex A.Juss.

Ricinus

1 species, probably native to NE. tropical Africa but now widely spread and often naturalised in warmer parts of the world. A glabrous annual or perennial tree-like herb to 5(-10) m with palmately lobed leaves and inflorescences terminating individual shoot-units. The oil has many uses, both medicinal (castor oil) and as a drying agent in paint, printers' ink and the like; the plants are also of ornamental value. These and other economic aspects are covered in Moshkin (1986), with additional summaries in van Welzen (1998) and in the PROSEA-series (group 6). The species is variable and numerous infraspecific taxa have been named; the most elaborate treatment is by Tavares de Carvalho (1956). None is accepted here, though all are listed in our synonymy. In the Webster system, the genus is the sole member of Ricininae (tribe Acalypheae). A possible close relative is the Australian *Adriana*. (Acalyphoideae)

Pax, F. & K. Hoffmann (1919). *Ricinus*. In A. Engler (ed.), Das Pflanzenreich, IV 147 XI (Euphorbiaceae-Acalypheae-Ricininae): 119-127. Berlin. (Heft 68.) La/Ge. — 1 species, given as native to Africa. Numerous varieties and forms recognized, described and keyed.

Tavares de Carvalho, J. A. (1956). *Ricinus communis* L. (estudo sistemático). 79, [1] pp., illus. (Anais Junta Investig. Ultramar 11(4), fasc. 1). Lisbon. Pt. — Elaborate treatment of infraspecific taxa, many of them new; key and descriptions. (French and English summaries included.)

Moshkin, V. A. (ed.) (1986). Castor. 315 pp., illus. Rotterdam: Balkema. (Originally publ. 1980, Moscow, as *Kleshchevina*.) En. — An 'economic monograph'; botanical aspects are in Chapter 3, pp. 17-27 (a single species with six subspecies is accepted, with photographs of the principal taxa); many references. [Written with special reference to the Commonwealth of Independent States, where at one time 200 000 ha. were under this crop.]

Welzen, P. C. van (1998). Revisions and phylogenies of Malesian Euphorbiaceae subtribe Lasiococcinae (*Homonoia, Lasiococca, Spathiostemon*) and *Clonostylis, Ricinus* and *Wetria*. Blumea 43: 131-164. (*Lasiococca* with Nguyen Nghia Thin & Vu Hoai Duc.) En. —

General introduction, with history of studies; a note of caution on the use as a character of monoecy vs. dioecy given the imperfect state of much material; phylogeny with character table and cladogram; revision of *Ricinus* in Malesia (pp. 151-154; 1 introduced species) with description, synonymy, types, literature citations, vernacular names, indication of distribution, ecology and habitat, and notes on anatomy, uses and properties, and systematics; identification list at end. [No infraspecific taxa recognised. The plants have high horticultural value as well as being an important source of oil, both for drying (e.g. in paints, printers' ink, etc.) and human consumption.]

Ricinus L., Sp. Pl.: 1007 (1753).
   NE. & E. Trop. Africa. 24 25.
      Cataputia Ludw., Ectypa Veg.: 81 (1760).

Ricinus communis L., Sp. Pl.: 1007 (1753). *Ricinus communis* var. *genuinus* Müll.Arg. in A.P.de Candolle, Prodr. 15(2): 1019 (1866), nom. inval. *Ricinus communis* var. *typicus* Fiori, Fl. Italia 2: 292 (1901), nom. inval.
   NE. Trop. Africa, cultivated and naturalised elsewere. (12) por spa 24 ETH SOM 25 KEN? (80) cos. Nanophan. or phan. – Oil (castor oil) is extracted from the seeds.
      Croton spinosus L., Sp. Pl.: 1005 (1753).
      Ricinus vulgaris Garsault, Fig. Pl. Méd.: 66 (1764).
      Ricinus africanus Mill., Gard. Dict. ed. 8: 5 (1768).
      Ricinus minor Mill., Gard. Dict. ed. 8: 7 (1768).
      Ricinus rugosus Mill., Gard. Dict. ed. 8: 4 (1768).
      Ricinus speciosus Burm.f., Fl. Indica: 207 (1768). *Ricinus communis* var. *speciosus* (Burm.f.) Müll.Arg. in A.P.de Candolle, Prodr. 15(2): 1021 (1866).
      Ricinus urens Mill., Gard. Dict. ed. 8: 3 (1768).
      Ricinus vulgaris Mill., Gard. Dict. ed. 8: 1 (1768).
      Ricinus medicus Forssk., Fl. Aegypt.-Arab.: 164 (1775).
      Ricinus inermis Mill., Gard. Dict. ed. 8: 6 (1776). *Ricinus communis* f. *inermis* (Mill.) Müll.Arg. in A.P.de Candolle, Prodr. 15(2): 1018 (1866). *Ricinus communis* var. *inermis* (Mill.) Pax & K.Hoffm. in H.G.A.Engler, Pflanzenr., IV, 147, XI: 122 (1919).
      Ricinus lividus Jacq., Misc. Austriac. 2: 360 (1781). *Ricinus communis* var. *lividus* (Jacq.) Müll.Arg. in A.P.de Candolle, Prodr. 15(2): 1018 (1866).
      Ricinus digitatus Noronha, Verh. Batav. Genootsch. Kunsten 5(4): 26 (1790).
      Ricinus peltatus Noronha, Verh. Batav. Genootsch. Kunsten 5(4): 2 (1790).
      Ricinus medius J.F.Gmel., Syst. Nat.: 1615 (1792).
      Ricinus armatus Andr., Bot. Repos. 6: t. 430 (1805). *Ricinus communis* var. *armatus* (Andr.) Müll.Arg. in A.P.de Candolle, Prodr. 15(2): 1018 (1866).
      Ricinus viridis Willd., Sp. Pl. 4: 564 (1805). *Ricinus communis* f. *viridis* (Willd.) Müll.Arg. in A.P.de Candolle, Prodr. 15(2): 1020 (1866). *Ricinus communis* subvar. *viridus* (Willd.) T.Carvalho, Anais Junta Invest. Ultramar 11(4): 67 (1956).
      Ricinus undulatus Besser, Cat. Jard. Bot. Krzemieniec: 90 (1811). *Ricinus communis* var. *undulatus* (Besser) Müll.Arg. in A.P.de Candolle, Prodr. 15(2): 1021 (1866).
      Ricinus nanus Bald., Cat. Pl., ed. 1813: 65 (1813).
      Ricinus hybridus Besser, Cat. Jard. Bot. Krzemieniec, Suppl. 2: 13 (1814). *Ricinus communis* f. *hybridus* (Besser) Müll.Arg. in A.P.de Candolle, Prodr. 15(2): 1020 (1866).
      Ricinus angulatus Thunb., Ricin.: 6 (1815).
      Ricinus japonicus Thunb., Ricin.: 4 (1815).
      Ricinus laevis DC., Rapp. [Not.] Pl. Rar. Genève 1: 31 (1823). *Ricinus communis* f. *laevis* (DC.) Müll.Arg. in A.P.de Candolle, Prodr. 15(2): 1021 (1866).
      Ricinus communis var. *minor* Steud., Nomencl. Bot. 2: 459 (1824).
      Ricinus leucocarpus Bertol., Giorn. Arcadico Sci. 21: 192 (1824). *Ricinus communis* var. *leucocarpus* (Bertol.) Müll.Arg. in A.P.de Candolle, Prodr. 15(2): 1018 (1866).
      Ricinus macrophyllus Bertol., Giorn. Arcadico Sci. 21: 192 (1824).

Ricinus glaucus Hoffmanns., Verz. Pfl. Nachtr. 2: 199 (1826). *Ricinus communis* f. *glaucus* (Hoffmanns.) Müll.Arg. in A.P.de Candolle, Prodr. 15(2): 1020 (1866).

Ricinus spectabilis Blume, Bijdr.: 623 (1826).

Ricinus megalospermus Delile in F.Cailliaud, Voy. Méroé 4: 89 (1827). *Ricinus communis* var. *megalospermus* (Delile) Müll.Arg. in A.P.de Candolle, Prodr. 15(2): 1017 (1866).

Ricinus scaber Bertol. ex Moris, Stirp. Sard. Elench. 1: 41 (1827). *Ricinus communis* f. *scaber* (Bertol. ex Moris) Müll.Arg. in A.P.de Candolle, Prodr. 15(2): 1019 (1866).

Ricinus badius Rchb., Iconogr. Bot. Exot. 2: 21 (1829). *Ricinus communis* var. *badius* (Rchb.) Müll.Arg. in A.P.de Candolle, Prodr. 15(2): 1019 (1866).

Ricinus tunisensis Desf., Tabl. École Bot., ed. 3: 340 (1829).

Ricinus europaeus T.Nees, Gen. Fl. Germ.: 38 (1833).

Ricinus krappa Steud., Nomencl. Bot., ed. 2, 2: 459 (1841).

Ricinus macrocarpus Steud., Nomencl. Bot., ed. 2, 2: 459 (1841).

Ricinus perennis Steud., Nomencl. Bot., ed. 2, 2: 459 (1841).

Ricinus purpurascens Bertol., Misc. Bot. 9: 4 (1851). *Ricinus communis* var. *purpurascens* (Bertol.) Müll.Arg. in A.P.de Candolle, Prodr. 15(2): 1018 (1866). *Ricinus communis* f. *purpurascens* (Bertol.) Pax in H.G.A.Engler, Pflanzenr., IV, 147, XI: 122 (1919). *Ricinus communis* subvar. *purpurascens* (Bertol.) T.Carvalho, Anais Junta Invest. Ultramar 11(4): 50 (1956).

Ricinus obermannii Groenl., Rev. Hort., IV, 7: 601 (1858).

Ricinus sanguineus Groenl., Rev. Hort., IV, 7: 601 (1858).

Ricinus ruber Miq., Fl. Ned. Ind. 1(2): 390 (1859).

Ricinus communis var. *sanguineus* Baill., Adansonia 1: 342 (1861).

Ricinus communis var. *benguelensis* Müll.Arg., J. Bot. 1: 337 (1864).

Ricinus compactus Huber, Cat. Print. 1865: 8 (1865).

Ricinus communis f. *americanus* Müll.Arg. in A.P.de Candolle, Prodr. 15(2): 1017 (1866). *Ricinus communis* subvar. *americanus* (Müll.Arg.) T.Carvalho, Anais Junta Invest. Ultramar 11(4): 44 (1956).

Ricinus communis f. *blumeanus* Müll.Arg. in A.P.de Candolle, Prodr. 15(2): 1018 (1866). *Ricinus communis* subvar. *blumeanus* (Müll.Arg.) T.Carvalho, Anais Junta Invest. Ultramar 11(4): 46 (1956).

Ricinus communis f. *denudatus* Müll.Arg. in A.P.de Candolle, Prodr. 15(2): 1020 (1866).

Ricinus communis f. *epiglaucus* Müll.Arg. in A.P.de Candolle, Prodr. 15(2): 1020 (1866).

Ricinus communis f. *erythrocladus* Müll.Arg. in A.P.de Candolle, Prodr. 15(2): 1020 (1866). *Ricinus communis* subvar. *erythrocladus* (Müll.Arg.) T.Carvalho, Anais Junta Invest. Ultramar 11(4): 64 (1956).

Ricinus communis f. *gracilis* Müll.Arg. in A.P.de Candolle, Prodr. 15(2): 1021 (1866). *Ricinus communis* subvar. *gracilis* (Müll.Arg.) T.Carvalho, Anais Junta Invest. Ultramar 11(4): 68 (1956).

Ricinus communis f. *intermedius* Müll.Arg. in A.P.de Candolle, Prodr. 15(2): 1020 (1866).

Ricinus communis f. *macrophyllus* Müll.Arg. in A.P.de Candolle, Prodr. 15(2): 1020 (1866). *Ricinus communis* subvar. *macrophyllus* (Müll.Arg.) T.Carvalho, Anais Junta Invest. Ultramar 11(4): 64 (1956).

Ricinus communis f. *oligacanthus* Müll.Arg. in A.P.de Candolle, Prodr. 15(2): 1020 (1866).

Ricinus communis f. *pruinosus* Müll.Arg. in A.P.de Candolle, Prodr. 15(2): 1017 (1866). *Ricinus communis* subvar. *pruinosus* (Müll.Arg.) T.Carvalho, Anais Junta Invest. Ultramar 11(4): 41 (1956).

Ricinus communis f. *rutilans* Müll.Arg. in A.P.de Candolle, Prodr. 15(2): 1017 (1866). *Ricinus communis* subvar. *rutilans* (Müll.Arg.) T.Carvalho, Anais Junta Invest. Ultramar 11(4): 40 (1956).

Ricinus communis f. *subpurpurascens* Müll.Arg. in A.P.de Candolle, Prodr. 15(2): 1020 (1866).

Ricinus communis f. *subviridus* Müll.Arg. in A.P.de Candolle, Prodr. 15(2): 1019 (1866).

Ricinus communis f. *zollingeri* Müll.Arg. in A.P.de Candolle, Prodr. 15(2): 1018 (1866).

Ricinus communis var. *africanus* Müll.Arg. in A.P.de Candolle, Prodr. 15(2): 1019 (1866).

Ricinus communis var. *amblyocalyx* Müll.Arg. in A.P.de Candolle, Prodr. 15(2): 1019 (1866).

Ricinus communis var. *microcarpus* Müll.Arg. in A.P.de Candolle, Prodr. 15(2): 1020 (1866).

Ricinus communis var. *reichenbachianus* Müll.Arg. in A.P.de Candolle, Prodr. 15(2): 1019 (1866).

Ricinus communis var. *rheedianus* Müll.Arg. in A.P.de Candolle, Prodr. 15(2): 1020 (1866).

Ricinus communis var. *rugosus* Müll.Arg. in A.P.de Candolle, Prodr. 15(2): 1018 (1866).

Ricinus rutilans Müll.Arg. in A.P.de Candolle, Prodr. 15(2): 1017 (1866).

Ricinus gibsonii auct., Gard. Chron., II, 4: 692 (1877).

Ricinus cambodgensis Benary, Gartenflora 36: 102 (1887).

Ricinus zanzibarensis auct., Ill. Hort. 41: 99 (1894).

Ricinus messeniacus Heldr., Exsicc. (Herb. Graec. Norm.): 1480 (1899).

Ricinus communis var. *bailundensis* Coult., Bull. Soc. Portug. Sci. Nat. 1918: 82 (1918).

Ricinus atropurpureus Pax & K.Hoffm. in H.G.A.Engler, Pflanzenr., IV, 147, IX: 126 (1919).

Ricinus borboniensis Pax & K.Hoffm. in H.G.A.Engler, Pflanzenr., IV, 147, IX: 126 (1919).

Ricinus communis var. *brasiliensis* Müll.Arg. ex Pax & K.Hoffm. in H.G.A.Engler, Pflanzenr., IV, 147, XI: 119 (1919).

Ricinus giganteus Pax & K.Hoffm. in H.G.A.Engler, Pflanzenr., IV, 147, IX: 126 (1919).

Ricinus metallicus Pax & K.Hoffm. in H.G.A.Engler, Pflanzenr., IV, 147, IX: 126 (1919).

Ricinus persicus Popova, Trudy Byuro Evgen. 16: 229 (1926).

Ricinus communis subsp. *persicus* Popova, Kleshchevina izd. VIR: 27 (1930).

Ricinus communis subsp. *sanguineus* Popova, Kleshchevina izd. VIR: 35 (1930).

Ricinus communis subsp. *zanzibarinus* Popova, Kleshchevina izd. VIR: 43 (1930).

Ricinus communis subsp. *mexicanus* Popova, Bukasov S.M. Vozdelyv. Rast. Meksiki: 409 (1931). *Ricinus communis* var. *mexicanus* (Popova) Moshkin, Kleshchevina (Nauch. Trudy Vasknil): 40 (1980).

Ricinus communis subsp. *sinensis* Hiltebr., Mirov. Rast. Res.: 63 (1935).

Ricinus communis subsp. *manshuricus* V.Bork., Descr. Var. Azcherkraya: 106 (1936).

Ricinus communis prol. *persicus* Popova, Kul't. Fl. URSS 7: 289 (1941).

Ricinus communis var. *caesius* Popova, Kul't. Fl. URSS 7: 287 (1941).

Ricinus communis var. *virens* Popova, Kul't. Fl. URSS 7: 290 (1941).

Ricinus macrocarpus Popova in N.I.Vavilov & al. (eds.), Kyltyrnaia Fl. SSR 7: 278 (1941), nom. illeg.

Ricinus macrocarpus prol. *indicus* Popova, Kul't. Fl. URSS 7: 283 (1941).

Ricinus macrocarpus prol. *japonicus* Popova, Kul't. Fl. URSS 7: 285 (1941).

Ricinus macrocarpus prol. *sanguineus* Popova, Kul't. Fl. URSS 7: 280 (1941).

Ricinus macrocarpus var. *nudus* Popova, Kul't. Fl. URSS 7: 287 (1941).

Ricinus microcarpus Popova in N.I.Vavilov & al. (eds.), Kyltyrnaia Fl. SSR 7: 287 (1941).

Ricinus microcarpus prol. *aegypticus* Popova, Kul't. Fl. URSS 7: 291 (1941). *Ricinus communis* var. *aegyptiaceus* (Popova) Moshkin, Kleshchevina (Nauch. Trudy Vasknil): 40 (1980).

Ricinus microcarpus prol. *indostanicus* Popova, Kul't. Fl. URSS 7: 290 (1941).

Ricinus microcarpus subsp. *spontaneus* Popova, Kul't. Fl. URSS 7: 288 (1941).

Ricinus microcarpus var. *atrovirens* Popova, Kul't. Fl. URSS 7: 290 (1941).

Ricinus zanzibarinus Popova in N.I.Vavilov & al. (eds.), Kyltyrnaia Fl. SSR 7: 292 (1941).

Ricinus communis f. *argentatus* T.Carvalho, Anais Junta Invest. Ultramar 11(4): 34 (1956).

Ricinus communis f. *argyratus* T.Carvalho, Anais Junta Invest. Ultramar 11(4): 36 (1956).

Ricinus communis f. *atratus* T.Carvalho, Anais Junta Invest. Ultramar 11(2): 32 (1956).

Ricinus communis f. *atrobrunneatus* T.Carvalho, Anais Junta Invest. Ultramar 11(4): 37 (1956).

Ricinus communis f. *atrofulvatus* T.Carvalho, Anais Junta Invest. Ultramar 11(4): 55 (1956).

Ricinus communis f. *atrofuscatus* T.Carvalho, Anais Junta Invest. Ultramar 11(4): 53 (1956).

Ricinus communis f. *atrophoeniceus* T.Carvalho, Anais Junta Invest. Ultramar 11(4): 66 (1956).

Ricinus communis f. *atropunicatus* T.Carvalho, Anais Junta Invest. Ultramar 11(4): 43 (1956).

Ricinus communis f. *atropurpureatus* T.Carvalho, Anais Junta Invest. Ultramar 11(4): 50 (1956).

Ricinus communis f. *avellanatus* T.Carvalho, Anais Junta Invest. Ultramar 11(4): 56 (1956).

Ricinus communis f. *canatus* T.Carvalho, Anais Junta Invest. Ultramar 11(4): 47 (1956).

Ricinus communis f. *canescens* T.Carvalho, Anais Junta Invest. Ultramar 11(4): 48 (1956).

Ricinus communis f. *carneatus* T.Carvalho, Anais Junta Invest. Ultramar 11(4): 44 (1956).

Ricinus communis f. *cervatus* T.Carvalho, Anais Junta Invest. Ultramar 11(4): 43 (1956).

Ricinus communis f. *cinerascens* T.Carvalho, Anais Junta Invest. Ultramar 11(4): 27 (1956).

Ricinus communis f. *cinereatus* T.Carvalho, Anais Junta Invest. Ultramar 11(4): 65 (1956).

Ricinus communis f. *exiguus* T.Carvalho, Anais Junta Invest. Ultramar 11(4): 68 (1956).

Ricinus communis f. *fulvatus* T.Carvalho, Anais Junta Invest. Ultramar 11(4): 68 (1956).

Ricinus communis f. *fumatus* T.Carvalho, Anais Junta Invest. Ultramar 11(4): 41 (1956).

Ricinus communis f. *fuscatus* T.Carvalho, Anais Junta Invest. Ultramar 11(4): 35 (1956).

Ricinus communis f. *gilvus* T.Carvalho, Anais Junta Invest. Ultramar 11(4): 53 (1956).

Ricinus communis f. *guttatus* T.Carvalho, Anais Junta Invest. Ultramar 11(4): 47 (1956).

Ricinus communis f. *incarnatus* T.Carvalho, Anais Junta Invest. Ultramar 11(4): 46 (1956).

Ricinus communis f. *maculatus* T.Carvalho, Anais Junta Invest. Ultramar 11(4): 29 (1956).

Ricinus communis f. *marmoreatus* T.Carvalho, Anais Junta Invest. Ultramar 11(4): 40 (1956).

Ricinus communis f. *murinatus* T.Carvalho, Anais Junta Invest. Ultramar 11(4): 66 (1956).

Ricinus communis f. *nigellus* T.Carvalho, Anais Junta Invest. Ultramar 11(4): 51 (1956).

Ricinus communis f. *nigrescens* T.Carvalho, Anais Junta Invest. Ultramar 11(4): 45 (1956).

Ricinus communis f. *niveatus* T.Carvalho, Anais Junta Invest. Ultramar 11(4): 39 (1956).

Ricinus communis f. *oblongus* T.Carvalho, Anais Junta Invest. Ultramar 11(4): 31 (1956).

Ricinus communis f. *obscurus* T.Carvalho, Anais Junta Invest. Ultramar 11(4): 42 (1956).

Ricinus communis f. *ostrinatus* T.Carvalho, Anais Junta Invest. Ultramar 11(4): 58 (1956).

Ricinus communis f. *pardalinus* T.Carvalho, Anais Junta Invest. Ultramar 11(4): 31 (1956).

Ricinus communis f. *picturatus* T.Carvalho, Anais Junta Invest. Ultramar 11(4): 30 (1956).

Ricinus communis f. *plumbeatus* T.Carvalho, Anais Junta Invest. Ultramar 11(4): 62 (1956).

Ricinus communis f. *pullatus* T.Carvalho, Anais Junta Invest. Ultramar 11(4): 49 (1956).

Ricinus communis f. *punctatus* T.Carvalho, Anais Junta Invest. Ultramar 11(4): 48 (1956).

Ricinus communis f. *punctulatus* T.Carvalho, Anais Junta Invest. Ultramar 11(4): 61 (1956).

Ricinus communis f. *punicans* T.Carvalho, Anais Junta Invest. Ultramar 11(4): 57 (1956).

Ricinus communis f. *radiatus* T.Carvalho, Anais Junta Invest. Ultramar 11(4): 64 (1956).

Ricinus communis f. *rufescens* T.Carvalho, Anais Junta Invest. Ultramar 11(4): 55 (1956).

Ricinus communis f. *russatus* T.Carvalho, Anais Junta Invest. Ultramar 11(4): 57 (1956).

Ricinus communis f. *scriptus* T.Carvalho, Anais Junta Invest. Ultramar 11(4): 59 (1956).

Ricinus communis f. *sordidus* T.Carvalho, Anais Junta Invest. Ultramar 11(4): 33 (1956).

Ricinus communis f. *stigmosus* T.Carvalho, Anais Junta Invest. Ultramar 11(4): 28 (1956).

Ricinus communis f. *striatus* T.Carvalho, Anais Junta Invest. Ultramar 11(4): 63 (1956).

Ricinus communis f. *subrotundus* T.Carvalho, Anais Junta Invest. Ultramar 11(4): 54 (1956).

Ricinus communis f. *sulcatus* T.Carvalho, Anais Junta Invest. Ultramar 11(4): 38 (1956).

Ricinus communis f. *tigrinus* T.Carvalho, Anais Junta Invest. Ultramar 11(4): 39 (1956).

Ricinus communis f. *umbrinus* T.Carvalho, Anais Junta Invest. Ultramar 11(4): 59 (1956).

Ricinus communis f. *venosus* T.Carvalho, Anais Junta Invest. Ultramar 11(4): 37 (1956).

Ricinus communis f. *vinatus* T.Carvalho, Anais Junta Invest. Ultramar 11(4): 52 (1956).

Ricinus communis f. *zebrinus* T.Carvalho, Anais Junta Invest. Ultramar 11(4): 60 (1956).

Ricinus communis f. *zonatus* T.Carvalho, Anais Junta Invest. Ultramar 11(4): 50 (1956).

Ricinus communis subvar. *almeidae* T.Carvalho, Anais Junta Invest. Ultramar 11(4): 58 (1956).

Ricinus communis subvar. *epruinosus* T.Carvalho, Anais Junta Invest. Umtramar 11(4): 43 (1956).

Ricinus communis subvar. *glauceus* T.Carvalho, Anais Junta Invest. Ultramar 11(4): 27 (1956).

Ricinus communis subvar. *griseus* T.Carvalho, Anais Junta Invest. Ultramar 11(4): 33 (1956).

Ricinus communis subvar. *roseus* T.Carvalho, Anais Junta Invest. Ultramar 11(4): 55 (1956).

Ricinus communis subvar. *subviridus* (Müll.Arg.) T.Carvalho, Anais Junta Invest. Ultramar 11(4): 65 (1956).

Ricinus communis subvar. *violaceus* T.Carvalho, Anais Junta Invest. Ultramar 11(4): 36 (1956).

Ricinus communis subvar. *violeus* T.Carvalho, Anais Junta Invest. Ultramar 11(4): 30 (1956).

Ricinus communis var. *macrocarpus* T.Carvalho, Anais Junta Invest. Ultramar 11(4): 27 (1956).

Ricinus communis var. *vasconcellosii* T.Carvalho, Anais Junta Invest. Ultramar 11(4): 33 (1956).

Ricinus communis subsp. *indicus* Popova & Moshkin, Kleshchevina (Nauch. Trudy Vasknil): 35 (1980).

Ricinus communis subsp. *ruderalis* Popova & Moshkin, Kleshchevina (Nauch. Trudy Vasknil): 38 (1980), nom. illeg.

Ricinus communis subsp. *sinensis* Popova & Moshkin, Kleshchevina (Nauch. Trudy Vasknil): 34 (1980).

Ricinus communis var. *brevinodis* Moshkin, Kleshchevina (Nauch. Trudy Vasknil): 32 (1980).

Ricinus communis var. *glaucus* Popova & Moshkin, Kleshchevina (Nauch. Trudy Vasknil): 37 (1980).

Ricinus communis var. *griseofolius* Moshkin, Kleshchevina (Nauch. Trudy Vasknil): 36 (1980).

Ricinus communis var. *indehiscens* Moshkin, Kleshchevina (Nauch. Trudy Vasknil): 33 (1980).

Ricinus communis var. *japonicus* Popova & Moshkin, Kleshchevina (Nauch. Trudy Vasknil): 35 (1980).

Ricinus communis var. *microspermus* Moshkin, Kleshchevina (Nauch. Trudy Vasknil): 31 (1980).

Ricinus communis var. *nanus* Moshkin, Kleshchevina (Nauch. Trudy Vasknil): 38 (1980).

Ricinus communis var. *roseus* Popova & Moshkin, Kleshchevina (Nauch. Trudy Vasknil): 31 (1980).

Ricinus communis var. *spontaneus* Popova & Moshkin, Kleshchevina (Nauch. Trudy Vasknil): 39 (1980).

Ricinus communis var. *violaceocaulis* Moshkin, Kleshchevina (Nauch. Trudy Vasknil): 34 (1980).

Ricinus communis var. *viridis* Popova & Moshkin, Kleshchevina (Nauch. Trudy Vasknil): 32 (1980).

Synonyms:

Ricinus africanus Mill. === **Ricinus communis** L.

Ricinus americanus Mill. === **Jatropha curcas** L.

Ricinus angulatus Thunb. === **Ricinus communis** L.

Ricinus apelta Lour. === **Mallotus apelta** (Lour.) Müll.Arg.

Ricinus armatus Andr. === **Ricinus communis** L.

Ricinus atropurpureus Pax & K.Hoffm. === **Ricinus communis** L.

Ricinus badius Rchb. === **Ricinus communis** L.

Ricinus borboniensis Pax & K.Hoffm. === **Ricinus communis** L.

Ricinus cambodgensis Benary === **Ricinus communis** L.

Ricinus chinensis Thunb. === **Mallotus paniculatus** (Lam.) Müll.Arg.

Ricinus communis var. *aegyptiaceus* (Popova) Moshkin === **Ricinus communis** L.

Ricinus communis var. *africanus* Müll.Arg. === **Ricinus communis** L.

Ricinus communis subvar. *almeidae* T.Carvalho === **Ricinus communis** L.

Ricinus communis var. *amblyocalyx* Müll.Arg. === **Ricinus communis** L.

Ricinus communis f. *americanus* Müll.Arg. === **Ricinus communis** L.

Ricinus communis subvar. *americanus* (Müll.Arg.) T.Carvalho === **Ricinus communis** L.

Ricinus communis f. *argentatus* T.Carvalho === **Ricinus communis** L.

Ricinus communis f. *argyratus* T.Carvalho === **Ricinus communis** L.

Ricinus communis var. *armatus* (Andr.) Müll.Arg. === **Ricinus communis** L.

Ricinus communis f. *atratus* T.Carvalho === **Ricinus communis** L.

Ricinus communis f. *atrobrunneatus* T.Carvalho === **Ricinus communis** L.

Ricinus communis f. *atrofulvatus* T.Carvalho === **Ricinus communis** L.

Ricinus communis f. *atrofuscatus* T.Carvalho === **Ricinus communis** L.

Ricinus communis f. *atrophoeniceus* T.Carvalho === **Ricinus communis** L.

Ricinus communis f. *atropunicatus* T.Carvalho === **Ricinus communis** L.

Ricinus communis f. *atropurpureatus* T.Carvalho === **Ricinus communis** L.

Ricinus communis f. *avellanatus* T.Carvalho === **Ricinus communis** L.

Ricinus communis var. *badius* (Rchb.) Müll.Arg. === **Ricinus communis** L.

Ricinus communis var. *bailundensis* Coult. === **Ricinus communis** L.

Ricinus communis var. *benguelensis* Müll.Arg. === **Ricinus communis** L.

Ricinus communis f. *blumeanus* Müll.Arg. === **Ricinus communis** L.

Ricinus communis subvar. *blumeanus* (Müll.Arg.) T.Carvalho === **Ricinus communis** L.

Ricinus communis var. *brasiliensis* Müll.Arg. ex Pax & K.Hoffm. === **Ricinus communis** L.

Ricinus communis var. *brevinodis* Moshkin === **Ricinus communis** L.

Ricinus communis var. *caesius* Popova === **Ricinus communis** L.

Ricinus communis f. *canatus* T.Carvalho === **Ricinus communis** L.

Ricinus communis f. *canescens* T.Carvalho === **Ricinus communis** L.

Ricinus communis f. *carneatus* T.Carvalho === **Ricinus communis** L.

Ricinus communis f. *cervatus* T.Carvalho === **Ricinus communis** L.

Ricinus communis f. *cinerascens* T.Carvalho === **Ricinus communis** L.

Ricinus communis f. *cinereatus* T.Carvalho === **Ricinus communis** L.

Ricinus communis f. *denudatus* Müll.Arg. === **Ricinus communis** L.

Ricinus communis f. *epiglaucus* Müll.Arg. === **Ricinus communis** L.

Ricinus communis subvar. *epruinosus* T.Carvalho === **Ricinus communis** L.

Ricinus communis subvar. *erythrocladus* (Müll.Arg.) T.Carvalho === **Ricinus communis** L.

Ricinus communis f. *erythrocladus* Müll.Arg. === **Ricinus communis** L.

Ricinus communis f. *exiguus* T.Carvalho === **Ricinus communis** L.

Ricinus communis f. *fulvatus* T.Carvalho === **Ricinus communis** L.

Ricinus communis f. *fumatus* T.Carvalho === **Ricinus communis** L.

Ricinus communis f. *fuscatus* T.Carvalho === **Ricinus communis** L.

Ricinus communis var. *genuinus* Müll.Arg. === **Ricinus communis** L.

Ricinus communis f. *gilvus* T.Carvalho === **Ricinus communis** L.

Ricinus communis subvar. *glauceus* T.Carvalho === **Ricinus communis** L.

Ricinus communis f. *glaucus* (Hoffmanns.) Müll.Arg. === **Ricinus communis** L.

Ricinus communis var. *glaucus* Popova & Moshkin === **Ricinus communis** L.

Ricinus communis subvar. *gracilis* (Müll.Arg.) T.Carvalho === **Ricinus communis** L.

Ricinus communis f. *gracilis* Müll.Arg. === **Ricinus communis** L.

Ricinus communis var. *griseofolius* Moshkin === **Ricinus communis** L.

Ricinus communis subvar. *griseus* T.Carvalho === **Ricinus communis** L.

Ricinus communis f. *guttatus* T.Carvalho === **Ricinus communis** L.

Ricinus communis f. *hybridus* (Besser) Müll.Arg. === **Ricinus communis** L.

Ricinus communis f. *incarnatus* T.Carvalho === **Ricinus communis** L.

Ricinus communis var. *indehiscens* Moshkin === **Ricinus communis** L.

Ricinus communis subsp. *indicus* Popova & Moshkin === **Ricinus communis** L.

Ricinus communis var. *inermis* (Mill.) Pax & K.Hoffm. === **Ricinus communis** L.

Ricinus communis f. *inermis* (Mill.) Müll.Arg. === **Ricinus communis** L.

Ricinus communis f. *intermedius* Müll.Arg. === **Ricinus communis** L.

Ricinus communis var. *japonicus* Popova & Moshkin === **Ricinus communis** L.

Ricinus communis f. *laevis* (DC.) Müll.Arg. === **Ricinus communis** L.

Ricinus communis var. *leucocarpus* (Bertol.) Müll.Arg. === **Ricinus communis** L.

Ricinus communis var. *lividus* (Jacq.) Müll.Arg. === **Ricinus communis** L.

Ricinus communis var. *macrocarpus* T.Carvalho === **Ricinus communis** L.

Ricinus communis subvar. *macrophyllus* (Müll.Arg.) T.Carvalho === **Ricinus communis** L.

Ricinus communis f. *macrophyllus* Müll.Arg. === **Ricinus communis** L.

Ricinus communis f. *maculatus* T.Carvalho === **Ricinus communis** L.

Ricinus communis subsp. *manshuricus* V.Bork. === **Ricinus communis** L.

Ricinus communis f. *marmoreatus* T.Carvalho === **Ricinus communis** L.

Ricinus communis var. *megalospermus* (Delile) Müll.Arg. === **Ricinus communis** L.

Ricinus communis var. *mexicanus* (Popova) Moshkin === **Ricinus communis** L.

Ricinus communis subsp. *mexicanus* Popova === **Ricinus communis** L.

Ricinus communis var. *microcarpus* Müll.Arg. === **Ricinus communis** L.

Ricinus communis var. *microspermus* Moshkin === **Ricinus communis** L.

Ricinus communis var. *minor* Steud. === **Ricinus communis** L.

Ricinus communis f. *murinatus* T.Carvalho === **Ricinus communis** L.

Ricinus communis var. *nanus* Moshkin === **Ricinus communis** L.

Ricinus communis f. *nigellus* T.Carvalho === **Ricinus communis** L.

Ricinus communis f. *nigrescens* T.Carvalho === **Ricinus communis** L.

Ricinus communis f. *niveatus* T.Carvalho === **Ricinus communis** L.

Ricinus communis f. *oblongus* T.Carvalho === **Ricinus communis** L.

Ricinus communis f. *obscurus* T.Carvalho === **Ricinus communis** L.

Ricinus communis f. *oligacanthus* Müll.Arg. === **Ricinus communis** L.

Ricinus communis f. *ostrinatus* T.Carvalho === **Ricinus communis** L.

Ricinus communis f. *pardalinus* T.Carvalho === **Ricinus communis** L.

Ricinus communis subsp. *persicus* Popova === **Ricinus communis** L.

Ricinus communis prol. *persicus* Popova === **Ricinus communis** L.

Ricinus communis f. *picturatus* T.Carvalho === **Ricinus communis** L.

Ricinus communis f. *plumbeatus* T.Carvalho === **Ricinus communis** L.

Ricinus communis f. *pruinosus* Müll.Arg. === **Ricinus communis** L.

Ricinus communis subvar. *pruinosus* (Müll.Arg.) T.Carvalho === **Ricinus communis** L.

Ricinus communis f. *pullatus* T.Carvalho === **Ricinus communis** L.

Ricinus communis f. *punctatus* T.Carvalho === **Ricinus communis** L.

Ricinus communis f. *punctulatus* T.Carvalho === **Ricinus communis** L.

Ricinus communis f. *punicans* T.Carvalho === **Ricinus communis** L.

Ricinus communis f. *purpurascens* (Bertol.) Pax === **Ricinus communis** L.

Ricinus communis subvar. *purpurascens* (Bertol.) T.Carvalho === **Ricinus communis** L.

Ricinus communis var. *purpurascens* (Bertol.) Müll.Arg. === **Ricinus communis** L.

Ricinus communis f. *radiatus* T.Carvalho === **Ricinus communis** L.

Ricinus communis var. *reichenbachianus* Müll.Arg. === **Ricinus communis** L.

Ricinus communis var. *rheedianus* Müll.Arg. === **Ricinus communis** L.

Ricinus communis var. *roseus* Popova & Moshkin === **Ricinus communis** L.

Ricinus communis subvar. *roseus* T.Carvalho === **Ricinus communis** L.

Ricinus communis subsp. *ruderalis* Popova & Moshkin === **Ricinus communis** L.

Ricinus communis f. *rufescens* T.Carvalho === **Ricinus communis** L.

Ricinus communis var. *rugosus* Müll.Arg. === **Ricinus communis** L.

Ricinus communis f. *russatus* T.Carvalho === **Ricinus communis** L.

Ricinus communis f. *rutilans* Müll.Arg. === **Ricinus communis** L.

Ricinus communis subvar. *rutilans* (Müll.Arg.) T.Carvalho === **Ricinus communis** L.

Ricinus communis subsp. *sanguineus* Popova === **Ricinus communis** L.

Ricinus communis var. *sanguineus* Baill. === **Ricinus communis** L.

Ricinus communis f. *scaber* (Bertol. ex Moris) Müll.Arg. === **Ricinus communis** L.

Ricinus communis f. *scriptus* T.Carvalho === **Ricinus communis** L.

Ricinus communis subsp. *sinensis* Popova & Moshkin === **Ricinus communis** L.

Ricinus communis subsp. *sinensis* Hiltebr. === **Ricinus communis** L.

Ricinus communis f. *sordidus* T.Carvalho === **Ricinus communis** L.

Ricinus communis var. *speciosus* (Burm.f.) Müll.Arg. === **Ricinus communis** L.

Ricinus communis var. *spontaneus* Popova & Moshkin === **Ricinus communis** L.

Ricinus communis f. *stigmosus* T.Carvalho === **Ricinus communis** L.

Ricinus communis f. *striatus* T.Carvalho === **Ricinus communis** L.

Ricinus communis f. *subpurpurascens* Müll.Arg. === **Ricinus communis** L.

Ricinus communis f. *subrotundus* T.Carvalho === **Ricinus communis** L.

Ricinus communis f. *subviridus* Müll.Arg. === **Ricinus communis** L.

Ricinus communis subvar. *subviridus* (Müll.Arg.) T.Carvalho === **Ricinus communis** L.

Ricinus communis f. *sulcatus* T.Carvalho === **Ricinus communis** L.

Ricinus communis f. *tigrinus* T.Carvalho === **Ricinus communis** L.

Ricinus communis var. *typicus* Fiori === **Ricinus communis** L.

Ricinus communis f. *umbrinus* T.Carvalho === **Ricinus communis** L.

Ricinus communis var. *undulatus* (Besser) Müll.Arg. === **Ricinus communis** L.

Ricinus communis var. *vasconcellosii* T.Carvalho === **Ricinus communis** L.

Ricinus communis f. *venosus* T.Carvalho === **Ricinus communis** L.

Ricinus communis f. *vinatus* T.Carvalho === **Ricinus communis** L.

Ricinus communis var. *violaceocaulis* Moshkin === **Ricinus communis** L.

Ricinus communis subvar. *violaceus* T.Carvalho === **Ricinus communis** L.

Ricinus communis subvar. *violeus* T.Carvalho === **Ricinus communis** L.

Ricinus communis var. *virens* Popova === **Ricinus communis** L.

Ricinus communis f. *viridis* (Willd.) Müll.Arg. === **Ricinus communis** L.

Ricinus communis var. *viridis* Popova & Moshkin === **Ricinus communis** L.

Ricinus communis subvar. *viridus* (Willd.) T.Carvalho === **Ricinus communis** L.

Ricinus communis subsp. *zanzibarinus* Popova === **Ricinus communis** L.

Ricinus communis f. *zebrinus* T.Carvalho === **Ricinus communis** L.

Ricinus communis f. *zollingeri* Müll.Arg. === **Ricinus communis** L.

Ricinus communis f. *zonatus* T.Carvalho === **Ricinus communis** L.

Ricinus compactus Huber === **Ricinus communis** L.

Ricinus dicoccus Roxb. === **Aleurites moluccanus**

Ricinus digitatus Noronha === **Ricinus communis** L.

Ricinus dioicus Wall. ex Roxb. === **Melanolepis multiglandulosa** (Reinw. ex Blume) Rchb. & Zoll. var. **multiglandulosa**

Ricinus dioicus Chev. ex Steud. === **Cordemoya integrifolia** (Willd.) Baill.

Ricinus dioicus G.Forst. === **Macaranga dioica** (G.Forst.) Müll.Arg.

Ricinus europaeus T.Nees === **Ricinus communis** L.

Ricinus floribundus Reinw. ex Müll.Arg. === **Mallotus floribundus** (Blume) Müll.Arg. var. **floribundus**

Ricinus furfuraceus Wall. === **Daphniphyllum majus** Müll.Arg. (Daphniphyllaceae)

Ricinus gibsonii auct. === **Ricinus communis** L.

Ricinus giganteus Pax & K.Hoffm. === **Ricinus communis** L.

Ricinus glaucus Hoffmanns. === **Ricinus communis** L.

Ricinus globosus (Sw.) Willd. === **Chaetocarpus globosus** (Sw.) Fawc. & Rendle

Ricinus hybridus Besser === **Ricinus communis** L.

Ricinus inermis Mill. === **Ricinus communis** L.

Ricinus integrifolius Willd. === **Cordemoya integrifolia** (Willd.) Baill.

Ricinus japonicus Thunb. === **Ricinus communis** L.

Ricinus jarak Thunb. === **Jatropha curcas** L.

Ricinus krappa Steud. === **Ricinus communis** L.

Ricinus laevis DC. === **Ricinus communis** L.

Ricinus lanceolatus Thouars ex Baill. === **Cordemoya integrifolia** (Willd.) Baill.

Ricinus leucocarpus Bertol. === **Ricinus communis** L.

Ricinus lividus Jacq. === **Ricinus communis** L.

Ricinus macrocarpus Popova === **Ricinus communis** L.

Ricinus macrocarpus Steud. === **Ricinus communis** L.

Ricinus macrocarpus prol. *indicus* Popova === **Ricinus communis** L.

Ricinus macrocarpus prol. *japonicus* Popova === **Ricinus communis** L.

Ricinus macrocarpus var. *nudus* Popova === **Ricinus communis** L.

Ricinus macrocarpus prol. *sanguineus* Popova === **Ricinus communis** L.

Ricinus macrophyllus Bertol. === **Ricinus communis** L.

Ricinus mappa L. === **Macaranga mappa** (L.) Müll.Arg.

Ricinus medicus Forssk. === **Ricinus communis** L.

Ricinus medius J.F.Gmel. === **Ricinus communis** L.

Ricinus megalospermus Delile === **Ricinus communis** L.

Ricinus messeniacus Heldr. === **Ricinus communis** L.

Ricinus metallicus Pax & K.Hoffm. === **Ricinus communis** L.

Ricinus microcarpus Popova === **Ricinus communis** L.

Ricinus microcarpus prol. *aegypticus* Popova === **Ricinus communis** L.

Ricinus microcarpus var. *atrovirens* Popova === **Ricinus communis** L.

Ricinus microcarpus prol. *indostanicus* Popova === **Ricinus communis** L.

Ricinus microcarpus subsp. *spontaneus* Popova === **Ricinus communis** L.

Ricinus minor Mill. === **Ricinus communis** L.

Ricinus montanus (Willd.) Wall. === **Baliospermum montanum** (Willd.) Müll.Arg.

Ricinus morifolius Noronha === **?**

Ricinus nanus Bald. === **Ricinus communis** L.

Ricinus obermannii Groenl. === **Ricinus communis** L.

Ricinus odoratus Noronha === **Commersonia echinata**

Ricinus peltatus Noronha === **Ricinus communis** L.

Ricinus perennis Steud. === **Ricinus communis** L.

Ricinus persicus Popova === **Ricinus communis** L.

Ricinus pictus Noronha === **Codiaeum variegatum** (L.) Blume

Ricinus portoricensis Juss. ex Baill. === **Jatropha hernandiifolia** var. **portoricensis** (Millsp.) Urb.

Ricinus pulchellus Noronha === **?**

Ricinus purpurascens Bertol. === **Ricinus communis** L.

Ricinus ruber Miq. === **Ricinus communis** L.

Ricinus rugosus Mill. === **Ricinus communis** L.

Ricinus rutilans Müll.Arg. === **Ricinus communis** L.

Ricinus salicinus Hassk. === **Homonoia riparia** Lour.

Ricinus sanguineus Groenl. === **Ricinus communis** L.

Ricinus scaber Bertol. ex Moris === **Ricinus communis** L.

Ricinus speciosus Burm.f. === **Ricinus communis** L.

Ricinus spectabilis Blume === **Ricinus communis** L.

Ricinus tanarius L. === **Macaranga tanarius** (L.) Müll.Arg.

Ricinus tomentosus Thunb. === **Adriana tomentosa** (Thunb.) Gaudich.

Ricinus tomentosus Gaudich. === **Adriana tomentosa** (Thunb.) Gaudich.

Ricinus trilobus Thunb. === **Macaranga triloba** (Thunb.) Müll.Arg.

Ricinus trilobus Reinw. ex Blume === **Macaranga triloba** (Thunb.) Müll.Arg.

Ricinus tunisensis Desf. === **Ricinus communis** L.

Ricinus undulatus Besser === **Ricinus communis** L.

Ricinus urens Mill. === **Ricinus communis** L.

Ricinus viridis Willd. === **Ricinus communis** L.

Ricinus vulgaris Mill. === **Ricinus communis** L.

Ricinus vulgaris Garsault === **Ricinus communis** L.

Ricinus zanzibarensis auct. === **Ricinus communis** L.

Ricinus zanzibarinus Popova === **Ricinus communis** L.

Riseleya

Synonyms:
Riseleya Hemsl. === **Drypetes** Vahl
Riseleya griffithii Hemsl. === **Drypetes riseleyi** Airy Shaw

Ritchieophyton

Introduced as a synonym of *Givotia* by Pax (1911; see that genus).

Rivinoides

An Afzelius name for *Erythrococca* published in synonymy by Prain (1911; see that genus).

Rockinghamia

2 species, E. Australia (Queensland); related to *Mallotus* and *Trewia*. Small to medium trees to 20 m in moist or wet forest with pseudo-verticillately or tightly spirally arranged leaves in flushes and terminal inflorescences. The two species are effectively altitudinal vicariants. *Ptychopyxis chrysantha* (New Guinea) appears superficially similar but that genus is in the Webster system only distantly related. (Acalyphoideae)

Airy-Shaw, H. K. (1966). Notes on Malaysian and other Asiatic Euphorbiaceae, LXI. A new genus from Queensland. Kew Bull. 20: 29-31. En. — Protologue of *Rockinghamia* and transfer of the single species from *Mallotus*.

Airy-Shaw, H. K. (1980). *Rockinghamia*. Kew Bull. 35: 667, 669, illus. (p. 668). (Euphorbiaceae-Platylobeae of Australia.) En. — The 2 species are endemic.

Rockinghamia Airy Shaw, Kew Bull. 20: 29 (1966).
　Trop. Australia. 50.

Rockinghamia angustifolia (Benth.) Airy Shaw, Kew Bull. 20: 29 (1966).
　N. & NE. Queensland. 50 QLD. Phan.
　　*Mallotus angustifolius Benth., Fl. Austral. 6: 141 (1873).

Rockinghamia brevipes Airy Shaw, Kew Bull. 31: 389 (1976). – FIGURE, p. 1413.
　Queensland (Cook). 50 QLD. Nanophan. or phan.

Roeperia

Synonyms:
Roeperia Spreng. === **Ricinocarpos** Desf.

Roigia

Synonyms:
Roigia Britton === **Phyllanthus** L.

Romanoa

1 species, S. America (Brazil, Paraguay); slender woody climbers or twiners reminiscent of Cucurbitaceae but without tendrils. Closely related to *Plukenetia* and possibly not distinct; however, the fruits are 3-angled. Some collections have been made in coastal restinga vegetation. (Acalyphoideae)

Pax, F. & K. Hoffmann (1919). *Anabaenella*. In A. Engler (ed.), Das Pflanzenreich, IV 147 IX (Euphorbiaceae-Acalypheae-Plukenetiinae): 27-28. Berlin. (Heft 68.) La/Ge. — 1 species, now included in *Romanoa*. [An earlier name, *Anabaena*, is not available.]

Radcliffe-Smith, A. (1980). A note on *Romanoa* (Euphorbiaceae). Kew Bull. 34: 589-590. En. —*Romanoa* Trevis. replaces *Anabaena* A. Juss. and *Anabaenella* Pax & Hoffm., nomina illeg.; new name for species and one variety along with complete synonymy.

Romanoa Trevis., Sagg. Algh. Coccot.: 99 (1848).
　E. & SE. Brazil, Paraguay. 84 85.
　　*Anabaena A.Juss., Euphorb. Gen.: 46 (1824), nom. illeg.
　　Anabaenella Pax & K.Hoffm. in H.G.A.Engler, Pflanzenr., IV, 147, IX: 27 (1919).

Rockinghamia brevipes Airy Shaw
Artist: Ann Davies
Kew Bull. 35: 668, fig. 6 (1980, lower left
KEW ILLUSTRATIONS COLLECTION

Romanoa tamnoides (A.Juss.) Radcl.-Sm., Kew Bull. 34: 589 (1980).
E. Brazil, Paraguay. 84 BZE BZL 85 PAR.
* *Anabaena tamnoides* A.Juss., Euphorb. Gen.: 46 (1824). *Plukenetia tamnoides* (A.Juss.)
Müll.Arg., Linnaea 34: 158 (1865). *Anabaenella tamnoides* (A.Juss.) Pax & K.Hoffm. in
H.G.A.Engler, Pflanzenr., IV, 147, IX: 27 (1919). *Anabaenella tamnoides* var. *genuina*
Pax & K.Hoffm. in H.G.A.Engler, Pflanzenr., IV, 147, IX: 27 (1919), nom. inval.

var. **sinuata** (Ule) Radcl.-Sm., Kew Bull. 34: 589 (1980).
Brazil (Bahia). 84 BZE.
* *Plukenetia sinuata* Ule, Bot. Jahrb. Syst. 42: 217 (1908). *Anabaenella tamnoides* var.
sinuata (Ule) Pax & K.Hoffm. in H.G.A.Engler, Pflanzenr., IV, 147, IX: 27 (1919).

var. **tamnoides**
SE. Brazil, Paraguay. 84 BZL 85 PAR.
Croton scandens Vell., Fl. Flumin. 10: 72 (1831).
Plukenetia occidentalis Leandro ex Baill., Étude Euphorb.: 484 (1858).

Ronnowia

Synonyms:
Ronnowia Buc'hoz === **Omphalea** L.

Rottlera

Rottlera Willd. has priority over *Rottlera* Roxb. but is synonymous with *Trewia*. *Rottlera* Roxb. was united by Müller with *Mallotus* but remains in use at sectional rank.

Synonyms:

Rottlera Willd. === **Trewia** L.

Rottlera Roxb. === **Mallotus** Lour.

Rottlera alba Roxb. === **Mallotus paniculatus** (Lam.) Müll.Arg.

Rottlera aureopunctata Dalzell === **Mallotus aureopunctatus** (Dalzell) Müll.Arg.

Rottlera barbata Wall. ex Baill. === **Mallotus barbatus** Müll.Arg.

Rottlera brasiliensis Spreng. === **Croton polyandrus** Spreng.

Rottlera dispar Blume === **Mallotus dispar** (Blume) Müll.Arg.

Rottlera eriocarpa Thwaites === **Mallotus eriocarpus** (Thwaites) Müll.Arg.

Rottlera ferruginea Roxb. === **Mallotus tetracoccus** (Roxb.) Kurz

Rottlera fuscescens Thwaites === **Mallotus fuscescens** (Thwaites) Müll.Arg.

Rottlera glaberrima Hassk. === **Macaranga glaberrima** (Hassk.) Airy Shaw
var. **glaberrima**

Rottlera hexandra Roxb. === **Macaranga hexandra** (Roxb.) Müll.Arg.

Rottlera longifolia Rchb.f. & Zoll. === **Mallotus peltatus** (Geiseler) Müll.Arg.

Rottlera longistipulata Kurz ex Teijsm. & Binn. === **Macaranga longistipulata** (Kurz ex Tejism. & Binn.) Müll.Arg.

Rottlera macrostachya Miq. === **Mallotus macrostachyus** (Miq.) Müll.Arg.

Rottlera miqueliana Scheff. === **Mallotus miquelianus** (Scheff.) Boerl.

Rottlera montana B.Heyne ex Baill. === **Macaranga heynei** I.M.Johnst.

Rottlera oblongifolia Miq. === **Mallotus oblongifolius** (Miq.) Müll.Arg.

Rottlera oppositifolia Blume === **Mallotus blumeanus** Müll.Arg.

Rottlera rufidula Miq. === **Mallotus rufidulus** (Miq.) Müll.Arg.

Rottlera scabrifolia A.Juss. === **Mallotus repandus** (Willd.) Müll.Arg.

Rottlera scandens Span. === **Mallotus repandus** (Willd.) Müll.Arg.

Rottlera sphaerocarpa Miq. === **Mallotus sphaerocarpus** (Miq.) Müll.Arg.

Rottlera subfalcata Rchb.f. & Zoll. === **Macaranga glaberrima** (Hassk.) Airy Shaw
var. **glaberrima**

Rottlera tetracocca Roxb. === **Mallotus tetracoccus** (Roxb.) Kurz

Rottlera tiliifolia Blume === **Mallotus tiliifolius** (Blume) Müll.Arg.

Rubina

Synonyms:

Rubina Noronha === **Antidesma** L.

Sagotia

2 species, S. America (Peru and Brazil northwards), extending disjunctly to Panama and Costa Rica. Small to medium forest trees to 18 m with clear or yellowish latex, elliptic leaves, terminal inflorescences and leaf-like perianth segments. The genus is regarded as related to *Codiaeum* and in the Webster system forms part of the Codiaeae. Most recently revised by Secco (1990). (Crotonoideae)

Pax, F. (with K. Hoffmann) (1911). *Sagotia*. In A. Engler (ed.), Das Pflanzenreich, IV 147 III (Euphorbiaceae-Cluytieae): 39-41. Berlin. (Heft 47.) La/Ge. — 1 species, N South America; four additional varieties recognized.

Jablonski, E. (1967). *Sagotia*. Euphorbiaceae, Guayana Highland (Mem. New York Bot. Gard. 17(1)): 151-152. New York. En. — 1 species.

Secco, R. de S. (1990). Revisão dos gêneros *Anomalocalyx* Ducke, *Dodecastigma* Ducke, *Pausandra* Radlk., *Pogonophora* Miers ex Benth. e *Sagotia* Baill. (Euphorbiaceae-Crotonoideae) para a América do Sul. 133 pp., illus., maps. Belém: Museu Paraense 'Emilio Goeldi'. Pt. — Pp. 95-108 a comprise revision of *Sagotia* in South America, with key, descriptions, *exsiccatae*, commentary, figures and map (p. 114).

Sagotia Baill., Adansonia 1: 53 (1860).
Trop. America. 80 82 83 84.

Sagotia brachysepala (Müll.Arg.) Secco, Acta Amazon., Supl. 15: 81 (1985).
N. South America to N. Brazil. 82 FRG GUY SUR VEN 84 BZN. Phan.
   * *Sagotia racemosa* var. *brachysepala* Müll.Arg., Flora 47: 516 (1864).

Sagotia racemosa Baill., Adansonia 1: 54 (1860). *Sagotia racemosa* var. *genuina* Müll.Arg., Flora 47: 516 (1864), nom. inval.
   Costa Rica, Panama, SE. Colombia, Venezuela, Guianas, Brazil. 80 COS PAN 82 FRG GUY SUR VEN 83 CLM 84 BZC BZE BZN. Phan.
   Sagotia racemosa var. *ligularis* Müll.Arg., Flora 47: 516 (1864).
   Sagotia racemosa var. *macrocarpa* Müll.Arg., Flora 47: 516 (1864).
   Sagotia racemosa var. *microsepala* Müll.Arg., Flora 47: 516 (1864).
   Sagotia tafelbergii Croizat, Bull. Torrey Bot. Club 75: 404 (1948).

Synonyms:
Sagotia gardenioides Vieill. ex Guillaumin === **Morierina montana** Vieill. (Rubiaceae)
Sagotia racemosa var. *brachysepala* Müll.Arg. === **Sagotia brachysepala** (Müll.Arg.) Secco
Sagotia racemosa var. *genuina* Müll.Arg. === **Sagotia racemosa** Baill.
Sagotia racemosa var. *ligularis* Müll.Arg. === **Sagotia racemosa** Baill.
Sagotia racemosa var. *macrocarpa* Müll.Arg. === **Sagotia racemosa** Baill.
Sagotia racemosa var. *microsepala* Müll.Arg. === **Sagotia racemosa** Baill.
Sagotia tafelbergii Croizat === **Sagotia racemosa** Baill.

Saipania

Synonyms:
Saipania Hosok. === **Croton** L.
Saipania glandulosa Hosok. === **Croton saipanensis** Hosok.

Sajorium

Synonyms:
Sajorium Endl. === **Pterococcus** Hassk.

Samaropyxis

Synonyms:
Samaropyxis Miq. === **Hymenocardia** Wall. ex Lindl.
Samaropyxis elliptica Miq. === **Hymenocardia punctata** Wall. ex Lindl.

Sampantaea

1 species, SE. Asia (Cambodia, Thailand); closely related to *Wetria* and possibly congeneric (Webster, Synopsis 1994). Shrubs or small trees with pinnately veined leaves in closed evergreen or semi-evergreen forest. The flushes are somewhat reminiscent of some species of *Terminalia*. (Acalyphoideae)

Airy-Shaw, H. K. (1966). Notes on Malaysian and other Asiatic Euphorbiaceae, LXIV. A remarkable new *Alchornea* from Siam. Kew Bull. 20: 45-47. En. — *A. amentiflora* described.

Airy-Shaw, H. K. (1968). Notes on Malesian and other Asiatic Euphorbiaceae, XCI. The female flower of *Alchornea amentiflora* Airy Shaw. Kew Bull. 21: 400-401. En. — Includes discussion; suggestion that species should be separated from *Alchornea* or that *Alchornea* be widened to include *Wetria* and *Lautembergia* (=*Orfilea*) as well as this plant.

Airy-Shaw, H. K. (1971). Notes on Malesian and other Asiatic Euphorbiaceae, CXXXIX. *Alchornea amentiflora* in Cambodia. Kew Bull. 25: 528. En. — In Battambang Province; previously known from Thailand.

Airy-Shaw, H. K. (1972). *Sampantaea*. Kew Bull. 26: 328-329. (Euphorbiaceae of Siam.) En. — Protologue of genus, with transfer of the single species from *Alchornea*.

Airy-Shaw, H. K. (1974). *Sampantaea amentiflora*. Ic. Pl. 38: pl. 3717. En. — Plant portrait with description and commentary.

Sampantaea Airy Shaw, Kew Bull. 26: 328 (1972).
 Indo-China. 41.

Sampantaea amentiflora (Airy Shaw) Airy Shaw, Kew Bull. 26: 328 (1972).
 – FIGURE, p. 1417.
 Thailand, Cambodia. 41 CBD THA. Nanophan. or phan.
 * Alchornea amentiflora Airy Shaw, Kew Bull. 20: 45 (1966).

Sandwithia

2 species, northern S. America (Venezuela, Guianas, Brazil); shrubs or small trees of forest understorey, sometimes common. Rather similar to *Sagotia*, a point made by Secco (1988) when describing *S. heterocalyx*; previously, Secco (1987) suggested that those two genera, along with *Anomalocalyx*, could well be accorded subtribal or tribal recognition in their own right. Webster (Synopsis, 1994), however, continues to refer *Sandwithia* and *Anomalocalyx* to subtribe Grosserinae (Aleuritideae) and not Codiaeae (*Sagotia*). (Crotonoideae)

Jablonski, E. (1967). *Sandwithia*. Euphorbiaceae, Guayana Highland (Mem. New York Bot. Gard. 17(1)): 152. New York. En. — 1 species, *S. guyanensis*.

Secco, R. de S. (1987). Aspectos sistemáticos e evolutivos do gênero *Sandwithia* Lanj. (Euphorbiaceae) em relação às suas afinidades. Bol. Mus. Paraense 'Emilio Goeldi', Bot. 3: 157-181, illus., maps. Pt. — Detailed study including an attempt to distinguish the genus from *Sagotia racemosa*; putative systematic relationships including *Anomalocalyx* and *Sagotia*; analysis of characters and discussion; conclusion (*Sandwithia* has a number of 'primitive' features and, along with *Anomalocalyx* and *Sagotia*, separate subtribal or tribal status within Crotonoideae may be merited).

Secco, R. de S. (1988). Dialissepalia no gênero *Sandwithia* Lanj. (Euphorbiaceae): uma novidade botânica do alto Rio Negro e da Venezuela. Bol. Mus. Paraense 'Emilio Goeldi', Bot. 4: 177-185, illus., maps. Pt. — Includes description of a novelty, *S. heterocalyx*, which exhibits discrete sepals in the female flowers (as in *Sagotia*) rather than the mostly united sepals of *S. guianensis*; key to the two species; discussion (with suggestion that the two genera are closely related).

Sandwithia Lanj., Bull. Misc. Inform. Kew. 1932: 185 (1932).
 S. Trop. America. 82 83 84.

Sandwithia guyanensis Lanj., Bull. Misc. Inform. Kew 1932: 185 (1932).
 Guiana, Guyana, S. Venezuela, Brazil (N. Amazonas, Rondônia, Amapá). 82 FRG GUY VEN 84 BZN. Nanophan. or phan.

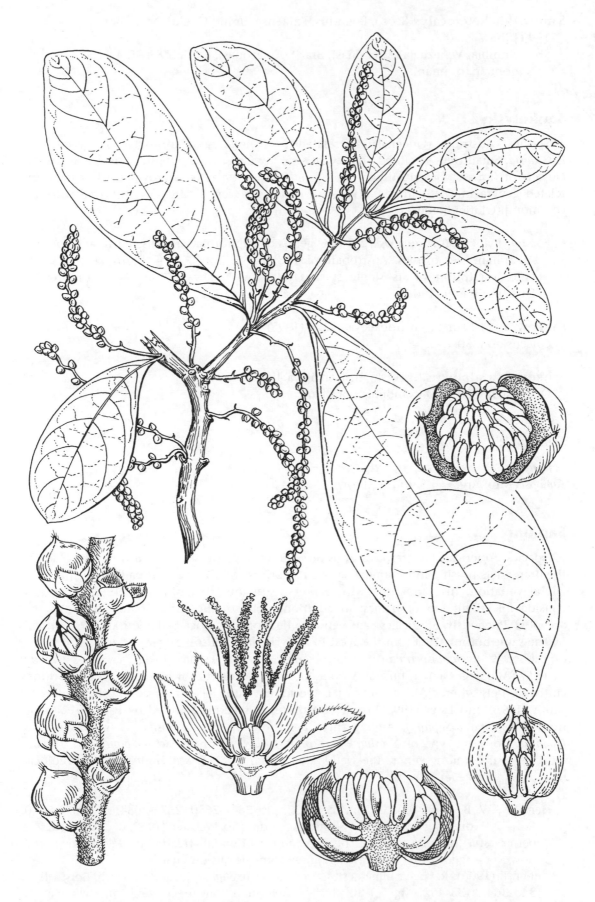

Sampantaea amentiflora (Airy Shaw) Airy Shaw

Artist: Mary Grierson
Ic. Pl. 38: pl. 3717 (1974)
KEW ILLUSTRATIONS COLLECTION

Sandwithia heterocalyx Secco, Bol. Mus. Paraense Emilio Goeldi, N. S., Bot. 4: 179 (1988).

    SE. Colombia, Venezuela (Amazonas), Brazil (Amazonas). 82 VEN 83 CLM 84 BZN. Nanophan. or phan.

Sankowskya

1 species, Australia (Queensland); related to *Longetia* and formerly associated with *Dissiliaria*. Small trees of closed, somewhat swampy forest with opposite leaves and axillary clusters of flowers. Described and revised by Forster (1995) who provides a key to it and related genera in Dissiliariinae. [The known range is very limited but the tree is locally common.] (Oldfieldioideae)

- Forster, P. I. (1995). *Sankowskya*, a new genus of Euphorbiaceae (Dissiliariinae) from the Australian wet tropics. Austrobaileya 4: 329-335, illus. En. — Protologue of genus and description of its single species (also new); key to the genus and related Dissiliariinae included.

Sankowskya P.I.Forst., Austrobaileya 4: 329 (1995).

    Queensland. 50.

Sankowskya stipularis P.I.Forst., Austrobaileya 4: 331 (1995).

    Queensland. 50 QLD. Nanophan. or phan.

Sapiopsis

Synonyms:
Sapiopsis Müll.Arg. === **Sapium** P.Browne

Sapium

22 species, Americas; laticiferous shrubs or small to large trees (*S. marmieri* reaching 35 m), the branchlets passing into terminal spikes of flowers and the leaves usually with spreading lateral venation. In comparison with earlier treatments, that of Kruijt (1996) features a considerable reduction of names; in addition, none of the earlier named sections was accepted. Previously, 58 species had been recorded for S. America by Jablonski (1967) and 27 by the same author (1968) for the West Indies. The latex of some species has been a source of rubber. *S. haematospermum* of south-central S. America has foliage very like *Nerium oleander* (Apocynaceae) and like it is often along or near rivers; it has been introduced into cultivation elsewhere. [African and Asian species have now been referred to other genera by Kruijt (1996) and Esser (1994). These constitute all sections as listed by Pax (1912) apart from *Americana* (=*Sapium*). Among them are those in sects. *Parasapium* (inclusive of *S. japonica*, now the type of *Neoshirakia*), *Armata* (now mostly in *Sclerocroton*) and *Triadica* (encompassing among others the well-known ornamental *S. sebiferum*).] (Euphorbioideae (except Euphorbieae))

    Hemsley, W. B. (1901-09). *Sapium*. Ic. Pl. 27: pls. 2647-2650, 2677-2684; 29: 2878-2900. En. — 35 drawings, with discussions. [According to Webster (1967; see **Americas**), a better 'starting point' than the Pax and Hoffmann treatment. Jablonski (1967, however, suggests that the drawings are in some details unreliable.]

    Huber, J. (1906). Revue critique des espèces du genre *Sapium* Jacq. Bull. Herb. Boiss., II, 6: 345-364, 433-452, figs. 1-50. Fr. — Regionally arranged revisions with keys, descriptions, selected synonymy, places of publication, indication of distribution, and commentary. ['The best work ever done on this genus' (Jablonski 1967). Both Huber's and Jablonski's works are now succeeded by that of Kruijt (1996).]

Pittier, H. (1908). The Mexican and Central American species of *Sapium*. Contr. U.S. Natl. Herb. 12: 159-169. En. — 9 species; key, descriptions and illustrations. [Superseded by Kruijt, 1996.]

Pax, F. (with K. Hoffmann) (1912). *Sapium*. In A. Engler (ed.), Das Pflanzenreich, IV 147 V (Euphorbiaceae-Hippomaneae): 199-258. Berlin. (Heft 52.) La/Ge. — c. 95 species in 3 subgenera with seven sections; widely distributed in warmer regions but most strongly represented in the Americas. [Sect. *Triadica* sometimes kept as a separate genus; more recently, subgen. *Conosapium* (Madagascar) has also been returned to generic rank but with the exclusion of *S. goudotianum*. The treatment is reportedly 'extremely difficult to use' (Webster 1967; see **Americas**).]

Léonard, J. (1959). Notulae systematicae XXVI. Notes sur les espèces africaines continentales des genres *Sapium* P. Br. et *Excoecaria* L. (Euphorbiacées). Bull. Jard. Bot. État 29: 133-146. Fr. — Discussion of generic limits, with comparative survey of past treatments; synopsis (without key) of *Sapium* (pp. 139-146; 12 species) with synonymy, selected exsiccatae, commentary, and descriptions of *SS. schmitzii* and *carterianum*. [*Sapium* is now limited to the New World and those in Africa are variously transferred (with some reductions) to *Sclerocroton*, *Sebastiania* and *Shirakia* among others (Kruijt, 1996).]

Jablonski, E. (1967). *Sapium*. Euphorbiaceae, Guayana Highland (Mem. New York Bot. Gard. 17(1)): 181-186. New York. En. — 13 species, one new; key. [Superseded by Kruijt 1996.]

Jablonski, E. (1967). Notes on neotropical Euphorbiaceae, 1. Synopsis of South American *Sapium*. Phytologia 14: 441-449. En. — Covers 58 species; comprises a mere key (without authorities for the species) along with a distribution map, index and brief history of research on the genus. [Now superseded; a number of taxa now in other genera.]

Jablonski, E. (1967). Notes on neotropical Euphorbiaceae, 2. New species and transfers. Phytologia 14: 450-456. En. — Mostly on *Sapium*.

Jablonski, E. (1968). Notes on neotropical Euphorbiaceae, 3. Synopsis of Caribbean *Sapium*. Phytologia 16: 393-435. En. — A synopsis of 27 species with keys, synonymy, citations of exsiccatae, commentary and maps; complements his South American treatment but is much more extensive. [Now superseded; some taxa in other genera.]

• Kruijt, R. C. (1996). A taxonomic monograph of *Sapium* Jacq., *Anomostachys* (Baill.) Hurus., *Duvigneaudia* J. Léonard and *Sclerocroton* Hochst. (Euphorbiaceae tribe Hippomaneae). 109 pp., illus. (Biblioth. Bot. 146). Stuttgart. En. — *Sapium*, pp. 27-91; treatment of 21 species with key, synonymy, descriptions, distribution and habitat, localities with exsiccatae, and notes and illustrations.

Esser, H.-J. (1999). A partial revision of the Hippomaneae (Euphorbiaceae) in Malesia. Blumea 44: 149-215, illus., maps. En. — *Sapium*, pp. 180-182; in Malesia all species introduced with no sign of any spreading. *S. glandulosum* produces a high quality latex suitable for rubber but this has proved difficult to extract.

Sapium P.Browne, Civ. Nat. Hist. Jamaica: 338 (1756).
Trop. & Subtrop. America. 76 78 79 80 81 82 83 84 85.
Seborium Raf., Sylva Tellur.: 63 (1838).
Gymnobothrys Wall. ex Baill., Étude Euphorb.: 526 (1858).
Sapiopsis Müll.Arg. in A.P.de Candolle, Prodr. 15(2): 1205 (1866).

Sapium adenodon Griseb., Mem. Amer. Acad. Arts, n.s. 8: 160 (1861). *Bonania adenodon* (Griseb.) Benth. & Hook.f., Gen. Pl. 3: 335 (1880).
E. Cuba. 81 CUB. Phan.
Sapium cubense Britton & P.Wilson, Mem. Torrey Bot. Club 16: 77 (1920).
Sapium maestrense Urb., Repert. Spec. Nov. Regni Veg. 28: 230 (1930).

Sapium allenii Huft, Phytologia 63: 441 (1987).
S. Costa Rica. 80 COS. Phan.

Sapium argutum (Müll.Arg.) Huber, Bull. Herb. Boissier, II, 6: 439 (1906).
Guiana, Surinam, NE. Brazil (incl. Fernando de Noronha). 82 FRG SUR 84 BZE. Phan.
  * Excoecaria arguta* Müll.Arg. in C.F.P.von Martius, Fl. Bras. 11(2): 614 (1874).
  Excoecaria tristis Müll.Arg. in C.F.P.von Martius, Fl. Bras. 11(2): 615 (1874). *Sapium triste*
    (Müll.Arg.) Huber, Bull. Herb. Boissier, II, 6: 451 (1906).
  Sapium sceleratum Ridl., J. Linn. Soc., Bot. 27: 60 (1890).
  Sapium cicatricosum Pax & K.Hoffm. in H.G.A.Engler, Pflanzenr., IV, 147, XVII: 203 (1924).
  Sapium montanum Lanj., Euphorb. Surinam: 47 (1931).

Sapium ciliatum Hemsl., Hooker's Icon. Pl. 27: t. 2683 (1901).
Guiana, Surinam, Brazil (Pará). 82 FRG SUR 84 BZN. Phan.
  Sapium patens Jabl., Phytologia 14: 450 (1967).

Sapium cuneatum Griseb., Fl. Brit. W. I.: 49 (1859). *Excoecaria cuneata* (Griseb.) Müll.Arg. in
A.P.de Candolle, Prodr. 15(2): 1208 (1866).
Jamaica. 81 JAM. Phan.
  Sapium harrisii Urb. ex Pax in H.G.A.Engler, Pflanzenr., IV, 147, V: 236 (1912).

Sapium daphnoides Griseb., Nachr. Königl. Ges. Wiss. Georg-Augusts-Univ. 1: 176 (1865).
Cuba, Hispaniola. 81 CUB DOM HAI. Phan.
  Sebastiania buchii Urb., Notizbl. Bot. Gart. Berlin-Dahlem 7: 497 (1921).
  Sapium buchii Urb., Ark. Bot. 20A(15): 65 (1926).

Sapium glandulosum (L.) Morong, Ann. New York Acad. Sci. 7: 227 (1893).
  – FIGURE, p. 1421.
Trop. America. 79 MXT 80 BLZ COS ELS GUA HON NIC PAN 81 JAM LEE TRT WIN 82
  FRG GUY SUR VEN 83 BOL CLM ECU PER 84 BZC BZE BZL BZN BZS 85 AGE AGW PAR
  URU. Phan.
  * Hippomane glandulosa* L., Sp. Pl.: 1191 (1753).
  Hippomane biglandulosa L., Sp. Pl. ed. 2: 1431 (1763). *Sapium hippomane* G.Mey., Prim.
    Fl. Esseq.: 275 (1818), nom. illeg. *Stillingia hippomane* (G.Mey.) Baill., Étude
    Euphorb.: 513 (1858). *Sapium aucuparium* var. *hippomane* (G.Mey.) Griseb., Fl. Brit.
    W. I.: 49 (1859). *Sapium biglandulosum* (L.) Müll.Arg., Linnaea 32: 116 (1863).
    Stillingia biglandulosa (L.) Baill., Adansonia 5: 320 (1865). *Excoecaria biglandulosa* (L.)
    Müll.Arg. in A.P.de Candolle, Prodr. 15(2): 1204 (1866).
  Sapium aucuparium Jacq., Enum. Syst. Pl.: 21 (1763). *Hippomane aucuparia* (Jacq.)
    Crantz, Inst. Rei Herb. 1: 169 (1766). *Stillingia aucuparia* (Jacq.) Oken, Allg.
    Naturgesch. 3(3): 1607 (1841). *Sapium biglandulosum* var. *aucuparium* (Jacq.)
    Müll.Arg., Linnaea 32: 119 (1863). *Excoecaria biglandulosa* var. *aucuparia* (Jacq.)
    Müll.Arg. in A.P.de Candolle, Prodr. 15(2): 1206 (1866).
  Sapium aucuparium Willd., Sp. Pl. 4: 572 (1805), nom. illeg.
  Sapium salicifolium Kunth in F.W.H.von Humboldt, A.J.A.Bonpland & C.S.Kunth, Nov.
    Gen. Sp. 2: 65 (1817). *Excoecaria biglandulosa* var. *salicifolia* (Kunth) Müll.Arg. in
    A.P.de Candolle, Prodr. 15(2): 1207 (1866).
  Omphalea glandulata Vell., Fl. Flumin. 10: 14 (1831). *Excoecaria biglandulosa* var.
    glandulata (Vell.) Müll.Arg. in C.F.P.von Martius, Fl. Bras. 11(2): 621 (1874). *Sapium*
    glandulatum (Vell.) Pax in H.G.A.Engler, Pflanzenr., IV, 147, V: 229 (1912).
  Sapium prunifolium Klotzsch, Hooker's J. Bot. Kew Gard. Misc. 2: 45 (1843). *Stillingia*
    prunifolia (Klotzsch) Baill., Étude Euphorb.: 513 (1858), nom. inval. *Excoecaria*
    biglandulosa var. *prunifolia* (Klotzsch) Müll.Arg. in A.P.de Candolle, Prodr. 15(2):
    1205 (1866).
  Sapium moritzianum Klotzsch in B.Seemann, Bot. Voy. Herald: 100 (1853). *Sapium*
    biglandulosum var. *moritzianum* (Klotzsch) Müll.Arg., Linnaea 32: 119 (1863).
    Excoecaria biglandulosa var. *moritziana* (Klotzsch) Müll.Arg. in A.P.de Candolle, Prodr.
    15(2): 1206 (1866).

Sapium glandulosum (L.) Morong (as *Sapium aucuparium* Jacq.)
Artist: Matilda Smith
Ic. Pl. 27: pl. 2650 (1900)
KEW ILLUSTRATIONS COLLECTION

Sapium aereum Klotzsch ex Müll.Arg., Linnaea 32: 119 (1863). *Excoecaria aerea* (Klotzsch ex Müll.Arg.) Müll.Arg. in A.P.de Candolle, Prodr. 15(2): 1207 (1866).

Sapium biglandulosum f. *minus* Müll.Arg., Linnaea 32: 117 (1863).

Sapium biglandulosum f. *oblongatum* Müll.Arg., Linnaea 32: 116 (1863). *Excoecaria biglandulosa* f. *oblongata* (Müll.Arg.) Müll.Arg. in A.P.de Candolle, Prodr. 15(2): 1205 (1866).

Sapium biglandulosum f. *obovatum* Müll.Arg., Linnaea 32: 116 (1863). *Excoecaria biglandulosa* f. *obovata* (Müll.Arg.) Müll.Arg. in A.P.de Candolle, Prodr. 15(2): 1205 (1866).

Sapium biglandulosum var. *aubletianum* Müll.Arg., Linnaea 32: 117 (1863). *Excoecaria biglandulosa* var. *aubletianum* (Müll.Arg.) Müll.Arg. in A.P.de Candolle, Prodr. 15(2): 1205 (1866). *Sapium aubletianum* (Müll.Arg.) Huber, Bull. Herb. Boissier, II, 6: 362 (1906).

Sapium biglandulosum var. *hamatum* Müll.Arg., Linnaea 32: 116 (1863). *Excoecaria biglandulosa* var. *hamata* (Müll.Arg.) Müll.Arg. in A.P.de Candolle, Prodr. 15(2): 1204 (1866). *Sapium hamatum* (Müll.Arg.) Pax & K.Hoffm. in H.G.A.Engler, Pflanzenr., IV, 147, V: 229 (1912).

Sapium biglandulosum var. *klotzschianum* Müll.Arg., Linnaea 32: 116 (1863). *Excoecaria biglandulosa* var. *klotzschiana* (Müll.Arg.) Müll.Arg. in C.F.P.von Martius, Fl. Bras. 11(2): 619 (1874). *Sapium klotzschianum* (Müll.Arg.) Huber, Bull. Herb. Boissier, II, 6: 438 (1906).

Sapium biglandulosum var. *lanceolatum* Müll.Arg., Linnaea 32: 118 (1863). *Excoecaria biglandulosa* var. *lanceolata* (Müll.Arg.) Müll.Arg. in A.P.de Candolle, Prodr. 15(2): 1206 (1866). *Sapium lanceolatum* (Müll.Arg.) Huber, Bull. Herb. Boissier, II, 6: 441 (1906).

Sapium biglandulosum var. *meyerianum* Müll.Arg., Linnaea 32: 116 (1863).

Sapium biglandulosum var. *pavonianum* Müll.Arg., Linnaea 32: 116 (1863). *Excoecaria biglandulosa* var. *pavoniana* (Müll.Arg.) Müll.Arg. in A.P.de Candolle, Prodr. 15(2): 1204 (1866). *Sapium pavonianum* (Müll.Arg.) Huber, Bull. Herb. Boissier, II, 6: 356 (1906).

Sapium biglandulosum var. *serratum* Müll.Arg., Linnaea 32: 118 (1863). *Sapium serratum* (Müll.Arg.) Klotzsch ex Baill., Adansonia 5: 320 (1865). *Excoecaria biglandulosa* var. *serrata* (Müll.Arg.) Müll.Arg. in A.P.de Candolle, Prodr. 15(2): 1206 (1866).

Sapium marginatum Müll.Arg., Linnaea 32: 120 (1863). *Stillingia marginata* (Müll.Arg.) Baill., Adansonia 5: 321 (1865). *Excoecaria marginata* (Müll.Arg.) Müll.Arg. in A.P.de Candolle, Prodr. 15(2): 1208 (1866).

Sapium marginatum f. *majus* Müll.Arg., Linnaea 32: 120 (1863). *Excoecaria marginata* f. *major* (Müll.Arg.) Müll.Arg. in A.P.de Candolle, Prodr. 15(2): 1208 (1866).

Sapium marginatum var. *lanceolatum* Müll.Arg., Linnaea 32: 120 (1863). *Excoecaria marginatum* var. *lanceolata* (Müll.Arg.) Müll.Arg. in A.P.de Candolle, Prodr. 15(2): 1208 (1866).

Sapium marginatum var. *spathulatum* Müll.Arg., Linnaea 34: 120 (1863). *Excoecaria marginata* var. *spathulata* (Müll.Arg.) Müll.Arg. in A.P.de Candolle, Prodr. 15(2): 1208 (1866).

Sapium obtusilobum Müll.Arg., Linnaea 32: 116 (1863). *Excoecaria obtusiloba* (Müll.Arg.) Müll.Arg. in A.P.de Candolle, Prodr. 15(2): 1203 (1866).

Sapium montevidense Klotzsch ex Baill., Adansonia 5: 320 (1865). *Excoecaria biglandulosa* var. *montevidensis* (Klotzsch ex Baill.) Müll.Arg. in C.F.P.von Martius, Fl. Bras. 11(2): 621 (1874).

Sapium suberosum Müll.Arg., Linnaea 34: 217 (1865).

Stillingia dracunculoides Baill., Adansonia 5: 321 (1865). *Excoecaria biglandulosa* var. *dracunculoides* (Baill.) Müll.Arg. in A.P.de Candolle, Prodr. 15(2): 1207 (1866).

Excoecaria biglandulosa var. *cuneata* Müll.Arg. in A.P.de Candolle, Prodr. 15(2): 1206 (1866).

Excoecaria suberosa Müll.Arg. in A.P.de Candolle, Prodr. 15(2): 1202 (1866).

Excoecaria biglandulosa var. *clausseniana* Müll.Arg. in C.F.P.von Martius, Fl. Bras. 11(2): 618 (1874). *Sapium claussenianum* (Müll.Arg.) Huber, Bull. Herb. Boissier, II, 6: 436 (1906).

Excoecaria biglandulosa var. *grandifolia* Müll.Arg. in C.F.P.von Martius, Fl. Bras. 11(2): 620 (1874).

Excoecaria biglandulosa var. *intercedens* Müll.Arg. in C.F.P.von Martius, Fl. Bras. 11(2): 618 (1874).

Excoecaria biglandulosa var. *leptadenia* Müll.Arg. in C.F.P.von Martius, Fl. Bras. 11(2): 620 (1874). *Sapium leptadenium* (Müll.Arg.) Huber, Bull. Herb. Boissier, II, 6: 436 (1906).

Excoecaria biglandulosa var. *petiolaris* Müll.Arg. in C.F.P.von Martius, Fl. Bras. 11(2): 621 (1874). *Sapium petiolare* (Müll.Arg.) Huber, Bull. Herb. Boissier, II, 6: 434 (1906).

Excoecaria marginata var. *conjungens* Müll.Arg. in C.F.P.von Martius, Fl. Bras. 11(2): 617 (1874). *Sapium marginatum* var. *conjungens* (Müll.Arg.) Pax in H.G.A.Engler, Pflanzenr., IV, 147, V: 224 (1912).

Excoecaria marginata var. *grandifolia* Müll.Arg. in C.F.P.von Martius, Fl. Bras. 11(2): 618 (1874). *Sapium marginatum* var. *grandifolium* (Müll.Arg.) Pax in H.G.A.Engler, Pflanzenr., IV, 147, V: 224 (1912).

Excoecaria marginata var. *longifolia* Müll.Arg. in C.F.P.von Martius, Fl. Bras. 11(2): 618 (1874). *Sapium marginatum* var. *longifolium* (Müll.Arg.) Pax in H.G.A.Engler, Pflanzenr., IV, 147, V: 224 (1912).

Excoecaria occidentalis Müll.Arg. in C.F.P.von Martius, Fl. Bras. 11(2): 615 (1874). *Sapium occidentale* (Müll.Arg.) Huber, Bull. Herb. Boissier, II, 6: 441 (1906).

Sapium aureum H.Buek, Gen. Sp. Candoll. 4: 364 (1874), sphalm.

Sapium poeppigii Hemsl., Hooker's Icon. Pl. 27: t. 2678 (1901).

Sapium caribaeum Urb., Symb. Antill. 3: 308 (1902).

Sapium taburu Ule, Bot. Jahrb. Syst. 35: 671 (1905).

Sapium bogotense Huber, Bull. Herb. Boissier, II, 6: 355 (1906).

Sapium hemsleyanum Huber, Bull. Herb. Boissier, II, 6: 362 (1906).

Sapium intercedens Huber, Bull. Herb. Boissier, II, 6: 437 (1906).

Sapium pittieri Huber, Bull. Herb. Boissier, II, 6: 350 (1906).

Sapium submarginatum Huber, Bull. Herb. Boissier, II, 6: 443 (1906).

Sapium oligoneurum K.Schum. & Pittier, Contr. U. S. Natl. Herb. 12: 168 (1908).

Sapium sulciferum Pittier, Contr. U. S. Natl. Herb. 12: 169 (1908).

Sapium fendleri Hemsl., Hooker's Icon. Pl. 29: t. 2888 (1909).

Sapium muelleri Hemsl., Hooker's Icon. Pl. 29: t. 2884 (1909).

Hippomane zeocca L. ex B.D.Jacks., Index Linn. Herb.: 86 (1912), nom. nud.

Sapium endlicherianum Klotzsch ex Pax in H.G.A.Engler, Pflanzenr., IV, 147, V: 228 (1912), pro syn.

Sapium obtusatum Klotzsch ex Pax in H.G.A.Engler, Pflanzenr., IV, 147, V: 224 (1912), pro syn.

Sapium pohlianum Klotzsch ex Pax in H.G.A.Engler, Pflanzenr., IV, 147, V: 228 (1912), pro syn.

Sapium punctatum Klotzsch ex Pax in H.G.A.Engler, Pflanzenr., IV, 147, V: 223 (1912), pro syn.

Sapium pycnostachys K.Schum. ex Pax in H.G.A.Engler, Pflanzenr., IV, 147, V: 208 (1912), nom. illeg.

Sapium subserratum Klotzsch ex Pax in H.G.A.Engler, Pflanzenr., IV, 147, V: 226 (1912), pro syn.

Sapium giganteum Pittier, Contr. U. S. Natl. Herb. 20: 128 (1918).

Sapium cremostachyum I.M.Johnst., Contr. Gray Herb. 68: 91 (1923).

Sapium albomarginatum Pax & K.Hoffm. in H.G.A.Engler, Pflanzenr., IV, 147, XVII: 203 (1924).

Sapium fragile Pax & K.Hoffm. in H.G.A.Engler, Pflanzenr., IV, 147, XVII: 202 (1924).

Sapium paranaense Pax & K.Hoffm. in H.G.A.Engler, Pflanzenr., IV, 147, XVII: 203 (1924).

Sapium tenellum Pax & K.Hoffm. in H.G.A.Engler, Pflanzenr., IV, 147, XVII: 202 (1924).

Sapium guaricense Pittier, J. Wash. Acad. Sci. 19: 355 (1929).

Sapium naiguatense Pittier, J. Wash. Acad. Sci. 19: 355 (1929).

Sapium paucistamineum Pittier, J. Wash. Acad. Sci. 19: 356 (1929).

Sapium integrifolium Splitg. ex Lanj., Euphorb. Surinam: 176 (1931), nom. inval.

Sapium klotzschianum var. *glaziovii* Lanj., Euphorb. Surinam: 47 (1931).

Stillingia haematantha Standl., Ann. Missouri Bot. Gard. 27: 314 (1940).

Sapium haematospermum var. *saltense* O'Donell & Lourteig, Lilloa 8: 586 (1942). *Sapium saltense* (O'Donell & Lourteig) Jabl., Phytologia 14: 451 (1967), nom. illeg.

Sapium schippii Croizat, Amer. Midl. Naturalist 29: 477 (1943).

Sapium contortum Croizat, J. Arnold Arbor. 26: 193 (1945).

Sapium nitidum Alain, Contr. Ocas. Mus. Hist. Nat. Colegio "De La Salle" 11: 8 (1952). *Sapium alainianum* P.T.Li, Guihaia 14: 130 (1994).

Sapium moaense Alain, Revista Soc. Cub. Bot. 10: 27 (1953).

Sapium ixiamasense Jabl., Phytologia 14: 450 (1967).

Sapium itzanum Lundell, Wrightia 5: 77 (1975).

Sapium izabalense Lundell, Wrightia 5: 346 (1977).

Sapium haematospermum Müll.Arg., Linnaea 34: 217 (1865). *Excoecaria haematosperma* (Müll.Arg.) Müll.Arg. in A.P.de Candolle, Prodr. 15(2): 1209 (1866).

Bolivia, Paraguay, Uruguay, N. Argentina, S. Brazil. 83 BOL 84 BZS 84 AGE AGW PAR URU. Nanophan. or phan.

Sapium biglandulosum f. *squarrosum* Müll.Arg., Linnaea 32: 118 (1863). *Excoecaria biglandulosa* var. *squarrosa* (Müll.Arg.) Müll.Arg. in C.F.P.von Martius, Fl. Bras. 11(2): 622 (1874).

Sapium biglandulosum var. *longifolium* Müll.Arg., Linnaea 32: 118 (1863). *Excoecaria biglandulosa* f. *longifolia* (Müll.Arg.) Müll.Arg. in A.P.de Candolle, Prodr. 15(2): 1206 (1866). *Excoecaria biglandulosa* var. *longifolia* (Müll.Arg.) Müll.Arg. in C.F.P.von Martius, Fl. Bras. 11(2): 622 (1874). *Sapium longifolium* (Müll.Arg.) Huber, Bull. Herb. Boissier, II, 6: 444 (1906).

Sapium biglandulosum var. *stenophyllum* Müll.Arg., Linnaea 32: 119 (1863). *Excoecaria biglandulosa* var. *stenophylla* (Müll.Arg.) Müll.Arg. in A.P.de Candolle, Prodr. 15(2): 1207 (1866). *Sapium stenophyllum* (Müll.Arg.) Huber, Bull. Herb. Boissier, II, 6: 360 (1906).

Sapium marginatum f. *stenophyllum* Müll.Arg., Linnaea 32: 120 (1863). *Excoecaria marginata* f. *stenophylla* (Müll.Arg.) Müll.Arg. in A.P.de Candolle, Prodr. 15(2): 1208 (1866). *Excoecaria marginata* var. *stenophylla* (Müll.Arg.) Müll.Arg. in C.F.P.von Martius, Fl. Bras. 11(2): 617 (1874). *Sapium marginatum* var. *stenophyllum* (Müll.Arg.) Pax in H.G.A.Engler, Pflanzenr., IV, 147, V: 224 (1912).

Stillingia salicifolia Klotzsch ex Baill., Adansonia 5: 320 (1865), nom. illeg.

Excoecaria biglandulosa var. *angustifolia* Müll.Arg. in C.F.P.von Martius, Fl. Bras. 11(2): 622 (1874).

Excoecaria biglandulosa var. *lanceolata* Müll.Arg. in C.F.P.von Martius, Fl. Bras. 11(2): 621 (1874).

Excoecaria tijueensis Müll.Arg. in C.F.P.von Martius, Fl. Bras. 11(2): 616 (1874).

Stillingia sylvatica var. *paraguayensis* Morong, Ann. New York Acad. Sci. 7: 226 (1893).

Sapium biglandulosum f. *longissimum* Chodat & Hassl., Bull. Herb. Boissier, II, 5: 677 (1905).

Sapium haematospermum f. *arboreum* Chodat & Hassl., Bull. Herb. Boissier, II, 5: 677 (1905).

Sapium tijucense Huber, Bull. Herb. Boissier, II, 6: 449 (1906).

Sapium gibertii Hemsl., Hooker's Icon. Pl. 29: t. 2886 (1909).

Sapium linearifolium Hemsl., Hooker's Icon. Pl. 29: t. 2881 (1909).

Sapium cupuliferum Herzog, Bot. Jahrb. Syst. 44: 322 (1910), nom. illeg.

Sapium bolivianum Pax & K.Hoffm. in H.G.A.Engler, Pflanzenr., IV, 147, V: 221 (1912).

Sapium squarrosum Klotzsch ex Pax in H.G.A.Engler, Pflanzenr., IV, 147, V: 219 (1912), pro syn.

Sapium rojasii H.Lév., Bull. Acad. Int. Géogr. Bot. 28: 162 (1918).

Sapium haitiense Urb., Ark. Bot. 17(7): 39 (1922).

Haiti (Massif de la Hotte). 81 HAI. Phan.

Sapium jenmannii Hemsl., Hooker's Icon. Pl. 27: t. 2649 (1900).
Colombia, S. Venezuela, Guyana, N. Brazil. 82 GUY VEN 83 CLM 84 BZN.
Sapium cladogyne Hutch., Bull. Misc. Inform. Kew 1912: 224 (1912).

Sapium lateriflorum Hemsl., Hooker's Icon. Pl. 27: t. 2680 (1901).
Mexico, Belize, Guatemala, Honduras, Costa Rica. 79 MXG MXN MXS MXT 80 BLZ COS GUA HON. Phan.
Sapium biglandulosum var. *nitidum* Monach., Bull. Torrey Bot. Club 67: 771 (1920).
Sapium nitidum (Monach.) Lundell, Amer. Midl. Naturalist 29: 477 (1943).
Sapium mammosum Lundell, Wrightia 5: 77 (1975).
Sapium ovalifolium Lundell, Wrightia 5: 78 (1975).

Sapium laurifolium (A.Rich.) Griseb., Fl. Brit. W. I.: 49 (1859). *Sapium laurocerasus* var. *laurifolium* (Griseb.) Müll.Arg., Linnaea 32: 116 (1863). *Excoecaria laurocerasus* var. *laurifolia* (Griseb.) Müll.Arg. in A.P.de Candolle, Prodr. 15(2): 1203 (1866).
Trop. America. 79 MXT 80 COS GUA HON PAN 81 CUB DOM HAI JAM PUE 83 CLM ECU PER 84 BZN. Phan.
Sapium jamaicense Sw., Adnot. Bot.: 62 (1829), nom. illeg.
* *Stillingia laurifolia* A.Rich. in R.de la Sagra, Hist. Fis. Cuba, Bot. 2: 201 (1850).
Sapium brownei Banks ex Griseb., Fl. Brit. W. I.: 49 (1859), pro syn.
Sapium laurocerasus var. *ellipticum* Müll.Arg., Linnaea 32: 116 (1863). *Excoecaria laurocerasus* var. *elliptica* (Müll.Arg.) Müll.Arg. in A.P.de Candolle, Prodr. 15(2): 1203 (1866).
Sapium decipiens Preuss, Exped. C.-Südamer.: 386 (1901).
Sapium utile Preuss, Exped. C.-Südamer.: 386 (1901).
Sapium eglandulosum Ule, Bot. Jahrb. Syst. 35: 673 (1905).
Sapium anadenum Pittier, Contr. U. S. Natl. Herb. 12: 164 (1908).
Sapium pleiostachys K.Schum. & Pittier, Contr. U. S. Natl. Herb. 12: 164 (1908).

Sapium laurocerasus Desf., Tabl. École Bot., ed. 3: 411 (1829). *Stillingia laurocerasus* (Desf.) Baill., Étude Euphorb.: 513 (1858). *Sapium laurocerasus* var. *genuinum* Müll.Arg., Linnaea 32: 116 (1863), nom. inval. *Excoecaria laurocerasus* var. *genuina* Müll.Arg. in A.P.de Candolle, Prodr. 15(2): 1202 (1866), nom. inval. *Excoecaria laurocerasus* (Desf.) Müll.Arg. in A.P.de Candolle, Prodr. 15(2): 1202 (1866).
Puerto Rico. 81 PUE. Phan.
Sapium latifolium Klotzsch ex Pax in H.G.A.Engler, Pflanzenr., IV, 147, V: 206 (1912), pro syn.

Sapium leucogynum C.Wright ex Griseb., Nachr. Königl. Ges. Wiss. Georg-Augusts-Univ. 1: 176 (1865). *Excoecaria leucogyna* (C.Wright ex Griseb.) Müll.Arg. in A.P.de Candolle, Prodr. 15(2): 1208 (1866).
W. Cuba. 81 CUB. Phan.
Sapium leucospermum Griseb., Nachr. Königl. Ges. Wiss. Georg-Augusts-Univ. 1: 177 (1865). *Excoecaria leucosperma* (Griseb.) Müll.Arg. in A.P.de Candolle, Prodr. 15(2): 1209 (1866).

Sapium macrocarpum Müll.Arg., Linnaea 32: 119 (1863). *Excoecaria macrocarpa* (Müll.Arg.) Müll.Arg. in A.P.de Candolle, Prodr. 15(2): 1207 (1866).
Mexico to C. Costa Rica. 79 MXC MXG MXN MXS MXT 80 COS ELS HON NIC. Phan.
Sapium mexicanum Hemsl., Hooker's Icon. Pl. 27: t. 2680 (1901).
Sapium pedicellatum Huber, Bull. Herb. Boissier, II, 6: 352 (1906).
Sapium thelocarpum K.Schum. & Pittier, Contr. U. S. Natl. Herb. 12: 166 (1908).
Sapium dolichostachys K.Schum. ex Pax in H.G.A.Engler, Pflanzenr., IV, 147, V: 233 (1912), pro syn.
Sapium bourgeaui Croizat, J. Arnold Arbor. 24: 171 (1943).

Sapium marmieri Huber, Bull. Soc. Bot. France 49: 49 (1902).
  Colombia, Ecuador, Peru, Bolivia, Brazil (Amazonas). 83 BOL CLM ECU PER 84 BZN. Phan.
   Sapium peloto Pax & K.Hoffm. in H.G.A.Engler, Pflanzenr., IV, 147, V: 210 (1912).
   Sapium leitera Gleason, Bull. Torrey Bot. Club 60: 364 (1933).

Sapium obovatum Klotzsch ex Müll.Arg., Linnaea 32: 120 (1863). *Stillingia obovata*
  (Klotzsch ex Müll.Arg.) Baill., Adansonia 5: 321 (1865). *Excoecaria obovata* (Klotzsch ex
  Müll.Arg.) Müll.Arg. in A.P.de Candolle, Prodr. 15(2): 1203 (1866). *Excoecaria marginata*
  var. *obovata* (Klotzsch ex Müll.Arg.) Müll.Arg. in C.F.P.von Martius, Fl. Bras. 11(2): 616
  (1874). *Sapium marginatum* var. *obovatum* (Klotzsch ex Müll.Arg.) Pax in H.G.A.Engler,
  Pflanzenr., IV, 147, V: 222 (1912).
  E. & SE. Brazil, Paraguay. 84 BZC BZE BZL 85 PAR. Nanophan.
   Excoecaria marginata var. *intermedia* Müll.Arg. in C.F.P.von Martius, Fl. Bras. 11(2): 617
    (1874). *Sapium marginatum* var. *intermedium* (Müll.Arg.) Pax in H.G.A.Engler,
    Pflanzenr., IV, 147, V: 223 (1912).
   Excoecaria martii Müll.Arg. in C.F.P.von Martius, Fl. Bras. 11(2): 614 (1874). *Sapium
    martii* (Müll.Arg.) Huber, Bull. Herb. Boissier, II, 6: 448 (1906).
   Sapium marginatum var. *paraguariense* Chodat & Hassl., Bull. Herb. Boissier, II, 5:
    677 (1905).
   Sapium obovatum var. *ellipticum* Chodat & Hassl., Bull. Herb. Boissier, II, 5: 676 (1905).
   Sapium hasslerianum Huber, Bull. Herb. Boissier, II, 6: 448 (1906), nom. illeg.

Sapium pachystachys K.Schum. & Pittier, Contr. U. S. Natl. Herb. 12: 168 (1908).
  Nicaragua, Costa Rica, W. Panama. 80 COS NIC PAN. Phan.
   Sapium caudatum Pittier, Contr. U. S. Natl. Herb. 20: 127 (1918).

Sapium pallidum (Müll.Arg.) Huber, Bull. Herb. Boissier, II, 6: 450 (1906).
  Brazil (Bahia, Minas Gerais). 84 BZE BZL. Nanophan.
  * *Sapium biglandulosum* var. *pallidum* Müll.Arg., Linnaea 32: 116 (1863). *Excoecaria
    biglandulosa* var. *pallida* (Müll.Arg.) Müll.Arg. in A.P.de Candolle, Prodr. 15(2): 1205
    (1866). *Excoecaria pallida* (Müll.Arg.) Müll.Arg. in C.F.P.von Martius, Fl. Bras. 11(2):
    623 (1874).
   Sapium duckei Huber ex Huft, Phytologia 63: 442 (1987).

Sapium paucinervium Hemsl., Hooker's Icon. Pl. 27: t. 2648 (1900).
  Guiana, Guyana, Surinam, Venezuela, N. & NE. Brazil. 82 FRG GUY SUR VEN 84 BZE BZN.
   Sapium microdentatum Lanj., Euphorb. Surinam: 46 (1931).

Sapium sellowianum (Müll.Arg.) Klotzsch ex Baill., Adansonia 5: 320 (1865).
  Brazil (Minas Gerais, São Paulo). 84 BZL. Phan.
  * *Sapium biglandulosum* var. *sellowianum* Müll.Arg., Linnaea 32: 118 (1863). *Excoecaria
    biglandulosa* var. *sellowiana* (Müll.Arg.) Müll.Arg. in A.P.de Candolle, Prodr. 15(2):
    1206 (1866).
   Excoecaria biglandulosa var. *longipes* Müll.Arg. in C.F.P.von Martius, Fl. Bras. 11(2): 619
    (1874). *Sapium longipes* (Müll.Arg.) Huber, Bull. Herb. Boissier, II, 6: 435 (1906).
   Excoecaria biglandulosa var. *sublanceolata* Müll.Arg. in C.F.P.von Martius, Fl. Bras.
    11(2): 621 (1874). *Sapium sublanceolatum* (Müll.Arg.) Huber, Bull. Herb. Boissier, II,
    6: 441 (1906).

Sapium stylare Müll.Arg., Linnaea 32: 119 (1863). *Excoecaria stylaris* (Müll.Arg.) Müll.Arg.
  in A.P.de Candolle, Prodr. 15(2): 1204 (1866).
  Costa Rica to Venezuela and Ecuador. 80 COS PAN 82 VEN 83 CLM ECU. Phan.
   Sapium thomsonii God.-Leb. ex Jum., Pl. Caoutch. Gutta: 151 (1898).
   Sapium tolimense Jum., Pl. Caoutch. Gutta: 151 (1898).
   Sapium verum Hemsl., Hooker's Icon. Pl. 27: t. 2647 (1900).
   Sapium cuatrecasasii Croizat, J. Arnold Arbor. 24: 172 (1943).

Sapium myrmecophilum Croizat, J. Arnold Arbor. 24: 172 (1943).

Sapium putumayense Croizat, Caldasia 2: 131 (1943).

Sapium rigidifolium Huft, Phytologia 63: 444 (1987).

Sapium solisii Huft, Phytologia 63: 446 (1987).

Synonyms:

Sapium abyssinicum (Müll.Arg.) Benth. === **Shirakiopsis elliptica** (Hochst.) Esser (ined.)

Sapium acetosella Milne-Redh. === **Microstachys acetosella** (Milne-Redh.) Esser

Sapium acetosella var. *elatius* Radcl.-Sm. === **Microstachys acetosella** (Milne-Redh.) Esser

Sapium acutifolium Benth. === **Stillingia acutifolia** (Benth.) Benth. & Hook.f. ex Hemsl.

Sapium aereum Klotzsch ex Müll.Arg. === **Sapium glandulosum** (L.) Morong

Sapium africanum (Sond.) Kuntze === **Spirostachys africana** Sond.

Sapium alainianum P.T.Li === **Sapium glandulosum** (L.) Morong

Sapium albomarginatum Pax & K.Hoffm. === **Sapium glandulosum** (L.) Morong

Sapium anadenum Pittier === **Sapium laurifolium** (A.Rich.) Griseb.

Sapium angulatum Klotzsch ex Pax === **Stillingia sanguinolenta** Müll.Arg.

Sapium angustifolium Alain === **Gymnanthes pallens** (Griseb.) Müll.Arg.

Sapium annuum Torr. === **Stillingia spinulosa** Torr.

Sapium annuum var. *dentatum* Torr. === **Stillingia treculiana** (Müll.Arg.) I.M.Johnst.

Sapium appendiculatum (Müll.Arg.) Pax & K.Hoffm. === **Sebastiania appendiculata** (Müll.Arg.) Jabl.

Sapium aquifolium Js.Sm. === **Alchornea aquifolia** (Js.Sm.) Domin

Sapium armatum Pax & K.Hoffm. === **Sclerocroton integerrimus** Hochst.

Sapium atrobadiomaculatum F.P.Metcalf === ?

Sapium aubletianum (Müll.Arg.) Huber === **Sapium glandulosum** (L.) Morong

Sapium aubrevillei Leandri === **Shirakiopsis aubrevillei** (Leandri) Esser (ined.)

Sapium aucuparium Willd. === **Sapium glandulosum** (L.) Morong

Sapium aucuparium Jacq. === **Sapium glandulosum** (L.) Morong

Sapium aucuparium var. *hippomane* (G.Mey.) Griseb. === **Sapium glandulosum** (L.) Morong

Sapium aureum H.Buek === **Sapium glandulosum** (L.) Morong

Sapium baccatum Roxb. === **Balakata baccata** (Roxb.) Esser

Sapium balansae Parodi === **Sebastiania** ?

Sapium berberifolium Meisn. === **Alchornea aquifolia** (Js.Sm.) Domin

Sapium biglandulosum (L.) Müll.Arg. === **Sapium glandulosum** (L.) Morong

Sapium biglandulosum var. *aubletianum* Müll.Arg. === **Sapium glandulosum** (L.) Morong

Sapium biglandulosum var. *aucuparium* (Jacq.) Müll.Arg. === **Sapium glandulosum** (L.) Morong

Sapium biglandulosum var. *hamatum* Müll.Arg. === **Sapium glandulosum** (L.) Morong

Sapium biglandulosum var. *klotzschianum* Müll.Arg. === **Sapium glandulosum** (L.) Morong

Sapium biglandulosum var. *lanceolatum* Müll.Arg. === **Sapium glandulosum** (L.) Morong

Sapium biglandulosum var. *longifolium* Müll.Arg. === **Sapium haematospermum** Müll.Arg.

Sapium biglandulosum f. *longissimum* Chodat & Hassl. === **Sapium haematospermum** Müll.Arg.

Sapium biglandulosum var. *meyerianum* Müll.Arg. === **Sapium glandulosum** (L.) Morong

Sapium biglandulosum f. *minus* Müll.Arg. === **Sapium glandulosum** (L.) Morong

Sapium biglandulosum var. *moritzianum* (Klotzsch) Müll.Arg. === **Sapium glandulosum** (L.) Morong

Sapium biglandulosum var. *nitidum* Monach. === **Sapium lateriflorum** Hemsl.

Sapium biglandulosum f. *oblongatum* Müll.Arg. === **Sapium glandulosum** (L.) Morong

Sapium biglandulosum f. *obovatum* Müll.Arg. === **Sapium glandulosum** (L.) Morong

Sapium biglandulosum var. *pallidum* Müll.Arg. === **Sapium pallidum** (Müll.Arg.) Huber

Sapium biglandulosum var. *pavonianum* Müll.Arg. === **Sapium glandulosum** (L.) Morong

Sapium biglandulosum var. *sellowianum* Müll.Arg. === **Sapium sellowianum** (Müll.Arg.) Klotzsch ex Baill.

Sapium biglandulosum var. *serratum* Müll.Arg. === **Sapium glandulosum** (L.) Morong

Sapium biglandulosum f. *squarrosum* Müll.Arg. === **Sapium haematospermum** Müll.Arg.

Sapium biglandulosum var. *stenophyllum* Müll.Arg. === **Sapium haematospermum** Müll.Arg.

Sapium biloculare (S.Watson) Pax === **Sebastiania bilocularis** S.Watson

Sapium biloculare var. *amplum* I.M.Johnst. === **Sebastiania ampla** (I.M.Johnst.) Jabl.

Sapium bingerium Roxb. ex Willd. === **Shirakiopsis indica** (Willd.) Esser

Sapium bingyricum Roxb. ex Baill. === **Shirakiopsis indica** (Willd.) Esser

Sapium bodenbenderi Kuntze === **Stillingia bodenbenderi** (Kuntze) D.J.Rogers

Sapium bogotense Huber === **Sapium glandulosum** (L.) Morong

Sapium bolanderi Kuntze ex Pax === **Stillingia bodenbenderi** (Kuntze) D.J.Rogers

Sapium bolivianum Pax & K.Hoffm. === **Sapium haematospermum** Müll.Arg.

Sapium bourgeaui Croizat === **Sapium macrocarpum** Müll.Arg.

Sapium brasiliense Spreng. ex Pax === **Croton** sp.

Sapium brownei Banks ex Griseb. === **Sapium laurifolium** (A.Rich.) Griseb.

Sapium buchii Urb. === **Sapium daphnoides** Griseb.

Sapium bussei Pax === **Excoecaria bussei** (Pax) Pax

Sapium caribaeum Urb. === **Sapium glandulosum** (L.) Morong

Sapium carterianum J.Léonard === **Sclerocroton carterianus** (J.Léonard) Kruijt & Roebers

Sapium cassinefolium Tausch ex Pax === **Stillingia lineata** (Lam.) Müll.Arg. subsp. **lineata**

Sapium caudatum Pittier === **Sapium pachystachys** K.Schum. & Pittier

Sapium chihsinianum S.K.Lee === **Triadica sebifera** (L.) Small

Sapium cicatricosum Pax & K.Hoffm. === **Sapium argutum** (Müll.Arg.) Huber

Sapium cladogyne Hutch. === **Sapium jenmannii** Hemsl.

Sapium claussenianum (Müll.Arg.) Huber === **Sapium glandulosum** (L.) Morong

Sapium cochinchinense (Lour.) Kuntze === **Excoecaria cochinchinensis** Lour.

Sapium cochinchinense (Lour.) Gagnep. === **Triadica cochinchinensis** Lour.

Sapium contortum Croizat === **Sapium glandulosum** (L.) Morong

Sapium cordifolium Roxb. === **Alchornea mollis** (Benth.) Müll.Arg.

Sapium cornutum Pax === **Sclerocroton cornutus** (Pax) Kruijt & Roebers

Sapium cornutum var. *coriaceum* Pax === **Sclerocroton cornutus** (Pax) Kruijt & Roebers

Sapium cornutum var. *genuinum* Pax === **Sclerocroton cornutus** (Pax) Kruijt & Roebers

Sapium cornutum var. *lineolatum* Pax === **Sclerocroton cornutus** (Pax) Kruijt & Roebers

Sapium cornutum var. *poggei* (Pax) Pax === **Sclerocroton cornutus** (Pax) Kruijt & Roebers

Sapium crassifolium Elmer === **Antidesma bunius** (L.) Spreng.

Sapium cremostachyum I.M.Johnst. === **Sapium glandulosum** (L.) Morong

Sapium cuatrecasasii Croizat === **Sapium stylare** Müll.Arg.

Sapium cubense Britton & P.Wilson === **Sapium adenodon** Griseb.

Sapium cupuliferum Hemsl. === **Stillingia salpingadenia** Huber

Sapium cupuliferum Herzog === **Sapium haematospermum** Müll.Arg.

Sapium dacdece Buch.-Ham. ex Wall. === **Balakata baccata** (Roxb.) Esser

Sapium dalzielii Hutch. === **Microstachys faradianensis** (Beille) Esser

Sapium decipiens Preuss === **Sapium laurifolium** (A.Rich.) Griseb.

Sapium diandrum (Müll.Arg.) Pax === **Sebastiania** sp.

Sapium dichotomum Klotzsch ex Pax === **Stillingia dichotoma** Müll.Arg.

Sapium discolor (Champ. ex Benth.) Müll.Arg. === **Triadica cochinchinensis** Lour.

Sapium discolor var. *wenhsienensis* S.B.Ho === **Triadica cochinchinensis** Lour.

Sapium diversifolium (Miq.) Pax === **Shirakiopsis indica** (Willd.) Esser

Sapium dolichostachys K.Schum. ex Pax === **Sapium macrocarpum** Müll.Arg.

Sapium drummondii Jacques === **Scolopia zeyheri** (Nees) Harv. (Flacourtiaceae)

Sapium duckei Huber ex Huft === **Sapium pallidum** (Müll.Arg.) Huber

Sapium eglandulosum Ule === **Sapium laurifolium** (A.Rich.) Griseb.

Sapium ellipticum (Hochst. ex Krauss) Pax === **Shirakiopsis elliptica** (Hochst.) Esser (ined.)

Sapium endlicherianum Klotzsch ex Pax === **Sapium glandulosum** (L.) Morong

Sapium erythrospermum (Griseb.) Müll.Arg. === **Bonania erythrosperma** (Griseb.) Benth. & Hook.f. ex B.D.Jacks.

Sapium eugeniifolium Buch.-Ham. ex Hook.f. === **Triadica cochinchinensis** Lour.

Sapium faradianense (Beille) Pax === **Microstachys faradianensis** (Beille) Esser

Sapium fendleri Hemsl. === **Sapium glandulosum** (L.) Morong

Sapium fragile Pax & K.Hoffm. === **Sapium glandulosum** (L.) Morong

Sapium gibertii Hemsl. === **Sapium haematospermum** Müll.Arg.

Sapium giganteum Pittier === **Sapium glandulosum** (L.) Morong

Sapium glandulatum (Vell.) Pax === **Sapium glandulosum** (L.) Morong

Sapium goudotianum (Baill.) Pax === **Taeniosapium goudotianum** (Baill.) Müll.Arg.

Sapium grahamii (Stapf) Prain === **Excoecaria grahamii** Stapf

Sapium guaricense Pittier === **Sapium glandulosum** (L.) Morong

Sapium guatemalense Lundell === **Tetrorchidium brevifolium** Standl. & Steyerm.

Sapium guineense (Benth.) Kuntze === **Excoecaria guineensis** (Benth.) Müll.Arg.

Sapium gymnogyne Leandri === **Anomostachys lastellei** (Müll.Arg.) Kruijt

Sapium haematospermum f. *arboreum* Chodat & Hassl. === **Sapium haematospermum** Müll.Arg.

Sapium haematospermum var. *saltense* O'Donell & Lourteig === **Sapium glandulosum** (L.) Morong

Sapium hamatum (Müll.Arg.) Pax & K.Hoffm. === **Sapium glandulosum** (L.) Morong

Sapium harrisii Urb. ex Pax === **Sapium cuneatum** Griseb.

Sapium hasslerianum Huber === **Sapium obovatum** Klotzsch ex Müll.Arg.

Sapium hemsleyanum Huber === **Sapium glandulosum** (L.) Morong

Sapium herbertiifolium Jacques === ?

Sapium hexandrum Wall. === **Balakata baccata** (Roxb.) Esser

Sapium hildebrandtii Pax === **Sclerocroton melanostictus** (Baill.) Kruijt & Roebers

Sapium hippomane G.Mey. === **Sapium glandulosum** (L.) Morong

Sapium hookeri Hook.f. === **Triadica cochinchinensis** Lour.

Sapium hurmais Buch.-Ham. === **Shirakiopsis indica** (Willd.) Esser

Sapium ilicifolium Willd. === **Hippomane spinosa** L.

Sapium indicum Willd. === **Shirakiopsis indica** (Willd.) Esser

Sapium insigne (Royle) Benth. === **Falconeria insignis** Royle

Sapium insigne var. *genuinum* Pax === **Falconeria insignis** Royle

Sapium insigne var. *malabaricum* (Wight) Hook.f. === **Falconeria insignis** Royle

Sapium integerrimum (Hochst. ex Krauss) J.Léonard === **Sclerocroton integerrimus** Hochst.

Sapium integrifolium Splitg. ex Lanj. === **Sapium glandulosum** (L.) Morong

Sapium intercedens Huber === **Sapium glandulosum** (L.) Morong

Sapium itzanum Lundell === **Sapium glandulosum** (L.) Morong

Sapium ixiamasense Jabl. === **Sapium glandulosum** (L.) Morong

Sapium izabalense Lundell === **Sapium glandulosum** (L.) Morong

Sapium jamaicense Sw. === **Sapium laurifolium** (A.Rich.) Griseb.

Sapium japonicum (Siebold & Zucc.) Pax & K.Hoffm. === **Neoshirakia japonica** (Siebold & Zucc.) Esser

Sapium kerstingii Pax === **Shirakiopsis elliptica** (Hochst.) Esser (ined.)

Sapium klotzschianum (Müll.Arg.) Huber === **Sapium glandulosum** (L.) Morong

Sapium klotzschianum var. *glaziovii* Lanj. === **Sapium glandulosum** (L.) Morong

Sapium laevifolium Thouars ex Baill. === **Stillingia lineata** (Lam.) Müll.Arg. subsp. **lineata**

Sapium laevigatum Lam. === **Stillingia lineata** (Lam.) Müll.Arg. subsp. **lineata**

Sapium lanceolatum (Müll.Arg.) Huber === **Sapium glandulosum** (L.) Morong

Sapium lateriflorum Merr. === **Balakata luzonica** (Vidal) Esser

Sapium latifolium Klotzsch ex Pax === **Sapium laurocerasus** Desf.

Sapium laui Croizat === **Triadica cochinchinensis** Lour.

Sapium laurocerasus var. *ellipticum* Müll.Arg. === **Sapium laurifolium** (A.Rich.) Griseb.

Sapium laurocerasus var. *genuinum* Müll.Arg. === **Sapium laurocerasus** Desf.

Sapium laurocerasus var. *laurifolium* (Griseb.) Müll.Arg. === **Sapium laurifolium** (A.Rich.) Griseb.

Sapium leitera Gleason === **Sapium marmieri** Huber

Sapium leonardii-crispi J.Léonard === **Duvigneaudia leonardii-crispi** (J.Léonard) Kruijt & Roebers

Sapium leonardii-crispi var. *pubescentifolium* J.Léonard === **Duvigneaudia leonardii-crispi** (J.Léonard) Kruijt & Roebers

Sapium leptadenium (Müll.Arg.) Huber === **Sapium glandulosum** (L.) Morong

Sapium leucospermum Griseb. === **Sapium leucogynum** C.Wright ex Griseb.

Sapium linearifolium Hemsl. === **Sapium haematospermum** Müll.Arg.

Sapium lineatum Lam. === **Stillingia lineata** (Lam.) Müll.Arg.

Sapium longifolium (Müll.Arg.) Huber === **Sapium haematospermum** Müll.Arg.

Sapium longipes (Müll.Arg.) Huber === **Sapium sellowianum** (Müll.Arg.) Klotzsch ex Baill.

Sapium loziense Leandri === **Anomostachys lastellei** (Müll.Arg.) Kruijt

Sapium luzonicum (Vidal) Merr. === **Balakata luzonica** (Vidal) Esser

Sapium macrophyllum Klotzsch ex Pax === **Caryodendron janeirense** Müll.Arg.

Sapium madagascariense (Baill.) Prain === **Excoecaria madagascariensis** (Baill.) Müll.Arg.

Sapium madagascariense (Müll.Arg.) Pax === **Conosapium madagascariensis** Müll.Arg.

Sapium maestrense Urb. === **Sapium adenodon** Griseb.

Sapium mammosum Lundell === **Sapium lateriflorum** Hemsl.

Sapium mannianum Hiern === **Shirakiopsis elliptica** (Hochst.) Esser (ined.)

Sapium mannianum (Müll.Arg.) Benth. === **Shirakiopsis elliptica** (Hochst.) Esser (ined.)

Sapium marginatum Müll.Arg. === **Sapium glandulosum** (L.) Morong

Sapium marginatum var. *conjungens* (Müll.Arg.) Pax === **Sapium glandulosum** (L.) Morong

Sapium marginatum var. *grandifolium* (Müll.Arg.) Pax === **Sapium glandulosum** (L.) Morong

Sapium marginatum var. *intermedium* (Müll.Arg.) Pax === **Sapium obovatum** Klotzsch ex Müll.Arg.

Sapium marginatum var. *lanceolatum* Müll.Arg. === **Sapium glandulosum** (L.) Morong

Sapium marginatum var. *longifolium* (Müll.Arg.) Pax === **Sapium glandulosum** (L.) Morong

Sapium marginatum f. *majus* Müll.Arg. === **Sapium glandulosum** (L.) Morong

Sapium marginatum var. *obovatum* (Klotzsch ex Müll.Arg.) Pax === **Sapium obovatum** Klotzsch ex Müll.Arg.

Sapium marginatum var. *paraguariense* Chodat & Hassl. === **Sapium obovatum** Klotzsch ex Müll.Arg.

Sapium marginatum var. *spathulatum* Müll.Arg. === **Sapium glandulosum** (L.) Morong

Sapium marginatum f. *stenophyllum* Müll.Arg. === **Sapium haematospermum** Müll.Arg.

Sapium marginatum var. *stenophyllum* (Müll.Arg.) Pax === **Sapium haematospermum** Müll.Arg.

Sapium martii (Müll.Arg.) Huber === **Sapium obovatum** Klotzsch ex Müll.Arg.

Sapium martii var. *peruvianum* J.F.Macbr. === **Sapium sp.**

Sapium melanostictum (Baill.) Pax & K.Hoffm. === **Sclerocroton melanostictus** (Baill.) Kruijt & Roebers

Sapium merrillianum Pax & K.Hoffm. === **Balakata luzonica** (Vidal) Esser

Sapium mexicanum Hemsl. === **Sapium macrocarpum** Müll.Arg.

Sapium microdentatum Lanj. === **Sapium paucinervium** Hemsl.

Sapium moaense Alain === **Sapium glandulosum** (L.) Morong

Sapium montanum Lanj. === **Sapium argutum** (Müll.Arg.) Huber

Sapium montevidense Klotzsch ex Baill. === **Sapium glandulosum** (L.) Morong

Sapium moritzianum Klotzsch === **Sapium glandulosum** (L.) Morong

Sapium muelleri Hemsl. === **Sapium glandulosum** (L.) Morong

Sapium myricifolium C.Wright ex Griseb. === **Bonania myricifolia** (Griseb.) Benth. & Hook.f.

Sapium myrmecophilum Croizat === **Sapium stylare** Müll.Arg.

Sapium naiguatense Pittier === **Sapium glandulosum** (L.) Morong

Sapium nitidum Alain === **Sapium glandulosum** (L.) Morong

Sapium nitidum (Monach.) Lundell === **Sapium lateriflorum** Hemsl.

Sapium oblongifolium De Wild. === **Sclerocroton cornutus** (Pax) Kruijt & Roebers

Sapium oblongifolium (Müll.Arg.) Pax === **Sclerocroton oblongifolius** (Müll.Arg.) Kruijt & Roebers

Sapium obovatum var. *ellipticum* Chodat & Hassl. === **Sapium obovatum** Klotzsch ex Müll.Arg.

Sapium obtusatum Klotzsch ex Pax === **Sapium glandulosum** (L.) Morong

Sapium obtusifolium Kunth === **Sebastiania obtusifolia** Pax & K.Hoffm.

Sapium obtusifolium Lam. === **Stillingia lineata** (Lam.) Müll.Arg. subsp. **lineata**

Sapium obtusilobum Müll.Arg. === **Sapium glandulosum** (L.) Morong

Sapium occidentale (Müll.Arg.) Huber === **Sapium glandulosum** (L.) Morong

Sapium oligoneurum K.Schum. & Pittier === **Sapium glandulosum** (L.) Morong

Sapium oppositifolium Klotzsch ex Baill. === **Stillingia oppositifolia** Baill. ex Müll.Arg.

Sapium ovalifolium Lundell === **Sapium lateriflorum** Hemsl.

Sapium pallens (Griseb.) Borhidi === **Gymnanthes pallens** (Griseb.) Müll.Arg.

Sapium pallens var. *tenax* (Griseb.) Borhidi === **Gymnanthes pallens** (Griseb.) Müll.Arg.

Sapium paranaense Pax & K.Hoffm. === **Sapium glandulosum** (L.) Morong

Sapium parvifolium Alain === **Gymnanthes** ?

Sapium patagonicum (Speg.) D.J.Rogers === **?**

Sapium patens Jabl. === **Sapium ciliatum** Hemsl.

Sapium paucistamineum Pittier === **Sapium glandulosum** (L.) Morong

Sapium pavonianum (Müll.Arg.) Huber === **Sapium glandulosum** (L.) Morong

Sapium pedicellatum Huber === **Sapium macrocarpum** Müll.Arg.

Sapium peloto Pax & K.Hoffm. === **Sapium marmieri** Huber

Sapium perrieri Leandri === **Anomostachys lastellei** (Müll.Arg.) Kruijt

Sapium peruvianum Steud. === **Sebastiania obtusifolia** Pax & K.Hoffm.

Sapium peruvianum (J.F.Macbr.) Jabl. === **?**

Sapium petiolare (Müll.Arg.) Huber === **Sapium glandulosum** (L.) Morong

Sapium pittieri Huber === **Sapium glandulosum** (L.) Morong

Sapium pleiocarpum Y.C.Tseng === **Triadica sebifera** (L.) Small

Sapium pleiostachys K.Schum. & Pittier === **Sapium laurifolium** (A.Rich.) Griseb.

Sapium plumerioides Croizat === **Stillingia lineata** subsp. **pacifica** (Müll.Arg.) Steenis

Sapium poeppigii Hemsl. === **Sapium glandulosum** (L.) Morong

Sapium poggei Pax === **Sclerocroton cornutus** (Pax) Kruijt & Roebers

Sapium pohlianum Klotzsch ex Pax === **Sapium glandulosum** (L.) Morong

Sapium populifolium Wall. ex Wight === **Balakata baccata** (Roxb.) Esser

Sapium prunifolium Klotzsch === **Sapium glandulosum** (L.) Morong

Sapium punctatum Klotzsch ex Pax === **Sapium glandulosum** (L.) Morong

Sapium putumayense Croizat === **Sapium stylare** Müll.Arg.

Sapium pycnostachys K.Schum. ex Pax === **Sapium glandulosum** (L.) Morong

Sapium reticulatum (Hochst.) Pax === **Sclerocroton integerrimus** Hochst.

Sapium rhombifolium Rusby === **Sebastiania brasiliensis** Spreng.

Sapium rigidifolium Huft === **Sapium stylare** Müll.Arg.

Sapium rojasii H.Lév. === **Sapium haematospermum** Müll.Arg.

Sapium rotundifolium Hemsl. === **Triadica** sp.

Sapium ruizii Hemsl. === **Stillingia** sp.

Sapium salicifolium Kunth === **Sapium glandulosum** (L.) Morong

Sapium salpingadenium Müll.Arg. === **Stillingia salpingadenia** Huber

Sapium saltense (O'Donell & Lourteig) Jabl. === **Sapium glandulosum** (L.) Morong

Sapium sanchezii Merr. === **Shirakiopsis sanchezii** (Merr.) Esser

Sapium sanguinolentum Klotzsch ex Pax === **Stillingia oppositifolia** Baill. ex Müll.Arg.

Sapium sceleratum Ridl. === **Sapium argutum** (Müll.Arg.) Huber

Sapium schippii Croizat === **Sapium glandulosum** (L.) Morong

Sapium schmitzii J.Léonard === **Sclerocroton schmitzii** (J.Léonard) Kruijt & Roebers

Sapium sebiferum (L.) Roxb. === **Triadica sebifera** (L.) Small

Sapium sebiferum var. *cordatum* S.Y.Wang === **Triadica sebifera** (L.) Small

Sapium sebiferum var. *dabeshanense* B.C.Ding & T.B.Chao === **Triadica sebifera** (L.) Small

Sapium sebiferum var. *multiracemosum* B.C.Ding & T.B.Chao === **Triadica sebifera** (L.) Small

Sapium sebiferum var. *pendulum* B.C.Ding & T.B.Chao === **Triadica sebifera** (L.) Small

Sapium serratum (Müll.Arg.) Klotzsch ex Baill. === **Sapium glandulosum** (L.) Morong

Sapium simii Kuntze === **Excoecaria simii** (Kuntze) Pax

Sapium simile Hemsl. === **Sebastiania** sp.

Sapium solisii Huft === **Sapium stylare** Müll.Arg.

Sapium squarrosum Klotzsch ex Pax === **Sapium haematospermum** Müll.Arg.

Sapium stenophyllum (Müll.Arg.) Huber === **Sapium haematospermum** Müll.Arg.

Sapium sterculiaceum Wall. === **Baccaurea bracteata** Müll.Arg. var. **bracteata**

Sapium steyermarkii Jabl. === **Tetrorchidium** sp.

Sapium suberosum Müll.Arg. === **Sapium glandulosum** (L.) Morong

Sapium sublanceolatum (Müll.Arg.) Huber === **Sapium sellowianum** (Müll.Arg.) Klotzsch ex Baill.

Sapium submarginatum Huber === **Sapium glandulosum** (L.) Morong

Sapium subrotundifolium Elmer === **Blumeodendron kurzii** (Hook.f.) J.J.Sm. ex Koord. & Valeton

Sapium subserratum Klotzsch ex Pax === **Sapium glandulosum** (L.) Morong

Sapium subsessile Chodat & Hassl. === **Sebastiania subsessilis** (Müll.Arg.) Pax

Sapium subsessile Hemsl. === **Stillingia bodenbenderi** (Kuntze) D.J.Rogers

Sapium subulatum (Müll.Arg.) Chodat & Hassl. === **Sebastiania subulata** (Müll.Arg.) Pax

Sapium suffruticosum Pax === **Sclerocroton oblongifolius** (Müll.Arg.) Kruijt & Roebers

Sapium sulciferum Pittier === **Sapium glandulosum** (L.) Morong

Sapium sylvaticum (L.) Torr. === **Stillingia sylvatica** L.

Sapium sylvaticum var. *linearifolium* Torr. === **Stillingia texana** I.M.Johnst.

Sapium taburu Ule === **Sapium glandulosum** (L.) Morong

Sapium tanguinum (Baill.) Müll.Arg. === **Stillingia lineata** (Lam.) Müll.Arg. subsp. **lineata**

Sapium tenellum Pax & K.Hoffm. === **Sapium glandulosum** (L.) Morong

Sapium terminale (Baill.) Müll.Arg. === **Stillingia terminalis** Baill.

Sapium thelocarpum K.Schum. & Pittier === **Sapium macrocarpum** Müll.Arg.

Sapium thomsonii God.-Leb. ex Jum. === **Sapium stylare** Müll.Arg.

Sapium tijucense Huber === **Sapium haematospermum** Müll.Arg.

Sapium tolimense Jum. === **Sapium stylare** Müll.Arg.

Sapium triloculare Pax & K.Hoffm. === **Shirakiopsis trilocularis** (Pax & K.Hoffm.) Esser (ined.)

Sapium triste (Müll.Arg.) Huber === **Sapium argutum** (Müll.Arg.) Huber

Sapium tuerckheimianum Pax & K.Hoffm. === **Sebastiania tuerckheimiana** (Pax & K.Hoffm.) Lundell

Sapium utile Preuss === **Sapium laurifolium** (A.Rich.) Griseb.

Sapium veracruzense Lundell === **Sapium sp.**

Sapium verum Hemsl. === **Sapium stylare** Müll.Arg.

Sapium virgatum (Zoll. & Moritzi ex Miq.) Hook.f. === **Shirakiopsis virgata** (Zoll. & Moritzi ex Miq.) Esser

Sapium warmingii Chodat & Hassl. === **Sebastiania warmingii** Pax

Sapium xylocarpum Pax === **Sclerocroton cornutus** (Pax) Kruijt & Roebers

Sapium xylocarpum var. *genuinum* Pax === **Sclerocroton cornutus** (Pax) Kruijt & Roebers

Sapium xylocarpum var. *lineolatum* Pax === **Sclerocroton cornutus** (Pax) Kruijt & Roebers

Sapium yutajense Jabl. === **Dendrothrix yutajensis** (Jabl.) Esser

Sapium zelayense Kunth === **Stillingia zelayensis** (Kunth) Müll.Arg.

Saragodra

Synonyms:
Saragodra Steud. === **Suregada** Roxb. ex Rottl.

Sarcoclinium

Synonyms:
Sarcoclinium Wight === **Agrostistachys** Dalzell

Sarcoclinium gaudichaudii Baill. === **Agrostistachys gaudichaudii** (Baill.) Müll.Arg.

Sarcoclinium hookeri Thwaites === **Agrostistachys hookeri** (Thwaites) Benth. & Hook.f.

Sarcoclinium longifolium Wight === **Agrostistachys borneensis** Becc.

Sarcoclinium sessilifolium Kurz === **Agrostistachys sessilifolia** (Kurz) Pax & K.Hoffm.

Sarothrostachys

Synonyms:
Sarothrostachys Klotzsch === **Gymnanthes** Sw.
Sarothrostachys multiramea Klotzsch === **Gymnanthes multiramea** (Klotzsch) Müll.Arg.

Sauropus

83 species, Asia, Malesia and Australia with *S. bacciformis*, a shore species, extending to the Mascarenes; includes the largely Australian *Synostemon*. Annual or perennial herbs or shrubs with distichously arranged leaves, some resembling species of *Phyllanthus*. The subgenera and sections accepted by Pax & Hoffmann (1922, for *Sauropus* s.s.; *Synostemon* was never treated) – some inherited from Mueller – were by Airy-Shaw (1980) considered to be all but valueless. The 20 accepted Australian species (revised by Airy-Shaw, 1980) are not closely related to those in SE. Asia and W. Malesia; apart from the very widely distributed *S. macranthus* the only link was thought to be between *S. calcareus* (Peninsular Malaysia) and *S. brunonis* (Northern Territory, Australia). Some species are cultivated for fragrant flowers (*SS. spatulifolius, thorelii*). Along with *Breynia*, *Sauropus* is separated from *Phyllanthus* (with which it has been confused by collectors) by an obscure or absent floral disk and usually ventrally invaginated seeds (Webster, Synopsis, 1994). Certain species were once included with *Agyneia* (Pax & Hoffmann 1922) but that name is properly a synonym of *Glochidion*. (Phyllanthoideae)

Pax, F. & K. Hoffmann (1922). *Agyneia*. In A. Engler (ed.), Das Pflanzenreich, IV 147 XV (Euphorbiaceae-Phyllanthoideae-Phyllantheae): 212-215. Berlin. (Heft 81.) La/Ge. — 1 species, *A. bacciformis*, in Madagascar, Asia and W Malesia. An additional species, *A. affinis*, thought to belong to *Glochidion*. [Both species now referable to *Sauropus bacciformis*.]

Pax, F. & K. Hoffmann (1922). *Sauropus*. In A. Engler (ed.), Das Pflanzenreich, IV 147 XV (Euphorbiaceae-Phyllanthoideae-Phyllantheae): 215-226. Berlin. (Heft 81.) La/Ge. — 29 species (1 doubtful) in 2 subgenera with 6 sections; Asia, Sunda and the Philippines.

Croizat, L. (1942b). Notes on the Euphorbiaceae, III. Bull. Bot. Gard. Buitenzorg, III, 17: 209-219. En. — Includes (among other topics) a proposal for the conservation of *Sauropus* over *Arachne*.

Leandri, J. (1958). *Agyneia*. Fl. Madag. Comores 111 (Euphorbiacées), I: 137-138. Paris. Fr. — Madagascar records of *Agyneia bacciformis*, a coastal species, are questionable and based only on old collections; the plant (now in *Sauropus*) is otherwise native from the Mascarenes eastwards to Borneo and southern China.

Airy-Shaw, H. K. (1960). Notes on Malaysian Euphorbiaceae, III. A noteworthy new species of *Sauropus* Bl. from Borneo. Kew Bull. 14: 354-355. En. — Description of *S. micrasterias*, from W. Sarawak limestone.

Webster, G. L. (1960). The status of *Agyneia* and *Glochidion* (Euphorbiaceae). Taxon 9: 25-26. En. — An outgrowth of the author's revision of West Indian *Phyllanthus* (1956-58). *Agyneia* L. synonymous with *Glochidion* (the latter here proposed for conservation); *Agyneia* auct. long used for *Phyllanthus bacciformis* L. but as not available a new combination in *Synostemon* is proposed.

Airy-Shaw, H. K. (1963). Notes on Malaysian and other Asiatic Euphorbiaceae, XXIII. Further collection of *Sauropus micrasterias*. Kew Bull. 16: 344. En. — New record.

Airy-Shaw, H. K. (1969). Notes on Malesian and other Asiatic Euphorbiaceae, CI. New or noteworthy species of *Sauropus* Bl. Kew Bull. 23: 42-55. En. — Notes and novelties, encompassing 12 species and 1 (new) section; a key to sect. *Hemisauropus* is also presented.

Airy-Shaw, H. K. (1971). Notes on Malesian and other Asiatic Euphorbiaceae, CXXVIII. *Sauropus suberosus* in Malaya. Kew Bull. 25: 500. En. —*S. suberosus*, of Thailand, in Peninsular Malaysia.

Airy-Shaw, H. K. (1978). Notes on Malesian and other Asiatic Euphorbiaceae, CCX. *Synostemon* F. Muell. Kew Bull. 33: 37-38. En. — Description of *S. sphenophyllus* from New Guinea. [*Synostemon*, centered in Australia, is now combined with *Sauropus*.]

Airy-Shaw, H. K. (1979). Notes on Malesian and other Asiatic Euphorbiaceae, CCXXIII. *Sauropus* Bl. Kew Bull. 33: 530-531. En. — Amplified description of the Vietnamese *S. spatulifolius* Beille, grown in Kuala Lumpur. [Like *S. thorelii*, the species is apparently grown for its fragrant flowers.]

Airy-Shaw, H. K. (1980). *Sauropus*. Kew Bull. 35: 669-686. (Euphorbiaceae-Platylobeae of Australia.) En. — Treatment of c. 20 species; viewed as hardly related to those of SE Asia.

Airy-Shaw, H. K. (1980). Notes on Euphorbiaceae from Indomalesia, Australia and the Pacific, CCXXXIII. *Sauropus* Bl. Kew Bull. 35: 389. En. — Description of *S. gramineus* from Thailand.

Brunel, J. F. (1987). *Synostemon*. In *idem*, Sur le genre *Phyllanthus* L. et quelques genres voisins de la tribu des Phyllantheae Dumort. Strasbourg. [See also *Phyllanthus*.] Fr. — Description of the genus; synopsis (1 species reported from Madagascar – but see Leandri 1958). [That species, *S. bacciformis* ranges along shores in the Indian Ocean and South China and Java Seas; the remainder are Australian. All have been transferred to *Sauropus*.]

Hunter, J. T. & J. J. Bruhl (1997). Four new rare species of *Sauropus* Blume (Euphorbiaceae: Phyllantheae) from north Queensland. Austrobaileya 4: 661-672, illus. En. — Descriptions with illustrations of four novelties; key to all species (12) in Queensland; some reductions.

Sauropus Blume, Bijdr.: 595 (1826).
W. Indian Ocean, Trop. Asia, Australia. 29 40 41 42 50.
Aalius Rumph. ex Lam., Encycl. 1: 1 (1793).
Ceratogynum Wight, Icon. Pl. Ind. Orient. 5(2): 26 (1852).
Diplomorpha Griff., Not. Pl. Asiat. 4: 479 (1854).
Synostemon F.Muell., Fragm. 1: 32 (1858).
Breyniopsis Beille, Bull. Soc. Bot. France 72: 157 (1925).
Heterocalymnantha Domin, Biblioth. Bot. 89: 313 (1927).

Sauropus albiflorus (F.Muell. ex Müll.Arg.) Airy Shaw, Kew Bull. 35: 671 (1980). – FIGURE, p. 1435.
Queensland, New South Wales. 50 NSW QLD. Nanophan.
 * *Phyllanthus albiflorus* F.Muell. ex Müll.Arg., Linnaea 34: 70 (1865). *Synostemon albiflorus* (F.Muell. ex Müll.Arg.) Airy Shaw, Kew Bull. 33: 37 (1978).

Sauropus amabilis Airy Shaw, Kew Bull. 23: 49 (1969).
N. Thailand. 41 THA. Hemicr. or cham.

Sauropus amoebiflorus Airy Shaw, Kew Bull. 23: 45 (1969).
Thailand. 41 THA. Cham. or nanophan.

Sauropus androgynus (L.) Merr., Forest. Bur. Philipp. Bull. 1: 30 (1903).
Trop. & Subtrop. Asia. 36 CHC CHH CHS 40 ASS EHM IND SRL 41 CBD LAO THA VIE 42 BOR JAW LSI MLY MOL NWG PHI SUL SUM. (Cl.) nanophan. or phan.
 * *Clutia androgyna* L., Mant. Pl. 1: 128 (1767). *Aalius androgyna* (L.) Kuntze, Revis. Gen. Pl. 2: 591 (1891).
Phyllanthus acidissimus Noronha, Verh. Batav. Genootsch. Kunsten 5(4): 22 (1790).
Phyllanthus speciosus Noronha, Verh. Batav. Genootsch. Kunsten 5(4): 22 (1790), nom. nud.
Agyneia ovata Poir. in J.B.A.M.de Lamarck, Encycl., Suppl. 1: 243 (1810).
Andrachne ovata Lam. ex Poir. in J.B.A.M.de Lamarck, Encycl., Suppl. 1: 243 (1810).
Sauropus albicans Blume, Bijdr.: 595 (1826). *Sauropus albicans* var. *genuinus* Müll.Arg. in A.P.de Candolle, Prodr. 15(2): 241 (1866), nom. inval.
Phyllanthus strictus Roxb., Fl. Ind. ed. 1832, 3: 670 (1832).
Sauropus gardnerianus Wight, Icon. Pl. Ind. Orient. 6: 6, t. 1951 (1853). *Sauropus albicans* var. *gardnerianus* (Wight) Müll.Arg., Linnaea 32: 72 (1863).
Sauropus indicus Wight, Icon. Pl. Ind. Orient. 6: 6, t. 1952 (1853).

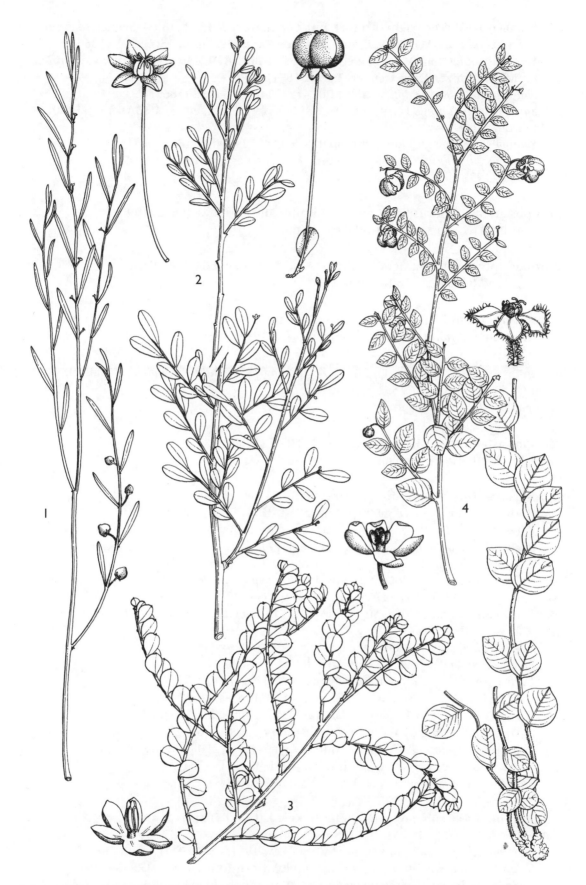

Sauropus trachyspermus (F. Muell.) Airy-Shaw (1, as *S. hubbardii* Airy-Shaw); *S. albiflorus* (F. Muell. ex Müll. Arg.) Airy-Shaw (2, as *S. albiflorus* subsp. *albiflorus*); *S. crassifolius* (Müll. Arg.) Airy-Shaw (3); *S. brunonis* (S. Moore) Airy-Shaw (4, as *S. brunonis* var. *ovatus* Airy-Shaw)

Artist: Ann Davies
Kew Bull. 35: 672, fig. 7 (1980)
KEW ILLUSTRATIONS COLLECTION

Sauropus zeylanicus Wight, Icon. Pl. Ind. Orient. 6: 6, t. 1952 (1853). *Sauropus albicans* var. *zeylanicus* (Wight) Müll.Arg. in A.P.de Candolle, Prodr. 15(2): 241 (1866).
Sauropus sumatranus Miq., Fl. Ned. Ind., Eerste Bijv.: 179, 446 (1861). *Aalius sumatrana* (Miq.) Kuntze, Revis. Gen. Pl. 2: 591 (1891).
Sauropus albicans var. *intermedius* Müll.Arg., Linnaea 32: 72 (1863).
Sauropus macranthus Fern.-Vill. in F.M.Blanco, Fl. Filip., ed. 3, 4(13A): 187 (1880), nom. illeg.
Sauropus scandens C.B.Rob., Philipp. J. Sci., C 4: 72 (1909).
Sauropus parviflorus Pax & K.Hoffm. in H.G.A.Engler, Pflanzenr., IV, 147, XV: 218 (1922).

Sauropus anemoniflorus J.T.Hunter & J.J.Bruhl, Austrobaileya 4: 662 (1997).
Queensland (North Kennedy). 50 QLD. Hemicr.

Sauropus aphyllus J.T.Hunter & J.J.Bruhl, Austrobaileya 4: 664 (1997).
Queensland (Cook). 50 QLD. Hemicr.

Sauropus arenosus J.T.Hunter & J.J.Bruhl, Nuytsia 11: 166 (1997).
N. Western Australia. 50 WAU. Nanophan.

Sauropus assimilis Thwaites, Enum. Pl. Zeyl.: 284 (1861). *Aalius assimilis* (Thwaites) Kuntze, Revis. Gen. Pl. 2: 591 (1891).
Sri Lanka. 40 SRL. Nanophan.

Sauropus asteranthos Airy Shaw, Kew Bull. 23: 47 (1969).
NE. Thailand. 41 THA. Cham.

Sauropus bacciformis (L.) Airy Shaw, Kew Bull. 35: 685 (1980).
W. Indian Ocean, Trop. & Subtrop. Asia. 29 MAU MDG REU 36 CHH CHS 38 TAI 40 BAN IND SRL 41 THA VIE 42 BOR JAW SUL SUM. Nanophan.
  * *Phyllanthus bacciformis* L., Mant. Pl. 2: 294 (1771). *Agyneia bacciformis* (L.) A.Juss., Euphorb. Gen.: 24 (1824). *Agyneia bacciformis* var. *genuina* Müll.Arg. in A.P.de Candolle, Prodr. 15(2): 238 (1866), nom. inval. *Diplomorpha bacciformis* (L.) Kuntze, Revis. Gen. Pl. 2: 603 (1891). *Synostemon bacciformis* (L.) G.L.Webster, Taxon 9: 26 (1960).
  Phyllanthus racemosus L.f., Suppl. Pl.: 415 (1782). *Emblica racemosa* (L.f.) Spreng., Syst. Veg. 3: 20 (1826).
  Agyneia phyllanthoides Spreng., Syst. Veg. 3: 19 (1826).
  Emblica annua Raf., Sylva Tellur.: 91 (1838).
  Diplomorpha herbacea Griff., Not. Pl. Asiat. 4: 479 (1854).
  Agyneia bacciformis var. *angustifolia* Müll.Arg., Linnaea 32: 72 (1863).
  Agyneia bacciformis var. *oblongifolia* Müll.Arg., Linnaea 32: 71 (1863).
  Agyneia affinis Kurz ex Teijsm. & Binn., Tijdschr. Ned.-Indië 27: 118 (1864).
  Phyllanthus goniocladus Merr. & Chun, Sunyatsenia 2: 260 (1935). *Agyneia gonioclada* (Merr. & Chun) H.Keng, J. Wash. Acad. Sci. 41: 201 (1951).
  Agyneia taiwaniana H.Keng, J. Wash. Acad. Sci. 41: 200 (1951).

Sauropus bicolor Craib, Bull. Misc. Inform. Kew 1914: 11 (1914).
Burma, N. Thailand. 41 BMA THA. Nanophan.
  Sauropus rigidus Craib, Bull. Misc. Inform. Kew 1911: 457 (1911), nom. illeg.
  Sauropus similis var. *microphylla* Craib, Bull. Misc. Inform. Kew 1912: 184 (1912).
    Sauropus bicolor var. *microphyllus* (Craib) Airy Shaw, Kew Bull. 26: 333 (1972).

Sauropus bonii Beille in H.Lecomte, Fl. Indo-Chine 5: 651 (1927).
China (Guangxi), Vietnam, Thailand. 36 CHS 41 THA VIE. Nanophan.

Sauropus brevipes Müll.Arg., Linnaea 32: 73 (1863). *Aalius brevipes* (Müll.Arg.) Kuntze, Revis. Gen. Pl. 2: 591 (1891).
Burma, Thailand, Cambodia, Pen. Malaysia. 41 BMA CBD THA 42 MLY. Nanophan.
Sauropus parvifolius Ridl., J. Straits Branch Roy. Asiat. Soc. 59: 175 (1911).

Sauropus brunonis (S.Moore) Airy Shaw, Kew Bull. 35: 672 (1980). – FIGURE, p. 1435.
Northern Territory. 50 NTA. Hemicr. or cham.
  * *Phyllanthus brunonis* S.Moore, J. Linn. Soc., Bot. 45: 213 (1920).
  Sauropus brunonis var. *ovatus* Airy Shaw, Kew Bull. 35: 673 (1980).

Sauropus calcareus M.R.Hend., Gard. Bull. Straits Settlem. 7: 121 (1933).
Pen. Malaysia. 42 MLY.

Sauropus concinnus Collett & Hemsl., J. Linn. Soc., Bot. 28: 123 (1890).
N. Burma (Shan Hills). 41 BMA. Cham. or nanophan.

Sauropus convallarioides J.T.Hunter & J.J.Bruhl, Austrobaileya 4: 666 (1997).
Queensland (Cook). 50 QLD. Cham.

Sauropus convexus J.J.Sm., Bull. Jard. Bot. Buitenzorg, III, 6: 82 (1924).
? 42 +. – Provisionally accepted.

Sauropus crassifolius (Müll.Arg.) Airy Shaw, Kew Bull. 35: 673 (1980). – FIGURE, p. 1435.
Western Australia. 50 WAU. Nanophan.
  * *Phyllanthus crassifolius* Müll.Arg., Flora 47: 513 (1864). *Glochidion crassifolium*
    (Müll.Arg.) Gardner ex Beard, W. Austr. Pl.: 58 (1967).

Sauropus decrescentifolia J.T.Hunter & J.J.Bruhl, Austrobaileya 4: 668 (1997).
Queensland (Cook, North Kennedy). 50 QLD. Hemicr.

Sauropus delavayi Croizat, J. Arnold Arbor. 21: 496 (1940).
China (Yunnan, Guangxi). 36 CHC CHS. Nanophan.

Sauropus ditassoides (Müll.Arg.) Airy Shaw, Kew Bull. 35: 674 (1980).
N. Northern Territory. 50 NTA. Cham.
  * *Phyllanthus ditassoides* Müll.Arg., Flora 47: 487 (1864).

Sauropus dunlopii J.T.Hunter & J.J.Bruhl, Nuytsia 11: 169 (1997).
Northern Territory. 50 NTA. Cham.

Sauropus elachophyllus (F.Muell. ex Benth.) Airy Shaw, Kew Bull. 35: 675 (1980).
Northern Territory, Queensland. 50 NTA QLD. Cham. or nanophan.
  * *Phyllanthus elachophyllus* F.Muell. ex Benth., Fl. Austral. 6: 101 (1873).
  Sauropus elachophyllus var. *glaber* Airy Shaw, Kew Bull. 35: 675 (1980).
  Sauropus elachophyllus var. *latior* Airy Shaw, Kew Bull. 35: 675 (1980).

Sauropus elegantissimus Ridl., Bull. Misc. Inform. Kew 1926: 476 (1926).
Pen. Malaysia. 42 MLY.

Sauropus filicinus J.T.Hunter & J.J.Bruhl, Nuytsia 11: 170 (1997).
Northern Territory. 50 NTA. Cham. or nanophan.

Sauropus garrettii Craib, Bull. Misc. Inform. Kew 1914: 284 (1914).
S. China, Hainan, Burma to Singapore. 36 CHC CHH CHS 41 BMA THA 42 MLY.
Nanophan. or phan.
  Sauropus yunnanensis Pax & K.Hoffm. in H.G.A.Engler, Pflanzenr., IV, 147, XV: 220 (1922).
  Sauropus chorisepalus Merr. & Chun, Sunyatsenia 2: 10 (1934).

Sauropus glaucus (F.Muell.) Airy Shaw, Kew Bull. 35: 676 (1980).
N. Northern Territory. 50 NTA. Cham.
* *Synostemon glaucus* F.Muell., Fragm. 1: 33 (1858).
Phyllanthus adamii Müll.Arg. in A.P.de Candolle, Prodr. 15(2): 327 (1866). *Glochidion adamii* (Müll.Arg.) Gardner ex Beard, W. Austr. Pl.: 58 (1967).
Phyllanthus bossiaeoides A.Cunn. ex Benth., Fl. Austral. 6: 98 (1873).

Sauropus gracilis J.T.Hunter & J.J.Bruhl, Nuytsia 11: 171 (1997).
Northern Territory. 50 NTA. Cham.

Sauropus gramineus Airy Shaw, Kew Bull. 35: 389 (1980).
Thailand. 41 THA.

Sauropus granulosus Airy Shaw, Kew Bull. 23: 53 (1969).
NE. Thailand. 41 THA. Cham.

Sauropus harmandii Beille in H.Lecomte, Fl. Indo-Chine 5: 657 (1927).
Cambodia. 41 CBD.

Sauropus hayatae Beille in H.Lecomte, Fl. Indo-Chine 5: 650 (1927).
Vietnam. 41 VIE.

Sauropus heteroblastus Airy Shaw, Kew Bull. 23: 48 (1969).
Thailand, S. Vietnam, Cambodia. 41 CBD THA VIE. Nanophan.

Sauropus hirsutus Beille in H.Lecomte, Fl. Indo-Chine 5: 657 (1927).
Indo-China. 41 CBD LAO THA. Cham.

Sauropus hirtellus (F.Muell.) Airy Shaw, Kew Bull. 35: 677 (1980).
SE. Queensland, NE. New South Wales. 50 NSW QLD. Cham.
* *Synostemon hirtellus* F.Muell., Fragm. 3: 89 (1863). *Phyllanthus thymoides* var. *hirtellus* (F.Muell.) Müll.Arg. in A.P.de Candolle, Prodr. 15(2): 372 (1866). *Phyllanthus hirtellus* (F.Muell.) Müll.Arg. in A.P.de Candolle, Prodr. 15(2): 326 (1866), nom. illeg. *Glochidion hirtellum* (F.Muell.) H.Eichler, Suppl. J. M. Black Fl. S. Austral. ed. 2: 210 (1965).

Sauropus huntii (Ewart & Davies) Airy Shaw, Kew Bull. 35: 679 (1980).
S. Northern Territory. 50 NTA. Nanophan.
* *Phyllanthus huntii* Ewart & O.B.Davies, Fl. N. Territory: 164 (1917).

Sauropus kerrii Airy Shaw, Kew Bull. 23: 52 (1969).
E. Thailand. 41 THA. Hemicr. or cham.

Sauropus kitanovii Thin, Euphorb. Vietnam: 49 (1995).
Vietnam. 41 VIE. Nanophan. or phan.

Sauropus lanceolatus Hook.f., Fl. Brit. India 5: 333 (1887). *Aalius lanceolata* (Hook.f.) Kuntze, Revis. Gen. Pl. 2: 591 (1891).
Arunachal Pradesh (Mishmi Hills). 40 EHM.

Sauropus macranthus Hassk., Retzia 1: 166 (1855). *Aalius macrantha* (Hassk.) Kuntze, Revis. Gen. Pl. 2: 591 (1891).
Trop. & Subtrop. Asia, NE. Queensland. 36 CHC CHH CHS 40 ASS EHM NEP 41 THA VIE 42 BOR JAW LSI MOL NWG SUL SUM 50 QLD. Nanophan. or phan.
Sauropus spectabilis Miq., Fl. Ned. Ind., Eerste Bijv.: 179, 446 (1861). *Aalius spectabilis* (Miq.) Kuntze, Revis. Gen. Pl. 2: 591 (1891).
Sauropus forcipatus Hook.f., Fl. Brit. India 5: 334 (1887). *Aalius forcipata* (Hook.f.) Kuntze, Revis. Gen. Pl. 2: 591 (1891).

Sauropus macrophyllus Hook.f., Fl. Brit. India 5: 333 (1887). *Aalius macrophylla* (Hook.f.)
Kuntze, Revis. Gen. Pl. 2: 591 (1891).

Glochidion umbratile Maiden & Betche, Proc. Linn. Soc. New South Wales, II, 30:
370 (1905).

Sauropus robinsonii Merr., Philipp. J. Sci., C 7: 407 (1912 publ. 1913).

Sauropus grandifolius Pax & K.Hoffm. in H.G.A.Engler, Pflanzenr., IV, 147, XV:
222 (1922).

Sauropus ruber Teijsm. & Binn. ex Pax & K.Hoffm. in H.G.A.Engler, Pflanzenr., IV, 147,
XV: 219 (1922).

Sauropus wichurae Müll.Arg. ex Pax & K.Hoffm. in H.G.A.Engler, Pflanzenr., IV, 147, XV:
220 (1922).

Sauropus longipedicellatus Merr. & Chun, Sunyatsenia 2: 34 (1934).

Sauropus maichauensis Thin, J. Biol. (Vietnam) 14(2): 24 (1992).
Vietnam. 41 VIE.

Sauropus micrasterias Airy Shaw, Kew Bull. 14: 354 (1960).
Borneo (Sarawak). 42 BOR. Nanophan.

Sauropus oblongifolius Hook.f., Fl. Brit. India 5: 333 (1887). *Aalius oblongifolia* (Hook.f.)
Kuntze, Revis. Gen. Pl. 2: 591 (1891).
Arunachal Pradesh (Mishmi Hills). 40 EHM.

Sauropus ochrophyllus (Benth.) Airy Shaw, Kew Bull. 35: 682 (1980).
N. Northern Territory. 50 NTA. Hemicr.
  * *Phyllanthus ochrophyllus* Benth., Fl. Austral. 6: 99 (1873).

Sauropus orbicularis Craib, Bull. Misc. Inform. Kew 1914: 284 (1914).
Thailand. 41 THA. Cham. or nanophan.
    Sauropus orbicularis var. *minor* Airy Shaw, Kew Bull. 23: 45 (1969).

Sauropus paucifolius J.T.Hunter & J.J.Bruhl, Nuytsia 11: 172 (1997).
Northern Territory. 50 NTA. Hemicr. or cham.

Sauropus pierrei (Beille) Croizat, J. Arnold Arbor. 21: 494 (1940).
China (Yunnan), Indo-China, N. Borneo. 36 CHC 41 CBD LAO VIE 42 BOR. Nanophan.
  or phan.
  * *Breyniopsis pierrei* Beille, Bull. Soc. Bot. France 72: 158 (1925).

Sauropus podenzanae (S.Moore) Airy Shaw, Kew Bull. 35: 682 (1980).
Queensland (near Cooktown). 50 QLD. Cham.
  * *Phyllanthus podenzanae* S.Moore, J. Linn. Soc., Bot. 45: 214 (1920).

Sauropus poilanei Beille in H.Lecomte, Fl. Indo-Chine 5: 653 (1927).
Vietnam. 41 VIE.

Sauropus pulchellus Airy Shaw, Kew Bull. 23: 54 (1969).
E. Thailand. 41 THA. Cham. or nanophan.

Sauropus quadrangularis (Willd.) Müll.Arg., Linnaea 32: 73 (1863). *Aalius quadrangularis*
(Müll.Arg.) Kuntze, Revis. Gen. Pl. 2: 591 (1891).
C. & S. India to Indo-China. 36 CHC CHS CHT 40 ASS EHM IND NEP 41 BMA CBD THA
VIE. Cham.
  * *Phyllanthus quadrangularis* Willd., Sp. Pl. 4: 585 (1805).
    Phyllanthus rhamnoides Roxb., Fl. Ind. ed. 1832, 3: 663 (1832). *Ceratogynum rhamnoides*
    (Roxb.) Wight, Icon. Pl. Ind. Orient. 5(2): 26 (1852).

Phyllanthus myrtillus Wall., Numer. List: 7892A (1847), nom. inval.

Sauropus ceratogynum Baill., Étude Euphorb.: 635 (1858), nom. nud.

Phyllanthus leschenaultii var. *tenellus* Müll.Arg., Linnaea 32: 38 (1863).

Sauropus compressus Müll.Arg. in A.P.de Candolle, Prodr. 15(2): 243 (1866). *Aalius compressa* (Müll.Arg.) Kuntze, Revis. Gen. Pl. 2: 591 (1891). *Sauropus quadrangularis* var. *compressus* (Müll.Arg.) Airy Shaw, Kew Bull. 26: 337 (1972).

Sauropus quadrangularis var. *puberulus* Kurz, Forest Fl. Burma 2: 350 (1877).

Sauropus pubescens Hook.f., Fl. Brit. India 5: 335 (1887). *Sauropus quadrangularis* var. *puberulus* Kurz, Forest Fl. Burma 2: 350 (1877). *Aalius pubescens* (Hook.f.) Kuntze, Revis. Gen. Pl. 2: 591 (1891).

Aalius ceratogynum Kuntze, Revis. Gen. Pl. 2: 591 (1891).

Sauropus racemosus Beille in H.Lecomte, Fl. Indo-Chine 5: 648 (1927).
Vietnam. 41 VIE.

Sauropus ramosissimus (F.Muell.) Airy Shaw, Kew Bull. 35: 682 (1980).
S. Northern Territory, Queensland, New South Wales, South Australia. 50 NSW NTA QLD SOA. Cham.
  * *Synostemon ramosissimus* F.Muell., Fragm. 1: 33 (1858). *Phyllanthus ramosissimus* (F.Muell.) Müll.Arg., Linnaea 34: 70 (1865).

Sauropus repandus Müll.Arg., Flora 55: 2 (1872).
Sikkim, China (Yunnan). 36 CHC 40 EHM. Nanophan.

Sauropus reticulatus X.L.Mo ex P.T.Li, Acta Phytotax. Sin. 25: 133 (1987).
China (Yunnan, Guangxi). 36 CHC CHS. Nanophan.

Sauropus retroversus Wight, Icon. Pl. Ind. Orient. 6: 6, t. 1951 (1853). *Aalius retroversa* (Wight) Kuntze, Revis. Gen. Pl. 2: 591 (1891).
Sri Lanka. 40 SRL. Nanophan.

Sauropus rhamnoides Blume, Bijdr.: 596 (1826). *Aalius rhamnodes* (Blume) Kuntze, Revis. Gen. Pl. 2: 591 (1891).
Jawa, Borneo, Philippines. 42 BOR JAW PHI. (Cl.) nanophan. or phan.
  Phyllanthus placentatus Noronha, Verh. Batav. Genootsch. Kunsten 5(4): 22 (1790).
  Phyllanthus placenta Hassk., Tijdschr. Natuurl. Gesch. Physiol. 11: 225 (1844), sphalm.

Sauropus rigens (F.Muell.) Airy Shaw, Kew Bull. 35: 683 (1980).
Northern Territory, Queensland, New South Wales, South Australia. 50 NSW NTA QLD SOA. Cham. or nanophan.
  * *Synostemon rigens* F.Muell., Fragm. 2: 153 (1861). *Phyllanthus rigens* (F.Muell.) Müll.Arg., Flora 47: 513 (1864). *Glochidion rigens* (F.Muell.) H.Eichler, Suppl. J. M. Black Fl. S. Austral. ed. 2: 210 (1965).
  Heterocalymnantha minutifolia Domin, Biblioth. Bot. 89: 313 (1927).

Sauropus rigidulus (F.Muell. ex Müll.Arg.) Airy Shaw, Kew Bull. 35: 684 (1980).
N. Northern Territory. 50 NTA. Nanophan.
  * *Phyllanthus rigidulus* F.Muell. ex Müll.Arg., Linnaea 34: 72 (1865).

Sauropus rigidus Thwaites, Enum. Pl. Zeyl.: 284 (1861). *Aalius rigida* (Thwaites) Kuntze, Revis. Gen. Pl. 2: 591 (1891).
Sri Lanka. 40 SRL. Nanophan.

Sauropus rimophilus J.T.Hunter & J.J.Bruhl, Nuytsia 11: 173 (1997).
Northern Territory. 50 NTA. Nanophan.

Sauropus rostratus Miq., Fl. Ned. Ind., Eerste Bijv.: 179, 447 (1861). *Aalius rostrata* (Miq.) Kuntze, Revis. Gen. Pl. 2: 591 (1891).
Sumatera. 42 SUM. Nanophan.

Sauropus saksenianus Manilal, Prasann. & Sivar., J. Indian Bot. Soc. 64: 294 (1985).
India. 40 IND.

Sauropus salignus J.T.Hunter & J.J.Bruhl, Nuytsia 11: 175 (1997).
N. Western Australia. 50 WAU. Nanophan.

Sauropus siamensis Chakrab. & M.G.Gangop., J. Econ. Taxon. Bot. 19: 452 (1995).
C. Thailand. 41 THA.

Sauropus similis Craib, Bull. Misc. Inform. Kew 1911: 457 (1911).
N. Thailand. 41 THA. Nanophan. or phan.

Sauropus spatulifolius Beille in H.Lecomte, Fl. Indo-Chine 5: 652 (1927).
N. Vietnam. (36) chs 41 VIE (42) mly phi. Cham. – Cultivated for medicine.
  Sauropus changianus S.Y.Hu, J. Arnold Arbor. 32: 393 (1951).

Sauropus sphenophyllus (Airy Shaw) Airy Shaw, Kew Bull., Addit. Ser. 8: 221 (1980).
Papua New Guinea. 42 NWG. Nanophan.
  * Synostemon sphenophyllus Airy Shaw, Kew Bull. 33: 37 (1978).

Sauropus stenocladus (Müll.Arg.) J.T.Hunter & J.J.Bruhl, Nuytsia 11: 176 (1997).
Northern Territory. 50 NTA. Cham. or nanophan.
  * Phyllanthus stenocladus Müll.Arg., Flora 47: 536 (1864).

  subsp. **pinifolius** J.T.Hunter & J.J.Bruhl, Nuytsia 11: 177 (1997).
  Northern Territory. 50 NTA. Cham. or nanophan.

  subsp. **stenocladus**
  Northern Territory. 50 NTA. Cham. or nanophan.
  Phyllanthus lissocarpus S.Moore, J. Linn. Soc., Bot. 45: 215 (1920). *Sauropus lissocarpus* (S.Moore) Airy Shaw, Kew Bull. 35: 680 (1980).
  Sauropus glaucus var. *glaber* Airy Shaw, Kew Bull. 35: 676 (1980).
  Sauropus latzii Airy Shaw, Kew Bull. 35: 680 (1980).

Sauropus stipitatus Hook.f., Fl. Brit. India 5: 333 (1887). *Aalius stipitata* (Hook.f.) Kuntze, Revis. Gen. Pl. 2: 591 (1891).
Sikkim. 40 EHM. Nanophan.

Sauropus suberosus Airy Shaw, Kew Bull. 23: 42 (1969).
Thailand, Pen. Malaysia. 41 THA 42 MLY. Nanophan.

Sauropus thesioides (Benth.) Airy Shaw, Kew Bull. 35: 684 (1980).
Queensland. 50 QLD. Hemicr.
  * Phyllanthus thesioides Benth., Fl. Austral. 6: 98 (1873). *Glochidion thesioides* H.Eichler, Suppl. J. M. Black Fl. S. Austral. ed. 2: 210 (1965).

Sauropus thoii Thin, Euphorb. Vietnam: 48 (1995).
Vietnam. 41 VIE. Nanophan.

Sauropus thorelii Beille in H.Lecomte, Fl. Indo-Chine 5: 649 (1927).
Laos. 41 LAO. Nanophan. – In Thailand cultivated for the scented flowers.

Sauropus tiepii Thin, Euphorb. Vietnam: 49 (1995).
Vietnam. 41 VIE. Nanophan. or phan.

Sauropus torridus J.T.Hunter & J.J.Bruhl, Nuytsia 11: 178 (1997).
N. Western Australia, Northern Territory. 50 NTA WAU. hemicr. or cham.

Sauropus trachyspermus (F.Muell.) Airy Shaw, Kew Bull. 35: 685 (1980). – FIGURE, p. 1435.
Australia. 50 NSW NTA QLD SOA WAU. Hemicr. or cham.
   * *Phyllanthus trachyspermus* F.Muell., Trans. Philos. Soc. Victoria 1: 14 (1855). *Glochidion trachyspermum* (F.Muell.) H.Eichler, Suppl. J. M. Black Fl. S. Austral. ed. 2: 210 (1965).
   Phyllanthus rhytidospermus F.Muell. ex Müll.Arg., Linnaea 34: 70 (1865). *Glochidion rhytidospermum* (F.Muell. ex Müll.Arg.) H.Eichler, Suppl. J. M. Black Fl. S. Austral. ed. 2: 210 (1965).
   Sauropus hubbardii Airy Shaw, Kew Bull. 35: 677 (1980).
   Phyllanthus minutiflorus F.Muell. & Tate, Trans. & Proc. Roy. Soc. South Australia 13: 98 (June 1890).

Sauropus trinervius Hook.f. & Thomson ex Müll.Arg., Linnaea 32: 72 (1863). *Aalius trinervia* (Hook.f. & Thomson ex Müll.Arg.) Kuntze, Revis. Gen. Pl. 2: 591 (1891).
Sikkim to China (Yunnan). 36 CHC 40 ASS BAN EHM. Nanophan.
   Phyllanthus trinervius Wall., Numer. List: 7922 (1847), nom. inval.

Sauropus tsiangii P.T.Li, Acta Phytotax. Sin. 25: 135 (1987).
China (Guangxi: Longzhou). 36 CHS. Nanophan.

Sauropus varieri Sivar. & Balach., J. Econ. Taxon. Bot. 5: 918 (1984).
SW. India. 40 IND. Nanophan.

Sauropus villosus (Blanco) Merr., Contr. Arnold Arbor. 8: 86 (1934).
S. Vietnam, Thailand, pen. Malaysia, Sumatera, Philippines. 41 THA VIE 42 MLY PHI SUM. Nanophan.
   * *Kirganelia villosa* Blanco, Fl. Filip.: 712 (1837). *Glochidion llanosii* Müll.Arg., Linnaea 32: 68 (1863). *Phyllanthus llanosii* (Müll.Arg.) Müll.Arg., Flora 48: 387 (1865). *Sauropus llanosii* (Müll.Arg.) Gage, Rec. Bot. Surv. India 9: 223 (1922).
   Phyllanthus pubescens Klotzsch, Nova Acta Acad. Caes. Leop-Carol. German. Nat. Cur. 19(Suppl. 1): 420 (1843).

Sauropus yanhuianus P.T.Li, Acta Phytotax. Sin. 25: 134 (1987).
China (Yunnan: Cangyuan). 36 CHC. Nanophan.

Synonyms:
Sauropus albicans Blume === **Sauropus androgynus** (L.) Merr.
Sauropus albicans var. *gardnerianus* (Wight) Müll.Arg. === **Sauropus androgynus** (L.) Merr.
Sauropus albicans var. *genuinus* Müll.Arg. === **Sauropus androgynus** (L.) Merr.
Sauropus albicans var. *intermedius* Müll.Arg. === **Sauropus androgynus** (L.) Merr.
Sauropus albicans var. *zeylanicus* (Wight) Müll.Arg. === **Sauropus androgynus** (L.) Merr.
Sauropus albiflorus subsp. *microcladus* (Müll.Arg.) Airy Shaw === **Phyllanthus microcladus** Müll.Arg.
Sauropus bicolor var. *microphyllus* (Craib) Airy Shaw === **Sauropus bicolor** Craib
Sauropus brunonis var. *ovatus* Airy Shaw === **Sauropus brunonis** (S.Moore) Airy Shaw
Sauropus ceratogynum Baill. === **Sauropus quadrangularis** (Willd.) Müll.Arg.
Sauropus changianus S.Y.Hu === **Sauropus spatulifolius** Beille
Sauropus chorisepalus Merr. & Chun === **Sauropus garrettii** Craib
Sauropus compressus Müll.Arg. === **Sauropus quadrangularis** (Willd.) Müll.Arg.
Sauropus elachophyllus var. *glaber* Airy Shaw === **Sauropus elachophyllus** (F.Muell. ex Benth.) Airy Shaw

Sauropus elachophyllus var. *latior* Airy Shaw === **Sauropus elachophyllus** (F.Muell. ex Benth.) Airy Shaw

Sauropus forcipatus Hook.f. === **Sauropus macranthus** Hassk.

Sauropus gardnerianus Wight === **Sauropus androgynus** (L.) Merr.

Sauropus glaucus var. *glaber* Airy Shaw === **Sauropus stenocladus** (Müll.Arg.) J.T.Hunter & J.J.Bruhl subsp. **stenocladus**

Sauropus grandifolius Pax & K.Hoffm. === **Sauropus macranthus** Hassk.

Sauropus hubbardii Airy Shaw === **Sauropus trachyspermus** (F.Muell.) Airy Shaw

Sauropus indicus Wight === **Sauropus androgynus** (L.) Merr.

Sauropus latzii Airy Shaw === **Sauropus stenocladus** (Müll.Arg.) J.T.Hunter & J.J.Bruhl subsp. **stenocladus**

Sauropus lissocarpus (S.Moore) Airy Shaw === **Sauropus stenocladus** (Müll.Arg.) J.T.Hunter & J.J.Bruhl subsp. **stenocladus**

Sauropus llanosii (Müll.Arg.) Gage === **Sauropus villosus** (Blanco) Merr.

Sauropus longipedicellatus Merr. & Chun === **Sauropus macranthus** Hassk.

Sauropus macranthus Fern.-Vill. === **Sauropus androgynus** (L.) Merr.

Sauropus macrophyllus Hook.f. === **Sauropus macranthus** Hassk.

Sauropus orbicularis var. *minor* Airy Shaw === **Sauropus orbicularis** Craib

Sauropus parviflorus Pax & K.Hoffm. === **Sauropus androgynus** (L.) Merr.

Sauropus parvifolius Ridl. === **Sauropus brevipes** Müll.Arg.

Sauropus pubescens Hook.f. === **Sauropus quadrangularis** (Willd.) Müll.Arg.

Sauropus quadrangularis var. *compressus* (Müll.Arg.) Airy Shaw === **Sauropus quadrangularis** (Willd.) Müll.Arg.

Sauropus quadrangularis var. *puberulus* Kurz === **Sauropus quadrangularis** (Willd.) Müll.Arg.

Sauropus rigidus Craib === **Sauropus bicolor** Craib

Sauropus robinsonii Merr. === **Sauropus macranthus** Hassk.

Sauropus ruber Teijsm. & Binn. ex Pax & K.Hoffm. === **Sauropus macranthus** Hassk.

Sauropus scandens C.B.Rob. === **Sauropus androgynus** (L.) Merr.

Sauropus similis var. *microphylla* Craib === **Sauropus bicolor** Craib

Sauropus spectabilis Miq. === **Sauropus macranthus** Hassk.

Sauropus sumatranus Miq. === **Sauropus androgynus** (L.) Merr.

Sauropus wichurae Müll.Arg. ex Pax & K.Hoffm. === **Sauropus macranthus** Hassk.

Sauropus yunnanensis Pax & K.Hoffm. === **Sauropus garrettii** Craib

Sauropus zeylanicus Wight === **Sauropus androgynus** (L.) Merr.

Savia

14 species, Americas (W. Indies, SE. Brazil) and, in addition (as presently conceived), Africa, Madagascar and the Comoros; includes *Kleinodendron*. Shrubs or small trees with distichously arranged foliage. In the Americas, a strong secondary centre of diversity exists in Cuba. Some species formerly here are now in *Andrachne*, *Petalodiscus*, *Blotia* and *Gonatogyne* but even with excision of these the genus still appears unnatural. Webster (1994, Classification) draws attention to the highly disjunct distribution (see his Fig. 1). No overall revision has appeared since 1922. [When current studies by Petra Hoffmann are complete the genus will be limited to the Americas. Those in Madagascar not already transferred will be in *Petalodiscus*. *S. fadenii* (=*Petalodiscus fadenii*; E. Africa) is projected to be removed to a new genus.] (Phyllanthoideae)

Pax, F. & K. Hoffmann (1922). *Savia*. In A. Engler (ed.), Das Pflanzenreich, IV 147 XV (Euphorbiaceae-Phyllanthoideae-Phyllantheae): 181-188. Berlin. (Heft 81.) La/Ge. — 18 species, Americas, Africa and Madagascar. [The species of sect. *Petalodiscus* are now at generic rank; those of sect. *Maschalanthus* (=*Savia*) outside the Americas are now, or will be, in other genera including *Blotia* and *Petalodiscus* with the southern African *S. ovalis* going to *Andrachne*; that of sect. *Gonatogyne* is also at generic rank.]

Leandri, J. (1958). *Savia*. Fl. Madag. Comores 111 (Euphorbiacées), I: 116-126. Paris. Fr. — Flora treatment. [All species are now, or will be, in *Petalodiscus*.]

Smith, L. B. & R. J. Downs (1964). *Kleinodendron*: novo gênero de Euforbiáceas. Sellowia 16: 175-178, illus. Pt. — Protologue of genus and description of *K. riosulense* (also new) from Santa Catarina.

Savia Willd., Sp. Pl. 4: 771 (1806).
  Mexico, Caribbean, Brazil. 79 81 84. Nanophan. or phan.
  Kleinodendron L.B.Sm. & Downs, Sellowia 16: 177 (1964).

Savia andringitrana Leandri, Bull. Soc. Bot. France 81: 587 (1934).
  C. & E. Madagascar. 29 MDG. Nanophan.
  Savia andringitrana var. *micrantha* Leandri, Notul. Syst. (Paris) 7: 190 (1939).

Savia bahamensis Britton, Torreya 4: 104 (1904).
  Bahamas, E. Cuba, Jamaica. 78 FLA? 81 BAH CUB JAM. Nanophan. or phan.
  Savia apiculata Urb., Repert. Spec. Nov. Regni Veg. 28: 209 (1930).

Savia bojeriana Baill., Adansonia 8: 345 (1868).
  C. & EC. Madagascar. 29 MDG. Nanophan.
  Savia bojeriana var. *perrieri* Leandri, Notul. Syst. (Paris) 7: 189 (1939).

Savia clementis Alain, Contr. Ocas. Mus. Hist. Nat. Colegio "De La Salle" 11: 2 (1952).
  E. Cuba (Sierra de Moa). 81 CUB. Nanophan.

Savia clusiifolia Griseb., Nachr. Königl. Ges. Wiss. Georg-Augusts-Univ. 1: 164 (1865). *Savia clusiifolia* var. *genuina* Müll.Arg. in A.P.de Candolle, Prodr. 15(2): 230 (1866), nom. inval.
  Cuba. 81 CUB. Nanophan.
  Savia clusiifolia var. *fallax* Müll.Arg. in A.P.de Candolle, Prodr. 15(2): 231 (1866).
  Savia clusiifolia var. *intermedia* Müll.Arg. in A.P.de Candolle, Prodr. 15(2): 231 (1866).
  Savia maculata Urb., Symb. Antill. 9: 181 (1924).

Savia cuneifolia Urb., Repert. Spec. Nov. Regni Veg. 28: 210 (1930).
  NE. Cuba. 81 CUB. Nanophan. or phan.

Savia danguyana Leandri, Mém. Inst. Sci. Madagascar, Sér. B, Biol. Vég. 8: 239 (1957).
  C. & E. Madagascar. 29 MDG. Nanophan. or phan.
  Blotia oblongifolia var. *louvelii* Leandri, Mém. Inst. Sci. Madagascar, Sér. B, Biol. Veg. 8: 243 (1957).

Savia dictyocarpa Müll.Arg. in C.F.P.von Martius, Fl. Bras. 11(2): 704 (1874).
  SE. Brazil. 84 BZL. Nanophan.
  Kleinodendron riosulense L.B.Sm. & Downs, Sellowia 16: 177 (1964).

Savia erythroxyloides Griseb., Mem. Amer. Acad. Arts, n.s. 8: 157 (1861).
  Cuba, Hispaniola, Jamaica. 81 CUB DOM HAI JAM. Nanophan. or phan.

  var. **erythroxyloides**
    Cuba, Hispaniola, Jamaica. 81 CUB DOM HAI JAM. Nanophan. or phan.

  var. **parvifolia** Urb., Repert. Spec. Nov. Regni Veg. 23: 209 (1930).
    Cuba (Villa Clara). 81 CUB. Nanophan. or phan.

Savia laurifolia Griseb., Nachr. Königl. Ges. Wiss. Georg-Augusts-Univ. 1: 163 (1865).
  C. & E. Cuba. 81 CUB. Nanophan.
  Savia clusiifolia var. *membranacea* Müll.Arg. in A.P.de Candolle, Prodr. 15(2): 231 (1866).
    Savia membranacea (Müll.Arg.) Urb., Repert. Spec. Nov. Regni Veg. 28: 209 (1930).
  Savia impressa Urb., Repert. Spec. Nov. Regni Veg. 28: 211 (1930).
  Savia longipes Urb., Repert. Spec. Nov. Regni Veg. 28: 211 (1930).

Savia perlucens Britton, Bull. Torrey Bot. Club 43: 464 (1916).
  Cuba (Pinar del Río, I. de la Juventud). 81 CUB. Nanophan.

Savia ranavalonae Leandri, Mém. Inst. Sci. Madagascar, Sér. B, Biol. Vég. 8: 239 (1957).
  C. Madagascar. 29 MDG. Nanophan. or phan.

Savia revoluta Scott-Elliot, J. Linn. Soc., Bot. 29: 48 (1891).
  Madagascar. 29 MDG. Nanophan.

Savia sessiliflora (Sw.) Willd., Sp. Pl. 4: 771 (1806).
  Caribbean. 81 CUB DOM HAI JAM? LEE PUE. Nanophan. or phan.
    * *Croton sessiliflorus* Sw., Prodr.: 100 (1788).
      Phyllanthus laurifolius A.Rich. in R.de la Sagra, Hist. Fis. Cuba, Bot. 2: 216 (1850).
      Phyllanthus pubigerus A.Rich. in R.de la Sagra, Hist. Fis. Cuba, Bot. 2: 216 (1850).

Synonyms:

Savia actephila Hassk. === **Actephila excelsa** var. **javanica** (Miq.) Pax & K.Hoffm.
Savia andringitrana var. *micrantha* Leandri === **Savia andringitrana** Leandri
Savia apiculata Urb. === **Savia bahamensis** Britton
Savia arida Warnock & M.C.Johnst. === **Andrachne arida** (Warnock & M.C.Johnst.)
  G.L.Webster
Savia bemarensis Leandri === **Blotia bemarensis** (Leandri) Leandri
Savia bojeriana var. *perrieri* Leandri === **Savia bojeriana** Baill.
Savia brasiliensis (Baill.) Pax & K.Hoffm. === **Gonatogyne brasiliensis** (Baill.) Müll.Arg.
Savia clusiifolia var. *fallax* Müll.Arg. === **Savia clusiifolia** Griseb.
Savia clusiifolia var. *genuina* Müll.Arg. === **Savia clusiifolia** Griseb.
Savia clusiifolia var. *intermedia* Müll.Arg. === **Savia clusiifolia** Griseb.
Savia clusiifolia var. *membranacea* Müll.Arg. === **Savia laurifolia** Griseb.
Savia decaryi Leandri === **Blotia oblongifolia** (Baill.) Leandri
Savia elegans (Baill.) Müll.Arg. === **Wielandia elegans** Baill.
Savia fadenii Radcl.-Sm. === **Petalodiscus fadenii** (Radcl.-Sm.) Radcl.-Sm.
Savia hildebrandtii Baill. === **Blotia mimosoides** (Baill.) Petra Hoffm. & McPherson
Savia impressa Urb. === **Savia laurifolia** Griseb.
Savia laureola Baill. === **Petalodiscus laureola** (Baill.) Pax
Savia longipes Urb. === **Savia laurifolia** Griseb.
Savia maculata Urb. === **Savia clusiifolia** Griseb.
Savia maroando Danguy === **Blotia mimosoides** (Baill.) Petra Hoffm. & McPherson
Savia membranacea (Müll.Arg.) Urb. === **Savia laurifolia** Griseb.
Savia mimosoides Baill. === **Blotia mimosoides** (Baill.) Petra Hoffm. & McPherson
Savia oblongifolia (Baill.) Baill. === **Blotia oblongifolia** (Baill.) Leandri
Savia ovalis (E.Mey. ex Sond.) Pax & K.Hoffm. === **Andrachne ovalis** (E.Mey. ex Sond.)
  Müll.Arg.
Savia phyllanthoides (Nutt.) Pax & K.Hoffm. === **Andrachne phyllanthoides** (Nutt.)
  Müll.Arg.
Savia phyllanthoides var. *reverchonii* (Coult.) Pax & K.Hoffm. === **Andrachne phyllanthoides**
  (Nutt.) Müll.Arg.
Savia phyllanthoides var. *roemeriana* (Scheele) Pax & K.Hoffm. === **Andrachne**
  phyllanthoides (Nutt.) Müll.Arg.
Savia platyrachis Baill. === **Petalodiscus platyrachis** (Baill.) Pax
Savia platyrachis var. *microphylla* Leandri === **Petalodiscus platyrachis** (Baill.) Pax
Savia pulchella Baill. === **Petalodiscus pulchella** (Baill.) Pax
Savia volubilis Raf. === **Amphicarpaea monoica** (L.) Fern. (Fabaceae)
Savia zeylanica Baill. === **Actephila excelsa** (Dalzell) Müll.Arg. var. **excelsa**

Scagea

2 species, New Caledonia; shrubs or small trees to 10 m with alternate leaves (in *S. oligostemon* relatively narrow) and axillary inflorescences in the upper parts of flushes. Webster (Synopsis, 1994) referred the genus to subtribe Pseudanthinae (tribe Caletieae) of which most other genera are in Australia including *Micrantheum* and *Pseudanthus*. (Oldfieldioideae)

McPherson, G. & C. Tirel (1987). *Scagea*. Fl. Nouvelle-Calédonie, 14 (Euphorbiacées, I): 90-95. Paris. Fr. — Flora treatment (2 species, mostly in peridotite areas); key. [*S. depauperata* previously treated in *Longetia* and *Austrobuxus*.]

Scagea McPherson, Bull. Mus. Natl. Hist. Nat., B, Adansonia 7: 247 (1985 publ. 1986). New Caledonia. 60. Nanophan.

Scagea depauperata (Baill.) McPherson, Bull. Mus. Natl. Hist. Nat., B, Adansonia 7: 248 (1985 publ. 1986).
SE. New Caledonia. 60 NWC. Nanophan.
* *Longetia depauperata* Baill., Adansonia 11: 100 (1874). *Austrobuxus depauperatus* (Baill.) Airy Shaw, Kew Bull. 25: 508 (1971).

Scagea oligostemon (Guillaumin) McPherson, Bull. Mus. Natl. Hist. Nat., B, Adansonia 7: 248 (1985 publ. 1986).
New Caledonia. 60 NWC. Nanophan.
* *Baloghia oligostemon* Guillaumin, Bull. Mus. Natl. Hist. Nat., II, 21: 263 (1949).
    Austrobuxus oligostemon (Guillaumin) Airy Shaw, Kew Bull. 33: 531 (1979).
    Longetia gynotricha Guillaumin, Bull. Mus. Natl. Hist. Nat., II, 28: 130 (1956).
    Austrobuxus gynotrichus (Guillaumin) Airy Shaw, Kew Bull. 25: 508 (1971).

Scepa

Synonyms:
Scepa Lindl. === **Aporusa** Blume
Scepa aurita Tul. === **Aporusa octandra** var. **malesiana** Schot
Scepa chinensis Champ. ex Benth. === **Aporusa octandra** var. **chinensis** (Champ. ex Benth.) Schot
Scepa lindleyana Wight === **Aporusa cardiosperma** (Gaertn.) Merr.
Scepa microstachya Tul. === **Aporusa microstachya** (Tul.) Müll.Arg.
Scepa stipulacea Lindl. === **Aporusa octandra** (Buch.-Ham. ex D.Don) Vickery var. **octandra**
Scepa villosa Lindl. === **Aporusa villosa** (Lindl.) Baill.

Scepasma

Synonyms:
Scepasma Blume === **Phyllanthus** L.
Scepasma buxifolia Blume === **Phyllanthus buxifolius** (Blume) Müll.Arg.
Scepasma longifolium Hassk. === **Phyllanthus acutissimus** Miq.
Scepasma parvifolia Rchb.f. & Zoll. ex Teijsm. & Binn. === ?

Schinziophyton

1 species, tropical Africa (Namibia and Mozambique to Tanzania, Zaire and Angola, mostly in a relatively limited band running through the Caprivi Strip in Namibia). Deciduous, sometimes multi-trunked trees to 25 m with spreading crowns similar to *Ricinodendron* but stipules deciduous and ovary mostly 1-locular; in addition, it is more usually found in open forest or woodland, usually in river basins or other low-lying land. Where trees occur, they

may become dominant. A habit photograph appears on *Flore du Congo Belge* 8(1)(1962). *S. rautanenii* was in the past conventionally included in *Ricinodendron*. The trees furnish a useful balsa-like light timber, the fruit pulp is made into porridge and the kernels yield an edible oil. (Crotonoideae)

Pax, F. (with K. Hoffmann) (1911). *Ricinodendron*. In A. Engler (ed.), Das Pflanzenreich, IV 147 III (Euphorbiaceae-Cluytieae): 45-49. Berlin. (Heft 47.) La/Ge. — 3 species in 2 subgenera; *R. rautanenii*, pp. 48-49. [Now in *Schinziophyton*.]

Peters, C. R. (1987). *Ricinodendron rautanenii*. Econ. Bot. 41: 494-502, illus., map. En. — Botanical description; discussion of distribution, preferred habitats (often associated with modern or fossil watercourses and sometimes dominant, usually on sandy soils), and yield; recommendations; literature. [The pulp of the fleshy fruits is nutrient-rich and the kernels are also edible. These are a major food resource for the nomadic !Kung San.]

Radcliffe-Smith, A. (1990). Notes on African Euphorbiaceae, XXII. The genus *Schinziophyton*. Kew Bull. 45: 157-160. En. — Description of genus, originally proposed by Hutchinson for *Ricinodendron rautanenii* of central Africa (including also *R. viticoides*); transfer.

Schinziophyton Hutch. ex Radcl.Sm., Kew Bull. 45: 157 (1990).
  Zaire and Tanzania to S. Africa. 23 25 26 27.

Schinziophyton rautanenii (Schinz) Radcl.-Sm., Kew Bull. 45: 157 (1990).
  Zaire, Tanzania, Mozambique, Malawi, Zambia, Zimbabwe, Botswana, Angola, Caprivi
    Strip, Namibia. 23 ZAI 25 TAN 26 ANG MLW MOZ ZAM ZIM 27 BOT CPV NAM. Phan.
  * *Ricinodendron rautanenii* Schinz, Bull. Herb. Boissier 6: 744 (1898).
    Ricinodendron viticoides Mildbr., Notizbl. Bot. Gart. Berlin-Dahlem 12: 517 (1935).

Schismatopera

Synonyms:
Schismatopera Klotzsch === **Pera** Mutis
Schismatopera laurina Benth. === **Pera distichophylla** (Mart.) Baill.

Schistostigma

Synonyms:
Schistostigma Lauterb. === **Cleistanthus** Hook.f. ex Planch.
Schistostigma papuanum Lauterb. === **Cleistanthus papuanus** (Lauterb.) Jabl.

Schizogyne

An Ehrenberg name introduced as a synonym of *Acalypha* by Pax (1924; see there); a later homonym of *Schizogyne* Cass. (Compositae)

Schorigeram

Synonyms:
Schorigeram Adans. === **Tragia** Plum. ex L.

Schousboea

Synonyms:
Schousboea Willd. === **Croton** L.

Schousboea

A later homonym of *Schousboea* Willd.

Synonyms:
Schousboea Schumach. & Thonn. === **Alchornea** Sw.
Schousboea cordifolia Schumach. & Thonn. === **Alchornea cordifolia** (Schumach. & Thonn.)
   Müll.Arg.

Schradera

Synonyms:
Schradera Willd. === **Croton** L.

Sclerocroton

6 species, tropical and subtropical Africa and Madagascar; segregated from *Sapium* s.l. by
Kruijt (1996). Subshrubs (*S. oblongifolius*) or shrubs or small trees to 12 m in secondary
forest, open woodland or bushland. *S. oblongifolius* is in particular fire-adapted. The genus
was by Kruijt (l.c.) thought to be most closely related to his *Shirakia* (= *Neoshirakia* and
Shirakiopsis). (Euphorbioideae (except Euphorbieae))

   Pax, F. (with K. Hoffmann) (1912). *Sapium*. In A. Engler (ed.), Das Pflanzenreich, IV 147 V
      (Euphorbiaceae-Hippomaneae): 199-258. Berlin. (Heft 52.) La/Ge. — Nos. 78-83 and 85
      (pp. 243-249), comprising subgen. *Sclerocroton* sect. *Armata*, pertain to *Sclerocroton* as
      now accepted. [These now comprise 4 species.]
 • Kruijt, R. C. (1996). A taxonomic monograph of *Sapium* Jacq., *Anomostachys* (Baill.) Hurus.,
      Duvigneaudia J. Léonard and *Sclerocroton* Hochst. (Euphorbiaceae tribe Hippomaneae).
      109 pp., illus. (Biblioth. Bot. 146). Stuttgart. En. —*Sclerocroton* (by Kruijt with G. J.
      Roebers), pp. 16-27; treatment of 5 species with key, descriptions, synonymy,
      distribution and habitat, localities with exsiccatae, and notes and illustrations

Sclerocroton Hochst., Flora 28: 85 (1845).
   Trop. & S. Africa, Madagascar. 22 23 25 26 27 29.

Sclerocroton carterianus (J.Léonard) Kruijt & Roebers, Biblioth. Bot. 146: 18 (1996).
   Liberia, Sierra Leone. 22 LBR SIE. Nanophan.
      * Sapium carterianum J.Léonard, Bull. Jard. Bot. État 29: 146 (1959).

Sclerocroton cornutus (Pax) Kruijt & Roebers, Biblioth. Bot. 146: 20 (1996).
   W. & C. Trop. Africa. 22 IVO 23 CMN CON EQG GAB ZAI 26 ANG ZAM ZIM. Nanophan.
      or phan.
      * Sapium cornutum Pax, Bot. Jahrb. Syst. 19: 114 (1894). *Sapium cornutum* var. *genuinum*
         Pax in H.G.A.Engler, Pflanzenr., IV, 147, V: 246 (1912), nom. inval.
      Sapium cornutum var. *coriaceum* Pax, Bot. Jahrb. Syst. 19: 115 (1894).
      Sapium poggei Pax, Bot. Jahrb. Syst. 19: 115 (1894). *Sapium cornutum* var. *poggei* (Pax)
         Pax in H.G.A.Engler, Pflanzenr., IV, 147, V: 246 (1912).
      Sapium xylocarpum Pax, Bot. Jahrb. Syst. 19: 115 (1894). *Sapium xylocarpum* var.
         genuinum Pax in II.G.A.Engler, Pflanzenr., IV, 147, V: 246 (1912), nom. inval.
      Sapium oblongifolium De Wild., Ann. Mus. Congo Belge, Bot., V, 1: 279 (1906), nom. illeg.
      Sapium cornutum var. *lineolatum* Pax in H.G.A.Engler, Pflanzenr., IV, 147, V:
         247 (1912).
      Sapium xylocarpum var. *lineolatum* Pax in H.G.A.Engler, Pflanzenr., IV, 147, V:
         247 (1912).

Sclerocroton integerrimus Hochst., Flora 28: 85 (1845). *Stillingia integerrima* (Hochst.)
Baill., Adansonia 3: 162 (1863). *Excoecaria integerrima* (Hochst.) Müll.Arg. in A.P.de
Candolle, Prodr. 15(2): 948 (1866). *Sapium integerrimum* (Hochst.) J.Léonard, Bull. Jard.
Bot. État 29: 142 (1959).
Trop. & S. Africa. 22 GUI 23 ZAI 25 TAN 26 MOZ ZIM 27 NAT SWZ TVL. Nanophan.
or phan.
Sclerocroton reticulatus Hochst., Flora 28: 85 (1845). *Excoecaria reticulata* (Hochst.)
Müll.Arg. in A.P.de Candolle, Prodr. 15(2): 1213 (1866). *Sapium reticulatum* (Hochst.)
Pax in H.G.A.Engler, Pflanzenr., IV, 147, V: 245 (1912).
Tragia integerrima Hochst., Flora 28: 85 (1845), pro syn.
Tragia natalensis Hochst., Flora 28: 85 (1845), pro syn.
Excoecaria hochstetteriana Müll.Arg., Linnaea 32: 122 (1863), nom. illeg.
Sapium armatum Pax & K.Hoffm. in H.G.A.Engler, Pflanzenr., IV, 147, V: 244 (1912).

Sclerocroton melanostictus (Baill.) Kruijt & Roebers, Biblioth. Bot. 146: 23 (1996).
Madagascar. 29 MDG. Nanophan. or phan.
Croton melanostictus Boivin ex Baill., Adansonia 1: 285 (1861).
* *Stillingia melanosticta* Baill., Adansonia 1: 285 (1861). *Excoecaria melanosticta* (Baill.)
Müll.Arg., Linnaea 32: 122 (1863). *Sapium melanostictum* (Baill.) Pax & K.Hoffm. in
H.G.A.Engler, Pflanzenr., IV, 147, V: 248 (1912).
Sapium hildebrandtii Pax, Bot. Jahrb. Syst. 19: 116 (1894).

Sclerocroton oblongifolius (Müll.Arg.) Kruijt & Roebers, Biblioth. Bot. 146: 24 (1996).
C. Trop. Africa. 23 ZAI 26 ANG ZAM ZIM. Cham. or nanophan.
* *Excoecaria oblongifolia* Müll.Arg., J. Bot. 2: 337 (1864). *Sapium oblongifolium* (Müll.Arg.)
Pax, Bot. Jahrb. Syst. 19: 114 (1894).
Sapium suffruticosum Pax in O.Warburg (ed.), Kunene-Sambesi Exped.: 284 (1903).

Sclerocroton schmitzii (J.Léonard) Kruijt & Roebers, Biblioth. Bot. 146: 26 (1996).
C. Trop. Africa. 23 BUR RWA ZAI 26 ZAM ZIM. Nanophan. or phan.
* *Sapium schmitzii* J.Léonard, Bull. Jard. Bot. État 29: 142 (1958).

Synonyms:
Sclerocroton ellipticus Hochst. === **Shirakiopsis elliptica** (Hochst.) Esser (ined.)
Sclerocroton reticulatus Hochst. === **Sclerocroton integerrimus** Hochst.

Sclerocyathium

A segregate of *Euphorbia*.

Synonyms:
Sclerocyathium Prokh. === **Euphorbia** L.

Scortechinia

Replaced by *Neoscortechinia* for nomenclatural reasons.

Synonyms:
Scortechinia Hook.f. === **Neoscortechinia** Pax
Scortechinia forbesii Hook.f. === **Neoscortechinia forbesii** (Hook.f.) S.Moore
Scortechinia kingii Hook.f. === **Neoscortechinia kingii** (Hook.f.) Pax & K.Hoffm.
Scortechinia nicobarica Hook.f. === **Neoscortechinia nicobarica** (Hook.f.) Pax & K.Hoffm.

Sebastiania

78 species, warmer parts of Americas (notably in Brazil with another, smaller centre in Mexico); includes *Dendrocousinsia*. Shrubs or small trees with pinnately veined, usually serrate leaves tending to cluster near the ends of branches. Malesian species were by some thought not properly to be in the genus (van Steenis 1948); indeed, one of them, *S. chamaelea* (in sect. 2, *Elachocroton*), is now referred to *Microstachys* and others have gone to *Gymnanthes* (Esser, 1999). *Microstachys* was, along with *Ditrysina*, removed by Esser (1994; see **Euphorbioideae (except Euphorbieae**)); the two correspond to sects. 1, 2, 3 and 5 in Pax & Hoffmann (1931). The genus is evidently closely related to *Gymnanthes*, differing mainly in the presence of terminal or leaf-opposed spikes. *Excoecaria* is related but features mainly ecarunculate seeds; recently, however, Esser et al. (1997; see **Euphorbioideae (except Euphorbieae**)) have proposed union of these two genera. The position of the three species formerly in *Dendrocousinsia* (sect. 4 in Pax & Hoffmann 1931), all more or less small-leaved Jamaican shrubs or trees to 5 m, requires further assessment. Esser (1994, l.c.) has tentatively proposed a return to generic rank; for him '*Dendrocousinsia* ist sicher nicht zu *Sebastiania* zu stellen'. This has here been followed. As the only representatives away from the mainland, they are if nothing else geographically anomalous. Two Mexican species, *SS. palmeri* and *pringlei* (possibly only forms of *S. pavoniana*), are the source of 'semillas brincadores' or 'Mexican jumping beans', a phenomenon caused by the activites of a microlepidopteran larva; further discussion and references appear in Webster (1967: 386-387; see **Americas**) as well as in Pax (1912). (Euphorbioideae (except Euphorbieae))

Pax, F. (with K. Hoffmann) (1912). *Sebastiania*. In A. Engler (ed.), Das Pflanzenreich, IV 147 V (Euphorbiaceae-Hippomaneae): 88-153. Berlin. (Heft 52.) La/Ge. — 79 species (of which 4 doubtful) in 7 sections; diagram of sectional and generic relationships. Sect. *Dendrocousinsia* is covered in Pax (1914; see below). The biology of 'jumping beans' is reviewed, with literature references, under *S. palmerii* (no. 66, pp. 147-148). [Some sections are now in, or likely to be referable to, other genera.]

Millspaugh, C. F. (1913). The genera *Pedilanthus* and *Cubanthus*, and other American Euphorbiaceae. Publ. Field Mus. Nat. Hist., Bot. Ser., 2: 353-377. En. — Includes (pp. 374-375) protologue of *Dendrocousinsia*; 2 species, both also new, were described and a key supplied. [*Dendrocousinsia* has in recent years been included in *Sebastiania*.]

Pax, F. (with K. Hoffmann) (1914). *Sebastiania*. In A. Engler (ed.), Das Pflanzenreich, IV 147 VII [Euphorbiaceae-Additamentum V]: 422. Berlin. (Heft 63.) La/Ge. — The 2 known species of *Dendrocousinsia* (Millspaugh 1913) here reduced to sectional rank.

Steenis, C. G. G. J. van (1948). Provisional note on the genus *Sebastiania* in Malaysia. Bull. Bot. Gard. Buitenzorg, III, 17: 409-410. (Miscellaneous botanical notes, I, 14.) En. — Synopsis of 4 species (including the coastal *S. chamaelea*) with key and descriptions of 2 novelties, *SS. lancifolia* and *remota*. [*S. chamaelea* is now referred to *Microstachys*, the remainder appear to be related to *Duvigneaudia* and, in turn, *Gymnanthes*.]

Airy-Shaw, H. K. (1960). Notes on Malaysian Euphorbiaceae, XV. Additional records for *Sebastiania borneënsis* Pax. Kew Bull. 14: 396-397. En. — Further localities, with exsiccatae; fruit described. [The species is now removed from *Sebastiania* and likely to be transferred to *Gymnanthes*.]

Oliveira, A. Souza de (1988). Taxinomia das espécies do gênero *Sebastiania* secção *Elachocroton* (Baill.) Pax (Imphorbiaceae)[sic!] ocorrentes no Brasil. Arq. Jard. Bot. Rio de Janeiro 27: 3-65, illus., map. Pt. — Detailed study of two species (now referable to *Microstachys* but formal transfers have yet to be made); for more details see that genus. The revision is preceded by a character review and a history of research in *Sebastiania*.

• McVaugh, R. (1995). Euphorbiacearum sertum Novo-Galicianarum revisarum. Contr. Univ. Michigan Herb. 20: 173-215, illus. En. — Revision precursory to treatment in *Flora Novo-Galiciana*; includes (pp. 204-210) a treatment of *Sebastiania*.

Esser, H.-J. (1999). A partial revision of the Hippomaneae (Euphorbiaceae) in Malesia. Blumea 44: 149-215, illus., maps. En. — *Sebastiania*, pp. 206; brief note regarding exclusion of the genus from Malesia (the two species concerned having been transferred to *Gymnanthes*).

Sebastiania Spreng., Neue Entd. 2: 118 (1821).
  Mexico, S. America. 79 80 81 82 83 84 85.
    Adenogyne Klotzsch, Arch. Naturgesch. 7: 183 (1841).
    Clonostachys Klotzsch, Arch. Naturgesch. 7: 185 (1841).
    Dendrocousinsia Millsp., Publ. Field Mus. Nat. Hist., Bot. Ser. 2: 374 (1913).

Sebastiania alpina (Fawc. & Rendle) Pax & K.Hoffm. in H.G.A.Engler, Nat. Pflanzenfam. ed.
  2, 19c: 193 (1931).
  Jamaica. 81 JAM.
    * *Dendrocousinsia alpina* Fawc. & Rendle, J. Bot. 57: 313 (1919).

Sebastiania ampla (I.M.Johnst.) Jabl., Phytologia 16: 423 (1968).
  Mexico (Baja California). 79 MXN.
    * *Sapium biloculare* var. *amplum* I.M.Johnst., Proc. Calif. Acad. Sci., IV, 12: 1077 (1924).

Sebastiania appendiculata (Müll.Arg.) Jabl., Phytologia 16: 423 (1968).
  Mexico (Sonaloa, Oaxaca). 79 MXN MXS. Phan. – Provisionally accepted.
    * *Stillingia appendiculata* Müll.Arg., Linnaea 32: 87 (1863). *Sapium appendiculatum*
      (Müll.Arg.) Pax & K.Hoffm. in H.G.A.Engler, Pflanzenr., IV, 147, V: 214 (1912).

Sebastiania argutidens Pax & K.Hoffm. in H.G.A.Engler, Pflanzenr., IV, 147, V:
  129 (1912).
  Brazil (Santa Catarina). 84 BZS. Nanophan. or phan.

Sebastiania bahiensis (Müll.Arg.) Müll.Arg. in A.P.de Candolle, Prodr. 15(2): 1183 (1866).
  Brazil (Bahia). 84 BZE. Nanophan.
    * *Gymnanthes bahiensis* Müll.Arg., Linnaea 32: 102 (1863). *Stillingia bahiensis* (Müll.Arg.)
      Baill., Adansonia 5: 329 (1865).

Sebastiania bicalcarata (Müll.Arg.) Pax in H.G.A.Engler, Pflanzenr., IV, 147, V:
  144 (1912).
  Brazil (Goiás). 84 BZC. Nanophan.
    * *Excoecaria bicalcarata* Müll.Arg., Linnaea 32: 125 (1863).

Sebastiania bilocularis S.Watson, Proc. Amer. Acad. Arts 20: 374 (1885). *Sapium biloculare*
  (S.Watson) Pax in H.G.A.Engler, Pflanzenr., IV, 147, V: 221 (1912).
  Mexico (NW. Sonora), Arizona. 76 ARI 79 MXN. Nanophan. or phan. – Provisionally
    accepted

Sebastiania boliviana Rusby, Descr. S. Amer. Pl.: 50 (1920).
  Bolivia. 83 BOL.

Sebastiania brasiliensis Spreng., Neue Entd. 2: 118 (1821). *Gymnanthes brasiliensis* (Spreng.)
  Müll.Arg., Linnaea 32: 104 (1863). *Stillingia brasiliensis* (Spreng.) Müll.Arg. in A.P.de
  Candolle, Prodr. 15(2): 1191 (1866). *Sebastiania brasiliensis* var. *genuina* Müll.Arg. in A.P.de
  Candolle, Prodr. 15(2): 1187 (1866), nom. inval. *Actinostemon brasiliensis* (Spreng.) Pax in
  H.G.A.Engler, Pflanzenr., IV, 147, V: 80 (1912).
  Brazil, Bolivia, Paraguay, Uruguay, N. Argentina. 83 BOL 84 BZC BZE BZL BZS 85 AGE PAR
    URU. Nanophan. or phan.
    Excoecaria brasiliensis Spreng., Neue Entd. 2: 117 (1821).
    Microstachys ramosissima A.St.-Hil., Hist. Pl. Remarq. Brésil: 242 (1825). *Stillingia*
      ramosissima (A.St.-Hil.) Baill., Adansonia 5: 328 (1865). *Sebastiania brasiliensis* var.
      ramosissima (A.St.-Hil.) Müll.Arg. in A.P.de Candolle, Prodr. 15(2): 1187 (1866).
    Sebastiania desertorum Klotzsch, Arch. Naturgesch. 7: 183 (1841), nom. nud.
    Sebastiania divaricata Klotzsch, Arch. Naturgesch. 7: 183 (1841), nom. nud.
    Sebastiania foveata Klotzsch, Arch. Naturgesch. 7: 183 (1841), nom. nud.

Sebastiania macrophylla Klotzsch, Arch. Naturgesch. 7: 183 (1841), nom. nud.

Sebastiania reticulata Klotzsch, Arch. Naturgesch. 7: 183 (1841), nom. nud.

Stillingia desertorum Klotzsch, Arch. Naturgesch. 7: 183 (1841), pro syn.

Gymnanthes brasiliensis f. *microphylla* Müll.Arg., Linnaea 32: 104 (1863). *Sebastiania brasiliensis* f. *microphylla* (Müll.Arg.) Müll.Arg. in A.P.de Candolle, Prodr. 15(2): 1187 (1866). *Sebastiania brasiliensis* var. *microphylla* (Müll.Arg.) Müll.Arg. in C.F.P.von Martius, Fl. Bras. 11(2): 585 (1874).

Gymnanthes brasiliensis f. *rufescens* Müll.Arg., Linnaea 32: 104 (1863). *Sebastiania brasiliensis* f. *rufescens* (Müll.Arg.) Müll.Arg. in A.P.de Candolle, Prodr. 15(2): 1187 (1866). *Sebastiania brasiliensis* var. *rufescens* (Müll.Arg.) Müll.Arg. in C.F.P.von Martius, Fl. Bras. 11(2): 585 (1874).

Gymnanthes brasiliensis var. *divaricata* Müll.Arg., Linnaea 32: 104 (1863). *Sebastiania brasiliensis* var. *divaricata* (Müll.Arg.) Müll.Arg. in A.P.de Candolle, Prodr. 15(2): 1187 (1866).

Gymnanthes brasiliensis var. *erythroxyloides* Müll.Arg., Linnaea 32: 104 (1863). *Sebastiania brasiliensis* var. *erythoxyloides* (Müll.Arg.) Müll.Arg. in A.P.de Candolle, Prodr. 15(2): 1188 (1866).

Gymnanthes brasiliensis var. *obovata* Müll.Arg., Linnaea 32: 104 (1863). *Sebastiania brasiliensis* var. *obovata* (Müll.Arg.) Müll.Arg. in A.P.de Candolle, Prodr. 15(2): 1187 (1866).

Gymnanthes brasiliensis var. *robusta* Müll.Arg., Linnaea 32: 104 (1863). *Sebastiania brasiliensis* var. *robusta* (Müll.Arg.) Müll.Arg. in A.P.de Candolle, Prodr. 15(2): 1186 (1866).

Sebastiania brasiliensis var. *anisophylla* Müll.Arg. in C.F.P.von Martius, Fl. Bras. 11(2): 141 (1873).

Sebastiania brasiliensis f. *viridis* Müll.Arg. in C.F.P.von Martius, Fl. Bras. 11(2): 586 (1874).

Sebastiania brasiliensis var. *eremophila* Müll.Arg. in C.F.P.von Martius, Fl. Bras. 11(2): 587 (1874).

Sebastiania brasiliensis var. *polymorpha* Müll.Arg. in C.F.P.von Martius, Fl. Bras. 11(2): 586 (1874).

Sebastiania brasiliensis var. *rigida* Müll.Arg. in C.F.P.von Martius, Fl. Bras. 11(2): 585 (1874).

Sapium rhombifolium Rusby, Bull. Torrey Bot. Club 28: 307 (1901). *Sebastiania rhombifolia* (Rusby) Jabl., Phytologia 14: 452 (1967), nom. illeg.

Sebastiania brasiliensis var. *brachystachys* Pax & K.Hoffm. in H.G.A.Engler, Pflanzenr., IV, 147, V: 142 (1912).

Sebastiania brasiliensis var. *brevispicata* Pax & K.Hoffm. in H.G.A.Engler, Pflanzenr., IV, 147, V: 140 (1912).

Sebastiania fiebrigii Pax in H.G.A.Engler, Pflanzenr., IV, 147, V: 142 (1912), nom. provis.

Sebastiania robusta Klotzsch ex Pax in H.G.A.Engler, Pflanzenr., IV, 147, V: 141 (1912), nom. nud.

Sebastiania anisandra Lillo, Seg. Contr. Arb. Argent.: 16 (1917).

Sebastiania brevifolia (Müll.Arg.) Müll.Arg. in A.P.de Candolle, Prodr. 15(2): 1186 (1866). Brazil (Bahia). 84 BZE.

* *Gymnanthes brevifolia* Müll.Arg., Linnaea 32: 104 (1863). *Stillingia brevifolia* (Müll.Arg.) Baill., Adansonia 5: 328 (1865).

Sebastiania bridgesii (Müll.Arg.) Pax in H.G.A.Engler, Pflanzenr., IV, 147, V: 143 (1912). Bolivia. 83 BOL. Nanophan.

* *Excoecaria bridgesii* Müll.Arg., Linnaea 32: 124 (1863).

Sebastiania catingae Ule, Bot. Jahrb. Syst. 42: 222 (1908). Colombia, Brazil (Bahia). 83 CLM 84 BZE. Nanophan. or phan.

Sebastiania chaetodonta Müll.Arg. in C.F.P.von Martius, Fl. Bras. 11(2): 577 (1874). Brazil (Minas Gerais). 84 BZL. Nanophan.

Sebastiania chahalana Lundell, Wrightia 4: 90 (1968).
  Guatemala. 80 GUA.

Sebastiania chiapensis Lundell, Wrightia 4: 35 (1968).
  Mexico (Chiapas). 79 MXT.

Sebastiania confusa Lundell, Lloydia 2: 99 (1939).
  Belize, Guatemala. 80 BLZ GUA.

Sebastiania cornuta McVaugh, Contrib. Univ. Mich. Herb. 20: 205 (1995).
  N. Mexico (to Nayarit). 79 MXE MXN MXS. Nanophan. or phan.

Sebastiania cruenta (Standl. & Steyerm.) Miranda, Anales Inst. Biol. Univ. Nac. México 24: 64 (1953).
  Guatemala. 80 GUA.
    * *Stillingia cruenta* Standl. & Steyerm., Publ. Field Mus. Nat. Hist., Bot. Ser. 23: 125 (1944).

Sebastiania daphniphylla (Baill.) Müll.Arg. in A.P.de Candolle, Prodr. 15(2): 1180 (1866).
  Brazil (Rio de Janeiro). 84 BZL. Nanophan.
    * *Stillingia daphniphylla* Baill., Adansonia 5: 326 (1865).

Sebastiania dimorphocalyx Müll.Arg. in C.F.P.von Martius, Fl. Bras. 11(2): 574 (1874).
  Brazil (Minas Gerais). 84 BZL. Nanophan.

Sebastiania echinocarpa Müll.Arg. in C.F.P.von Martius, Fl. Bras. 11(2): 584 (1874).
  Brazil (Bahia). 84 BZE.

Sebastiania edwalliana Pax & K.Hoffm. in H.G.A.Engler, Pflanzenr., IV, 147, V: 134 (1912).
  Brazil (São Paulo), Paraguay. 84 BZL 85 PAR. Nanophan. or phan.
    Sebastiania edwalliana var. *acuminata* Pax & K.Hoffm. in H.G.A.Engler, Pflanzenr., IV, 147, V: 134 (1912).
    Sebastiania edwalliana var. *minor* Pax & K.Hoffm. in H.G.A.Engler, Pflanzenr., IV, 147, V: 134 (1912).
    Sebastiania edwalliana var. *vestita* Pax & K.Hoffm. in H.G.A.Engler, Pflanzenr., IV, 147, V: 134 (1912).

Sebastiania eglandulata (Vell.) Pax in H.G.A.Engler, Pflanzenr., IV, 147, V: 151 (1912).
  Brazil (Rio de Janeiro). 84 BZL. Nanophan.
    * *Omphalea eglandulata* Vell., Fl. Flumin. 10: 12 (1831). *Excoecaria eglandulata* (Vell.) Müll.Arg. in C.F.P.von Martius, Fl. Bras. 11(2): 628 (1874).

Sebastiania fasciculata (Millsp.) Pax & K.Hoffm. in H.G.A.Engler, Pflanzenr., IV, 147, VII: 422 (1914).
  Jamaica. 81 JAM.
    * *Dendrocousinsia fasciculata* Millsp., Publ. Field Mus. Nat. Hist., Bot. Ser. 2: 375 (1913).

Sebastiania glabrescens Pax & K.Hoffm. in H.G.A.Engler, Pflanzenr., IV, 147, XVII: 200 (1924).
  Brazil (Minas Gerais). 84 BZL.

Sebastiania gracilis Pax & K.Hoffm. in H.G.A.Engler, Pflanzenr., IV, 147, XVII: 201 (1924).
  Brazil (Rio de Janeiro). 84 BZL.

Sebastiania granatensis (Müll.Arg.) Müll.Arg. in A.P.de Candolle, Prodr. 15(2): 1189 (1866).
  Colombia. 83 CLM.
    * *Gymnanthes granatensis* Müll.Arg., Linnaea 32: 107 (1863).
    Stillingia arborescens Pittier, Arb. Arbust. Venez.: 9 (1921).

Sebastiania haploclada Briq., Annuaire Conserv. Jard. Bot. Genève 4: 231 (1900).
　Peru. 83 PER. Nanophan. or phan.

Sebastiania heteroica Müll.Arg. in C.F.P.von Martius, Fl. Bras. 11(2): 708 (1874).
　Brazil (Rio de Janeiro). 84 BZL. Nanophan.

Sebastiania hintonii Lundell, Wrightia 2: 105 (1960).
　Mexico (Guerrero). 79 MXS. Phan.

Sebastiania hippophaifolia (Griseb.) Pax in H.G.A.Engler, Pflanzenr., IV, 147, V: 152 (1912).
　NE. Argentina. 85 AGE.
　　* *Excoecaria hippophaifolia* Griseb., Abh. Königl. Ges. Wiss. Göttingen 24: 61 (1879).

Sebastiania huallagensis Croizat, J. Arnold Arbor. 24: 177 (1943).
　Peru. 83 PER.

Sebastiania jacobinensis (Müll.Arg.) Müll.Arg. in A.P.de Candolle, Prodr. 15(2): 1188 (1866).
　Brazil (Bahia). 84 BZE. Nanophan.
　　* *Gymnanthes jacobinensis* Müll.Arg., Linnaea 32: 106 (1863). *Stillingia jacobinensis*
　　　(Müll.Arg.) Baill., Adansonia 5: 329 (1865).

Sebastiania jaliscensis McVaugh, Brittonia 13: 200 (1961).
　Mexico (Jalisco). 79 MXS.

Sebastiania klotzschiana (Müll.Arg.) Müll.Arg. in A.P.de Candolle, Prodr. 15(2): 1178 (1866).
　SE. & S. Brazil, Uruguay, Paraguay, N. Argentina. 84 BZL BZS 85 AGE PAR URU.
　　Nanophan. or phan.
　　Excoecaria discolor Spreng., Syst. Veg. 3: 24 (1826).
　　Adenogyne discolor Klotzsch, Arch. Naturgesch. 7: 184 (1841), nom. nud.
　　Adenogyne marginata Klotzsch, Arch. Naturgesch. 7: 184 (1841), nom. nud.
　　Gymnanthes discolor Baill., Étude Euphorb.: 531 (1858), nom. nud.
　　Gymnanthes marginata Baill., Étude Euphorb.: 531 (1858), nom. nud.
　　Gymnanthes brachyclada Müll.Arg., Linnaea 32: 98 (1863). *Sebastiania brachyclada*
　　　(Müll.Arg.) Müll.Arg. in A.P.de Candolle, Prodr. 15(2): 1178 (1866). *Sebastiania*
　　　klotzschiana var. *brachyclada* (Müll.Arg.) Pax & K.Hoffm. in H.G.A.Engler, Pflanzenr.,
　　　IV, 147, V: 129 (1912).
　　* *Gymnanthes klotzschiana* Müll.Arg., Linnaea 32: 98 (1863). *Sebastiania klotzschiana* var.
　　　genuina Müll.Arg. in A.P.de Candolle, Prodr. 15(2): 1178 (1866), nom. inval.
　　Stillingia commersoniana Baill., Adansonia 5: 330 (1865). *Sebastiania commersoniana*
　　　(Baill.) L.B.Sm. & Downs, Fl. Ilustr. Catar. 1(Euforbiac.): 308 (1988).
　　Stillingia cremostachys Baill., Adansonia 5: 322 (1865).
　　Excoecaria marginata Griseb., Abh. Königl. Ges. Wiss. Göttingen 19: 97 (1874).
　　Sebastiania klotzschiana var. *trichoneura* Müll.Arg. in C.F.P.von Martius, Fl. Bras. 11(2):
　　　574 (1874).
　　Excoecaria glauca Parodi, Anales Soc. Ci. Argent. 11: 54 (1881).

Sebastiania laureola (Baill.) Müll.Arg. in A.P.de Candolle, Prodr. 15(2): 1180 (1866).
　Brazil (Rio de Janeiro). 84 BZL. Nanophan.
　　* *Stillingia laureola* Baill., Adansonia 5: 327 (1865).

Sebastiania leptopoda Lundell, Wrightia 5: 164 (1975).
　Guatemala. 80 GUA.

Sebastiania longicuspis Standl., Publ. Field Mus. Nat. Hist., Bot. Ser. 11: 134 (1932).
　Belize, Guatemala. 80 BLZ GUA. – Provisionally accepted.

Sebastiania longispicata Pax & K.Hoffm. in H.G.A.Engler, Pflanzenr., IV, 147, V: 142 (1912).
Paraguay. 85 PAR. Nanophan.

Sebastiania lottiae McVaugh, Contrib. Univ. Mich. Herb. 20: 208 (1995).
C. & SW. Mexico. 79 MXC MXS.

Sebastiania macrocarpa Müll.Arg. in A.P.de Candolle, Prodr. 15(2): 1188 (1866).
Brazil (Ceará). 84 BZE. Nanophan.
   Gymnanthes macrocarpa Müll.Arg. in H.G.A.Engler, Pflanzenr., IV, 147, V: 144 (1912), pro syn.
   Sebastiania ovata Klotzsch ex Pax in H.G.A.Engler, Pflanzenr., IV, 147, V: 144 (1912), pro syn.

Sebastiania membranifolia Müll.Arg. in C.F.P.von Martius, Fl. Bras. 11(2): 579 (1874).
Brazil (Goiás, Minas Gerais). 84 BZC BZL. Nanophan. or phan.

Sebastiania mosenii Pax & K.Hoffm. in H.G.A.Engler, Pflanzenr., IV, 147, XVII: 201 (1924).
Brazil (Minas Gerais). 84 BZL.

Sebastiania obtusifolia Pax & K.Hoffm. in H.G.A.Engler, Pflanzenr., IV, 147, V: 149 (1912).
Peru. 83 PER. Nanophan.
   * *Sapium obtusifolium* Kunth in F.W.H.von Humboldt, A.J.A.Bonpland & C.S.Kunth, Nov. Gen. Sp. 2: 65 (1817), nom. illeg. *Excoecaria obtusifolia* (Kunth) Müll.Arg., Linnaea 32: 125 (1863).
   Sapium peruvianum Steud., Nomencl. Bot., ed. 2, 2: 512 (1841).

Sebastiania pachyphylla Pax & K.Hoffm. in H.G.A.Engler, Pflanzenr., IV, 147, V: 142 (1912).
Paraguay. 85 PAR. Cham.

Sebastiania pachystachys (Klotzsch) Müll.Arg. in A.P.de Candolle, Prodr. 15(2): 1182 (1866).
Brazil (São Paulo, Paraná). 84 BZL BZS. Nanophan.
   * *Adenogyne pachystachya* Klotzsch, Arch. Naturgesch. 7: 184 (1841). *Gymnanthes pachystachys* (Klotzsch) Baill., Étude Euphorb.: 531 (1858). *Stillingia pachystachya* (Klotzsch) Müll.Arg., Linnaea 32: 102 (1863). *Sebastiania pachystachys* var. *genuina* Müll.Arg. in A.P.de Candolle, Prodr. 15(2): 1182 (1866), nom. inval.
   Gymnanthes pachystachys var. *pubescens* Müll.Arg., Linnaea 32: 101 (1863). *Sebastiania pachystachys* var. *pubescens* (Müll.Arg.) Müll.Arg. in A.P.de Candolle, Prodr. 15(2): 1182 (1866).
   Sebastiania pachystachys var. *glabra* Pax in H.G.A.Engler, Pflanzenr., IV, 147, V: 131 (1912).

Sebastiania panamensis G.L.Webster, Ann. Missouri Bot. Gard. 75: 1128 (1988).
W. Panama. 80 PAN. Nanophan. or phan. – Provisionally accepted.

Sebastiania pavoniana (Müll.Arg.) Müll.Arg. in A.P.de Candolle, Prodr. 15(2): 1189 (1866).
Mexico, NW. Costa Rica. 79 MXG MXN MXS 80 COS. Nanophan. or phan.
   * *Gymnanthes pavoniana* Müll.Arg., Linnaea 32: 106 (1863).
   Sebastiania palmeri Rose, Contr. U. S. Natl. Herb. 1: 112 (1891).
   Sebastiania ramirezii Maury, La Naturaleza, II, 2: 406 (1894).

Sebastiania potamophila (Müll.Arg.) Pax in H.G.A.Engler, Pflanzenr., IV, 147, V: 149 (1912).
Brazil (Bahia). 84 BZE. Nanophan.
   * *Excoecaria potamophila* Müll.Arg. in C.F.P.von Martius, Fl. Bras. 11(2): 627 (1874).
   Sebastiania martii Müll.Arg. in H.G.A.Engler, Pflanzenr., IV, 147, V: 149 (1912), pro syn.

Sebastiania pringlei S.Watson, Proc. Amer. Acad. Arts 26: 149 (1891).
Mexico (San Luis Potosí). 79 MXE. Nanophan. – Provisionally accepted.
Gymnanthes pringlei S.Watson ex Pax in H.G.A.Engler, Pflanzenr., IV, 147, V: 147 (1912).

Sebastiania pteroclada (Müll.Arg.) Müll.Arg. in A.P.de Candolle, Prodr. 15(2): 1190 (1866).
Brazil (Rio de Janeiro). 84 BZL. Nanophan.
* *Gymnanthes pteroclada* Müll.Arg., Linnaea 32: 107 (1863). *Stillingia pteroclada*
(Müll.Arg.) Baill., Adansonia 5: 329 (1865).

Sebastiania pubescens Pax & K.Hoffm. in H.G.A.Engler, Pflanzenr., IV, 147, V: 128 (1912).
Brazil (Minas Gerais). 84 BZL. Nanophan.

Sebastiania pubiflora Lundell, Wrightia 5: 80 (1975).
Guatemala. 80 GUA.

Sebastiania pusilla Croizat, J. Arnold Arbor. 26: 193 (1945).
Uruguay. 85 URU.

Sebastiania ramulosa Pax & K.Hoffm. in H.G.A.Engler, Pflanzenr., IV, 147, XVI: 200 (1924).
Brazil (São Paulo). 84 BZL.

Sebastiania rhombifolia Müll.Arg. in C.F.P.von Martius, Fl. Bras. 11(2): 590 (1874).
Brazil (Minas Gerais). 84 BZL. Nanophan.

Sebastiania riedelii Müll.Arg. in C.F.P.von Martius, Fl. Bras. 11(2): 578 (1874).
Brazil (Minas Gerais). 84 BZL. Nanophan.
* *Gymnanthes serrata* var. *pubescens* Müll.Arg., Linnaea 32: 99 (1863). *Sebastiania serrata*
var. *pubescens* (Müll.Arg.) Müll.Arg. in A.P.de Candolle, Prodr. 15(2): 1179 (1866).

Sebastiania rigida (Müll.Arg.) Müll.Arg. in A.P.de Candolle, Prodr. 15(2): 1180 (1866).
Brazil (Minas Gerais, Rio de Janeiro). 84 BZL. Nanophan.
* *Gymnanthes rigida* Müll.Arg., Linnaea 32: 99 (1863). *Stillingia rigida* (Müll.Arg.) Baill.,
Adansonia 5: 330 (1865).

Sebastiania riparia Schrad., Gött. Gel. Anz. 1: 713 (1821).
Brazil (Espírito Santo). 84 BZL. Nanophan.
Sebastiania viminea Nees, Flora 4: 328 (1821).

Sebastiania rotundifolia Pax & K.Hoffm. in H.G.A.Engler, Pflanzenr., IV, 147, V: 134 (1912).
S. Brazil. 84 BZS.

Sebastiania rupicola Pax & K.Hoffm. in H.G.A.Engler, Pflanzenr., IV, 147, V: 151 (1912).
Brazil (Rio de Janeiro). 84 BZL. Nanophan.

Sebastiania sarmentosa M.E.Jones, Contr. W. Bot. 18: 49 (1933).
Mexico (Baja California). 79 MXN.

Sebastiania schottiana (Müll.Arg.) Müll.Arg. in A.P.de Candolle, Prodr. 15(2): 1176 (1866).
WC., SE. & S. Brazil, Paraguay, Uruguay. 84 BZC BZL BZS 85 PAR URU. Nanophan.
Gymnanthes angustifolia Müll.Arg., Linnaea 32: 99 (1863). *Sebastiania angustifolia*
(Müll.Arg.) Müll.Arg. in A.P.de Candolle, Prodr. 15(2): 1179 (1866). *Sebastiania*
schottiana var. *angustifolia* (Müll.Arg.) Pax & K.Hoffm. in H.G.A.Engler, Pflanzenr., IV,
147, V: 127 (1912).
* *Gymnanthes schottiana* Müll.Arg., Linnaea 32: 96 (1863). *Stillingia schottiana* (Müll.Arg.)
Baill., Adansonia 5: 331 (1865). *Sebastiania schottiana* var. *genuina* Müll.Arg. in
C.F.P.von Martius, Fl. Bras. 11(2): 568 (1874), nom. inval.

Stillingia phyllanthiformis Baill., Adansonia 5: 331 (1865). *Sebastiania phyllanthiformis* (Baill.) Müll.Arg. in A.P.de Candolle, Prodr. 15(2): 1176 (1866). *Sebastiania schottiana* var. *phyllanthiformis* (Baill.) Pax & K.Hoffm. in H.G.A.Engler, Pflanzenr., IV, 147, V: 127 (1912).

Sebastiania mucronata Müll.Arg. in A.P.de Candolle, Prodr. 15(2): 1180 (1866). *Sebastiania schottiana* var. *mucronata* (Müll.Arg.) Müll.Arg. in C.F.P.von Martius, Fl. Bras. 11(2): 568 (1874).

Sebastiania serrata (Baill. ex Müll.Arg.) Müll.Arg. in A.P.de Candolle, Prodr. 15(2): 1179 (1866). Brazil (Minas Gerais, São Paulo, Rio Grande do Sul), Paraguay. 84 BZL BZS 85 PAR. Nanophan.
* *Gymnanthes serrata* Baill. ex Müll.Arg., Linnaea 32: 99 (1863). *Stillingia serrata* (Baill. ex Müll.Arg.) Klotzsch ex Baill., Adansonia 5: 329 (1865). *Sebastiania serrata* var. *genuina* Müll.Arg. in A.P.de Candolle, Prodr. 15(2): 1179 (1866), nom. inval. *Sebastiania serrata* var. *typica* Pax & K.Hoffm. in H.G.A.Engler, Pflanzenr., IV, 147, V: 133 (1912), nom. inval.
 Sebastiania serrata var. *grandifolia* Chodat & Hassl., Bull. Herb. Boissier, II, 5: 674 (1905). *Sebastiania grandifolia* (Chodat & Hassl.) Pax & K.Hoffm. in H.G.A.Engler, Pflanzenr., IV, 147, V: 133 (1912).
 Sebastiania serrata var. *major* Pax & K.Hoffm. in H.G.A.Engler, Pflanzenr., IV, 147, V: 133 (1912).

Sebastiania spicata (Millsp.) Pax & K.Hoffm. in H.G.A.Engler, Pflanzenr., IV, 147, VII: 422 (1914).
Jamaica. 81 JAM.
* *Dendrocousinsia spicata* Millsp., Publ. Field Mus. Nat. Hist., Bot. Ser. 2: 374 (1913).

Sebastiania standleyana Lundell, Lloydia 2: 97 (1939).
Belize, Guatemala. 80 BLZ GUA.

Sebastiania stipulacea (Müll.Arg.) Müll.Arg. in A.P.de Candolle, Prodr. 15(2): 1176 (1866).
S. Brazil. 84 BZL. Hemicr.
* *Gymnanthes stipulacea* Müll.Arg., Linnaea 32: 96 (1863). *Stillingia stipulacea* (Müll.Arg.) Baill., Adansonia 5: 325 (1865). *Microstachys stipulacea* (Müll.Arg.) Klotzsch ex Baill., Adansonia 5: 325 (1865).

Sebastiania subsessilis (Müll.Arg.) Pax in H.G.A.Engler, Pflanzenr., IV, 147, V: 145 (1912).
S. Brazil. 84 BZS. Nanophan.
* *Excoecaria subsessilis* Müll.Arg. in A.P.de Candolle, Prodr. 15(2): 1223 (1866).
 Sapium subsessile Chodat & Hassl., Bull. Herb. Boissier, II, 5: 677 (1905).

Sebastiania subulata (Müll.Arg.) Pax in H.G.A.Engler, Pflanzenr., IV, 147, V: 143 (1912).
Paraguay. 85 PAR. Nanophan. or phan.
* *Excoecaria subulata* Müll.Arg. in C.F.P.von Martius, Fl. Bras. 11(2): 627 (1874). *Sapium subulatum* (Müll.Arg.) Chodat & Hassl., Bull. Herb. Boissier, II, 5: 678 (1905).
 Sebastiania subulata var. *ramosa* Pax in H.G.A.Engler, Pflanzenr., IV, 147, V: 143 (1912).
 Sebastiania subulata var. *virgata* Pax in H.G.A.Engler, Pflanzenr., IV, 147, V: 143 (1912).

Sebastiania tikalana Lundell, Wrightia 2: 54 (1960).
Guatemala. 80 GUA.

Sebastiania trichogyne Pax & K.Hoffm. in H.G.A.Engler, Pflanzenr., IV, 147, XVII: 201 (1924).
Brazil (Minas Gerais). 84 BZL.

Sebastiania trinervia (Müll.Arg.) Müll.Arg. in A.P.de Candolle, Prodr. 15(2): 1182 (1866).
Brazil (Bahia). 84 BZE.

Gymnanthes trinervia Müll.Arg., Linnaea 32: 101 (1863). *Stillingia trinervia* (Müll.Arg.) Baill., Adansonia 5: 328 (1865).

Sebastiania tuerckheimiana (Pax & K.Hoffm.) Lundell, Wrightia 5: 80 (1975). Guatemala. 80 GUA.
Sapium tuerckheimianum Pax & K.Hoffm. in H.G.A.Engler, Pflanzenr., IV, 147, XIV: 61 (1919).

Sebastiania venezolana Pax & K.Hoffm. in H.G.A.Engler, Pflanzenr., IV, 147, XVII: 201 (1924). Venezuela. 82 VEN.

Sebastiania vestita Müll.Arg. in C.F.P.von Martius, Fl. Bras. 11(2): 575 (1874). Brazil (Minas Gerais). 84 BZL. Nanophan.

Sebastiania warmingii (Müll.Arg.) Pax in H.G.A.Engler, Pflanzenr., IV, 147, V: 143 (1912). Brazil (Minas Gerais). 84 BZL. Nanophan.
Excoecaria warmingii Müll.Arg. in C.F.P.von Martius, Fl. Bras. 11(2): 626 (1874). *Sapium warmingii* (Müll.Arg.) Chodat & Hassl., Bull. Herb. Boissier, II, 5: 677 (1905).

Sebastiania weddelliana (Baill.) Müll.Arg. in A.P.de Candolle, Prodr. 15(2): 1188 (1866). Brazil (Mato Grosso). 84 BZC. Nanophan.
Stillingia weddelliana Baill., Adansonia 5: 329 (1865).

Sebastiania ypanemensis (Müll.Arg.) Müll.Arg. in A.P.de Candolle, Prodr. 15(2): 1179 (1866). Brazil (Minas Gerais, São Paulo), Paraguay. 84 BZL 85 PAR. Nanophan.
Gymnanthes ypanemensis Müll.Arg., Linnaea 32: 100 (1863). *Stillingia ypanemensis* (Müll.Arg.) Baill., Adansonia 5: 330 (1865).

Synonyms:

Sebastiania acetosella (Milne-Redh.) Kruijt === **Microstachys acetosella** (Milne-Redh.) Esser
Sebastiania actinostemoides (Müll.Arg.) Müll.Arg. === **Gymnanthes actinostemoides** Müll.Arg.
Sebastiania adenophora Pax & K.Hoffm. === ?
Sebastiania albicans (Griseb.) C.Wright === **Gymnanthes albicans** (Griseb.) Urb.
Sebastiania angustifolia (Müll.Arg.) Müll.Arg. === **Sebastiania schottiana** (Müll.Arg.) Müll.Arg.
Sebastiania anisandra Lillo === **Sebastiania brasiliensis** Spreng.
Sebastiania anisodonta Müll.Arg. === **Microstachys** sp.
Sebastiania bidentata (Mart. & Zucc.) Pax === **Microstachys bidentata** (Mart. & Zucc.) Esser
Sebastiania bidentata var. *genuina* Pax === **Microstachys bidentata** (Mart. & Zucc.) Esser
Sebastiania bidentata var. *odontococca* (Müll.Arg.) Pax === **Microstachys bidentata** (Mart. & Zucc.) Esser
Sebastiania bidentata var. *pilgeri* Pax & K.Hoffm. === **Microstachys bidentata** (Mart. & Zucc.) Esser
Sebastiania bidentata var. *scoparia* (Mart.) Pax === **Microstachys bidentata** (Mart. & Zucc.) Esser
Sebastiania borneensis Pax & K.Hoffm. === **Gymnanthes borneensis** (Pax & K. Hoffm.) Esser
Sebastiania brachyclada (Müll.Arg.) Müll.Arg. === **Sebastiania klotzschiana** (Müll.Arg.) Müll.Arg.
Sebastiania brachypoda (Griseb.) C.Wright === **Actinostemon brachypodus** (Griseb.) Urb.
Sebastiania brasiliensis var. *anisophylla* Müll.Arg. === **Sebastiania brasiliensis** Spreng.
Sebastiania brasiliensis var. *brachystachys* Pax & K.Hoffm. === **Sebastiania brasiliensis** Spreng.
Sebastiania brasiliensis var. *brevispicata* Pax & K.Hoffm. === **Sebastiania brasiliensis** Spreng.
Sebastiania brasiliensis var. *divaricata* (Müll.Arg.) Müll.Arg. === **Sebastiania brasiliensis** Spreng.
Sebastiania brasiliensis var. *eremophila* Müll.Arg. === **Sebastiania brasiliensis** Spreng.
Sebastiania brasiliensis var. *erythoxyloides* (Müll.Arg.) Müll.Arg. === **Sebastiania brasiliensis** Spreng.
Sebastiania brasiliensis var. *genuina* Müll.Arg. === **Sebastiania brasiliensis** Spreng.
Sebastiania brasiliensis var. *microphylla* (Müll.Arg.) Müll.Arg. === **Sebastiania brasiliensis** Spreng.

Sebastiania brasiliensis f. *microphylla* (Müll.Arg.) Müll.Arg. === **Sebastiania brasiliensis** Spreng.

Sebastiania brasiliensis var. *obovata* (Müll.Arg.) Müll.Arg. === **Sebastiania brasiliensis** Spreng.

Sebastiania brasiliensis var. *polymorpha* Müll.Arg. === **Sebastiania brasiliensis** Spreng.

Sebastiania brasiliensis var. *ramosissima* (A.St.-Hil.) Müll.Arg. === **Sebastiania brasiliensis** Spreng.

Sebastiania brasiliensis var. *rigida* Müll.Arg. === **Sebastiania brasiliensis** Spreng.

Sebastiania brasiliensis var. *robusta* (Müll.Arg.) Müll.Arg. === **Sebastiania brasiliensis** Spreng.

Sebastiania brasiliensis var. *rufescens* (Müll.Arg.) Müll.Arg. === **Sebastiania brasiliensis** Spreng.

Sebastiania brasiliensis f. *rufescens* (Müll.Arg.) Müll.Arg. === **Sebastiania brasiliensis** Spreng.

Sebastiania brasiliensis f. *viridis* Müll.Arg. === **Sebastiania brasiliensis** Spreng.

Sebastiania buchii Urb. === **Sapium daphnoides** Griseb.

Sebastiania chamaelea (L.) Müll.Arg. === **Microstachys chamaelea** (L.) Müll.Arg.

Sebastiania chamaelea var. *africana* Pax & K.Hoffm. === **Microstachys chamaelea** (L.) Müll.Arg.

Sebastiania chamaelea var. *asperococca* (F.Muell.) Pax === **Microstachys chamaelea** (L.) Müll.Arg.

Sebastiania chamaelea var. *chariensis* Beille === **Microstachys chamaelea** (L.) Müll.Arg.

Sebastiania commersoniana (Baill.) L.B.Sm. & Downs === **Sebastiania klotzschiana** (Müll.Arg.) Müll.Arg.

Sebastiania corniculata (Vahl) Müll.Arg. === **Microstachys corniculata** (Vahl) Griseb.

Sebastiania corniculata var. *acalyphoides* (Mart.) Müll.Arg. === **Microstachys corniculata** (Vahl) Griseb.

Sebastiania corniculata var. *blepharophylla* Müll.Arg. === **Microstachys corniculata** (Vahl) Griseb.

Sebastiania corniculata var. *campestris* (Baill.) Müll.Arg. === **Microstachys corniculata** (Vahl) Griseb.

Sebastiania corniculata f. *crotonoides* Müll.Arg. === **Microstachys corniculata** (Vahl) Griseb.

Sebastiania corniculata f. *discopoda* Müll.Arg. === **Microstachys corniculata** (Vahl) Griseb.

Sebastiania corniculata var. *egensis* Müll.Arg. === **Microstachys corniculata** (Vahl) Griseb.

Sebastiania corniculata var. *fallax* (Müll.Arg.) Müll.Arg. === **Microstachys corniculata** (Vahl) Griseb.

Sebastiania corniculata f. *fallax* Müll.Arg. === **Microstachys corniculata** (Vahl) Griseb.

Sebastiania corniculata var. *ferruginea* Müll.Arg. === **Microstachys hispida** (Mart.) Govaerts

Sebastiania corniculata var. *fischeri* Müll.Arg. === **Microstachys corniculata** (Vahl) Griseb.

Sebastiania corniculata var. *genuina* Müll.Arg. === **Microstachys corniculata** (Vahl) Griseb.

Sebastiania corniculata var. *glabrata* (Mart.) Müll.Arg. === **Microstachys corniculata** (Vahl) Griseb.

Sebastiania corniculata f. *glabrata* Müll.Arg. === **Microstachys corniculata** (Vahl) Griseb.

Sebastiania corniculata f. *glandulosa* (Mart.) Müll.Arg. === **Microstachys corniculata** (Vahl) Griseb.

Sebastiania corniculata var. *guianensis* (Klotzsch) Pax === **Microstachys corniculata** (Vahl) Griseb.

Sebastiania corniculata var. *hassleriana* Chodat === **Microstachys corniculata** (Vahl) Griseb.

Sebastiania corniculata var. *heterophylla* Müll.Arg. === **Microstachys corniculata** (Vahl) Griseb.

Sebastiania corniculata var. *hispida* (Mart.) Müll.Arg. === **Microstachys corniculata** (Vahl) Griseb.

Sebastiania corniculata var. *incana* Müll.Arg. === **Microstachys hispida** (Mart.) Govaerts

Sebastiania corniculata var. *intercedens* Müll.Arg. === **Microstachys hispida** (Mart.) Govaerts

Sebastiania corniculata var. *klotzschiana* Müll.Arg. === **Microstachys hispida** (Mart.) Govaerts

Sebastiania corniculata var. *laeta* Müll.Arg. === **Microstachys hispida** (Mart.) Govaerts

Sebastiania corniculata var. *lagoensis* Müll.Arg. === **Microstachys hispida** (Mart.) Govaerts

Sebastiania corniculata var. *leptoclada* Müll.Arg. === **Microstachys corniculata** (Vahl) Griseb.

Sebastiania corniculata var. *leucoblepharis* Müll.Arg. === **Microstachys hispida** (Mart.) Govaerts

Sebastiania corniculata f. *longifolia* (Mart.) Müll.Arg. === **Microstachys corniculata** (Vahl) Griseb.

Sebastiania corniculata var. *longifolia* (Mart.) Müll.Arg. === **Microstachys corniculata** (Vahl) Griseb.

Sebastiania corniculata var. *lurida* Müll.Arg. === **Microstachys corniculata** (Vahl) Griseb.

Sebastiania corniculata var. *macrophylla* Müll.Arg. === **Microstachys hispida** (Mart.) Govaerts

Sebastiania corniculata var. *major* Müll.Arg. === **Microstachys hispida** (Mart.) Govaerts

Sebastiania corniculata var. *mansoana* Müll.Arg. === **Microstachys hispida** (Mart.) Govaerts

Sebastiania corniculata var. *megapontica* Müll.Arg. === **Microstachys hispida** (Mart.) Govaerts

Sebastiania corniculata var. *micrantha* (Benth.) Müll.Arg. === **Microstachys corniculata** (Vahl) Griseb.

Sebastiania corniculata var. *microdendron* Müll.Arg. === **Microstachys corniculata** (Vahl) Griseb.

Sebastiania corniculata var. *obtusifolia* Müll.Arg. === **Microstachys corniculata** (Vahl) Griseb.

Sebastiania corniculata var. *occidentalis* Müll.Arg. === **Microstachys hispida** (Mart.) Govaerts

Sebastiania corniculata f. *olfersiana* Müll.Arg. === **Microstachys corniculata** (Vahl) Griseb.

Sebastiania corniculata var. *oligophylla* Müll.Arg. === **Microstachys hispida** (Mart.) Govaerts

Sebastiania corniculata f. *ovata* Müll.Arg. === **Microstachys corniculata** (Vahl) Griseb.

Sebastiania corniculata var. *paraguayensis* Chodat === **Microstachys hispida** (Mart.) Govaerts

Sebastiania corniculata var. *parvifolia* Müll.Arg. === **Microstachys corniculata** (Vahl) Griseb.

Sebastiania corniculata var. *patula* (Mart.) Müll.Arg. === **Microstachys corniculata** (Vahl) Griseb.

Sebastiania corniculata var. *petiolaris* Müll.Arg. === **Microstachys corniculata** (Vahl) Griseb.

Sebastiania corniculata var. *poeppingii* Müll.Arg. === **Microstachys corniculata** (Vahl) Griseb.

Sebastiania corniculata var. *pohlii* Müll.Arg. === **Microstachys corniculata** (Vahl) Griseb.

Sebastiania corniculata var. *potamophila* Müll.Arg. === **Microstachys corniculata** (Vahl) Griseb.

Sebastiania corniculata var. *prostrata* (Mart.) Müll.Arg. === **Microstachys corniculata** (Vahl) Griseb.

Sebastiania corniculata var. *psilophylla* Müll.Arg. === **Microstachys corniculata** (Vahl) Griseb.

Sebastiania corniculata var. *purpurella* Müll.Arg. === **Microstachys hispida** (Mart.) Govaerts

Sebastiania corniculata var. *regnellii* Müll.Arg. === **Microstachys hispida** (Mart.) Govaerts

Sebastiania corniculata var. *riedelii* Müll.Arg. === **Microstachys hispida** (Mart.) Govaerts

Sebastiania corniculata var. *rufescens* Müll.Arg. === **Microstachys corniculata** (Vahl) Griseb.

Sebastiania corniculata var. *salicifolia* (Mart.) Müll.Arg. === **Microstachys corniculata** (Vahl) Griseb.

Sebastiania corniculata f. *salicifolia* (Mart.) Müll.Arg. === **Microstachys corniculata** (Vahl) Griseb.

Sebastiania corniculata var. *scabra* Müll.Arg. === **Microstachys corniculata** (Vahl) Griseb.

Sebastiania corniculata var. *schuechiana* Müll.Arg. === **Microstachys hispida** (Mart.) Govaerts

Sebastiania corniculata var. *sclerophylla* Müll.Arg. === **Microstachys hispida** (Mart.) Govaerts

Sebastiania corniculata var. *sellowiana* Müll.Arg. === **Microstachys corniculata** (Vahl) Griseb.

Sebastiania corniculata var. *speciosa* Müll.Arg. === **Microstachys hispida** (Mart.) Govaerts

Sebastiania corniculata var. *subglabrata* Müll.Arg. === **Microstachys hispida** (Mart.) Govaerts

Sebastiania corniculata var. *tomentosa* Müll.Arg. === **Microstachys hispida** (Mart.) Govaerts

Sebastiania corniculata var. *tragioides* (Mart.) Pax === **Microstachys corniculata** (Vahl) Griseb.

Sebastiania corniculata var. *transiens* Müll.Arg. === **Microstachys corniculata** (Vahl) Griseb.

Sebastiania corniculata f. *velutina* Müll.Arg. === **Microstachys corniculata** (Vahl) Griseb.

Sebastiania corniculata var. *villaricensis* Müll.Arg. === **Microstachys corniculata** (Vahl) Griseb.

Sebastiania corniculata var. *weddelliana* Müll.Arg. === **Microstachys hispida** (Mart.) Govaerts

Sebastiania daphnoides (Mart.) Müll.Arg. === **Microstachys daphnoides** (Mart.) Müll.Arg.

Sebastiania daphnoides var. *genuina* Müll.Arg. === **Microstachys daphnoides** (Mart.) Müll.Arg.

Sebastiania daphnoides var. *incana* (Müll.Arg.) Müll.Arg. === **Microstachys daphnoides** (Mart.) Müll.Arg.

Sebastiania daphnoides var. *intermedia* Müll.Arg. === **Microstachys daphnoides** (Mart.) Müll.Arg.

Sebastiania daphnoides var. *major* Müll.Arg. === **Microstachys daphnoides** (Mart.) Müll.Arg.

Sebastiania daphnoides var. *myrtilloides* (Mart.) Müll.Arg. === **Microstachys daphnoides** (Mart.) Müll.Arg.

Sebastiania daphnoides var. *oleoides* (Mart.) Müll.Arg. === **Microstachys daphnoides** (Mart.) Müll.Arg.

Sebastiania desertorum Klotzsch === **Sebastiania brasiliensis** Spreng.

Sebastiania diamantinensis Glaz. === ?

Sebastiania discolor (Spreng.) Müll.Arg. === **Gymnanthes discolor** (Spreng.) Müll.Arg.

Sebastiania discolor var. *fiebrigii* Pax & K.Hoffm. === **Gymnanthes discolor** (Spreng.) Müll.Arg.

Sebastiania discolor var. *genuina* Müll.Arg. === **Gymnanthes discolor** (Spreng.) Müll.Arg.

Sebastiania discolor var. *subconcolor* (Müll.Arg.) Müll.Arg. === **Gymnanthes discolor** (Spreng.) Müll.Arg.

Sebastiania ditassoides (Didr.) Müll.Arg. === **Microstachys ditassoides** (Didr.) Esser

Sebastiania ditassoides f. *apiculata* Müll.Arg. === **Microstachys ditassoides** (Didr.) Esser

Sebastiania ditassoides var. *discolor* Pax & K.Hoffm. === **Microstachys ditassoides** (Didr.) Esser

Sebastiania ditassoides var. *genuina* Müll.Arg. === **Microstachys ditassoides** (Didr.) Esser

Sebastiania ditassoides var. *glabrata* (Müll.Arg.) Müll.Arg. === **Microstachys ditassoides** (Didr.) Esser

Sebastiania ditassoides f. *hastata* Müll.Arg. === **Microstachys ditassoides** (Didr.) Esser

Sebastiania ditassoides var. *ledifolia* (Müll.Arg.) Müll.Arg. === **Microstachys ditassoides** (Didr.) Esser

Sebastiania ditassoides var. *parvifolia* (Müll.Arg.) Müll.Arg. === **Microstachys ditassoides** (Didr.) Esser

Sebastiania ditassoides var. *vellerifolia* (Müll.Arg.) Pax === **Microstachys ditassoides** (Didr.) Esser

Sebastiania divaricata Klotzsch === **Sebastiania brasiliensis** Spreng.

Sebastiania edwalliana var. *acuminata* Pax & K.Hoffm. === **Sebastiania edwalliana** Pax & K.Hoffm.

Sebastiania edwalliana var. *minor* Pax & K.Hoffm. === **Sebastiania edwalliana** Pax & K.Hoffm.

Sebastiania edwalliana var. *vestita* Pax & K.Hoffm. === **Sebastiania edwalliana** Pax & K.Hoffm.

Sebastiania elliptica (Sw.) Müll.Arg. === **Gymnanthes elliptica** Sw.

Sebastiania faradianensis (Beille) Kruijt === **Microstachys faradianensis** (Beille) Esser

Sebastiania fiebrigii Pax === Sebastiania brasiliensis Spreng.

Sebastiania foveata Klotzsch === **Sebastiania brasiliensis** Spreng.

Sebastiania fruticosa (Bartram) Fernald === **Ditrysinia fruticosa** (Bartram) Govaerts & Frodin

Sebastiania gaudichaudii (Müll.Arg.) Müll.Arg. === **Gymnanthes gaudichaudii** Müll.Arg.

Sebastiania glandulosa (Sw.) Müll.Arg. === **Gymnanthes glandulosa** (Sw.) Müll.Arg.

Sebastiania glandulosa Pax === **Microstachys corniculata** (Vahl) Griseb.

Sebastiania glandulosa f. *calvescens* Pax === **Microstachys corniculata** (Vahl) Griseb.

Sebastiania glandulosa var. *campestris* (Baill.) Pax === **Microstachys corniculata** (Vahl) Griseb.

Sebastiania glandulosa var. *fallax* (Müll.Arg.) Pax === **Microstachys corniculata** (Vahl) Griseb.

Sebastiania glandulosa var. *microdendron* (Müll.Arg.) Pax === **Microstachys corniculata** (Vahl) Griseb.

Sebastiania glandulosa var. *obtusifolia* (Müll.Arg.) Pax === **Microstachys corniculata** (Vahl) Griseb.

Sebastiania glandulosa f. *olfersiana* (Müll.Arg.) Pax === **Microstachys corniculata** (Vahl) Griseb.

Sebastiania glandulosa f. *ovata* (Müll.Arg.) Pax === **Microstachys corniculata** (Vahl) Griseb.

Sebastiania glandulosa var. *parvifolia* (Müll.Arg.) Pax === **Microstachys corniculata** (Vahl) Griseb.

Sebastiania glandulosa var. *pohlii* (Müll.Arg.) Pax === **Microstachys corniculata** (Vahl) Griseb.

Sebastiania glandulosa var. *psilophylla* (Müll.Arg.) Pax === **Microstachys corniculata** (Vahl) Griseb.

Sebastiania glandulosa var. *scabra* (Müll.Arg.) Pax === **Microstachys corniculata** (Vahl) Griseb.

Sebastiania glandulosa var. *sellowiana* (Müll.Arg.) Pax === **Microstachys corniculata** (Vahl) Griseb.

Sebastiania glandulosa var. *transiens* (Müll.Arg.) Pax === **Microstachys corniculata** (Vahl) Griseb.

Sebastiania glandulosa f. *velutina* (Müll.Arg.) Pax === **Microstachys corniculata** (Vahl) Griseb.

Sebastiania glaucophylla Müll.Arg. === **Gymnanthes** sp.

Sebastiania grandifolia (Chodat & Hassl.) Pax & K.Hoffm. === **Sebastiania serrata** (Baill. ex Müll.Arg.) Müll.Arg.

Sebastiania grisebachiana Müll.Arg. === **Gymnanthes albicans** (Griseb.) Urb.

Sebastiania guyanensis (Müll.Arg.) Müll.Arg. === **Gymnanthes guyanensis** Müll.Arg.

Sebastiania heterodoxa (Müll.Arg.) Benth. === **Microstachys heterodoxa** (Müll.Arg.) Esser

Sebastiania hexaptera Urb. === ?

Sebastiania hispida (Mart.) Pax === **Microstachys hispida** (Mart.) Govaerts

Sebastiania hispida var. *ambigua* Pax & K.Hoffm. === **Microstachys hispida** (Mart.) Govaerts

Sebastiania hispida var. *aspera* Pax & K.Hoffm. === **Microstachys hispida** (Mart.) Govaerts

Sebastiania hispida f. *brevipila* Pax === **Microstachys hispida** (Mart.) Govaerts

Sebastiania hispida var. *crotonoides* (Mart.) Pax === **Microstachys hispida** (Mart.) Govaerts

Sebastiania hispida var. *euhispida* Pax === **Microstachys hispida** (Mart.) Govaerts

Sebastiania hispida var. *ferruginea* (Müll.Arg.) Pax === **Microstachys hispida** (Mart.) Govaerts

Sebastiania hispida f. *glabrescens* Pax === **Microstachys hispida** (Mart.) Govaerts

Sebastiania hispida var. *graciliramea* Pax & K.Hoffm. === **Microstachys hispida** (Mart.) Govaerts

Sebastiania hispida var. *incana* (Müll.Arg.) Pax === **Microstachys hispida** (Mart.) Govaerts

Sebastiania hispida var. *intercedens* (Müll.Arg.) Pax === **Microstachys hispida** (Mart.) Govaerts

Sebastiania hispida var. *klotzschiana* (Müll.Arg.) Pax === **Microstachys hispida** (Mart.) Govaerts

Sebastiania hispida var. *laeta* (Müll.Arg.) Pax === **Microstachys hispida** (Mart.) Govaerts

Sebastiania hispida var. *lagoensis* (Müll.Arg.) Pax === **Microstachys hispida** (Mart.) Govaerts

Sebastiania hispida var. *leucoblepharis* (Müll.Arg.) Pax === **Microstachys hispida** (Mart.) Govaerts

Sebastiania hispida var. *macrophylla* (Müll.Arg.) Pax === **Microstachys hispida** (Mart.) Govaerts

Sebastiania hispida var. *major* (Müll.Arg.) Pax === **Microstachys hispida** (Mart.) Govaerts

Sebastiania hispida var. *mansoana* (Müll.Arg.) Pax === **Microstachys hispida** (Mart.) Govaerts

Sebastiania hispida var. *megapontica* (Müll.Arg.) Pax === **Microstachys hispida** (Mart.) Govaerts

Sebastiania hispida var. *occidentalis* (Müll.Arg.) Pax === **Microstachys hispida** (Mart.) Govaerts

Sebastiania hispida var. *oligophylla* (Müll.Arg.) Pax === **Microstachys hispida** (Mart.) Govaerts

Sebastiania hispida var. *paraguayensis* (Chodat) Pax === **Microstachys hispida** (Mart.) Govaerts

Sebastiania hispida var. *patula* (Mart.) Pax === **Microstachys corniculata** (Vahl) Griseb.

Sebastiania hispida var. *purpurella* (Müll.Arg.) Pax === **Microstachys hispida** (Mart.) Govaerts

Sebastiania hispida var. *regnellii* (Müll.Arg.) Pax === **Microstachys hispida** (Mart.) Govaerts

Sebastiania hispida var. *riedelii* (Müll.Arg.) Pax === **Microstachys hispida** (Mart.) Govaerts

Sebastiania hispida var. *scandens* Pax & K.Hoffm. === **Microstachys hispida** (Mart.) Govaerts

Sebastiania hispida var. *schuechiana* (Müll.Arg.) Pax === **Microstachys hispida** (Mart.) Govaerts

Sebastiania hispida var. *sclerophylla* (Müll.Arg.) Pax === **Microstachys hispida** (Mart.) Govaerts

Sebastiania hispida var. *speciosa* (Müll.Arg.) Pax === **Microstachys hispida** (Mart.) Govaerts

Sebastiania hispida var. *stenophylla* Pax & K.Hoffm. === **Microstachys hispida** (Mart.) Govaerts

Sebastiania hispida var. *subglabrata* (Müll.Arg.) Pax === **Microstachys hispida** (Mart.) Govaerts

Sebastiania hispida var. *subpatula* Pax & K.Hoffm. === **Microstachys hispida** (Mart.) Govaerts

Sebastiania hispida var. *tomentosa* (Müll.Arg.) Pax === **Microstachys hispida** (Mart.) Govaerts

Sebastiania hispida f. *villosula* Pax === **Microstachys hispida** (Mart.) Govaerts

Sebastiania hispida var. *weddeliana* (Müll.Arg.) Pax === **Microstachys hispida** (Mart.) Govaerts

Sebastiania hypoleuca (Benth.) Müll.Arg. === **Gymnanthes hypoleuca** Benth.

Sebastiania inopinata Prain === **Duvigneaudia inopinata** (Prain) J.Léonard

Sebastiania klotzschiana var. *brachyclada* (Müll.Arg.) Pax & K.Hoffm. === **Sebastiania klotzschiana** (Müll.Arg.) Müll.Arg.

Sebastiania klotzschiana var. *genuina* Müll.Arg. === **Sebastiania klotzschiana** (Müll.Arg.) Müll.Arg.

Sebastiania klotzschiana var. *trichoneura* Müll.Arg. === **Sebastiania klotzschiana** (Müll.Arg.) Müll.Arg.

Sebastiania lanceifolia Steenis === **Gymnanthes** sp.

Sebastiania larensis Croizat & Tamayo === **Bonania** sp.

Sebastiania ligustrina (Michx.) Müll.Arg. === **Ditrysinia fruticosa** (Bartram) Govaerts & Frodin

Sebastiania linearifolia Lanj. === **Microstachys corniculata** (Vahl) Griseb.

Sebastiania longipes (Müll.Arg.) Müll.Arg. === **Gymnanthes longipes** Müll.Arg.

Sebastiania lucida (Sw.) Müll.Arg. === **Gymnanthes lucida** Sw.

Sebastiania macrophylla Klotzsch === **Sebastiania brasiliensis** Spreng.

Sebastiania marginata (Mart.) Müll.Arg. === **Microstachys marginata** (Mart.) Klotzsch ex Müll.Arg.

Sebastiania marginata var. *coriacea* (Mart.) Pax === **Microstachys marginata** (Mart.) Klotzsch ex Müll.Arg.

Sebastiania martii Müll.Arg. === **Sebastiania potamophila** (Müll.Arg.) Pax

Sebastiania mexicana Brandegee === **Microstachys corniculata** (Vahl) Griseb.

Sebastiania micrantha (Benth.) Lanj. === **Microstachys corniculata** (Vahl) Griseb.

Sebastiania mucronata Müll.Arg. === **Sebastiania schottiana** (Müll.Arg.) Müll.Arg.

Sebastiania multiramea (Klotzsch) Mart. === **Gymnanthes glabrata** (Mart.) Govaerts

Sebastiania multiramea var. *genuina* Müll.Arg. === **Gymnanthes glabrata** (Mart.) Govaerts

Sebastiania multiramea var. *glabrata* (Mart.) Pax === **Gymnanthes glabrata** (Mart.) Govaerts

Sebastiania multiramea var. *luschnathiana* (Baill.) Müll.Arg. === **Gymnanthes glabrata** (Mart.) Govaerts

Sebastiania myricifolia (Griseb.) C.Wright === **Bonania myricifolia** (Griseb.) Benth. & Hook.f.

Sebastiania myrtilloides (Mart.) Pax === **Microstachys daphnoides** (Mart.) Müll.Arg.

Sebastiania myrtilloides var. *daphnoides* (Mart.) Pax === **Microstachys daphnoides** (Mart.) Müll.Arg.

Sebastiania myrtilloides var. *incana* (Müll.Arg.) Pax === **Microstachys daphnoides** (Mart.) Müll.Arg.

Sebastiania myrtilloides var. *intermedia* (Müll.Arg.) Pax === **Microstachys daphnoides** (Mart.) Müll.Arg.

Sebastiania myrtilloides var. *major* (Müll.Arg.) Pax === **Microstachys daphnoides** (Mart.) Müll.Arg.

Sebastiania myrtilloides var. *martiana* Pax === **Microstachys daphnoides** (Mart.) Müll.Arg.

Sebastiania nervosa (Müll.Arg.) Müll.Arg. === **Gymnanthes nervosa** Müll.Arg.

Sebastiania nummularifolia Cordeiro === **Microstachys nummularifolia** (Cordeiro) Esser

Sebastiania oleoides (Mart.) Müll.Arg. === **Microstachys daphnoides** (Mart.) Müll.Arg.

Sebastiania ovata Klotzsch ex Pax === **Sebastiania macrocarpa** Müll.Arg.

Sebastiania pachystachys var. *genuina* Müll.Arg. === **Sebastiania pachystachys** (Klotzsch) Müll.Arg.

Sebastiania pachystachys var. *glabra* Pax === **Sebastiania pachystachys** (Klotzsch) Müll.Arg.

Sebastiania pachystachys var. *pubescens* (Müll.Arg.) Müll.Arg. === **Sebastiania pachystachys** (Klotzsch) Müll.Arg.

Sebastiania pallens (Griseb.) Müll.Arg. === **Gymnanthes pallens** (Griseb.) Müll.Arg.

Sebastiania palmeri Rose === **Sebastiania pavoniana** (Müll.Arg.) Müll.Arg.

Sebastiania phyllanthiformis (Baill.) Müll.Arg. === **Sebastiania schottiana** (Müll.Arg.) Müll.Arg.

Sebastiania picardae Urb. === **?**

Sebastiania ramirezii Maury === **Sebastiania pavoniana** (Müll.Arg.) Müll.Arg.

Sebastiania remota Steenis === **Gymnanthes remota** (Steenis) Esser

Sebastiania reticulata Klotzsch === **Sebastiania brasiliensis** Spreng.

Sebastiania revoluta Ule === **Microstachys revoluta** (Ule) Esser

Sebastiania rhombifolia (Rusby) Jabl. === **?**

Sebastiania robusta Klotzsch ex Pax === **Sebastiania brasiliensis** Spreng.

Sebastiania salicifolia (Mart.) Pax === **Microstachys** sp.

Sebastiania salicifolia var. *fischeri* (Müll.Arg.) Pax === **Microstachys corniculata** (Vahl) Griseb.

Sebastiania salicifolia var. *genuina* Pax === **Microstachys** sp.

Sebastiania salicifolia var. *glaziovii* Pax & K.Hoffm. === **Microstachys** sp.

Sebastiania salicifolia var. *leptoclada* (Müll.Arg.) Pax === **Microstachys** sp.

Sebastiania salicifolia var. *longifolia* (Mart.) Pax === **Microstachys** sp.

Sebastiania salicifolia var. *lurida* (Müll.Arg.) Pax === **Sebastiania salicifolia**

Sebastiania salicifolia var. *similis* Pax & K.Hoffm. === **Microstachys** sp.

Sebastiania schlechtendaliana (Müll.Arg.) Müll.Arg. === **Gymnanthes riparia** (Schltdl.) Klotzsch

Sebastiania schottiana var. *angustifolia* (Müll.Arg.) Pax & K.Hoffm. === **Sebastiania schottiana** (Müll.Arg.) Müll.Arg.

Sebastiania schottiana var. *genuina* Müll.Arg. === **Sebastiania schottiana** (Müll.Arg.) Müll.Arg.

Sebastiania schottiana var. *mucronata* (Müll.Arg.) Müll.Arg. === **Sebastiania schottiana** (Müll.Arg.) Müll.Arg.

Sebastiania schottiana var. *phyllanthiformis* (Baill.) Pax & K.Hoffm. === **Sebastiania schottiana** (Müll.Arg.) Müll.Arg.

Sebastiania serrata var. *genuina* Müll.Arg. === **Sebastiania serrata** (Baill. ex Müll.Arg.) Müll.Arg.

Sebastiania serrata var. *grandifolia* Chodat & Hassl. === **Sebastiania serrata** (Baill. ex Müll.Arg.) Müll.Arg.

Sebastiania serrata var. *major* Pax & K.Hoffm. === **Sebastiania serrata** (Baill. ex Müll.Arg.) Müll.Arg.

Sebastiania serrata var. *pubescens* (Müll.Arg.) Müll.Arg. === **Sebastiania riedelii** Müll.Arg.

Sebastiania serrata var. *typica* Pax & K.Hoffm. === **Sebastiania serrata** (Baill. ex Müll.Arg.) Müll.Arg.

Sebastiania serrulata (Mart.) Müll.Arg. === **Microstachys serrulata** (Mart.) Müll.Arg.

Sebastiania serrulata var. *fastigiata* Pax & K.Hoffm. === **Microstachys serrulata** (Mart.) Müll.Arg.

Sebastiania serrulata var. *genuina* Müll.Arg. === **Microstachys serrulata** (Mart.) Müll.Arg.

Sebastiania serrulata var. *hispida* Müll.Arg. === **Microstachys serrulata** (Mart.) Müll.Arg.

Sebastiania serrulata var. *klotzschiana* (Müll.Arg.) Müll.Arg. === **Microstachys serrulata** (Mart.) Müll.Arg.

Sebastiania serrulata var. *oblongifolia* Müll.Arg. === **Microstachys serrulata** (Mart.) Müll.Arg.

Sebastiania serrulata var. *oncoblepharis* (Müll.Arg.) Müll.Arg. === Microstachys serrulata (Mart.) Müll.Arg.

Sebastiania singularis Rizzini === ?

Sebastiania subulata var. *ramosa* Pax === **Sebastiania subulata** (Müll.Arg.) Pax

Sebastiania subulata var. *virgata* Pax === **Sebastiania subulata** (Müll.Arg.) Pax

Sebastiania treculiana (Müll.Arg.) Müll.Arg. === **Stillingia treculiana** (Müll.Arg.) I.M.Johnst.

Sebastiania uleana Pax & K.Hoffm. === **Microstachys uleana** (Pax & Hoffm.) Esser

Sebastiania viminea Nees === **Sebastiania riparia** Schrad.

Sebastiania virgata (Müll.Arg.) Müll.Arg. === **Microstachys bidentata** (Mart. & Zucc.) Esser

Sebastiania virgata var. *bidentata* (Mart. & Zucc.) Müll.Arg. === **Microstachys bidentata** (Mart. & Zucc.) Esser

Sebastiania virgata var. *odontococca* Müll.Arg. === **Microstachys bidentata** (Mart. & Zucc.) Esser

Sebastiania virgata var. *scoparia* (Mart.) Müll.Arg. === **Microstachys bidentata** (Mart. & Zucc.) Esser

Sebastiania widgrenii (Müll.Arg.) Müll.Arg. === **Gymnanthes widgrenii** Müll.Arg.

Seborium

Synonyms:
Seborium Raf. === **Sapium** P.Browne

Securinega

5 species, Madagascar and Mascarenes. Much-branched, sometimes deciduous twiggy shrubs or small to medium trees (*S. durissima*, to 8 m; *S. perrieri* to 20 m) of more or less dry forest or scrub with distichously arranged leaves. *S. capuronii* is recorded from bush with cactoids (*Allaudia*); *S. durissima*, the Mascarene *bois dur*, exhibits the heteromorphy frequent in those islands (the leaves in seedlings being much smaller than those of adults). Never revised for *Pflanzenreich* but studies since 1931 have shown that species outside this region are best referred to *Flueggea* and *Meineckia*. *Margaritaria* is another possible ally. Webster (Synopsis, 1994), on the basis of pollen and seeds, assigns the genus to a unique subtribe within Phyllantheae and has suggested affinities with Oldfieldioideae. (Phyllanthoideae)

Pax, F. & K. Hoffmann (1931). *Securinega*. In A. Engler (ed.), Die natürlichen Pflanzenfamilien, 2. Aufl., 19c: 60. Leipzig. Ge. — Synopsis with description of genus; accounts for some 20 species (including *Chascotheca*, *Flueggea* and *Meineckia*). [Never revised for *Pflanzenreich*.]

Leandri, J. (1958). *Securinega*. Fl. Madag. Comores 111 (Euphorbiacées), I: 107-116. Paris. Fr. — Flora treatment (4 species); key. [Two other doubtful species are also listed; one, *S. durissima*, otherwise native in the Mascarenes, the other, *S. trichogynis*, only once collected. The latter is now properly in *Meineckia*.]

Brunel, J. F. (1987). *Securinega*. In *idem*, Sur le genre *Phyllanthus* L. et quelques genres voisins de la tribu des Phyllantheae Dumort. Strasbourg. Fr. — Description of genus, with illustrations of key characters; list of species (1 only in Africa, 5 in Madagascar and Mascarenes). [See also *Phyllanthus*. Recent work suggests the genus properly is limited to Madagascar and the Mascarenes.]

Securinega Comm. ex A.Juss., Gen. Pl.: 388 (1789).
Madagascar, Mascarenes. 29. Nanophan. or phan.

Securinega antsingyensis Leandri, Mém. Inst. Sci. Madagascar, Sér. B, Biol. Vég. 8: 234 (1957).
W. Madagascar. 29 MDG. Nanophan.

Securinega capuronii Leandri, Mém. Inst. Sci. Madagascar, Sér. B, Biol. Vég. 8: 235 (1957).
W. & SW. Madagascar. 29 MDG. Nanophan. or phan.

Securinega durissima J.F.Gmel., Syst. Nat.: 1008 (1792). *Acidoton durissimus* (J.F.Gmel.) Kuntze, Revis. Gen. Pl. 2: 592 (1891).
Réunion, Mauritius, Rodrigues. 29 MAU REU ROD. Phan.
Securinega nitida Willd., Sp. Pl. 4: 761 (1806), nom. illeg.
Securinega commersonii G.Don in R.Sweet, Hort. Brit., ed. 3: 768 (1839), nom. nud.

Securinega perrieri Leandri, Bull. Soc. Bot. France 84: 65 (1937).
W. & SW. Madagascar. 29 MDG. Phan.

Securinega seyrigii Leandri, Mém. Inst. Sci. Madagascar, Sér. B, Biol. Vég. 8: 236 (1957).
W. & SW. Madagascar. 29 MDG. Phan.

Synonyms:
Securinega abeggii Urb. & Ekman === **Andrachne brittonii** Urb.
Securinega abyssinica A.Rich. === **Flueggea virosa** (Roxb. ex Willd.) Voigt
Securinega acicularis Croizat === **Flueggea acicularis** (Croizat) G.L.Webster
Securinega acidothamnus (Griseb.) Müll.Arg. === **Flueggea acidoton** (L.) G.L.Webster
Securinega acidoton (L.) Fawc. === **Flueggea acidoton** (L.) G.L.Webster
Securinega acuminatissima (C.B.Rob.) C.B.Rob. === **Flueggea flexuosa** Müll.Arg.
Securinega bailloniana Müll.Arg. === **Margaritaria discoidea** var. **triplosphaera** Radcl.-Sm.
Securinega buxifolia (Reut.) Müll.Arg. === **Flueggea tinctoria** (L.) G.L.Webster
Securinega capensis I.M.Johnst. === **Tetracoccus capensis** (I.M.Johnst.) Croizat
Securinega commersonii G.Don === **Securinega durissima** J.F.Gmel.
Securinega congesta (Benth. ex Müll.Arg.) Müll.Arg. === **Jablonskia congesta** (Benth. ex Müll.Arg.) G.L.Webster
Securinega elliptica (Spreng.) Müll.Arg. === **Flueggea elliptica** (Spreng.) Baill.
Securinega fasciculata (S.Watson) I.M.Johnst. === **Tetracoccus fasciculatus** (S.Watson) Croizat
Securinega flexuosa (Müll.Arg.) Müll.Arg. === **Flueggea flexuosa** Müll.Arg.
Securinega fluggeoides (Müll.Arg.) Müll.Arg. === **Flueggea suffruticosa** (Pall.) Baill.
Securinega grisea Müll.Arg. === **Flueggea virosa** (Roxb. ex Willd.) Voigt
Securinega guaraiuva Kuhlm. === **?**
Securinega hallii (Brandegee) I.M.Johnst. === **Tetracoccus fasciculatus** var. **hallii** (Brandegee) Dressler

Securinega hilariana (Baill.) Müll.Arg. === **Meineckia neogranatensis** subsp. **hilariana** (Baill.) G.L.Webster

Securinega hysterantha Bojer === **Margaritaria anomala** (Baill.) Fosberg

Securinega japonica Miq. === **Flueggea suffruticosa** (Pall.) Baill.

Securinega keyensis Warb. === **Flueggea virosa** subsp. **melanthesoides** (F.Muell.) G.L.Webster

Securinega leucopyrus Brandis === **Flueggea virosa** (Roxb. ex Willd.) Voigt

Securinega leucopyrus (Willd.) Müll.Arg. === **Flueggea leucopyrus** Willd.

Securinega melanthesoides (F.Muell.) Airy Shaw === **Flueggea virosa** subsp. **melanthesoides** (F.Muell.) G.L.Webster

Securinega melanthesoides var. *aridicola* (Domin) Airy Shaw === **Flueggea virosa** subsp. **melanthesoides** (F.Muell.) G.L.Webster

Securinega microcarpa (Blume) Müll.Arg. === **Flueggea virosa** (Roxb. ex Willd.) Voigt

Securinega muelleriana Baill. === **Neoroepera buxifolia** Müll.Arg.

Securinega multiflora S.B.Liang === **Flueggea suffruticosa** (Pall.) Baill.

Securinega neopeltandra (Griseb.) Urb. ex Pax & K.Hoffm. === **Chascotheca neopeltandra** (Griseb.) Urb.

Securinega nitida Willd. === **Securinega durissima** J.F.Gmel.

Securinega nitida W.T.Aiton === **Actephila lindleyi** (Steud.) Airy Shaw

Securinega obovata (Willd.) Müll.Arg. === **Flueggea virosa** (Roxb. ex Willd.) Voigt

Securinega phyllanthoides (Baill.) Müll.Arg. === **Meineckia phyllanthoides** Baill.

Securinega ramiflora (Aiton) Müll.Arg. === **Flueggea suffruticosa** (Pall.) Baill.

Securinega samoana Croizat === **Flueggea flexuosa** Müll.Arg.

Securinega schlechteri Pax === **Cleistanthus schlechteri** (Pax) Hutch.

Securinega schuechiana Müll.Arg. === **Flueggea schuechiana** (Müll.Arg.) G.L.Webster

Securinega schweinfurthii Balf.f. === **Andrachne schweinfurthii** (Balf.f.) Radcl.-Sm.

Securinega spirei (Beille) Croizat === **Flueggea spirei** Beille

Securinega suffruticosa (Pall.) Rehder === **Flueggea suffruticosa** (Pall.) Baill.

Securinega tinctoria (L.) Rothm. === **Flueggea tinctoria** (L.) G.L.Webster

Securinega trichogynis (Baill.) Müll.Arg. === **Meineckia trichogynis** (Baill.) G.L.Webster

Securinega verrucosa (Thunb.) Benth. ex Pax === **Flueggea verrucosa** (Thunb.) G.L.Webster

Securinega virgata (Poir.) Maire === **Flueggea tinctoria** (L.) G.L.Webster

Securinega virosa (Roxb. ex Willd.) Baill. === **Flueggea virosa** (Roxb. ex Willd.) Voigt

Securinega virosa var. *australiana* Baill. === **Flueggea virosa** subsp. **melanthesoides** (F.Muell.) G.L.Webster

Seidelia

2 species, southern Africa (Namibia and South Africa); small, delicate, annual much-branched herbs of dry habitats closely related to *Leidesia*; both are in turn related to the northern hemisphere *Mercurialis*. The local distribution of *Seidelia* is in general more southern and western than *Leidesia*. (Acalyphoideae)

Prain, D. (1913). Mercurialineae and Adenoclineae of South Africa. Ann. Bot. 27: 371-410. En. — Includes (pp. 398-399) a synopsis of *Seidelia* (2 species), with key, synonymy, references and citations, localities with exsiccatae, and brief comments; biogeographical review at end of paper.

Pax, F. (with K. Hoffmann) (1914). *Seidelia*. In A. Engler (ed.), Das Pflanzenreich, IV 147 VII (Euphorbiaceae-Acalypheae-Mercurialinae): 282-284, illus. Berlin. (Heft 63.) La/Ge. — 2 species, S Africa; *SS. pumila* and *triandra* distinct only as *formae*.

Seidelia Baill., Étude Euphorb.: 465 (1858).
S. Africa. 27.

Seidelia firmula (Prain) Pax & K.Hoffm. in H.G.A.Engler, Pflanzenr., IV, 147, VII: 282 (1914).
Namibia, Cape Prov. 27 CPP NAM. Ther.
* *Leidesia firmula* Prain, Bull. Misc. Inform. Kew 1912: 337 (1912).

Seidelia triandra (E.Mey.) Pax, Bot. Jahrb. Syst. 10: 35 (1889).

Namibia, Cape Prov., Free State. 27 CPP NAM OFS. Ther. – *S. pumila* sometimes separated, following *Flora capensis*.

* *Mercurialis triandra* E.Mey., Linnaea 4: 237 (1829). *Tragia triandra* var. *genuina* Müll.Arg. in A.P.de Candolle, Prodr. 15(2): 947 (1866), nom. inval.

Mercurialis pumila Sond., Linnaea 23: 112 (1850). *Seidelia pumila* (Sond.) Baill., Étude Euphorb.: 465 (1858). *Tragia triandra* var. *pumila* (Sond.) Müll.Arg. in A.P.de Candolle, Prodr. 15(2): 947 (1866). *Seidelia triandra* f. *pumila* (Sond.) Pax & K.Hoffm. in H.G.A.Engler, Pflanzenr., IV, 147, VII: 284 (1914).

Seidelia mercurialis Baill., Étude Euphorb.: 465 (1858). *Seidelia triandra* f. *mercurialis* (Baill.) Pax & K.Hoffm. in H.G.A.Engler, Pflanzenr., IV, 147, VII: 284 (1914).

Tragia triandra Müll.Arg. in A.P.de Candolle, Prodr. 15(2): 947 (1866).

Synonyms:
Seidelia mercurialis Baill. === **Seidelia triandra** (E.Mey.) Pax
Seidelia pumila (Sond.) Baill. === **Seidelia triandra** (E.Mey.) Pax
Seidelia triandra f. *mercurialis* (Baill.) Pax & K.Hoffm. === **Seidelia triandra** (E.Mey.) Pax
Seidelia triandra f. *pumila* (Sond.) Pax & K.Hoffm. === **Seidelia triandra** (E.Mey.) Pax

Semilta

Synonyms:
Semilta Raf. === **Croton** L.

Senefeldera

5 species, Americas (mainly in Amazonia, but in scattered fashion reaching north to Panama); shrubs or small to large trees (*S. verticillata* to 25 m) of primary or (*S. verticillata*) also in secondary forest with sometimes pseudo-verticillately arranged leaves surmounted by terminal inflorescences. The genus is related to *Mabea*; along with *Senefelderopsis*, they comprise the Mabeinae (tribe Hippomaneae). Preliminary research by Esser (1994) indicates that the genus will be reduced to the latter 3 species in the list, the two others going to as-yet unpublished segregate genera (one of them monotypic). (Euphorbioideae (except Euphorbieae))

Pax, F. (with K. Hoffmann) (1912). *Senefeldera*. In A. Engler (ed.), Das Pflanzenreich, IV 147 V (Euphorbiaceae-Hippomaneae): 23-25. Berlin. (Heft 52.) La/Ge. — 4-5 species, Americas.

Jablonski, E. (1965). *Senefeldera*. Euphorbiaceae, Guayana Highland (Mem. New York Bot. Gard. 12(3)): 171-174. New York. En. — 9 species, with key. 'The delimitation of *Senefeldera* .. appears to be artificial'. [Now superseded; Esser (1994) has proposed several transfers and reductions.]

Esser, H.-J. (1994). Systematische Studien an den Hippomaneae Adr. Jussieu ex Bartling (Euphorbiaceae) insbesondere den Mabeinae Pax et K. Hoffm. 305 pp., pp., illus., maps. Hamburg. (Unpubl. Ph.D. dissertation, Univ. of Hamburg.) Ge. —*Senefeldera*, pp. 245-259 (3 species); includes introduction, key, descriptions, synonymy, types, distribution and habitat, commentary, and localities with exsiccatae. Appendices A-C include for all genera an index to specimens seen, collections in herb. HBG, and illustrations, cladograms and distribution maps.

Senefeldera Mart., Flora 24(2): 29 (1841).
Panama to N. Brazil. 80 82 83 84.

Senefeldera inclinata Müll.Arg. in C.F.P.von Martius, Fl. Bras. 11(2): 530 (1874).
Panama to Peru, Venezuela (Amazonas), Brazil (Amazonas). 80 PAN 82 VEN 83 CLM ECU PER 84 BZN. Phan.

Senefeldera karsteniana Pax & K.Hoffm. in H.G.A.Engler, Pflanzenr., IV, 147, V: 25 (1912).
Senefeldera nitida Croizat, J. Wash. Acad. Sci. 33: 18 (1943).
Senefeldera skutchiana Croizat, J. Wash. Acad. Sci. 33: 18 (1943).
Senefeldera contracta R.E.Schult., Bot. Mus. Leafl. 17: 72 (1953).

Senefeldera macrophylla Ducke, Arch. Jard. Bot. Rio de Janeiro 4: 113 (1925).
Colombia (Amazonas), Surinam, Brazil (Pará, Amazonas), Guyana, Peru. 82 GUY SUR 83 CLM PER 84 BZN. Phan.

Senefeldera testiculata Pittier, Arb. Arbust. Venez.: 31 (1923).
Colombia, Venezuela. 82 VEN 83 CLM. Phan.

Senefeldera triandra Pax & K.Hoffm. in H.G.A.Engler, Pflanzenr., IV, 147, XIV: 55 (1919).
Brazil (Amazonas), Peru. 83 PER 84 BZN. Phan.

Senefeldera verticillata (Vell.) Croizat, J. Wash. Acad. Sci. 33: 18 (1943).
Brazil (Bahia to São Paulo). 84 BZE BZL. Phan.
* *Omphalea verticillata* Vell., Fl. Flumin. 10: 15 (1831).
Senefeldera angustifolia Klotzsch, Arch. Naturgesch. 7: 184 (1841), nom. nud.
Senefeldera latifolia Klotzsch, Arch. Naturgesch. 7: 184 (1841), nom. nud.
Senefeldera multiflora Mart., Flora 24(2): 29 (1841). *Senefeldera multiflora* var. *genuina* Müll.Arg. in A.P.de Candolle, Prodr. 15(2): 1153 (1866), nom. inval.
Senefeldera grandifolia Baill., Étude Euphorb.: 536 (1858).
Senefeldera multiflora var. *angustifolia* Müll.Arg. in A.P.de Candolle, Prodr. 15(2): 1154 (1866).
Senefeldera multiflora var. *intermedia* Müll.Arg. in A.P.de Candolle, Prodr. 15(2): 1154 (1866).
Senefeldera multiflora var. *obovata* Müll.Arg. in A.P.de Candolle, Prodr. 15(2): 1154 (1866).
Senefeldera dodecandra Müll.Arg. in C.F.P.von Martius, Fl. Bras. 11(2): 529 (1874).
Senefeldera multiflora var. *acutifolia* Müll.Arg. in C.F.P.von Martius, Fl. Bras. 11(2): 530 (1874).

Synonyms:
Senefeldera angustifolia Klotzsch === **Senefeldera verticillata** (Vell.) Croizat
Senefeldera chiribiquetensis R.E.Schult. & Croizat === **Senefelderopsis chiribiquetensis** (R.E.Schult. & Croizat) Steyerm.
Senefeldera contracta R.E.Schult. === **Senefeldera inclinata** Müll.Arg.
Senefeldera dodecandra Müll.Arg. === **Senefeldera verticillata** (Vell.) Croizat
Senefeldera grandifolia Baill. === **Senefeldera verticillata** (Vell.) Croizat
Senefeldera karsteniana Pax & K.Hoffm. === **Senefeldera inclinata** Müll.Arg.
Senefeldera latifolia Klotzsch === **Senefeldera verticillata** (Vell.) Croizat
Senefeldera multiflora Mart. === **Senefeldera verticillata** (Vell.) Croizat
Senefeldera multiflora var. *acutifolia* Müll.Arg. === **Senefeldera verticillata** (Vell.) Croizat
Senefeldera multiflora var. *angustifolia* Müll.Arg. === **Senefeldera verticillata** (Vell.) Croizat
Senefeldera multiflora var. *genuina* Müll.Arg. === **Senefeldera verticillata** (Vell.) Croizat
Senefeldera multiflora var. *intermedia* Müll.Arg. === **Senefeldera verticillata** (Vell.) Croizat
Senefeldera multiflora var. *obovata* Müll.Arg. === **Senefeldera verticillata** (Vell.) Croizat
Senefeldera nitida Croizat === **Senefeldera inclinata** Müll.Arg.
Senefeldera skutchiana Croizat === **Senefeldera inclinata** Müll.Arg.
Senefeldera yutajensis (Jabl.) G.L.Webster === **Dendrothrix yutajensis** (Jabl.) Esser

Senefelderopsis

2 species, N. South America (tepui areas of Colombia, Venezuela and Guyana); laticiferous shrubs to small or large trees (*S. croizatii* to 30 m; *S. chiribiquetensis* not over 8 m) with rigid, spirally arranged, clustered leaves and terminal, partly leafy inflorescences. The genus is confined to the Guayana Shield, where *S. chiribiquetensis* is sometimes dominant in formations on sandstones. In comparison with the closely related *Senefeldera* the flowers have only 2-3 rather than 5-12 stamens. (Euphorbioideae (except Euphorbieae))

> Jablonski, E. (1965). *Senefelderopsis*. Euphorbiaceae, Guayana Highland (Mem. New York Bot. Gard. 12(3)): 174-176. New York. En. — 4 species, with key (accounting for whole genus). [Now superseded.]
>
> Jablonski, E. (1967). *Senefelderopsis*. Euphorbiaceae, Guayana Highland (Mem. New York Bot. Gard. 17(1)): 186. New York. En. — Note on *Senefelderopsis sipapoensis*. [Now reduced to *S. croizatii*.]
>
> Esser, H.-J. (1994). Systematische Studien an den Hippomaneae Adr. Jussieu ex Bartling (Euphorbiaceae) insbesondere den Mabeinae Pax et K. Hoffm. 305 pp., pp., illus., maps. Hamburg. (Unpubl. Ph.D. dissertation, Univ. of Hamburg.) Ge. —*Senefelderopsis*, pp. 233-244 (2 species); includes introduction, key, descriptions, synonymy, types, distribution and habitat, commentary, and localities with exsiccatae. Appendices A-C include for all genera an index to specimens seen, collections in herb. HBG, and illustrations, cladograms and distribution maps. [Published as Esser, 1995.]
>
> • Esser, H.-J. (1995). A taxonomic revision of *Senefelderopsis* Steyerm. (Euphorbiaceae), including additional notes on *Dendrothrix* Esser. Mitt. Inst. Allg. Bot. Hamburg 25: 121-133. En. — Detailed treatment with taxonomic history, character survey, geography and ecology, relationships of the genus, key, synonymy, references and citations, types, descriptions, indication of distribution and habitat, localities with exsiccatae, and commentary; one distribution map; list of references. 2 species accepted.

Senefelderopsis Steyerm., Bot. Mus. Leafl. 15: 45 (1951).
S. Trop. America. 82 83.

Senefelderopsis chiribiquetensis (R.E.Schult. & Croizat) Steyerm., Bot. Mus. Leafl. 15: 47 (1951).
Colombia (Vaupés), Venezuela (Amazonas) ? 82 VEN? 83 CLM. Nanophan. or phan.
  * Senefeldera chiribiquetensis R.E.Schult. & Croizat, Caldasia 3: 122 (1944).

Senefelderopsis croizatii Steyerm., Bot. Mus. Leafl. 15: 45 (1951).
Guiana, Venezuela (Bolívar, Amazonas). 82 FRG VEN. Nanophan. or phan.
  Senefelderopsis sipapoensis Jabl., Mem. New York Bot. Gard. 12: 175 (1965).
  Senefelderopsis venamoensis Jabl., Mem. New York Bot. Gard. 12: 176 (1965).

Synonyms:
Senefelderopsis sipapoensis Jabl. === **Senefelderopsis croizatii** Steyerm.
Senefelderopsis venamoensis Jabl. === **Senefelderopsis croizatii** Steyerm.
Senefelderopsis yutajensis (Jabl.) Mennega === **Dendrothrix yutajensis** (Jabl.) Esser

Serophyton

Synonyms:
Serophyton Benth. === **Ditaxis** Vahl ex A.Juss.
Serophyton lanceolatum Benth. === **Ditaxis lanceolata** (Benth.) Pax & K.Hoffm.
Serophyton pilosissimum Benth. === **Ditaxis pilosissima** (Benth.) A.Heller

Shirakia

Replaced by *Neoshirakia* for nomenclatural reasons, with some species in addition referred to *Shirakiopsis*.

Synonyms:

Shirakia Hurus. === **Neoshirakia** Esser

Shirakia aubrevillei (Leandri) Kruijt === **Shirakiopsis aubrevillei** (Leandri) Esser

Shirakia cochinchinensis (Lour.) Hurus. === **Triadica cochinchinensis** Lour.

Shirakia elliptica (Hochst.) Kruijt === **Shirakiopsis elliptica** (Hochst.) Esser

Shirakia indica (Willd.) Hurus. === **Shirakiopsis indica** (Willd.) Esser

Shirakia japonica (Siebold & Zucc.) Hurus. === **Neoshirakia japonica** (Siebold & Zucc.) Esser

Shirakia sanchezii (Merr.) Kruijt === **Shirakiopsis sanchezii** (Merr.) Esser

Shirakia trilocularis (Pax & K.Hoffm.) Kruijt === **Shirakiopsis trilocularis** (Pax & K.Hoffm.) Esser

Shirakia virgata (Zoll. & Moritzi ex Miq.) Kruijt === **Shirakiopsis virgata** (Zoll. & Moritzi ex Miq.) Esser

Shirakiopsis

6 species, Africa, Asia and Malesia east to the Solomon Is.; formerly the major (but non-typical) part of *Sapium* sect. *Parasapium* (Muell. Arg.) Hook. f. Small to large laticiferous trees to 30 m or more in forest, sometimes much-branched and spreading; inflorescences terminal, sometimes on short shoots. The former *Sapium triloculare* is found in Tanzanian coastal forest and *S. virgatum* similarly so in Jawa. The former *S. indicum*, usually a small to medium tree but once recorded as 27 m in height, is widespread in Asia and Malesia in delta country, along estuarine rivers and behind mangrove. The former *S. ellipticum* is commonly in edges, secondary formations or along rivers where light is available. The genus was formerly thought to be related to *Excoecaria*, within which *Sapium indicum* has in the past sometimes been accomodated; more recent thinking suggests an affinity with *Gymnanthes* and *Neoshirakia*. (Euphorbioideae (except Euphorbieae))

> Pax, F. (with K. Hoffmann) (1912). *Sapium.* In A. Engler (ed.), Das Pflanzenreich, IV 147 V (Euphorbiaceae-Hippomaneae): 199-258. Berlin. (Heft 52.) La/Ge. — The species now referable to *Shirakiopsis* are covered in the authors' sects. *Parasapium* (nos. 86-88 and 90-91) and *Triadica* (no. 75, *S. diversifolium* = *Shirakiopsis indica*).
>
> Esser, H.-J. (1999). A partial revision of the Hippomaneae (Euphorbiaceae) in Malesia. Blumea 44: 149-215, illus., maps. En. — *Shirakiopsis*, pp. 182-190; protologue of genus and treatment (with new combinations) of 3 Malesian species with key, descriptions, synonymy, references, types, distribution, habitat, vernacular names, commentary (including uses), figure and map; all general references, identification list and index to botanical names at end of paper.

Shirakiopsis Esser, Blumea 44: 184 (1999).
  Trop. Old World. 22 23 24 25 26 27 40 41 42 60.

Shirakiopsis aubrevillei (Leandri) Esser (ined.).
  W. Trop. Africa. 22 IVO. Phan.
    * Sapium aubrevillei Leandri, Bull. Soc. Bot. France 81: 449 (1934). *Shirakia aubrevillei* (Leandri) Kruijt, Biblioth. Bot. 146: 92 (1996).

Shirakiopsis elliptica (Hochst.) Esser (ined.).
  Trop. & S. Africa. 22 BEN GHA GUI IVO NGA SIE TOG 23 CAF CMN CON GGI ZAI 24 ETH 25 KEN TAN UGA 26 ANG MLW MOZ ZAM ZIM 27 CPP NAT SWZ TVL? (82) guy. Phan.
    * Sclerocroton ellipticus Hochst., Flora 28: 85 (1845). *Stillingia elliptica* (Hochst.) Baill., Adansonia 3: 162 (1863). *Sapium ellipticum* (Hochst.) Pax in H.G.A.Engler, Pflanzenr., IV, 147, V: 253 (1912). *Shirakia elliptica* (Hochst.) Kruijt, Biblioth. Bot. 146: 93 (1996).
    Tragia elliptica Hochst., Flora 28: 85 (1845), pro syn.
    Excoecaria manniana Müll.Arg., Flora 47: 433 (1864). *Sapium mannianum* (Müll.Arg.) Benth. in G.Bentham & J.D.Hooker, Gen. Pl. 3: 335 (1880).

Excoecaria abyssinica Müll.Arg., Linnaea 34: 217 (1865). *Sapium abyssinicum* (Müll.Arg.)
Benth. in G.Bentham & J.D.Hooker, Gen. Pl. 3: 335 (1880).
Aporusa somalensis Mattei, Boll. Reale Orto Bot. Palermo 7: 101 (1908).
Sapium kerstingii Pax, Bot. Jahrb. Syst. 43: 85 (1909).

Shirakiopsis indica (Willd.) Esser, Blumea 44: 185 (1999).
Trop. Asia to Solomon Is. 40 IND 41 BMA THA VIE 42 BIS BOR JAW LSI MOL SUL SUM
60 SOL. Phan.
Sapium bingerium Roxb. ex Willd., Sp. Pl. 4: 572 (1805), pro syn.
*   *Sapium indicum* Willd., Sp. Pl. 4: 572 (1805). *Stillingia indica* (Willd.) Oken, Allg.
Naturgesch. 3(3): 1606 (1841). *Excoecaria indica* (Willd.) Müll.Arg., Linnaea 32:
123 (1863). *Shirakia indica* (Willd.) Hurus., J. Fac. Sci. Univ. Tokyo, Sect. 3, Bot. 6:
317 (1954).
Sapium hurmais Buch.-Ham., Trans. Linn. Soc. London 17: 229 (1835).
Sapium bingyricum Roxb. ex Baill., Étude Euphorb. Atlas: t. 6 (1858).
Stillingia bingyrica Baill., Étude Euphorb.: 513 (1858), nom. inval.
Stillingia diversifolia Miq., Fl. Ned. Ind., Eerste Bijv.: 461 (1861). *Excoecaria diversifolia*
(Miq.) Müll.Arg. in A.P.de Candolle, Prodr. 15(2): 1211 (1866). *Sapium diversifolium*
(Miq.) Pax in H.G.A.Engler, Pflanzenr., IV, 147, V: 241 (1912).

Shirakiopsis sanchezii (Merr.) Esser, Blumea 44: 189 (1999).
Philippines. 42 PHI. Phan.
*   *Sapium sanchezii* Merr., Philipp. J. Sci., C 7: 406 (1912 publ. 1913). *Shirakia sanchezii*
(Merr.) Kruijt, Biblioth. Bot. 146: 93 (1996).

Shirakiopsis trilocularis (Pax & K.Hoffm.) Esser (ined.).
Tanzania. 25 TAN.
Shirakia trilocularis (Pax & K.Hoffm.) Kruijt, Biblioth. Bot. 146: 93 (1996).

Shirakiopsis virgata (Zoll. & Moritzi ex Miq.) Esser, Blumea 44: 189 (1999).
Jawa. 42 JAW. Phan.
Excoecaria virgata Zoll. ex Baill., Étude Euphorb: 518 (1858), nom. nud. *Stillingia virgata*
Baill., Étude Euphorb.: 518 (1858), nom. nud.
*   *Excoecaria virgata* Zoll. & Moritzi ex Miq., Fl. Ned. Ind. 1(1): 416 (1855). *Sapium
virgatum* (Zoll. & Moritzi ex Miq.) Hook.f., Fl. Brit. India 5: 471 (1888). *Shirakia
virgata* (Zoll. & Moritzi ex Miq.) Kruijt, Biblioth. Bot. 146: 93 (1996).

Sibangea

3 species, WC. and E. Africa (Nigeria to Gabon and in Tanzania); large shrubs or small to
medium forest trees (*S. pleioneura* to 21 m) with large distichously arranged leaves (reminiscent
of some Annonaceae) and flowers in axillary clusters. Each of the three species is of relatively
limited distribution; at least *S. pleioneura* is montane. *S. similis* is usually in forest understorey.
The genus is a segregate of *Drypetes* differing in having narrower and non-imbricate sepals in
the female flowers; like that genus it is a member of Drypeteae (which in the past has been
given family status as Putranjivaceae; for further discussion, see *Drypetes*). (Phyllanthoideae)

Oliver, D. (1883). *Sibangea arborescens.* Ic. Pl. 15: 9, pl. 1411. En. — Plant portrait with
description and commentary; includes protologue of genus.
Pax, F. & K. Hoffmann (1922). *Drypetes.* In A. Engler (ed.), Das Pflanzenreich, IV 147 XV
(Euphorbiaceae-Phyllanthoideae-Phyllantheae): 229-279. Berlin. (Heft 81.) La/Ge. —
Nos. 99 (*D. similis*) and 107 (*D. arborescens*) now are referred to *Sibangea*.
Radcliffe-Smith, A. (1978). Notes on African Euphorbiaceae, VII. Kew Bull. 32: 475-481.
En. — Includes a treatment of *Sibangea* (a revived segregate of *Drypetes*); 3 species, one
new. [Precursory to FTEA.]
Radcliffe-Smith, A. (1986). Notes on African Euphorbiaceae, XVIII. Kew Bull. 41: 963-964.
En. — Additions; includes note on *Sibangea*.

Sibangea Oliv., Hooker's Icon. Pl. 15: t. 1411 (1883).
　　Trop. Africa. 22 23 25. Phan.

Sibangea arborescens Oliv., Hooker's Icon. Pl. 15: t. 1411 (1883). *Drypetes arborescens* (Oliv.)
　　Hutch. in D.Oliver, Fl. Trop. Afr. 6(1): 680 (1912). – FIGURE, p. 1473.
　　Cameroon, Gabon. 23 CMN GAB. Phan.

Sibangea pleioneura Radcl.-Sm., Kew Bull. 32: 481 (1978).
　　Tanzania. 25 TAN. Phan.

Sibangea similis (Hutch.) Radcl.-Sm., Kew Bull. 32: 481 (1978).
　　Nigeria, Cameroon. 22 NGA 23 CMN. Phan.
　　* *Drypetes similis* Hutch. in D.Oliver, Fl. Trop. Afr. 6(1): 679 (1912).

Silvaea

Synonyms:
Silvaea Hook. & Arn. === **Trigonostemon** Blume
Silvaea hookeriana Baill. === **Trigonostemon semperflorens** (Roxb.) Müll.Arg.
Silvaea semperflorens (Roxb.) Hook. & Arn. === **Trigonostemon semperflorens** (Roxb.)
　　Müll.Arg.

Sinopimelodendron

Synonyms:
Sinopimelodendron Tsiang === **Cleidiocarpon** Airy Shaw

Siphonanthus

Synonyms:
Siphonanthus Schreb. ex Baill. === **Hevea** Aubl.

Siphonia

Synonyms:
Siphonia Rich. === **Hevea** Aubl.
Siphonia brasiliensis Willd. ex A.Juss. === **Hevea brasiliensis** (Willd. ex A.Juss.) Müll.Arg.
Siphonia discolor Spruce ex Benth. === **Hevea spruceana** (Benth.) Müll.Arg.
Siphonia kunthiana Baill. === **Hevea pauciflora** (Spruce ex Benth.) Müll.Arg. var. pauciflora
Siphonia lutea Spruce ex Benth. === **Hevea guianensis** var. **lutea** (Spruce ex Benth.) Ducke &
　　R.E.Schult.
Siphonia pauciflora Spruce ex Benth. === **Hevea pauciflora** (Spruce ex Benth.) Müll.Arg.
Siphonia rigidifolia Spruce ex Benth. === **Hevea rigidifolia** (Spruce ex Benth.) Müll.Arg.
Siphonia spruceana Benth. === **Hevea spruceana** (Benth.) Müll.Arg.

Spathiostemon

2 species, Malesia (W. Malesia to Philippines and New Guinea); formerly a subgenus of *Homonoia*. Small to medium trees to 20 m with spirally arranged leaves and axillary inflorescences, *S. javensis* in or at edges of forest (and quickly appearing along roads therein). Both species can be very common. The genus differs from *Homonoia* in the absence of lepidote scales; it also is not associated with rivers. A revision has lately been published by van Welzen (1998), with *S. moniliformis* retained but *Clonostylis* (q.v.) excluded. (Acalyphoideae)

Sibangea arborescens Oliv.
Artist: Matilda Smith
Ic. Pl. 15: pl. 1411 (1883)

Pax, F. & K. Hoffmann (1919). *Homonoia*. In A. Engler (ed.), Das Pflanzenreich, IV 147 XI (Euphorbiaceae-Acalypheae-Ricininae): 114-118. Berlin. (Heft 68.) La/Ge. — Includes subgen. *Spathiostemon* (1 sp., W. Malesia, Philippines and Papuasia.).

Airy-Shaw, H. K. (1963). Notes on Malaysian and other Asiatic Euphorbiaceae, XXXVIII. A second species of *Spathiostemon* Bl. Kew Bull. 16: 357-358. En. —*S. moniliformis* described.

Airy-Shaw, H. K. (1966). Notes on Malaysian and other Asiatic Euphorbiaceae, LXXVIII. Further collections of *Spathiostemon moniliformis*. Kew Bull. 20: 408-409. En. — Additional records.

Airy-Shaw, H. K. (1974). *Spathiostemon moniliformis*. Ic. Pl. 38: pl. 3720. En. — Plant portrait with description and commentary.

Airy-Shaw, H. K. (1978). Notes on Malesian and other Asiatic Euphorbiaceae, CC. *Clonostylis* S. Moore reduced to *Spathiostemon* Bl. Kew Bull. 32: 407-408. En. — Reduction of genus and transfer of *C. forbesii*; key to 3 species.

Airy-Shaw, H. K. (1980). Notes on Malesian and other Asiatic Euphorbiaceae, CCXXXVIII. *Spathiostemon* Bl. Kew Bull. 35: 395. En. — Fruiting material of *S. moniliformis* described.

- Welzen, P. C. van (1998). Revisions and phylogenies of Malesian Euphorbiaceae subtribe Lasiococcinae (*Homonoia, Lasiococca, Spathiostemon*) and *Clonostylis, Ricinus* and *Wetria*. Blumea 43: 131-164. (*Lasiococca* with Nguyen Nghia Thin & Vu Hoai Duc.) En. — General introduction, with history of studies; a note of caution on the use as a character of monoecy vs. dioecy given the imperfect state of much material; phylogeny with character table and cladogram; revision of Malesian *Spathiostemon* (pp. 145-150; 2 species) with key, descriptions, synonymy, types, literature citations, vernacular names, indication of distribution, ecology and habitat, and notes on anatomy, uses and properties (where known), and systematics; identification list at end. [*S. moniliformis* is retained, but shorn of *Clonostylis forbesii*.]

Spathiostemon Blume, Bijdr.: 621 (1826).
Trop. Asia. 41 42.
> *Polydragma* Hook.f., Hooker's Icon. Pl. 18: t. 1701 (1887).

Spathiostemon javensis Blume, Bijdr.: 622 (1826). *Homonoia javensis* (Blume) Müll.Arg., Linnaea 34: 200 (1865).
Malesia. 42 BIS BOR JAW LSI MLY MOL NWG PHI SUL. Nanophan. or phan.
> *Adelia javanica* Miq., Fl. Ned. Ind. 1(2): 388 (1859).
> *Mallotus eglandulosus* Elmer, Leafl. Philipp. Bot. 1: 313 (1908).
> *Homonoia javensis* var. *ciliata* Merr., Philipp. J. Sci., C 7: 391 (1912 publ. 1913).
> *Mallotus calvus* Pax & K.Hoffm. in H.G.A.Engler, Pflanzenr., IV, 147, VII: 195 (1914).
> *Spathiostemon javensis* var. *nimae* Airy Shaw, Kew Bull., Addit. Ser. 8: 202 (1980).

Spathiostemon moniliformis Airy Shaw, Kew Bull. 16: 357 (1963).
S. Thailand. 41 THA. Nanophan. or phan.

Synonyms:
Spathiostemon forbesii (S.Moore) Airy Shaw === **Clonostylis forbesii** S.Moore
Spathiostemon javensis var. *nimae* Airy Shaw === **Spathiostemon javensis** Blume
Spathiostemon salicinus (Hassk.) Hassk. === **Homonoia riparia** Lour.
Spathiostemon salicinus var. *angustifolius* Miq. === **Homonoia riparia** Lour.

Speranskia

3 species, China; perennials or subshrubs with soft, coarse-looking serrate leaves and terminal racemes with tricoccous capsules which stick to socks or stock on contact. The genus is related to those in Ditaxinae (*Argythamnia* and allies) but differs in free stamens as well as the terminal inflorescence. (Acalyphoideae)

Pax, F. (with K. Hoffmann) (1912). *Speranskia*. In A. Engler (ed.), Das Pflanzenreich, IV 147 VI (Euphorbiaceae-Acalypheae-Chrozophorinae): 14-17. Berlin. (Heft 57.) La/Ge. — 3 mutually closely related species, China.

Hwang Shu-mei (1989). (A note on genus *Speranskia* in China (Euphorbiaceae).) Bull. Bot. Res., Harbin 9(4): 37-40, illus. Ch. — Concise treatment, with key, synonymy, references and citations, localities with exsiccatae, and summary of distribution; 3 species accepted (one new, *S. yunnanensis*).

Speranskia Baill., Étude Euphorb.: 388 (1858).
China. 36.

Speranskia cantonensis (Hance) Pax & K.Hoffm. in H.G.A.Engler, Pflanzenr., IV, 147, VI: 15 (1912). – FIGURE, p. 1476.
China (Hubei, Guangdong). 36 CHC CHS. Cham. or nanophan.
* *Argythamnia cantonensis* Hance, J. Bot. 16: 14 (1878).
Speranskia henryi Oliv., Hooker's Icon. Pl. 16: t. 1577 (1887).
Mercurialis acanthocarpa H.Lév. & Vaniot, Repert. Spec. Nov. Regni Veg. 3: 21 (1906).

Speranskia tuberculata (Bunge) Baill., Étude Euphorb.: 389 (1858).
NC. China. 36 CHN. Hemicr. or cham.
* *Croton tuberculatus* Bunge, Enum. Pl. China Bor.: 59 (1833). *Argythamnia tuberculata* (Bunge) Müll.Arg., Linnaea 34: 144 (1865).

var. **pekinensis** (Pax & K.Hoffm.) Hurus., J. Fac. Sci. Univ. Tokyo, Sect. 3, Bot. 6: 310 (1954).
NC. China. 36 CHN. Hemicr. or cham.
* *Speranskia pekinensis* Pax & K.Hoffm. in H.G.A.Engler, Pflanzenr., IV, 147, VI: 15 (1912).

var. **tuberculata**
NC. China. 36 CHN. Hemicr. or cham.

Speranskia yunnanensis S.M.Hwang, Bull. Bot. Res., Harbin 9: 38 (1989).
China (Yunnan). 36 CHC.

Synonyms:
Speranskia henryi Oliv. === **Speranskia cantonensis** (Hance) Pax & K.Hoffm.
Speranskia pekinensis Pax & K.Hoffm. === **Speranskia tuberculata** var. **pekinensis** (Pax & K.Hoffm.) Hurus.

Sphaerostylis

2 species, Madagascar; related to *Tragia* but staminate sepals forming a pseudodisk. Shrubby climbers with cordate leaves in humid forest at lower elevations. The Malesian species formerly referred here is now in *Megistostigma*. (Acalyphoideae)

Pax, F. & K. Hoffmann (1919). *Sphaerostylis*. In A. Engler (ed.), Das Pflanzenreich, IV 147 IX (Euphorbiaceae-Acalypheae-Plukenetiinae): 106-107. Berlin. (Heft 68.) La/Ge. — 2 species, Madagascar (1), W Malesia (1). [Malesian species now in *Megistostigma*.]

Sphaerostylis Baill., Étude Euphorb.: 466 (1858).
Madagascar. 29.

Sphaerostylis perrieri Leandri, Bull. Soc. Bot. France 85: 527 (1938 publ. 1939).
Madagascar. 29 MDG.

Sphaerostylis tulasneana Baill., Étude Euphorb.: 466 (1858).
N. Madagascar. 29 MDG.

Speranskia cantonensis (Hance) Pax & K. Hoffm. (as *S. henryi* Oliv.)

Artist: Matilda Smith
Ic. Pl. 16: pl. 1577 (1887)
KEW ILLUSTRATIONS COLLECTION

Synonyms:

Sphaerostylis anomala (Prain) Croizat === **Tragiella anomala** (Prain) Pax & K.Hoffm.
Sphaerostylis cordata (Merr.) Pax & K.Hoffm. === **Megistostigma cordatum** Merr.
Sphaerostylis frieseana (Prain) Croizat === **Tragiella frieseana** (Prain) Pax & K.Hoffm.
Sphaerostylis glabrata (Kurz) Merr. === **Megistostigma malaccense** Hook.f.
Sphaerostylis malaccensis (Hook.f.) Pax & K.Hoffm. === **Megistostigma malaccense** Hook.f.
Sphaerostylis natalensis (Sond.) Croizat === **Tragiella natalensis** (Sond.) Pax & K.Hoffm.

Sphragidia

Synonyms:

Sphragidia Thwaites === **Drypetes** Vahl
Sphragidia zeylanica Thwaites === **Drypetes longifolia** (Blume) Pax & K.Hoffm.

Sphyranthera

2 species, Andaman and Nicobar (Great Nicobar) Islands; shrubs or small trees to 5 m of sunny places in forest and scrub, the leaves spirally arranged and relatively nondescript. Its affinities are disputed; they remained uncertain for Airy-Shaw (1984) but Hooker had thought it acalyphoid and Webster (Synopsis, 1994) gave it a place next to *Codiaeum*. For a recent treatment (as part of a revision of Andamans Euphorbiaceae) see Chakraborty & Balakrishnan (1992, under **Asia**). (Crotonoideae)

Airy-Shaw, H. K. (1984). A note on the (presumed) female plant of *Sphyranthera lutescens* (Euphorbiaceae) in the Nicobar Islands. Kew Bull. 39: 807-808. En. — First record of genus for Nicobar Islands and additional morphological data.

Chakraborty, T. & M. K. Vasudeva Rao (1984). A new species of *Sphyranthera* (Euphorbiaceae) from North Andaman I. J. Econ. Taxon. Bot. 5: 959-961. En. —*S. airyshawii* described; further notes on *S. lutescens*. [Male plant of the latter described in ibid., 6: 429-430.]

Sphyranthera Hook.f., Hooker's Icon. Pl. 18: t. 1702 (1887).
Andaman Is., Nicobar Is. 41.

Sphyranthera airyshawii Chakrab. & Vasudeva Rao, J. Econ. Taxon. Bot. 5: 960 (1984).
N. Andaman Is. 41 AND. Nanophan. or phan.

Sphyranthera lutescens (Kurz) Pax & K.Hoffm. in H.G.A.Engler, Nat. Pflanzenfam. ed. 2, 19c: 231 (1931).
C. Andaman Is., Nicobar Is. 41 AND NCB. Nanophan. or phan.
  * *Codiaeum lutescens* Kurz, J. Asiat. Soc. Bengal, Pt. 2, Nat. Hist. 42(2): 246 (1873).
  Sphyranthera capitellata Hook.f., Hooker's Icon. Pl. 18: t. 1702 (1887).

Synonyms:

Sphyranthera capitellata Hook.f. === **Sphyranthera lutescens** (Kurz) Pax & K.Hoffm.

Spirostachys

2 species, NE. Africa and southwards; much-branched deciduous laticiferous shrubs or small to medium trees to 12 m (rarely more) of woodland, bushland and savanna (sometimes along rivers) with *Excoecaria*-like leaves and axillary inflorescences with crowded flowers. The genus is related to *Excoecaria*. In *S. africana*, with a more southerly distribution, the flowers appear before the leaves. *S. venenifera* is very poisonous, and particularly harmful to

eyes. *S. africana*, possessing a good timber, is additionally treated in non-systematic literature (cf. Alfaro Cardoso 1964; other references in Lebrun and Stork, *Énumération des plantes à fleurs d'Afrique tropicale*, 1. 1991). [Some specialists do not accept *S. venenifera* as distinct.] (Euphorbioideae (except Euphorbieae))

> Pax, F. (with K. Hoffmann) (1912). *Spirostachys*. In A. Engler (ed.), Das Pflanzenreich, IV 147 V (Euphorbiaceae-Hippomaneae): 153-156. Berlin. (Heft 52.) La/Ge. — 4 species in 2 sections. [Now reduced to 2 species.]
> Alfaro Cardoso, J. (1964). *Spirostachys africana*. 53 pp. (Publ. Serv. Agric. Moçambique, A, 18). Lourenço Marques. Pt. — Economic.

Spirostachys Sond., Linnaea 23: 106 (1850).
> Trop. & S. Africa. 24 25 26 27.
>> *Excoecariopsis* Pax, Bot. Jahrb. Syst. 45: 239 (1910).

Spirostachys africana Sond., Linnaea 23: 106 (1850). *Stillingia africana* (Sond.) Baill., Adansonia 3: 163 (1863). *Excoecaria africana* (Sond.) Müll.Arg., Linnaea 32: 123 (1863). *Sapium africanum* (Sond.) Kuntze, Revis. Gen. Pl. 3(2): 293 (1898).
> Kenya to S. Africa. 25 KEN TAN 26 ANG MOZ ZIM 27 BOT CPP NAM NAT SWZ TVL. Nanophan. or phan. – Furnishes a good timber.
>> *Excoecaria synandra* Pax, Bot. Jahrb. Syst. 43: 223 (1909). *Spirostachys synandra* (Pax) Pax in H.G.A.Engler, Pflanzenr., IV, 147, V: 155 (1912).

Spirostachys venenifera (Pax) Pax in H.G.A.Engler, Pflanzenr., IV, 147, V: 151 (1912).
> – FIGURE, p. 1479.
> S. Somalia, Kenya, Tanzania. 24 SOM 25 KEN TAN. Nanophan. or phan.
>> * *Excoecaria venenifera* Pax, Bot. Jahrb. Syst. 19: 113 (1894).
>> *Excoecaria glomeriflora* Pax in H.G.A.Engler, Pflanzenw. Ost-Afrikas C: 241 (1895). *Spirostachys glomeriflora* (Pax) Pax in H.G.A.Engler, Pflanzenr., IV, 147, V: 156 (1912).

Synonyms:
Spirostachys glomeriflora (Pax) Pax === **Spirostachys venenifera** (Pax) Pax
Spirostachys madagascarienis Baill. === **Excoecaria madagascariensis** (Baill.) Müll.Arg.
Spirostachys synandra (Pax) Pax === **Spirostachys africana** Sond.

Spixia

Synonyms:
Spixia Leandro === Pera Mutis
Spixia barbinervis Klotzsch === **Pera barbinervis** (Klotzsch) Pax & K.Hoffm.
Spixia distichophylla Mart. === **Pera distichophylla** (Mart.) Baill.
Spixia heteranthera Schrank === **Pera heteranthera** (Schrank) I.M.Johnst.
Spixia leandrii Mart. === **Pera heteranthera** (Schrank) I.M.Johnst.

Spondianthus

1 species, tropical Africa, with two varieties each widely distributed (Léonard & Nkounkou 1989). This genus of mainly lowland riverside, swamp and gallery forest trees (to 25 m) with tufted, sometimes rounded leaves and terminal, cluster-forming inflorescences was at one time thought to be anacardiaceous. Webster (Synopsis, 1994) includes it in his Antidesmieae but in its own subtribe, Spondianthinae; the latter may be an ancestral 'sister group' to other members of that tribe. Stuppy (1995; see **Phyllanthoideae**) has suggested that inclusion in one of the uniovulate subfamilies would be preferable but more study was needed. The two infraspecific taxa are distinguished partly on preferred habitat. (Phyllanthoideae)

Spirostachys venenifera (Pax) Pax
Artist: Mary Millar Watt
Fl. Trop. East Africa, Euphorbiaceae 1: 388, fig. 73 (1987)
KEW ILLUSTRATIONS COLLECTION

Hutchinson, J. (1913). *Spondianthus preussii*. Ic. Pl. 30: pl. 2986. La/En. — Plant portrait with description and text.

Pax, F. & K. Hoffmann (1922). *Spondianthus*. In A. Engler (ed.), Das Pflanzenreich, IV 147 XV (Euphorbiaceae-Phyllanthoideae-Phyllantheae): 13-15, illus. Berlin. (Heft 81.) La/Ge. — 1 species, Africa.

• Léonard, J. & J. Nkounkou (1989). Révision du genre *Spondianthus* Engl. (Euphorbiacée africaine). Bull. Jard. Bot. Natl. Belg. 59: 133-149. Fr. — Complete treatment, with map, citations of exsiccatae, synonymy, notes on uses, and discussion; a single species, *S. preussii*, accepted. Chorology is considered on pp. 147-148; the two subspecies have different 'affinities' with subsp. *preusii* being more 'oceanic' and at lower altitudes (see map, p. 141).

Spondianthus Engl., Bot. Jahrb. Syst. 36: 215 (1905).
Trop. Africa. 22 23 24 25 26. Phan.
Megabaria Pierre ex Hutch., Bull. Misc. Inform. Kew 1910: 56 (1910).

Spondianthus preussii Engl., Bot. Jahrb. Syst. 36: 216 (1905). *Spondianthus preussii* var. *genuinus* Pax & K.Hoffm. in H.G.A.Engler, Pflanzenr., IV, 147, XV: 15 (1922), nom. inval.
Trop. Africa. 22 BEN GHA GUI IVO LBR NGA 23 CAF CMN CON EQG GAB GGI ZAI 24 SUD 25 TAN UGA 26 MOZ. Phan.

var. **glaber** (Engl.) Engl., Notizbl. Bot. Gart. Berlin-Dahlem 5: 242 (1911).
Trop. Africa. 22 GUI IVO NGA 23 CAF CMN CON GAB ZAI 24 SUD 25 TAN UGA 26 ANG. Phan.
* Spondianthus glaber Engl., Bot. Jahrb. Syst. 36: 216 (1905). *Spondianthus preussii* subsp. *glaber* (Engl.) J.Léonard & Nkounkou, Bull. Jard. Bot. Belg. 59: 143 (1989).
Megabaria ugandensis Hutch., Bull. Misc. Inform. Kew 1910: 57 (1910). *Spondianthus ugandensis* (Hutch.) Hutch. in D.Oliver, Fl. Trop. Afr. 6(1): 1044 (1913).
Thecacoris trillesii Beille, Bull. Soc. Bot. France 57(8): 120 (1910).

var. **preussii**
W. & WC. Trop. Africa. 22 BEN GHA IVO LBR NGA 23 CMN EQG GAB GGI ZAI. Phan.
Megabaria trillesii Pierre ex Hutch., Bull. Misc. Inform. Kew 1910: 57 (1910).

Synonyms:
Spondianthus glaber Engl. === **Spondianthus preussii** var. glaber (Engl.) Engl.
Spondianthus obovatus Engl. === **Protomegabaria stapfiana** (Beille) Hutch.
Spondianthus preussii var. *genuinus* Pax & K.Hoffm. === **Spondianthus preussii** Engl.
Spondianthus preussii subsp. *glaber* (Engl.) J.Léonard & Nkounkou === **Spondianthus preussii** var. **glaber** (Engl.) Engl.
Spondianthus ugandensis (Hutch.) Hutch. === **Spondianthus preussii** var. **glaber** (Engl.) Engl.

Stachyandra

4 species, Madagascar (largely in the N.); allied to *Androstachys* and questionably distinct. Trees to 35 m with monopodial branch growth, the leaves opposite and palmately 3-7-foliolate, the blades with contrasting surfaces (the under being whitish-lanuginose). The petioles arise from and are mostly adnate to stipular sheaths at the end of each internode, which later falls; they are best seen as protection for the shoot apex but may have an additional function. The rather twiggy flushes are on the whole reminiscent of species of *Vitex* (Labiatae). The trees sometimes form pure stands, as reported by Capuron (Radcliffe-Smith, 1990). The wood is similar to that of *Androstachys*: hard and termite-resistant. Only *S. merana* is tolerably known. [Airy-Shaw (1965; see **General**) proposed for *Androstachys* s.l. separate family status as Androstachydaceae but this generally has not found favour.] (Oldfieldioideae)

Airy-Shaw, H. K. (1970). A review of *Androstachys* Prain in Madagascar. Adansonia, II, 10: 519-524. En. — Account of 5 species with descriptions of 4 novelties (these latter now in *Stachyandra*), key, synonymy, references and citations (mostly for *A. johnsonii*), localities with exsiccatae, and commentary. [Some of the localities of $$A. johnsonii in Madagascar are from ravines in the isolated southern Isalo massif.]

Radcliffe-Smith, A. (1990). Notes on Madagascan Euphorbiaceae, III: *Stachyandra*. Kew Bull. 45: 561-568. En. — Protologue of a new Malagasy genus (segregated from *Androstachys*); 3 new combinations.

Stachyandra J.-F.Leroy ex Radcl.-Sm., Kew Bull. 45: 562 (1990).
Madagascar. 29. Phan.

Stachyandra imberbis (Airy Shaw) Radcl.-Sm., Kew Bull. 45: 567 (1990).
NW. Madagascar. 29 MDG. Phan.
Androstachys imberbis Airy Shaw, Adansonia, n.s. 10: 521 (1971).

Stachyandra merana (Airy Shaw) J.-F.Leroy ex Radcl.-Sm., Kew Bull. 45: 567 (1990).
NW. Madagascar. 29 MDG. Phan.
Androstachys merana Airy Shaw, Adansonia, n.s. 10: 520 (1971).

var. **merana**
NW. Madagascar. 29 MDG. Phan.

var. **obovalifoliola** Radcl.-Sm., Kew Bull. 45: 567 (1990).
NW. Madagascar. 29 MDG. Phan.

Stachyandra rufibarbis (Airy Shaw) Radcl.-Sm., Kew Bull. 45: 568 (1990).
NE. Madagascar. 29 MDG. Phan.
Androstachys rufibarbis Airy Shaw, Adansonia, n.s. 10: 523 (1971).

Stachyandra viticifolia (Airy Shaw) Radcl.-Sm., Kew Bull. 45: 568 (1990).
− FIGURE, p. 1482.
NW. Madagascar. 29 MDG. Phan.
Androstachys viticifolia Airy Shaw, Adansonia, n.s. 10: 522 (1971).

Stachystemon

Now combined with *Pseudanthus* (Radcliffe-Smith, 1996; see that genus).

Synonyms:
Stachystemon Planch. === **Pseudanthus** Sieber ex Spreng.
Stachystemon axillaris George === **Pseudanthus axillaris** (A.S.George) Radcl.-Sm.
Stachystemon brachyphyllus Müll.Arg. === **Pseudanthus brachyphyllus** (Müll.Arg.) F.Muell.
Stachystemon brevifolius Planch. ex Grüning === **Pseudanthus brachyphyllus** (Müll.Arg.) F.Muell.
Stachystemon polyandrus (F.Muell.) Benth. === **Pseudanthus polyandrus** F.Muell.
Stachystemon vermicularis Planch. === **Pseudanthus vermicularis** (Planch.) F.Muell.

Staphysora

Synonyms:
Staphysora Pierre === **Maesobotrya** Benth.
Staphysora albida Pierre ex Pax === **Maesobotrya longipes** (Pax) Hutch.
Staphysora dusenii Pax === **Maesobotrya klaineana** (Pierre) J.Léonard
Staphysora klaineana Pierre === **Maesobotrya klaineana** (Pierre) J.Léonard
Staphysora sapinii De Wild. === **Maesobotrya bertramiana** Büttner

Stachyandra viticifolia (Airy Shaw) Radcl.-Sm.

Artist: Christine Grey-Wilson
Kew Bull. 45: 566, fig. 4 (1990)
KEW ILLUSTRATIONS COLLECTION

Staurothyrax

A posthumously published and never-used name of Griffith, now reduced to *Phyllanthus*.

Steigeria

Synonyms:
Steigeria Müll.Arg. === **Baloghia** Endl.
Steigeria montana Müll.Arg. === **Baloghia montana** (Müll.Arg.) Pax

Stelechanteria

Synonyms:
Stelechanteria Thouars ex Baill. === **Drypetes** Vahl
Stelechanteria thouarsiana Baill. === **Drypetes thouarsiana** (Baill.) Capuron

Stenadenium

Synonyms:
Stenadenium Pax === **Monadenium** Pax
Stenadenium spinescens Pax === **Monadenium spinescens** (Pax) Bally

Stenonia

Synonyms:
Stenonia Didr. === **Ditaxis** Vahl ex A.Juss.
Stenonia montevidensis Didr. === **Ditaxis montevidensis** (Didr.) Pax

Stenonia

A later homonym of *Stenonia* Didr.

Synonyms:
Stenonia Baill. === **Cleistanthus** Hook.f. ex Planch.

Stenoniella

Synonyms:
Stenoniella Kuntze === **Cleistanthus** Hook.f. ex Planch.

Sterigmanthe

A Klotzsch segregate of *Euphorbia*.

Synonyms:
Sterigmanthe Klotzsch & Garcke === **Euphorbia** L.

Stilaginella

Synonyms:
Stilaginella Tul. === **Hieronyma** Allemão
Stilaginella amazonica Tul. === **Hieronyma alchorneoides** Alleão var. **alchorneoides**
Stilaginella benthamii Tul. === **Hieronyma oblonga** (Tul.) Müll.Arg.

Stilaginella blanchetiana Tul. === **Hieronyma oblonga** (Tul.) Müll.Arg.
Stilaginella clusioides Tul. === **Hieronyma clusioides** (Tul.) Griseb.
Stilaginella ferruginea Tul. === **Hieronyma alchorneoides** Alleão var. **alchorneoides**
Stilaginella laxiflora Tul. === **Hieronyma alchorneoides** Alleão var. **alchorneoides**
Stilaginella oblonga Tul. === **Hieronyma oblonga** (Tul.) Müll.Arg.
Stilaginella scabrida Tul. === **Hieronyma scabrida** (Tul.) Müll.Arg.

Stilago

Synonyms:
Stilago L. === **Antidesma** L.
Stilago bunius L. === **Antidesma bunius** (L.) Spreng.
Stilago diandra Roxb. === **Antidesma acidum** Retz.
Stilago lanceolaria Roxb. === **Antidesma acidum** Retz.
Stilago tomentosa Roxb. === **Antidesma roxburghii** Wall. ex Tul.

Stillingfleetia

Synonyms:
Stillingfleetia Bojer === **Triadica** Lour.

Stillingia

29 species, warmer regions but mostly in Middle and South America; 2 (*SS. terminalis*, *thouarsiana*) occur in Madagascar and 1 (*S. lineata*) disjunctly from the Mascarenes to Fiji. Laticiferous annuals, perennials, subshrubs or shrubs or (rarely) sparely branching trees to 12 m (*S. lineata*) with simple alternate or opposite leaves and terminal spiciform thyrses. In the Pax system (1912) six sections were recognised, based partly on habit. Pax's revision has been largely succeeded by that of Rogers (1951); he recognised 23 species in 2 subgenera, *Stillingia* and *Gymnostillingia*, and 5 series (for distribution maps see pp. 216-217 therein). Esser (1994; see **Euphorbioideae (except Euphorbieae**)) credits the genus with three groups: subgen. *Stillingia* (exclusive of ser. *Dichotomae* Rogers) with 14 species; ser. *Dichotomae* with 8; and subgen. *Gymnostillingia* with 5. *Dichotomae* encompasses the extra-American species; in the Americas it is limited to Atlantic Brazil. Here the leaves and, often, the stems are 'succulent'; some species at least are more or less coastal. Subgen. *Gymnostillingia* ranges disjunctly from SW. North America to Guatemala, while subgen. *Stillingia* s.s. ranges widely (but also disjunctly) in the Americas. *S. lineata* exhibits considerable variability in the Mascarenes, including distinctive nearly-linear juvenile foliage. The genus is distinguished from both *Sapium* s.l. and *Excoecaria* in particular by a trilobed carpophore. *S. patagonica* has to be excluded (see **Unplaced taxa**). (Euphorbioideae (except Euphorbieae))

Pax, F. (with K. Hoffmann) (1912). *Stillingia*. In A. Engler (ed.), Das Pflanzenreich, IV 147 V (Euphorbiaceae-Hippomaneae): 180-199. Berlin. (Heft 52.) La/Ge. — 26 species in 6 sections, mostly in Middle and S America.

Rogers, D. J. (1951). A revision of *Stillingia* in the New World. Ann. Missouri Bot. Gard. 38: 207-259, illus., maps. (Based on a Ph.D. dissertation, Washington University, St. Louis.) En. — Revision (23 species in 2 subgenera and 5 series, a further species being doubtful) with keys, descriptions, synonymy, references, indication of distribution, localities with exsiccatae, taxonomic commentary, figures and maps; list of collections seen and index to all names at end.

Airy-Shaw, H. K. (1963). Notes on Malaysian and other Asiatic Euphorbiaceae, XLVIII. The genus *Stillingia* Garden in E. Malaysia. Kew Bull. 16: 372. En. — Extension of range; *S. pacifica*, described from Fiji, was found in Timor and Babar. The species is closely related to *S. lineata* in the Mascarene Islands.

Airy-Shaw, H. K. (1972). Notes on Malesian and other Asiatic Euphorbiaceae, CLXIX. Note on *Stillingia lineata*. Kew Bull. 27: 93. En. —*Stillingia lineata*, with its subspecies *pacifica*, ranges disjunctly from Mauritius and Réunion through Nusa Tenggara (Indonesia) to Fiji.

- McVaugh, R. (1995). Euphorbiacearum sertum Novo-Galicianarum revisarum. Contr. Univ. Michigan Herb. 20: 173-215, illus. En. — Revision precursory to treatment in *Flora Novo-Galiciana*; includes (pp. 210-215) a treatment of *Stillingia* (4 species), with key.

Esser, H.-J. (1999). A partial revision of the Hippomaneae (Euphorbiaceae) in Malesia. Blumea 44: 149-215, illus., maps. En. — *Stillingia*, pp. 190-197; treatment of 1 species (*S. lineata* subsp. *pacifica*) with description, synonymy, references, types, indication of distribution and habitat, illustration, map and commentary; all general references, identification list and index to botanical names at end of paper. [Found primarily on small islands.]

Stillingia L., Mant. Pl. 1: 19 (1767).
> Trop. & Subtrop. America, Madagascar, Mascarenes, Trop. Asia, Fiji. 29 42 60 76 77 78 79 80 82 83 84 85.
> *Gymnostillingia* Müll.Arg., Linnaea 32: 89 (1863).

Stillingia acutifolia (Benth.) Benth. & Hook.f. ex Hemsl., Biol. Cent.-Amer., Bot. 3: 135 (1883).
> Mexico (Chiapas), Guatemala. 79 MXT 80 GUA. Nanophan. or phan.
> * *Sapium acutifolium* Benth., Pl. Hartw.: 90 (1842).
> *Gymnostillingia macrantha* Müll.Arg., Linnaea 32: 90 (1863). *Stillingia macrantha* (Müll.Arg.) Benth. & Hook.f. ex Hemsl., Biol. Cent.-Amer., Bot. 3: 135 (1883).
> *Stillingia propria* Brandegee, Univ. Calif. Publ. Bot. 6: 185 (1915).

Stillingia aquatica Chapm., Fl. South U.S.: 405 (1860).
> Georgia, Florida. 76 FLA GEO. Hydronanophan.

Stillingia argutedentata Jabl., Phytologia 14: 451 (1967).
> Brazil (Minas Gerais). 84 BZL. Nanophan.

Stillingia bicarpellaris S.Watson, Proc. Amer. Acad. Arts 21: 455 (1886).
> Mexico (Coahuila, Hidlago, Guanajuato). 79 MXE. Nanophan.

Stillingia bodenbenderi (Kuntze) D.J.Rogers, Ann. Missouri Bot. Gard. 38: 222 (1951).
> Brazil (São Paulo), Argentina (Córdoba). 84 BZL 85 AGE. Nanophan. or phan.
> * *Sapium bodenbenderi* Kuntze, Revis. Gen. Pl. 3(2): 292 (1898). *Excoecaria bodenbenderi* (Kuntze) K.Schum., Just's Bot. Jahresber. 26(1): 349 (1900).
> *Sapium subsessile* Hemsl., Hooker's Icon. Pl. 27: t. 2684 (1901).
> *Sapium bolanderi* Kuntze ex Pax in H.G.A.Engler, Pflanzenr., IV, 147, V: 225 (1912), nom. illeg.

Stillingia dichotoma Müll.Arg., Linnaea 32: 88 (1863).
> Brazil (Rio de Janeiro). 84 BZL. Nanophan.
> *Sapium dichotomum* Klotzsch ex Pax in H.G.A.Engler, Pflanzenr., IV, 147, V: 184 (1912), nom. inval.

Stillingia diphtherina D.J.Rogers, Ann. Missouri Bot. Gard. 38: 226 (1951).
> Guatemala, Honduras. 80 GUA HON. Nanophan.

Stillingia dusenii Pax & K.Hoffm. in H.G.A.Engler, Pflanzenr., IV, 147, XVII: 202 (1924).
> Brazil (Paraná). 84 BZS. Cham.

Stillingia linearifolia S.Watson, Proc. Amer. Acad. Arts 14: 297 (1879). *Stillingia gymnogyna* Pax & K.Hoffm. in H.G.A.Engler, Pflanzenr., IV, 147, V: 196 (1912).
> S. California, W. Arizona, S. Nevada, Guadalupe I., Mexico (Baja California, Sonora). 76 ARI CAL NEV 79 MXI MXN. Hemicr.

Stillingia lineata (Lam.) Müll.Arg. in A.P.de Candolle, Prodr. 15(2): 1157 (1866).
Réunion, Mauritius, Malesia, Fiji. 29 MAU REU 42 LSI MOL PHI 60 FIJ. Nanophan.
* *Sapium lineatum* Lam., Encycl. 2: 734 (1788).

subsp. **lineata**
Réunion, Mauritius. 29 MAU REU. Nanophan.
Sapium laevigatum Lam., Encycl. 2: 735 (1788). *Stillingia lineata* var. *laevigata* (Lam.)
Baill., Adansonia 2: 27 (1861).
Sapium obtusifolium Lam., Encycl. 2: 735 (1790). *Stillingia obtusifolia* (Lam.) Baill.,
Étude Euphorb.: 513 (1858). *Stillingia lineata* var. *obtusifolia* (Lam.) Baill., Adansonia
2: 27 (1861).
Sapium laevifolium Thouars ex Baill., Étude Euphorb.: 513 (1858).
Stillingia mauritiana Baill., Adansonia 2: 27 (1861). *Excoecaria mauritiana* (Baill.) Baill.,
Traité Bot. Méd. Phan. 2: 946 (1884).
Stillingia tanguinia Baill., Adansonia 2: 28 (1861). *Sapium tanguinum* (Baill.) Müll.Arg.,
Linnaea 32: 116 (1863). *Stillingia lineata* var. *tanguina* (Baill.) Pax in H.G.A.Engler,
Pflanzenr., IV, 147, V: 183 (1912).
Sapium cassinefolium Tausch ex Pax in H.G.A.Engler, Pflanzenr., IV, 147, V: 183 (1912),
nom. illeg.

subsp. **pacifica** (Müll.Arg.) Steenis, Blumea, Suppl. 5: 302 (1966).
Lesser Sunda Is. (Timor), Malaku (Babar), Philippines, Fiji. 42 LSI MOL PHI 60 FIJ.
* *Stillingia pacifica* Müll.Arg. in A.P.de Candolle, Prodr. 15(2): 1156 (1866).
Sapium plumerioides Croizat, J. Arnold Arbor. 23: 507 (1942).

Stillingia oppositifolia Baill. ex Müll.Arg. in A.P.de Candolle, Prodr. 15(2): 1219 (1866).
Brazil (Minas Gerais, Rio Grande do Sul). 84 BZL BZS. Nanophan. or phan.
Sapium oppositifolium Klotzsch ex Baill., Étude Euphorb.: 513 (1858), nom. nud.
Sapium sanguinolentum Klotzsch ex Pax in H.G.A.Engler, Pflanzenr., IV, 147, V: 191
(1912), pro syn.

Stillingia parvifolia Sánchez Vega, Sagást. & Huft, Ann. Missouri Bot. Gard. 75: 1666 (1988
publ. 1989).
Peru. 83 PER.

Stillingia paucidentata S.Watson, Proc. Amer. Acad. Arts 14: 298 (1879). *Stillingia
linearifolia* var. *paucidentata* (S.Watson) Jeps., Fl. Calif. 2(1): 422 (1936).
California. 76 CAL. Hemicr.

Stillingia peruviana D.J.Rogers, Ann. Missouri Bot. Gard. 38: 223 (1951).
Peru (Huancavelica). 83 PER. Nanophan. – Fruit edible.

Stillingia pietatis McVaugh, Contrib. Univ. Mich. Herb. 20: 210 (1995).
Mexico (Michoacán). 79 MXS.

Stillingia querceticola McVaugh, Contrib. Univ. Mich. Herb. 20: 213 (1995).
Mexico (Nayarit). 79 MXS.

Stillingia salpingadenia Huber, Bull. Herb. Boissier, II, 6: 452 (1906). *Excoecaria
salpingadenia* (Huber) Müll.Arg. in A.P.de Candolle, Prodr. 15(2): 1209 (1866).
Bolivia, Paraguay. 83 BOL 85 PAR. Cham. or nanophan.
* *Sapium salpingadenium* Müll.Arg., Linnaea 32: 121 (1863).
Sapium cupuliferum Hemsl., Hooker's Icon. Pl. 27: t. 2679 (1901). *Stillingia salpingadenia*
var. *cupulifera* (Hemsl.) Pax in H.G.A.Engler, Pflanzenr., IV, 147, V: 190 (1912).
Stillingia saxatilis var. *salicina* Chodat & Hassl., Bull. Herb. Boissier, II, 5: 677 (1905).
Stillingia salpingadenia var. *salicina* (Chodat & Hassl.) Pax in H.G.A.Engler, Pflanzenr.,
IV, 147, V: 190 (1912).
Stillingia salpingadenia subsp. *anadena* Pax & K.Hoffm. in H.G.A.Engler, Pflanzenr., IV,
147, V: 190 (1912).

Stillingia sanguinolenta Müll.Arg., Linnaea 32: 88 (1863).
  Mexico. 79 MXE MXC MXT. Nanophan.
    Stillingia sanguinolenta var. *angustifolia* Müll.Arg., Linnaea 32: 88 (1863).
    Stillingia sanguinolenta var. *lanceolata* Müll.Arg., Linnaea 32: 88 (1863).
    Sapium angulatum Klotzsch ex Pax in H.G.A.Engler, Pflanzenr., IV, 147, V: 191 (1912),
      pro syn.

Stillingia saxatilis Müll.Arg. in C.F.P.von Martius, Fl. Bras. 11(2): 539 (1874). *Stillingia*
  salpingadenia subsp. *saxatilis* (Müll.Arg.) Pax & K.Hoffm. in H.G.A.Engler, Pflanzenr., IV,
  147, V: 189 (1912).
  Brazil (Bahia, Minas Gerais). 84 BZE BZL. Nanophan.
    Gymnostillingia loranthacea Müll.Arg. in C.F.P.von Martius, Fl. Bras. 11(2): 541 (1874).
      Stillingia loranthacea (Müll.Arg.) Pax in H.G.A.Engler, Pflanzenr., IV, 147, V: 185 (1912).

Stillingia scutellifera D.J.Rogers, Ann. Missouri Bot. Gard. 38: 235 (1951).
  Paraguay, NE. Argentina. 85 AGE PAR. Cham. or nanophan.
    Stillingia saxatilis f. *angustior* Chodat & Hassl., Bull. Herb. Boissier, II, 5: 676 (1905).
      Stillingia salpingadenia var. *angustior* (Chodat & Hassl.) Pax & K.Hoffm. in
        H.G.A.Engler, Pflanzenr., IV, 147, V: 190 (1912).
    Stillingia saxatilis f. *latior* Chodat & Hassl., Bull. Herb. Boissier, II, 5: 676 (1905).
    Stillingia saxatilis var. *grandifolia* Chodat & Hassl., Bull. Herb. Boissier, II, 5: 676 (1905).
      Stillingia salpingadenia var. *grandifolia* (Chodat & Hassl.) Pax & K.Hoffm. in
        H.G.A.Engler, Pflanzenr., IV, 147, V: 190 (1912).
  * *Stillingia saxatilis* var. *salicifolia* Chodat & Hassl., Bull. Herb. Boissier, II, 5: 676 (1905).
    Stillingia salpingadenia var. *elliptica* Pax & K.Hoffm. in H.G.A.Engler, Pflanzenr., IV, 147,
      V: 189 (1912).

Stillingia spinulosa Torr. in W.H.Emory, Not. Milit. Reconn.:151 (1848).
  SE. California, SW. Arizona, Mexico (Baja California del Norte, Sonora). 76 ARI CAL 79
  MXN. Hemicr.
    Sapium annuum Torr. in W.H.Emory, Rep. U.S. Mex. Bound. 2(1): 201 (1858). *Stillingia*
      annua (Torr.) Müll.Arg. in A.P.de Candolle, Prodr. 15(2): 1160 (1866).

Stillingia sylvatica L., Mant. Pl. 1: 126 (1767). *Sapium sylvaticum* (L.) Torr. in W.H.Emory,
  Rep. U.S. Mex. Bound. 2(1): 201 (1858). *Stillingia sylvatica* var. *genuina* Müll.Arg. in A.P.de
  Candolle, Prodr. 15(2): 1158 (1866), nom. inval. *Excoecaria sylvatica* (L.) Baill., Traité Bot.
  Méd. Phan. 2: 946 (1884). – FIGURE, p. 1488.
  C. & SE. USA. 74 KAN OKL 77 NWM TEX 78 ALA ARK FLA GEO LOU NCA SCA VRG.
  Cham. or nanophan.

subsp. **sylvatica**
    C. & SE. USA. 74 KAN OKL 77 NWM TEX 78 ALA ARK FLA GEO LOU NCA SCA VRG.
      Cham. or nanophan.
    Stillingia sylvatica var. *salicifolia* Torr., Ann. Lyceum Nat. Hist. New York 2: 245 (1826).
      Stillingia salicifolia (Torr.) Raf., Atlantic J.: 146 (1832).
    Stillingia lanceolata Nutt., Trans. Amer. Philos. Soc., n.s., 5: 176 (1837).
    Stillingia sylvatica var. *angustifolia* Müll.Arg. in A.P.de Candolle, Prodr. 15(2): 1158
      (1866). *Stillingia angustifolia* (Müll.Arg.) Engelm. ex S.Watson, Proc. Amer. Acad. Arts
      18: 154 (1888).
    Stillingia sylvatica var. *spathulata* Müll.Arg. in A.P.de Candolle, Prodr. 15(2): 1158
      (1866). *Stillingia spathulata* (Müll.Arg.) Small, Fl. S.E. U.S.: 704 (1903).
    Stillingia smallii Wooton & Standl., Contr. U. S. Natl. Herb. 19: 405 (1915).

subsp. **tenuis** (Small) D.J.Rogers, Ann. Missouri Bot. Gard. 38: 241 (1951).
  Florida (Dale Co.). 78 FLA. Cham.
  * *Stillingia tenuis* Small, Bull. New York Bot. Gard. 3: 429 (1905).

Stillingia sylvatica L. subsp. *sylvatica*

Artist: David Blair
Bentley & Trimen, Medicinal Plants, 4: Pl. 241 (1880), uncoloured version
KEW ILLUSTRATIONS COLLECTION

Stillingia terminalis Baill., Adansonia 2: 29 (1861). *Sapium terminale* (Baill.) Müll.Arg., Linnaea 32: 116 (1863).
Madagascar. 29 MDG.
  Stillingia mauritiana var. *salicina* Baill., Adansonia 2: 27 (1861).

Stillingia texana I.M.Johnst., Contr. Gray Herb. 68: 91 (1923).
Oklahoma, New Mexico, Texas, Mexico (Coahuila). 74 OKL 77 NWM TEX 79 MXE.
Cham. or nanophan.
  * *Sapium sylvaticum* var. *linearifolium* Torr. in W.H.Emory, Rep. U.S. Mex. Bound. 2(1): 201 (1858). *Stillingia sylvatica* var. *linearifolia* (Torr.) Müll.Arg. in A.P.de Candolle, Prodr. 15(2): 1158 (1866). *Stillingia linearifolia* (Torr.) Small, Fl. S.E. U.S.: 704 (1903), nom. illeg. *Stillingia texana* var. *typica* Waterf., Rhodora 50: 95 (1948), nom. inval.
  Stillingia texana var. *latifolia* Waterf., Rhodora 50: 95 (1948).

Stillingia thouarsiana Baill., Adansonia 2: 28 (1861). *Excoecaria thouarsiana* (Baill.) Müll.Arg. in A.P.de Candolle, Prodr. 15(2): 1218 (1866).
Madagascar. 29 MDG. Nanophan.

Stillingia trapezoidea Ule, Bot. Jahrb. Syst. 42: 223 (1908).
Brazil (Piauí). 84 BZE. Nanophan. or phan.

Stillingia treculiana (Müll.Arg.) I.M.Johnst., Contr. Gray Herb. 68: 91 (1923).
S. Texas, Mexico (Coahuila, Nuevo León, Tamaulipas). 77 TEX 79 MXE. Hemicr.
  Sapium annuum var. *dentatum* Torr. in W.H.Emory, Rep. U.S. Mex. Bound. 2(1): 201 (1858). *Stillingia torreyana* S.Watson, Proc. Amer. Acad. Arts 14: 298 (1879). *Stillingia dentata* (Torr.) Britton & Rusby, Trans. New York Acad. Sci. 7: 14 (1887).
  * *Gymnanthes treculiana* Müll.Arg., Linnaea 34: 216 (1865). *Sebastiania treculiana* (Müll.Arg.) Müll.Arg. in A.P.de Candolle, Prodr. 15(2): 1165 (1866).

Stillingia uleana Pax & K.Hoffm. in H.G.A.Engler, Pflanzenr., IV, 147, V: 187 (1912).
Brazil (Bahia). 84 BZE. Nanophan. or phan.

Stillingia zelayensis (Kunth) Müll.Arg., Linnaea 32: 87 (1863).
Mexico to W. Panama. 79 MXC MXE MXS MXT 80 BLZ COS GUA HON PAN. Nanophan.
  * *Sapium zelayense* Kunth in F.W.H.von Humboldt, A.J.A.Bonpland & C.S.Kunth, Nov. Gen. Sp. 2: 65 (1817).
  Stillingia microsperma Pax & K.Hoffm. in H.G.A.Engler, Pflanzenr., IV, 147, V: 187 (1912).

Synonyms:
Stillingia africana (Sond.) Baill. === **Spirostachys africana** Sond.
Stillingia agallocha (L.) Baill. === **Excoecaria agallocha** L.
Stillingia angustifolia (Müll.Arg.) Engelm. ex S.Watson === **Stillingia sylvatica** L. subsp. **sylvatica**
Stillingia annua (Torr.) Müll.Arg. === **Stillingia spinulosa** Torr.
Stillingia appendiculata Müll.Arg. === **Sebastiania appendiculata** (Müll.Arg.) Jabl.
Stillingia arborescens Pittier === **Sebastiania granatensis** (Müll.Arg.) Müll.Arg.
Stillingia asperococca (F.Muell.) Baill. === **Microstachys chamaelea** (L.) Müll.Arg.
Stillingia aucuparia (Jacq.) Oken === **Sapium glandulosum** (L.) Morong
Stillingia baccata (Roxb.) Baill. === **Balakata baccata** (Roxb.) Esser
Stillingia bahiensis (Müll.Arg.) Baill. === **Sebastiania bahiensis** (Müll.Arg.) Müll.Arg.
Stillingia bidentata Baill. === **Microstachys bidentata** (Mart. & Zucc.) Esser
Stillingia bidentata (Mart. & Zucc.) Baill. === **Microstachys bidentata** (Mart. & Zucc.) Esser
Stillingia biglandulosa (L.) Baill. === **Sapium glandulosum** (L.) Morong
Stillingia bingyrica Baill. === **Shirakiopsis indica** (Willd.) Esser
Stillingia brasiliensis (Spreng.) Müll.Arg. === **Sebastiania brasiliensis** Spreng.
Stillingia brevifolia (Müll.Arg.) Baill. === **Sebastiania brevifolia** (Müll.Arg.) Müll.Arg.
Stillingia campestris Baill. === **Microstachys corniculata** (Vahl) Griseb.

Stillingia chamaelea (L.) Müll.Arg. === **Microstachys chamaelea** (L.) Müll.Arg.

Stillingia cochinchinensis (Lour.) Baill. === **Triadica cochinchinensis** Lour.

Stillingia commersoniana Baill. === **Sebastiania klotzschiana** (Müll.Arg.) Müll.Arg.

Stillingia concolor (Spreng.) Baill. === **Actinostemon concolor** (Spreng.) Müll.Arg.

Stillingia coriacea (Mart.) Baill. === **Microstachys marginata** (Mart.) Klotzsch ex Müll.Arg.

Stillingia corniculata (Vahl) Baill. === **Microstachys corniculata** (Vahl) Griseb.

Stillingia cremostachys Baill. === **Sebastiania klotzschiana** (Müll.Arg.) Müll.Arg.

Stillingia crotonoides (Mart.) Baill. === **Microstachys corniculata** ?

Stillingia cruenta Standl. & Steyerm. === **Sebastiania cruenta** (Standl. & Steyerm.) Miranda

Stillingia cubana (A.Rich.) Baill. === **Bonania cubana** A.Rich.

Stillingia daphniphylla Baill. === **Sebastiania daphniphylla** (Baill.) Müll.Arg.

Stillingia dentata (Torr.) Britton & Rusby === **Stillingia treculiana** (Müll.Arg.) I.M.Johnst.

Stillingia desertorum Klotzsch === **Sebastiania brasiliensis** Spreng.

Stillingia discolor Champ. ex Benth. === **Triadica cochinchinensis** Lour.

Stillingia discolor (Spreng.) Baill. === **Gymnanthes discolor** (Spreng.) Müll.Arg.

Stillingia diversifolia Miq. === **Shirakiopsis indica** (Willd.) Esser

Stillingia dracunculoides Baill. === **Sapium glandulosum** (L.) Morong

Stillingia eglandulosa A.Rich. === **Grimmeodendron eglandulosum** (A.Rich.) Urb.

Stillingia elliptica (Hochst.) Balll. === **Shirakiopsis elliptica** (Hochst.) Esser

Stillingia frutescens Bosc ex Steud. === **Ditrysinia fruticosa** (Bartram) Govaerts & Frodin

Stillingia fruticosa Bartram === **Ditrysinia fruticosa** (Bartram) Govaerts & Frodin

Stillingia fruticosa Spreng. === **Ditrysinia fruticosa** (Bartram) Govaerts & Frodin

Stillingia gaudichaudii (Müll.Arg.) Baill. === **Gymnanthes gaudichaudii** Müll.Arg.

Stillingia glabrata (Mart.) Baill. === **Gymnanthes glabrata** (Mart.) Govaerts

Stillingia glandulosa Dombey ex Baill. === **Adenopeltis serrata** (Aiton) I.M.Johnst.

Stillingia goudotiana Baill. === **Taeniosapium goudotianum** (Baill.) Müll.Arg.

Stillingia guianensis (Aubl.) Baill. === **Maprounea guianensis** Aubl.

Stillingia guineensis Benth. === **Excoecaria guineensis** (Benth.) Müll.Arg.

Stillingia gymnogyna Pax & K.Hoffm. === **Stillingia linearifolia** S.Watson

Stillingia haematantha Standl. === **Sapium glandulosum** (L.) Morong

Stillingia hastata Klotzsch ex Baill. === **Microstachys ditassoides** (Didr.) Esser

Stillingia heterodoxa Müll.Arg. === **Microstachys heterodoxa** (Müll.Arg.) Esser

Stillingia hilariana Baill. === **Maprounea guianensis** Aubl.

Stillingia himalayensis Klotzsch === **Excoecaria acerifolia** Didr.

Stillingia hippomane (G.Mey.) Baill. === **Sapium glandulosum** (L.) Morong

Stillingia hypoleuca (Benth.) Baill. === **Gymnanthes hypoleuca** Benth.

Stillingia indica (Willd.) Oken === **Shirakiopsis indica** (Willd.) Esser

Stillingia integerrima (Hochst.) Baill. === **Sclerocroton integerrimus** Hochst.

Stillingia jacobinensis (Müll.Arg.) Baill. === **Sebastiania jacobinensis** (Müll.Arg.) Müll.Arg.

Stillingia japonica Siebold & Zucc. === **Neoshirakia japonica** (Siebold & Zucc.) Esser

Stillingia lanceolaria Miq. === **Triadica cochinchinensis** Lour.

Stillingia lanceolata Nutt. === **Stillingia sylvatica** L. subsp. **sylvatica**

Stillingia lastellei Baill. === **Anomostachys lastellei** (Müll.Arg.) Kruijt

Stillingia laureola Baill. === **Sebastiania laureola (Baill.)** Müll.Arg.

Stillingia laurifolia A.Rich. === **Sapium laurifolium** (A.Rich.) Griseb.

Stillingia laurocerasus (Desf.) Baill. === **Sapium laurocerasus** Desf.

Stillingia ligustrina Michx. === **Ditrysinia fruticosa** (Bartram) Govaerts & Frodin

Stillingia linearifolia (Torr.) Small === **Stillingia texana** I.M.Johnst.

Stillingia linearifolia var. *paucidentata* (S.Watson) Jeps. === **Stillingia paucidentata** S.Watson

Stillingia lineata var. *densiflora* Baker === **Excoecaria benthamiana** Hemsl.

Stillingia lineata var. *laevigata* (Lam.) Baill. === **Stillingia lineata** (Lam.) Müll.Arg.
   subsp. **lineata**

Stillingia lineata var. *obtusifolia* (Lam.) Baill. === **Stillingia lineata (Lam.) Müll.Arg.**
   subsp. **lineata**

Stillingia lineata var. *tanguina* (Baill.) Pax === **Stillingia lineata** (Lam.) Müll.Arg.
   subsp. **lineata**

Stillingia loranthacea (Müll.Arg.) Pax === **Stillingia saxatilis** Müll.Arg.

Stillingia luschnathiana Baill. === **Gymnanthes glabrata** (Mart.) Govaerts

Stillingia macrantha (Müll.Arg.) Benth. & Hook.f. ex Hemsl. === **Stillingia acutifolia** (Benth.) Benth. & Hook.f. ex Hemsl.

Stillingia madagascariensis Baill. === **Excoecaria madagascariensis** (Baill.) Müll.Arg.

Stillingia marginata (Müll.Arg.) Baill. === **Sapium glandulosum** (L.) Morong

Stillingia mauritiana Baill. === **Stillingia lineata** (Lam.) Müll.Arg. subsp. **lineata**

Stillingia mauritiana var. *salicina* Baill. === **Stillingia terminalis** Baill.

Stillingia melanosticta Baill. === **Sclerocroton melanostictus** (Baill.) Kruijt & Roebers

Stillingia microsperma Pax & K.Hoffm. === **Stillingia zelayensis** (Kunth) Müll.Arg.

Stillingia multiramea (Klotzsch) Baill. === **Gymnanthes glabrata** (Mart.) Govaerts

Stillingia myrtilloides (Mart.) Baill. === **Microstachys daphnoides** (Mart.) Müll.Arg.

Stillingia nervosa (Müll.Arg.) Baill. === **Gymnanthes nervosa** Müll.Arg.

Stillingia nutans (G.Forst.) Geiseler === **Homalanthus nutans** (G.Forst.) Guill.

Stillingia obovata (Klotzsch ex Müll.Arg.) Baill. === **Sapium obovatum** Klotzsch ex Müll.Arg.

Stillingia obtusifolia (Lam.) Baill. === **Stillingia lineata** (Lam.) Müll.Arg. subsp. **lineata**

Stillingia pachystachya (Klotzsch) Müll.Arg. === **Sebastiania pachystachys** (Klotzsch) Müll.Arg.

Stillingia pacifica Müll.Arg. === **Stillingia lineata** subsp. **pacifica** (Müll.Arg.) Steenis

Stillingia paniculata Miq. === **Balakata baccata** (Roxb.) Esser

Stillingia patagonica (Speg.) Pax & K.Hoffm. === ?

Stillingia patula Baill. === **Microstachys corniculata** ?

Stillingia phyllanthiformis Baill. === **Sebastiania schottiana** (Müll.Arg.) Müll.Arg.

Stillingia populnea Geiseler === **Homalanthus populneus** (Geiseler) Pax

Stillingia propria Brandegee === **Stillingia acutifolia** (Benth.) Benth. & Hook.f. ex Hemsl.

Stillingia prostrata (Mart.) Baill. === **Microstachys corniculata** (Vahl) Griseb.

Stillingia prunifolia (Klotzsch) Baill. === **Sapium glandulosum** (L.) Morong

Stillingia pteroclada (Müll.Arg.) Baill. === **Sebastiania pteroclada** (Müll.Arg.) Müll.Arg.

Stillingia ramosissima (A.St.-Hil.) Baill. === **Sebastiania brasiliensis** Spreng.

Stillingia rigida (Müll.Arg.) Baill. === **Sebastiania rigida** (Müll.Arg.) Müll.Arg.

Stillingia salicifolia Small === **Stillingia sylvatica** L. subsp. **sylvatica**

Stillingia salicifolia (Torr.) Raf. === **Stillingia sylvatica** L. subsp. **sylvatica**

Stillingia salicifolia Klotzsch ex Baill. === **Sapium haematospermum** Müll.Arg.

Stillingia salpingadenia subsp. *anadena* Pax & K.Hoffm. === **Stillingia salpingadenia** Huber

Stillingia salpingadenia var. *angustior* (Chodat & Hassl.) Pax & K.Hoffm. === **Stillingia scutellifera** D.J.Rogers

Stillingia salpingadenia var. *cupulifera* (Hemsl.) Pax === **Stillingia salpingadenia** Huber

Stillingia salpingadenia var. *elliptica* Pax & K.Hoffm. === **Stillingia scutellifera** D.J.Rogers

Stillingia salpingadenia var. *grandifolia* (Chodat & Hassl.) Pax & K.Hoffm. === **Stillingia scutellifera** D.J.Rogers

Stillingia salpingadenia var. *salicina* (Chodat & Hassl.) Pax === **Stillingia salpingadenia** Huber

Stillingia salpingadenia subsp. *saxatilis* (Müll.Arg.) Pax & K.Hoffm. === **Stillingia saxatilis** Müll.Arg.

Stillingia sanguinolenta var. *angustifolia* Müll.Arg. === **Stillingia sanguinolenta** Müll.Arg.

Stillingia sanguinolenta var. *lanceolata* Müll.Arg. === **Stillingia sanguinolenta** Müll.Arg.

Stillingia saxatilis f. *angustior* Chodat & Hassl. === **Stillingia scutellifera** D.J.Rogers

Stillingia saxatilis var. *grandifolia* Chodat & Hassl. === **Stillingia scutellifera** D.J.Rogers

Stillingia saxatilis f. *latior* Chodat & Hassl. === **Stillingia scutellifera** D.J.Rogers

Stillingia saxatilis var. *salicifolia* Chodat & Hassl. === **Stillingia scutellifera** D.J.Rogers

Stillingia saxatilis var. *salicina* Chodat & Hassl. === **Stillingia salpingadenia** Huber

Stillingia schottiana (Müll.Arg.) Baill. === **Sebastiania schottiana** (Müll.Arg.) Müll.Arg.

Stillingia sebifera (L.) Michx. === **Triadica sebifera** (L.) Small

Stillingia sellowiana Baill. === **Microstachys corniculata** (Vahl) Griseb.

Stillingia serrata (Baill. ex Müll.Arg.) Klotzsch ex Baill. === **Sebastiania serrata** (Baill. ex Müll.Arg.) Müll.Arg.

Stillingia serrulata (Mart.) Baill. === **Microstachys serrulata** (Mart.) Müll.Arg.

Stillingia sinensis (Lour.) Baill. === **Triadica sebifera** (L.) Small

Stillingia smallii Wooton & Standl. === **Stillingia sylvatica** L. subsp. **sylvatica**

Stillingia spathulata (Müll.Arg.) Small === **Stillingia sylvatica** L. subsp. **sylvatica**

Stillingia stipulacea (Müll.Arg.) Baill. === **Sebastiania stipulacea** (Müll.Arg.) Müll.Arg.

Stillingia sylvatica var. *angustifolia* Müll.Arg. === **Stillingia sylvatica** L. subsp. **sylvatica**

Stillingia sylvatica var. *genuina* Müll.Arg. === **Stillingia sylvatica** L.

Stillingia sylvatica var. *linearifolia* (Torr.) Müll.Arg. === **Stillingia texana** I.M.Johnst.

Stillingia sylvatica var. *paraguayensis* Morong === **Sapium haematospermum** Müll.Arg.

Stillingia sylvatica var. *salicifolia* Torr. === **Stillingia sylvatica** L. subsp. **sylvatica**

Stillingia sylvatica var. *spathulata* Müll.Arg. === **Stillingia sylvatica** L. subsp. **sylvatica**

Stillingia tanguinia Baill. === **Stillingia lineata** (Lam.) Müll.Arg. subsp. **lineata**

Stillingia tenuis Small === **Stillingia sylvatica** subsp. **tenuis** (Small) D.J.Rogers

Stillingia texana var. *latifolia* Waterf. === **Stillingia texana** I.M.Johnst.

Stillingia texana var. *typica* Waterf. === **Stillingia texana** I.M.Johnst.

Stillingia torreyana S.Watson === **Stillingia treculiana** (Müll.Arg.) I.M.Johnst.

Stillingia trinervia (Müll.Arg.) Baill. === **Sebastiania trinervia** (Müll.Arg.) Müll.Arg.

Stillingia velutina Baill. === **Microstachys corniculata** (Vahl) Griseb.

Stillingia virgata Baill. === **Shirakiopsis virgata** (Zoll. & Moritzi ex Miq.) Esser

Stillingia weddelliana Baill. === **Sebastiania weddelliana** (Baill.) Müll.Arg.

Stillingia widgrenii (Müll.Arg.) Baill. === **Gymnanthes widgrenii** Müll.Arg.

Stillingia ypanemensis (Müll.Arg.) Baill. === **Sebastiania ypanemensis** (Müll.Arg.) Müll.Arg.

Stipellaria

Synonyms:

Stipellaria Benth. === **Alchornea** Sw.

Stipellaria mollis Benth. === **Alchornea mollis** (Benth.) Müll.Arg.

Stipellaria parviflora Benth. === **Alchornea sicca** (Blanco) Merr.

Stipellaria tiliifolia Benth. === **Alchornea tiliifolia** (Benth.) Müll.Arg.

Stipellaria trewioides Benth. === **Alchornea trewioides** (Benth.) Müll.Arg.

Stipellaria villosa Benth. === **Alchornea villosa** (Benth.) Müll.Arg.

Stomatocalyx

Synonyms:

Stomatocalyx Müll.Arg. === **Pimelodendron** Hassk.

Stomatocalyx griffithianus Müll.Arg. === **Pimelodendron griffithianum** (Müll.Arg.) Benth.

Strophioblachia

1 species, S. China and SE. Asia to Malesia (Philippines, Sulawesi). Undershrubs or shrubs of dry forest or scrub to 2 m with thin, variously shaped (sometimes cordate or pandurate) foliage; may occur on limestone. A revision by Nguyen Nghia Thin and Vu Hoai Duc appeared in 1998 (see below); only one species has been accepted (cf. also Thin 1995: 15; under **Asia**). The genus is related to *Blachia* and to *Pantadenia*, both also Asiatic; Webster (Synopsis, 1994), however, also suggests an affinity with *Sagotia* which seems on the face of it unlikely. (Crotonoideae)

> Pax, F. (with K. Hoffmann) (1911). *Strophioblachia*. In A. Engler (ed.), Das Pflanzenreich, IV 147 III (Euphorbiaceae-Cluytieae): 35-36. Berlin. (Heft 47.) La/Ge. — 2 species, Sulawesi and SE Asia.

Airy-Shaw, H. K. (1971). Notes on Malesian and other Asiatic Euphorbiaceae, CXLIII. Notes on *Strophioblachia* Boerl. Kew Bull. 25: 544-545. En. — 2 species in SE. Asia and S. China.
- Nguyen Nghia Thin, Vu Hoai Duc & P. C. van Welzen (1998). A revision of the Indochinese-Malesian genus *Strophioblachia* (Euphorbiaceae). Blumea 43: 479-487, illus., map. En. — Treatment of 1 species with description, synonymy, references and citations, types, indication of distribution, habitat and vernacular names, and discussion (particularly of variability); list of specimens seen at end (without specific localities or herbaria where housed). No consideration of affinities is presented.

Strophioblachia Boerl., Handl. Fl. Ned. Ind. 3: 235 (1900).
Hainan, S. China, Indo-China, C. Malesia. 36 41 42.

Strophioblachia fimbricalyx Boerl., Handl. Fl. Ned. Ind. 3: 236 (1900).
S. China (S. Yunnan, SW. Guangxi), Hainan, Vietnam, Philippines, Sulawesi. 36 CHC CHH CHS 41 CBD VIE 42 PHI SUL. Nanophan. or phan. – A dry-monsoonal distribution. Dried seeds are used in fermented drinks (Mindoro, Philippines).
Blachia glandulosa Pierre ex Pax in H.G.A.Engler, Pflanzenr., IV, 147, III: 36 (1911), pro syn.
Strophioblachia glandulosa Pax in H.G.A.Engler, Pflanzenr., IV, 147, III: 36 (1911).
Strophioblachia glandulosa var. *tonkinensis* Gagnep. in H.Lecomte, Fl. Indo-Chine 5: 410 (1926).
Strophioblachia fimbricalyx var. *efimbriata* Airy Shaw, Kew Bull. 25: 544 (1971).
Strophioblachia glandulosa var. *cordifolia* Airy Shaw, Kew Bull. 25: 545 (1971).

Synonyms:
Strophioblachia fimbricalyx var. *efimbriata* Airy Shaw === **Strophioblachia fimbricalyx** Boerl.
Strophioblachia glandulosa Pax === **Strophioblachis fimbricalyx** Boerl.
Strophioblachia glandulosa var. *cordifolia* Airy Shaw === **Strophioblachia glandulosa** Boerl.
Strophioblachia glandulosa var. *tonkinensis* Gagnep. === **Strophioblachia fimbricalyx** Boerl.

Stylanthus

Synonyms:
Stylanthus Rchb. & Zoll. === **Mallotus** Lour.
Stylanthus thwaitesii Baill. === **Podadenia sapida** Thwaites

Stylodiscus

Synonyms:
Stylodiscus Benn. === **Bischofia** Blume
Stylodiscus trifoliatus (Roxb.) Benn. === **Bischofia javanica** Blume

Sumbavia

Synonyms:
Sumbavia Baill. === **Doryxylon** Zoll.
Sumbavia macrophylla Müll.Arg. === **Sumbaviopsis albicans** (Blume) J.J.Sm.
Sumbavia rottlerioides Baill. === **Doryxylon spinosum** Zoll.

Sumbaviopsis

1 species, S. China, SE. Asia and Malesia; shrubs or small trees to 15 m of dry forest with mallotoid foliage, several times recorded from limestone. The leaves are peltate and in the flowers a torus develops under the stamens. Some authorities include the genus with *Doryxylon*, but here the inflorescences are terminal rather than axillary. Along with *Melanolepis* and *Thyrsanthera*, the genera comprise subtribe Doryxylinae (tribe Chrozophoreae). A relationship with *Mallotus* has also been suggested but there petals are absent. *Sumbaviopsis*, along with relatives in Doryxylinae, is under study by P. van Welzen

(Leiden) for *Flora Malesiana*; preliminary results suggest that, while all genera in the subtribe are only weakly distinguishable among themselves and from *Chrozophora* (Chrozophorinae), for pragmatic reasons they are best kept mutually distinct. (Acalyphoideae)

Pax, F. (with K. Hoffmann) (1912). *Sumbaviopsis*. In A. Engler (ed.), Das Pflanzenreich, IV 147 VI (Euphorbiaceae-Acalypheae-Chrozophorinae): 13-14. Berlin. (Heft 57.) La/Ge. — 1 species, *S. albicans*, Java.

Airy-Shaw, H. K. (1960). Notes on Malaysian Euphorbiaceae, V. On *Sumbavia macrophylla* Muell. Arg. Kew Bull. 14: 357-358. En. —*Sumbavia macrophylla*, one of two previously recognized species, referred to *Sumbaviopsis albicans*.

Balakrishnan, N. P. (1967). Studies in Indian Euphorbiaceae, II. The genus *Doryxylon* Zoll. Bull. Bot. Surv. India 9: 56-58, illus. En. — Detailed treatment of *D. albicans*, with descriptions, type, synonymy, references and citations, indication of overall distribution, and Indian and Myanmarian localities with exsiccatae, and illustration. [The genus is considered to have 2 species, *D. spinosum* and *D. albicans*; however, the reduction here of *Sumbaviopsis*, erected by J. J. Smith for *S. albicans*, is not accepted by Webster (Synopsis, 1994) and others.]

Sumbaviopsis J.J.Sm., Meded. Dept. Landb. Ned.-Indië 10: 356 (1910).
S. China, Trop. Asia. 36 40 41 42.

Sumbaviopsis albicans (Blume) J.J.Sm., Meded. Dept. Landb. Ned.-Indië 10: 357 (1910).
S. China, Assam, Indo-China to Philippines (Palawan). 36 CHC 40 ASS 41 BMA THA VIE 42 BOR PHI SUM. Phan.
* Adisca albicans Blume, Bijdr.: 611 (1826). *Croton albicans* (Blume) Rchb.f. & Zoll., Acta Soc. Regiae Sci. Indo-Neerl. 1: 21 (1856). *Cephalocroton albicans* (Blume) Müll.Arg. in A.P.de Candolle, Prodr. 15(2): 760 (1866). *Doryxylon albicans* (Blume) N.P.Balakr., Bull. Bot. Surv. India 9: 58 (1968).
Sumbavia macrophylla Müll.Arg., Flora 47: 482 (1864).
Coelodiscus speciosus Müll.Arg., Linnaea 34: 154 (1865). *Mallotus speciosus* (Müll.Arg.) Pax & K.Hoffm. in H.G.A.Engler, Pflanzenr., IV, 147, VII: 205 (1914).
Sumbaviopsis albicans var. *disperma* Gagnep. in ?, : 418 (1926).

Synonyms:
Sumbaviopsis albicans var. *disperma* Gagnep. === **Sumbaviopsis albicans** (Blume) J.J.Sm.

Suregada

31 species, Africa (8 in tropical Africa), Madagascar (13) and eastern tropics; glabrous shrubs or small trees with early fugacious stipules which leave a ring-like scar. The inflorescences are leaf-opposed. Species distinctions in the Old World 'are extremely uncertain' (Airy-Shaw 1963), and no infrageneric subdivisions have been proposed. Some species are mainly coastal or nearly so; *S. glomerulata* is widely distributed from Asia across Malesia to Australia. The genus, long known as *Gelonium* and so treated by Pax (1912), was considered the sole representative of tribe Gelonieae by Webster (Synopsis, 1994), a marked restriction from previous practice; other genera are now largely in Adenoclineae. (Crotonoideae)

Pax, F. (with K. Hoffmann) (1912). *Gelonium*. In A. Engler (ed.), Das Pflanzenreich, IV 147 IV (Euphorbiaceae-Gelonieae): 14-24. Berlin. (Heft 52.) La/Ge. — c. 18 species, Africa, Madagascar, Asia, Sunda, Philippines, Papuasia. [Now known as *Suregada*.]

Croizat, L. (1942). Notes on the Euphorbiaceae, III. Bull. Bot. Gard. Buitenzorg, III, 17: 209-219. En. — Includes (pp. 212-217) the substitution of *Suregada* for *Gelonium* on nomenclatural grounds, with necessary combinations.

Leandri, J. (1944). Contribution à l'étude des Euphorbiacées de Madagascar, VIII: Bridéliées, Géloniées. Notul. Syst. (Paris) 11: 151-159. Fr. — Includes key to 7 known Malagasy *Gelonium* (=*Suregada*) along with novelties and notes; in effect a regional revision. [Succeeded by Radcliffe-Smith, 1991.]

Merrill, E. D. (1951). On certain nomenclatural errors in the Euphorbiaceae. J. Arnold Arbor. 32: 79-81. En. — A reply to Croizat (1942); errors pointed out (Croizat having made new combinations indiscriminately, some for species not even euphorbiaceous).

Léonard, J. (1958). Notulae systematicae XXIV. Note sur les genres *Suregada* Roxb. ex Rottl. et *Gelonium* Roxb. ex Willd. (Euphorbiaceae). Bull. Jard. Bot. État 28: 443-450. Fr. — A follow-on to Croizat (1942) covering 8 African species (one new); includes synonymy, selected exsiccatae, and commentary with description of *S. croizatiana* from Congo.

Airy-Shaw, H. K. (1963). Notes on Malaysian and other Asiatic Euphorbiaceae, XLIV. An unexpected synonym in *Gelonium*. Kew Bull. 16: 367. En. — *Doryalis macrodendron* Gilg, first described as flacourtiaceous, is a synonym of *Gelonium papuanum* (now in *Suregada*).

Airy-Shaw, H. K. (1969). Notes on Malesian and other Asiatic Euphorbiaceae, CXVIII. New or noteworthy species of *Suregada* Roxb. ex Rottl. Kew Bull. 23: 128-130. En. — 2 species, one new (from W. Sarawak limestone).

Airy-Shaw, H. K. (1971). Notes on Malesian and other Asiatic Euphorbiaceae, CXLV. A remarkable variety of *Suregada multiflora*. Kew Bull. 25: 550. En. — *S. multiflora* var. *lamellata* from Peninsular Malaysia.

• Radcliffe-Smith, A. (1991). Notes on Madagascan Euphorbiaceae, IV. The genus *Suregada* in Madagascar and the Comoro Is. Kew Bull. 46: 711-726. En. — Synopsis of 13 species, with key; maps and illustrations included.

Suregada Roxb. ex Rottl., Ges. Naturf. Freunde Berlin Neue Schriften 4: 206 (1803).
Trop. & S. Africa, W. Indian Ocean, S. China, Taiwan, Trop. Asia. 22 23 24 25 26 27 29 36 38 40 41 42 43 50.
Gelonium Roxb. ex Willd., Sp. Pl. 4: 831 (1806).
Erythrocarpus Blume, Bijdr.: 604 (1826).
Saragodra Steud., Nomencl. Bot., ed. 2, 2: 513 (1841).
Ceratophorus Sond., Linnaea 23: 120 (1850).
Owataria Matsum., Bot. Mag. (Tokyo) 14: 1 (1900).

Suregada adenophora Baill., Adansonia 1: 253 (1861).
Madagascar. 29 MDG.
Suregada crenulata Baill., Adansonia 1: 252 (1861).

Suregada aequoreum (Hance) Seem., J. Bot. 4: 403 (1866).
S. Taiwan. 38 TAI. Nanophan.
* *Gelonium aequoreum* Hance, J. Bot. 4: 173 (1866).

Suregada africana (Sond.) Müll.Arg. in A.P.de Candolle, Prodr. 15(2): 1129 (1866).
Mozambique, S. Africa. 26 MOZ 27 CPP NAT SWZ TVL. Phan.
* *Ceratophorus africanus* Sond., Linnaea 23: 121 (1850). *Suregada ceratophorus* Baill., Adansonia 3: 154 (1863), nom. illeg.

Suregada bifaria (Roxb. ex Willd.) Baill., Hist. Pl. 5: 120 (1874).
Andaman Is., Pen. Malaysia. 41 AND 42 MLY. Phan.
* *Gelonium bifarium* Roxb. ex Willd., Sp. Pl. 4: 831 (1806).
Suregada dicocca Roxb. ex Pax in H.G.A.Engler, Pflanzenr., IV, 147, IV: 19 (1912), pro syn.

Suregada boiviniana Baill., Adansonia 1: 252 (1861).
Madagascar. 29 MDG.

var. **boiviniana**
Madagascar. 29 MDG.
Gelonium pycnantherum Pax & K.Hoffm. in H.G.A.Engler, Pflanzenr., IV, 147, IV: 21 (1912). *Suregada pycnanthera* (Pax & K.Hoffm.) Croizat, Bull. Jard. Bot. Buitenzorg, III, 17: 216 (1942).
Gelonium baronii S.Moore, J. Bot. 64: 42 (1926). *Suregada baronii* (S.Moore) Croizat, Bull. Jard. Bot. Buitenzorg, III, 17: 217 (1942).

var. **grandifolia** Leandri ex Radcl.-Sm., Kew Bull. 46: 712 (1991).
　　NW. Madagascar. 29 MDG.

var. **meridionalis** Leandri ex Radcl.-Sm., Kew Bull. 46: 713 (1991).
　　S. Madagascar. 29 MDG.

Suregada borbonica (Pax & K.Hoffm.) Croizat, Bull. Jard. Bot. Buitenzorg, III, 17: 216 (1942).
　Réunion. 29 REU.
　　* *Gelonium borbonicum* Pax & K.Hoffm. in H.G.A.Engler, Pflanzenr., IV, 147, IV: 19 (1912).

Suregada bracteata Radcl.-Sm., Kew Bull. 46: 713 (1991).
　N. Madagascar. 29 MDG. Phan.

Suregada calcicola Airy Shaw, Kew Bull. 23: 129 (1969).
　Borneo (Sarawak). 42 BOR. Phan.

Suregada capuronii Leandri, Bull. Soc. Bot. France 103: 607 (1957).
　C. & S. Madagascar. 29 MDG.

Suregada cicerosperma (Gagnep.) Croizat, Bull. Jard. Bot. Buitenzorg, III, 17: 216 (1942).
　Vietnam. 41 VIE.
　　* *Gelonium cicerospermum* Gagnep., Bull. Soc. Bot. France 70: 875 (1923 publ. 1924).

Suregada comorensis Baill., Bull. Mens. Soc. Linn. Paris 2: 978 (1891).
　Comoros. 29 COM.
　　Gelonium comorense S.Moore, J. Bot. 64: 42 (1926). *Suregada comorensis* (S.Moore)
　　　Croizat, Bull. Jard. Bot. Buitenzorg, III, 17: 217 (1942), nom. illeg.

Suregada croizatiana J.Léonard, Bull. Jard. Bot. État 28: 449 (1958).
　Zaire. 23 ZAI. Nanophan.

Suregada decidua Radcl.-Sm., Kew Bull. 46: 717 (1991).
　W. Madagascar. 29 MDG. Phan.

Suregada eucleoides Radcl.-Sm., Kew Bull. 46: 720 (1991).
　SC. Madagascar. 29 MDG. Phan.

Suregada gaultheriifolia Radcl.-Sm., Kew Bull. 46: 715 (1991).
　EC. Madagascar. 29 MDG. Nanophan.

Suregada glomerulata (Blume) Baill., Étude Euphorb.: 396 (1858). – FIGURE, p. 1497.
　S. China to Northern Territory. 36 CHC CHH CHS 41 VIE 42 BOR JAW LSI MOL NWG
　　PHI SUL SUM 50 NTA. Nanophan. or phan.
　　* *Erythrocarpus glomerulatus* Blume, Bijdr.: 605 (1826).
　　Suregada spicata Baill., Étude Euphorb.: 396 (1858).
　　Gelonium meliocarpum Elmer, Leafl. Philipp. Bot. 3: 919 (1910). *Suregada meliocarpa*
　　　(Elmer) Croizat, Bull. Jard. Bot. Buitenzorg, III, 17: 216 (1942).
　　Gelonium pulgarense Elmer, Leafl. Philipp. Bot. 4: 1293 (1911). *Suregada pulgarensis*
　　　(Elmer) Croizat, Bull. Jard. Bot. Buitenzorg, III, 17: 216 (1942).
　　Gelonium subglomerulatum Elmer, Leafl. Philipp. Bot. 4: 1292 (1911). *Suregada*
　　　subglomerata (Elmer) Croizat, Bull. Jard. Bot. Buitenzorg, III, 17: 216 (1942).
　　Gelonium microcarpum Pax & K.Hoffm. in H.G.A.Engler, Pflanzenr., IV, 147, IV: 19
　　　(1912). *Suregada microcarpa* (Pax & K.Hoffm.) Croizat, Bull. Jard. Bot. Buitenzorg, III,
　　　17: 216 (1942).
　　Gelonium papuanum Pax in H.G.A.Engler, Pflanzenr., IV, 147, IV: 20 (1912). *Suregada*
　　　papuana (Pax) Croizat, Bull. Jard. Bot. Buitenzorg, III, 17: 216 (1942).
　　Gelonium philippinense Pax & K.Hoffm. in H.G.A.Engler, Pflanzenr., IV, 147, IV: 20
　　　(1912). *Suregada philippinensis* (Pax & K.Hoffm.) Croizat, Bull. Jard. Bot. Buitenzorg,
　　　III, 17: 216 (1942).

Gelonium mindanaense Elmer, Leafl. Philipp. Bot. 7: 2640 (1915). *Suregada mindanaensis* (Elmer) Croizat, Bull. Jard. Bot. Buitenzorg, III, 17: 216 (1942).

Gelonium rubrum Ridl., Bull. Misc. Inform. Kew 1926: 81 (1926). *Suregada rubra* (Ridl.) Croizat, Bull. Jard. Bot. Buitenzorg, III, 17: 217 (1942).

Gelonium borneense Pax & K.Hoffm., Mitt. Inst. Bot. Hamburg 7: 230 (1931). *Suregada borneensis* (Pax & K.Hoffm.) Croizat, Bull. Jard. Bot. Buitenzorg, III, 17: 217 (1942).

Suregada glomerulata (Blume) Baill.
Artist: Ann Davies
Kew Bull. 35: 668, fig. 6 (1980), upper right
KEW ILLUSTRATIONS COLLECTION

Suregada gossweileri (S.Moore) Croizat, Bull. Jard. Bot. Buitenzorg, III, 17: 217 (1942).
Zaire, Angola. 23 ZAI 26 ANG.
> *Gelonium congoense* S.Moore, J. Bot. 64: 42 (1926). *Suregada congoensis* (S.Moore) Croizat, Bull. Jard. Bot. Buitenzorg, III, 17: 217 (1942).
> * *Gelonium gossweileri* S.Moore, J. Bot. 64: 43 (1926).

Suregada grandiflora Radcl.-Sm., Kew Bull. 46: 725 (1991).
E. Madagascar. 29 MDG. Phan.

Suregada humbertii (Leandri) Radcl.-Sm., Kew Bull. 46: 718 (1991).
EC. Madagascar. 29 MDG.
Gelonium humbertii Leandri, Notul. Syst. (Paris) 11: 156 (1944).

Suregada ivorensis (Aubrév. & Pellegr.) J.Léonard, Bull. Jard. Bot. État 28: 449 (1958).
Ivory Coast. 22 IVO.
Gelonium ivorense Aubrév. & Pellegr., Bull. Soc. Bot. France 85: 290 (1938).

Suregada lanceolata (Willd.) Kuntze, Revis. Gen. Pl. 2: 619 (1891).
S. India, Sri Lanka. 40 IND SRL. Phan.
Gelonium lanceolatum Willd., Sp. Pl. 4: 832 (1806).
Suregada angustifolia Baill., Étude Euphorb.: 396 (1858), nom. nud.
Gelonium angustifolium Müll.Arg. in A.P.de Candolle, Prodr. 15(2): 1128 (1866).
Suregada angustifolia (Müll.Arg.) Airy Shaw, Kew Bull. 23: 128 (1969).

Suregada laurina Baill., Adansonia 1: 253 (1861).
E. Madagascar. 29 MDG.

Suregada lithoxyla (Pax & K.Hoffm.) Croizat, Bull. Jard. Bot. Buitenzorg, III, 17: 216 (1942).
Tanzania. 25 TAN. Nanophan.
Gelonium lithoxylon Pax & K.Hoffm. in H.G.A.Engler, Pflanzenr., IV, 147, IV: 22 (1912).

Suregada multiflora (A.Juss.) Baill., Étude Euphorb.: 396 (1858).
India, Burma, Thailand, Pen. Malaysia, Vietnam, Sumatera, Lesser Sunda Is. 40 IND 41
BMA MLY THA VIA 42 LSI SUM. Nanophan. or phan.
Gelonium multiflorum A.Juss., Euphorb. Gen.: 111 (1824).
Suregada glabra Roxb., Fl. Ind. ed. 1832, 3: 831 (1832).
Gelonium oxyphyllum Miq., Fl. Ned. Ind., Eerste Bijv.: 452 (1861). *Suregada oxyphylla*
(Miq.) Kuntze, Revis. Gen. Pl. 2: 619 (1891).
Gelonium tenuifolium Ridl., J. Straits Branch Roy. Asiat. Soc. 59: 181 (1911). *Suregada*
tenuifolia (Ridl.) Croizat, Bull. Jard. Bot. Buitenzorg, III, 17: 216 (1942).
Gelonium affine S.Moore, J. Bot. 63(Suppl.): 104 (1925). *Suregada affinis* (S.Moore)
Croizat, Bull. Jard. Bot. Buitenzorg, III, 17: 217 (1942).
Gelonium sumatranum S.Moore, J. Bot. 63(Suppl.): 104 (1925). *Suregada sumatrana*
(S.Moore) Croizat, Bull. Jard. Bot. Buitenzorg, III, 17: 217 (1942).
Suregada multiflora var. *lamellata* Airy Shaw, Kew Bull. 25: 550 (1971).
Suregada multiflora var. *verrucigera* Airy Shaw, Kew Bull. 32: 81 (1977).

Suregada nigricaulis Radcl.-Sm., Kew Bull. 46: 722 (1991).
EC. Madagascar. 29 MDG. Phan.

Suregada occidentalis (Hoyle) Croizat, Bull. Jard. Bot. Buitenzorg, III, 17: 217 (1942).
Ghana, Nigeria. 22 GHA NGA.
Gelonium occidentale Hoyle, Bull. Misc. Inform. Kew 1935: 258 (1935).

Suregada perrieri (Leandri) Radcl.-Sm., Kew Bull. 46: 720 (1991).
E. Madagascar. 29 MDG.
Gelonium perrieri Leandri, Notul. Syst. (Paris) 11: 156 (1944).

Suregada procera (Prain) Croizat, Bull. Jard. Bot. Buitenzorg, III, 17: 216 (1942).
– FIGURE, p. 1499.
Trop. & S. Africa. 23 ZAI 24 ETH SUD 25 KEN TAN UGA 26 MLW MOZ ZAM ZIM 27 NAT
TVL. Nanophan.
Gelonium procerum Prain, Bull. Misc. Inform. Kew 1911: 233 (1911).

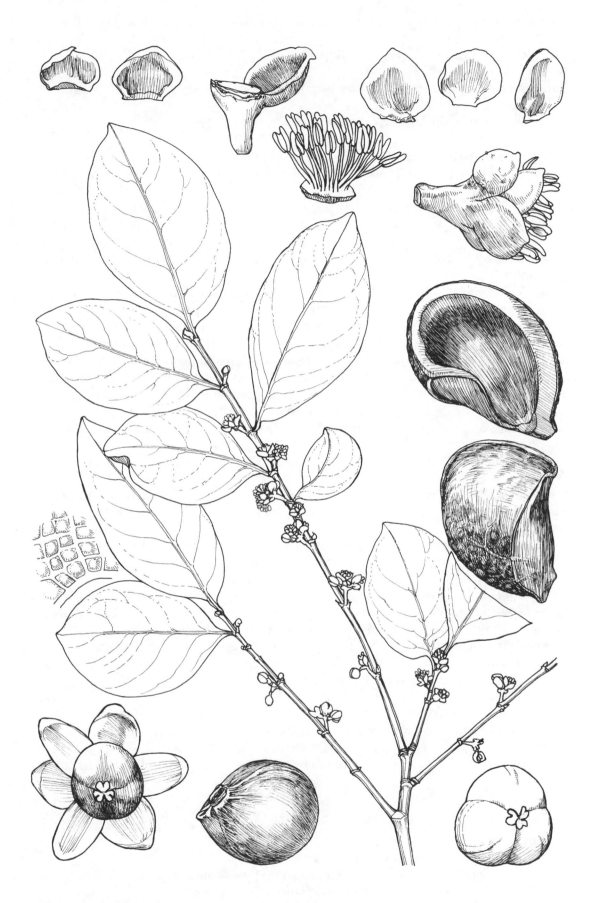

Suregada procera (Prain) Croizat

Artist: Mary Millar Watt

Fl. Trop. East Africa, Euphorbiaceae 1: 379, fig. 71 (1987)

Suregada racemulosa (Merr.) Croizat, Bull. Jard. Bot. Buitenzorg, III, 17: 216 (1942).
Philippines. 42 PHI.
* *Gelonium racemulosum* Merr., Philipp. J. Sci., C 4: 281 (1909).

Suregada stenophylla (Merr.) Croizat, Bull. Jard. Bot. Buitenzorg, III, 17: 216 (1942).
Philippines. 42 PHI.
* *Gelonium stenophyllum* Merr., Philipp. J. Sci. 27: 30 (1925).

Suregada zanzibariensis Baill., Adansonia 1: 254 (1861).
Somalia to S. Africa, Madagascar. 24 SOM 25 KEN TAN 26 MOZ ZIM 27 NAT 29 MDG.
Nanophan.

Synonyms:
Suregada affinis (S.Moore) Croizat === **Suregada multiflora** (A.Juss.) Baill.
Suregada angolensis (Prain) Croizat === **Tetrorchidium didymostemon** (Baill.) Pax &
K.Hoffm.
Suregada angustifolia (Müll.Arg.) Airy Shaw === **Suregada lanceolata** (Willd.) Kuntze
Suregada angustifolia Baill. === **Suregada lanceolata** (Willd.) Kuntze
Suregada baronii (S.Moore) Croizat --- **Suregada boiviniana** Baill. var. **boiviniana**
Suregada borneensis (Pax & K.Hoffm.) Croizat === **Suregada glomerulata** (Blume) Baill.
Suregada ceratophorus Baill. === **Suregada africana** (Sond.) Müll.Arg.
Suregada comorensis (S.Moore) Croizat === **Suregada comorensis** Baill.
Suregada congoensis (S.Moore) Croizat === **Suregada gossweileri** (S.Moore) Croizat
Suregada crenulata Baill. === **Suregada adenophora** Baill.
Suregada dicocca Roxb. ex Pax === **Suregada bifaria** (Roxb. ex Willd.) Baill.
Suregada glabra Roxb. === **Suregada multiflora** (A.Juss.) Baill.
Suregada glandulosa (Elmer) Croizat === **Rinorea bengalensis** (Wall.) Kuntze (Violaceae)
Suregada meliocarpa (Elmer) Croizat === **Suregada glomerulata** (Blume) Baill.
Suregada microcarpa (Pax & K.Hoffm.) Croizat === **Suregada glomerulata** (Blume) Baill.
Suregada mindanaensis (Elmer) Croizat === **Suregada glomerulata** (Blume) Baill.
Suregada multiflora var. *lamellata* Airy Shaw === **Suregada multiflora** (A.Juss.) Baill.
Suregada multiflora var. *verrucigera* Airy Shaw === **Suregada multiflora** (A.Juss.) Baill.
Suregada oxyphylla (Miq.) Kuntze === **Suregada multiflora** (A.Juss.) Baill.
Suregada papuana (Pax) Croizat === **Suregada glomerulata** (Blume) Baill.
Suregada philippinensis (Pax & K.Hoffm.) Croizat === **Suregada glomerulata** (Blume) Baill.
Suregada pinatubensis (Elmer) Croizat === **Casearia trivalvis** (Blanco) Merr. (Flacourtiaceae)
Suregada pulgarensis (Elmer) Croizat === **Suregada glomerulata** (Blume) Baill.
Suregada pycnanthera (Pax & K.Hoffm.) Croizat === **Suregada boiviniana** Baill.
var. **boiviniana**
Suregada rubra (Ridl.) Croizat === **Suregada glomerulata** (Blume) Baill.
Suregada spicata Baill. === **Suregada glomerulata** (Blume) Baill.
Suregada subglomerata (Elmer) Croizat === **Suregada glomerulata** (Blume) Baill.
Suregada sumatrana (S.Moore) Croizat === **Suregada multiflora** (A.Juss.) Baill.
Suregada tenuifolia (Ridl.) Croizat === **Suregada multiflora** (A.Juss.) Baill.
Suregada trifida (Elmer) Croizat === **Rinorea bengalensis** (Wall.) Kuntze (Violaceae)

Symphyllia

Synonyms:
Symphyllia Baill. === **Epiprinus** Griff.
Symphyllia mallotiformis Müll.Arg. === **Epiprinus mallotiformis** (Müll.Arg.) Croizat
Symphyllia siletiana Baill. === **Epiprinus siletianus** (Baill.) Croizat
Symphyllia siletiana var. *genuina* Müll.Arg. === **Epiprinus siletianus** (Baill.) Croizat
Symphyllia siletiana var. *trichantha* Müll.Arg. === **Epiprinus siletianus** (Baill.) Croizat
Symphyllia silhetense Benth. === **Epiprinus siletianus** (Baill.) Croizat

Synadenium

15 species, E. & S. tropical Africa; related to *Monadenium* and *Euphorbia*. Laticiferous shrubs or small trees to 10 m of dry country with 'succulent' cylindrical branches, fleshy leaves, and divaricating, ultimately cyatheoid inflorescences. The cyathia are (unlike *Monadenium*) radially symmetrical but have a continuous involucral gland with (also unlike *Monadenium*) no gaps. Never revised for *Pflanzenreich*; remains a taxonomically difficult genus (S. Carter, personal communication). Some species, including the widely distributed *S. grantii*, are grown as garden ornamentals in warmer regions. (Euphorbioideae (Euphorbieae))

- Pax, F. & K. Hoffmann (1931). *Synadenium*. In A. Engler (ed.), Die natürlichen Pflanzenfamilien, 2. Aufl., 19c: 221-222. Leipzig. Ge. — Synopsis with description of genus; 13 species listed.
  Jones, K. & J. B. Smith (1969). The chromosome identity of *Monadenium* Pax and *Synadenium* Pax (Euphorbiaceae). Kew Bull. 23: 491-498. En. — Karyological account; evidence for distinction of the genera.
  Carter, S. (1987). Taxonomic changes in *Synadenium* (Euphorbiaceae) from East Africa. Kew Bull. 42: 667-671. En. — Some novelties described. [Precursory to *Flora of Tropical East Africa* account.]

Synadenium Boiss. in A.P.de Candolle, Prodr. 15(2): 187 (1862).
Trop. & S. Africa. 23 24 26 27.

Synadenium angolense N.E.Br. in D.Oliver, Fl. Trop. Afr. 6(1): 469 (1911).
Angola. 26 ANG.

Synadenium calycinum S.Carter, Kew Bull. 42: 670 (1987).
Tanzania. 25 TAN. Nanophan. or phan.

Synadenium cameronii N.E.Br., Bull. Misc. Inform. Kew 1901: 133 (1901).
Malawi. 26 MLW.

Synadenium compactum N.E.Br. in D.Oliver, Fl. Trop. Afr. 6(1): 465 (1911).
Rwanda, Kenya. 23 RWA 25 KEN.

  var. **compactum**
    Rwanda, Kenya. 23 RWA 25 KEN.

  var. **rubrum** S.Carter, Kew Bull. 42: 669 (1987).
    Kenya (near Embu). 25 KEN. Nanophan. or phan. – Cultivated for ornament.

Synadenium cupulare (Boiss.) Wheeler ex A.C.White, R.A.Dyer & B.Sloane, Succ. Euphorb. 2: 953, App. A (1941). – FIGURE, p. 1502.
Natal, Swaziland, Transvaal, Malawi ? 26 MLW? 27 NAT SWZ TVL.
  Euphorbia arborescens E.Mey. in J.F.Drège, Zwei Pflanzengeogr. Dokum.: 184 (1843), nom. nud.
  *Euphorbia cupularis Boiss., Cent. Euphorb.: 23 (1860).
  Synadenium arborescens Boiss. in A.P.de Candolle, Prodr. 15(2): 187 (1862).
  Euphorbia synadenia Baill., Adansonia 3: 142 (1863), nom. illeg.

Synadenium cymosum N.E.Br. in D.Oliver, Fl. Trop. Afr. 6(1): 469 (1911).
Tanzania, Uganda. 25 TAN UGA.

Synadenium gazense N.E.Br., J. Linn. Soc., Bot. 40: 190 (1911).
Zambia, Zimbabwe, Mozambique. 26 MOZ ZAM ZIM.

Synadenium glabratum S.Carter, Kew Bull. 42: 667 (1987).
Tanzania, Zambia. 25 TAN 26 ZAM. Nanophan.

Synadenium cupulare (Boiss.) Wheeler ex A.C. White, R.A. Dyer & B. Sloane (as *S. arborescens* Boiss.)
Artist: Matilda Smith
Bot. Mag. 117: pl. 7184 (1891), uncoloured version
KEW ILLUSTRATIONS COLLECTION

Synadenium glaucescens Pax, Bot. Jahrb. Syst. 33: 289 (1903).
  Tanzania. 25 TAN. Nanophan. or phan.

Synadenium grantii Hook.f., Bot. Mag. 83: t. 5633 (1867).
  E. Zaire to E. Trop. Africa, Malawi ? 23 BUR RWA ZAI 25 KEN TAN UGA 26 MLW?
    Nanophan. or phan.
    Synadenium umbellatum Pax, Bot. Jahrb. Syst. 19: 125 (1894).
    Synadenium umbellatum var. *puberulum* N.E.Br. in D.Oliver, Fl. Trop. Afr. 6(1): 464 (1911).

Synadenium halipedicola L.C.Leach, Garcia de Orta, Sér. Bot. 6: 47 (1984).
  Mozambique. 26 MOZ.

Synadenium kirkii N.E.Br. in D.Oliver, Fl. Trop. Afr. 6(1): 466 (1911).
  Malawi ?, Mozambique ?, Zimbabwe. 26 MLW? MOZ? ZIM.

Synadenium molle Pax, Bot. Jahrb. Syst. 43: 88 (1909).
  Kenya, Tanzania. 25 KEN TAN. Nanophan. or phan.

Synadenium pereskiifolium (Houllet ex Baill.) Guillaumin, Bull. Mus. Natl. Hist. Nat., II, 7:
  135 (1935).
  Tanzania (incl. Zanzibar), Kenya. 25 KEN TAN. Nanophan. or phan.
   * *Euphorbia pereskiifolia* Houllet ex Baill., Adansonia 1: 105 (1861).
    Euphorbia sulcata Lem. ex Boiss. in A.P.de Candolle, Prodr. 15(2): 176 (1862), pro syn.
    Synadenium carinatum Boiss. in A.P.de Candolle, Prodr. 15(2): 187 (1862).
    Synadenium piscatorium Pax, Bot. Jahrb. Syst. 19: 125 (1894).

Synadenium volkensii Pax in H.G.A.Engler, Pflanzenw. Ost-Afrikas C: 242 (1895).
  Kenya (Masai), Tanzania (Mt. Kilimanjaro). 25 KEN TAN. Phan.

Synonyms:
Synadenium arborescens Boiss. === **Synadenium cupulare** (Boiss.) Wheeler ex A.C.White,
  R.A.Dyer & B.Sloane
Synadenium ballyi Werderm. ex Ball. === ?
Synadenium carinatum Boiss. === **Synadenium pereskiifolium** (Houllet ex Baill.) Guillaumin
Synadenium piscatorium Pax === **Synadenium pereskiifolium** (Houllet ex Baill.) Guillaumin
Synadenium umbellatum Pax === **Synadenium grantii** Hook.f.
Synadenium umbellatum var. *puberulum* N.E.Br. === **Synadenium grantii** Hook.f.

Synaspisma

Synonyms:
Synaspisma Endl. === **Codiaeum** Rumph. ex A.Juss.

Synastemon

An orthographic variant of *Synostemon*.

Syndyophyllum

2 species, Malesia (Sumatra, Borneo and New Guinea). The distribution of these medium to large, sympodially branching forest trees (which may reach 30 m) is seemingly disjunct; in New Guinea *S. excelsum* is so far known only from the Kepala Burung and in the Gogol/Naru basins of Madang Province (Papua New Guinea) with one intermediate locality (near Sarmi on the north coast of Irian Jaya). In Borneo *S. occidentale*, an 'ant-plant', is

known mostly from Sabah. Like its relatives *Erismanthus* and *Moultonianthus*, the leaves in this genus are opposite. A recent treatment is by van Welzen (1995) but I would, contrary to that author, conclude that those two genera are more closely related to one another than to *Syndyophyllum*. (Acalyphoideae)

Pax, F. (with K. Hoffmann) (1911). *Syndyophyllum*. In A. Engler (ed.), Das Pflanzenreich, IV 147 III (Euphorbiaceae-Cluytieae): 104-105. Berlin. (Heft 47.) La/Ge. — 1 species, New Guinea; a 30 m tall forest tree, the lvs subopposite.

Airy-Shaw, H. K. (1960). Notes on Malaysian Euphorbiaceae, XIII. The genus *Syndyophyllum* in Sumatra and Borneo. Kew Bull. 14: 392-394. En. — Range extension; genus (and species) formerly known only from New Guinea. Includes localities with exsiccatae.

Aity-Shaw, H.K. (1974). *Syndyophyllum excelsum* subsp. *occidentale*. Ic. Pl. 38: pl. 3722. La/En. — Plant portrait with description and commentary. [This subspecies is now given species rank.]

- Welzen, P. C. van (1995). Taxonomy and phylogeny of the Euphorbiaceae tribe Erismantheae G. L. Webster (*Erismanthus*, *Moultonianthus*, and *Syndyophyllum*). Blumea 40: 375-396, illus., maps. En. — Treatment of *Syndyophyllum*, pp. 388-394, with key, descriptions, synonymy, references, citations, indication of distribution and habitat, and vernacular names; 2 species covered with a further one removed to *Mallotus*.

Syndyophyllum K.Schum. & Lauterb., Fl. Schutzgeb. Südsee: 403 (1900).
Malesia. 42.

Syndyophyllum excelsum K.Schum. & Lauterb., Fl. Schutzgeb. Südsee: 404 (1900).
N. New Guinea. 42 NWG. Phan.

Syndyophyllum occidentale (Airy Shaw) Welzen, Blumea 40: 393 (1995).
– FIGURE, p. 1505.
N. Sumatera, Borneo. 42 BOR SUM. Phan.
 * *Syndyophyllum excelsum* subsp. *occidentale* Airy Shaw, Hooker's Icon. Pl. 38: t. 3722 (1974).

Synonyms:
Syndyophyllum excelsum subsp. *occidentale* Airy Shaw === **Syndyophyllum occidentale** (Airy Shaw) Welzen
Syndyophyllum trinervium K.Schum. & Lauterb. === **Mallotus trinervius** (K.Schum. & Lauterb.) Pax & K.Hoffm.

Synema

Synonyms:
Synema Dulac === **Mercurialis** L.

Synexemia

Synonyms:
Synexemia Raf. === **Phyllanthus** L.
Synexemia pumila Raf. === **Phyllanthus caroliniensis** Walter subsp. **caroliniensis**

Synostemon

Synonyms:
Synostemon F.Muell. === **Sauropus** Blume
Synostemon albiflorus (F.Muell. ex Müll.Arg.) Airy Shaw === **Sauropus albiflorus** (F.Muell. ex Müll.Arg.) Airy Shaw

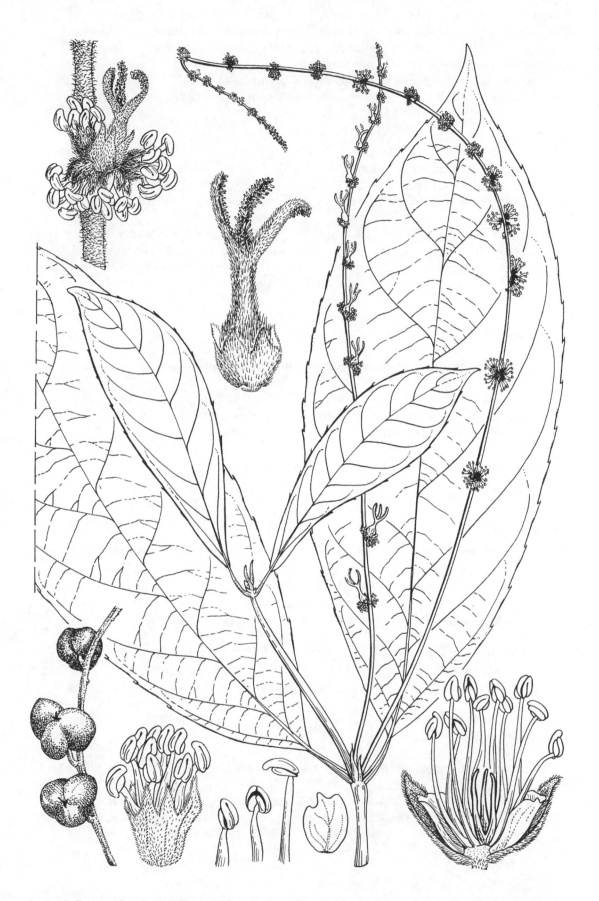

Syndyophyllum occidentale (Airy Shaw) Welzen (as *S. excelsum* K. Schum. & Lauterb. subsp. *occidentale* Airy Shaw)

Artist: Mary Grierson
Ic. Pl. 38: pl. 3722 (1974)

Synostemon bacciformis (L.) G.L.Webster === **Sauropus bacciformis** (L.) Airy Shaw
Synostemon glaucus F.Muell. === **Sauropus glaucus** (F.Muell.) Airy Shaw
Synostemon hirtellus F.Muell. === **Sauropus hirtellus** (F.Muell.) Airy Shaw
Synostemon ramosissimus F.Muell. === **Sauropus ramosissimus** (F.Muell.) Airy Shaw
Synostemon rigens F.Muell. === **Sauropus rigens** (F.Muell.) Airy Shaw
Synostemon sphenophyllus Airy Shaw === **Sauropus sphenophyllus** (Airy Shaw) Airy Shaw

Tacarcuna

3 species, South America (Amazonian Peru, western Venezuela and northwestern Colombia); small forest trees to 13 m. The species are evidently all more or less restricted in distribution. [No material in Kew Herbarium.] (Subfamily not established)

> Huft, M. J. (1989). New and critical taxa of Euphorbiaceae from South America. Ann. Missouri Bot. Gard. 76: 1077-1086, illus. En. — Pp. 1080-1085 contain the generic protologue of *Tacarcuna* along with a key to and descriptions of 3 new species; no concrete suggestions regarding affinities are given.

Tacarcuna Huft, Ann. Missouri Bot. Gard. 76: 1080 (1989).
Panama, Venezuela, Peru. 80 82 83.

Tacarcuna amanoifolia Huft, Ann. Missouri Bot. Gard. 76: 1082 (1989).
Peru. 83 PER. Phan.

Tacarcuna gentryi Huft, Ann. Missouri Bot. Gard. 76: 1081 (1989).
Panama. 80 PAN 83 CLM? Phan.

Tacarcuna tachirensis Huft, Ann. Missouri Bot. Gard. 76: 1083 (1989).
Venezuela. 82 VEN. Phan.

Taeniosapium

1 species, Madagascar; glabrous shrubs of forest remnants at higher elevations with small, closely spaced leaves and persistent stipules when mature; inflorescences small, leaf-opposed. Formerly reduced to *Sapium* s.l. but revived by Esser (in press). Esser (1994) previously included it with *Conosapium* but in habit *T. goudotianum* may more closely resemble *Sclerocroton melanostictus*. (Euphorbioideae (except Euphorbieae))

> Pax, F. (with K. Hoffmann) (1912). *Sapium.* In A. Engler (ed.), Das Pflanzenreich, IV 147 V (Euphorbiaceae-Hippomaneae): 199-258. Berlin. (Heft 52.) La/Ge. — No. 93 (p. 255), *S. goudotianum*, is the type species of *Taeniosapium*.

Taeniosapium Müll.Arg. in A.P.de Candolle, Prodr. 15(2): 1200 (1866).
Madagascar. 29.

Taeniosapium goudotianum (Baill.) Müll.Arg. in A.P.de Candolle, Prodr. 15(2): 1201 (1866).
Madagascar. 29 MDG. Nanophan.
> * *Stillingia goudotiana* Baill., Adansonia 2: 30 (1861). *Excoecaria goudotiana* (Baill.) Müll.Arg., Linnaea 32: 122 (1863). *Sapium goudotianum* (Baill.) Pax in H.G.A.Engler, Pflanzenr., IV, 147, V: 255 (1912).

Tanarius

Synonyms:
Tanarius Rumph. ex Kuntze === **Macaranga** Thouars
Tanarius kurzii Kuntze === **Macaranga kurzii** (Kuntze) Pax & K.Hoffm.

Tandonia

Synonyms:
Tandonia Baill. === **Tannodia** Baill.
Tandonia cordifolia Baill. === **Tannodia cordifolia** (Baill.) Baill.

Tannodia

9 species, E., SE., SC. & C. Africa (3, Zaïre and Kenya to Zimbabwe) and Madagascar & Comoro Is. (6); includes *Domohinea* and *Neoholstia* (Radcliffe-Smith 1998). Glabrous or sparely hairy 'crotonoid' shrubs or small or medium (occasionally large) trees of forest, riverine habitats or rocky open places to 20 m with slender branches, thin, distichously arranged leaves and weak terminal inflorescences. *T. congolensis* is in forest understorey. Tanzanian material referred to *T. swynnertonii* has larger leaves than the main population in Zimbabwe and Mozambique. Three of the Malagasy species are inhabitants of hill and submontane forest to 1100 m. In the Webster system (Synopsis, 1994) the genus is allied to *Grossera*. (Crotonoideae)

> Pax, F. (with K. Hoffmann) (1912). *Holstia*. In A. Engler (ed.), Das Pflanzenreich, IV 147 VI (Euphorbiaceae-Acalypheae-Chrozophorinae): 108-109. Berlin. (Heft 57.) La/Ge. — 2 species, E. Africa. [Only 1 of these species now accepted.]
>
> Pax, F. (with K. Hoffmann) (1912). *Tannodia*. In A. Engler (ed.), Das Pflanzenreich, IV 147 VI (Euphorbiaceae-Acalypheae-Chrozophorinae): 110-111. Berlin. (Heft 57.) La/Ge. — 2 species, Africa and Comoro Is.
>
> Prain, D. (1912). The genus *Tannodia* Baill. J. Bot. 50: 125-128. En. — Synoptic revision with diagnoses but no key; 4 species in 2 sections, *Tannodia* ('Eutannodia') and *Holstia*.
>
> Leandri, J. (1940(1941)). Sur un genre malgache nouveau d'Euphorbiacées. Bull. Soc. Bot. France 87: 279-285. Fr. — Protologue of *Domohinea* with 1 species, *D. perrieri*. [Now combined with *Tannodia*.]
>
> • Radcliffe-Smith, A. (1998). A synopsis of *Tannodia* Baill. (Crotonoïdeae-Aleuritideae-Grosserinae) with especial reference to Madagascar, and the subsumption of *Domohinea* Leandri. Kew Bull. 53: 173-186, illus. En. — General remarks; synoptic treatment of 9 species with key, descriptions of novelties, types, synonymy, commentary, and (for Madagascar) localities with exsiccatae. *Neoholstia* as well as *Domohinea* are reduced, in the former case representing a return to the opinion of Prain (1912). Two subgenera, each with two sections, are accepted (but without a distinct, separate list).

Tannodia Baill., Adansonia 1: 251 (1861).
  C. & E. Trop. Africa, W. Indian Ocean. 23 25 26 29.
    Tandonia Baill., Adansonia 1: 184 (1861).
    Holstia Pax, Bot. Jahrb. Syst. 43: 220 (1909), nom. illeg. *Neoholstia* Rauschert, Taxon 31: 559 (1982).
    Domohinea Leandri, Bull. Soc. Bot. France 87: 285 (1941).

Tannodia congolensis J.Léonard, Bull. Jard. Bot. État 25: 300 (1955).
  Zaire. 23 ZAI.

Tannodia cordifolia (Baill.) Baill., Adansonia 1: 251 (1861).
  Comoros, Madagascar. 29 COM MDG. Nanophan. or phan.
    * *Tandonia cordifolia* Baill., Adansonia 1: 251 (1861).
    Agrostistachys comorensis Pax, Bot. Jahrb. Syst. 23: 523 (1897).

Tannodia grandiflora Radcl.-Sm., Kew Bull. 53: 180 (1998).
  Madagascar (Antsiranana). 29 MDG. Phan.

var. **grandiflora**
Madagascar (Antsiranana). 29 MDG. Phan.

var. **myrtifolia** Radcl.-Sm., Kew Bull. 53: 184 (1998).
Madagascar (Antsiranana). 29 MDG. Phan.

Tannodia nitida Radcl.-Sm., Kew Bull. 53: 178 (1998).
Madagascar (Toamasina). 29 MDG. Phan.

Tannodia obovata Radcl.-Sm., Kew Bull. 53: 180 (1998).
Madagascar (Toamasina). 29 MDG. Phan.

Tannodia pennivenia Radcl.-Sm., Kew Bull. 53: 184 (1998).
Madagascar (Antsiranana). 29 MDG. Nanophan. or phan.

Tannodia perrieri (Leandri) Radcl.-Sm., Kew Bull. 53: 177 (1998).
Madagascar. 29 MDG. Nanophan. or phan.
   * *Domohinea perrieri* Leandri, Bull. Soc. Bot. France 87: 285 (1941).

var. **latifolia** Radcl.-Sm., Kew Bull. 53: 178 (1998).
Madagascar. 29 MDG. Nanophan. or phan.

var. **ludiifolia** Radcl.-Sm., Kew Bull. 53: 178 (1998).
Madagascar (Fianarantsoa). 29 MDG. Nanophan. or phan.

var. **perrieri**
Madagascar. 29 MDG. Nanophan. or phan.

Tannodia swynnertonii (S.Moore) Prain, J. Bot. 50: 127 (1912).
Tanzania, Mozambique, Zimbabwe. 25 TAN 26 MOZ ZIM. Phan.
   * *Croton swynnertonii* S.Moore, J. Linn. Soc., Bot. 40: 194 (1911).

Tannodia tenuifolia (Pax) Prain, J. Bot. 50: 128 (1912). – FIGURE, p. 1509.
SC. Trop. Africa to Kenya. 25 KEN TAN 26 MLW MOZ ZAM ZIM. Nanophan. or phan.
   * *Holstia tenuifolia* Pax, Bot. Jahrb. Syst. 43: 220 (1909). *Neoholstia tenuifolia* (Pax)
      Rauschert, Taxon 31: 559 (1982).

var. **glabrata** Prain, J. Bot. 50: 128 (1912). *Holstia tenuifolia* var. *glabrata* (Prain) Pax in
H.G.A.Engler, Pflanzenr., IV, 147, VI: 109 (1912). *Neoholstia tenuifolia* var. *glabrata*
(Prain) Radcl.-Sm., Kew Bull. 39: 791 (1984).
Kenya to Mozambique. 25 KEN TAN 26 MOZ. Nanophan. or phan.
Holstia sessiliflora Pax, Bot. Jahrb. Syst. 43: 220 (1909). *Tannodia sessiliflora* (Pax) Prain,
   J. Bot. 50: 128 (1912). *Neoholstia sessiliflora* (Pax) Rauschert, Taxon 31: 559 (1982).

var. **tenuifolia**
SC. & E. Trop. Africa. 25 KEN TAN 26 MLW MOZ ZAM ZIM. Nanophan. or phan.

Synonyms:
Tannodia sessiliflora (Pax) Prain === **Tannodia tenuifolia** var. **glabrata** Prain

Tapoides

1 species, Malesia (Borneo: mainly in Sabah); forest trees to 30 m with foliar dimorphism (leaves and foliaceous floral bracts). Initially referred to Pax's tribe Cluytieae (Acalyphoideae) but Airy-Shaw (1967) later concluded that a place in the mainly American Jatropheae was justified. He also indicated, moreover, that the whole group was 'primitive' with links with Cluytieae as well as other Crotonoideae. In the Webster system the genus is in subtribe Grosserinae (Aleuritideae). (Crotonoideae)

Tannodia tenuifolia (Pax) Prain var *glabrata* Prain (as *Neoholstia tenuifolia* var *glabrata* (Prain) Radcl.-Sm.)

Artist: G. Papadopoulos
Fl. Trop East Africa, Euphorbiaceae 1: 170, fig. 31 1987

Airy-Shaw, H. K. (1960). Notes on Malaysian Euphorbiaceae, XIX. A new genus from Borneo. Kew Bull. 14: 473-475. En. —*Ostodes villamilii* transferred to a new genus *Tapoides*, placed questionably in Pax's tribe Cluytieae. [For illustration, see Ic. Pl. 37: pl. 3632 (1967).]

Airy-Shaw, H. K. (1967). *Tapoïdes villamilii*. Ic. Pl. 37: pl. 3632. La/En. — Plant portrait with description and commentary (including speculations on affinities and on Jatropheae in general).

Tapoides Airy Shaw, Kew Bull. 14: 473 (1960).
Borneo. 42.

Tapoides villamilii (Merr.) Airy Shaw, Kew Bull. 14: 473 (1960). – FIGURE, p. 1511.
Borneo (NE. Sarawak, Sabah). 42 BOR. Phan.
    * Ostodes villamilii Merr., J. Straits Branch Roy. Asiat. Soc. 76: 92 (1917).

Telephioides

Synonyms:
Telephioides Tourn. ex Moench === **Andrachne** L.
Telephioides procumbens Moench === **Andrachne telephioides** L.

Telogyne

Synonyms:
Telogyne Baill. === **Trigonostemon** Blume
Telogyne indica Baill. === **Trigonostemon verticillatus** (Jack) Pax

Telopea

A later homonym of *Telopea* R. Br. (Proteaceae).

Synonyms:
Telopea Sol. ex Baill. === **Aleurites** J.R.Forst. & G.Forst.

Tephranthus

Synonyms:
Tephranthus Neck. === **Phyllanthus** L.

Tetracarpidium

1 species, W. & WC. tropical Africa (scattered from Sierra Leone to Zaire). An oleiferous woody climber of partly shady places to 5-7 m, closely related to *Plukenetia* and formerly within as its section *Angostylidium*. The oil is edible and the vine has been cultivated in village-forest (Chevalier 1948). Continuing research on *Plukenetia* will likely show that maintenanace of Pax's segregate is not justified (cf. Gillespie 1993; see there). (Acalyphoideae)

Pax, F. & K. Hoffmann (1919). *Angostylidium*. In A. Engler (ed.), Das Pflanzenreich, IV 147 IX (Euphorbiaceae-Acalypheae-Plukenetiinae): 17-19. Berlin. (Heft 68.) La/Ge. — 1 species, *A. conophorum*. (Now in *Tetracarpidium*).

Hédin, L. (1929). Une plante oléagineuse peu connue de l'Ouest Africain: le *Tetracarpidium conophorum*. Rev. Bot. Appl. Agric. Trop. 9: 752-753. Fr. — Botanical and ecological treatment of *T. conophorum* along with notes on its cultivation and management and the properties of the oil (with particular reference to Cameroon).

Tapoides villamilii (Merr.) Airy Shaw

Artist: E. Margaret Stones
Ic. Pl. 37: pl. 3632 (1967)

Chevalier, A. (1948). Une plante oléagineuse africaine (*Tetracarpidium conophorum* Hutch. et Dalziel). Rev. Bot. Appl. Agric. Trop. 28: 465-466. Fr. — Botanical and economic treatment of *T. conophorum*; the oil is edible and the vine has been cultivated in village-forest.

Tetracarpidium Pax, Bot. Jahrb. Syst. 26: 329 (1899).
W. & WC. Trop. Africa. 22 23.
Angostylidium Pax & K.Hoffm. in H.G.A.Engler, Pflanzenr., IV, 147, IX: 17 (1919).

Tetracarpidium conophorum (Müll.Arg.) Hutch. & Dalziel, Fl. W. Trop. Afr. 1: 307 (1928).
W. & WC. Trop. Africa. 22 BEN NGA SIE 23 CMN EQG GAB GGI ZAI. Cl. nanophan.
* Plukenetia conophora Müll.Arg., Flora 47: 530 (1864).
Mallotus preussii Pax, Bot. Jahrb. Syst. 23: 525 (1897). *Cleidion preussii* (Pax) Baker, Bull. Misc. Inform. Kew 1910: 343 (1910).
Tetracarpidium staudtii Pax, Bot. Jahrb. Syst. 26: 329 (1899).
Cleidion mannii Baker, Bull. Misc. Inform. Kew 1910: 58 (1910).

Synonyms:
Tetracarpidium staudtii Pax === **Tetracarpidium conophorum** (Müll.Arg.) Hutch. & Dalziel

Tetracoccus

4 species, SW. North America (United States and Mexico). A genus of more or less xerophytic shrubs with small, sometimes narrow leaves, arguably now relictual in its distribution. Covered by Pax and Hoffmann only in *Pflanzenfamilien* (1931: 74) but fully revised by Dressler (1954). Differences of opinion exist with regard to its placement and Webster (1994; see **General**) believes that separate tribal status may be in order. (Oldfieldioideae)

• Dressler, R. (1954). The genus *Tetracoccus*. Rhodora 56: 45-61, map. En. — Introduction covering general biology, taxonomy, geography (including map with distribution of all taxa) and putative phylogeny along with previous studies; descriptive treatment with key, synonymy, references, types, localities with exsiccatae, and sometimes extensive taxonomic, biological, ecological and geographical commentary. [4 species and 1 additional variety accepted.]

Tetracoccus Engelm. ex Parry, W. Amer. Sci. 1: 13 (1885).
SW. U.S.A., N. Mexico. 76 79. Nanophan.
Halliophytum I.M.Johnst., Contr. Gray Herb. 68: 88 (1923).

Tetracoccus capensis (I.M.Johnst.) Croizat, Bull. Torrey Bot. Club 69: 457 (1942).
Mexico (S. Baja California Sur). 79 MXN. Nanophan.
* Securinega capensis I.M.Johnst., Univ. Calif. Publ. Bot. 7: 441 (1922). *Halliophytum capense* (I.M.Johnst.) I.M.Johnst., Contr. Gray Herb. 68: 89 (1923).

Tetracoccus dioicus Parry, W. Amer. Sci. 1: 13 (1885).
California (San Diego Co.), Mexico (Baja California del Norte). 76 CAL 79 MXN. Nanophan.
Tetracoccus engelmannii S.Watson, Proc. Amer. Acad. Arts 20: 373 (1885).

Tetracoccus fasciculatus (S.Watson) Croizat, Bull. Torrey Bot. Club 69: 456 (1942).
SW. U.S.A., Mexico (Chihuahua, Durango, Coahuila). 76 ARI CAL 79 MXE. Nanophan.
* Bernardia fasciculata S.Watson, Proc. Amer. Acad. Arts 18: 153 (1883). *Securinega fasciculata* (S.Watson) I.M.Johnst., Univ. Calif. Publ. Bot. 7: 441 (1922). *Halliophytum fasciculatum* (S.Watson) I.M.Johnst., Contr. Gray Herb. 68: 88 (1923).

var. **fasciculatus**
  Mexico (Chihuahua, Durango, Coahuila). 79 MXE. Nanophan.

var. **hallii** (Brandegee) Dressler, Rhodora 56: 57 (1954).
  California, W. Arizona. 76 ARI CAL. Nanophan.
  * *Tetracoccus hallii* Brandegee, Zoe 5: 229 (1906). *Securinega hallii* (Brandegee) I.M.Johnst.,
    Univ. Calif. Publ. Bot. 7: 442 (1922). *Halliophytum hallii* (Brandegee) I.M.Johnst.,
    Contr. Gray Herb. 68: 88 (1923). *Halliophytum fasciculatum* var. *hallii* (Brandegee)
    McMinn, Man. Calif. Shrubs: 249 (1939).

Tetracoccus ilicifolius Coville & Gilman, J. Wash. Acad. Sci. 26: 531 (1936).
  California (Death Valley). 76 CAL. Nanophan.

Synonyms:
Tetracoccus engelmannii S.Watson === **Tetracoccus dioicus** Parry
Tetracoccus hallii Brandegee === **Tetracoccus fasciculatus** var. **hallii** (Brandegee) Dressler

Tetractinostigma

Synonyms:
Tetractinostigma Hassk. === **Aporusa** Blume
Tetractinostigma lucidum Miq. === **Aporusa lucida** (Miq.) Airy Shaw
Tetractinostigma microcalyx Hassk. === **Aporusa octandra** var. **malesiana** Schot

Tetraglochidion

Synonyms:
Tetraglochidion K.Schum. === **Glochidion** J.R.Forst. & G.Forst.
Tetraglochidion gimi K.Schum. === **Glochidion gimi** (K.Schum.) Pax & K.Hoffm.

Tetraglossa

Synonyms:
Tetraglossa Bedd. === **Cleidion** Blume

Tetraplandra

8 species, South America (Brazil); laticiferous glabrous shrubs or trees mainly of the *mata atlantica*, the twigs sometimes thorny and the oblanceolate, more or less finely-veined leaves sometimes coarsely toothed. Closely related to *Algernonia* and perhaps not really separable from it. Other possible relatives include *Ophthalmoblapton*, as suggested by Pax, and *Pachystroma*. (Euphorbioideae (except Euphorbieae))

  Pax, F. (with K. Hoffmann) (1912). *Tetraplandra*. In A. Engler (ed.), Das Pflanzenreich, IV
    147 V (Euphorbiaceae-Hippomaneae): 274-276. Berlin. (Heft 52.) La/Ge. — 5 species, E.
    Brazil (described from around Rio de Janeiro where information available). [4 species
    retained; *T. gibbosa* now in *Algernonia*.]
• Emmerich, M. (1981). Revisão taxinômica dos gêneros *Algernonia* Baill. e *Tetraplandra*
    Baill. (Euphorbiaceae-Hippomaneae). Arq. Mus. Nac. Rio de Janeiro 56: 91-110, 11 pls.
    in text. Pt. —*Tetraplandra*, pp. 94-98; revision (8 species, 4 new) with key, descriptions,
    synonymy, references and citations, types, indication of distribution and habitat, and
    localities with exsiccatae; phenological and ecological data, literature and well-drawn
    plates at end of paper.

Tetraplandra Baill., Ann. Sci. Nat., Bot., IV, 9: 200 (1858).
 Brazil. 84.

Tetraplandra amazonica Emmerich, Arq. Mus. Nac. Rio de Janeiro 56: 95 (1981).
 Brazil (Amazonas). 84 BZN.

Tetraplandra anomala Pax & K.Hoffm. in H.G.A.Engler, Pflanzenr., IV, 147, V: 276 (1912).
 Brazil (Rio de Janeiro). 84 BZL. Phan.

Tetraplandra bahiensis Emmerich, Arq. Mus. Nac. Rio de Janeiro 56: 95 (1981).
 Brazil (Bahia). 84 BZE.

Tetraplandra dimitrii Emmerich, Arq. Mus. Nac. Rio de Janeiro 56: 96 (1981).
 Brazil. 84 BZL.

Tetraplandra kuhlmannii Emmerich, Arq. Mus. Nac. Rio de Janeiro 56: 96 (1981).
 Brazil. 84 BZL.

Tetraplandra leandrii Baill., Ann. Sci. Nat., Bot., IV, 9: 202 (1858).
 SE. Brazil. 84 BZL. Phan.

Tetraplandra longipetiolata Pax & K.Hoffm. in H.G.A.Engler, Pflanzenr., IV, 147, V: 275 (1912).
 Brazil (Rio de Janeiro). 84 BZL. Nanophan. or phan.

Tetraplandra riedelii Müll.Arg. in C.F.P.von Martius, Fl. Bras. 11(2): 534 (1874).
 Brazil (Rio de Janeiro). 84 BZL. Nanophan. or phan.
  Tetraplandra riedelii var. *subcuneata* Müll.Arg. in C.F.P.von Martius, Fl. Bras. 11(2): 535 (1874).

Synonyms:
Tetraplandra gibbosa Pax & K.Hoffm. === **Algernonia gibbosa** (Pax & K.Hoffm.) Emmerich
Tetraplandra grandifolia Glaz. === ?
Tetraplandra riedelii var. *subcuneata* Müll.Arg. === **Tetraplandra riedelii** Müll.Arg.

Tetrorchidiopsis

Synonyms:
Tetrorchidiopsis Rauschert === **Tetrorchidium** Poepp.

Tetrorchidium

23 species, Americas and Africa; laticiferous shrubs or small to large forest trees (*T. rotundatum* can reach 35 m), the leaves alternate or spirally arranged or (in Africa) opposite and inflorescences axillary. Some species at least have thick twigs. A number grow at elevations over 1000 m, with *T. microphyllum* reaching 2300 m. *Hasskarlia*, comprising the African species, was finally united with *Tetrorchidium* by Pax & Hoffmann (1931; see **General**). Webster (Synopsis, 1994) considers the African *Klaineanthus* to be a close relative; in that genus, however, the leaves are without glands and the inflorescences may be only partly subtended by leaves. The only moderately recent treatment (apart from floras) is for Colombia (Cuatrecasas 1957). (Crotonoideae)

  Pax, F. (with K. Hoffmann) (1912). *Tetrorchidium*. In A. Engler (ed.), Das Pflanzenreich, IV 147 IV (Euphorbiaceae-Gelonieae): 29-32. Berlin. (Heft 52.) La/Ge. — 4 species, Americas; 3 in Andean region.

Pax, F. (with K. Hoffmann) (1914). *Hasskarlia*. In A. Engler (ed.), Das Pflanzenreich, IV 147 VII [Euphorbiaceae-Additamentum V]: 416-418. Berlin. (Heft 63.) La/Ge. — 4 species, Africa. [Now in *Tetrorchidium*].

Cuatrecasas, J. (1957). The Colombian species of *Tetrorchidium*. Brittonia 9: 76-82, illus. En. — Synoptic revision (8 species of which 6 endemic) with key, illustrations, descriptions of novelties, synonymy, references, localities with exsiccatae and commentary.

Jablonski, E. (1967). *Tetrorchidium*. Euphorbiaceae, Guayana Highland (Mem. New York Bot. Gard. 17(1)): 162. New York. En. — 1 species, *T. rubrivenum*.

Tetrorchidium Poepp. in E.F.Poeppig & S.L.Endlicher, Nov. Gen. Sp. Pl. 3: 23 (1841).
Mexico, Trop. America, Trop. Africa. 22 23 25 26 79 80 81 82 83 84.
Hasskarlia Baill., Adansonia 1: 51 (1860).
Tetrorchidiopsis Rauschert, Taxon 31: 559 (1982).

Tetrorchidium andinum Müll.Arg., Flora 47: 538 (1864).
Peru. 83 PER. Phan.

Tetrorchidium boyacanum Croizat, J. Arnold Arbor. 24: 170 (1943).
Colombia. 83 CLM. Phan.

Tetrorchidium brevifolium Standl. & Steyerm., Publ. Field Mus. Nat. Hist., Bot. Ser. 23: 126 (1944).
Guatemala. 80 GUA.
Sapium guatemalense Lundell, Wrightia 5: 76 (1975).

Tetrorchidium bulbipilosum Cuatrec., Brittonia 9: 81 (1957).
Colombia. 83 CLM. Phan.

Tetrorchidium costaricense Huft, Ann. Missouri Bot. Gard. 75: 1112 (1988).
Costa Rica, W. Panama. 80 COS PAN. Phan.

Tetrorchidium didymostemon (Baill.) Pax & K.Hoffm. in H.G.A.Engler, Pflanzenr., IV, 147, XIV: 53 (1919).
Trop. Africa. 22 GNB IVO LBZ NGA SIE TOG 23 CMN EQG GAB GGI ZAI 25 TAN UGA 26 ANG. (Cl.) nanophan. or phan.
  * *Hasskarlia didymostemon* Baill., Adansonia 1: 52 (1860).
  Gelonium angolense Prain, Bull. Misc. Inform. Kew 1911: 233 (1911). *Suregada angolensis* (Prain) Croizat, Bull. Jard. Bot. Buitenzorg, III, 17: 216 (1942).
  Hasskarlia minor Prain, Bull. Misc. Inform. Kew 1912: 231 (1912). *Tetrorchidium minus* (Prain) Pax & K.Hoffm. in H.G.A.Engler, Pflanzenr., IV, 147, XIV: 53 (1919).
  Tetrorchidium congolense J.Léonard, Bull. Jard. Bot. État 29: 197 (1959).
  Tetrorchidium congolense var. *lenifolium* J.Léonard, Bull. Jard. Bot. État 29: 201 (1959).

Tetrorchidium duckei Radcl.-Sm. & Govaerts, Kew Bull. 52: 189 (1997).
Brazil (Amazonas). 84 BZN. Nanophan. or phan.
  * *Adenophaedra minor* Ducke, Arq. Inst. Biol. Veg. 2: 66 (1935). *Tetrorchidium minus* (Ducke) Ducke, Bol. Técn. Inst. Agron. N. 19: 45 (1950), nom. illeg.

Tetrorchidium dusenii Pax & K.Hoffm. in H.G.A.Engler, Pflanzenr., IV, 147, XVI: 197 (1924).
Brazil (Paraná). 84 BZS.

Tetrorchidium euryphyllum Standl., Publ. Field Mus. Nat. Hist., Bot. Ser. 4: 219 (1929).
Costa Rica, Panama, N. Colombia. 80 COS PAN 83 CLM. Phan.

Tetrorchidium gorgonae Croizat, J. Arnold Arbor. 24: 170 (1943).
W. Ecuador, Colombia (Gorgona I.). 83 CLM ECU. Phan.

Tetrorchidium jamaicense Croizat, J. Arnold Arbor. 24: 171 (1943).
Jamaica. 81 JAM. Phan.

Tetrorchidium macrophyllum Müll.Arg. in A.P.de Candolle, Prodr. 15(2): 1133 (1866).
Peru, Ecuador, Colombia. 83 CLM ECU PER. Phan.

Tetrorchidium microphyllum Huft, Ann. Missouri Bot. Gard. 75: 1110 (1988).
Panama (NE. of Boquete). 80 PAN. Phan.

Tetrorchidium molinae L.O.Williams, Fieldiana, Bot. 29: 348 (1961).
Honduras. 80 HON.

Tetrorchidium ochroleucum Cuatrec., Brittonia 9: 79 (1957).
Colombia. 83 CLM. Phan.

Tetrorchidium oppositifolium (Pax) Pax in H.G.A.Engler, Pflanzenr., IV, 147, XIV: 53 (1919).
Liberia, Ivory Coast, Nigeria, Cameroon. 22 IVO LBR NGA 23 CMN. Phan.
  * *Hasskarlia oppositifolia* Pax, Bot. Jahrb. Syst. 43: 81 (1909).

Tetrorchidium parvulum Müll.Arg. in C.F.P.von Martius, Fl. Bras. 11(2): 513 (1874).
Brazil (Rio de Janeiro). 84 BZL. Phan.

Tetrorchidium popayanense Croizat, J. Arnold Arbor. 24: 171 (1943).
Colombia (Cauca). 83 CLM. Phan.

Tetrorchidium robledoanum Cuatrec., Brittonia 9: 81 (1957).
Colombia. 83 CLM. Phan.

Tetrorchidium rotundatum Standl., Trop. Woods 16: 44 (1928).
Mexico to C. Costa Rica. 79 MXG 80 COS HON NIC. Phan.

Tetrorchidium rubrivenium Poepp. in E.F.Poeppig & S.L.Endlicher, Nov. Gen. Sp. Pl. 3: 23 (1841). *Tetrorchidium rubrivenium* var. *genuinum* Müll.Arg. in A.P.de Candolle, Prodr. 15(2): 1133 (1866), nom. inval.
Costa Rica, Windward Is., Colombia, Peru, Venezuela, Brazil (Amazonas, Rio de Janeiro). 80 COS 81 WIN 82 VEN 83 CLM PER 84 BZN BZL. Phan.
  Tetrorchidium trigynum Baill., Étude Euphorb.: 440 (1858). *Tetrorchidium rubrivenium* var. *trigynum* (Baill.) Baill., Adansonia 5: 225 (1865).
  Tetrorchidium rubrivenium var. *fendleri* Müll.Arg. in A.P.de Candolle, Prodr. 15(2): 1133 (1866).
  Tetrorchidium rubrivenium var. *integrifolium* Müll.Arg. in A.P.de Candolle, Prodr. 15(2): 1133 (1866).

Tetrorchidium tenuifolium (Pax & K.Hoffm.) Pax & K.Hoffm. in H.G.A.Engler, Pflanzenr., IV, 147, XIV: 53 (1919).
Cameroon, Zaire. 23 CMN ZAI. Nanophan.
  * *Hasskarlia tenuifolia* Pax & K.Hoffm., Bot. Jahrb. Syst. 45: 238 (1910).

Tetrorchidium ulugurense Verdc., Kew Bull. 12: 347 (1957).
Tanzania (Morogoro). 25 TAN. Nanophan. or phan. – Very close to *T. tenuifolium*.

Thecacoris

19 species, W. to SC. tropical Africa and in Madagascar (5); includes *Cyathogyne*. Perennials, subshrubs or shrubs (sometimes scandent) or small to medium trees of wet or dry forest and bushland with alternate leaves and flowers in axillary spikes. The foliage in some species is reminiscent of the related genus *Antidesma* but here the fruit are dehiscent. Along with that genus it is referred to subtribe Antidesminae (Antidesmeae). The smaller shrubby species are twiggy and recall the Malagasy *Leptonema*. No full revision of either genus has been published since 1922. [Though retained by some authorities, including Léonard (1995) and Stuppy (1995; see **Phyllanthoideae**) *Cyathogyne* is here included in synonymy following Radcliffe-Smith who argues that *T. usambarensis* connects the two concepts.] (Phyllanthoideae)

Pax, F. & K. Hoffmann (1922). *Cyathogyne*. In A. Engler (ed.), Das Pflanzenreich, IV 147 XV (Euphorbiaceae-Phyllanthoideae-Phyllantheae): 40-43. Berlin. (Heft 81.) La/Ge. — 5 species, Africa.

Pax, F. & K. Hoffmann (1922). *Thecacoris*. In A. Engler (ed.), Das Pflanzenreich, IV 147 XV (Euphorbiaceae-Phyllanthoideae-Phyllantheae): 8-13. Berlin. (Heft 81.) La/Ge. — 11 species; Africa, Madagascar.

Leandri, J. (1958). *Thecacoris*. Fl. Madag. Comores 111 (Euphorbiacées), I: 4-12. Paris. Fr. — Flora treatment (5 species); key.

- Léonard, J. (1995). Révision des espèces zaïroises des genres *Thecacoris* A. Juss. et *Cyathogyne* Mull.-Arg. (Euphorbiaceae). Bull. Jard. Bot. Natl. Belg. 64: 13-52, illus., maps. Fr. — Includes descriptive revisions of *Cyathogyne* (pp. 43-52; 4 species and some infraspecific taxa) and *Thecacoris* (pp. 26-42; 5 species) with keys, synonymy, references and citations, types, vernacular names, indication of distribution, habitat and chorology, localities with exsiccatae, notes on uses, and sometimes detailed taxonomic commentary; index at end. The transitional *C. usambarensis* is covered on pp. 22-25 where it is transferred from *Thecacoris*. A key to the two genera appears on pp. 25-26. The general part includes a checklist of *Thecacoris* in mainland Africa (pp. 15-16; 11 recorded) with indication of distinguishing features.

Thecacoris A.Juss., Euphorb. Gen.: 12 (1824).
  Trop. Africa, Madagascar. 22 23 25 26 29.
    Cyathogyne Müll.Arg., Flora 47: 536 (1864).
    Henribaillonia Kuntze, Revis. Gen. Pl. 2: 606 (1891).
    Baccaureopsis Pax, Bot. Jahrb. Syst. 43: 318 (1909).

Thecacoris annobonae Pax & K.Hoffm. in H.G.A.Engler, Pflanzenr., IV, 147, XV: 9 (1922).
  Annobon, Cameroon ? 23 CMN? GGI. Phan.

Thecacoris batesii Hutch., Bull. Misc. Inform. Kew 1910: 58 (1910).
  Cameroon. 23 CMN.
    Thecacoris reticulata Pax in H.G.A.Engler & C.G.O.Drude, Veg. Erde 9, III(2): 14 (1921).

Thecacoris cometia Leandri, Mém. Inst. Sci. Madagascar, Sér. B, Biol. Vég. 8: 211 (1957).
  Madagascar (incl. Nosi Bé I.). 29 MDG. Nanophan. or phan.
    * *Cometia lucida* Baill., Adansonia 2: 55 (1861).

Thecacoris glabroglandulosa (J.Léonard) J.Léonard, Bull. Jard. Bot. Belg. 64: 40 (1995).
  Zaire (Mayombe). 23 ZAI. Nanophan.
    * *Thecacoris gymnogyne* var. *glabroglandulosa* J.Léonard, Bull. Soc. Roy. Bot. Belgique 84: 52 (1951).

Thecacoris grandifolia (Pax & K.Hoffm.) Govaerts in R.Govaerts, D.G.Frodin & A.Radcliffe-Smith, World Checklist Bibliogr. Euphorbiaceae: 1517 (2000).
  WC. Trop. Africa. 23 CMN ZAI.
    * *Cyathogyne grandifolia* Pax & K.Hoffm. in H.G.A.Engler, Pflanzenr., IV, 147, XV: 41 (1922).

Thecacoris humbertii Leandri, Mém. Inst. Sci. Madagascar, Sér. B, Biol. Vég. 8: 211 (1957).
Madagascar. 29 MDG. Phan.

var. **anjanaharibes** Leandri, Mém. Inst. Sci. Madagascar, Sér. B, Biol. Vég. 8: 211 (1957).
E. Madagascar. 29 MDG. Phan.

var. **humbertii**
NW. Madagascar. 29 MDG. Phan.

Thecacoris lancifolia Pax & K.Hoffm. in H.G.A.Engler, Pflanzenr., IV, 147, XV: 12 (1922).
Equatorial Guinea. 23 EQG. Nanophan.

Thecacoris latistipula J.Léonard, Bull. Jard. Bot. État 17: 254 (1945).
Zaire. 23 ZAI. Nanophan.

Thecacoris leptobotrya (Müll.Arg.) Brenan, Kew Bull. 7: 446 (1953).
Nigeria, Cameroon, Congo, Zaire, Equatorial Guinea, Gabon. 22 NGA 23 CMN CON EQG GAB ZAI. Nanophan. or phan.
  * *Antidesma leptobotryum* Müll.Arg., Flora 47: 529 (1864).
    Thecacoris gymnogyne Pax, Bot. Jahrb. Syst. 28: 20 (1899).
    Thecacoris klainei Pierre ex Pax & K.Hoffm. in H.G.A.Engler, Pflanzenr., IV, 147, XV: 12 (1922), pro syn.
    Thecacoris obanensis Hutch. in J.Hutchinson & J.M.Dalziel, Fl. W. Trop. Afr. 1: 283 (1928).
    Thecacoris talbotae Hutch. in J.Hutchinson & J.M.Dalziel, Fl. W. Trop. Afr. 1: 283 (1928).

Thecacoris lucida (Pax) Hutch. in D.Oliver, Fl. Trop. Afr. 6(1): 660 (1912).
  – FIGURE, p. 1519.
Gabon, Congo, Zaire, W. Uganda, Tanzania, N. Angola. 23 CON GAB ZAI 25 TAN UGA 26 ANG. (Cl.) nanophan. or phan.
  * *Baccaureopsis lucida* Pax, Bot. Jahrb. Syst. 43: 319 (1909).

Thecacoris madagascariensis A.Juss., Euphorb. Gen.: 105 (1824).
Madagascar. 29 MDG. Nanophan.

var. **madagascariensis**
E. Madagascar. 29 MDG. Nanophan.
Acalypha glabrata Vahl ex A.Juss., Euphorb. Gen.: 105 (1824).

var. **montana** Leandri, Notul. Syst. (Paris) 6: 19 (1937).
C. & E. Madagascar. 29 MDG. Nanophan.

Thecacoris manniana (Müll.Arg.) Müll.Arg. in A.P.de Candolle, Prodr. 15(2): 246 (1866).
São Tomé. 23 GGI. Phan.
  * *Antidesma mannianum* Müll.Arg., Flora 47: 519 (1864).

Thecacoris membranacea Pax, Bol. Soc. Brot. 10: 158 (1892).
São Tomé. 23 GGI. Phan.

Thecacoris perrieri Leandri, Notul. Syst. (Paris) 6: 19 (1937).
C. Madagascar. 29 MDG. Phan.

Thecacoris spathulifolia (Pax) Leandri, Mém. Inst. Sci. Madagascar, Sér. B, Biol. Vég. 8: 211 (1957).

Thecacoris lucida (Pax) Hutch. in D. Oliver
Artist: G. Papadopoulos
Fl. Trop. East Africa, Euphorbiaceae 1: 109, fig. 18 (1987)
KEW ILLUSTRATIONS COLLECTION

S. Somalia, Kenya, Tanzania, Mozambique, Madagascar. 24 SOM 25 KEN TAN 26 MOZ 29
   MDG. Nanophan.
  * *Cyathogyne spathulifolia* Pax, Bot. Jahrb. Syst. 33: 281 (1903).

var. **greveana** (Leandri) Leandri, Mém. Inst. Sci. Madagascar, Sér. B, Biol. Vég. 8:
   211 (1957).
   W. Madagascar. 29 MDG. Nanophan.
  * *Cyathogyne spathulifolia* var. *greveana* Leandri, Notul. Syst. (Paris) 6: 19 (1937).

var. **spathulifolia** – FIGURE, p. 1521.
S. Somalia, Kenya, Tanzania, Mozambique. 24 SOM 25 KEN TAN 26 MOZ. Nanophan.
Cyathogyne bussei Pax, Bot. Jahrb. Syst. 33: 280 (1903).

Thecacoris stenopetala (Müll.Arg.) Müll.Arg. in A.P.de Candolle, Prodr. 15(2): 246 (1866).
Sierra Leone, Liberia, Ivory Coast, Ghana, Cameroon, Principe, São Tomé, Bioko. 22 GHA IVO LBR SIE 23 CMN GGI. Nanophan.
* *Antidesma stenopetalum* Müll.Arg., Flora 47: 520 (1864).
Thecacoris gymnogyne var. *reticulata* Pax, Bot. Jahrb. Syst. 28: 21 (1899).
Antidesma comoense Beille, Bull. Soc. Bot. France 57(8): 122 (1910).
Thecacoris chevalieri Beille, Bull. Soc. Bot. France 57(8): 119 (1910).

Thecacoris trichogyne Müll.Arg., J. Bot. 2: 328 (1864).
Angola, Zambia, Zaire, Congo, Cameroon. 23 CMN CON ZAI 26 ANG ZAM. Nanophan. or phan.
Thecacoris lenifolia J.Léonard, Bull. Jard. Bot. État 17: 253 (1945).

Thecacoris usambarensis Verdc., Kew Bull. 7: 357 (1953).
Kenya, Tanzania. 25 KEN TAN. Nanophan.

Thecacoris viridis (Müll.Arg.) G.L.Webster, Ann. Missouri Bot. Gard. 81: 52 (1994).
W. & WC. Trop. Africa. 22 NGA 23 CMN CON EQG GAB ZAI.Hemicr. or cham.
* *Cyathogyne viridis* Müll.Arg., Flora 47: 536 (1864).

subsp. **dewevrei** (Pax) Govaerts in R.Govaerts, D.G.Frodin & A.Radcliffe-Smith, World Checklist Bibliogr. Euphorbiaceae: 1520 (2000).
Zaire. 23 ZAI. Cham. or nanophan.
* *Cyathogyne dewevrei* Pax ex De Wild. & T.Durand, Ann. Mus. Congo Belge, Bot., II, 1: 49 (1899). *Cyathogyne viridis* subsp. *dewevrei* (Pax ex De Wild. & T.Durand) J.Léonard, Bull. Jard. Bot. Belg. 64: 47 (1995).

subsp. **glabra** (J.Léonard) Govaerts in R.Govaerts, D.G.Frodin & A.Radcliffe-Smith, World Checklist Bibliogr. Euphorbiaceae: 1520 (2000).
E. Zaire. 23 ZAI. Cham. or nanophan.
* *Cyathogyne viridis* subsp. *glabra* J.Léonard, Bull. Jard. Bot. Belg. 64: 51 (1995).

subsp. **viridis**
W. & WC. Trop. Africa. 22 NGA 23 CMN CON EQG GAB ZAI. Hemicr. or cham.
Cyathogyne preussii Pax, Bot. Jahrb. Syst. 23: 521 (1897). *Cyathogyne viridis* var. *preussii* (Pax) Pax in H.G.A.Engler, Pflanzenr., IV, 147, XV: 42 (1922).
Cyathogyne viridis var. *subintegra* Pax & K.Hoffm. in H.G.A.Engler, Pflanzenr., IV, 147, XV: 41 (1922).

Synonyms:
Thecacoris chevalieri Beille === **Thecacoris stenopetala** (Müll.Arg.) Müll.Arg.
Thecacoris glabrata Hutch. === **Maesobotrya glabrata** (Hutch.) Exell
Thecacoris gymnogyne Pax === **Thecacoris leptobotrya** (Müll.Arg.) Brenan
Thecacoris gymnogyne var. *glabroglandulosa* J.Léonard === **Thecacoris glabroglandulosa** (J.Léonard) J.Léonard
Thecacoris gymnogyne var. *reticulata* Pax === **Thecacoris stenopetala** (Müll.Arg.) Müll.Arg.
Thecacoris klainei Pierre ex Pax & K.Hoffm. === **Thecacoris leptobotrya** (Müll.Arg.) Brenan
Thecacoris lenifolia J.Léonard === **Thecacoris trichogyne** Müll.Arg.
Thecacoris obanensis Hutch. === **Thecacoris leptobotrya** (Müll.Arg.) Brenan
Thecacoris reticulata Pax === **Thecacoris batesii** Hutch.
Thecacoris talbotae Hutch. === **Thecacoris leptobotrya** (Müll.Arg.) Brenan
Thecacoris trillesii Beille === **Spondianthus preussii** var. **glaber** (Engl.) Engl.

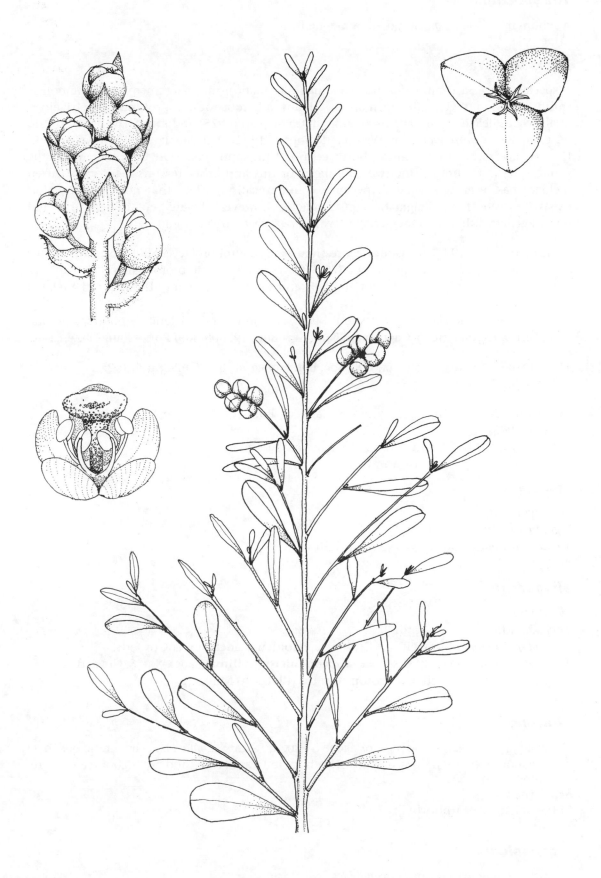

Thecacoris spathulifolia (Pax) Leandri var. *spathulifolia* (as *Cyathogyne bussei* Pax)

Artist: G. Papadopoulos
Fl. Trop. East Africa, Euphorbiaceae 1: 110, fig. 19 (1987)

Thelypotzium

A synonym of *Andrachne* proposed by Gagnepain.

Thyrsanthera

1 species, SE. Asia (Cambodia, Laos, Vietnam, Thailand); multistemmed white-tomentose shrubs 0.25-2 m of generally open areas with a woody rootstock and rotund-ovate triveined 'malvaceous' leaves. The stamens are united. Gagnepain (1925) and Airy-Shaw believed it to be related to *Chrozophora* but Webster (Synopsis, 1994) includes it with *Sumbaviopsis* in Doryxylinae rather than Chrozophorinae (tribe Chrozophoreae). On the other hand, the genus appears to 'bridge' the two subtribes. For this and other reasons Peter van Welzen (Leiden) has recently suggested (personal communication, 1998) that no real distinction exists between them. Pragmatic considerations, however, dictate retention of this and related genera rather than merger of all with *Chrozophora*. (Acalyphoideae)

> Gagnepain, F. (1925). Quelques genres nouveaux d'Euphorbiacées. Bul. Soc. Bot. France 71: 864-879. Fr. — Includes (pp. 878-879) generico-specific protologue of *Thyrsanthera suborbicularis* (described from Cambodia and southern Vietnam); thought to be related to *Chrozophora*.
>
> Airy-Shaw, H. K. (1965). Notes on Malaysian and other Asiatic Euphorbiaceae, LI. *Thyrsanthera* Pierre ex Gagnep. in Siam. Kew Bull. 19: 308-309. En. — Range extension.

Thyrsanthera Pierre ex Gagnep., Bull. Soc. Bot. France 71: 878 (1924 publ. 1925).
  Indo-China. 41.

Thyrsanthera suborbicularis Pierre ex Gagnep., Bull. Soc. Bot. France 71: 878 (1924 publ. 1925).
  Indo-China. 41 CBD LAO THA VIE. Cham. or nanophan.

Tiglium

Synonyms:
Tiglium Klotzsch === **Croton** L.
Tiglium officinale Klotzsch === **Croton tiglium** L.

Timandra

Synonyms:
Timandra Klotzsch === **Croton** L.
Timandra dichotoma Klotzsch === **Croton serratoideus** Radcl.-Sm. & Govaerts
Timandra erythroxyloides Klotzsch === **Croton microphyllinus** Radcl.-Sm. & Govaerts
Timandra serrata Klotzsch === **Croton serratus** (Klotzsch) Müll.Arg.

Tirucalia

Generally now regarded as an infrageneric taxon within *Euphorbia*. The numerous new combinations recently proposed by P. V. Heath are in Heath (1996) and not accounted for here.

Synonyms:
Tirucalia Raf. === **Euphorbia** L.

Tithymalodes

An orthographic variant of *Tithymaloides*.

Synonyms:
Tithymalodes Ludw. ex Kuntze === **Pedilanthus** Neck.

Tithymaloides

Rejected against *Pedilanthus*.

Synonyms:
Tithymaloides Ortega === **Pedilanthus** Neck.

Tithymalopsis

A Klotzsch segregate of *Euphorbia*, now usually regarded as infrageneric.

Synonyms:
Tithymalopsis Klotzsch & Garcke === **Euphorbia** L.
Tithymalopsis exserta Small === **Euphorbia exserta** (Small) Coker
Tithymalopsis ipecacuanhae f. *linearis* Moldenke === **Euphorbia ipecacuanhae** L.

Tithymalus

of Gaertner; usually included within *Euphorbia* and more or less inclusive of the non-succulent members of subgen. *Esula*. Some twentieth-century Eastern European and CIS authors have, however, maintained it given – like Klotzsch and others in disagreement with Boissier – their belief in a narrower generic concept within Euphorbiinae. It is typified by *E. peplus*. The name is conserved against that of Miller (= *Pedilanthus*).

Synonyms:
Tithymalus Gaertn., nom. cons. === **Euphorbia** L.
Tithymalus acalyphoides (Hochst. ex Boiss.) Schweinf. === **Euphorbia acalyphoides** Hochst. ex Boiss.
Tithymalus acanthothamnos (Heldr. & Sart. ex Boiss.) Soják === **Euphorbia acanthothamnos** Heldr. & Sart. ex Boiss.
Tithymalus acicularis Dulac === **Euphorbia cyparissias** L.
Tithymalus acutifolius Lam. === **Euphorbia pithyusa** L.
Tithymalus adenensis (Deflers) Soják === **Euphorbia balsamifera** subsp. **adenensis** (Deflers) Govaerts
Tithymalus adenochlorus (C.Morren & Decne.) Hara === **Euphorbia adenochlora** C.Morren & Decne.
Tithymalus adrianus Klotzsch & Garcke === **Euphorbia papillosa** A.St.-Hil.
Tithymalus aellenii (Rech.f.) Soják === **Euphorbia aellenii** Rech.f.
Tithymalus agowensis (Hochst. ex Boiss.) Schweinf. === **Euphorbia agowensis** Hochst. ex Boiss.
Tithymalus agrarius (M.Bieb.) Klotzsch & Garcke === **Euphorbia agraria** M.Bieb.
Tithymalus akenocarpus (Guss.) Klotzsch & Garcke === **Euphorbia akenocarpa** Guss.
Tithymalus alaicus Prokh. === **Euphorbia alaica** (Prokh.) Prokh.
Tithymalus alatavicus (Boiss.) Prokh. === **Euphorbia alatavica** Boiss.
Tithymalus aleppicus (L.) Klotzsch & Garcke === **Euphorbia aleppica** L.
Tithymalus alpigena (A.Kern.) Woerl. === **Euphorbia dulcis** L. subsp. **dulcis**
Tithymalus alpinus (C.A.Mey. ex Ledeb.) Klotzsch & Garcke === **Euphorbia alpina** C.A.Mey. ex Ledeb.
Tithymalus altaicus (C.A.Mey.) Klotzsch & Garcke === **Euphorbia altaica** C.A.Mey.
Tithymalus altissimus (Boiss.) Klotzsch & Garcke === **Euphorbia altissima** Boiss.
Tithymalus altotibeticus (Paulsen) Prokh. === **Euphorbia altotibetica** Paulsen
Tithymalus altus (Norton) Wooton & Standl. === **Euphorbia alta** Norton
Tithymalus amplexicaulis Klotzsch & Garcke === **Euphorbia condylocarpa** M.Bieb.
Tithymalus amygdaloides (L.) Garsault === **Euphorbia amygdaloides** L.
Tithymalus amygdaloides subsp. *arbuscula* (Meusel) Soják === **Euphorbia amygdaloides** subsp. **arbuscula** Meusel

Tithymalus amygdaloides subsp. *semiperfoliatus* (Viv.) Soják === **Euphorbia amygdaloides** subsp. **semiperfoliata** (Viv.) Radcl.-Sm.

Tithymalus anacampseros (Boiss.) Klotzsch === **Euphorbia anacampseros** Boiss.

Tithymalus ancyrensis (Azn. ex M.S.Khan) Soják === **Euphorbia coniosperma** Boiss. & Buhse

Tithymalus andrachnoides (Schrenk) Klotzsch & Garcke === **Euphorbia andrachnoides** Schrenk

Tithymalus androsaemifolius (Schousb. ex Willd.) Samp. === **Euphorbia esula** L. subsp. **esula**

Tithymalus angulatus (Jacq.) Raf. === **Euphorbia angulata** Jacq.

Tithymalus × *angustatus* (Rochel) Soják === **Euphorbia** × **angustata** (Rochel) Borza

Tithymalus angustifolius Gilib. === **Euphorbia cyparissias** L.

Tithymalus × *angustifrons* (Borbás) Holub === **Euphorbia** × **angustifrons** Borbás

Tithymalus antiquorus (L.) Moench === **Euphorbia antiquorum** L.

Tithymalus aphyllus (Brouss. ex Willd.) Klotzsch & Garcke === **Euphorbia aphylla** Brouss. ex Willd.

Tithymalus apiculatus Klotzsch & Garcke === **Euphorbia erythrina** Link

Tithymalus apios (L.) Hill === **Euphorbia apios** L.

Tithymalus arboreus Tourn. ex Lam. === **Euphorbia dendroides** L.

Tithymalus ardonensis (Galushko) Galushko === **Euphorbia ardonensis** Galushko

Tithymalus argutus (Banks & Sol.) Soják === **Euphorbia arguta** Banks & Sol.

Tithymalus aristatus (Schmalh.) Prokh. === **Euphorbia aristata** Schmalh.

Tithymalus arkansanus (Engelm. & A.Gray) Klotzsch & Garcke === **Euphorbia spathulata** Lam.

Tithymalus arvalis (Boiss. & Heldr.) Klotzsch & Garcke === **Euphorbia arvalis** Boiss. & Heldr.

Tithymalus asper (M.Bieb.) Klotzsch & Garcke === **Euphorbia squamosa** Willd.

Tithymalus atlantis (Maire) Soják === **Euphorbia clementei** Boiss. subsp. **clementei**

Tithymalus atropurpureus (Brouss. ex Willd.) Klotzsch & Garcke === **Euphorbia atropurpurea** Brouss. ex Willd.

Tithymalus atrosanguineus Klotzsch & Garcke === **Euphorbia portulacoides** L. subsp. **portulacoides**

Tithymalus attenuatus Klotzsch & Garcke === **Euphorbia silenifolia** (Haw.) Sweet

Tithymalus aucheri (Boiss.) Klotzsch & Garcke === **Euphorbia aucheri** Boiss.

Tithymalus aulacospermus (Boiss.) Klotzsch === **Euphorbia aulacosperma** Boiss.

Tithymalus auriculatus Lam. === **Euphorbia peplis** L.

Tithymalus austriacus (A.Kern.) Á.Löve & D.Löve === **Euphorbia illirica** Lam. subsp. **illirica**

Tithymalus austrinus Small === **Euphorbia commutata** Engelm. ex A.Gray

Tithymalus austroanatolicus (Hub.-Mor. & M.S.Khan) Soják === **Euphorbia glabriflora** Vis.

Tithymalus azoricus (Hochst.) Klotzsch & Garcke === **Euphorbia azorica** Hochst.

Tithymalus aztecus Croizat === **Pedilanthus bracteatus** (Jacq.) Boiss.

Tithymalus bahamensis (Millsp.) Croizat === **Pedilanthus tithymaloides** subsp. **bahamensis** (Millsp.) Dressler

Tithymalus balsamifer (Aiton) Haw. === **Euphorbia balsamifera** Aiton

Tithymalus barrelieri (Savi) Soják === **Euphorbia barrelieri** Savi

Tithymalus barrelieri subsp. *carnicus* (Boiss.) Soják === **Euphorbia triflora** Schott, Nyman & Kotschy subsp. **triflora**

Tithymalus barrelieri subsp. *hercegovinus* (Beck) Soják === **Euphorbia barrelieri** subsp. **hercegovina** (Beck) Kuzmanov

Tithymalus barrelieri subsp. *thessalus* (Formánek) Soják === **Euphorbia barrelieri** subsp. **thessala** (Formánek) Bornm.

Tithymalus basarabicus (Prodán) Soják === **Euphorbia basarabica** Prodán

Tithymalus baxanicus (Galushko) Galushko === **Euphorbia baxanica** Galushko

Tithymalus × *bazargicus* (Prodán) Soják === **Euphorbia** × **bazargica** Prodán

Tithymalus beamanii (M.C.Johnst.) Soják === **Euphorbia beamanii** M.C.Johnst.

Tithymalus bergii Klotzsch & Garcke === **Euphorbia silenifolia** (Haw.) Sweet

Tithymalus berytheus (Boiss. & Blanche) Soják === **Euphorbia berythea** Boiss. & Blanche

Tithymalus biglandulosus (Desf.) Haw. === **Euphorbia rigida** M.Bieb.

Tithymalus biumbellatus (Poir.) Klotzsch & Garcke === **Euphorbia biumbellata** Poir.

Tithymalus bivonae (Steud.) Soják === **Euphorbia bivonae** Steud.

Tithymalus blepharophyllus (C.A.Mey.) Klotzsch & Garcke === **Euphorbia blepharophylla** C.A.Mey.

Tithymalus boeticus (Boiss.) Samp. === **Euphorbia boetica** Boiss.

Tithymalus boissierianus Woronow === **Euphorbia boissieriana** (Woronow) Prokh.

Tithymalus bounophilus (Boiss.) Prokh. === **Euphorbia aucheri** Boiss.

Tithymalus brachycerus (Engelm.) Small === **Euphorbia brachycera** Engelm.

Tithymalus brachypus Klotzsch & Garcke === **Euphorbia mauritanica** L.

Tithymalus bracteatus (Jacq.) Haw. === **Pedilanthus bracteatus** (Jacq.) Boiss.

Tithymalus braunii Schweinf. === **Euphorbia longituberculosa** Hochst. ex Boiss.

Tithymalus bridgesii Klotzsch & Garcke === **Euphorbia portulacoides** L. subsp. **portulacoides**

Tithymalus briquetii (Emb. & Maire) Soják === **Euphorbia briquetii** Emb. & Maire

Tithymalus brittingeri (Opiz ex Samp.) Holub === **Euphorbia verrucosa** L.

Tithymalus broteroi (Daveau) Soják === **Euphorbia oxyphylla** Boiss.

Tithymalus buchtormensis (C.A.Mey.) Klotzsch & Garcke === **Euphorbia buchtormensis** C.A.Mey.

Tithymalus × *budensis* (T.Simon) Chrtek & Krísa === **Euphorbia** × **budensis** T.Simon

Tithymalus buhsei (Boiss.) Prokh. === **Euphorbia buhsei** Boiss.

Tithymalus bungei (Boiss.) Prokh. === **Euphorbia bungei** Boiss.

Tithymalus bupleurifolius (Jacq.) Haw. === **Euphorbia bupleurifolia** Jacq.

Tithymalus bupleuroides (Desf.) Soják === **Euphorbia bupleuroides** Desf.

Tithymalus buschianus (Grossh.) Soják === **Euphorbia buschiana** Grossh.

Tithymalus cadrilateri (Prodán) Soják === **Euphorbia nicaeensis** subsp. **cadrilateri** (Prodán) Kuzmanov

Tithymalus caeladenius (Boiss.) Soják === **Euphorbia caeladenia** Boiss.

Tithymalus caesalpinii Bubani === **Euphorbia hirsuta** L.

Tithymalus caesius (Kar. & Kir.) Klotzsch & Garcke === **Euphorbia esula** var. **caesia** (Kar. & Kir.) Ledeb.

Tithymalus caespitosus (Lam.) Soják === **Euphorbia caespitosa** Lam.

Tithymalus caiogalus (Ehrh.) Moench === **Euphorbia seguieriana** Neck. subsp. **seguieriana**

Tithymalus calcaratus (Schltdl.) Croizat === **Pedilanthus calcaratus** Schltdl.

Tithymalus calendulifolius (Delile) Raf. === **Euphorbia arguta** Banks & Sol.

Tithymalus calyculatus (Kunth) Klotzsch & Garcke === **Euphorbia calyculata** Kunth

Tithymalus calyptratus (Coss. & Kralik) Soják === **Euphorbia calyptrata** Coss. & Kralik

Tithymalus campestris (S.Geuns) Klotzsch & Garcke === **Euphorbia seguieriana** Neck. subsp. **seguieriana**

Tithymalus canariensis (L.) H.Karst. === **Euphorbia canariensis** L.

Tithymalus capensis Klotzsch & Garcke === **Euphorbia striata** Thunb.

Tithymalus capitulatus (Rchb.) Soják === **Euphorbia capitulata** Rchb.

Tithymalus cardiophyllus (Boiss. & Heldr.) Klotzsch & Garcke === **Euphorbia cardiophylla** Boiss. & Heldr.

Tithymalus carniolicus (Jacq.) Raf. === **Euphorbia carniolica** Jacq.

Tithymalus carpaticus (Wol.) Á.Löve & D.Löve === **Euphorbia carpatica** Wol.

Tithymalus cartaginiensis (Porta & Rigo) Soják === **Euphorbia squamigera** Loisel.

Tithymalus cassia (Boiss.) Klotzsch & Garcke === **Euphorbia cassia** Boiss.

Tithymalus cataputia Garsault === **Euphorbia lathyris** L.

Tithymalus caudiculosus (Boiss.) Prokh. === **Euphorbia caudiculosa** Boiss.

Tithymalus cebrinus Klotzsch & Garcke === **Euphorbia petitiana** A.Rich.

Tithymalus celerieri (Emb.) Soják === **Euphorbia celerieri** (Emb.) Emb. ex Vindt

Tithymalus ceratocarpus (Ten.) Soják === **Euphorbia ceratocarpa** Ten.

Tithymalus cerebrinus (Hochst. ex Boiss.) Schweinf. === **Euphorbia petitiana** A.Rich.

Tithymalus chaborasius (Gomb.) Soják === **Euphorbia chaborasia** Gomb.

Tithymalus chamaebuxus (Bernard ex Gren. & Godr.) Soják === **Euphorbia pyrenaica** Jord.

Tithymalus chamaepeplus (Boiss. & Gaill.) Soják === **Euphorbia chamaepeplus** Boiss. & Gaill.

Tithymalus chamaesula (Boiss.) Wooton & Standl. === **Euphorbia chamaesula** Boiss.

Tithymalus chamaesyce (L.) Moench === **Euphorbia chamaesyce** L.

Tithymalus chankoanus (Vorosch.) Soják === **Euphorbia esula** L. subsp. **esula**

Tithymalus characias (L.) Hill === **Euphorbia characias** L.

Tithymalus cheiradenius (Boiss. & Hohen.) Klotzsch & Garcke === **Euphorbia cheiradenia** Boiss. & Hohen.

Tithymalus cheirolepioides (Rech.f.) Soják === **Euphorbia cheirolepioides** Rech.f.

Tithymalus chesneyi Klotzsch & Garcke === **Euphorbia cuspidata** Bertol.

Tithymalus chilensis (Gay) Klotzsch & Garcke === **Euphorbia portulacoides** L. subsp. **portulacoides**

Tithymalus chimaerus (Lipsky) Galushko === **Euphorbia chimaera** Lipsky

Tithymalus chrysophyllus Klotzsch & Garcke === **Euphorbia portulacoides** subsp. **collina** (Phil.) Croizat

Tithymalus cinerascens Moench === **Euphorbia segetalis** L. var. **segetalis**

Tithymalus cinerea Raf. === **?**

Tithymalus clementei (Boiss.) Klotzsch & Garcke === **Euphorbia clementei** Boiss.

Tithymalus clusii Bubani === **Euphorbia isatidifolia** Lam.

Tithymalus cognatus Klotzsch ex Klotzsch & Garcke === **Euphorbia cashmeriana** Royle

Tithymalus collinus (Phil.) Klotzsch & Garcke === **Euphorbia portulacoides** subsp. **collina** (Phil.) Croizat

Tithymalus commutatus (Engelm. ex A.Gray) Klotzsch & Garcke === **Euphorbia commutata** Engelm. ex A.Gray

Tithymalus condylocarpus (M.Bieb.) Klotzsch & Garcke === **Euphorbia condylocarpa** M.Bieb.

Tithymalus confertus Klotzsch & Garcke === **Euphorbia ericoides** Lam.

Tithymalus connata Raf. === **Euphorbia aleppica** L.

Tithymalus consanguineus (Schrenk) Klotzsch & Garcke === **Euphorbia consanguinea** Schrenk

Tithymalus corallioides (L.) Raf. === **Euphorbia corallioides** L.

Tithymalus cornutus (Pers.) Schweinf. ex Asch. === **Euphorbia retusa** Forssk.

Tithymalus correllii (M.C.Johnst.) Soják === **Euphorbia correllii** M.C.Johnst.

Tithymalus corsicus (Req.) Soják === **Euphorbia corsica** Req.

Tithymalus cossonianus (Boiss.) Soják === **Euphorbia cossoniana** Boiss.

Tithymalus cotinifolius (L.) Haw. === **Euphorbia cotinifolia** L.

Tithymalus craspedius (Boiss.) Klotzsch & Garcke === **Euphorbia craspedia** Boiss.

Tithymalus crebrifoliatus Bubani === **Euphorbia segetalis** var. **pinea** (L.) Lanke

Tithymalus crenulatus (Engelm.) A.Heller === **Euphorbia crenulata** Engelm.

Tithymalus cressoides (M.C.Johnst.) Soják === **Euphorbia cressoides** M.C.Johnst.

Tithymalus crispus Haw. === **Euphorbia crispa** (Haw.) Sweet

Tithymalus cristatus Raf. === **Euphorbia retusa** Forssk.

Tithymalus cuneifolius (Guss.) Klotzsch & Garcke === **Euphorbia cuneifolia** Guss.

Tithymalus cupanii (Guss. ex Bertol.) Klotzsch & Garcke === **Euphorbia pithyusa** subsp. **cupanii** (Guss. ex Bertol.) Radcl.-Sm.

Tithymalus cyatophorus (Murray) Moench === **Euphorbia cyathophora** Murr.

Tithymalus cybirensis (Boiss.) Klotzsch & Garcke === **Euphorbia valerianifolia** Lam.

Tithymalus cymbifer (Schltdl.) Croizat === **Pedilanthus cymbiferus** Schltdl.

Tithymalus cyparissias (L.) Hill === **Euphorbia cyparissias** L.

Tithymalus cyrtophyllus Prokh. === **Euphorbia cyrtophylla** (Prokh.) Prokh.

Tithymalus damascenus (Boiss.) Klotzsch & Garcke === **Euphorbia macroclada** Boiss.

Tithymalus darlingtonii (A.Gray) Small === **Euphorbia purpurea** (Raf.) Fernald

Tithymalus davisii (M.S.Khan) Soják === **Euphorbia davisii** M.S.Khan

Tithymalus deamii (Millsp.) Croizat === **Pedilanthus tithymaloides** (L.) Poit. subsp. **tithymaloides**

Tithymalus decipiens (Boiss. & Buhse) Soják === **Euphorbia decipiens** Boiss. & Buhse

Tithymalus declinatus Moench === **Euphorbia portlandica** L.

Tithymalus deflexus (Sibth. & Sm.) Klotzsch & Garcke === **Euphorbia deflexa** Sibth. & Sm.

Tithymalus deltobracteatus Prokh. === **Euphorbia deltobracteata** (Prokh.) Prokh.

Tithymalus dendroides (L.) Hill === **Euphorbia dendroides** L.

Tithymalus densiusculiformis (Pazij) Pazij === **Euphorbia densiusculiformis** (Pazij) Botsch.

Tithymalus densiusculus (Popov) Prokh. === **Euphorbia densiuscula** Popov

Tithymalus densus (Schrenk) Klotzsch & Garcke === **Euphorbia densa** Schrenk

Tithymalus denticulatus Moench === **Euphorbia serrata** L.

Tithymalus depauperatus (Hochst. ex A.Rich.) Klotzsch & Garcke === **Euphorbia depauperata** Hochst. ex A.Rich.

Tithymalus deseglisei (Boreau ex Boiss.) Soják === **Euphorbia dulcis** L. subsp. **dulcis**

Tithymalus desertorum (Weinm.) Klotzsch & Garcke === **Euphorbia villosa** subsp. **semivillosa**

Tithymalus dictyospermus (Fisch. & C.A.Mey.) A.Heller === **Euphorbia spathulata** Lam.

Tithymalus diffusus Lam. === **Euphorbia spinosa** L. subsp. **spinosa**

Tithymalus dilatatus (Hochst. ex A.Rich.) Klotzsch & Garcke === **Euphorbia schimperiana** Scheele var. **schimperiana**

Tithymalus dimorphocaulon (P.H.Davis) Soják === **Euphorbia dimorphocaulon** P.H.Davis

Tithymalus discolor (Ledeb.) Klotzsch & Garcke === **Euphorbia esula** L. subsp. **esula**

Tithymalus divergens Klotzsch === **Euphorbia rothiana** Spreng.

Tithymalus djimilensis (Boiss.) Soják === **Euphorbia djimilensis** Boiss.

Tithymalus dobrogensis (Prodán) Soják === **Euphorbia nicaeensis** subsp. **dobrogensis** (Prodán) Kuzmanov

Tithymalus dominii (Rohlena) Chrtek & Krísa === **Euphorbia helioscopia** L. subsp. **helioscopia**

Tithymalus dracunculoides (Lam.) Klotzsch & Garcke === **Euphorbia dracunculoides** Lam.

Tithymalus dracunculoides subsp. *flamandii* (Batt.) Soják === **Euphorbia dracunculoides** subsp. **flamandii** (Batt.) Maire

Tithymalus dracunculoides subsp. *glebulosus* (Coss. & Durieu) Soják === **Euphorbia dracunculoides** subsp. **glebulosa** (Coss. & Durieu) Maire

Tithymalus dracunculoides subsp. *hesperius* (Maire) Soják === **Euphorbia dracunculoides** subsp. **hesperia** Maire

Tithymalus dracunculoides subsp. *inconspicuus* (Ball) Soják === **Euphorbia dracunculoides** subsp. **inconspicua** (Ball) Maire

Tithymalus dracunculoides subsp. *intermedius* (Maire) Soják === **Euphorbia dracunculoides** subsp. **intermedia** (Maire) Maire

Tithymalus dracunculoides subsp. *volutianus* (Maire) Soják === **Euphorbia dracunculoides** subsp. **volutiana** Maire

Tithymalus dulcis Scop. === **Euphorbia dulcis** L. subsp. **dulcis**

Tithymalus dulcis subsp. *ellipticus* (Pers.) Soják === **Euphorbia dulcis** L. subsp. **dulcis**

Tithymalus dulcis subsp. *incomptus* (Ces.) Soják === **Euphorbia dulcis** L. subsp. **dulcis**

Tithymalus dulcis subsp. *purpuratus* (Thuill.) Holub === **Euphorbia dulcis** subsp. **purpurata** (Thuill.) Rothm.

Tithymalus dumosus (Boiss.) Klotzsch & Garcke === **Euphorbia hierosolymitana** Boiss. var. **hierosolymitana**

Tithymalus duvalii (Lecoq & Lamotte) Soják === **Euphorbia duvalii** Lecoq & Lamotte

Tithymalus ebracteolatus (Hayata) Hara === **Euphorbia ebracteolata** Hayata

Tithymalus ecklonii Klotzsch & Garcke === **Euphorbia ecklonii** (Klotzsch & Garcke) Baill.

Tithymalus edgeworthii (Boiss.) Soják === **Euphorbia edgeworthii** Boiss.

Tithymalus edulis (Lour.) H.Karst. === **Euphorbia neriifolia** L.

Tithymalus elliottii Klotzsch & Garcke === **Euphorbia gracilior** Cronquist

Tithymalus ellipticus Klotzsch & Garcke === **Euphorbia silenifolia** (Haw.) Sweet

Tithymalus elwendicus (Stapf) Soják === **Euphorbia elwendica** Stapf

Tithymalus eochlorus Croizat === **Pedilanthus bracteatus** (Jacq.) Boiss.

Tithymalus epicyparissias E.Mey. ex Klotzsch & Garcke === **Euphorbia epicyparissias** (E.Mey ex Klotzsch & Garcke) Boiss.

Tithymalus epithymoides (L.) Klotzsch & Garcke === **Euphorbia epithymoides** L.

Tithymalus epithymoides var. *microspermus* (Murb.) Soják === **Euphorbia epithymoides** L.

Tithymalus ericetorum (Zumagl.) Soják === **Euphorbia verrucosa** L.

Tithymalus erinaceus (Boiss. & Kotschy) Soják === **Euphorbia erinacea** Boiss. & Kotschy

Tithymalus eriophorus (Boiss.) Klotzsch & Garcke === **Euphorbia eriophora** Boiss.

Tithymalus erubescens (Boiss.) Klotzsch & Garcke === **Euphorbia erubescens** Boiss.

Tithymalus erythradenius (Boiss.) Klotzsch & Garcke === **Euphorbia erythradenia** Boiss.

Tithymalus erythraeus (Hemsl.) Soják === **Euphorbia sieboldiana** Morris & Decne.

Tithymalus erythrinus (Link) Klotzsch & Garcke === **Euphorbia erythrina** Link

Tithymalus erythrodon (Boiss. & Heldr.) Klotzsch & Garcke === **Euphorbia erythrodon** Boiss. & Heldr.

Tithymalus erythrorrhizus Klotzsch & Garcke === **Euphorbia papillosa** A.St.-Hil.

Tithymalus esula (L.) Hill === **Euphorbia esula** L.

Tithymalus esula subsp. *mosanus* (Lej.) Soják === **Euphorbia esula** L. subsp. **esula**

Tithymalus esula subsp. *pseudocyparissias* (Jord.) Soják === **Euphorbia esula** L. subsp. **esula**

Tithymalus esula subsp. *riparia* (Jord.) Chrtek & Krísa === **Euphorbia esula** L. subsp. **esula**

Tithymalus esula subsp. *tristis* (Besser) Chrtek & Krísa === **Euphorbia esula** L. subsp. **esula**

Tithymalus esuliformis (Schauer ex Nees & Schauer) Klotzsch & Garcke === **Euphorbia esuliformis** Schauer ex Nees & Schauer

Tithymalus exiguus (L.) Hill === **Euphorbia exigua** L.

Tithymalus falcatus (L.) Klotzsch & Garcke ex Garcke === **Euphorbia falcata** L.

Tithymalus falcatus subsp. *acuminatus* (Lam.) Soják === **Euphorbia falcata** L. subsp. **falcata**

Tithymalus falcatus subsp. *macrostegius* (Bornm.) Soják === **Euphorbia falcata** subsp. **macrostegia** (Bornm.) O.Schwarz

Tithymalus fendleri (Torr. & A.Gray) Klotzsch & Garcke === **Euphorbia fendleri** Torr. & A.Gray

Tithymalus ferganensis (B.Fedtsch.) Prokh. === **Euphorbia ferganensis** B.Fedtsch.

Tithymalus filicinus (Port.) Soják === **Euphorbia esula** L. subsp. **esula**

Tithymalus firmus (Ledeb.) Klotzsch & Garcke === **Euphorbia seguieriana** Neck. subsp. **seguieriana**

Tithymalus fischerianus (Steud.) Soják === **Euphorbia fischeriana** Steud.

Tithymalus fistulosus (M.S.Khan) Soják === **Euphorbia fistulosa** M.S.Khan

Tithymalus flaccidus Moench === **Euphorbia mauritanica** L.

Tithymalus flavicomus (DC.) Bubani === **Euphorbia flavicoma** DC.

Tithymalus flavicomus subsp. *brittingeri* (Opiz ex Samp.) Soják === **Euphorbia verrucosa** L.

Tithymalus flavicomus subsp. *costeanus* (Rouy) Soják === **Euphorbia flavicoma** subsp. **costeana** (Rouy) Vindt & Guin. ex Greuter & Burdet

Tithymalus floridana Raf. === ?

Tithymalus foliosus Klotzsch & Garcke === **Euphorbia foliosa** (Klotzsch & Garcke) N.E.Br.

Tithymalus fontquerianus (Greuter) Soják === **Euphorbia fontqueriana** Greuter

Tithymalus fragifer Klotzsch & Garcke === **Euphorbia epithymoides** L.

Tithymalus franchetii (B.Fedtsch.) Prokh. === **Euphorbia franchetii** B.Fedtsch.

Tithymalus franciscanus (Norton) A.Heller === **Euphorbia crenulata** Engelm.

Tithymalus fruticosus Klotzsch & Garcke === **Euphorbia bivonae** Steud. subsp. **bivonae**

Tithymalus fruticosus Gilib. === **Euphorbia palustris** L.

Tithymalus furcillatus (Kunth) Klotzsch & Garcke === **Euphorbia furcillata** Kunth

Tithymalus gaillardotii (Boiss. & Blanche) Klotzsch & Garcke === **Euphorbia gaillardotii** Boiss. & Blanche

Tithymalus galilaeus (Boiss.) Klotzsch & Garcke === **Euphorbia falcata** var. **galilaea** (Boiss.) Boiss.

Tithymalus gasparrinii (Boiss.) Soják === **Euphorbia gasparrinii** Boiss.

Tithymalus gaubae Soják === **Euphorbia gaubae** (Soják) Radcl.-Sm.

Tithymalus × gayeri (Borbás & Soó) Holub === **Euphorbia × gayeri** Borbás & So$O

Tithymalus gayi (Salis) Klotzsch & Garcke === **Euphorbia gayi** Salis

Tithymalus gedrosiacus (Rech.f., Aellen & Esfand.) Soják === **Euphorbia gedrosiaca** Rech.f., Aellen & Esfand.

Tithymalus genistoides (Bergius) Klotzsch & Garcke === **Euphorbia genistoides** Bergius

Tithymalus gerardianus (Jacq.) Steud. === **Euphorbia seguieriana** Neck. subsp. **seguieriana**

Tithymalus gerardii Raf. === **Euphorbia seguieriana** Neck. subsp. **seguieriana**

Tithymalus gibellianus (Peola) Soják === **Euphorbia hyberna** subsp. **gibelliana** (Peola) Raffaelli

Tithymalus glaberrimus (K.Koch) Klotzsch & Garcke === **Euphorbia glaberrima** K.Koch

Tithymalus glabriflorus (Vis.) Soják === **Euphorbia glabriflora** Vis.

Tithymalus glaucus Moench === **Euphorbia moenchiana** === ?

Tithymalus glomerulans Prokh. === **Euphorbia esula** L. subsp. **esula**

Tithymalus gmelinii (Steud.) Prokh. === **Euphorbia esula** L. subsp. **esula**

Tithymalus gracilis Klotzsch & Garcke === **Euphorbia esula** L. subsp. **esula**

Tithymalus graecus (Boiss. & Spruner) Klotzsch & Garcke === **Euphorbia taurinensis** All.

Tithymalus graminifolius (Vill.) Soják === **Euphorbia graminifolia** Vill.

Tithymalus graminifolius subsp. *jaxarticus* (Prokh.) Soják === **Euphorbia esula** subsp. **tommasiniana** (Bertol.) Kuzmanov

Tithymalus graminifolius subsp. *orientalis* (Boiss.) Soják === **Euphorbia esula** subsp. **tommasiniana** (Bertol.) Kuzmanov

Tithymalus graminifolius subsp. *tommasiniana* (Bertol.) Soják === **Euphorbia esula** subsp. **tommasiniana** (Bertol.) Kuzmanov

Tithymalus graminifolius subsp. *waldsteinii* (Soják) Soják === **Euphorbia esula** subsp. **tommasiniana** (Bertol.) Kuzmanov

Tithymalus graminifolius subsp. *zhiguliensis* (Prokh.) Soják === **Euphorbia esula** L. subsp. **esula**

Tithymalus granulatus (Forssk.) Raf. === **Euphorbia granulata** Forssk.

Tithymalus gregersenii (K.Maly ex Beck) Soják === **Euphorbia gregersenii** K.Maly ex Beck

Tithymalus greggii (Millsp.) Croizat === **Pedilanthus bracteatus** (Jacq.) Boiss.

Tithymalus griffithii (Hook.f.) Hara === **Euphorbia griffithii** Hook.f.

Tithymalus grisebachii (Millsp. & Britton) Croizat === **Pedilanthus tithymaloides** subsp. **jamaicensis** (Millsp. & Britton) Dressler

Tithymalus grisophyllus (M.S.Khan) Soják === **Euphorbia grisophylla** M.S.Khan

Tithymalus grossheimii Prokh. === **Euphorbia grossheimii** (Prokh.) Prokh.

Tithymalus guestii (Blakelock) Soják === **Euphorbia gaillardotii** Boiss. & Blanche

Tithymalus gulestanicus (Podlech) Soják === **Euphorbia gulestanica** Podlech

Tithymalus guntensis Prokh. === **Euphorbia guntensis** (Prokh.) Prokh.

Tithymalus guyonianus (Boiss. & Reut.) Klotzsch & Garcke === **Euphorbia guyoniana** Boiss. & Reut.

Tithymalus gypsicola (Rech.f. & Aellen) Soják === **Euphorbia gypsicola** Rech.f. & Aellen

Tithymalus haussknechtii (Boiss.) Soják === **Euphorbia haussknechtii** Boiss.

Tithymalus hebecarpus (Boiss.) Klotzsch & Garcke === **Euphorbia hebecarpa** Boiss.

Tithymalus heldreichii (Orph. ex Boiss.) Soják === **Euphorbia amygdaloides** subsp. **heldreichii** (Orph. ex Boiss.) Aldén

Tithymalus heldreichii subsp. *semiverticillatus* (Halácsy) Soják === **Euphorbia amygdaloides** subsp. **heldreichii** (Orph. ex Boiss.) Aldén

Tithymalus helioscopioides (Loscos & Pardo) Holub === **Euphorbia helioscopia** L.

Tithymalus helioscopius (L.) Hill === **Euphorbia helioscopia** L.

Tithymalus helioscopius subsp. *dominii* (Röhl.) Soják === **Euphorbia helioscopia** L. subsp. **helioscopia**

Tithymalus helioscopius subsp. *helioscopioides* (Loscos & Pardo) Soják === **Euphorbia helioscopia** L.

Tithymalus helleri (Millsp.) Small === **Euphorbia helleri** Millsp.

Tithymalus henryi (Hemsl.) Soják === **Euphorbia sieboldiana** Morris & Decne.

Tithymalus herniariifolius (Willd.) Klotzsch & Garcke === **Euphorbia herniariifolia** Willd.

Tithymalus herpetorrhizus Prokh. === **Euphorbia aucheri** Boiss.

Tithymalus heteradenus (Jaub. & Spach) Soják === **Euphorbia heteradena** Jaub. & Spach

Tithymalus heterophyllus (L.) Haw. === **Euphorbia heterophylla** L.

Tithymalus hieroglyphicus (Coss. & Durieu ex Boiss.) Soják === **Euphorbia hieroglyphica** Coss. & Durieu ex Boiss.

Tithymalus hieronymi (Subils) Soják === **Euphorbia hieronymi** Subils

Tithymalus himalayensis Klotzsch === **Euphorbia himalayensis** (Klotzsch) Boiss.

Tithymalus hinkleyorum (I.M.Johnst.) Soják === **Euphorbia hinkleyorum** I.M.Johnst.

Tithymalus hippocrepicus (Hemsl.) Soják === **Euphorbia sieboldiana** Morris & Decne.

Tithymalus hirsutus Lam. === **Euphorbia illirica** Lam. subsp. **illirica**

Tithymalus hochstetterianus Klotzsch & Garcke === **Euphorbia schimperiana** Scheele var. **schimperiana**

Tithymalus hohenackeri Klotzsch & Garcke === **Euphorbia orphanidis** Boiss.

Tithymalus huanchahanus Klotzsch & Garcke === **Euphorbia huanchahana** (Klotzsch & Garcke) Boiss.

Tithymalus humifusus (Willd.) Bubani === **Euphorbia humifusa** Willd.

Tithymalus humilis (C.A.Mey.) Klotzsch & Garcke === **Euphorbia humilis** C.A.Mey.

Tithymalus hybernus (L.) Hill === **Euphorbia hyberna** L.

Tithymalus hybernus subsp. *canuti* (Parl.) Soják === **Euphorbia hyberna** subsp. **canuti** (Parl.) Tutin

Tithymalus hybernus subsp. *insularis* (Boiss.) Soják === **Euphorbia hyberna** subsp. **insularis** (Boiss.) Briq.

Tithymalus hypoleucus Prokh. === **Euphorbia boissieriana** (Woronow) Prokh.

Tithymalus ibericus (Boiss.) Prokh. === **Euphorbia iberica** Boiss.

Tithymalus ierensis (Britton) Croizat === **Pedilanthus tithymaloides** (L.) Poit. subsp. **tithymaloides**

Tithymalus imbricatus (Vahl) Klotzsch & Garcke === **Euphorbia portlandica** L.

Tithymalus imperfoliatus (Vis.) Soják === **Euphorbia esula** L. subsp. **esula**

Tithymalus incisus (Engelm.) W.A.Weber === **Euphorbia schizoloba** Engelm.

Tithymalus inderiensis (Less. ex Kar. & Kir.) Klotzsch & Garcke === **Euphorbia inderiensis** Less. ex Kar. & Kir.

Tithymalus inundatus (Torr. ex Chapm.) Small === **Euphorbia inundata** Torr. ex Chapm.

Tithymalus involucratus Klotzsch & Garcke === **Euphorbia epicyparissias** (E.Mey ex Klotzsch & Garcke) Boiss.

Tithymalus ipecacuanhae (L.) Klotzsch & Garcke === **Euphorbia ipecacuanhae** L.

Tithymalus irgisensis (Litv.) Prokh. === **Euphorbia irgisensis** Litv.

Tithymalus isatidifolius (Lam.) Soják === **Euphorbia isatidifolia** Lam.

Tithymalus isauricus (M.S.Khan) Soják === **Euphorbia isaurica** M.S.Khan

Tithymalus ispahanicus (Boiss.) Klotzsch & Garcke === **Euphorbia heteradena** Jaub. & Spach

Tithymalus issykkulensis Prokh. === **Euphorbia tranzschelii** (Prokh.) Prokh.

Tithymalus isthmius (Täckh.) Soják === **Euphorbia grossheimii** (Prokh.) Prokh.

Tithymalus itzaeus (Millsp.) Croizat === **Pedilanthus tithymaloides** subsp. **parasiticus** (Klotzsch & Garcke) Dressler

Tithymalus ivanjohnstonii (M.C.Johnst.) Soják === **Euphorbia ivanjohnstonii** M.C.Johnst.

Tithymalus × *jablonskianus* (Polatschek) Holub === **Euphorbia** × **jablonskiana** Polatschek

Tithymalus jacquemontii (Boiss.) Soják === **Euphorbia jacquemontii** Boiss.

Tithymalus jacquinii (Fenzl ex Boiss.) Soják === **Euphorbia jacquinii** Fenzl ex Boiss.

Tithymalus jamaicensis (Millsp. & Britton) Croizat === **Pedilanthus tithymaloides** subsp. **jamaicensis** (Millsp. & Britton) Dressler

Tithymalus jasiewiczii Chrtek & Krísa === **Euphorbia jasiewiczii** (Chrtek & Krísa) Radcl.-Sm.

Tithymalus jaxarticus Prokh. === **Euphorbia esula** subsp. **tommasiniana** (Bertol.) Kuzmanov

Tithymalus jolkinii (Boiss.) Hara === **Euphorbia jolkinii** Boiss.

Tithymalus kanaoricus (Boiss.) Prokh. === **Euphorbia kanaorica** Boiss.

Tithymalus kansuensis (Prokh.) Prokh. === **Euphorbia kansuensis** Prokh.

Tithymalus kerneri (Huter ex A.Kern.) Pacher === **Euphorbia triflora** subsp. **kerneri** (Huter ex A.Kern.) Poldini

Tithymalus khasyanus (Boiss.) Soják === **Euphorbia khasyana** Boiss.

Tithymalus klokovianus (Railyan) Holub === **Euphorbia bessarabica** Klokov

Tithymalus koilopremnos Croizat === **Pedilanthus palmeri** Millsp.

Tithymalus kopetdaghi Prokh. === **Euphorbia kopetdaghi** (Prokh.) Prokh.

Tithymalus kotschyanus (Fenzl) Klotzsch & Garcke === **Euphorbia kotschyana** Fenzl

Tithymalus kozlovii (Prokh.) Prokh. === **Euphorbia kozlovii** Prokh.

Tithymalus kurdicus (Boiss.) Prokh. === **Euphorbia aucheri** Boiss.

Tithymalus laetus (Aiton) Haw. === **Euphorbia dendroides** L.

Tithymalus lagascae (Spreng.) Klotzsch & Garcke === **Euphorbia lagascae** Spreng.

Tithymalus lamprocarpus Prokh. === **Euphorbia soongarica** subsp. **lamprocarpa** (Prokh.) Prokh.

Tithymalus lateriflorus Klotzsch & Garcke === **Euphorbia osyridea** Boiss.

Tithymalus lathyris (L.) Hill === **Euphorbia lathyris** L.

Tithymalus latifolius (C.A.Mey. ex Ledeb.) Klotzsch & Garcke === **Euphorbia latifolia** C.A.Mey. ex Ledeb.

Tithymalus laurocerasifolius Mill. === **Pedilanthus tithymaloides** subsp. **padifolius** (L.) Dressler

Tithymalus leiococcus (Engelm.) Small === **Euphorbia spathulata** Lam.

Tithymalus leptaleus (Schauer) Klotzsch & Garcke === **Euphorbia graminea** Jacq. var. **graminea**

Tithymalus leptocerus (Engelm. ex Boiss.) Arthur === **Euphorbia crenulata** Engelm.

Tithymalus leptophyllus Bubani === **Euphorbia segetalis** L. var. **segetalis**

Tithymalus linearis (Retz.) Raf. === **Euphorbia articulata** Burm.

Tithymalus lingulatus (Heuff.) Soják === **Euphorbia lingulata** Heuff.

Tithymalus linifolius (Nath. ex L.) Klotzsch & Garcke === **Euphorbia segetalis** var. **pinea** (L.) Lanke

Tithymalus linifolius Lam. === **Euphorbia esula** L. subsp. **esula**

Tithymalus lipskyi Prokh. === **Euphorbia lipskyi** (Prokh.) Prokh.

Tithymalus literatus (Jacq.) Raf. === **Euphorbia platyphyllos** subsp. **literata** (Jacq.) Holub

Tithymalus litwinowii Prokh. === **Euphorbia aucheri** Boiss.

Tithymalus longicorniculatus (Kitam.) Soják === **Euphorbia longicorniculata** Kitam.

Tithymalus longicruris (Scheele) Small === **Euphorbia longicruris** Scheele

Tithymalus longifolius Hurus. & Yu.Tanaka === **Euphorbia donii** Oudejans

Tithymalus longipetiolatus Klotzsch & Garcke === **Euphorbia silenifolia** (Haw.) Sweet

Tithymalus longiradiatus (Lapeyr.) Bubani === **Euphorbia segetalis** var. **pinea** (L.) Lanke

Tithymalus louisii (Thiébaut) Soják === **Euphorbia hierosolymitana** Boiss. var. **hierosolymitana**

Tithymalus lucidus (Waldst. & Kit.) Klotzsch & Garcke ex Garcke === **Euphorbia lucida** Waldst. & Kit.

Tithymalus lunulatus (Bunge) Soják === **Euphorbia esula** L. subsp. **esula**

Tithymalus luridus (Engelm.) Wooton & Standl. === **Euphorbia lurida** Engelm.

Tithymalus lutescens (C.A.Mey.) Klotzsch & Garcke === **Euphorbia pilosa** L.

Tithymalus × *macinensis* (Prodán) Soják === **Euphorbia** × **macinensis** Prodán

Tithymalus macradenius (Donn.Sm.) Croizat === **Pedilanthus calcaratus** Schltdl.

Tithymalus macrocarpus (Benth.) Croizat === **Pedilanthus macrocarpus** Benth.

Tithymalus macroceras (Fisch. & C.A.Mey.) Klotzsch & Garcke === **Euphorbia macroceras** Fisch. & C.A.Mey.

Tithymalus macrocladus (Boiss.) Klotzsch & Garcke === **Euphorbia macroclada** Boiss.

Tithymalus macrorrhizus (C.A.Mey. ex Ledeb.) Klotzsch & Garcke === **Euphorbia macrorrhiza** C.A.Mey. ex Ledeb.

Tithymalus macrostegius Soják === **Euphorbia erubescens** Boiss.

Tithymalus maculatus (L.) Moench === **Euphorbia maculata** L.

Tithymalus macvaughianus (M.C.Johnst.) Soják === **Euphorbia macvaughiana** M.C.Johnst.

Tithymalus maddenii (Boiss.) Soják === **Euphorbia maddenii** Boiss.

Tithymalus malleatus (Boiss.) Soják === **Euphorbia malleata** Boiss.

Tithymalus malurensis (Rech.f.) Soják === **Euphorbia malurensis** Rech.f.

Tithymalus malvanus (Maire) Soják === **Euphorbia malvana** Maire

Tithymalus mancus (A.Nelson) A.Heller === **Euphorbia crenulata** Engelm.

Tithymalus mandshuricus (Maxim.) Soják === **Euphorbia esula** L. subsp. **esula**

Tithymalus maresii (Knoche) Soják === **Euphorbia maresii** Knoche

Tithymalus mariolensis (Rouy) Holub === **Euphorbia flavicoma** DC. subsp. **flavicoma**

Tithymalus maritimus Lam. === **Euphorbia paralias** L.

Tithymalus marschallianus (Boiss.) Klotzsch & Garcke === **Euphorbia marschalliana** Boiss.

Tithymalus marschallianus subsp. *armenus* (Prokh.) Soják === **Euphorbia marschalliana** subsp. **armena** (Prokh.) Oudejans

Tithymalus marschallianus subsp. *woronowii* (Grossh.) Soják === **Euphorbia woronowii** Grossh.

Tithymalus matritensis (Boiss.) Samp. === **Euphorbia matritensis** Boiss.

Tithymalus matthiolii Bubani === **Euphorbia nicaeensis** All. subsp. **nicaeensis**

Tithymalus mauritanicus (L.) Haw. === **Euphorbia mauritanica** L.

Tithymalus mazicum (Emb. & Maire) Soják === **Euphorbia mazicum** Emb. & Maire

Tithymalus medicagineus (Boiss.) Klotzsch & Garcke === **Euphorbia medicaginea** Boiss.

Tithymalus medicagineus subsp. *arsenariensis* (Batt.) Soják === **Euphorbia medicaginea** var. **arsenariensis** Batt.

Tithymalus megalanthus (Boiss.) Klotzsch & Garcke === **Euphorbia heteradena** Jaub. & Spach

Tithymalus megalatlanticus (Ball) Soják === **Euphorbia megalatlantica** Ball

Tithymalus megalocarpus (Rech.f.) Soják === **Euphorbia megalocarpa** Rech.f.

Tithymalus melanopotamicus Croizat === **Pedilanthus tithymaloides** subsp. **retusus** (Benth.) Dressler

Tithymalus melapetalus (Gasp. ex Guss.) Klotzsch & Garcke === **Euphorbia characias** L. subsp. **characias**

Tithymalus melliferus (Aiton) Moench === **Euphorbia longifolia** Lam. var. **longifolia**

Tithymalus mesopotamicus (M.S.Khan) Soják === **Euphorbia chaborasia** Gomb.

Tithymalus mexicanus (Engelm.) Wooton & Standl. === **Euphorbia spathulata** Lam.

Tithymalus meyeri Klotzsch & Garcke === **Euphorbia erubescens** Boiss.

Tithymalus meyerianus (Galushko) Galushko === **Euphorbia meyeriana** Galushko

Tithymalus micractinus (Boiss.) Soják === **Euphorbia micractina** Boiss.

Tithymalus micranthus (Stephan ex Willd.) Raf. === **Euphorbia stricta** L.

Tithymalus microcarpus Prokh. === **Euphorbia microcarpa** (Prokh.) Krylov

Tithymalus microsciadius (Boiss.) Klotzsch & Garcke === **Euphorbia microsciadia** Boiss.

Tithymalus microsphaerus (Boiss.) Klotzsch & Garcke === **Euphorbia microsphaera** Boiss.

Tithymalus millspaughii (Pax & K.Hoffm.) Croizat === **Pedilanthus millspaughii** Pax & K.Hoffm.

Tithymalus minutus (Loscos & Pardo) Soják === **Euphorbia minuta** Loscos & Pardo

Tithymalus missouriensis (Norton) Small === **Euphorbia spathulata** Lam.

Tithymalus monchiquensis (Franco & P.Silva) Soják === **Euphorbia paniculata** subsp. **monchiquensis** (Franco & P.Silva) Vicens, Molero & C.Blanché

Tithymalus mongolicus Prokh. === **Euphorbia mongolica** (Prokh.) Prokh.

Tithymalus monocyathium Prokh. === **Euphorbia monocyathium** (Prokh.) Prokh.

Tithymalus montenegrinus (Bald.) Soják === **Euphorbia montenegrina** (Bald.) K.Maly

Tithymalus monticola (Boiss.) Klotzsch & Garcke === **Euphorbia aucheri** Boiss.

Tithymalus mucronata Raf. === ?

Tithymalus mucronatus Bubani === **Euphorbia falcata** L. subsp. **falcata**

Tithymalus mucronulatus Prokh. === **Euphorbia mucronulata** (Prokh.) Pavlov

Tithymalus multicaulis Klotzsch & Garcke === **Euphorbia sclerophylla** Boiss.

Tithymalus multifurcatus (Rech.f., Aellen & Esfand.) Soják === **Euphorbia caeladenia** Boiss.

Tithymalus myrsinites (L.) Hill === **Euphorbia myrsinites** L.

Tithymalus myrtifolius (L.) Mill. === **Pedilanthus tithymaloides** (L.) Poit. subsp. **tithymaloides**

Tithymalus nakaianus (H.Lév.) Hara === **Euphorbia esula** L. subsp. **esula**

Tithymalus neilmulleri (M.C.Johnst.) Soják === **Euphorbia neilmulleri** M.C.Johnst.

Tithymalus nereidum (Jahand. & Maire) Soják === **Euphorbia nereidum** Jahand. & Maire

Tithymalus nevadensis (Boiss. & Reut.) Klotzsch & Garcke === **Euphorbia nevadensis** Boiss. & Reut.

Tithymalus nicaeensis (All.) Klotzsch & Garcke === **Euphorbia nicaeensis** All.

Tithymalus nicaeensis subsp. *cadrilateri* (Prodán) Soják === **Euphorbia nicaeensis** subsp. **cadrilateri** (Prodán) Kuzmanov

Tithymalus nicaeensis subsp. *glareosus* (Pall. ex M.Bieb.) Soják === **Euphorbia nicaeensis** subsp. **glareosa** (Pall. ex M.Bieb.) Radcl.-Sm.

Tithymalus nicaeensis subsp. *goldei* (Prokh.) Soják === **Euphorbia nicaeensis** subsp. **goldei** (Prokh.) Greuter & Burdet

Tithymalus nicaeensis subsp. *japygicus* (Ten.) Soják === **Euphorbia nicaeensis** All. subsp. **nicaeensis**

Tithymalus nicaeensis subsp. *pannonicus* (Host) Soják === **Euphorbia nicaeensis** All. subsp. **nicaeensis**

Tithymalus nicaeensis subsp. *stepposus* (Zoz) Soják === **Euphorbia nicaeensis** subsp. **stepposa** (Zoz) Greuter & Burdet

Tithymalus nicicianus (Borbás ex Novák) Soják === **Euphorbia seguieriana** subsp. **niciciana** (Borbás ex Novák) Rchb.f.

Tithymalus nodiflorus (Millsp.) Croizat === **Pedilanthus nodiflorus** Millsp.

Tithymalus notabilis Soják === **Euphorbia macrocarpa** Boiss. & Buhse

Tithymalus notadenius (Boiss. & Hohen.) Klotzsch & Garcke === **Euphorbia orientalis** L.

Tithymalus nudiflorus Haw. === **Euphorbia cymosa** Poir.

Tithymalus nudus (Velen.) Prokh. === **Euphorbia palustris** L.

Tithymalus nummularius Lam. === **Euphorbia chamaesyce** L.

Tithymalus nutans (Lag.) Samp. === **Euphorbia nutans** Lag.

Tithymalus oblongatus (Griseb.) Soják === **Euphorbia oblongata** Griseb.

Tithymalus oblongifolius (K.Koch) Klotzsch & Garcke === **Euphorbia oblongifolia** (K.Koch) K.Koch

Tithymalus obovata Raf. === **Euphorbia helioscopia** L. subsp. **helioscopia**

Tithymalus obtusatus (Pursh) Klotzsch & Garcke === **Euphorbia spathulata** Lam.

Tithymalus obtusifolius Klotzsch & Garcke === **Euphorbia obtusifolia** Poir.

Tithymalus oerstedii (Klotzsch) Croizat === **Pedilanthus oerstedii** Klotzsch

Tithymalus officinarum (L.) H.Karst. === **Euphorbia officinarum** L.

Tithymalus oidorrhizus (Pojark.) Soják === **Euphorbia oidorrhiza** Pojark.

Tithymalus olsson-sefferi (Millsp.) Croizat === **Pedilanthus bracteatus** (Jacq.) Boiss.

Tithymalus orientalis (L.) Hill === **Euphorbia orientalis** L.

Tithymalus orphanidis (Boiss.) Soják === **Euphorbia orphanidis** Boiss.

Tithymalus oschtenicus (Galushko) Galushko === **Euphorbia oschtenica** Galushko

Tithymalus osyrideus (Boiss.) Soják === **Euphorbia osyridea** Boiss.

Tithymalus ovatus E.Mey. ex Klotzsch & Garcke === **Euphorbia ovata** (E.Mey. ex Klotzsch & Garcke) Boiss.

Tithymalus oxyodontus (Boiss.) Soják === **Euphorbia oxyodonta** Boiss.

Tithymalus pachyrrhizus (Kar. & Kir.) Klotzsch & Garcke === **Euphorbia pachyrrhiza** Kar. & Kir.

Tithymalus padifolius (L.) Croizat === **Pedilanthus tithymaloides** subsp. **padifolius** (L.) Dressler

Tithymalus pallasii (Turcz.) Klotzsch & Garcke === **Euphorbia fischeriana** Steud.

Tithymalus palmeri (Engelm. ex S.Watson) Dayton === **Euphorbia palmeri** Engelm. ex S.Watson

Tithymalus palmyrenus (Mouterde) Soják === **Euphorbia cuspidata** Bertol.

Tithymalus palustris (L.) Garsault === **Euphorbia palustris** L.

Tithymalus pamiricus Prokh. === **Euphorbia pamirica** (Prokh.) Prokh.

Tithymalus pampeanus (Speg.) Soják === **Euphorbia pampeana** Speg.

Tithymalus panaceus (Webb & Berthel.) Klotzsch & Garcke === **Euphorbia terracina** L.

Tithymalus pancicii (Beck) Soják === **Euphorbia esula** L. subsp. **esula**

Tithymalus paniculata Raf. === **Euphorbia corollata** var. **paniculata** Boiss.

Tithymalus panjutinii (Grossh.) Soják === **Euphorbia panjutinii** Grossh.

Tithymalus pannonicus (Host) Á.Löve & D.Löve === **Euphorbia nicaeensis** All. subsp. **nicaeensis**

Tithymalus papillosus Klotzsch & Garcke === **Euphorbia duvalii** Lecoq & Lamotte

Tithymalus papillosus (A.St.-Hil.) Soják === **Euphorbia papillosa** A.St.-Hil.

Tithymalus × paradoxus (Schur) Holub === **Euphorbia × paradoxa** (Schur) Simonk.

Tithymalus paralias (L.) Hill === **Euphorbia paralias** L.

Tithymalus parasiticus (Boiss. ex Klotzsch) Croizat === **Pedilanthus tithymaloides** subsp. **parasiticus** (Klotzsch & Garcke) Dressler

Tithymalus parvulus Klotzsch & Garcke === **Euphorbia arvalis** Boiss. & Heldr. subsp. **arvalis**

Tithymalus parvulus (Delile) Schweinf. ex Schweinf. & Asch. === **Euphorbia parvula** Delile

Tithymalus pauciflorus Bubani === **Euphorbia minuta** Loscos & Pardo

Tithymalus pedersenii (Subils) Soják === **Euphorbia pedersenii** Subils

Tithymalus × peisonis (Rech.f.) Soják === **Euphorbia × peisonis** Rech.f.

Tithymalus pekinensis (Rupr.) Hara === **Euphorbia pekinensis** Rupr.

Tithymalus peltatus (Roxb.) Soják === **Euphorbia peltata** Roxb.

Tithymalus pendulus Haw. === **Sarcostemma sp.** (Asclepiadaceae)

Tithymalus penicillatus Millsp. === **Euphorbia huanchahana** (Klotzsch & Garcke) Boiss. subsp. **huanchahana**

Tithymalus peplidion (Engelm.) Small === **Euphorbia peplidion** Engelm.

Tithymalus peplis (L.) Scop. === **Euphorbia peplis** L.

Tithymalus peploides (Gouan) Klotzsch & Garcke === **Euphorbia peplus** var. **minima** DC.

Tithymalus peplus (L.) Hill === **Euphorbia peplus** L.

Tithymalus peplus subsp. *calabricus* (Huter, Porta & Rigo) Soják === **Euphorbia peplus** L. var. **peplus**

Tithymalus peplus subsp. *maritimus* (Boiss.) Soják === **Euphorbia peplus** var. **minima** DC.

Tithymalus peplus subsp. *peploides* (Gouan) Soják === **Euphorbia peplus** var. **minima** DC.

Tithymalus perbracteatus (Gage) Soják === **Euphorbia perbracteata** Gage

Tithymalus peritropoides (Millsp.) Croizat === **Pedilanthus palmeri** Millsp.

Tithymalus persepolitanus (Boiss.) Klotzsch & Garcke === **Euphorbia microsciadia** Boiss.

Tithymalus pestalozzae (Boiss.) Klotzsch & Garcke === **Euphorbia pestalozzae** Boiss.

Tithymalus petanophyllus Croizat === **Pedilanthus tithymaloides** subsp. **parasiticus** (Klotzsch & Garcke) Dressler

Tithymalus petiolaris (Sims) Haw. === **Euphorbia petiolaris** Sims

Tithymalus petiolatus (Banks & Sol.) Soják === **Euphorbia petiolata** Banks & Sol.

Tithymalus petitianus (A.Rich.) Klotzsch & Garcke === **Euphorbia petitiana** A.Rich.

Tithymalus petraeus (Brandegee) Croizat === **Pedilanthus tithymaloides** (L.) Poit. subsp. **tithymaloides**

Tithymalus petrophilus (C.A.Mey.) Soják === **Euphorbia petrophila** C.A.Mey.

Tithymalus philippianus Klotzsch & Garcke === **Euphorbia philippiana** (Klotzsch & Garcke) Boiss.

Tithymalus philorus Cockerell === **Euphorbia brachycera** Engelm.

Tithymalus philorus f. *dichotomus* Daniels === **Euphorbia brachycera** Engelm.

Tithymalus phymatospermus (Boiss.) Soják === **Euphorbia phymatosperma** Boiss.

Tithymalus phymatospermus subsp. *cernuus* (Coss. & Durieu ex Boiss.) Soják === **Euphorbia phymatosperma** subsp. **cernua** (Coss. & Durieu ex Boiss.) Vindt

Tithymalus physocaulos (Mouterde) Soják === **Euphorbia physocaulos** Mouterde

Tithymalus pictus (Jacq.) Haw. === **Euphorbia graminea** Jacq. var. **graminea**

Tithymalus pilosus (L.) Hill === **Euphorbia pilosa** L.

Tithymalus piluliferus (L.) Moench === **Euphorbia hirta** L.

Tithymalus pineus (L.) Klotzsch & Garcke === **Euphorbia segetalis** subsp. **linifolia**

Tithymalus pinkavanus (M.C.Johnst.) Soják === **Euphorbia pinkavana** M.C.Johnst.

Tithymalus piscatorius (Aiton) Haw. === **Euphorbia piscatoria** Aiton

Tithymalus pisidicus (Hub.-Mor. & M.S.Khan) Soják === **Euphorbia pisidica** Hub.-Mor. & M.S.Khan

Tithymalus pithyusa (L.) Scop. === **Euphorbia pithyusa** L.

Tithymalus pithyusa subsp. *cupanii* (Guss. ex Bertol.) Soják === **Euphorbia pithyusa** subsp. **cupanii** (Guss. ex Bertol.) Radcl.-Sm.

Tithymalus platyphyllos (L.) Hill === **Euphorbia platyphyllos** L.

Tithymalus platyphyllos subsp. *literatus* (Jacq.) Chrtek & Krísa === **Euphorbia platyphyllos** subsp. **literata** (Jacq.) Holub

Tithymalus platyphylos (L.) Raf. === **Euphorbia platyphyllos** L.

Tithymalus plebeius (Boiss.) Klotzsch & Garcke === **Euphorbia plebeia** Boiss.

Tithymalus poecilophyllus Prokh. === **Euphorbia poecilophylla** (Prokh.) Prokh.

Tithymalus polycaulis (Boiss.) Klotzsch & Garcke === **Euphorbia decipiens** Boiss. & Buhse

Tithymalus polygalifolius (Boiss. & Reut.) Soják === **Euphorbia polygalifolia** Boiss. & Reut.

Tithymalus polygalifolius subsp. *mariolensis* (Rouy) Soják === **Euphorbia flavicoma** DC. subsp. **flavicoma**

Tithymalus polytimeticus Prokh. === **Euphorbia kanaorica** Boiss.

Tithymalus × *popovii* (Rotschild) Soják === **Euphorbia** × **popovii** Rotschild

Tithymalus portlandicus (L.) Hill === **Euphorbia portlandica** L.

Tithymalus portulacoides (L.) Standl. === **Euphorbia portulacoides** L.

Tithymalus potaninii (Prokh.) Prokh. === **Euphorbia potaninii** Prokh.

Tithymalus pouzolzii Bubani === **Euphorbia duvalii** Lecoq & Lamotte

Tithymalus pringlei (Rob.) Croizat === **Pedilanthus tithymaloides** (L.) Poit. subsp. **tithymaloides**

Tithymalus procerus (M.Bieb.) Klotzsch & Garcke ex Garcke === **Euphorbia illirica** Lam. subsp. **illirica**

Tithymalus × procopianii (Savul. & Rayss) Soják === **Euphorbia × procopianii** Savul. & Rayss

Tithymalus proliferus (Buch.-Ham. ex D.Don) Soják === **Euphorbia prolifera** Buch.-Ham. ex D.Don

Tithymalus promecocarpus (Davis) Soják === **Euphorbia promecocarpa** Davis

Tithymalus prostratus (Aiton) Samp. === **Euphorbia prostrata** Aiton

Tithymalus prunifolius Haw. === **Euphorbia heterophylla** L.

Tithymalus przewalskii (Prokh.) Soják === **Euphorbia altotibetica** Paulsen

Tithymalus pseudoapios (Maire & Weiller) Soják === **Euphorbia pseudoapios** Maire & Weiller

Tithymalus × pseudoesula (Schur) Dostál === **Euphorbia × pseudoesula** Schur

Tithymalus pseudofalcatus (Chiov.) Soják === **Euphorbia pseudofalcata** Chiov.

Tithymalus × pseudolucidus (Schur) Dostál === **Euphorbia × pseudolucida** Schur

Tithymalus pseudopeplus (Speg.) Soják === **Euphorbia spathulata** Lam.

Tithymalus pseudosikkimensis Hurus. & Yu.Tanaka === **Euphorbia pseudosikkimensis** (Hurus. & Yu.Tanaka) Radcl.-Sm.

Tithymalus pseudosororius Prokh. === **Euphorbia aserbajdzhanica** Bordz.

Tithymalus × pseudovirgatus (Schur) Soják === **Euphorbia esula** nothosubsp. **pseudovirgata** (Schur) Govaerts

Tithymalus pterococcus (Brot.) Klotzsch & Garcke === **Euphorbia pterococca** Brot.

Tithymalus pubescens (Vahl) Samp. === **Euphorbia hirsuta** L.

Tithymalus pumilus (Sibth. & Sm.) Klotzsch & Garcke === **Euphorbia herniariifolia** Willd. var. **herniariifolia**

Tithymalus punctatus (Delile) Soják === **Euphorbia punctata** Delile

Tithymalus puniceus (Sw.) Haw. === **Euphorbia punicea** Sw.

Tithymalus purpureus Lam. === **Euphorbia characias** L. subsp. **characias**

Tithymalus purpusii (Brandegee) Croizat === **Pedilanthus calcaratus** Schltdl.

Tithymalus pygmaeus (Ledeb.) Klotzsch & Garcke === **Euphorbia pygmaea** Ledeb.

Tithymalus pyrenaeus Bubani === **Euphorbia pyrenaica** Jord.

Tithymalus raphanorrhizus Millsp. === **Euphorbia raphanorrhiza** (Millsp.) J.F.Macbr.

Tithymalus rapulum (Kar. & Kir.) Klotzsch & Garcke === **Euphorbia rapulum** Kar. & Kir.

Tithymalus reboudianus (Coss. ex Batt. & Trab.) Soják === **Euphorbia reboudiana** Coss. ex Batt. & Trab.

Tithymalus rechingeri (Greuter) Soják === **Euphorbia myrsinites** subsp. **rechingeri** (Greuter) Aldén

Tithymalus regis-jubae (J.Gay) Klotzsch & Garcke === **Euphorbia regis-jubae** J.Gay

Tithymalus repandus Haw. === **Euphorbia repanda** (Haw.) Sweet

Tithymalus repens (K.Koch) Klotzsch & Garcke === **Euphorbia repens** K.Koch

Tithymalus repetitus (Hochst. ex A.Rich.) Klotzsch & Garcke === **Euphorbia repetita** Hochst. ex A.Rich.

Tithymalus resiniferus (O.Berg) H.Karst. === **Euphorbia resinifera** O.Berg

Tithymalus retusus (Forssk.) Klotzsch & Garcke === **Euphorbia retusa** Forssk.

Tithymalus reuterianus (Boiss.) Klotzsch & Garcke === **Euphorbia reuteriana** Boiss.

Tithymalus revolutus Klotzsch & Garcke === **Euphorbia genistoides** var. **corifolia** (Lam.) N.E.Br.

Tithymalus rhabdotospermus (Radcl.-Sm.) Holub === **Euphorbia rhabdotosperma** Radcl.-Sm.

Tithymalus rhytidospermus (Boiss. & Balansa) Soják === **Euphorbia rhytidosperma** Boiss. & Balansa

Tithymalus riae (Pax & K.Hoffm.) Soják === **Euphorbia stracheyi** Boiss.

Tithymalus rigidus (M.Bieb.) Soják === **Euphorbia rigida** M.Bieb.

Tithymalus rimarum (Coss. & Balansa) Soják === **Euphorbia rimarum** Coss. & Balansa

Tithymalus robbiae (Turrill) Soják === **Euphorbia amygdaloides** subsp. **robbiae** (Turrill) Stace

Tithymalus robustus (Engelm.) Small === **Euphorbia brachycera** Engelm.

Tithymalus roemerianus (Scheele) Small === **Euphorbia roemeriana** Scheele
Tithymalus roschanicus Ikonn. === **Euphorbia roschanica** (Ikonn.) Czerepan.
Tithymalus rosea (Retz.) Raf. === **Euphorbia rosea** Retz.
Tithymalus rosularis (Fed.) Pazij === **Euphorbia rosularis** Fed.
Tithymalus rothianus (Spreng.) Klotzsch & Garcke === **Euphorbia rothiana** Spreng.
Tithymalus rotundifolius Lam. === **Euphorbia peplus** L. var. **peplus**
Tithymalus ruber (Cav.) Bubani === **Euphorbia falcata** L. subsp. **falcata**
Tithymalus rupestris (C.A.Mey.) Klotzsch & Garcke === **Euphorbia rupestris** C.A.Mey.
Tithymalus rupestris Lam. === **Euphorbia seguieriana** Neck. subsp. **seguieriana**
Tithymalus rupicola (Boiss.) Klotzsch & Garcke === **Euphorbia squamigera** Loisel.
Tithymalus ruscinonensis (Boiss.) Soják === **Euphorbia duvalii** Lecoq & Lamotte
Tithymalus sahendi (Bornm.) Soják === **Euphorbia sahendi** Bornm.
Tithymalus salicifolius (Host) Klotzsch & Garcke === **Euphorbia salicifolia** Host
Tithymalus sanasunitensis (Hand.-Mazz.) Soják === **Euphorbia sanasunitensis** Hand.-Mazz.
Tithymalus sarissophyllus Croizat === **Pedilanthus tithymaloides** subsp. **angustifolius**
    (Poit.) Dressler
Tithymalus saxatilis (Jacq.) Raf. === **Euphorbia saxatilis** Jacq.
Tithymalus saxicola (Radcl.-Sm.) Soják === **Euphorbia saxicola** Radcl.-Sm.
Tithymalus schickendantzii (Hieron.) Soják === **Euphorbia schickendantzii** Hieron.
Tithymalus schimperianus (Scheele) Klotzsch & Garcke === **Euphorbia schimperiana**
    Scheele
Tithymalus schizoceras (Boiss.) Klotzsch & Garcke === **Euphorbia macroclada** Boiss.
Tithymalus schizolobus (Engelm.) Norton === **Euphorbia schizoloba** Engelm.
Tithymalus schottii Klotzsch & Garcke === **Euphorbia schottiana** Boiss.
Tithymalus schugnanicus (B.Fedtsch.) Prokh. === **Euphorbia schugnanica** B.Fedtsch.
Tithymalus segetalis (L.) Hill === **Euphorbia segetalis** L.
Tithymalus segetalis subsp. *pineus* (L.) Soják === **Euphorbia segetalis** var. **pinea** (L.) Lanke
Tithymalus seguieri Scop. === **Euphorbia seguieriana** Neck. subsp. **seguieriana**
Tithymalus seguierianus (Neck.) Prokh. === **Euphorbia seguieriana** Neck.
Tithymalus seguierianus subsp. *loiseleurii* (Rouy) Soják === **Euphorbia seguieriana** Neck.
    subsp. **seguieriana**
Tithymalus seguierianus subsp. *minor* (Duby) Chrtek & Krísa === **Euphorbia seguieriana**
    Neck. subsp. **seguieriana**
Tithymalus seguierianus subsp. *nicicianus* (Borbás ex Novák) Chrtek & Krísa === **Euphorbia**
    seguieriana subsp. **niciciana** (Borbás ex Novák) Rchb.f.
Tithymalus seguierianus subsp. *saxicola* (Velen.) Chrtek & Krísa === **Euphorbia seguieriana**
    Neck. subsp. **seguieriana**
Tithymalus semiperfoliatus (Viv.) Klotzsch & Garcke === **Euphorbia amygdaloides** subsp.
    semiperfoliata (Viv.) Radcl.-Sm.
Tithymalus semivillosus Prokh. === **Euphorbia illirica** subsp. **semivillosa** (Prokh.) Govaerts
Tithymalus sendaicus (Makino) Hara === **Euphorbia sendaica** Makino
Tithymalus sera-comans (Bubani) Bubani === **Euphorbia sera-comans** Bubani
Tithymalus serotina Raf. === **Euphorbia characias** L. subsp. **characias**
Tithymalus serratus (L.) Hill === **Euphorbia serrata** L.
Tithymalus serratus Gilib. === **Euphorbia helioscopia** L. subsp. **helioscopia**
Tithymalus serrulatus (Thuill.) Holub === **Euphorbia stricta** L.
Tithymalus sessiliflorus (Roxb.) Klotzsch & Garcke === **Euphorbia sessiliflora** Roxb.
Tithymalus sewerzowii Prokh. === **Euphorbia sewerzowii** (Prokh.) Pavlov
Tithymalus shetoensis (Pax & K.Hoffm.) Soják === **Euphorbia stracheyi** Boiss.
Tithymalus shouanensis (H.Keng) S.S.Ying === **Euphorbia shouanensis** H.Keng
Tithymalus sibthorpii (Boiss.) Soják === **Euphorbia characias** subsp. **wulfenii** (Hoppe ex
    W.Koch) Radcl.-Sm.
Tithymalus sieboldianus (Morris & Decne.) Hara === **Euphorbia sieboldiana** Morris & Decne.
Tithymalus sikkimensis (Boiss.) Hurus. & Yu.Tanaka === **Euphorbia sikkimensis** Boiss.
Tithymalus silenifolius Haw. === **Euphorbia silenifolia** (Haw.) Sweet
Tithymalus sintenisii (Boiss. ex Freyn) Soják === **Euphorbia sintenisii** Boiss. ex Freyn

Tithymalus smallii (Millsp.) Small === **Pedilanthus tithymaloides** subsp. **smallii** (Millsp.) Dressler

Tithymalus sogdianus (Popov) Prokh. === **Euphorbia sogdiana** Popov

Tithymalus sojakii (Chrtek & Krísa) Chrtek & Krísa === **Euphorbia sojakii** (Chrtek & Krísa) Dubovik

Tithymalus × *somboriensis* (Prodán) Dostál === **Euphorbia** × **somboriensis** Prodán

Tithymalus × *sooi* (Simon) Soják === **Euphorbia** × **sooi** Simon

Tithymalus soongaricus (Boiss.) Prokh. === **Euphorbia soongarica** Boiss.

Tithymalus soongaricus subsp. *aristatus* (Schmalh.) Soják === **Euphorbia aristata** Schmalh.

Tithymalus soongaricus subsp. *lamprocarpus* (Prokh.) Soják === **Euphorbia soongarica** subsp. **lamprocarpa** (Prokh.) Prokh.

Tithymalus sororius (Schrenk) Klotzsch & Garcke === **Euphorbia sororia** Schrenk

Tithymalus spathulatus (Lam.) W.A.Weber === **Euphorbia spathulata** Lam.

Tithymalus spectabilis (Rob.) Croizat === **Pedilanthus bracteatus** (Jacq.) Boiss.

Tithymalus sphaerospermus (Shuttlew. ex Boiss.) Small === **Euphorbia floridana** Chapm.

Tithymalus spinidens Prokh. === **Euphorbia spinidens** (Prokh.) Prokh.

Tithymalus spinosus (L.) Raf. === **Euphorbia spinosa** L.

Tithymalus splendens (Bojer ex Hook.) M.Gómez === **Euphorbia milii** var. **splendens** (Bojer ex Hook.) Ursch & Leandri

Tithymalus squamigerus (Loisel.) Soják === **Euphorbia squamigera** Loisel.

Tithymalus squamigerus subsp. *rupicola* (Boiss.) Soják === **Euphorbia squamigera** Loisel.

Tithymalus squamosus (Willd.) Klotzsch & Garcke === **Euphorbia squamosa** Willd.

Tithymalus stenophyllus Klotzsch & Garcke === **Euphorbia stenophylla** (Klotzsch & Garcke) Boiss.

Tithymalus stocksianus (Boiss.) Soják === **Euphorbia caeladenia** Boiss.

Tithymalus stracheyi (Boiss.) Hurus. & Yu.Tanaka === **Euphorbia stracheyi** Boiss.

Tithymalus striatus (Thunb.) Klotzsch & Garcke === **Euphorbia striata** Thunb.

Tithymalus strictus (L.) Klotzsch & Garcke ex Garcke === **Euphorbia stricta** L.

Tithymalus stygianus (H.C.Watson) Soják === **Euphorbia stygiana** H.C.Watson

Tithymalus subamplexicaulis (Kar. & Kir.) Klotzsch & Garcke === **Euphorbia buchtormensis** C.A.Mey.

Tithymalus subcordatus Klotzsch & Garcke === **Euphorbia illirica** Lam. subsp. **illirica**

Tithymalus subcrenatus Klotzsch & Garcke === **Euphorbia portulacoides** L. subsp. **portulacoides**

Tithymalus subpavonianus Croizat === **Pedilanthus bracteatus** (Jacq.) Boiss.

Tithymalus subpubens (Engelm.) Norton === **Euphorbia palmeri** var. **subpubens** (Engelm.) L.C.Wheeler

Tithymalus sulcatus (Lens ex Loisel.) Bubani === **Euphorbia sulcata** Lens ex Loisel.

Tithymalus sylvaticus (L.) Hill === **Euphorbia amygdaloides** L. subsp. **amygdaloides**

Tithymalus syspirensis (K.Koch) Klotzsch & Garcke === **Euphorbia macroclada** Boiss.

Tithymalus szechuanicus (Pax & K.Hoffm.) Soják === **Euphorbia sieboldiana** Morris & Decne.

Tithymalus szovitsii (Fisch. & C.A.Mey.) Klotzsch & Garcke === **Euphorbia szovitsii** Fisch. & C.A.Mey.

Tithymalus talastavicus Prokh. === **Euphorbia talastavica** (Prokh.) Prokh.

Tithymalus tanaiticus (Pacz.) Galushko === **Euphorbia sareptana** Becker

Tithymalus tanguticus (Prokh.) Prokh. === **Euphorbia micractina** Boiss.

Tithymalus tarokoensis (Hayata) Soják === **Euphorbia esula** L. subsp. **esula**

Tithymalus tauricola (Prokh.) Holub === **Euphorbia illirica** Lam. subsp. **illirica**

Tithymalus taurinensis (All.) Klotzsch & Garcke === **Euphorbia taurinensis** All.

Tithymalus tchen-ngoi Soják === **Euphorbia pekinensis** Rupr. subsp. **pekinensis**

Tithymalus teheranicus (Boiss.) Soják === **Euphorbia teheranica** Boiss.

Tithymalus tehuacanus (Brandegee) Croizat === **Pedilanthus tehuacanus** Brandegee

Tithymalus telephioides (Chapm.) Small === **Euphorbia telephioides** Chapm.

Tithymalus tenuifolius (Lam.) Klotzsch & Garcke === **Euphorbia graminifolia** Vill.

Tithymalus terracinus (L.) Klotzsch & Garcke === **Euphorbia terracina** L.

Tithymalus tetraporus (Engelm.) Small === **Euphorbia tetrapora** Engelm.

Tithymalus thamnoides (Boiss.) Soják === **Euphorbia hierosolymitana** Boiss. var. **hierosolymitana**

Tithymalus thompsonii (Holmboe) Soják === **Euphorbia thompsonii** Holmboe

Tithymalus thomsonianus (Boiss.) Soják === **Euphorbia thomsoniana** Boiss.

Tithymalus thwaitesii Klotzsch & Garcke === **Euphorbia rothiana** Spreng.

Tithymalus thyrsoideus (Boiss.) Soják === **Euphorbia thyrsoidea** Boiss.

Tithymalus tianshanicus Prokh. === **Euphorbia thomsoniana** Boiss.

Tithymalus tibeticus (Boiss.) Prokh. === **Euphorbia tibetica** Boiss.

Tithymalus tithymaloides (L.) Croizat === **Pedilanthus tithymaloides** (L.) Poit.

Tithymalus togakusensis (Hayata) Hara === **Euphorbia togakusensis** Hayata

Tithymalus tomentellus (B.L.Rob. & Greenm.) Croizat === **Pedilanthus tomentellus** B.L.Rob. & Greenm.

Tithymalus tommasinianus (Bertol.) Soják === **Euphorbia esula** subsp. **tommasiniana** (Bertol.) Kuzmanov

Tithymalus tommasinianus subsp. *virgultosus* (Klokov) Soják === **Euphorbia esula** subsp. **tommasiniana** (Bertol.) Kuzmanov

Tithymalus tommasinianus subsp. *waldsteinii* (Soják) Soják === **Euphorbia esula** subsp. **tommasiniana** (Bertol.) Kuzmanov

Tithymalus transoxanus Prokh. === **Euphorbia transoxana** (Prokh.) Prokh.

Tithymalus transtaganus (Boiss.) Samp. === **Euphorbia transtagana** Boiss.

Tithymalus tranzschelii Prokh. === **Euphorbia tranzschelii** (Prokh.) Prokh.

Tithymalus trapezoidalis (Viv.) Klotzsch & Garcke === **Euphorbia terracina** L.

Tithymalus trichotomus (Kunth) Klotzsch & Garcke === **Euphorbia trichotoma** Kunth

Tithymalus triflorus (Schott, Nyman & Kotschy) Soják === **Euphorbia triflora** Schott, Nyman & Kotschy

Tithymalus triflorus subsp. *kerneri* (Huter ex A.Kern.) Soják === **Euphorbia triflora** subsp. **kerneri** (Huter ex A.Kern.) Poldini

Tithymalus trinervius Klotzsch & Garcke === **Euphorbia boetica** Boiss.

Tithymalus triodontus Prokh. --- **Euphorbia triodonta** (Prokh.) Prokh.

Tithymalus tristis (Besser) Klotzsch & Garcke === **Euphorbia esula** L. subsp. **esula**

Tithymalus truncatus Klotzsch & Garcke === **Euphorbia kraussiana** Bernh. ex Krauss

Tithymalus tshuiensis Prokh. === **Euphorbia tshuiensis** (Prokh.) Serg. ex Krylov

Tithymalus tuberosus (L.) Hill === **Euphorbia tuberosa** L.

Tithymalus tuckeyanus (Steud. ex Webb) Bolle ex Klotzsch & Garcke === **Euphorbia tuckeyana** Steud. ex Webb

Tithymalus turczaninowii (Kar. & Kir.) Klotzsch & Garcke === **Euphorbia turczaninowii** Kar. & Kir.

Tithymalus turkestanicus (Regel) Prokh. === **Euphorbia turkestanica** Regel

Tithymalus uliginosus (Welw. ex Boiss.) Samp. === **Euphorbia uliginosa** Welw. ex Boiss.

Tithymalus undulatus (M.Bieb.) Klotzsch & Garcke === **Euphorbia undulata** M.Bieb.

Tithymalus uniflorus Haw. ===?

Tithymalus unilateralis (Blakelock) Holub === **Euphorbia microsphaera** Boiss.

Tithymalus uralensis (Fisch. ex Link) Prokh. === **Euphorbia esula** subsp. **tommasiniana** (Bertol.) Kuzmanov

Tithymalus vallinianus (Belli) Soják === **Euphorbia valliniana** Belli

Tithymalus variabilis (Ces.) Klotzsch & Garcke === **Euphorbia variabilis** Ces.

Tithymalus variegatus (Sims) Haw. === **Euphorbia marginata** Pursh

Tithymalus velenovskyi (Bornm.) Soják === **Euphorbia palustris** L.

Tithymalus veneris (M.S.Khan) Soják === **Euphorbia veneris** M.S.Khan

Tithymalus venetus (Willd.) Klotzsch & Garcke === **Euphorbia characias** L. subsp. **characias**

Tithymalus verrucosus (L.) Hill === **Euphorbia verrucosa** L.

Tithymalus villicus Croizat === **Pedilanthus tithymaloides** (L.) Poit. subsp. **tithymaloides**

Tithymalus villosus (Waldst. & Kit. ex Willd.) Pacher === **Euphorbia illirica** Lam. subsp. **illirica**

Tithymalus villosus var. *trichocarpus* (Koch) Dostál === **Euphorbia illirica** Lam. subsp. **illirica**

Tithymalus villosus var. *tuberculatus* (Koch) Dostál === **Euphorbia illirica** Lam. subsp. **illirica**

Tithymalus virgatus (Haw.) Klotzsch & Garcke ex Garcke === **Euphorbia esula** subsp.
    tommasiniana (Bertol.) Kuzmanov
Tithymalus virgatus (Desf.) Haw. === **Euphorbia obtusifolia** Poir.
Tithymalus virgultosus (Klokov) Holub === **Euphorbia esula** subsp. **tommasiniana**
    (Bertol.) Kuzmanov
Tithymalus volgensis (Krysht.) Prokh. === **Euphorbia nicaeensis** subsp. **volgensis**
    (Krysht.) Oudejans
Tithymalus volhynicus (Besser ex Raiborski) Holub === **Euphorbia illirica** Lam. subsp. **illirica**
Tithymalus volkii (Rech.f.) Soják === **Euphorbia volkii** Rech.f.
Tithymalus × *wagneri* (Soó) Holub === **Euphorbia** × **wagneri** Soó
Tithymalus waldsteinii Soják === **Euphorbia esula** subsp. **tommasiniana** (Bertol.) Kuzmanov
Tithymalus wallichii (Hook.f.) Soják === **Euphorbia wallichii** Hook.f.
Tithymalus welwitschii (Boiss. & Reut.) Klotzsch & Garcke === **Euphorbia paniculata** subsp.
    welwitschii (Boiss. & Reut.) Vicens, Molero & C.Blanché
Tithymalus × *wimmerianus* (J.Wagner) Dostál === **Euphorbia** × **wimmeriana** J.Wagner
Tithymalus woronowii (Grossh.) Prokh. === **Euphorbia woronowii** Grossh.
Tithymalus wulfenii (Hoppe ex W.Koch) Soják === **Euphorbia characias** subsp. **wulfenii**
    (Hoppe ex W.Koch) Radcl.-Sm.
Tithymalus yamashitae (Kitam.) Soják === **Euphorbia microsciadia** Boiss.
Tithymalus yaroslavii (Poljakov) Soják === **Euphorbia yaroslavii** Poljakov
Tithymalus yemenicus (Boiss.) Soják === **Euphorbia yemenica** Boiss.
Tithymalus zeravschanicus (Regel) Prokh. === **Euphorbia sarawschanica** Regel
Tithymalus zeyheri Klotzsch & Garcke === **Euphorbia mauritanica** L.

Tithymalus

of Miller (1754); a synonym of *Pedilanthus*, typified by *Tithymalus myrtifolius* (= *Pedilanthus tithymaloides*). Rejected against *Tithymalus* Gaertn., under which all names are listed.

Synonyms:
Tithymalus Mill., nom. rejic. === **Pedilanthus** Neck.

Tithymalus

of Séguier (1754); a synonym of *Euphorbia*, typified by *E. pithyusa*.

Synonyms:
Tithymalus Ség. === **Euphorbia** L.

Tithymalus

of Adanson em. Hill (1768); a synonym of *Euphorbia*, typified by *E. peplus*.

Torfasadis

A synonym of *Euphorbia*, proposed by Rafinesque.

Tournesol

Rejected against *Chrozophora*.

Synonyms:
Tournesol Adans. === **Chrozophora** Neck. ex A.Juss.

Tournesolia

A Latinised orthographic variant of *Tournesol*.

Synonyms:
Tournesolia Nissole ex Scop. === **Chrozophora** Neck. ex A.Juss.

Toxicodendrum

Synonyms:
Toxicodendrum Thunb. === **Hyaenanche** Lamb. & Vahl
Toxicodendrum acutifollium Benth. === **Xymalos monospora** (Harv.) Baill. (Monimiaceae)
Toxicodendrum capense Thunb. === **Hyaenanche globosa** (Gaertn.) Lamb. & Vahl
Toxicodendrum globosum (Gaertn.) Pax & K.Hoffm. === **Hyaenanche globosa** (Gaertn.)
    Lamb. & Vahl

Trachycaryon

Synonyms:
Trachycaryon Klotzsch === **Adriana** Gaudich.
Trachycaryon cunninghamii F.Muell. === **Adriana tomentosa** Gaudich. var. **tomentosa**
Trachycaryon cunninghamii var. *glabrum* F.Muell. === **Adriana tomentosa** Gaudich.
    var. **tomentosa**
Trachycaryon cunninghamii var. *tomentosum* F.Muell. === **Adriana tomentosa** Gaudich.
    var. **tomentosa**
Trachycaryon hookeri F.Muell. === **Adriana tomentosa** var. **hookeri** (F.Muell.) C.L.Gross &
    M.A.Whalen
Trachycaryon hookeri var. *glabriusculum* F.Muell. === **Adriana tomentosa** var. **hookeri**
    (F.Muell.) C.L.Gross & M.A.Whalen
Trachycaryon hookeri var. *velutinum* F.Muell. === **Adriana tomentosa** var. **hookeri** (F.Muell.)
    C.L.Gross & M.A.Whalen
Trachycaryon klotzschii F.Muell. === **Adriana quadripartita** (Labill.) Gaudich.

Traganthus

Synonyms:
Traganthus Klotzsch === **Bernardia** Houst. ex Mill.
Traganthus sidoides Klotzsch === **Bernardia sidoides** (Klotzsch) Müll.Arg.

Tragia

168 species, mainly in Africa, Madagascar and the Americas but also extending from Arabia across Asia to Australia; includes *Ctenomeria*. This genus of creeping, twining, scandent or erect herbs or subshrubs, with sometimes urticating hairs, is mainly found in drier areas; some species have woody rootstocks. Pax & Hoffmann (1919) gave a total for the genus of 125 species. Gillespie (1994) briefly reviewed the genus, giving an estimate of 130 species. She also presented a key to five tropical American sections; of these *Tragia* at 40-50 species is much the biggest. The largest concentration of species in the Americas is in Brazil. In Africa and Asia the largest section is *Tagira*, also with 40-50 species. For tropical Africa (54 spp.), the only overall survey following Pax and Hoffmann is that in Lebrun and Stork, *Énumération des plantes à fleurs* 1 (1991). 27 species are covered in *Flora of Tropical East Africa*, Euphorbiaceae 1 (1987). Malagasy species were revised by Leandri (1938) with additions in 1971 (including description of a new subgenus, *Mauroya*). Those in Argentina were last revised by Mulgura & Gutierrez (1989) and North American species north of Mexico were re-surveyed by Miller & Webster (1967). More recently, the two Australian species were treated by Forster (1994); a third, *T. finalis*, was added in 1997. For subgen. *Mauroya* Leandri claimed affinities with *Tragiella*, *Sphaerostylis* and *Ramelia* (=

Bocquillonia), but the last-named is in the Webster system rather distant. The only Malesian species, *T. luzoniensis*, is now excluded from Euphorbiaceae. (Acalyphoideae)

Prain, D. (1913). The genus *Ctenomeria*. J. Bot. 51: 168-172. En. — 2 species, southern Africa. [Included in *Tragia* by Pax and Hoffmann (1919) as their section VII.]

Pax, F. & K. Hoffmann (1919). *Tragia*. In A. Engler (ed.), Das Pflanzenreich, IV 147 IX (Euphorbiaceae-Acalypheae-Plukenetiinae): 32-101. Berlin. (Heft 68.) La/Ge. — 125 species in 9 sections, Americas, Africa and Madagascar; *Ctenomeria* included as sect. 7.

Leandri, J. (1938). Le genre *Tragia* (Euphorbiacées) à Madagascar. Bull. Acad. Malgache, n.s., 21: 65-68. Fr. — Regional treatment with synoptic keys, descriptions of novelties, indication of distribution, and notes; 8 species in all accounted for.

Lourteig, A. & C. A. O'Donell (1941). *Tragiae* argentinae. Lilloa 6: 347-380, illus., map. Sp. — Treatment of 9 species with key, descriptions, synonymy, exsiccatae, notes, illustrations and map. [Precursor to account in *Genera et species plantarum argentinarum*.]

Miller, K. I. (1964). A taxonomic study of the species of *Tragia* in the United States. 161 pp. Lafayette, Ind. (Unpubl. Ph.D. dissertation, Purdue University.) En. — Substantively included in Miller & Webster (1967).

Miller, K. I. & G. L. Webster (1967). A preliminary revision of *Tragia* (Euphorbiaceae) in the United States. Rhodora 69: 241-305, illus., maps. Based on the senior author's Ph.D. dissertation, Purdue University (Miller 1964). En. — Background, morphology and systematics; key to and descriptive treatment with synonymy, references, types, indication of distribution and habitat, selected localities with exsiccatae, and commentary; all species illustrated and mapped. [Covers 11 species in sect. *Tragia* and 2 in sect. *Leptobotrys*, the latter nearly confined to SE US.]

Airy-Shaw, H. K. (1969). Notes on Malesian and other Asiatic Euphorbiaceae, CXII. Notes on the subtribe Plukenetiinae Pax. Kew Bull. 23: 114-121. En. —*Tragia*, pp. 114-115; several species excluded to *Cnesmone*, *Megistostigma* and *Pachystylidium*. The genus is considered by the author not to occur in SE Asia and Malesia.

Leandri, J. (1971). Un sous-genre malgache nouveau de *Tragia* (Euphorbiacées). Adansonia, II, 11: 437-439, illus. Fr. — Protologue of a new subgenus *Mauroya* (after the artist of the illustration) and description of *T. ivohibeensis*.

Radcliffe-Smith, A. (1983). Notes on African Euphorbiaceae, XIII. *Tragia*, *Tragiella*, etc. Kew Bull. 37: 683-691. En. — The largest part of this paper (pp. 684-690) is on *Tragia*, with notes and novelties. [Precursory to *Flora of Tropical East Africa*.]

Radcliffe-Smith, A. (1985). Notes on African Euphorbiaceae, XV. *Tragia* [ii]. Kew Bull. 40: 231-234. En. — A further precursor to *Flora of Tropical East Africa* with 4 new species, 1 new variety and 1 new record.

Radcliffe-Smith, A. (1987). Notes on African Euphorbiaceae, XIX. *Tragia* (iii) and notes on *Croton* and *Erythrococca*. Kew Bull. 42: 395-399. En. — Precursory to *Flora Zambesiaca* treatment.

Mulgura de Romera, M.E. & M.M. Gutierrez de Sanguinetti (1989). Actualización taxonómica de *Tragia* (Euphorbiaceae) para Argentina y regiones limítrofes. Darwiniana (29: 77-138, illus., maps. Sp. — Descriptive revision with key, synonymy, types, references to illustrations, distribution, habitat and phenology, localities with exsiccatae, commentary, figures and maps; doubtful and excluded names, bibliography and index at end. [Accounts for 14 species and one additional variety.]

Forster, P. I. (1994). A taxonomic revision of *Tragia* (Euphorbiaceae) in Australia. Austral. Syst. Bot. 7: 377-383, illus., map. En. — General remarks including past studies; descriptive treatment (2 species, one new) with key, synonymy, types, localities with (for *T. novae-hollandiae* selected) exsiccatae, indication of distribution and ecology, and commentary; references at end. [The Australian species are twining, wiry perennial herbs of wetter tropical and subtropical habitats. A third species appeared in 1997.]

Gillespie, L. J. (1994). A new section and two new species of *Tragia* (Euphorbiaceae) from the Venezuelan Guayana and French Guiana. Novon 4: 330-338, illus. En. — Novelties and notes along with a concise review of the genus; includes key to tropical American sections (5 in all).

Tragia Plum. ex L., Sp. Pl.: 980 (1753).

Trop. America, Trop. Africa to Sri Lanka, Australia. 22 23 24 25 26 27 29 36 40 50 74 76 77 78 79 80 81 82 84 85.

Schorigeram Adans., Fam. Pl. 2: 355 (1763).

Allosandra Raf., Autik. Bot.: 51 (1840).

Bia Klotzsch, Arch. Naturgesch. 7: 189 (1841).

Leptorhachis Klotzsch, Arch. Naturgesch. 7: 189 (1841).

Leucandra Klotzsch, Arch. Naturgesch. 7: 188 (1841).

Ctenomeria Harv., Hooker's J. Bot. Kew Gard. Misc. 1: 29 (1842).

Agirta Baill., Étude Euphorb.: 463 (1858).

Lassia Baill., Étude Euphorb.: 464 (1858).

Leptobotrys Baill., Étude Euphorb.: 478 (1858).

Zuckertia Baill., Étude Euphorb.: 495 (1858).

Tragia abortiva M.G.Gilbert, Nordic J. Bot. 12: 396 (1992).
Ethiopia. 24 ETH.

Tragia acalyphoides Radcl.-Sm., Kew Bull. 37: 684 (1983).
Tanzania (Uzaramo). 25 TAN. Cl. hemicr.

Tragia adenanthera Baill., Adansonia 1: 275 (1861).
Tanzania (incl. Zanzibar), Malawi. 25 TAN 26 MLW. Cl. hemicr.

Tragia adenophila Pax & K.Hoffm. in H.G.A.Engler, Pflanzenr., IV, 147, IX: 55 (1919).
Tragia adenophila var. *mollis* Pax & K.Hoffm. in H.G.A.Engler, Pflanzenr., IV, 147, IX: 55 (1919), nom. inval.
Paraguay. 85 PAR. Cham.
Tragia adenophila var. *ferruginea* Pax & K.Hoffm. in H.G.A.Engler, Pflanzenr., IV, 147, IX: 55 (1919).

Tragia affinis Rob. & Greenm., Proc. Amer. Acad. Arts 29: 393 (1894).
C. Mexico. 79 MXC. Cl. hemicr.

Tragia aliena Pax & K.Hoffm. in H.G.A.Engler, Pflanzenr., IV, 147, IX: 52 (1919).
? 8 +. Cl. cham. – Related to *T. gracilis*.

Tragia alienata (Didr.) Múlgura & M.M.Gut., Candollea 46: 523 (1991).
S. & SE. Brazil, E. Paraguay, N. Argentina, Bolivia (Yungas). 83 BOL 84 BZL BZS 85 AGE AGW PAR. Cl. nanophan.
Bia sellowiana Klotzsch, Arch. Naturgesch. 7: 190 (1841), nom. nud.
* *Bia alienata* Didr., Vidensk. Meddel. Dansk Naturhist. Foren. Kjøbenhavn 1857: 131 (1857).
Tragia sellowiana Müll.Arg., Linnaea 34: 179 (1865).
Tragia cissoides Müll.Arg. in C.F.P.von Martius, Fl. Bras. 11(2): 406 (1874).

Tragia amblyodonta (Müll.Arg.) Pax & K.Hoffm. in H.G.A.Engler, Pflanzenr., IV, 147, IX: 51 (1919).
SE. Arizona, New Mexico, Texas, NE. Mexico. 76 ARI 77 NWM TEX 79 MXE. (Cl.) hemicr.
* *Tragia nepetifolia* var. *amblyodonta* Müll.Arg. in A.P.de Candolle, Prodr. 15(2): 934 (1866).
Tragia dentata Klotzsch ex Pax & K.Hoffm. in H.G.A.Engler, Pflanzenr., IV, 147, IX: 51 (1919), nom. illeg.

Tragia angolensis Mull.Arg., J. Bot. 2: 333 (1864).
Angola, Zambia. 26 ANG ZAM. Hemicr.

Tragia arabica Baill., Étude Euphorb.: 461 (1858).
Ethiopia, Arabian Pen. 24 ETH 35 SAU YEM. Cl. cham.
Tragia parvifolia Pax, Bot. Jahrb. Syst. 19: 102 (1894). *Tragia arabica* var. *parvifolia* (Pax) Prain in D.Oliver, Fl. Trop. Afr. 6(1): 982 (1913).

Tragia arechavaletae Herter ex Arechav., Anales Mus. Nac. Montevideo, II, 1: 82 (1910).
Uruguay. 85 URU. Cham.

Tragia arnhemica P.I.Forst., Austral. Syst. Bot. 7: 381 (1994).
Northern Territory. 50 NTA.

Tragia ashiae M.G.Gilbert, Nordic J. Bot. 12: 395 (1992).
Ethiopia. 24 ETH.

Tragia aurea Rusby, Bull. New York Bot. Gard. 4: 444 (1907).
Bolivia. 83 BOL. Cl. nanophan.

Tragia bahiensis Müll.Arg., Linnaea 34: 182 (1865).
Brazil (Bahia). 84 BZE. Hemicr.
  Tragia nepetifolia Wawra, Bot. Ergebn.: 34 (1866).

Tragia bailloniana Müll.Arg., Linnaea 34: 178 (1865).
Mexico, C. America. 79 MXG MXS MXT 80 COS GUA PAN. Cl. nanophan.

Tragia balfourii Prain in D.Oliver, Fl. Trop. Afr. 6(1): 983 (1913).
Socotra. 24 SOC. Cl. cham.
  * *Tragia dioica* Balf.f., Proc. Roy. Soc. Edinburgh 12: 95 (1884), nom. illeg. *Tragia balfouriana* J.B.Gillett ex G.B.Popov, J. Linn. Soc., Bot. 55: 719 (1957), nom. illeg.

Tragia ballyi Radcl.-Sm., Kew Bull. 37: 685 (1983).
Kenya (Fort Hall). 25 KEN. Hemicr.

Tragia bangii Rusby, Bull. New York Bot. Gard. 4: 445 (1907).
Bolivia. 83 BOL. Hemicr.

Tragia baroniana Prain, Bull. Misc. Inform. Kew 1912: 235 (1912).
C. Madagascar. 29 MDG. Nanophan.

Tragia benthamii Baker, Bull. Misc. Inform. Kew 1910: 128 (1910).
Trop. & S. Africa. 22 BEN GHA IVO NGA TOG 23 CAR CMN CON EQG GAB GGI ZAI 24
  ETH SUD 25 UGA 26 ANG MLW MOZ ZAM ZIM 27 BOT. Cl. hemicr. or nanophan.
  * *Tragia cordifolia* Benth. in W.J.Hooker, Niger Fl.: 501 (1849), nom. illeg.
  Tragia mitis var. *kirkii* Müll.Arg., Flora 47: 435 (1864).
  Tragia kassiliensis Beille, Bull. Soc. Bot. France 57(8): 126 (1910).
  Tragia keniensis Rendle, J. Bot. 70: 164 (1932).

Tragia betonicifolia Nutt., Trans. Amer. Philos. Soc., n.s., 5: 173 (1837).
Kansas, Missouri, Oklahoma, Arkansas, Texas. 74 KAN MSO OKL 77 TEX 78 ARK.
  (Cl.) hemicr.
  Tragia urticifolia var. *texana* Shinners, Field & Lab. 19: 183 (1951).

Tragia bicolor Miq., Linnaea 26: 222 (1853). *Tragia miqueliana* var. *bicolor* (Miq.) Müll.Arg.
in A.P.de Candolle, Prodr. 15(2): 943 (1866).
  S. India. 40 IND. Cl. hemicr.
  Tragia miqueliana Müll.Arg. in A.P.de Candolle, Prodr. 15(2): 942 (1866).

Tragia biflora Urb. & Ekman, Ark. Bot. 22A(8): 62 (1929).
Hispaniola. 81 DOM HAI. Cl.

Tragia boiviniana Müll.Arg., Linnaea 34: 183 (1865).
Madagascar (Nosi Bé I.). 29 MDG. Cham.

Tragia bongolana Prain, Bull. Misc. Inform. Kew 1912: 236 (1912).
Sudan. 24 SUD. Cl. cham.

Tragia brevipes Pax, Bot. Jahrb. Syst. 19: 103 (1894).
Cameroon to Somalia and Zimbabwe. 23 CMN ZAI 24 ETH SOM 25 KEN TAN UGA 26
MLW ZAM ZIM. Cl. hemicr. or nanophan.
Tragia velutina Pax, Bot. Jahrb. Syst. 19: 104 (1894).
Tragia volkensii Pax in H.G.A.Engler, Pflanzenw. Ost-Afrikas C: 240 (1895).

Tragia brevispica Engelm. & A.Gray, Boston J. Nat. Hist. 4: 262 (1845).
Texas, NE. Mexico. 77 TEX 79 MXE. (Cl.) hemicr.
Tragia teucriifolia Scheele, Linnaea 25: 586 (1852). *Tragia nepetifolia* var. *teucriifolia*
(Scheele) Müll.Arg. in A.P.de Candolle, Prodr. 15(2): 934 (1866).

Tragia brouniana Prain, Bull. Misc. Inform. Kew 1909: 51 (1909). *Tragia cannabina* var.
brouniana (Prain) Prain in D.Oliver, Fl. Trop. Afr. 6(1): 977 (1913).
NE. Trop. Africa. 24 ETH SOM SUD. Ther. or cham.
Tragia involucrata var. *intermedia* Müll.Arg. in A.P.de Candolle, Prodr. 15(2): 944
(1866). *Tragia cannabina* var. *intermedia* (Müll.Arg.) Prain in D.Oliver, Fl. Trop. Afr.
6(1): 976 (1913).

Tragia capensis Thunb., Prodr. Pl. Cap.: 14 (1794). *Ctenomeria capensis* (Thunb.) Harv. ex
Sond., Linnaea 23: 109 (1850).
S. Africa. 27 CPP NAT TVL. Cl. cham.
Ctenomeria cordata Harv., Hooker's J. Bot. Kew Gard. Misc. 1: 29 (1842). *Tragia cordata*
(Harv.) Burtt Davy, Ann. Transvaal Mus. 3: 122 (1912), nom. illeg.
Ctenomeria kraussiana Hochst., Flora 28: 85 (1845).

Tragia caperonioides Pax & K.Hoffm. in H.G.A.Engler, Pflanzenr., IV, 147, IX: 40 (1919).
Paraguay. 85 PAR. Cham.

Tragia catamarcensis Pax & K.Hoffm. in H.G.A.Engler, Pflanzenr., IV, 147, IX: 42 (1919).
Argentina (Catamaraca). 85 AGW. – Provisionally accepted.

Tragia ceanothifolia Radcl.-Sm., Kew Bull. 40: 231 (1985).
C. Kenya. 25 KEN. Hemicr.

Tragia cearensis Pax & K.Hoffm. in H.G.A.Engler, Pflanzenr., IV, 147, XVII: 186 (1924).
Brazil (Ceará). 84 BZE.

Tragia chevalieri Beille, Bull. Soc. Bot. France 57(8): 126 (1910).
Ivory Coast, Nigeria. 22 IVO NGA. Cl. cham.

Tragia chlorocaulon Baill., Étude Euphorb.: 461 (1858).
S. Brazil. 84 BZS. Cl. cham.

Tragia cinerea (Pax) M.G.Gilbert & Radcl.-Sm., Kew Bull. 40: 394 (1985).
Eritrea to S. Kenya. 24 ERI ETH SOM 25 KEN. Cl. hemicr.
* *Tragia mitis* var. *cinerea* Pax, Bot. Jahrb. Syst. 19: 103 (1894). *Tragia cordifolia* var. *cinerea*
(Pax) Prain in D.Oliver, Fl. Trop. Afr. 6(1): 981 (1913). *Tragia pungens* var. *cinerea*
(Pax) Pax in H.G.A.Engler, Pflanzenr., IV, 147, IX: 79 (1919).

Tragia cocculifolia Prain, Bull. Misc. Inform. Kew 1912: 335 (1912).
C. Madagascar. 29 MDG. Nanophan.

Tragia collina Prain, Bull. Misc. Inform. Kew 1912: 335 (1912).
KwaZulu-Natal. 27 NAT. Cham.

Tragia cordata Michx., Fl. Bor.-Amer. 2: 176 (1803). *Tragia macrocarpa* Willd., Sp. Pl. 4: 323 (1805), nom. illeg. *Tragia michauxii* Baill., Étude Euphorb.: 460 (1858), nom. illeg.
EC. & E. U.S.A. 74 ILL MSO OKL 75 INI 77 TEX 78 ALA ARK FLA GEO KTY TEN. Cl. hemicr.

Tragia correae Huft, Ann. Missouri Bot. Gard. 75: 1107 (1988).
S. Costa Rica, Panama. 80 COS PAN. Cl. nanophan.

Tragia cubensis Urb., Repert. Spec. Nov. Regni Veg. 28: 226 (1930).
C. & E. Cuba. 81 CUB.

Tragia cuneata Müll.Arg., Linnaea 34: 180 (1865).
Brazil (Bahia). 84 BZE. Cl. cham.

Tragia depauperata Pax & K.Hoffm. in H.G.A.Engler, Pflanzenr., IV, 147, IX: 56 (1919).
Paraguay. 85 PAR. Hemicr.

Tragia descampsii De Wild., Ann. Mus. Congo Belge, Bot., IV, 1: 207 (1903).
Zaire (Shaba), W. Tanzania, Zambia. 23 ZAI 25 TAN 36 ZAM. Hemicr.

Tragia dinteri Pax, Bot. Jahrb. Syst. 43: 82 (1909).
Namibia. 27 NAM. Hemicr. or cham.

Tragia dioica Sond., Linnaea 23: 109 (1850).
Zimbabwe, S. Africa. 26 ZIM 27 BOT CPP NAM TVL. Hemicr. or cham.

var. **dioica**
Zimbabwe, S. Africa. 26 ZIM 27 BOT CPP NAM TVL. Hemicr. or cham.
Tragia schinzii Pax, Bull. Herb. Boissier 6: 734 (1898).

var. **lobata** (Müll.Arg.) Pax & K.Hoffm. in H.G.A.Engler, Pflanzenr., IV, 68: 88 (1919).
Namibia. 27 NAM. Hemicr. or cham.
* *Tragia rupestris* var. *lobata* Müll.Arg. in A.P.de Candolle, Prodr. 15(2): 941 (1866).
Tragia recta Dinter ex Pax & K.Hoffm. in H.G.A.Engler, Pflanzenr., IV, 147, IX: 88 (1919), pro syn.

Tragia dodecandra Griseb., Abh. Königl. Ges. Wiss. Göttingen 19: 98 (1874).
Argentina (Córdoba). 85 AGE. (Cl.) hemicr. or cham.

Tragia doryodes M.G.Gilbert, Nordic J. Bot. 12: 391 (1992).
Ethiopia. 24 ETH.

Tragia durbanensis Kuntze, Revis. Gen. Pl. 3(2): 293 (1898).
Mozambique to Cape Prov. 26 MOZ 27 CPP NAT. Cl. Hemicr.
Tragia capensis E.Mey. ex Sond., Linnaea 23: 110 (1850), nom. illeg.
* *Tragia meyeriana* var. *glabrata* Müll.Arg. in A.P.de Candolle, Prodr. 15(2): 938 (1866).
Tragia glabrata (Müll.Arg.) Pax & K.Hoffm. in H.G.A.Engler, Pflanzenr., IV, 147, IX: 94 (1919), nom. illeg.
Tragia glabrata var. *hispida* Radcl.-Sm., Kew Bull. 42: 396 (1987).

Tragia emilii Pax & K.Hoffm. in H.G.A.Engler, Pflanzenr., IV, 147, IX: 43 (1919).
Paraguay, Argentina (Misiones). 85 AGE PAR. Cl. hemicr.

Tragia emrichii Herter, Revista Sudamer. Bot. 8: 26 (1949).
Brazil (Rio Grande do Sul). 84 BZS.

Tragia fallacina Pax & K.Hoffm. in H.G.A.Engler, Pflanzenr., IV, 147, IX: 39 (1919).
Uruguay. 85 URU. Cham.

Tragia fallax Müll.Arg., Linnaea 34: 179 (1865).
Peru. 83 PER. Cl. cham.

Tragia fasciculata Beille, Bull. Soc. Bot. France 55(8): 83 (1908).
Central African Rep. 23 CAF. Hemicr. or cham.

Tragia fendleri Müll.Arg., Linnaea 34: 179 (1865).
Guyana, Venezuela, Brazil (Amazonas). 82 GUY VEN 84 BZN. Cl. cham.

Tragia finalis P.I.Forst., Austral. Syst. Bot. 10: 863 (1997).
N. & NE. Queensland. 50 QLD. Cl. cham. or nanophan.

Tragia friesii Pax & K.Hoffm. in H.G.A.Engler, Pflanzenr., IV, 147, XVII: 186 (1924).
Bolivia. 83 BOL.

Tragia furialis Bojer, Hortus Maurit.: 286 (1837). *Tragia angustifolia* var. *furialis* (Bojer)
Müll.Arg. in A.P.de Candolle, Prodr. 15(2): 939 (1866). *Tragia furialis* var. *eufurialis* Pax &
K.Hoffm. in H.G.A.Engler, Pflanzenr., IV, 147, IX: 94 (1919), nom. inval.
Kenya to Mozambique, Comoros, Madagascar. 25 KEN TAN 26 MOZ ZIM 29 COM MDG.
Cl. hemicr.
 Tragia scheffleri Baker, Bull. Misc. Inform. Kew 1908: 440 (1908). *Tragia furialis* var.
  scheffleri (Baker) Pax & K.Hoffm. in H.G.A.Engler, Pflanzenr., IV, 147, IX: 94 (1919).

Tragia gardneri Prain, Bull. Misc. Inform. Kew 1909: 52 (1909).
Zimbabwe. 26 ZIM. Hemicr. or cham.

Tragia geraniifolia Klotzsch ex Müll.Arg. in A.P.de Candolle, Prodr. 15(2): 933 (1866).
E. Bolivia, S. Brazil, Paraguay, Uruguay, N. Argentina. 83 BOL 84 BZS 85 AGE AGW PAR
URU. Cl. hemicr. or cham.
 Tragia geraniifolia var. *multifida* Griseb., Symb. Fl. Argent.: 60 (1879).

Tragia giardelliae M.M.Gut. & Múlgura, Darwiniana 27: 491 (1986).
Argentina (Misiones). 85 AGE. Cl. hemicr. or cham.

Tragia glabrescens Pax, Bot. Jahrb. Syst. 19: 104 (1894).
SE. Kenya, NE. Tanzania. 25 KEN TAN. Cl. hemicr.

Tragia glanduligera Pax & K.Hoffm. in H.G.A.Engler, Pflanzenr., IV, 147, IX: 55 (1919).
SE. Texas, E. Mexico. 77 TEX 79 MXE MXG MXT. Cl. hemicr.

Tragia gracilis Griseb., Nachr. Königl. Ges. Wiss. Georg-Augusts-Univ. 1: 176 (1865).
E. Cuba. 81 CUB. Cham.

Tragia guatemalensis Lotsy, Bot. Gaz. 20: 354 (1895).
Guatemala. 80 GUA. Cl. cham.

Tragia guyanensis Gillespie, Novon 4: 331 (1994).
Venezuela (Amazonas). 82 VEN. Cl. nanophan.

Tragia hassleriana Chodat, Bull. Herb. Boissier, II, 5: 606 (1905).
Paraguay. 85 PAR. Cham.

Tragia hieronymi Pax & K.Hoffm. in H.G.A.Engler, Pflanzenr., IV, 147, IX: 62 (1919).
N. Argentina, Paraguay. 85 AGE AGW PAR. Cham.
Tragia tenella Pax & K.Hoffm. in H.G.A.Engler, Pflanzenr., IV, 147, IX: 51 (1919).

Tragia hildebrandtii Müll.Arg., Bremen Abh. 7: 26 (1880). *Tragia cannabina* var. *hildebrandtii*
(Müll.Arg.) Pax & K.Hoffm. in H.G.A.Engler, Pflanzenr., IV, 147, IX: 86 (1919).
Ethiopia to Malawi. 24 ETH 25 KEN TAN 26 MLW. Ther. or cham.
Tragia hildebrandtii subsp. *glaucescens* Pax, Bot. Jahrb. Syst. 19: 103 (1894).
Tragia gallabatensis Prain, Bull. Misc. Inform. Kew 1909: 51 (1909).
Tragia mombassana Vatke ex Prain in D.Oliver, Fl. Trop. Afr. 6(1): 978 (1913), pro syn.

Tragia hispida Willd., Sp. Pl. 4: 323 (1805). *Tragia involucrata* var. *hispida* (Willd.) Müll.Arg.
in A.P.de Candolle, Prodr. 15(2): 943 (1866).
Bangladesh. 40 BAN. Cl. hemicr. – Perhaps identical with *T. involucrata*.
Tragia involucrata var. *angustifolia* Hook.f., Fl. Brit. India 5: 465 (1888).

Tragia imerinica Prain, Bull. Misc. Inform. Kew 1912: 336 (1912).
N. & C. Madagascar. 29 MDG. Nanophan.

Tragia impedita Prain, Bull. Misc. Inform. Kew 1909: 52 (1909).
Kenya, Tanzania. 25 KEN TAN. Cl. hemicr. or nanophan. – Intermediate between *T.*
brevipes & *T. benthamii*.

Tragia incana Klotzsch ex Müll.Arg. in A.P.de Candolle, Prodr. 15(2): 935 (1866).
Uruguay. 85 URU.

Tragia incisifolia Prain, Bull. Misc. Inform. Kew 1912: 237 (1912).
Zimbabwe to S. Africa. 26 MOZ ZIM 27 NAT TVL.

Tragia insuavis Prain, Bull. Misc. Inform. Kew 1912: 237 (1912).
Kenya, Tanzania. 25 KEN TAN. Cl. hemicr.

Tragia involucrata L., Sp. Pl.: 980 (1753). *Tragia involucrata* var. *genuina* Müll.Arg. in A.P.de
Candolle, Prodr. 15(2): 943 (1866), nom. inval.
India, Bangladesh, Sri Lanka. 40 BAN IND SRL. Cl. hemicr.
Croton urens L., Sp. Pl.: 1005 (1753).
Tragia cordata B.Heyne ex Wall., Numer. List: 7791A (1847), nom. inval.
Tragia trifida Wall., Numer. List: 7795 (1847), nom. inval.
Tragia involucrata var. *rheediana* Müll.Arg. in A.P.de Candolle, Prodr. 15(2): 943 (1866).

Tragia ivohibeensis Leandri, Adansonia, n.s., 11: 439 (1971).
Madagascar. 29 MDG.

Tragia japurensis Müll.Arg. in C.F.P.von Martius, Fl. Bras. 11(2): 404 (1874).
Brazil (Amazonas). 84 BZN. Cl. cham.

Tragia jonesii Radcl.-Sm. & Govaerts, Kew Bull. 52: 480 (1997).
W. Mexico. 79 MXN MXS. Cl. Hemicr.
* Tragia scandens M.E.Jones, Contr. W. Bot. 18: 49 (1933), nom. illeg.

Tragia karsteniana Pax & K.Hoffm. in H.G.A.Engler, Pflanzenr., IV, 147, IX: 65 (1919).
Colombia. 83 CLM. Cl. cham.

Tragia kirkiana Müll.Arg., Flora 47: 538 (1864).
Kenya to Northern Prov. 25 KEN TAN 26 MLW MOZ ZAM ZIM 27 TVL. Cl. hemicr.
Tragia angustifolia var. *hastata* Müll.Arg., Flora 47: 435 (1864).
Tragia stolziana Pax & K.Hoffm. in H.G.A.Engler, Pflanzenr., IV, 147, IX: 78 (1919).

Tragia laciniata (Torr.) Müll.Arg., Linnaea 34: 182 (1865).
SE. Arizona, Mexico (Chihuahua, Sonora). 76 ARI 79 MXE MXN. (Cl.) hemicr.
* *Tragia urticifolia* var. *laciniata* Torr., Bot. U.S. Mex. Bound.: 200 (1858).

Tragia lagoensis Müll.Arg. in C.F.P.von Martius, Fl. Bras. 11(2): 409 (1874).
Brazil (Minas Gerais). 84 BZL. Cham.

Tragia laminularis Müll.Arg., Linnaea 34: 183 (1865).
Sierra Leone. 22 SIE. Cl. cham.

Tragia lancifolia Dinter ex Pax & K.Hoffm in H.G.A.Engler, Pflanzenr., IV, 147, IX: 91 (1919).
Namibia. 27 NAM.

Tragia lasiophylla Pax & K.Hoffm. in H.G.A.Engler, Pflanzenr., IV, 147, IX: 86 (1919).
Tanzania to Zambia. 25 TAN 26 MLW ZAM. Hemicr.

Tragia lassia Radcl.-Sm. & Govaerts, Kew Bull. 52: 480 (1997).
Madagascar. 29 MDG.
* *Lassia scandens* Baill., Étude Euphorb.: 464 (1858). *Tragia scandens* (Baill.) Müll.Arg., Linnaea 34: 183 (1865), nom. illeg.

Tragia leptorhachis Radcl.-Sm. & Govaerts, Kew Bull. 52: 480 (1997).
S. Brazil. 84 BZS.
* *Leptorhachis hastata* Klotzsch, Arch. Naturgesch. 7: 189 (1841). *Tragia hastata* (Klotzsch) Müll.Arg. in C.F.P.von Martius, Fl. Bras. 11(2): 407 (1874), nom. illeg.

Tragia lessertiana (Baill.) Müll.Arg., Linnaea 34: 178 (1865).
Guianas, Brazil (Amapá, Maranhão). 82 FRG GUY SUR 84 BZE BZN. Cl. cham.
* *Bia lessertiana* Baill., Étude Euphorb.: 502 (1858).

Tragia leucandra Pax & K.Hoffm. in H.G.A.Engler, Pflanzenr., IV, 147, IX: 38 (1919).
S. Brazil. 84 BZS. Cham.
* *Leucandra betonicifolia* Klotzsch, Arch. Naturgesch. 7: 188 (1841). *Tragia betonicifolia* (Klotzsch) Müll.Arg., Linnaea 34: 181 (1865), nom. illeg.

Tragia lippiifolia Radcl.-Sm., Kew Bull. 40: 232 (1985).
Tanzania (Masai). 25 TAN. Cham.

Tragia lukafuensis De Wild., Ann. Mus. Congo Belge, Bot., IV, 1: 206 (1903).
Zaire (Shaba), Zambia. 23 ZAI 26 ZAM. Cl. cham.

Tragia mansfeldiana Herter, Revista Sudamer. Bot. 5: 35 (1936 publ. 1937).
Uruguay. 85 URU.

Tragia mazoensis Radcl.-Sm., Kew Bull. 42: 396 (1987).
Zimbabwe. 26 ZIM. Cl. hemicr.

Tragia melochioides Griseb., Abh. Königl. Ges. Wiss. Göttingen 24: 60 (1879).
NW. Argentina (Jujuy to Córdoba). 85 AGE AGW. Hemicr. or cham.
Tragia micrococca Pax & K.Hoffm. in H.G.A.Engler, Pflanzenr., IV, 147, IX: 54 (1919).
Tragia pseudomelochioides Pax & K.Hoffm. in H.G.A.Engler, Pflanzenr., IV, 147, IX: 55 (1919).

Tragia mexicana Müll.Arg., Linnaea 34: 181 (1865).
Mexico, Guatemala. 79 MXS 80 GUA. Cl. cham.

Tragia meyeriana Müll.Arg. in A.P.de Candolle, Prodr. 15(2): 938 (1866).
KwaZulu-Natal, Cape Prov. 27 CPP NAT.
Tragia bolusii Kuntze, Revis. Gen. Pl. 3(2): 293 (1898).

Tragia micromenes Radcl.-Sm., Kew Bull. 47: 681 (1992).
Zambia. 26 ZAM.

Tragia mildbraediana Pax & K.Hoffm. in H.G.A.Engler, Pflanzenr., IV, 147, IX: 96 (1919).
W. Trop. Africa ?, Cameroon. 22 LBR? NGA? SIE? 23 CMN. Cl. cham.

Tragia minor Sond., Linnaea 23: 108 (1850).
Mozambique to S. Africa. 26 MOZ 27 NAT SWZ TVL.

Tragia mitis Hochst. ex A.Rich., Tent. Fl. Abyss. 2: 244 (1850). *Tragia mitis* var. *genuina*
Müll.Arg. in A.P.de Candolle, Prodr. 15(2): 942 (1866), nom. inval.
Sudan, Ethiopia. 24 ETH SUD. Cl. cham.
Tragia cordata A.Rich., Tent. Fl. Abyss. 2: 244 (1850).

Tragia mixta M.G.Gilbert, Nordic J. Bot. 12: 396 (1992).
NE. & E. Trop. Africa. 24 ETH SOM 25 KEN TAN UGA.

Tragia moammarensis Baill., Étude Euphorb.: 461 (1858).
Yemen. 35 YEM. Cl. hemicr. or nanophan.

Tragia monadelpha Schumach. & Thonn. in C.F.Schumacher, Beskr. Guin. Pl.:
404 (1827).
Guinea. 22 GUI.

Tragia montana (Thwaites) Müll.Arg., Linnaea 34: 183 (1865).
Sri Lanka. 40 SRL. (Cl.) hemicr.
* *Tragia involucrata* var. *montana* Thwaites, Enum. Pl. Zeyl.: 270 (1861).

Tragia muelleriana Pax & K.Hoffm. in H.G.A.Engler, Pflanzenr., IV, 147, IX: 80 (1919).
S. India, Sri Lanka. 40 IND SRL. Cl. ther. or hemicr.
Tragia involucrata var. *cordata* Müll.Arg. in A.P.de Candolle, Prodr. 15(2): 943 (1866).
Tragia muelleriana var. *cordata* (Müll.Arg.) Pax & K.Hoffm. in H.G.A.Engler,
Pflanzenr., IV, 147, IX: 81 (1919).
Tragia miqueliana var. *unicolor* Müll.Arg. in A.P.de Candolle, Prodr. 15(2): 943 (1866).
Tragia muelleriana var. *unicolor* (Müll.Arg.) Pax & K.Hoffm. in H.G.A.Engler,
Pflanzenr., IV, 147, IX: 81 (1919).

Tragia negeliensis M.G.Gilbert, Nordic J. Bot. 12: 398 (1992).
Ethiopia. 24 ETH. Cham. or nanophan.

Tragia nepetifolia Cav., Icon. 6: 37 (1800). *Tragia nepetifolia* var. *genuina* Müll.Arg. in A.P.de
Candolle, Prodr. 15(2): 934 (1866), nom. inval.
Arizona, N. Mexico. 76 ARI 79 MXE MXN. (Cl.) hemicr.
Tragia urticifolia Benth., Pl. Hartw.: 14 (1839).

Tragia nigricans Bush, Rep. (Annual) Missouri Bot. Gard. 1906: 122 (1906).
SC. Texas. 77 TEX. Hemicr.

Tragia novae-hollandiae Müll.Arg., Linnaea 34: 180 (1865).
E. Queensland, NE. New South Wales. 50 NSW QLD. (Cl.) cham.
Tragia setosa A.Cunn. ex Pax & K.Hoffm. in H.G.A.Engler, Pflanzenr., IV, 147, IX: 44 (1919), pro syn.

Tragia okanyua Pax, Bull. Herb. Boissier 6: 735 (1898).
Tanzania to Northern Prov. 25 TAN 26 ANG MLW MOZ ZAM ZIM 27 BOT NAM TVL. Cl. hemicr. or nanophan.
Tragia madandensis S.Moore, J. Linn. Soc., Bot. 40: 203 (1911).

Tragia oligantha Pax & K.Hoffm. in H.G.A.Engler, Pflanzenr., IV, 147, XVII: 187 (1924).
Bolivia. 83 BOL.

Tragia pacifica McVaugh, Brittonia 13: 203 (1961).
Mexico (Sinaloa, Nayarit, Jalisco, Colima). 79 MXN MXS. Cl. nanophan.

Tragia paraguariensis Pax & K.Hoffm. in H.G.A.Engler, Pflanzenr., IV, 147, IX: 53 (1919).
S. Brazil, Paraguay, Argentina (Misiones). 84 BZS 85 AGE PAR. Hemicr.
Tragia bahiensis var. *subsessilis* Chodat & Hassl., Bull. Herb. Boissier, II, 5: 607 (1905). *Tragia paraguariensis* var. *subsessilis* (Chodat & Hassl.) Pax & K.Hoffm. in H.G.A.Engler, Pflanzenr., IV, 147, IX: 54 (1919).
Tragia uberabana var. *discolor* Chodat & Hassl., Bull. Herb. Boissier, II, 5: 607 (1905). *Tragia paraguariensis* var. *discolor* (Chodat & Hassl.) Pax & K.Hoffm. in H.G.A.Engler, Pflanzenr., IV, 147, IX: 54 (1919).
Tragia uberabana var. *macrophylla* Chodat & Hassl., Bull. Herb. Boissier, II, 5: 607 (1905). *Tragia paraguariensis* var. *macrophylla* (Chodat & Hassl.) Pax & K.Hoffm. in H.G.A.Engler, Pflanzenr., IV, 147, IX: 54 (1919).
Tragia paraguariensis var. *canescens* Pax & K.Hoffm. in H.G.A.Engler, Pflanzenr., IV, 147, IX: 54 (1919).
Tragia paraguariensis var. *glabrescens* Pax & K.Hoffm. in H.G.A.Engler, Pflanzenr., IV, 147, IX: 54 (1919).

Tragia paxii Lourteig & O'Donell, Lilloa 6: 351 (1941).
Argentina (Misiones). 85 AGE. Cl. cham.

Tragia peltata Vell., Fl. Flumin. 10: 6 (1831).
Brazil (Rio de Janeiro). 84 BZL. Cl. cham.

Tragia perrieri Leandri, Bull. Trimestriel Acad. Malgache, n.s., 21: 67 (1938 publ. 1939).
Madagascar. 29 MDG.

Tragia petiolaris Radcl.-Sm., Kew Bull. 37: 687 (1983).
Tanzania, Zambia. 25 TAN 26 ZAM. (Cl.) hemicr.

Tragia physocarpa Prain, Bull. Misc. Inform. Kew 1912: 238 (1912).
Namibia, Northern Prov. 27 NAM TVL. Cham.

Tragia pinnata (Poir.) A.Juss., Euphorb. Gen.: 48 (1824).
NE. Argentina, Uruguay. 85 AGE URU. (Cl.) hemicr. or cham.
*Acalypha pinnata Poir. in J.B.A.M.de Lamarck, Encycl. 6: 205 (1804).

Tragia platycalyx Radcl.-Sm., Kew Bull. 40: 232 (1985).
Tanzania (Singida). 25 TAN. Cham.

Tragia plukenetii Radcl.-Sm., Kew Bull. 37: 688 (1983).
Nigeria to Somalia and Zimbabwe, C. & S. India, Sri Lanka. 22 NGA 23 CAF CMN ZAI 24 ETH SOM SUD 25 KEN TAN UGA 26 MOZ ZAM ZIM 40 IND SRL. Ther. or cham.

* *Croton hastatus* L., Sp. Pl.: 1005 (1753). *Tragia cannabina* var. *hastata* (L.) Pax &
    K.Hoffm. in H.G.A.Engler, Pflanzenr., IV, 147, IX: 85 (1919).
    Tragia cannabina L.f., Suppl. Pl.: 415 (1782), nom. illeg. *Tragia involucrata* var. *cannabina*
    Müll.Arg. in A.P.de Candolle, Prodr. 15(2): 944 (1866).
    Tragia tripartita Beille, Bull. Soc. Bot. France 55(8): 83 (1908), nom. illeg.

Tragia plumieri Lourteig, Bradea 5: 353 (1990).
    Haiti. 81 HAI. Cl. cham.
    * *Tragia scandens* Aubl., Hist. Pl. Guiane 1: 97 (1775), nom. illeg.

Tragia pogostemonoides Radcl.-Sm., Kew Bull. 40: 233 (1985).
    Tanzania (Uzaramo). 25 TAN. Cham.

Tragia pohlii Müll.Arg., Linnaea 34: 81 (1865).
    Brazil (Goiás). 84 BZC. Cl. cham.
    Tragia mollis Klotzsch ex Pax & K.Hoffm. in H.G.A.Engler, Pflanzenr., IV, 147, IX: 64
        (1919), nom. inval.

Tragia polyandra Vell., Fl. Flumin. 10: 7 (1831).
    SE. Brazil. 84 BZL. Cl. cham.

Tragia polygonoides Prain, Bull. Misc. Inform. Kew 1912: 194 (1912).
    Ivory Coast. 22 IVO. Cl. cham.

Tragia potosina Lundell, Phytologia 1: 370 (1940).
    Mexico (San Luis Potosí). 79 MXE.

Tragia preussii Pax, Bot. Jahrb. Syst. 19: 102 (1894).
    Nigeria to Zaire. 22 NGA 23 CMN ZAI. Cl. cham.
    Tragia winkleri Pax, Bot. Jahrb. Syst. 43: 82 (1909).

Tragia prionoides Radcl.-Sm., Kew Bull. 42: 397 (1987).
    Zimbabwe, Northern Prov. 26 ZIM 27 TVL. Cl. hemicr.

Tragia prostrata Radcl.-Sm., Kew Bull. 42: 398 (1987).
    Zambia. 26 ZAM. Hemicr.

Tragia pungens (Forssk.) Müll.Arg. in A.P.de Candolle, Prodr. 15(2): 941 (1866).
    Eritrea, Ethiopia, Somalia, Yemen. 24 ERI ETH SOM 35 YEM. Cl. cham.
    * *Jatropha pungens* Forssk., Fl. Aegypt.-Arab.: 163 (1775).
    Tragia cordifolia Vahl, Symb. Bot. 1: 76 (1790).
    Tragia cordata Willd., Sp. Pl. 4: 322 (1805), nom. illeg.

Tragia ramosa Torr., Ann. Lyceum Nat. Hist. New York 2: 245 (1828). *Tragia nepetifolia* var.
    ramosa (Torr.) Müll.Arg. in A.P.de Candolle, Prodr. 15(2): 934 (1866).
    WC. to S. U.S.A., N. Mexico. 73 COL 74 KAN MSO NEB OKL 76 ARI CAL NEV UTA 77
    NWM TEX 78 ARK 79 MXE MXN. (Cl.) hemicr.
    Tragia angustifolia Nutt., Trans. Amer. Philos. Soc., n.s., 5: 172 (1837).
    Tragia scutellariifolia Scheele, Linnaea 25: 587 (1852). *Tragia nepetifolia* var. *scutellarifolia*
        (Scheele) Müll.Arg. in A.P.de Candolle, Prodr. 15(2): 934 (1866).
    Tragia ramosa var. *leptophylla* Torr. in W.H.Emory, Rep. U.S. Mex. Bound. 2(1): 201
        (1858). *Tragia stylaris* var. *leptophylla* (Torr.) Müll.Arg., Linnaea 34: 181 (1865). *Tragia*
        leptophylla (Torr.) I.M.Johnst., Contr. Gray Herb. 68: 91 (1923). *Tragia nepetifolia* var.
        leptophylla (Torr.) Shinners, SouthW. Naturalist 6: 101 (1961).
    Tragia stylaris Müll.Arg., Linnaea 34: 180 (1865).

Tragia stylaris var. *angustifolia* Müll.Arg., Linnaea 34: 180 (1865). *Tragia nepetifolia* var. *angustifolia* (Müll.Arg.) Müll.Arg. in A.P.de Candolle, Prodr. 15(2): 934 (1866).

Tragia stylaris var. *latifolia* Müll.Arg., Linnaea 34: 180 (1865). *Tragia nepetifolia* var. *latifolia* (Müll.Arg.) Müll.Arg. in A.P.de Candolle, Prodr. 15(2): 934 (1866). *Tragia ramosa* var. *latifolia* (Müll.Arg.) Pax & K.Hoffm. in H.G.A.Engler, Pflanzenr., IV, 147, IX: 40 (1919).

Tragia rhodesiae Pax, Bot. Jahrb. Syst. 39: 665 (1907).
SC. Trop. Africa. 26 MLW ZAM ZIM. Cham.

Tragia rhoicifolia Chiov., Fl. Somala 1: 310 (1929).
Somalia. 24 SOM.

Tragia rogersii Prain, Bull. Misc. Inform. Kew 1912: 238 (1912).
Northern Prov. 27 TVL.

Tragia rubiginosa Huft, Ann. Missouri Bot. Gard. 76: 1085 (1989).
Peru. 83 PER.

Tragia rupestris Sond., Linnaea 23: 108 (1850). *Tragia rupestris* var. *genuina* Müll.Arg. in A.P.de Candolle, Prodr. 15(2): 940 (1866), nom. inval.
Mozambique, S. Africa. 26 MOZ 27 NAT SWZ TVL.
Tragia rupestris var. *glabrata* Sond., Linnaea 23: 108 (1850).

Tragia saxicola Small, Fl. S.E. U.S.: 702 (1903).
Florida (incl. Key West). 78 FLA. Hemicr.

Tragia schlechteri Pax, Bull. Herb. Boissier 6: 735 (1898). *Ctenomeria schlechteri* (Pax) Prain, J. Bot. 51: 171 (1913).
KwaZulu-Natal. 27 NAT. Cl. cham.

Tragia schweinfurthii Baker, Bull. Misc. Inform. Kew 1908: 308 (1908).
Ethiopia. 24 ETH. Hemicr. or cham.

Tragia senegalensis Müll.Arg., Linnaea 34: 182 (1865).
W. Trop. Africa. 22 BEN GAM GHA GUI SIE TOG. Cl. hemicr. or cham.

Tragia shirensis Prain, Bull. Misc. Inform. Kew 1912: 239 (1912).
Malawi, Mozambique. 26 MLW MOZ. Hemicr.

var. **glabriuscula** Radcl.-Sm., Kew Bull. 42: 399 (1987).
Mozambique. 26 MOZ.

var. **shirensis**
Malawi, Mozambique. 26 MLW MOZ. Hemicr.

Tragia smallii Shinners, Field & Lab. 24: 37 (1956).
SE. U.S.A. to Texas. 77 TEX 78 ALA FLA GEO LOU MSI. Hemicr.

Tragia sonderi Prain, Bull. Misc. Inform. Kew 1912: 337 (1912).
KwaZulu-Natal, Swaziland. 27 NAT SWZ.

Tragia spathulata Benth. in W.J.Hooker, Niger Fl.: 502 (1894).
W. Trop. Africa. 22 GHA LBR NGA SIE TOG. Cl. cham.

Tragia stipularis Radcl.-Sm., Kew Bull. 37: 688 (1983).
Uganda, Tanzania, Zambia. 25 TAN UGA 26 ZAM. Hemicr.

Tragia subhastata Poepp. & Endl., Nov. Gen. Sp. Pl. 3: 20 (1841).
Peru. 83 PER. Cl. cham.

Tragia subsessilis Pax, Bot. Jahrb. Syst. 19: 101 (1894).
Kenya, N. Tanzania. 25 KEN TAN. (Cl.) hemicr.

Tragia tabulaemontana Gillespie, Novon 4: 333 (1994).
Guiana. 82 FRG. Cl. cham.

Tragia tenuifolia Benth. in W.J.Hooker, Niger Fl.: 502 (1849).
Trop. Africa. 22 GHA GUI IVO LBR NGA SIE TOG 23 CMN GAB GGI ZAI 24 SUD 25 UGA
26 ZIM. Cl. hemicr.
Tragia manniana Müll.Arg., Flora 47: 436 (1864).
Tragia klingii Pax, Bot. Jahrb. Syst. 19: 105 (1894).
Tragia zenkeri Pax, Bot. Jahrb. Syst. 23: 528 (1897).
Tragia calvescens Pax, Bot. Jahrb. Syst. 43: 324 (1909).

Tragia tiverneana Leandri, Bull. Trimestriel Acad. Malgache, n.s., 21: 65 (1938 publ. 1939).
Madagascar. 29 MDG.

Tragia tripartita Schweinf., Reliq. Kotschy.: 34 (1868).
Sudan. 24 SUD.

Tragia tristis Müll.Arg. in C.F.P.von Martius, Fl. Bras. 11(2): 410 (1874).
WC. & SE. Brazil, Paraguay. 84 BZC BZL 85 PAR. Cl. cham.

Tragia triumfetoides M.G.Gilbert, Nordic J. Bot. 12: 393 (1992).
Ethiopia. 24 ETH.

Tragia uberabana Müll.Arg. in C.F.P.von Martius, Fl. Bras. 11(2): 417 (1874).
S. & SE. Brazil, Uruguay. 84 BZL BZS 85 URU. Cham.

Tragia ukambensis Pax, Bot. Jahrb. Syst. 19: 105 (1894).
E. Trop. Africa. 25 KEN TAN? UGA. Cl. hemicr. or nanophan.

var. **ugandensis** Radcl.-Sm., Kew Bull. 40: 234 (1985).
Uganda (Karamoja). 25 UGA. Cl. hemicr. or nanophan.

var. **ukambensis**
Kenya, Tanzania ? 25 KEN TAN? Cl. hemicr. or nanophan. – Close to *T. impedita*.

Tragia uncinata M.G.Gilbert, Nordic J. Bot. 12: 399 (1992).
Ethiopia. 24 ETH.

Tragia urens L., Sp. Pl. ed. 2: 1391 (1763).
SE. U.S.A. to Texas. 77 TEX 78 ALA FLA GEO LOU MSI NCA SCA VRG. Hemicr.
Tragia innocua Walter, Fl. Carol.: 229 (1788). *Tragia urens* var. *lanceolata* Michx., Fl. Bor.-
Amer. 2: 175 (1803). *Tragia discolor* f. *lanceolata* (Michx.) Müll.Arg.in A.P.de
Candolle, Prodr. 15(2): 946 (1866). *Tragia urens* var. *innocua* (Walter) Pax & K.Hoffm.
in H.G.A.Engler, Pflanzenr., IV, 147, IX: 58 (1919).
Tragia urens var. *linearis* Michx., Fl. Bor.-Amer. 2: 175 (1803). *Tragia discolor* var. *linearis*
(Michx.) Müll.Arg. in A.P.de Candolle, Prodr. 15(2): 947 (1866).
Tragia urens var. *subovalis* Michx., Fl. Bor.-Amer. 2: 175 (1803). *Tragia discolor* var.
subovalis (Michx.) Müll.Arg. in A.P.de Candolle, Prodr. 15(2): 946 (1866).
Tragia linearifolia Elliott, Sketch Bot. S. Carolina 2: 563 (1824).
Leptobotrys discolor Baill., Étude Euphorb.: 479 (1858). *Tragia discolor* (Baill.) Müll.Arg.
in A.P.de Candolle, Prodr. 15(2): 946 (1866).
Tragia discolor f. *latifolia* Müll.Arg. in A.P.de Candolle, Prodr. 15(2): 946 (1866).

Tragia urticifolia Michx., Fl. Bor.-Amer. 2: 176 (1803).
SE. U.S.A. to Texas. 77 TEX 78 ALA ARK FLA GEO LOU MSI NCA SCA. (Cl.) hemicr.

Tragia vogelii Keay, Kew Bull. 10: 139 (1955).
Ivory Coast to Uganda. 22 IVO NGA 23 CAF CMN 25 UGA. Cl. hemicr.
Tragia angustifolia Benth. in W.J.Hooker, Niger Fl.: 502 (1849), nom. illeg. *Tragia angustifolia* var. *genuina* Müll.Arg. in A.P.de Candolle, Prodr. 15(2): 939 (1866), nom. inval.

Tragia volubilis L., Sp. Pl.: 980 (1753). *Tragia volubilis* var. *genuina* Müll.Arg. in A.P.de Candolle, Prodr. 15(2): 936 (1866), nom. inval.
S. Mexico, Trop. America, Trop. Africa. 22 GHA IVO NGA SIE 23 CAF CMN CON GAB GGI ZAI 24 SUD 25 UGA 26 ANG 79 MXS 80 COS GUA PAN 81 CUB DOM HAI JAM LEE WIN 82 FRG SUR VEN 83 BOL CLM ECU PER 84 BZC BZE BZL BZN BZS 85 AGE AGW PAR URU. Cl. cham. or cl. nanophan.
Tragia plumosa Desf., Tabl. École Bot.: 207 (1804).
Tragia pedunculata P.Beauv., Fl. Oware 1: 90 (1807).
Tragia virgata Lam., Tabl. Encycl. 3: 347 (1823).
Tragia diffusa Vell., Fl. Flumin. 10: 10 (1831).
Tragia triangularis Vell., Fl. Flumin. 10: 8 (1831). *Tragia volubilis* var. *triangularis* (Vell.) Müll.Arg. in A.P.de Candolle, Prodr. 15(2): 936 (1866).
Tragia serra Poepp. & Endl., Nov. Gen. Sp. Pl. 3: 20 (1841). *Tragia volubilis* var. *serra* (Poepp. & Endl.) Müll.Arg. in A.P.de Candolle, Prodr. 15(2): 936 (1866).
Croton scandens Sieber ex C.Presl, Abh. Königl. Böhm. Ges. Wiss., V, 3: 539 (1845).
Tragia infesta Mart. ex D.Dietr., Syn. Pl. 5: 256 (1852).
Tragia gayana Baill., Étude Euphorb.: 461 (1858).
Tragia haguensis Goudot ex Baill., Étude Euphorb.: 461 (1858).
Tragia monandra Baill., Étude Euphorb.: 461 (1858).
Tragia ibaguensis Goudot ex Müll.Arg. in A.P.de Candolle, Prodr. 15(2): 936 (1866), nom. illeg.
Tragia volubilis var. *grandifolia* Müll.Arg. in A.P.de Candolle, Prodr. 15(2): 936 (1866).
Tragia volubilis var. *tenuifolia* Müll.Arg. in A.P.de Candolle, Prodr. 15(2): 936 (1866).
Tragia amoena Müll.Arg. in C.F.P.von Martius, Fl. Bras. 11(2): 414 (1874).
Tragia pedicillaris Müll.Arg. in C.F.P.von Martius, Fl. Bras. 11(2): 415 (1874). *Tragia volubilis* var. *pedicellaris* (Müll.Arg.) Pax & K.Hoffm. in H.G.A.Engler, Pflanzenr., IV, 147, IX: 50 (1919).
Tragia volubilis var. *lanceolata* Müll.Arg. in C.F.P.von Martius, Fl. Bras. 11(2): 414 (1874).
Tragia ovata Parodi, Anales Soc. Ci. Argent. 11: 51 (1881).
Tragia volubilis var. *longifolia* Pax & K.Hoffm. in H.G.A.Engler, Pflanzenr., IV, 147, IX: 49 (1919).

Tragia wahlbergiana Prain, J. Bot. 51: 169 (1913).
Mozambique, Northern Prov. 26 MOZ 27 TVL.
Tragia affinis Müll.Arg. ex Prain, Bull. Misc. Inform. Kew 1912: 334 (1912), nom. illeg.

Tragia websteri Allem & J.L.Wächt., Revista Brasil. Biol. 37: 83 (1977).
Brazil (Rio Grande do Sul). 84 BZS. Hemicr. or cham.

Tragia wildemanii Beille, Bull. Soc. Bot. France 55(8): 82 (1908).
Guinea, Mali. 22 GUI MLI. Hemicr.
Tragia akwapimensis Prain, Bull. Misc. Inform. Kew 1912: 235 (1912).

Tragia wingei Allem & J.L.Wächt., Revista Brasil. Biol. 37: 85 (1977).
Brazil (Rio Grande do Sul). 84 BZS. Hemicr. or cham.

Tragia yucatanensis Millsp., Publ. Field Mus. Nat. Hist., Bot. Ser. 2: 420 (1916).
Mexico (Yucatán). 79 MXT.

Synonyms:

Tragia adenophila var. *ferruginea* Pax & K.Hoffm. === **Tragia adenophila** Pax & K.Hoffm.

Tragia adenophila var. *mollis* Pax & K.Hoffm. === **Tragia adenophila** Pax & K.Hoffm.

Tragia affinis Müll.Arg. ex Prain === **Tragia wahlbergiana** Prain

Tragia akwapimensis Prain === **Tragia wildemanii** Beille

Tragia ambigua S.Moore === **Tragiella natalensis** (Sond.) Pax & K.Hoffm.

Tragia ambigua var. *urticans* S.Moore === **Tragiella natalensis** (Sond.) Pax & K.Hoffm.

Tragia amoena Müll.Arg. === **Tragia volubilis** L.

Tragia angustifolia Benth. === **Tragia vogelii** Keay

Tragia angustifolia Nutt. === **Tragia ramosa** Torr.

Tragia angustifolia var. *furialis* (Bojer) Müll.Arg. === **Tragia furialis** Bojer

Tragia angustifolia var. *genuina* Müll.Arg. === **Tragia vogelii** Keay

Tragia angustifolia var. *hastata* Müll.Arg. === **Tragia kirkiana** Müll.Arg.

Tragia anisopetala Merr. & Chun === **Cnesmone anisopetala** (Merr. & Chun) Croizat

Tragia anomala Prain === **Tragiella anomala** (Prain) Pax & K.Hoffm.

Tragia arabica var. *parvifolia* (Pax) Prain === **Tragia arabica** Baill.

Tragia arborea Comm. ex Baill. === **Acalypha filiformis** Poir. subsp. **filiformis**

Tragia bahiensis var. *subsessilis* Chodat & Hassl. === **Tragia paraguariensis** Pax & K.Hoffm.

Tragia balfouriana J.B.Gillett ex G.B.Popov === **Tragia balfourii** Prain

Tragia betonicifolia (Klotzsch) Müll.Arg. === **Tragia leucandra** Pax & K.Hoffm.

Tragia bicornis Vahl ex A.Juss. === **Microstachys corniculata** (Vahl) Griseb.

Tragia bolusii Kuntze === **Tragia meyeriana** Müll.Arg.

Tragia bracteata Blanco === **Omphalea bracteata** (Blanco) Merr.

Tragia buettneri Pax === **Dalechampia ipomoeifolia** Benth.

Tragia burmanica Kurz === **Megistostigma burmanicum** (Kurz) Airy Shaw

Tragia calvescens Pax === **Tragia tenuifolia** Benth.

Tragia cannabina L.f. === **Tragia plukenetii** Radcl.-Sm.

Tragia cannabina var. *brouniana* (Prain) Prain === **Tragia brouniana** Prain

Tragia cannabina var. *hastata* (L.) Pax & K.Hoffm. === **Tragia plukenetii** Radcl.-Sm.

Tragia cannabina var. *hildebrandtii* (Müll.Arg.) Pax & K.Hoffm. === **Tragia hildebrandtii** Müll.Arg.

Tragia cannabina var. *intermedia* (Müll.Arg.) Prain === **Tragia brouniana** Prain

Tragia capensis E.Mey. ex Sond. === **Tragia durbanensis** Kuntze

Tragia castaneifolia Juss. ex Baill. === **Acalypha integrifolia** subsp. **marginata** (Poir.) Coode

Tragia chamaelea L. === **Microstachys chamaelea** (L.) Müll.Arg.

Tragia cissoides Müll.Arg. === **Tragia alienata** (Didr.) Múlgura & M.M.Gut.

Tragia colorata Poir. === **Acalypha integrifolia** Willd. subsp. **integrifolia**

Tragia cordata A.Rich. === **Tragia mitis** Hochst. ex A.Rich.

Tragia cordata (Harv.) Burtt Davy === **Tragia capensis** Thunb.

Tragia cordata Willd. === **Tragia pungens** (Forssk.) Müll.Arg.

Tragia cordata B.Heyne ex Wall. === **Tragia involucrata** L.

Tragia cordifolia Benth. === **Tragia benthamii** Baker

Tragia cordifolia Vahl === **Tragia pungens** (Forssk.) Müll.Arg.

Tragia cordifolia var. *cinerea* (Pax) Prain === **Tragia cinerea** (Pax) M.G.Gilbert & Radcl.-Sm.

Tragia corniculata Vahl === **Microstachys corniculata** (Vahl) Griseb.

Tragia delpyana Gagnep. === **Pachystylidium hirsutum** (Blume) Pax & K.Hoffm.

Tragia dentata Klotzsch ex Pax & K.Hoffm. === **Tragia amblyodonta** (Müll.Arg.) Pax & K.Hoffm.

Tragia dentata (Alain) Alain === **Platygyna dentata** Alain

Tragia diffusa Vell. === **Tragia volubilis** L.

Tragia dioica Balf.f. === **Tragia balfourii** Prain

Tragia discolor (Baill.) Müll.Arg. === **Tragia urens** L.

Tragia discolor f. *lanceolata* (Michx.) Müll.Arg. === **Tragia urens** L.

Tragia discolor f. *latifolia* Müll.Arg. === **Tragia urens** L.

Tragia discolor var. *linearis* (Michx.) Müll.Arg. === **Tragia urens** L.

Tragia discolor var. *subovalis* (Michx.) Müll.Arg. === **Tragia urens** L.
Tragia elliptica Hochst. === **Shirakiopsis elliptica** (Hochst.) Esser
Tragia erosa Hochst. === **Pyrenacantha scandens** Planch. ex Harv. (Icacinaceae)
Tragia filiformis Poir. === **Cleidion spiciflorum** (Burm.f.) Merr. var. **spiciflorum**
Tragia frieseana Prain === **Tragiella frieseana** (Prain) Pax & K.Hoffm.
Tragia fruticosa Comm. ex Baill. === **Acalypha integrifolia** Willd. subsp. **integrifolia**
Tragia furialis var. *eufurialis* Pax & K.Hoffm. === **Tragia furialis** Bojer
Tragia furialis var. *scheffleri* (Baker) Pax & K.Hoffm. === **Tragia furialis** Bojer
Tragia gagei Haines === **Pachystylidium hirsutum** (Blume) Pax & K.Hoffm.
Tragia gallabatensis Prain === **Tragia hildebrandtii** Müll.Arg.
Tragia gayana Baill. === **Tragia volubilis** L.
Tragia geraniifolia var. *multifida* Griseb. === **Tragia geraniifolia** Klotzsch ex Müll.Arg.
Tragia glabrata (Müll.Arg.) Pax & K.Hoffm. === **Tragia durbanensis** Kuntze
Tragia glabrata var. *hispida* Radcl.-Sm. === **Tragia durbanensis** Kuntze
Tragia grandifolia Klotzsch === **Adenophaedra grandifolia** (Klotzsch) Müll.Arg.
Tragia haguensis Goudot ex Baill. === **Tragia volubilis** L.
Tragia hastata (Klotzsch) Müll.Arg. === **Tragia leptorhachis** Radcl.-Sm. & Govaerts
Tragia hastata Reinw. ex Hassk. === **Cnesmone javanica** Blume var. **javanica**
Tragia hexandra Jacq. === **Platygyna hexandra** (Jacq.) Müll.Arg.
Tragia hildebrandtii subsp. *glaucescens* Pax === **Tragia hildebrandtii** Müll.Arg.
Tragia hirsuta Blume === **Pachystylidium hirsutum** (Blume) Pax & K.Hoffm.
Tragia howardii Alain === **Platygyna volubilis** Howard
Tragia ibaguensis Goudot ex Müll.Arg. === **Tragia volubilis** L.
Tragia infesta Mart. ex D.Dietr. === **Tragia volubilis** L.
Tragia innocua Blanco === **Alchornea rugosa** (Lour.) Müll.Arg.
Tragia innocua Walter === **Tragia urens** L.
Tragia integerrima Hochst. === **Sclerocroton integerrimus** Hochst.
Tragia integrifolia Willd. ex Müll.Arg. === **Acalypha integrifolia** Willd. subsp. **integrifolia**
Tragia interrupta Spreng. === **Laportea interrupta** (L.) Chew (Urticaceae)
Tragia involucrata var. *angustifolia* Hook.f. === **Tragia hispida** Willd.
Tragia involucrata var. *cannabina* Müll.Arg. === **Tragia plukenetii** Radcl.-Sm.
Tragia involucrata var. *cordata* Müll.Arg. === **Tragia muelleriana** Pax & K.Hoffm.
Tragia involucrata var. *genuina* Müll.Arg. === **Tragia involucrata** L.
Tragia involucrata var. *hispida* (Willd.) Müll.Arg. === **Tragia hispida** Willd.
Tragia involucrata var. *intermedia* Müll.Arg. === **Tragia brouniana** Prain
Tragia involucrata var. *montana* Thwaites === **Tragia montana** (Thwaites) Müll.Arg.
Tragia involucrata var. *rheediana* Müll.Arg. === **Tragia involucrata** L.
Tragia irritans Merr. === **Pachystylidium hirsutum** (Blume) Pax & K.Hoffm.
Tragia kassiliensis Beille === **Tragia benthamii** Baker
Tragia keniensis Rendle === **Tragia benthamii** Baker
Tragia klingii Pax === **Tragia tenuifolia** Benth.
Tragia laevis Ridl. === **Cnesmone laevis** (Ridl.) Airy Shaw
Tragia leonis (Alain) Alain === **Platygyna leonis** Alain
Tragia leptophylla (Torr.) I.M.Johnst. === **Tragia ramosa** Torr.
Tragia linearifolia Elliott === **Tragia urens** L.
Tragia lobata Wall. === **Acalypha integrifolia** Willd. subsp. **integrifolia**
Tragia luzoniensis Merr. === **Pyrenacantha** sp. (Icacinaceae)
Tragia macrocarpa Willd. === **Tragia cordata** Michx.
Tragia macrophylla Wall. === **Cnesmone javanica** Blume var. **javanica**
Tragia madandensis S.Moore === **Tragia okanyua** Pax
Tragia mairei (II.Lév.) Rehder === **Cnesmone mairei** (H.Lév.) Croizat
Tragia manniana Müll.Arg. === **Tragia tenuifolia** Benth.
Tragia marginata Poir. === **Acalypha integrifolia** subsp. **marginata** (Poir.) Coode
Tragia mercurialis L. === **Micrococca mercurialis** (L.) Benth.
Tragia meyeriana var. *glabrata* Müll.Arg. === **Tragia durbanensis** Kuntze
Tragia michauxii Baill. === **Tragia cordata** Michx.

Tragia micrococca Pax & K.Hoffm. === **Tragia melochioides** Griseb.

Tragia miqueliana Müll.Arg. === **Tragia bicolor** Miq.

Tragia miqueliana var. *bicolor* (Miq.) Müll.Arg. === **Tragia bicolor** Miq.

Tragia miqueliana var. *unicolor* Müll.Arg. === **Tragia muelleriana** Pax & K.Hoffm.

Tragia mitis var. *cinerea* Pax === **Tragia cinerea** (Pax) M.G.Gilbert & Radcl.-Sm.

Tragia mitis var. *genuina* Müll.Arg. === **Tragia mitis** Hochst. ex A.Rich.

Tragia mitis var. *kirkii* Müll.Arg. === **Tragia benthamii** Baker

Tragia mitis var. *oblongifolia* Müll.Arg. === **Tragiella natalensis** (Sond.) Pax & K.Hoffm.

Tragia mollis Klotzsch ex Pax & K.Hoffm. === **Tragia pohlii** Müll.Arg.

Tragia mombassana Vatke ex Prain === **Tragia hildebrandtii** Müll.Arg.

Tragia monandra Baill. === **Tragia volubilis** L.

Tragia muelleriana var. *cordata* (Müll.Arg.) Pax & K.Hoffm. === **Tragia muelleriana** Pax & K.Hoffm.

Tragia muelleriana var. *unicolor* (Müll.Arg.) Pax & K.Hoffm. === **Tragia muelleriana** Pax & K.Hoffm.

Tragia natalensis Sond. === **Tragiella natalensis** (Sond.) Pax & K.Hoffm.

Tragia natalensis Hochst. === **Sclerocroton integerrimus** Hochst.

Tragia nepetifolia Wawra === **Tragia bahiensis** Müll.Arg.

Tragia nepetifolia var. *amblyodonta* Müll.Arg. === **Tragia amblyodonta** (Müll.Arg.) Pax & K.Hoffm.

Tragia nepetifolia var. *angustifolia* (Müll.Arg.) Müll.Arg. === **Tragia ramosa** Torr.

Tragia nepetifolia var. *genuina* Müll.Arg. === **Tragia nepetifolia** Cav.

Tragia nepetifolia var. *latifolia* (Müll.Arg.) Müll.Arg. === **Tragia ramosa** Torr.

Tragia nepetifolia var. *leptophylla* (Torr.) Shinners === **Tragia ramosa** Torr.

Tragia nepetifolia var. *ramosa* (Torr.) Müll.Arg. === **Tragia ramosa** Torr.

Tragia nepetifolia var. *scutellarifolia* (Scheele) Müll.Arg. === **Tragia ramosa** Torr.

Tragia nepetifolia var. *teucriifolia* (Scheele) Müll.Arg. === **Tragia brevispica** Engelm. & A.Gray

Tragia obovata (Borhidi) Borhidi === **Platygyna obovata** Borhidi

Tragia obtusata Vahl ex Baill. === **Acalypha integrifolia** Willd. subsp. **integrifolia**

Tragia odorata Steud. === **Acalypha integrifolia** Willd. subsp. **integrifolia**

Tragia ovata Parodi === **Tragia volubilis** L.

Tragia paraguariensis var. *canescens* Pax & K.Hoffm. === **Tragia paraguariensis** Pax & K.Hoffm.

Tragia paraguariensis var. *discolor* (Chodat & Hassl.) Pax & K.Hoffm. === **Tragia paraguariensis** Pax & K.Hoffm.

Tragia paraguariensis var. *glabrescens* Pax & K.Hoffm. === **Tragia paraguariensis** Pax & K.Hoffm.

Tragia paraguariensis var. *macrophylla* (Chodat & Hassl.) Pax & K.Hoffm. === **Tragia paraguariensis** Pax & K.Hoffm.

Tragia paraguariensis var. *subsessilis* (Chodat & Hassl.) Pax & K.Hoffm. === **Tragia paraguariensis** Pax & K.Hoffm.

Tragia parvifolia Pax === **Tragia arabica** Baill.

Tragia parvifolia (Alain) Alain === **Platygyna parvifolia** Alain

Tragia pedicillaris Müll.Arg. === **Tragia volubilis** L.

Tragia pedunculata P.Beauv. === **Tragia volubilis** L.

Tragia philippinensis Merr. === **Cnesmone philippinensis** (Merr.) Airy Shaw

Tragia pilosa Vell. === **Microstachys corniculata** (Vahl) Griseb.

Tragia plumosa Desf. === **Tragia volubilis** L.

Tragia pruricus Willd. ex Endl. === **Platygyna hexandra** (Jacq.) Müll.Arg.

Tragia pseudomelochioides Pax & K.Hoffm. === **Tragia melochioides** Griseb.

Tragia pubescens Glaz. === **?**

Tragia pungens var. *cinerea* (Pax) Pax === **Tragia cinerea** (Pax) M.G.Gilbert & Radcl.-Sm.

Tragia ramosa var. *latifolia* (Müll.Arg.) Pax & K.Hoffm. === **Tragia ramosa** Torr.

Tragia ramosa var. *leptophylla* Torr. === **Tragia ramosa** Torr.

Tragia recta Dinter ex Pax & K.Hoffm. === **Tragia dioica** var. **lobata** (Müll.Arg.) Pax & K.Hoffm.

Tragia reticulata Poir. === **Acalypha filiformis** Poir. subsp. **filiformis**

Tragia rugosa Wall. === **Cnesmone javanica** Blume var. **javanica**

Tragia rupestris var. *genuina* Müll.Arg. === **Tragia rupestris** Sond.

Tragia rupestris var. *glabrata* Sond. === **Tragia rupestris** Sond.

Tragia rupestris var. *lobata* Müll.Arg. === **Tragia dioica** var. **lobata** (Müll.Arg.) Pax & K.Hoffm.

Tragia salviifolia Bojer ex Baill. === **Acalypha salviifolia** Baill.

Tragia saxatilis Bojer ex Pax === **Acalypha spachiana** Baill.

Tragia scandens Aubl. === **Tragia plumieri** Lourteig

Tragia scandens L. === **Tetracera scandens** (L.) Merr. (Dilleniaceae)

Tragia scandens (Baill.) Müll.Arg. === **Tragia lassia** Radcl.-Sm. & Govaerts

Tragia scandens M.E.Jones === **Tragia jonesii** Radcl.-Sm. & Govaerts

Tragia scheffleri Baker === **Tragia furialis** Bojer

Tragia schinzii Pax === **Tragia dioica** Sond. var. **dioica**

Tragia schultzeana Dinter ex Pax & K.Hoffm. === **Pterococcus africanus** (Sond.) Pax & K.Hoffm.

Tragia scutellariifolia Scheele === **Tragia ramosa** Torr.

Tragia sellowiana Müll.Arg. === **Tragia alienata** (Didr.) Múlgura & M.M.Gut.

Tragia serra Poepp. & Endl. === **Tragia volubilis** L.

Tragia setosa A.Cunn. ex Pax & K.Hoffm. === **Tragia novae-hollandiae** Müll.Arg.

Tragia shankii A.Molina === **Dalechampia shankii** (Molina) Huft

Tragia stolziana Pax & K.Hoffm. === **Tragia kirkiana** Müll.Arg.

Tragia stylaris Müll.Arg. === **Tragia ramosa** Torr.

Tragia stylaris var. *angustifolia* Müll.Arg. === **Tragia ramosa** Torr.

Tragia stylaris var. *latifolia* Müll.Arg. === **Tragia ramosa** Torr.

Tragia stylaris var. *leptophylla* (Torr.) Müll.Arg. === **Tragia ramosa** Torr.

Tragia tenella Pax & K.Hoffm. === **Tragia hieronymi** Pax & K.Hoffm.

Tragia tenuis Wall. === **Acalypha supera** Forssk.

Tragia teucriifolia Scheele === **Tragia brevispica** Engelm. & A.Gray

Tragia triandra (Borhidi) Borhidi === **Platygyna triandra** Borhidi

Tragia triandra Müll.Arg. === **Seidelia triandra** (E.Mey.) Pax

Tragia triandra var. *genuina* Müll.Arg. === **Seidelia triandra** (E.Mey.) Pax

Tragia triandra var. *pumila* (Sond.) Müll.Arg. === **Seidelia triandra** (E.Mey.) Pax

Tragia triangularis Vell. === **Tragia volubilis** L.

Tragia trifida Wall. === **Tragia involucrata** L.

Tragia tripartita Beille === **Tragia plukenetii** Radcl.-Sm.

Tragia uberabana var. *discolor* Chodat & Hassl. === **Tragia paraguariensis** Pax & K.Hoffm.

Tragia uberabana var. *macrophylla* Chodat & Hassl. === **Tragia paraguariensis** Pax & K.Hoffm.

Tragia urens var. *innocua* (Walter) Pax & K.Hoffm. === **Tragia urens** L.

Tragia urens var. *lanceolata* Michx. === **Tragia urens** L.

Tragia urens var. *linearis* Michx. === **Tragia urens** L.

Tragia urens var. *subovalis* Michx. === **Tragia urens** L.

Tragia urticifolia Benth. === **Tragia nepetifolia** Cav.

Tragia urticifolia var. *laciniata* Torr. === **Tragia laciniata** (Torr.) Müll.Arg.

Tragia urticifolia var. *texana* Shinners === **Tragia betonicifolia** Nutt.

Tragia velutina Pax === **Tragia brevipes** Pax

Tragia villosa Thunb. === **Acalypha capensis** (L.f.) Prain

Tragia virgata Lam. === **Tragia volubilis** L.

Tragia volkensii Pax === **Tragia brevipes** Pax

Tragia volubilis var. *genuina* Müll.Arg. === **Tragia volubilis** L.

Tragia volubilis var. *grandifolia* Müll.Arg. === **Tragia volubilis** L.

Tragia volubilis var. *lanceolata* Müll.Arg. === **Tragia volubilis** L.

Tragia volubilis var. *longifolia* Pax & K.Hoffm. === **Tragia volubilis** L.

Tragia volubilis var. *pedicellaris* (Müll.Arg.) Pax & K.Hoffm. === **Tragia volubilis** L.

Tragia volubilis var. *serra* (Poepp. & Endl.) Müll.Arg. === **Tragia volubilis** L.

Tragia volubilis var. *tenuifolia* Müll.Arg. === **Tragia volubilis** L.

Tragia volubilis var. *triangularis* (Vell.) Müll.Arg. === **Tragia volubilis** L.
Tragia winkleri Pax === **Tragia preussii** Pax
Tragia zenkeri Pax === **Tragia tenuifolia** Benth.

Tragiella

4 species, E. to S. Africa; very close to *Sphaerostylis*. Twining or erect perennials with urticating hairs, leaf-opposed inflorescences and conspicuous bracts, generally in deciduous or dry evergreen forest and developing quickly after rains. No full revision has appeared since 1919; however, since then only one good species has been added. The genus resembles *Tragia* but the sepals in male flowers do not form a pseudodisk and the styles are massive. All these genera are assigned to subtribe Tragiinae (Plukenetieae) in the Webster system. (Acalyphoideae)

Pax, F. & K. Hoffmann (1919). *Tragiella*. In A. Engler (ed.), Das Pflanzenreich, IV 147 IX (Euphorbiaceae-Acalypheae-Plukenetiinae): 104-106. Berlin. (Heft 68.) La/Ge. — 3 species, Africa.

Radcliffe-Smith, A. (1983). Notes on African Euphorbiaceae, XIII. *Tragia, Tragiella*, etc. Kew Bull. 37: 683-691. En. —*Tragiella*, pp. 690-691; description of *T. pyxostigma*. [Precursory to *Flora of Tropical East Africa*.]

Tragiella Pax & K.Hoffm. in H.G.A.Engler, Pflanzenr., IV, 147, IX: 104 (1919).
Trop. & S. Africa. 24 25 26 27.

Tragiella anomala (Prain) Pax & K.Hoffm. in H.G.A.Engler, Pflanzenr., IV, 147, IX: 106 (1919).
Tanzania, Malawi, E. Zambia. 25 TAN 26 MLW ZAM. Cl. hemicr.
* *Tragia anomala* Prain, Bull. Misc. Inform. Kew 1912: 194 (1912). *Sphaerostylis anomala* (Prain) Croizat, J. Arnold Arbor. 22: 430 (1941).

Tragiella frieseana (Prain) Pax & K.Hoffm. in H.G.A.Engler, Pflanzenr., IV, 147, IX: 106 (1919).
Zambia. 26 ZAM.
* *Tragia frieseana* Prain in R.E.Fries, Wiss. Erg. Schwed. Rhod.-Kongo Exped. 1: 125 (1914). *Sphaerostylis frieseana* (Prain) Croizat, J. Arnold Arbor. 22: 430 (1941).

Tragiella natalensis (Sond.) Pax & K.Hoffm. in H.G.A.Engler, Pflanzenr., IV, 147, IX: 105 (1919).
E. & S. Africa. 24 SUD 25 KEN TAN UGA 26 MLW MOZ ZIM 27 CPP NAT SWZ TVL. Cl. hemicr.
* *Tragia natalensis* Sond., Linnaea 23: 107 (1850). *Sphaerostylis natalensis* (Sond.) Croizat, J. Arnold Arbor. 22: 430 (1941).
Tragia mitis var. *oblongifolia* Müll.Arg., Flora 47: 435 (1864).
Tragia ambigua S.Moore, J. Linn. Soc., Bot. 40: 202 (1911).
Tragia ambigua var. *urticans* S.Moore, J. Linn. Soc., Bot. 40: 203 (1911).

Tragiella pyxostigma Radcl.-Sm., Kew Bull. 37: 690 (1983).
Tanzania. 25 TAN. Cl. nanophan.

Synonyms:
Tragiella pavoniifolia Chiov. === **Dalechampia pavoniifolia** (Chiov.) M.G.Gilbert

Tragiopsis

Synonyms:
Tragiopsis H.Karst. === **Microstachys** A.Juss.

Treisia

An early succulent segregate of *Euphorbia*; the name remains in infrageneric use.

Synonyms:
Treisia Haw. === **Euphorbia** L.

Trevia

An orthographic variant of *Trewia*.

Trewia

1 species, warmer parts of Asia and in W. Malesia (including the Philippines) but evidently everywhere rare or local (Airy-Shaw 1966); rapidly-growing, light-demanding small to medium-sized deciduous trees to 20 m, the flowers appearing before the thin, broadly ovate, triveined leaves. The genus is in the Webster system grouped together with its allegedly near relative *Neotrewia*, the large genus *Mallotus*, and other Rottlerinae in the Acalypheae. (Acalyphoideae)

> Pax, F. (with K. Hoffmann) (1914). *Trewia*. In A. Engler (ed.), Das Pflanzenreich, IV 147 VII (Euphorbiaceae-Acalypheae-Mercurialinae): 140-142. Berlin. (Heft 63.) La/Ge. — Monotypic, warmer parts of Asia and in Malesia to Java and the Philippines.
> Airy-Shaw, H. K. (1966). Notes on Malaysian and other Asiatic Euphorbiaceae, LXXVI. *Trewia nudiflora* in Borneo. Kew Bull. 20: 405-406. En. — Extension of range; commentary.
> Airy-Shaw, H. K. (1969). Notes on Malesian and other Asiatic Euphorbiaceae, CVII. *Trewia nudiflora* in the Malay Peninsula. Kew Bull. 23: 79-80. En. — Range extension.

Trewia L., Sp. Pl.: 1193 (1753).
> S. China, Trop. Asia. 36 40 41 42.
> *Canschi* Adans., Fam. Pl. 2: 443 (1763).
> *Rottlera* Willd., Gött. J. Naturwiss. 1: 7 (1797).

Trewia nudiflora L., Sp. Pl.: 1193 (1753). – FIGURE, p. 1561.
> India, Sri Lanka, N. Thailand, Vietnam, SC. China, Hainan, Sumatera, Jawa, Borneo (E. Kalimantan), Philippines. 36 CHC CHH 40 IND SRL 41 THA VIE 42 BOR JAW PHI SUM. Phan.
> *Trewia integerrima* Stokes, Bot. Mat. Med. 4: 570 (1812).
> *Trewia macrophylla* Roth, Nov. Pl. Sp.: 373 (1821).
> *Trewia macrostachya* Klotzsch, Bot. Ergebn. Reise Waldemar: 117 (1862).
> *Trewia polycarpa* Benth. & Hook.f., Gen. Pl. 3: 318 (1880).
> *Mallotus cardiophyllus* Merr., Philipp. J. Sci., C 7: 398 (1912 publ. 1913).

Synonyms:
Trewia africana Baill. === **Erythrococca africana** (Baill.) Prain
Trewia ambigua Merr. === **Neotrewia cumingii** (Müll.Arg.) Pax & K.Hoffm.
Trewia discolor Sm. === **Mallotus** sp.
Trewia glabrata Banks & Sol. ex Pax & K.Hoffm. === **Mallotus polyadenos** F.Muell.
Trewia hernandifolia Roth === **Macaranga indica** Wight
Trewia inophyllum (G.Forst.) Spreng. === **Baloghia inophylla** (G.Forst.) P.S.Green
Trewia insignis Steud. === **Wetria insignis** (Steud.) Airy Shaw
Trewia integerrima Stokes === **Trewia nudiflora** L.
Trewia macrophylla Roth === **Trewia nudiflora** L.
Trewia macrophylla Blume === **Wetria insignis** (Steud.) Airy Shaw
Trewia macrostachya Klotzsch === **Trewia nudiflora** L.

Trewia nudiflora L. (as *T. macrostachya* Klotzsch)
Artist: C.F. Schmidt
Klotzch & Garcke, Bot. Ergebn. Reise Waldemar: pl. 23 (1862)
KEW ILLUSTRATIONS COLLECTION

Trewia polycarpa Benth. & Hook.f. === **Trewia nudiflora** L.
Trewia pubescens Sm. === **Macaranga** sp.
Trewia rusciflora B.Heyne ex Roth === **Mallotus** sp.
Trewia tricuspidata Willd. === **Mallotus paniculatus** (Lam.) Müll.Arg.

Triadica

2 species, E. & SE. Asia, west to Assam; conventionally recognised as a section of *Sapium* but restoration of generic rank, recommended by Esser (1994), is followed here. Laticiferous fast-growing trees to 24 m (but usually smaller), the thin leaves turning red before falling. *S. sebifera*, native of E. Asia, is now widely cultivated in warmer parts of the world. [Research on this genus is still in progress.] (Euphorbioideae (except Euphorbieae))

Pax, F. (with K. Hoffmann) (1912). *Sapium*. In A. Engler (ed.), Das Pflanzenreich, IV 147 V (Euphorbiaceae-Hippomaneae): 199-258. Berlin. (Heft 52.) La/Ge. — Nos. 70-73 (pp. 237-241), the majority of sect. *Triadica*, and 90 (pp. 252-253) pertain to *Triadica*.

Esser, H.-J. (1994). Systematische Studien an den Hippomaneae Adr. Jussieu ex Bartling (Euphorbiaceae) insbesondere den Mabeinae Pax et K. Hoffm. 305 pp., pp., illus., maps. Hamburg. (Unpubl. Ph.D. dissertation, Univ. of Hamburg.) Ge. — Includes justification for resurrection of *Triadica* (cf. pp. 61-62).

Esser, H.-J. (1999). A partial revision of the Hippomaneae (Euphorbiaceae) in Malesia. Blumea 44: 149-215, illus., maps. En. — *Triadica*, pp. 197-206; treatment of 2 species with key, descriptions, synonymy, references, citations, types, distribution, habitat, vernacular names, commentary (including uses), figures and maps; all general references, identification list and index to botanical names at end of paper. [Relates to the genus only with respect to Malesia.]

Triadica Lour., Fl. Cochinch.: 610 (1790).
Trop. & Subtrop. Asia. 36 38 40 41 42.
Stillingfleetia Bojer, Hortus Maurit.: 284 (1837).
Carumbium Kurz, Forest Fl. Burma 2: 411 (1877), nom. illeg.

Triadica cochinchinensis Lour., Fl. Cochinch.: 610 (1790). *Stillingia cochinchinensis* (Lour.) Baill., Adansonia 1: 351 (1861). *Excoecaria loureiroana* Müll.Arg. in A.P.de Candolle, Prodr. 15(2): 1217 (1866). *Sapium cochinchinense* (Lour.) Gagnep. in H.Lecomte, Fl. Indo-Chine 5: 401 (1926), nom. illeg. *Shirakia cochinchinensis* (Lour.) Hurus., J. Fac. Sci. Univ. Tokyo, Sect. 3, Bot. 6: 318 (1954).
China and India to Philippines. 36 CHC 40 ASS 41 VIE 42 BOR PHI. Phan.
Stillingia discolor Champ. ex Benth., Hooker's J. Bot. Kew Gard. Misc. 6: 1 (1854). *Sapium discolor* (Champ. ex Benth.) Müll.Arg., Linnaea 32: 121 (1863). *Excoecaria discolor* (Champ. ex Benth.) Müll.Arg. in A.P.de Candolle, Prodr. 15(2): 1210 (1866), nom. illeg.
Stillingia lanceolaria Miq., Fl. Ned. Ind., Eerste Bijv.: 461 (1861). *Excoecaria lanceolaria* (Miq.) Müll.Arg. in A.P.de Candolle, Prodr. 15(2): 1221 (1866).
Sapium eugeniifolium Buch.-Ham. ex Hook.f., Fl. Brit. India 5: 470 (1888).
Sapium hookeri Hook.f., Fl. Brit. India 5: 470 (1888).
Sapium laui Croizat, J. Arnold Arbor. 21: 505 (1940).
Sapium discolor var. *wenhsienensis* S.B.Ho, Fl. Tsinlingensis 1(3): 451 180 (1981).

Triadica sebifera (L.) Small, Florida Trees: 59 (1913).
China, Taiwan, Japan. 36 CHC CHN CHS 38 JAP KOR. Phan. – Widely grown in warmer parts of the world. The leaves in autumn turn a beautiful colour.
* *Croton sebiferum* L., Sp. Pl.: 1004 (1753). *Stillingia sebifera* (L.) Michx., Fl. Bor.-Amer. 2: 213 (1803). *Sapium sebiferum* (L.) Roxb., Fl. Ind. ed. 1832, 3: 693 (1832). *Excoecaria sebifera* (L.) Müll.Arg. in A.P.de Candolle, Prodr. 15(2): 1210 (1866).
Triadica sinensis Lour., Fl. Cochinch.: 610 (1790). *Stillingia sinensis* (Lour.) Baill., Étude Euphorb.: 512 (1858).
Triadica chinensis Spreng., Syst. Veg. 1: 93 (1824), sphalm.
Croton macrocarpus Rchb. ex Müll.Arg. in A.P.de Candolle, Prodr. 15(2): 698 (1866).
Sapium chihsinianum S.K.Lee, Acta Phytotax. Sin. 5: 12 (1956).
Sapium pleiocarpum Y.C.Tseng, Acta Phytotax. Sin. 20: 105 (1982).

Sapium sebiferum var. *cordatum* S.Y.Wang, in Fl. Henan 2: 480 (1988).
Sapium sebiferum var. *dabeshanense* B.C.Ding & T.B.Chao, in Fl. Henan 2: 481 (1988).
Sapium sebiferum var. *multiracemosum* B.C.Ding & T.B.Chao, in Fl. Henan 2: 480 (1988).
Sapium sebiferum var. *pendulum* B.C.Ding & T.B.Chao, in Fl. Henan 2: 481 (1988).

Synonyms:
Triadica chinensis Spreng. === **Triadica sebifera** (L.) Small
Triadica japonica (Siebold & Zucc.) Baill. === **Neoshirakia japonica** (Siebold & Zucc.) Esser
Triadica sinensis Lour. === **Triadica sebifera** (L.) Small

Tricarium

Synonyms:
Tricarium Lour. === **Phyllanthus** L.
Tricarium cochinchinense Lour. === **Phyllanthus acidus** (L.) Skeels

Tricherostigma

An orthographic variant of *Trichosterigma*.

Trichosterigma

A Klotzsch segregate of *Euphorbia*; the name remains in use within subgen. *Agaloma*.

Synonyms:
Trichosterigma Klotzsch & Garcke === **Euphorbia** L.

Tridesmis

Synonyms:
Tridesmis Lour. === **Croton** L.
Tridesmis tomentosa Lour. === **Croton crassifolius** Geiseler

Trigonopleura

3 species, Sumatra through Peninsular Malaysia to Borneo and the Philippines, with one record of *T. malayana* in central Sulawesi; forest trees to 30 m on well-drained ground with distichously arranged leaves and axillary inflorescences. *T. macrocarpa* is limited to the Semengoh Forest Reserve near Kuching, Sarawak and *T. dubia* to the eastern Visayas and NE Mindanao. The genus is closely related to the more widely distributed *Chaetocarpus* with one of its species (or its ancestor) possibly being ancestral (van Welzen, 1995). (Acalyphoideae)

Pax, F. (with K. Hoffmann) (1911). *Trigonopleura*. In A. Engler (ed.), Das Pflanzenreich, IV 147 III (Euphorbiaceae-Cluytieae): 95-96. Berlin. (Heft 47.) La/Ge. — 1 species, Peninsular Malaysia. Additions in ibid., XIV (Additamentum VI): 42 (1919).

Airy-Shaw, H. K. (1981). Notes on Asiatic, Malesian and Melanesian Euphorbiaceae, CCLI. *Trigonopleura* Hook. f. Kew Bull. 36: 610-611. En. — Description of *T. macrocarpa* from Sarawak.

• Welzen, P. C. van, L. J. Bulalacao & Tran Van Ôn (1995). A taxonomic revision of the Malesian genus *Trigonopleura* Hook. f. (Euphorbiaceae). Blumea 40: 363-374, illus., maps. En. — Descriptive revision (3 species) with key, synonymy, references, types, indication of distribution, habitat and ecology, vernacular names, and notes; identification list and list of references at end but no separate index. The general part includes a comparison of the genus with *Chaetocarpus* and phylogenetic analyses (with one conclusion pointing towards their possible union under *Chaetocarpus*). The species are mutually very closely related.

Trigonopleura Hook.f., Fl. Brit. India 5: 399 (1887).
   Malesia. 42.
      Peniculifera Ridl., J. Straits Branch Roy. Asiat. Soc. 82: 173 (1920).

Trigonopleura dubia (Elmer) Merr., Philipp. J. Sci., C 11: 77 (1916).
   EC. Philippines. 42 PHI. Phan.
      * *Alsodeia dubia* Elmer, Leafl. Philipp. Bot. 8: 2875 (1915).
      Trigonopleura philippinensis Merr., Philipp. J. Sci., C 10: 275 (1915).

Trigonopleura macrocarpa Airy Shaw, Kew Bull. 36: 610 (1981).
   Borneo (S. Sarawak). 42 BOR. Phan.

Trigonopleura malayana Hook.f., Fl. Brit. India 5: 399 (1887).
   Pen. Malaysia, Sumatera (incl. Bangka), Borneo, Sulawesi. 42 BOR MLY SUL SUM. Phan.
      Trigonopleura borneensis Merr., Philipp. J. Sci., C 11: 76 (1916).

Synonyms:
Trigonopleura borneensis Merr. === **Trigonopleura malayana** Hook.f.
Trigonopleura philippinensis Merr. === **Trigonopleura dubia** (Elmer) Merr.

Trigonostemon

92 species, S. China and S. and SE. Asia to Malesia (including the Philippines and New Guinea), New Caledonia and northern Australia; includes *Poilaniella* and *Tritaxis*. Shrubs or small trees of forest, some at least with red sap, the leaves spirally arranged to almost opposite and often clustered at branchlet ends. At least two species are litter-accumulators, with adventitious roots: *TT. detritiferus* and *wetriifolius*. A new revision of these interesting rosette-trees is much needed; the present list merely represents a status quo and it is quite possible several names will be reduced in future. Subsequent to the revisions by Pax (1911) and (in part) by Jablonski (1963) many additions were made by Airy-Shaw; more recently, further additions and changes have come from R. Milne. Pax & Hoffmann (1931) were of the opinion that the genus 'sicherlich ist .. in mehrere Gattungen aufzuspalten' but their argument for dismemberment has not found favour. Milne (1995) has indicated that the time-delay between development of female and male flowers varies both within and between species and may cause an individual collection to appear unisexual when actually this is but an artefact. The generic name is conserved against *Enchidium* Jack. (Crotonoideae)

   Pax, F. (1910). *Tritaxis*. In A. Engler (ed.), Das Pflanzenreich, IV 147 [I] (Euphorbiaceae-Jatropheae): 113-114. Berlin. (Heft 42.) La/Ge. — 3 species, India, SE Asia, Philippines. [Now merged with *Trigonostemon*.]
   Pax, F. (with K. Hoffmann) (1911). *Trigonostemon*. In A. Engler (ed.), Das Pflanzenreich, IV 147 III (Euphorbiaceae-Cluytieae): 85-95. Berlin. (Heft 47.) La/Ge. — 18 species in 4 sections, SE Asia and Malesia; two additional species of uncertain position. [No. 20 may be in *Cleidion*.]
   Pax, F. (with K. Hoffmann) (1914). *Trigonostemon*. In A. Engler (ed.), Das Pflanzenreich, IV 147 VII [Euphorbiaceae-Additamentum V]: 406-408. Berlin. (Heft 63.) La/Ge. — 6 species added.
   Pax, F. & K. Hoffmann (1931). *Trigonostemon*. In A. Engler (ed.), Die natürlichen Pflanzenfamilien, 2. Aufl., 19c: 169-170. Leipzig. Ge. — Synopsis with description of genus; 5 sections with indication of selected species. [Grouped with the genus, but not reduced, were *Actephilopsis* Ridl., *Prosartema* Gagnep. and *Poilaniella* Gagnep. All of these have since been reduced, the last by Airy-Shaw (1978).]
   Jablonski, E. (1963). Revision of *Trigonostemon* (Euphorbiaceae) of Malaya, Sumatra and Borneo. Brittonia 15: 151-168. En. — Synoptic treatment of 27 species (15 imperfectly known; one doubtfully in genus) with descriptions of novelties and localities with exsiccatae; one additional species, *T. asahanensis*, almost certainly to be excluded. A

section key is provided but not species keys. [Based only on material in NY and from SING. The species are distributed across three sections.]

Airy-Shaw, H. K. (1966). Notes on Malaysian and other Asiatic Euphorbiaceae, LXV. Two new species of *Trigonostemon*. Kew Bull. 20: 47-49. En. — 2 novelties described.

Airy-Shaw, H. K. (1968). Notes on Malesian and other Asiatic Euphorbiaceae, XCV. A new *Trigonostemon* from Borneo. Kew Bull. 21: 407-408. En. — Description of *T. ionthocarpus*.

Airy-Shaw, H. K. (1969). Notes on Malesian and other Asiatic Euphorbiaceae, CXVI. Notes on *Tritaxis* Baill. and *Dimorphocalyx* Thw. Kew Bull. 23: 123-126. En. —*Tritaxis* reduced to *Trigonostemon* with transfer of *T. gaudichaudii*; transfers of three other *Tritaxis* species to *Dimorphocalyx*; key differentiating *Dimorphocalyx* and *Trigonostemon*.

Airy-Shaw, H. K. (1969). Notes on Malesian and other Asiatic Euphorbiaceae, CXVII. Notes on *Trigonostemon aurantiacus* Boerl. Kew Bull. 23: 126-128. En. — Clarification of species; reductions of some other names and citation of all material at K with distribution.

Airy-Shaw, H. K. (1971). Notes on Malesian and other Asiatic Euphorbiaceae, CXLIV. New or noteworthy species of *Trigonostemon* Bl. Kew Bull. 25: 545-550. En. — Treatment of 5 species, including novelties.

Airy-Shaw, H. K. (1978). Notes on Malesian and other Asiatic Euphorbiaceae, CCV. New taxa and new names in *Trigonostemon* Bl. Kew Bull. 32: 415-418. En. — Treatment of 6 species (one new); includes reductions of *Poilaniella* and *Prosartema*.

Airy-Shaw, H. K. and F. S. P. Ng (1978). *Trigonostemon wetriifolius*, a new species from Endau-Rompin. Malays. Forester 41: 237-240. En. — Novelty; from southern Peninsular Malaysia.

Airy-Shaw, H. K. (1979). Notes on Malesian and other Asiatic Euphorbiaceae, CCXXVI. *Trigonostemon* Bl. Kew Bull. 33: 534-536. En. — Descriptions of 2 new species from New Guinea.

Milne, R. (1994). New species of, and notes on, Bornean *Trigonostemon*, *Cleistanthus* and *Macaranga* (Euphorbiaceae). Kew Bull. 50: 25-49, illus. En. — Includes descriptions of two new *Trigonostemon*, one, *T. detritiferus*, a litter-accumulator.

Milne, R. (1995). *Trigonostemon magnificum* (Euphorbiaceae), a new species from Sumatra. Kew Bull. 50: 51-53, illus. En. — Novelty.

Milne, R. (1995). Notes on Bornean and other West Malesian *Trigonostemon* (Euphorbiaceae). Kew Bull. 50: 25-49, illus. En. — General remarks on the genus, mainly on inflorescence development with the comment that inflorescences always have potential to produce flowers of both sexes; descriptions of 2 new subspecies and 3 new varieties; key to Bornean species; miscellaneous species notes (Borneo and elsewhere). [A somewhat miscellaneous paper.]

Trigonostemon Blume, Bijdr.: 600 (1826), nom. cons.
S. China, Trop. Asia, NE. Queensland, SW. Pacific. 36 40 41 42 43 50 60.
Enchidium Jack, Malayan Misc. 2(7): 89 (1822), nom. rejic.
Silvaea Hook. & Arn., Bot. Beechey Voy.: 211 (1837).
Athroisma Griff., Not. Pl. Asiat. 4: 477 (1854).
Telogyne Baill., Étude Euphorb.: 327 (1858).
Tritaxis Baill., Étude Euphorb.: 342 (1858).
Tylosepalum Kurz ex Teijsm. & Binn., Tijdschr. Ned.-Indië 27: 50 (1864).
Nepenthandra S.Moore, J. Bot. 43: 149 (1905).
Actephilopsis Ridl., Bull. Misc. Inform. Kew 1923: 360 (1923).
Prosartema Gagnep., Bull. Soc. Bot. France 71: 875 (1924 publ. 1925).
Poilaniella Gagnep., Bull. Soc. Bot. France 72: 467 (1925).
Neotrigonostemon Pax & K.Hoffm., Notizbl. Bot. Gart. Berlin-Dahlem 10: 385 (1928).
Kurziodendron N.P.Balakr., Bull. Bot. Surv. India 8: 68 (1966).

Trigonostemon adenocalyx Gagnep., Bull. Soc. Bot. France 69: 747 (1922 publ. 1923).
Vietnam. 41 VIE.

Trigonostemon albiflorus Airy Shaw, Kew Bull. 25: 547 (1971).
NW. Thailand. 41 THA. Nanophan.

Trigonostemon angustifolius Merr., Philipp. J. Sci. 20: 396 (1922).
Philippines. 42 PHI.

Trigonostemon apetalogyne Airy Shaw, Kew Bull. 33: 534 (1979).
Irian Jaya. 42 NWG. Nanophan.

Trigonostemon aurantiacus (Kurz ex Teijsm. & Binn.) Boerl., Handl. Fl. Ned. Ind. 3(1): 284 (1900).
Andaman Is., S. Thailand, Pen. Malaysia, Sumatera (Bangka), Jawa. 41 AND THA 42 JAW MLY SUM. Nanophan.
* *Tylosepalum aurantiacum* Kurz ex Teijsm. & Binn., Tijdschr. Ned.-Indië 27: 50 (1864).
  Codiaeum aurantiacum (Kurz ex Teijsm. & Binn.) Müll.Arg. in A.P.de Candolle, Prodr. 15(2): 1118 (1866).

var. **aurantiacus**
S. Thailand, Pen. Malaysia, Sumatera (Bangka), Jawa. 41 THA 42 JAW MLY SUM. Nanophan.
Actephilopsis malayana Ridl., Bull. Misc. Inform. Kew 1923: 361 (1923). *Trigonostemon malayanus* (Ridl.) Airy Shaw, Kew Bull. 20: 413 (1966).

var. **rubriflorus** N.P.Balakr. & Chakrab., J. Econ. Taxon. Bot. 5: 169 (1984).
Andaman Is. 41 AND. Nanophan.

Trigonostemon beccarii Ridl., Bull. Misc. Inform. Kew 1925: 89 (1925).
Sumatera (incl. Sib I.). 42 SUM. Nanophan.
Trigonostemon longisepalus Ridl., Bull. Misc. Inform. Kew 1925: 89 (1925).

Trigonostemon birmanicus Chakrab. & N.P.Balakr., J. Econ. Taxon. Bot. 5: 175 (1984).
Burma. 41 BMA.

Trigonostemon bonianus Gagnep., Bull. Soc. Bot. France 69: 747 (1922 publ. 1923).
Vietnam. 41 VIE.

Trigonostemon capillipes (Hook.f.) Airy Shaw, Kew Bull. 20: 413 (1966).
S. Thailand, Pen. Malaysia (Perlis, incl. Singapore), Borneo (Sarawak), Philippines (Surigao). 41 THA 42 BOR MLY PHI. Phan.
* *Dimorphocalyx capillipes* Hook.f., Fl. Brit. India 5: 405 (1887).
  Trigonostemon pachyphyllus Airy Shaw, Kew Bull. 25: 546 (1971).
  Trigonostemon diffusus subsp. *condensus* R.Milne, Kew Bull. 50: 47 (1995).

Trigonostemon capitellatum Gagnep., Bull. Soc. Bot. France 69: 748 (1922 publ. 1923).
Laos, Vietnam. 41 LAO VIE.

Trigonostemon carnosulus Airy Shaw, Kew Bull. 32: 415 (1978).
Pen. Malaysia. 42 MLY. Nanophan. or phan.

Trigonostemon cherrieri Veillon, Bull. Mus. Natl. Hist. Nat., B, Adansonia 14: 55 (1992).
New Caledonia. 60 NWC.

Trigonostemon chinensis Merr., Philipp. J. Sci. 21: 498 (1922).
China (S. Guangxi, Hainan). 36 CHH CHS. Nanophan. or phan.
Trigonostemon fungii Merr., Lingnan Sci. J. 11: 47 (1932). *Trigonostemon chinensis* f. *fungii* (Merr.) Y.T.Chang, Acta Phytotax. Sin. 27: 149 (1989).

Trigonostemon cochinchinensis Gagnep., Bull. Soc. Bot. France 69: 748 (1922 publ. 1923).
Vietnam. 41 VIE.

Trigonostemon croceus B.C.Stone, Malaysian Forester 43: 249 (1980).
Pen. Malaysia. 42 MLY.

Trigonostemon detritiferus R.Milne, Kew Bull. 49: 446 (1994).
Borneo (Brunei). 42 BOR. Nanophan.

Trigonostemon diffusus Merr., Sarawak Mus. J. 3: 525 (1928).
Borneo (Sarawak). 42 BOR.

Trigonostemon diplopetalus Thwaites, Enum. Pl. Zeyl.: 277 (1861). – FIGURE, p. 1568.
Sri Lanka. 40 SRL. Nanophan. or phan.

Trigonostemon dipteranthus Airy Shaw, Kew Bull. 20: 47 (1966).
W. Sumatera. 42 SUM. Phan.

Trigonostemon eberhardtii Gagnep., Bull. Soc. Bot. France 69: 749 (1922 publ. 1923).
Vietnam. 41 VIE.

Trigonostemon elegantissimus Airy Shaw, Kew Bull. 20: 48 (1966). *Trigonostemon viridissimus* var. *elegantissimus* (Airy Shaw) Airy Shaw, Kew Bull., Addit. Ser. 4: 206 (1975).
Borneo. 42 BOR.

Trigonostemon elmeri Merr., Univ. Calif. Publ. Bot. 15: 162 (1929).
Borneo (Sabah). 42 BOR. Phan.

Trigonostemon everettii Merr., Philipp. J. Sci., C 7: 408 (1912 publ. 1913).
Philippines. 42 PHI.

Trigonostemon filiformis Quisumb., Philipp. J. Sci. 41: 328 (1930).
Philippines (Luzon). 42 PHI.

Trigonostemon filipes Y.T.Chang & X.L.Mo, Acta Phytotax. Sin. 27: 149 (1989).
China (E. Yunnan, SW. Guangxi). 36 CHC CHS. Nanophan.

Trigonostemon flavidus Gagnep., Bull. Soc. Bot. France 69: 749 (1922 publ. 1923).
Laos. 41 LAO.

Trigonostemon fragilis (Gagnep.) Airy Shaw, Kew Bull. 32: 415 (1978).
Vietnam. 41 VIE.
   Poilaniella fragilis Gagnep., Bull. Soc. Bot. France 72: 467 (1925).

Trigonostemon gagnepainianus Airy Shaw, Kew Bull. 32: 415 (1978).
Vietnam. 41 VIE.
   Prosartema gaudichaudii Gagnep., Bull. Soc. Bot. France 72: 468 (1925).

Trigonostemon gaudichaudii Müll.Arg., Linnaea 34: 213 (1865).
Vietnam. 41 VIE.

Trigonostemon harmandii Gagnep., Bull. Soc. Bot. France 69: 750 (1922 publ. 1923).
Cambodia. 41 CBD.

Trigonostemon hartleyi Airy Shaw, Kew Bull. 33: 535 (1979).
NE. Papua New Guinea. 42 NWG. Nanophan. or phan.

Trigonostemon heteranthus Wight, Icon. Pl. Ind. Orient. 5: t. 1890 (1852).
Indo-China, W. Malesia. 41 BMA THA 42 BOR MLY SUM. Phan.
      Trigonostemon longifolius Baill., Étude Euphorb.: 341 (1858), nom. nud.
      Pseudotrewia cuneifolia Miq., Fl. Ned. Ind., Eerste Bijv.: 462 (1861). *Alchornea cuneifolia*
         (Miq.) Müll.Arg. in A.P.de Candolle, Prodr. 15(2): 900 (1866). *Wetria cuneifolia* (Miq.)
         Pax in H.G.A.Engler & K.A.E.Prantl, Nat. Pflanzenfam. 3(5): 57 (1890).

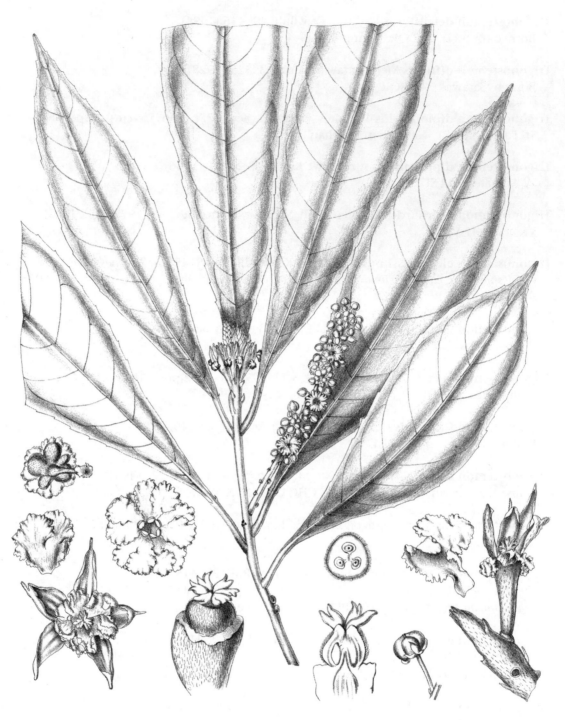

Trigonostemon diplopetalus Thwaites
Artist: W. de Alwis
Trimen, Handb. Fl. Ceylon, pl. 83 (1898), uncoloured state
KEW ILLUSTRATIONS COLLECTION

Alchornea cuneata Miq., Ann. Mus. Bot. Lugduno-Batavi 4: 122 (1869). *Trigonostemon sanguineus* Ridl., Bull. Misc. Inform. Kew 1926: 80 (1926). *Trigonostemon ridleyi* Merr. ex Jabl., Brittonia 15: 165 (1963).

Trigonostemon heterophyllus Merr., Lingnan Sci. J. 9: 38 (1930).
Hainan. 36 CHH. Nanophan.

Trigonostemon hirsutus C.B.Rob., Philipp. J. Sci., C 6: 335 (1911).
Philippines, Borneo (Sabah, Sarawak). 42 BOR PHI. Phan.

Trigonostemon oblongifolius Merr., Philipp. J. Sci., C 7: 409 (1912 publ. 1913).
Trigonostemon luzoniensis Merr., Philipp. J. Sci. 16: 568 (1920).

Trigonostemon howii Merr. & Chun, Sunyatsenia 2: 262 (1935).
Hainan. 36 CHH. Nanophan. or phan.

Trigonostemon huangmosu Y.T.Chang, Guihaia 3: 174 (1983).
China (SE. Yunnan). 36 CHC. Nanophan.

Trigonostemon hybridus Gagnep., Bull. Soc. Bot. France 69: 750 (1922 publ. 1923).
Cambodia. 41 CBD.

Trigonostemon inopinatus Airy Shaw, Kew Bull. 31: 396 (1976).
Queensland (South Kennedy). 50 QLD. Nanophan. or phan.

Trigonostemon ionthocarpus Airy Shaw, Kew Bull. 31: 407 (1968).
Borneo (Sarawak ?, Sabah). 42 BOR. Nanophan. or phan.

Trigonostemon kerrii Craib, Bull. Misc. Inform. Kew 1924: 97 (1924).
NC. Thailand. 41 THA. Nanophan. or phan. – Related to *T. murtonii*.

Trigonostemon kwangsiensis Hand.-Mazz., Sinensia 2: 130 (1932).
China (Guangxi). 36 CHS.
  Trigonostemon kwangsiensis var. *viridulus* H.S.Kiu, Guihaia 12: 210 (1992).

Trigonostemon laetus Baill., Étude Euphorb.: 341 (1858).
Burma. 41 BMA. Nanophan.

Trigonostemon laevigatus Müll.Arg., Flora 47: 538 (1864).
Andaman Is., S. Thailand, Pen. Malaysia, Sumatera, Borneo, Philippines. 41 AND THA 42
  BOR MLY PHI SUM. Nanophan. or phan.
  Trigonostemon anomalus Merr., Philipp. J. Sci. 16: 569 (1920).
  Trigonostemon petelotii Merr., Univ. Calif. Publ. Bot. 10: 425 (1924).
  Trigonostemon laevigatus var. *petiolaris* Airy Shaw, Kew Bull. 32: 417 (1978).

Trigonostemon lanceolatus (S.Moore) Pax in H.G.A.Engler, Pflanzenr., IV, 147, III:
92 (1911).
Burma. 41 BMA.
  * Nepenthandra lanceolata S.Moore, J. Bot. 43: 149 (1905).

Trigonostemon laoticus Gagnep., Bull. Soc. Bot. France 69: 751 (1922 publ. 1923).
Laos. 41 LAO.

Trigonostemon laxiflorus Merr., Philipp. J. Sci. 16: 567 (1920).
Philippines. 42 PHI.

Trigonostemon leucanthus Airy Shaw, Kew Bull. 25: 548 (1971).
S. China, Hainan, Thailand. 36 CHH CHS 41 THA. Nanophan.

var. **hainanensis** H.S.Kiu, Guihaia 12: 211 (1992).
  Hainan. 36 CHH.

var. **leucanthus**
  China (S. Guangxi). 36 CHS. Nanophan.

var. **siamensis** H.S.Kiu, Guihaia 12: 211 (1992).
  Thailand. 41 THA.

Trigonostemon lii Y.T.Chang, Guihaia 3: 175 (1983).
China (S. Yunnan). 36 CHC. Nanophan. or phan.

Trigonostemon longipedunculatus (Elmer) Elmer, Leafl. Philipp. Bot. 4: 1306 (1911).
Philippines, Borneo. 42 BOR PHI. Nanophan. or phan.
  * *Croton longipedunculatus* Elmer, Leafl. Philipp. Bot. 1: 311 (1908).

  var. **longipedunculatus**
    Philippines, Borneo. 42 BOR PHI. Nanophan. or phan.

  var. **mollis** R.I.Milne, Kew Bull. 50: 40 (1995).
    Borneo (Sabah), Philippines. 42 BOR PHI. Nanophan. or phan.

Trigonostemon longipes (Merr.) Merr., Philipp. J. Sci., C 11: 191 (1916).
Philippines. 42 PHI. Phan.
  * *Dimorphocalyx longipes* Merr., Philipp. J. Sci. 1(Suppl.): 82 (1906).

Trigonostemon lutescens Y.T.Chang & J.Y.Liang, Guihaia 3: 173 (1983).
China (S. Guangxi). 36 CHS. Nanophan.

Trigonostemon magnificus R.I.Milne, Kew Bull. 50: 51 (1995).
N. Sumatera. 42 SUM. Nanophan. or phan.

Trigonostemon malaccanus Müll.Arg., Flora 47: 482 (1864).
Pen. Malaysia, N. Sumatera. 42 MLY SUM. Nanophan. or phan.

Trigonostemon matangensis R.I.Milne, Kew Bull. 49: 445 (1994).
Borneo (SW. Sarawak). 42 BOR. Nanophan. or phan.

Trigonostemon matanginsu R.Milne, Kew Bull. 49: 445 (1994).
Borneo (Sarawak). 42 BOR.

Trigonostemon merrillii Elmer, Leafl. Philipp. Bot. 4: 1304 (1911).
Philippines, Borneo (NE. Sabah). 42 BOR PHI.

Trigonostemon murtonii Craib, Bull. Misc. Inform. Kew 1911: 464 (1911).
SE. Thailand, Vietnam. 41 THA VIE. Nanophan. or phan.
    Trigonostemon pinnatus Gagnep., Bull. Soc. Bot. France 64: 752 (1922 publ. 1923).

Trigonostemon nemoralis Thwaites, Enum. Pl. Zeyl.: 277 (1861).
S. India, Sri Lanka. 40 IND SRL. Nanophan. or phan.

Trigonostemon nigrifolius N.P.Balakr. & Chakrab., J. Econ. Taxon. Bot. 5: 173 (1984).
Burma. 41 BMA.

Trigonostemon oblanceolatus C.B.Rob., Philipp. J. Sci., C 6: 337 (1911).
Philippines. 42 PHI.

Trigonostemon pentandrus Pax & K.Hoffm. in H.G.A.Engler, Pflanzenr., IV, 147, VII: 406 (1914).
Pen. Malaysia. 42 MLY. Nanophan.

Trigonostemon philippinensis Stapf, Leafl. Philipp. Bot. 1: 206 (1907).
Sumatera, Borneo, Philippines. 42 BOR PHI SUM. Phan.

Trigonostemon phyllocalyx Gagnep., Bull. Soc. Bot. France 72: 469 (1925).
SE. Thailand, Vietnam. 41 THA VIE. Phan.

Trigonostemon pierrei Gagnep., Bull. Soc. Bot. France 64: 752 (1922 publ. 1923).
Cambodia (Phu-quoc I.), Borneo (Sarawak ?, Brunei ?). 41 CBD 42 BOR? Phan.

Trigonostemon poilanei Gagnep., Bull. Soc. Bot. France 64: 753 (1922 publ. 1923).
Vietnam. 41 VIE.

Trigonostemon polyanthus Merr., Philipp. J. Sci., C 9: 492 (1914 publ. 1915).
Philippines, Borneo. 42 BOR PHI. Nanophan. or phan.

    var. **calcicola** R.I.Milne, Kew Bull. 50: 33 (1995).
      Borneo (Sarawak). 42 BOR. Nanophan.
      Trigonostemon borneensis Merr., Univ. Calif. Publ. Bot. 15: 162 (1929).

    var. **lychnos** R.Milne, Kew Bull. 50: 35 (1995).
      Borneo (Brunei). 42 BOR.

    var. **polyanthus**
      Philippines, Borneo. 42 BOR PHI. Nanophan. or phan.
      Dimorphocalyx borneensis Merr., Philipp. J. Sci., C 11: 73 (1916). *Trigonostemon villosus*
        var. *borneensis* (Merr.) Airy Shaw, Kew Bull., Addit. Ser. 4: 205 (1975).
      Trigonostemon acuminatus Merr., Philipp. J. Sci., C 11: 190 (1916).
      Trigonostemon merrillianus Airy Shaw, Kew Bull. 25: 549 (1971).

Trigonostemon praetervisus Airy Shaw, Kew Bull. 37: 121 (1982).
Assam. 40 ASS.

Trigonostemon quocensis Gagnep., Bull. Soc. Bot. France 69: 753 (1922 publ. 1923).
Indo-China. 41 CBD THA VIE. Nanophan. or phan.

Trigonostemon reidioides (Kurz) Craib, Bull. Misc. Inform. Kew 1911: 464 (1911).
Burma, Thailand, Cambodia, Laos, Vietnam. 41 BMA CBD LAO THA VIE. Nanophan.
   * *Baliospermum reidioides* Kurz, Flora 58: 32 (1875).

Trigonostemon rubescens Gagnep., Bull. Soc. Bot. France 69: 754 (1922 publ. 1923).
Laos, Cambodia. 41 CBD LAO.

Trigonostemon rufescens Jabl., Brittonia 15: 152 (1963).
Pen. Malaysia. 42 MLY.

Trigonostemon salicifolius Ridl., Bull. Misc. Inform. Kew 1923: 366 (1923). *Trigonostemon*
verticillatus var. *salicifolius* (Ridl.) Whitmore, Gard. Bull. Singapore 26: 52 (1972).
Pen. Malaysia (Selangor). 42 MLY. Nanophan. or phan.

Trigonostemon sandakanensis Jabl., Brittonia 15: 159 (1963).
Borneo (Sabah). 42 BOR. Nanophan.

Trigonostemon sanguineus Gagnep., Bull. Soc. Bot. France 72: 470 (1925).
Vietnam. 41 VIE.

Trigonostemon semperflorens (Roxb.) Müll.Arg. in A.P.de Candolle, Prodr. 15(2): 1110 (1866).
Assam. 40 ASS. Nanophan.
  * *Clutia semperflorens* Roxb., Fl. Ind. ed. 1832, 3: 730 (1832). *Silvaea semperflorens* (Roxb.)
    Hook. & Arn., Bot. Beechey Voy.: 211 (1837).
  Agyneia ciliata Wall., Numer. List: 7852 (1847).
  Agyneia tetrandra Wall., Numer. List: 7951 (1847), nom. nud.
  Silvaea hookeriana Baill., Étude Euphorb.: 342 (1858). *Trigonostemon hookerianus* (Baill.)
    Müll.Arg. in A.P.de Candolle, Prodr. 15(2): 1109 (1866).
  Clutia sempervirens Müll.Arg. in A.P.de Candolle, Prodr. 15(2): 764 (1866).

Trigonostemon serratus Blume, Bijdr.: 600 (1826).
Jawa, Lesser Sunda Is. (Bali). 42 JAW LSI.

Trigonostemon sinclairii Jabl., Brittonia 15: 154 (1963).
Pen. Malaysia. 42 MLY.

Trigonostemon stellaris (Gagnep.) Airy Shaw, Kew Bull. 32: 415 (1978).
Vietnam. 41 VIE.
    * *Prosartema stellare* Gagnep., Bull. Soc. Bot. France 71: 875 (1924).

Trigonostemon stenophyllus Quisumb., Philipp. J. Sci. 41: 330 (1930).
Philippines (Luzon). 42 PHI.

Trigonostemon sunirmalii Chakrab. & N.P.Balakr., J. Econ. Taxon. Bot. 5: 179 (1984).
Burma. 41 BMA.

Trigonostemon thorelii Gagnep., Bull. Soc. Bot. France 69: 755 (1922 publ. 1923).
Laos. 41 LAO.

Trigonostemon thyrsoidcus Stapf, Bull. Misc. Inform. Kew 1909: 264 (1909).
NW. Thailand, Vietnam, China (Guizhou, S. Yunnan, Guangxi). 36 CHC CHS 41 THA
    VIE. Nanophan. or phan.

Trigonostemon verrucosus J.J.Sm., Bull. Jard. Bot. Buitenzorg, III, 6: 97 (1924).
Jawa ? 42 JAW? – Described from plants cultivated in Hort. Bog.; shrub to 2 m.

Trigonostemon verticillatus (Jack) Pax in H.G.A.Engler, Pflanzenr., IV, 147, III: 87 (1911).
S. Thailand, Pen. Malaysia. 41 THA 42 MLY. Nanophan. or phan.
    * *Enchidium verticillatum* Jack, Malayan Misc. 2(7): 89 (1822).
    Telogyne indica Baill., Étude Euphorb.: 328 (1858). *Trigonostemon indicus* (Baill.)
        Müll.Arg., Linnaea 34: 214 (1865).

Trigonostemon villosus Hook.f., Fl. Brit. India 5: 397 (1887).
Nicobar Is., Pen. Malaysia, N. Sumatera ?, Borneo. 41 NCB 42 BOR MLY SUM? Nanophan.

subsp. **caesius** R.I.Milne, Kew Bull. 50: 37 (1995).
    Borneo (Sabah, Sarawak). 42 BOR.

var. **nicobaricus** (Chakrab.) N.P.Balakr. & Chakrab., Candollea 46: 629 (1991).
    Nicobar Is. 41 NCB.
    * *Trigonostemon nicobaricus* Chakrab., J. Econ. Taxon. Bot. 5: 203 (1984).

subsp. **villosus**
    Pen. Malaysia, N. Sumatera ? 42 MLY SUM? Nanophan.
    Trigonostemon tomentellus Pax & K.Hoffm. in H.G.A.Engler, Pflanzenr., IV, 147, III:
        89 (1911).

Trigonostemon viridissimus (Kurz) Airy Shaw, Kew Bull. 25: 545 (1971).
Andaman Is. to Lesser Sunda Is. 41 AND BMA 42 LSI MLY SUM. Nanophan. or phan.
    * *Sabia viridissima* Kurz, J. Asiat. Soc. Bengal, Pt. 2, Nat. Hist. 41(2): 304 (1872). *Blachia
        viridissima* (Kurz) King, J. Asiat. Soc. Bengal, Pt. 2, Nat. Hist. 65: 455 (1896).
    Trigonostemon ovatifolius J.J.Sm. ex Koord. & Valeton, Meded. Dept. Landb. Ned.-Indië
        10: 583 (1910).
    Trigonostemon membranaceus Pax & K.Hoffm. in H.G.A.Engler, Pflanzenr., IV, 147, III:
        91 (1911).
    Trigonostemon sumatranus Pax & K.Hoffm. in H.G.A.Engler, Pflanzenr., IV, 147, III:
        90 (1911).
    Trigonostemon macgregorii Merr., Philipp. J. Sci. 16: 566 (1920).

Trigonostemon chatterjii Deb & G.K.Deka, Indian Forester 91: 577 (1965). *Trigonostemon viridissimus* var. *chatterjii* (Deb & G.K.Deka) N.P.Balakr. & Chakrab., J. Econ. Taxon. Bot. 5: 967 (1984).

Trigonostemon viridissimus var. *confertifolius* N.P.Balakr. & N.G.Nair, Bull. Bot. Surv. India 24: 36 (1982 publ. 1983).

Trigonostemon voratus Croizat, Sargentia 1: 52 (1942).
Fiji. 60 FIJ.

Trigonostemon wenzelii Merr., Philipp. J. Sci., C 8: 380 (1913).
Philippines (Leyte). 42 PHI.

Trigonostemon wetriifolius Airy Shaw & Ng, Malaysian Forester 41: 237 (1978).
S. Pen. Malaysia. 42 MLY.

Trigonostemon whiteanus (Croizat) Airy Shaw, Kew Bull. 38: 68 (1983).
Philippines. 42 PHI.
* *Cheilosa whiteana* Croizat, J. Arnold Arbor. 23: 507 (1942).

Trigonostemon wui H.S.Kiu, J. Trop. Subtrop. Bot. 3: 19 (1995).
China (W. Guangdong). 36 CHS.

Trigonostemon xyphophylloides (Croizat) L.K.Dai & T.L.Wu, Acta Phytotax. Sin. 8: 278 (1963).
Hainan. 36 CHH. Nanophan.
* *Cleidion xyphophylloides* Croizat, J. Arnold Arbor. 21: 503 (1940).

Synonyms:

Trigonostemon acuminatus Merr. === **Trigonostemon polyanthus** Merr. var. **polyanthus**
Trigonostemon anomalus Merr. === **Trigonostemon laevigatus** Müll.Arg.
Trigonostemon arboreus Ridl. === **Omphalea malayana** Merr.
Trigonostemon asahanensis Croizat === **Dimorphocalyx muricatus** (Hook.f.) Airy Shaw
Trigonostemon beddomei (Benth.) N.P.Balakr. === **Dimorphocalyx beddomei** (Benth.) Airy Shaw
Trigonostemon borneensis Merr. === **Trigonostemon polyanthus** var. **calcicola** R.I.Milne
Trigonostemon bulusanensis Elmer === **Dimorphocalyx bulusanensis** (Elmer) Airy Shaw
Trigonostemon chatterjii Deb & G.K.Deka === **Trigonostemon viridissimus** (Kurz) Airy Shaw
Trigonostemon chinensis f. *fungii* (Merr.) Y.T.Chang === **Trigonostemon chinensis** Merr.
Trigonostemon cumingii Müll.Arg. === **Dimorphocalyx cumingii** (Müll.Arg.) Airy Shaw
Trigonostemon diffusus subsp. *condensus* R.Milne === **Trigonostemon capillipes** (Hook.f.) Airy Shaw
Trigonostemon forbesii Pax === **Wetria insignis** (Steud.) Airy Shaw
Trigonostemon fungii Merr. === **Trigonostemon chinensis** Merr.
Trigonostemon hookerianus (Baill.) Müll.Arg. === **Trigonostemon semperflorens** (Roxb.) Müll.Arg.
Trigonostemon indicus (Baill.) Müll.Arg. === **Trigonostemon verticillatus** (Jack) Pax
Trigonostemon kwangsiensis var. *viridulus* H.S.Kiu === **Trigonostemon kwangsiensis** Hand.-Mazz.
Trigonostemon laevigatus var. *petiolaris* Airy Shaw === **Trigonostemon laevigatus** Müll.Arg.
Trigonostemon lawianus Müll.Arg. === **Dimorphocalyx glabellus** var. **lawianus** (Müll.Arg.) Chakrab. & N.P.Balakr.
Trigonostemon longifolius Baill. === **Trigonostemon heteranthus** Wight
Trigonostemon longisepalus Ridl. === **Trigonostemon beccarii** Ridl.
Trigonostemon luzoniensis Merr. === **Trigonostemon hirsutus** C.B.Rob.
Trigonostemon macgregorii Merr. === **Trigonostemon viridissimus** (Kurz) Airy Shaw
Trigonostemon macrophyllus (Müll.Arg.) Müll.Arg. === **Paracroton pendulus** (Hassk.) Miq. subsp. **pendulus**

Trigonostemon malayanus (Ridl.) Airy Shaw === **Trigonostemon aurantiacus** (Kurz ex Teijsm. & Binn.) Boerl. var. **aurantiacus**

Trigonostemon membranaceus Pax & K.Hoffm. === **Trigonostemon viridissimus** (Kurz) Airy Shaw

Trigonostemon merrillianus Airy Shaw === **Trigonostemon polyanthus** Merr. var. **polyanthus**

Trigonostemon nicobaricus Chakrab. === **Trigonostemon villosus** var. **nicobaricus** (Chakrab.) N.P.Balakr. & Chakrab.

Trigonostemon oblongifolius Merr. === **Trigonostemon hirsutus** C.B.Rob.

Trigonostemon oliganthus K.Schum. === **Cleidion papuanum** Lauterb.

Trigonostemon ovatifolius J.J.Sm. ex Koord. & Valeton === **Trigonostemon viridissimus** (Kurz) Airy Shaw

Trigonostemon pachyphyllus Airy Shaw === **Trigonostemon capillipes** (Hook.f.) Airy Shaw

Trigonostemon petelotii Merr. === **Trigonostemon laevigatus** Müll.Arg.

Trigonostemon pinnatus Gagnep. === **Trigonostemon murtonii** Craib

Trigonostemon ridleyi Merr. ex Jabl. === **Trigonostemon heteranthus** Wight

Trigonostemon sanguineus Ridl. === **Trigonostemon heteranthus** Wight

Trigonostemon sumatranus Pax & K.Hoffm. === **Trigonostemon viridissimus** (Kurz) Airy Shaw

Trigonostemon tomentellus Pax & K.Hoffm. === **Trigonostemon villosus** Hook.f. subsp. **villosus**

Trigonostemon verticillatus var. *salicifolius* (Ridl.) Whitmore === **Trigonostemon salicifolius** Ridl.

Trigonostemon villosus var. *borneensis* (Merr.) Airy Shaw === **Trigonostemon polyanthus** Merr. var. **polyanthus**

Trigonostemon viridissimus var. *chatterjii* (Deb & G.K.Deka) N.P.Balakr. & Chakrab. === **Trigonostemon viridissimus** (Kurz) Airy Shaw

Trigonostemon viridissimus var. *confertifolius* N.P.Balakr. & N.G.Nair === **Trigonostemon viridissimus** (Kurz) Airy Shaw

Trigonostemon viridissimus var. *elegantissimus* (Airy Shaw) Airy Shaw === **Trigonostemon elegantissimus** Airy Shaw

Trigonostemon zeylanicus (Thwaites) Müll.Arg. === **Paracroton pendulus** subsp. **zeylanicus** (Thwaites) N.P.Balakr. & Chakrab.

Triplandra

Synonyms:
Triplandra Raf. === **Croton** L.

Tritaxis

Synonyms:
Tritaxis Baill. === **Trigonostemon** Blume
Tritaxis beddomei Benth. === **Dimorphocalyx beddomei** (Benth.) Airy Shaw
Tritaxis macrophylla Müll.Arg. === **Paracroton pendulus** (Hassk.) Miq. subsp. **pendulus**
Tritaxis zeylanica Müll.Arg. === **Paracroton zeylanicus** (Müll.Arg.) N.P.Balakr. & Chakrab.

Tumalis

Synonyms:
Tumalis Raf. === **Euphorbia** L.

Tylosepalum

Synonyms:
Tylosepalum Kurz ex Teijsm. & Binn. === **Trigonostemon** Blume
Tylosepalum aurantiacum Kurz ex Teijsm. & Binn. === **Trigonostemon aurantiacus** (Kurz ex Teijsm. & Binn.) Boerl.

Tyria

Synonyms:
Tyria Klotzsch === **Bernardia** Houst. ex Mill.
Tyria myricifolia Scheele === **Bernardia myricifolia** (Scheele) S.Watson

Tzellemtinia

Synonyms:
Tzellemtinia Chiov. === **Bridelia** Willd.
Tzellemtinia nervosa Chiov. === **Bridelia scleroneura** Müll.Arg. subsp. **scleroneura**

Uapaca

52 species, Africa and Madagascar (with c. 12 in the latter); 'pachycaulous' shrubs or small to large trees to 35 m or more occupying a diversity of wooded or savannoid habitats, but relatively prominent in seasonally dry (e.g. *Brachystegia*) forests. No full revision has appeared since 1922 when Pax and Hoffmann accepted 27 species. De Wildeman (1936) recognised 47 species for mainland Africa and essayed two keys as well as a full (but in part compiled) enumeration; for SE. Zaïre a further revision was made by Duvigneaud (1950). For Madagascar the most recent treatment is that of Leandri (1958). In Madagascar the tapia, *U. bojeri*, is a characteristic and symbolic highlands tree of considerable economic importance. Other species are isolated big forest trees with females bearing fruits near the base. Many are also of economic value, including timber. Adventitous (stilt) roots are sometimes present (though not in *U. bojeri*); De Wildeman (1936) furnishes 5 plates of different species exhibiting this feature. Buttresses in the genus (and in general) were also extensively discussed by De Wildeman (1936), with 13 African species featuring them or possessing a fluted trunk. Some authors (e.g. Airy-Shaw 1965; Meeuse 1990; see **General**) have proposed separate family status. Airy-Shaw singled out the habit, leaves (reminiscent of Anacardiaceae) and involucres; he also drew attention to an anatomy not known elsewhere in Euphorbiaceae s.l. (though foliar anatomy was used by Webster to justify its retention in the family). In any case the genus has no or but few close relatives in the Phyllanthoideae as traditionally circumscribed, being monotypic at subtribal or tribal rank. The species, however, are more often than not poorly understood. (Phyllanthoideae)

Perrier de la Bâthie, H. (1912). Le tapia. J. Agric. Trop. 12: 300-302. Fr. — Botany, ecology and biology in relation to long-term management and enhancement of stands; conclusion that bushfires are over a longer term harmful (serious damage may result wherever there has been a substantial buildup of herbage and underbrush).

Pax, F. & K. Hoffmann (1922). *Uapaca*. In A. Engler (ed.), Das Pflanzenreich, IV 147 XV (Euphorbiaceae-Phyllanthoideae-Phyllantheae): 298-311. Berlin. (Heft 81.) La/Ge. — 27 species; Africa (1-21) and Madagascar (22-27).

Denis, M. (1927). Observations sur les *Uapaca* malgaches et diagnoses d'espèces nouvelles. Arch. Bot. (Caen), Bull. Mens. 1: 223-231. Fr. — Novelties (7) and notes, additional to Pax and Hoffmann (1922); localities with exsiccatae included for all 12 species recognised but no key for identification. [Treatment succeeded by that of Leandri, 1958.]

• De Wildeman, É. (1936). Contributions à l'étude des espèces du genre *Uapaca* Baill. (Euphorbiacées). 192, [1] pp., illus., 5 pls. (Mém. Inst. Roy. Colon. Belge, Sci. Nat. Med., coll. 8°, 4(5)). Brussels. Fr. — Substantial descriptive treatment covering all of Africa; includes different kinds of keys (pp. 71-90), descriptions, synonymy, references and citations, localities with exsiccatae (Congo only), and usually extended commentary; index to all names at end. The revision is preceded by a rambling introduction ranging over many issues in tropical botany and ecology (not limited to *Uapaca*) along with a review of characters used in classification. A phytogeographical table for the genus covering all of Africa appears on pp. 58-60, followed by a lexicon of known vernacular names.

- Duvigneaud, P. (1950). Les «Uapaca» (Euphorbiacées) des forêts claires du Congo méridional. Bull. Inst. Roy. Colon. Belge 20: 863-892. Fr. — Introduction (with a note that established vernacular groupings have here some taxonomic value); running notes on Congo species with documentation; phytogeographical review (pp. 881-886) and formal synopsis (pp. 886-891) with key, names and synonymy with references and citations, Latin diagnoses of new taxa, and types; literature, p. 892. [An extension and revision of De Wildeman, 1936.]
- Leandri, J. (1958). *Uapaca*. Fl. Madag. Comores 111 (Euphorbiacées), I: 163-181. Paris. Fr. — Flora treatment (12 species); key. (Species limits ill-understood, according to the author.)
  Radcliffe-Smith, A. (1993). Notes on African Euphorbiaceae, XXIX: *Uapaca*. Kew Bull. 48: 611-617, illus. En. — Description of *U. lissopyrena* from SC. Africa; notes on *U. kirkiana*; historical survey of genus.

Uapaca Baill., Étude Euphorb.: 595 (1858).
  Trop. Africa, Madagascar. 22 23 24 25 26 29. Nanophan. or phan.
  Gymnocarpus Thouars ex Baill., Étude Euphorb.: 595 (1858).
  Aapaca Metzdorff, Just's Bot. Jahresber. 1885: 212 (1888).

Uapaca acuminata (Hutch.) Pax & K.Hoffm. in H.G.A.Engler, Pflanzenr., IV, 147, XV: 308 (1922).
  Nigeria, Cameroon. 22 NGA 23 CMN.
  * *Uapaca heudelotii* var. *acuminata* Hutch. in D.Oliver, Fl. Trop. Afr. 6(1): 636 (1912).

Uapaca ambanjensis Leandri, Mém. Inst. Sci. Madagascar, Sér. B, Biol. Vég. 8: 256 (1957).
  N. Madagascar. 29 MDG. Nanophan. or phan.

Uapaca amplifolia Denis, Arch. Bot. Bull. Mens. 1: 224 (1927).
  N. & W. Madagascar. 29 MDG. Phan.

Uapaca angustipyrena De Wild., Ann. Soc. Sci. Bruxelles, Sér. B 52: 205 (1932).
  Zaire. 23 ZAI.

Uapaca benguelensis Müll.Arg., J. Bot. 2: 332 (1864). *Uapaca benguelensis* f. *pilosa* P.A.Duvign., Bull. Inst. Roy. Colon. Belge 20: 890 (1949), nom. illeg.
  Angola, Zaire. 23 ZAI 26 ANG.
  Uapaca teusczii Pax, Bot. Jahrb. Syst. 19: 79 (1894).
  Uapaca angolensis Hutch. ex Pax & K.Hoffm. in H.G.A.Engler, Pflanzenr., IV, 147, XV: 301 (1922), pro syn.
  Uapaca benguelensis f. *glabra* P.A.Duvign., Bull. Inst. Roy. Colon. Belge 20: 890 (1949).
  Uapaca benguelensis var. *pedunculata* P.A.Duvign., Bull. Inst. Roy. Colon. Belge 20: 890 (1949).

Uapaca betamponensis Leandri, Mém. Inst. Sci. Madagascar, Sér. B, Biol. Vég. 8: 256 (1957).
  EC. Madagascar. 29 MDG. Phan.

Uapaca bojeri Baill., Adansonia 11: 176 (1874).
  Madagascar. 29 MDG. Phan. – A significant tree ('tapia') in the central highlands.
  Uapaca clusiacea Baker, J. Linn. Soc., Bot. 18: 278 (1881).
  Chorizotheca macrophylla Heckel, Ann. Inst. Bot.-Géol. Colon. Marseille: 215 (1910).

Uapaca bossenge De Wild., Ann. Mus. Congo Belge, Bot., V, 2: 271 (1908).
  – FIGURE, p. 1577.
  Zaire. 23 ZAI.

Uapaca brevipedunculata De Wild., Ann. Soc. Sci. Bruxelles, Sér. B 53: 145 (1933).
  Zaire. 23 ZAI.

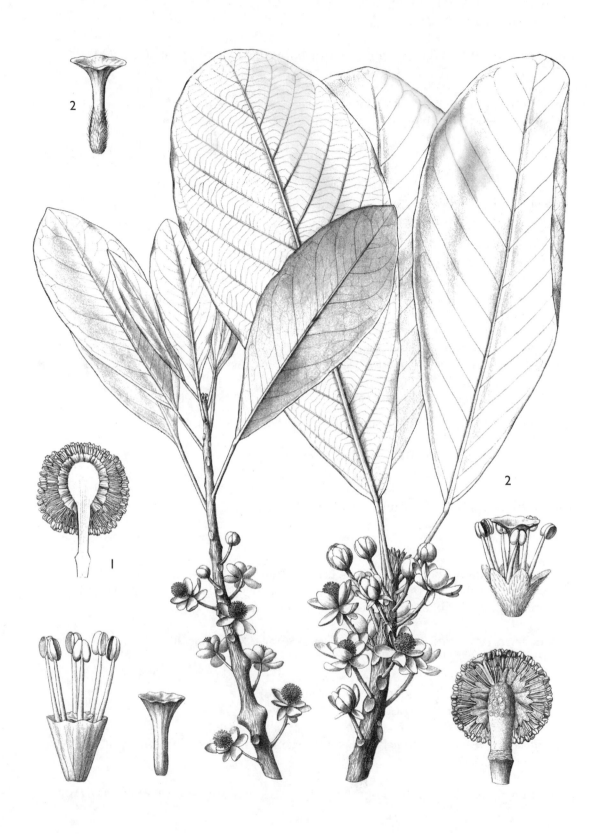

Uapaca bossenge De Wild. (1), *U. ealaensis* De Wild. (2)

Artist: A d'Apreval
Ann. Mus. Congo Belge, Bot., V, 2: pl. 70 (1908) (Études Fl. Bas-Moyen-Congo)
KEW ILLUSTRATIONS COLLECTION

Uapaca brieyi De Wild., Bull. Jard. Bot. État 4: 362 (1914).
Zaire (Mayumbe). 23 ZAI.

Uapaca casteelsii De Wild., Compt.-Rend. Hebd. Séances Mém. Soc. Biol. 96: 700 (1927).
Zaire. 23 ZAI.

Uapaca chevalieri Beille, Bull. Soc. Bot. France 55(8): 68 (1908).
Guinea, Sierra Leone, Ivory Coast. 22 GUI IVO SIE. Phan.

Uapaca corbisieri De Wild., Compt.-Rend. Hebd. Séances Mém. Soc. Biol. 96: 118 (1927).
Zaire. 23 ZAI. Phan.

Uapaca densifolia Baker, J. Linn. Soc., Bot. 20: 252 (1883).
Madagascar. 29 MDG. Nanophan. or phan.

> **var. densifolia**
> C. & E. Madagascar. 29 MDG. Nanophan. or phan.
> *Uapaca myricifolia* Baker, J. Linn. Soc., Bot. 21: 440 (1885).
> *Uapaca angustifolia* Denis, Arch. Bot. Bull. Mens. 1: 227 (1927). *Uapaca densifolia* var.
> *angustifolia* (Denis) Leandri, in Fl. Madag. 111: 176 (1958).
> *Uapaca perrieri* Denis, Arch. Bot. Bull. Mens. 1: 228 (1927). *Uapaca densifolia* var. *perrieri*
> (Denis) Leandri, in Fl. Madag. 111: 176 (1958).
> *Uapaca densifolia* var. *humbertii* Leandri, Mém. Inst. Sci. Madagascar, Sér. B, Biol. Vég. 8:
> 257 (1957).
> *Uapaca densifolia* f. *orientalis* Leandri, in Fl. Madag. 111: 174 (1958).

> **var. sambiranensis** (Denis) Leandri, Mém. Inst. Sci. Madagascar, Sér. B, Biol. Vég. 8:
> 356 (1957).
> NW. Madagascar. 29 MDG. Nanophan. or phan.
> * *Uapaca sambiranensis* Denis, Arch. Bot. Bull. Mens. 1: 226 (1927).

Uapaca ealaensis De Wild., Ann. Mus. Congo Belge, Bot., V, 2: 272 (1908).
  – FIGURE, p. 1577.
Zaire. 23 ZAI.

Uapaca esculenta A.Chev. ex Aubrév. & Leandri, Bull. Soc. Bot. France 82: 52 (1935).
Sierra Leone, Liberia, Ivory Coast, Ghana, Nigeria. 22 GHA IVO LBR NGA SIE. Phan. –
Fruits edible. – Probably identical with *U. corbisieri*.

Uapaca ferrarii De Wild., Compt.-Rend. Hebd. Séances Mém. Soc. Biol. 96: 117 (1927).
Zaire (Kasaï). 23 ZAI.

Uapaca ferruginea Baill., Étude Euphorb.: 596 (1858).
C. & E. Madagascar. 29 MDG. Phan.
> *Uapaca clusioides* Baker, J. Linn. Soc., Bot. 21: 441 (1885).
> *Uapaca ferruginea* var. *carnotiana* Leandri, Mém. Inst. Sci. Madagascar, Sér. B, Biol. Vég.
> 8: 257 (1957).

Uapaca goossensii De Wild., Compt.-Rend. Hebd. Séances Mém. Soc. Biol. 96: 118 (1927).
Zaire (Eala). 23 ZAI.

Uapaca gossweileri Hutch., Bull. Misc. Inform. Kew 1912: 101 (1912).
Angola, Zaire. 23 ZAI 26 ANG. Phan. – Close to *U. benguelensis*.

Uapaca guineensis Müll.Arg., Flora 47: 517 (1864).
Trop. Africa. 22 GHA IVO LBR NGA SIE 23 CAF CMN GAB GGI ZAI 24 SUD 25 TAN 26
MLW ZAM ZIM. Phan.

var. **guineensis**

  Trop. Africa. 22 GHA IVO LBR NGA SIE 23 CMN GAB GGI ZAI 25 TAN. Phan.

  Antidesma guineensis G.Don ex Hook., Niger Fl.: 515 (1849), nom. nud.

  Uapaca mole Pax, Bot. Jahrb. Syst. 19: 79 (1894).

  Uapaca bingervillensis Beille, Bull. Soc. Bot. France 55(8): 66 (1908).

  Uapaca laurentii De Wild., Ann. Mus. Congo Belge, Bot., V, 2: 272 (1908).

  Uapaca perrotii Beille, Bull. Soc. Bot. France 55(8): 67 (1908).

  Uapaca seretii De Wild., Ann. Mus. Congo Belge, Bot., V, 2: 274 (1908).

  Uapaca gabonensis Pierre ex Pax & K.Hoffm. in H.G.A.Engler, Pflanzenr., IV, 147, XV:
    306 (1922), pro syn.

var. **sudanica** (Beille) Hutch. in D.Oliver, Fl. Trop. Afr. 6(1): 641 (1912).

  Central African Rep., Sudan, Malawi, Zambia, Zimbabwe. 23 CAF 24 SUD 26 MLW
    ZAM ZIM. Phan.

  * *Uapaca guignardii* var. *sudanica* Beille, Bull. Soc. Bot. France 55(8): 67 (1908).

Uapaca heudelotii Baill., Adansonia 1: 81 (1860).

  W. & WC. Trop. Africa. 22 BEN GHA GUI IVO LBR MLI NGA SIE TOG 23 CMN EQG GAB
    ZAI. Phan.

  Uapaca marquesii Pax, Bot. Jahrb. Syst. 23: 522 (1897).

Uapaca katentaniensis De Wild., Ann. Soc. Sci. Bruxelles, Sér. B 53: 58 (1933).

  Zaire (Shaba). 23 ZAI. Phan. – Provisionally accepted.

Uapaca kibuatii De Wild., Ann. Soc. Sci. Bruxelles, Sér. B 53: 309 (1933).

  W. Zaire. 23 ZAI.

Uapaca kirkiana Müll.Arg., Flora 47: 517 (1864).

  Angola, Tanzania, Burundi, Zaire, Malawi, Zambia, Zimbabwe, Mozambique. 23 BUR ZAI
    25 TAN 26 ANG MLW MOZ ZAM ZIM. Phan.

  Uapaca goetzei Pax, Bot. Jahrb. Syst. 28: 418 (1900).

  Uapaca kirkiana var. *goetzei* Pax, Bot. Jahrb. Syst. 34: 370 (1904).

  Uapaca homblei De Wild., Ann. Soc. Sci. Bruxelles 45: 309 (1926).

  Uapaca munamensis De Wild., Ann. Soc. Sci. Bruxelles, Sér. B 52: 206 (1932).

  Uapaca albida De Wild., Ann. Soc. Sci. Bruxelles, Sér. B 53: 57 (1933).

  Uapaca dubia De Wild., Ann. Soc. Sci. Bruxelles, Sér. B 53: 307 (1933). *Uapaca kirkiana*
    var. *dubia* (De Wild.) P.A.Duvign., Bull. Inst. Roy. Colon. Belge 20: 891 (1949).

  Uapaca neomasuku De Wild., Ann. Soc. Sci. Bruxelles, Sér. B 52: 207 (1933).

  Uapaca kirkiana var. *kwangoensis* P.A.Duvign., Bull. Inst. Roy. Colon. Belge 20:
    891 (1949).

  Uapaca kirkiana var. *sessilifolia* P.A.Duvign., Bull. Inst. Roy. Colon. Belge 20:
    891 (1949).

  Uapaca greenwayi Suess., Trans. Rhodesia Sci. Assoc. 43: 85 (1951).

Uapaca lebrunii De Wild., Ann. Soc. Sci. Bruxelles, Sér. B 53: 146 (1933).

  Zaire (Eala). 23 ZAI.

Uapaca letestuana A.Chev., Veg. Ut. Afr. Trop. Franc. 9: 304 (1917).

  Gabon. 23 GAB.

Uapaca lissopyrena Radcl.-Sm., Kew Bull. 48: 612 (1993).

  SC. Trop. Africa. 26 MLW MOZ ZAM ZIM.

Uapaca littoralis Denis, Arch. Bot. Bull. Mens. 1: 228 (1927).

  E. Madagascar. 29 MDG. Phan.

Uapaca louvelii Denis, Arch. Bot. Bull. Mens. 1: 230 (1927).
E. Madagascar. 29 MDG. Phan.
Uapaca louvelii var. *paralia* Leandri, Mém. Inst. Sci. Madagascar, Sér. B, Biol. Vég. 8: 258 (1957).

Uapaca macrostipulata De Wild., Ann. Soc. Sci. Bruxelles, Sér. B 47: 36 (1927).
Zaire (Eala). 23 ZAI.

Uapaca mangorensis Leandri, Mém. Inst. Sci. Madagascar, Sér. B, Biol. Vég. 8: 258 (1957).
EC. Madagascar. 29 MDG. Nanophan. or phan.

Uapaca multinervata De Wild., Compt.-Rend. Hebd. Séances Mém. Soc. Biol. 96: 118 (1927).
Zaire (Eala). 23 ZAI.

Uapaca nitida Müll.Arg., Flora 47: 517 (1864).
Kenya, Tanzania, Malawi, Zambia, Zimbabwe, Mozambique, Angola, Burundi, Zaire. 23 BUR ZAI 25 KEN TAN 26 ANG MLW MOZ ZAM ZIM. Nanophan. or phan.

var. **longifolia** (P.A.Duvign.) Radcl.-Sm., Kew Bull. 48: 616 (1993).
Zaire to Mozambique. 23 ZAI 26 MLW MOZ ZIM.
* *Uapaca nitida* f. *longifolia* P.A.Duvign., Bull. Inst. Roy. Colon. Belge 20: 890 (1949).

var. **nitida**
Kenya, Tanzania, Malawi, Zambia, Zimbabwe, Mozambique, Angola, Burundi, Zaire. 23 BUR ZAI 25 KEN TAN 26 ANG MLW MOZ ZAM ZIM. Nanophan. or phan.
Uapaca microphylla Pax, Bot. Jahrb. Syst. 23: 523 (1897).
Uapaca microphylla var. *hendrickxii* De Wild., Études Fl. Bas- Moyen-Congo 2: 273 (1908).
Uapaca similis Pax & K.Hoffm. in H.G.A.Engler, Pflanzenr., IV, 147, XV: 305 (1922).
Uapaca nitida f. *aucta* P.A.Duvign., Bull. Inst. Roy. Colon. Belge 20: 888 (1949).
Uapaca nitida f. *bianoensis* P.A.Duvign., Bull. Inst. Roy. Colon. Belge 20: 889 (1949).
Uapaca nitida f. *latiuscula* P.A.Duvign., Bull. Inst. Roy. Colon. Belge 20: 888 (1949).
Uapaca nitida f. *scalarinervosa* P.A.Duvign., Bull. Inst. Roy. Colon. Belge 20: 889 (1949).
Uapaca nitida var. *mulengo* P.A.Duvign., Bull. Inst. Roy. Colon. Belge 20: 889 (1949).
Uapaca nitida var. *nsambi* P.A.Duvign., Bull. Inst. Roy. Colon. Belge 20: 888 (1949).
Uapaca nitida var. *sokolobe* P.A.Duvign., Bull. Inst. Roy. Colon. Belge 20: 889 (1949).
Uapaca nitida var. *suffrutescens* P.A.Duvign., Bull. Inst. Roy. Colon. Belge 20: 889 (1949).

Uapaca nymphaeantha Pax & K.Hoffm. in H.G.A.Engler, Pflanzenr., IV, 147, XV: 306 (1922).
Cameroon. 23 CMN. Phan. – Close to *U. pynaertii*.

Uapaca paludosa Aubrév. & Leandri, Bull. Soc. Bot. France 82: 50 (1935).
Liberia, Ivory Coast, Ghana, Nigeria, Cameroon, Zaire, Uganda, Tanzania. 22 GHA IVO LBR NGA 23 CMN ZAI 25 TAN UGA. Phan.

Uapaca pilosa Hutch., Bull. Misc. Inform. Kew 1912: 102 (1912). *Uapaca pilosa* f. *hirsuta* P.A.Duvign., Bull. Inst. Roy. Colon. Belge 20: 891 (1949), nom. illeg.
Cameroon, Tanzania, Zambia, Malawi, Zaire. 23 CMN ZAI 25 TAN 26 MLW ZAM. Nanophan.

var. **petiolata** P.A.Duvign., Bull. Inst. Roy. Colon. Belge 20: 890 (1949).
Zaire, Tanzania, Malawi, Zambia. 23 ZAI 25 TAN 26 MLW ZAM. Nanophan.
Uapaca macrocephala Pax & K.Hoffm. in H.G.A.Engler, Pflanzenr., IV, 147, XV: 305 (1922).
Uapaca sapinii De Wild., Proc. Internat. Congr. Pl. Sc. Ithaca: 1420 (1929).
Uapaca pilosa f. *subglabra* P.A.Duvign., Bull. Inst. Roy. Colon. Belge 20: 891 (1949).

var. **pilosa**
Cameroon, Tanzania, Zambia, Malawi, Zaire. 23 CMN ZAI 25 TAN 26 MLW ZAM.
Nanophan.
Uapaca masuku De Wild., Ann. Soc. Sci. Bruxelles, Sér. B 45: 311 (1926).

Uapaca prominenticarinata De Wild., Ann. Soc. Sci. Bruxelles, Sér. B 47: 37 (1927).
W. Zaire. 23 ZAI.

Uapaca pynaertii De Wild., Ann. Mus. Congo Belge, Bot., V, 2: 274 (1908).
Zaire. 23 ZAI. Phan.

Uapaca rivularis Denis, Arch. Bot. Bull. Mens. 1: 225 (1927).
W. Madagascar. 29 MDG. Phan.

Uapaca robynsii De Wild., Ann. Soc. Sci. Bruxelles, Sér. B 53: 60 (1933).
Zaire (Shaba), Zambia, Malawi. 23 ZAI 26 MLW ZAM. Nanophan.

Uapaca rufopilosa (De Wild.) P.A.Duvign., Bull. Séances Inst. Roy. Colon. Belge 20: 890 (1949).
Zaire, Angola, Zambia. 23 ZAI 26 ANG ZAM.
   * *Uapaca nitida* var. *rufopilosa* De Wild., Mém. Inst. Roy. Colon. Belge, Sect. Sci. Nat.
   (8vo) 4(5): 161 (1936).

Uapaca samfii De Wild., Ann. Soc. Sci. Bruxelles, Sér. B 53: 310 (1933).
Zaire (Kivu). 23 ZAI.

Uapaca sansibarica Pax, Bot. Jahrb. Syst. 34: 370 (1904).
Zaire (Shaba), Tanzania, Uganda, Burundi, Sudan, Malawi, Zambia, Zimbabwe,
Mozambique, Angola. 23 BUR ZAI 24 SUD 25 TAN UGA 26 ANG MLW MOZ ZAM
ZIM. Phan.
Uapaca sansibarica var. *cuneata* Pax, Bot. Jahrb. Syst. 34: 370 (1904).

Uapaca silvestris Leandri, Mém. Inst. Sci. Madagascar, Sér. B, Biol. Vég. 8: 259 (1957).
E. Madagascar. 29 MDG. Nanophan. or phan.

Uapaca staudtii Pax, Bot. Jahrb. Syst. 23: 522 (1897).
Nigeria, Cameroon, Bioko. 22 NGA 23 CMN GGI. Phan.

Uapaca stipularis Pax & K.Hoffm. in H.G.A.Engler, Pflanzenr., IV, 147, XV: 303 (1922).
Cameroon, Equatorial Guinea. 23 CMN EQG. Phan.

Uapaca thouarsii Baill., Étude Euphorb.: 596 (1858).
E. Madagascar. 29 MDG. Phan.

Uapaca togoensis Pax, Bot. Jahrb. Syst. 34: 371 (1904).
Guinea-Bissau to Central African Rep. 22 BEN GHA GNB GUI IVO MLI NGA SIE TOG 23
CAF CMN. Phan.
Uapaca guignardii A.Chev. ex Beille, Bull. Soc. Bot. France 55(8): 66 (1908).
Uapaca somon Aubrév. & Leandri, Bull. Soc. Bot. France 82: 50 (1935).

Uapaca vanderystii De Wild., Compt.-Rend. Hebd. Séances Mém. Soc. Biol. 96: 701 (1927).
W. Zaire. 23 ZAI.

Uapaca vanhouttei De Wild., Ann. Mus. Congo Belge, Bot., V, 2: 275 (1908).
S. Nigeria, Zaire. 22 NGA 23 CMN? EQG? GAB? GGI? ZAI. Phan.

Uapaca verruculosa De Wild., Ann. Soc. Sci. Bruxelles, Sér. B 53: 148 (1933).
W. Zaire. 23 ZAI.

Synonyms:

Uapaca albida De Wild. === **Uapaca kirkiana** Müll.Arg.

Uapaca angolensis Hutch. ex Pax & K.Hoffm. === **Uapaca benguelensis** Müll.Arg.

Uapaca angustifolia Denis === **Uapaca densifolia** Baker var. **densifolia**

Uapaca benguelensis f. *glabra* P.A.Duvign. === **Uapaca benguelensis** Müll.Arg.

Uapaca benguelensis var. *pedunculata* P.A.Duvign. === **Uapaca benguelensis** Müll.Arg.

Uapaca benguelensis f. *pilosa* P.A.Duvign. === **Uapaca benguelensis** Müll.Arg.

Uapaca bingervillensis Beille === **Uapaca guineensis** Müll.Arg. var. **guineensis**

Uapaca clusiacea Baker === **Uapaca bojeri** Baill.

Uapaca clusioides Baker === **Uapaca ferruginea** Baill.

Uapaca densifolia var. *angustifolia* (Denis) Leandri === **Uapaca densifolia** Baker var. **densifolia**

Uapaca densifolia var. *humbertii* Leandri === **Uapaca densifolia** Baker var. **densifolia**

Uapaca densifolia f. *orientalis* Leandri === **Uapaca densifolia** Baker var. **densifolia**

Uapaca densifolia var. *perrieri* (Denis) Leandri === **Uapaca densifolia** Baker var. **densifolia**

Uapaca dubia De Wild. === **Uapaca kirkiana** Müll.Arg.

Uapaca ferruginea var. *carnotiana* Leandri === **Uapaca ferruginea** Baill.

Uapaca gabonensis Pierre ex Pax & K.Hoffm. === **Uapaca guineensis** Müll.Arg. var. **guineensis**

Uapaca goetzei Pax === **Uapaca kirkiana** Müll.Arg.

Uapaca greenwayi Suess. === **Uapaca kirkiana** Müll.Arg.

Uapaca griffithii Hemsl. === **Drypetes riseleyi** Airy Shaw

Uapaca guignardii A.Chev. ex Beille === **Uapaca togoensis** Pax

Uapaca guignardii var. *sudanica* Beille === **Uapaca guineensis** var. **sudanica** (Beille) Hutch.

Uapaca heudelotii var. *acuminata* Hutch. === **Uapaca acuminata** (Hutch.) Pax & K.Hoffm.

Uapaca homblei De Wild. === **Uapaca kirkiana** Müll.Arg.

Uapaca kirkiana var. *dubia* (De Wild.) P.A.Duvign. === **Uapaca kirkiana** Müll.Arg.

Uapaca kirkiana var. *goetzei* Pax === **Uapaca kirkiana** Müll.Arg.

Uapaca kirkiana var. *kwangoensis* P.A.Duvign. === **Uapaca kirkiana** Müll.Arg.

Uapaca kirkiana var. *sessilifolia* P.A.Duvign. === **Uapaca kirkiana** Müll.Arg.

Uapaca laurentii De Wild. === **Uapaca guineensis** Müll.Arg. var. **guineensis**

Uapaca louvelii var. *paralia* Leandri === **Uapaca louvelii** Denis

Uapaca macrocephala Pax & K.Hoffm. === **Uapaca pilosa** var. **petiolata** P.A.Duvign.

Uapaca marquesii Pax === **Uapaca heudelotii** Baill.

Uapaca masuku De Wild. === **Uapaca pilosa** Hutch. var. **pilosa**

Uapaca microphylla Pax === **Uapaca nitida** Müll.Arg. var. **nitida**

Uapaca microphylla var. *hendrickxii* De Wild. === **Uapaca nitida** Müll.Arg. var. **nitida**

Uapaca mole Pax === **Uapaca guineensis** Müll.Arg. var. **guineensis**

Uapaca munamensis De Wild. === **Uapaca kirkiana** Müll.Arg.

Uapaca myricifolia Baker === **Uapaca densifolia** Baker var. **densifolia**

Uapaca neomasuku De Wild. === **Uapaca kirkiana** Müll.Arg.

Uapaca nitida f. *aucta* P.A.Duvign. === **Uapaca nitida** Müll.Arg. var. **nitida**

Uapaca nitida f. *bianoensis* P.A.Duvign. === **Uapaca nitida** Müll.Arg. var. **nitida**

Uapaca nitida f. *latiuscula* P.A.Duvign. === **Uapaca nitida** Müll.Arg. var. **nitida**

Uapaca nitida f. *longifolia* P.A.Duvign. === **Uapaca nitida** var. **longifolia** (P.A.Duvign.) Radcl.-Sm.

Uapaca nitida var. *mulengo* P.A.Duvign. === **Uapaca nitida** Müll.Arg. var. **nitida**

Uapaca nitida var. *nsambi* P.A.Duvign. === **Uapaca nitida** Müll.Arg. var. **nitida**

Uapaca nitida var. *rufopilosa* De Wild. === **Uapaca rufopilosa** (De Wild.) P.A.Duvign.

Uapaca nitida f. *scalarinervosa* P.A.Duvign. === **Uapaca nitida** Müll.Arg. var. **nitida**

Uapaca nitida var. *sokolobe* P.A.Duvign. === **Uapaca nitida** Müll.Arg. var. **nitida**

Uapaca nitida var. *suffrutescens* P.A.Duvign. --- **Uapaca nitida** Müll.Arg. var. **nitida**

Uapaca perrieri Denis === **Uapaca densifolia** Baker var. **densifolia**

Uapaca perrotii Beille === **Uapaca guineensis** Müll.Arg. var. **guineensis**

Uapaca pilosa f. *hirsuta* P.A.Duvign. === **Uapaca pilosa** Hutch.

Uapaca pilosa f. *subglabra* P.A.Duvign. === **Uapaca pilosa** var. **petiolata** P.A.Duvign.

Uapaca sambiranensis Denis === **Uapaca densifolia** var. **sambiranensis** (Denis) Leandri

Uapaca sansibarica var. *cuneata* Pax === **Uapaca sansibarica** Pax

Uapaca sapinii De Wild. === **Uapaca pilosa** var. **petiolata** P.A.Duvign.
Uapaca seretii De Wild. === **Uapaca guineensis** Müll.Arg. var. **guineensis**
Uapaca similis Pax & K.Hoffm. === **Uapaca nitida** Müll.Arg. var. **nitida**
Uapaca somon Aubrév. & Leandri === **Uapaca togoensis** Pax
Uapaca teusczii Pax === **Uapaca benguelensis** Müll.Arg.

Uranthera

A part of the *Phyllanthodendron* group now within *Phyllanthus*.

Synonyms:
Uranthera Pax & K.Hoffm. === **Phyllanthus** L.
Uranthera siamensis Pax & K.Hoffm. === **Phyllanthus roseus** (Craib & Hutch.) Beille

Urinaria

An early segregate of *Phyllanthus*.

Synonyms:
Urinaria Medik. === **Phyllanthus** L.

Usteria

Synonyms:
Usteria Dennst. === **Acalypha** L.

Vallaris

Synonyms:
Vallaris Raf. === **Euphorbia** L.

Vandera

Synonyms:
Vandera Raf. === **Croton** L.

Vaupesia

1 species, S. America; trees to 30 m or so with a large crown, simple leaves and red or white latex produced only in the branches. Originally allied by Schultes to *Micrandra*, *Hevea* and *Joannesia* but in the Webster system thought to be (along with *Joannesia*) closer to *Jatropha* on account of presumed pollen affinities. (Crotonoideae)

Schultes, R. E. (1955). A new generic concept in the Euphorbiaceae. Bot. Mus. Leafl. 17: 27-36, 3 pls. (incl. map). En. — Along with particulars on *Vaupesia* (newly described here, with a new species, *V. cataractorum*), this paper includes ecological and ethnobiological considerations touching also on several other Euphorbiaceae. Illustrations (drawing, pl. 12) and a map are included.

Vaupesia R.E.Schult., Bot. Mus. Leafl. 17: 27 (1955).
Colombia, N. Brazil. 83 84.

Vaupesia cataractarum R.E.Schult., Bot. Mus. Leafl. 17: 28 (1955).
Colombia (Rio Negro), NW. Brazil. 83 CLM 84 BZN.

Veconcibea

Synonyms:
Veconcibea Pax & K.Hoffm. === **Conceveiba** Aubl.

Ventenatia

Synonyms:
Ventenatia Tratt. === **Pedilanthus** Neck.

Vernicia

3 species, E. and SE. Asia from Japan to Indochina; deciduous trees to 20 m. *V. fordii* furnishes the tung-oil of commerce and the timber is also important. These and other economic aspects have been covered in several works, among them Langeron (1902), Wilson (1913), Motte (1933), Hinkul (1935) and Hoh (1939). Formerly not distinguished from *Aleurites*, the genera remain associated within subtribe Aleuritinae in tribe Aleuritideae. [Several additional references, the greater part specific to *V. fordii*, are given in Webster, 1967 (see **Americas**). Further information is incorporated in the PROSEA database and series (group 6).] (Crotonoideae)

Langeron, M. (1902). See *Aleurites*.

Hemsley, W. B. (1906). *Aleurities fordii*. Ic. Pl. 29: pls. 2801-2802. En. — Plant portrait (2 figures) with description, distribution and commentary; the first formal botanical account of this well-known wood-oil tree.

Pax, F. (1910). *Aleurites*. In A. Engler (ed.), Das Pflanzenreich, IV 147 [I] (Euphorbiaceae-Jatropheae): 128-133. Berlin. (Heft 42.) La/Ge. — 4 species, of which 2 in sect. *Dryandra* (=*Vernicia*). [Now superseded.]

Wilson, E. H. (1913). Revision of the synonymy of the species of *Aleurites*. Bull. Imperial Inst. 11: 441-461. En. — General introduction; identification of Chinese provenances and their botany and agronomy; extraction of the oil and its uses in China and export therefrom; composition of the oil; uses outside China; notes on the Japanese species (now *Vernicia cordata*); conclusions; formal botanical treatment (Appendix, pp. 460-461) with synonymised accounts of 3 species (now all in *Vernicia*) but no key.

Hemsley, W. B. (1914). The wood-oil trees of China and Japan. Kew Bull. Misc. Inform. 1914: 1-4. En. — Written in support of the detailed survey by Wilson (1913) establishing that recognition of three species was merited; summary of his taxonomic treatment, with synonymy, references, citations, and indication of distribution.

Motte, J. (1933). Les *Aleurites* de la sect. *Dryandra*, à huiles de bois. 36 pp., map (Ann. Inst. Bot.-Géol. Colon. Marseille, V 1(2)). Marseille. Fr. — A historical, botanical and economic survey of the three species of *Aleurites* sect. *Dryandra* (=*Vernicia*); includes key (pp. 8-9) and extensive accounts of each, particularly *V. cordata*. The author concludes that of the species *V. fordii* should be chosen for plantations.

Hinkul, S. G. (1935). Tungovoe derevo [Tung-oil tree]. Trudy Prikl. Bot. Genet. Selekts. (Bull. Appl. Bot.), × (Dendrol. Dekor. Sadovodst.), 2: 137-153, illus. Ru. — Historical treatment; descriptions, with illustrations, of *V. cordata*, *V. fordii* and *V. montana* (as species of *Aleurites*) along with synonymy, references, citations, distinguishing features, variability, and properties as well as trial records in the then-USSR. [An English summary appears on p. 153. There had been much previous confusion as to the proper identity and names for these trees.]

Hoh Hin-cheung (1939). Genus *Aleurites* in Kwantung and Kwangsi. Lingnan Sci. J. 18: 303-327, 513-524, illus., maps. En/Ch. — Key to 3 species (p. 308) with descriptions of each; discussion of distributions with maps; all exsiccatae cited with localities and individual notes (in table 11); literature and personal records (table 12). The second part of the paper covers cultural and environmental requirements, practices and proposals along with a list of literature consulted. [All these species are now referable to *Vernicia*.]

Stockar, A. (1947). Comunicación preliminar sobre hibridaciones entre varias especies de *Aleurites*. Revista Argent. Agron. 14: 33-38. Sp. — Report of trials among the three known species, with 6 crosses and backcrosses in all studied; descriptions of features in each; some suggestions for continued work. The hybrids were all reported as fertile (all parent species having the same chromosome number, 2n=22). [A lead time of 3-5 years was required between generations.]

Airy-Shaw, H. K. (1966). Notes on Malaysian and other Asiatic Euphorbiaceae, LXXII. Generic segregation in the affinity of *Aleurites* J. R. et G. Forst. Kew Bull. 20: 393-395. En. —*Vernicia* separated from *Aleurites*; 3 species, with 2 new combinations.

Wang, Z. (1980). (A preliminary study on local varieties of *Aleurites fordii* Hemsl. in Zhejiang.) J. Nanjing Techn. Coll. Forest Prod. 2: 55-66. Ch. — Related to provenance studies; no formal taxonomy (save for a key, p. 57). English summary, p. 66. [5 forms recognised, based on a variety of morphological features.]

Stuppy, W. et al. (1999). Revision of the genera *Aleurites*, *Reutealis* and *Vernicia* (Euphorbiaceae). Blumea 44: 73-98, illus. En. — *Vernicia*, pp. 89-95; treatment of 3 species with key, descriptions, synonymy, references, types, distribution, habitat, commentary, and notes on uses, etc.; all general references, identification list and index to botanical names at end of paper. A general survey and key to genera appear on pp. 73-78.

Vernicia Lour., Fl. Cochinch.: 586 (1790).
Burma to Japan. 36 38 41.
Dryandra Thunb., Nov. Gen. Pl.: 60 (1783).
Elaeococca Comm. ex A.Juss., Euphorb. Gen.: 38 (1824).

Vernicia cordata (Thunb.) Airy Shaw, Kew Bull. 20: 394 (1966).
S. Japan. 38 JAP. Phan.
* *Dryandra cordata* Thunb., Nova Acta Regiae Soc. Sci. Upsal. 4: 38 (1783). *Aleurites cordata* (Thunb.) R.Br. ex Steud., Nomencl. Bot., ed. 2, 1: 49 (1840).
Aleurites japonica Blume, Ann. Mus. Bot. Lugduno-Batavi 4: 120 (1869).
Aleurites verniciflua Baill., Hist. Pl. 5: 116 (1874).

Vernicia fordii (Hemsl.) Airy Shaw, Kew Bull. 20: 394 (1966).
S. China, Hainan, Vietnam. 36 CHC CHH CHS 41 VIE. Phan.
* *Aleurites fordii* Hemsl., Hooker's Icon. Pl. 29: t. 2801 (1906).

Vernicia montana Lour., Fl. Cochinch.: 586 (1790). *Aleurites montana* (Lour.) E.H.Wilson, Bull. Imp. Inst. Gr. Brit. 11: 460 (1913). – FIGURE, p. 1586.
S. China, Hainan, Taiwan, Indo-China. 36 CHC CHH CHS 38 jap TAI 41 BMA CBD THA VIE. Phan.
Dryandra vernicia Corrêa, Ann. Mus. Natl. Hist. Nat. 8: 69 (1822). *Aleurites vernicia* (Corrêa) Hassk., Flora 25(2): 40 (1842).

Victorinia

Synonyms:
Victorinia Léon === **Cnidoscolus** Pohl

Vigia

Now united with *Plukenetia* (Gillespie 1993; see there).

Synonyms:
Vigia Vell. === **Plukenetia** L.
Vigia serrata Vell. === **Plukenetia serrata** (Vell.) L.J.Gillespie

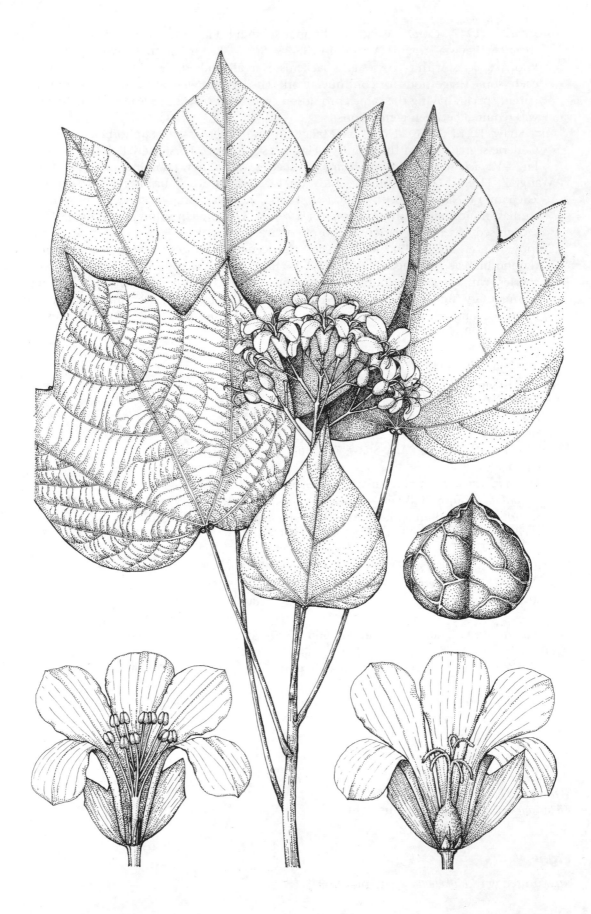

Vernicia montana Lour.

Artist: Judy C. Dunkley
Fl. Trop. East Africa, Euphorbiaceae, 1: 180, fig. 35 (1987)

Villanova

Synonyms:
Villanova Pourr. ex Cutanda === **Flueggea** Willd.

Voatamalo

2 species, Madagascar. Small to medium trees to 20 m in well-drained forest with foliage resembling some species of *Eugenia*, the leaves likewise being opposite; fruits subtended by a persistent perianth. In the Webster system the leaf arrangment distinguishes this genus from *Mischodon* (with whorled leaves) and *Aristogeitonia* (with spirally arranged leaves). *Androstachys* and *Stachyandra* are more distantly related, although Bosser (1975) thought that an affinity did exist. (Oldfieldioideae)

Bosser, J. (1975(1976)). *Voatamalo*, nouveau genre d'Euphorbiaceae de Madagascar. Adansonia, II, 15: 333-340, illus. Fr. — Protologue of genus, extensive discussion, and descriptions of two species (both also new); no key. Some consideration on the homogeneity or otherwise of Euphorbiaceae s.l. is also presented, without resolution.

Voatamalo Capuron ex Bosser, Adansonia, n.s. 15: 333 (1975 publ. 1976).
Madagascar. 29. Nanophan. or phan.

Voatamalo capuronii Bosser, Adansonia, n.s. 15: 339 (1975 publ. 1976).
NE. Madagascar. 29 MDG. Nanophan. or phan.

Voatamalo eugenioides Capuron ex Bosser, Adansonia, n.s. 15: 337 (1975 publ. 1976).
– FIGURE, p. 1588.
E. Madagascar. 29 MDG. Phan.

Wartmannia

Synonyms:
Wartmannia Müll.Arg. === **Homalanthus** A.Juss.

Wetria

2 species, SE. Asia, Malesia (extending to New Guinea) and Australia (NE. Queensland); shrubs or small to medium trees to 27 m with many-veined leaves, found at least sometimes in somewhat dry or seasonally dry forest and here and there recorded from limestone. Along with *Spathiostemon* and *Homonoia*, in the Webster system it forms subtribe Lasiococcinae in the Acalypheae; it is not considered related to *Alchornea* despite superficial similarity. A new revision has been published by van Welzen (1998); among other features is the transfer of New Guinean records of *W. insignis* to *W. australiensis*. (Acalyphoideae)

Pax, F. (with K. Hoffmann) (1914). *Wetria*. In A. Engler (ed.), Das Pflanzenreich, IV 147 VII (Euphorbiaceae-Acalypheae-Mercurialinae): 219-220. Berlin. (Heft 63.) La/Ge. — 2 species (1 doubtfully here), W Malesia.
Airy-Shaw, H. K. (1960). Notes on Malaysian Euphorbiaceae, XVIII. The genus *Wetria* Baill. recorded from Burma. Kew Bull. 14: 472-473. En. — Range extension; *Wetria macrophylla* (now *W. insignis*) in SE Myanmar.
Airy-Shaw, H. K. (1963). Notes on Malaysian and other Asiatic Euphorbiaceae, XXXIV. New records for *Wetria macrophylla*. Kew Bull. 16: 353. En. — Additional records for what is now *Wetria insignis*.

Voatamalo eugenioides Capuron ex Bosser
Artist: D. Godot de Mauroy
Adansonia, n.s. 15: 335 (1975 publ. 1976)

Airy-Shaw, H. K. (1972). Notes on Malesian and other Asiatic Euphorbiaceae, CLXIII: *Wetria insignis* in New Guinea. Kew Bull. 27: 87-88. En. — Range extension, with first record of the genus for New Guinea; includes reduction of *Trigonostemon forbesii*. [These records are now included in *Wetria australiensis*.]

Forster, P. I. (1994). *Wetria australiensis* sp. nov. (Euphorbiaceae), a new generic record for Australia. Austrobaileya 4: 139-143, illus. En. — Range extension for genus with a new species, *W. australiensis*, from near Cairns in northern Queensland. No revised key is, however, presented.

- Welzen, P. C. van (1998). Revisions and phylogenies of Malesian Euphorbiaceae subtribe Lasiococcinae (*Homonoia, Lasiococca, Spathiostemon*) and *Clonostylis, Ricinus* and *Wetria*. Blumea 43: 131-164. (*Lasiococca* with Nguyen Nghia Thin & Vu Hoai Duc.) En. — General introduction, with history of studies; a note of caution on the use as a character of monoecy vs. dioecy given the imperfect state of much material; phylogeny with character table and cladogram; revision of Malesian *Wetria* (pp. 155-160; 2 species) with key, descriptions, synonymy, types, literature citations, vernacular names, indication of distribution, ecology and habitat, and notes on anatomy, uses and properties (where known), and systematics; identification list at end.

Wetria Baill., Étude Euphorb.: 409 (1858).
> Trop. Asia, N. Australia. 41 42 50.
> > *Pseudotrewia* Miq., Fl. Ned. Ind. 1(2): 414 (1859).

Wetria australiensis P.I.Forst., Austrobaileya 4: 141 (1994).
> Papua New Guinea, Queensland (Cook). 50 QLD. Phan.

Wetria insignis (Steud.) Airy Shaw, Kew Bull. 26: 350 (1972).
> Burma, Thailand, W. & C. Malesia. 41 BMA THA 42 BOR JAW LSI MLY PHI SUM. Phan.
> > * *Trewia macrophylla* Blume, Bijdr.: 612 (1826), nom. illeg. *Trewia insignis* Steud., Nomencl. Bot., ed. 2, 2: 698 (1841). *Wetria macrophylla* J.J.Sm. ex Koord. & Valeton, Meded. Dept. Landb. Ned.-Indië 10: 471 (1910).
> > *Wetria trewioides* Baill., Étude Euphorb.: 409 (1858).
> > *Alchornea blumeana* Müll.Arg., Linnaea 34: 167 (1865).
> > *Agrostistachys pubescens* Merr., Philipp. J. Sci., C 4: 274 (1909).
> > *Trigonostemon forbesii* Pax in H.G.A.Engler, Pflanzenr., IV, 147, III: 88 (1911).

Synonyms:
Wetria cuneifolia (Miq.) Pax === **Trigonostemon heteranthus** Wight
Wetria macrophylla J.J.Sm. ex Koord. & Valeton === **Wetria insignis** (Steud.) Airy Shaw
Wetria rufescens (Franch.) Pamp. === **Discocleidion rufescens** (Franch.) Pax & K.Hoffm.
Wetria trewioides Baill. === **Wetria insignis** (Steud.) Airy Shaw

Wetriaria

Synonyms:
Wetriaria (Müll.Arg.) Kuntze === **Argomuellera** Pax

Whyanbeelia

1 species, Australia (NE. Queensland); medium trees of the Daintree rainforests to 20 m with opposite leaves, axillary inflorescences and large fruits with conspicuous styles. Related to *Canaca* and *Dissiliaria* but monoecious; more distantly, there is a relationship with *Austrobuxus*. (Oldfieldioideae)

Airy-Shaw, H. K. (1980). *Whyanbeelia*. Kew Bull. 35: 691-692. (Euphorbiaceae-Platylobeae of Australia.) En. — Treatment of 1 species. [For protologue see Airy-Shaw 1976 under **Australasia**.]

Whyanbeelia Airy Shaw & B.Hyland, Kew Bull. 31: 375 (1976).
  Queensland. 50. Phan.

Whyanbeelia terrae-reginae Airy Shaw & B.Hyland, Kew Bull. 31: 376 (1976).
  – FIGURE, p. 1591.
  Queensland (Cook). 50 QLD. Phan.

Wielandia

1 species, Madagascar, Comoros, Aldabra and Seychelles; glabrous shrubs or understorey trees with slender, flexible branches to 20 m in dry forest (in Madagascar deciduous), sometimes in rocky situations. The sole species, apparently a relict, is considered to be among the less specialised members of the subfamily; its pentamerous flowers with a five-locular ovary are distinctive. Webster (Synopsis, 1994) considers it closest to *Petalodiscus*, though that has a dissected staminate disk. (Phyllanthoideae)

Hemsley, W. B. (1906). *Wielandia elegans*. Ic. Pl. 29: pl. 2813. En. — Plant portrait with description and commentary; in Seychelles.

Pax, F. & K. Hoffmann (1922). *Wielandia*. In A. Engler (ed.), Das Pflanzenreich, IV 147 XV (Euphorbiaceae-Phyllanthoideae-Phyllantheae): 181. Berlin. (Heft 81.) La/Ge. — 1 species (*W. elegans*), Seychelles.

Leandri, J. (1958). *Wielandia*. Fl. Madag. Comores 111 (Euphorbiacées), I: 135-137. Paris. Fr. — A variety of *Wielandia elegans* (otherwise in Seychelles) present.

• Hoffmann, P. (1998). Revision of the genus *Wielandia* (Euphorbiaceae-Phyllanthoideae). Adansonia, III, 20: 333-340, illus., map. — Brief introduction; treatment of 1 species with description, synonymy, references, literature citations, vernacular names, indication of distribution and habitat, references to figures, commentary, and localities with exsiccatae; short bibliography at end. [*W. elegans* var. *perrieri* no longer recognised as distinct.]

Wielandia Baill., Étude Euphorb.: 568 (1858).
  Madagascar to Seychelles. 29.

Wielandia elegans Baill., Étude Euphorb.: 569 (1858). *Savia elegans* (Baill.) Müll.Arg.,
  Linnaea 32: 78 (1863). – FIGURE, p. 1592.
  Madagascar, Comoros, Aldabra Seychelles. 29 ALD COM MDG SEY. Nanophan. or phan.
    Wielandia elegans var. *perrieri* Leandri, Notul. Syst. (Paris) 7: 190 (1939).

Synonyms:
Wielandia oblongifolia Baill. === **Blotia oblongifolia** (Baill.) Leandri

Williamia

A segregate of *Phyllanthus*.

Synonyms:
Williamia Baill. === **Phyllanthus** L.

Wurtzia

Synonyms:
Wurtzia Baill. === **Margaritaria** L.f.
Wurtzia tetracocca Baill. === **Margaritaria tetracocca** (Baill.) G.L.Webster

Whyanbeelia terrae-reginae Airy Shaw & B. Hyland
Artist: Ann Davies
Kew Bull. 35: 668, Fig. 6 (1980) lower right

Xamesike

Synonyms:
Xamesike Raf. === **Euphorbia** L.

Wielandia elegans Baill.

Artist: Matilda Smith
Ic. Pl. 29: pl. 2813 (1906)
KEW ILLUSTRATIONS COLLECTION

Xylococcus

An unpublished name of Robert Brown proposed by Britton and S. Moore, but a later homonym of *Xylococcus* Nutt. (Ericaceae); synonymous with *Petalostigma*.

Xylophylla

An American segregate of *Phyllanthus*, early recognised on account of distinctive foliage but not revived in the twentieth century except by Britton and his 'school'.

Synonyms:
Xylophylla L. === **Phyllanthus** L.
Xylophylla angustifolia Sw. === **Phyllanthus angustifolius** (Sw.) Sw.
Xylophylla angustifolia var. *linearis* Sw. === **Phyllanthus arbuscula** (Sw.) J.F.Gmel.
Xylophylla arbuscula Sw. === **Phyllanthus arbuscula** (Sw.) J.F.Gmel.
Xylophylla artensis Montrouz. === **Exocarpos sp.** (Santalaceae)
Xylophylla asplenifolia (L'Hér.) Salisb. === **Phyllanthus latifolius** (L.) Sw.
Xylophylla contorta Britton === **Phyllanthus angustifolius** (Sw.) Sw.
Xylophylla × *elongata* Jacq. === **Phyllanthus 'Elongatus'**
Xylophylla ensifolia Bojer ex Drake === **Phylloxylon ensifolius** Baill. (Fabaceae)
Xylophylla ensifolia Bojer ex Baker === **Exocarpos xylophylloides** Baker (Santalaceae)
Xylophylla epiphyllanthus (L.) Hornem. === **Phyllanthus epiphyllanthus** L.
Xylophylla epiphyllanthus Britton === **Phyllanthus epiphyllanthus** L. subsp. **epiphyllanthus**
Xylophylla falcata Sw. === **Phyllanthus epiphyllanthus** L. subsp. **epiphyllanthus**
Xylophylla latifolia L. === **Phyllanthus latifolius** (L.) Sw.
Xylophylla linearis (Sw.) Steud. === **Phyllanthus arbuscula** (Sw.) J.F.Gmel.
Xylophylla longifolia L. === **Exocarpos ceramicus** (L.) Endl. (Santalaceae)
Xylophylla lucena Roth === **Flueggea leucopyrus** Willd.
Xylophylla media Lodd. ex G.Don === **Phyllanthus** ?
Xylophylla montana Sw. === **Phyllanthus montanus** (Sw.) Sw.
Xylophylla obovata Willd. === **Flueggea virosa** (Roxb. ex Willd.) Voigt
Xylophylla obtusata Billb. === **Phyllanthus obtusatus** (Billb.) Müll.Arg.
Xylophylla parviflora Bellardi ex Colla === **Flueggea suffruticosa** (Pall.) Baill.
Xylophylla ramiflora Aiton === **Flueggea suffruticosa** (Pall.) Baill.

Zalitea

Precedes *Zygophyllidium* if that taxon be ever separated from *Euphorbia*.

Synonyms:
Zalitea Raf. === **Euphorbia** L.

Zarcoa

Synonyms:
Zarcoa Llanos === **Glochidion** J.R.Forst. & G.Forst.
Zarcoa phillipica Llanos === **Glochidion album** (Blanco) Boerl.

Zenkerodendron

An unpublished name of Gilg included by Jablonski in the synonymy of *Cleistanthus* (Jablonszky 1915; see that genus).

Zimmermannia

Now united with *Meineckia* (Radcliffe Smith, 1997; see there).

Synonyms:
Zimmermannia Pax === **Meineckia** Baill.
Zimmermannia acuminata Verdc. === **Meineckia acuminata** (Verdc.) Brunel ex Radcl.-Sm.
Zimmermannia capillipes Pax === **Meineckia paxii** Brunel ex Radcl.-Sm.
Zimmermannia decaryi (Leandri) G.L.Webster === **Meineckia decaryi** (Leandri) Brunel ex
  Radcl.-Sm.
Zimmermannia decaryi var. *occidentalis* (Leandri) Poole === **Meineckia decaryi** var.
  occidentalis (Leandri) Radcl.-Sm.
Zimmermannia grandiflora Verdc. === **Meineckia grandiflora** (Verdc.) Brunel ex Radcl.-Sm.
Zimmermannia nguruensis Radcl.-Sm. === **Meineckia nguruensis** (Radcl.-Sm.) Brunel ex
  Radcl.-Sm.
Zimmermannia ovata E.A.Bruce === **Meineckia ovata** (E.A.Bruce) Brunel ex Radcl.-Sm.
Zimmermannia stipularis Radcl.-Sm. === **Meineckia stipularis** (Radcl.-Sm.) Brunel ex Radcl.-Sm.

Zimmermanniopsis

Now united with *Meineckia* (Radcliffe Smith, 1997; see there).

Synonyms:
Zimmermanniopsis Radcl.-Sm. === **Meineckia** Baill.
Zimmermanniopsis uzungwaensis Radcl.-Sm. === **Meineckia uzungwaensis** (Radcl.-Sm.)
  Radcl.-Sm.

Zuckertia

Synonyms:
Zuckertia Baill. === **Tragia** Plum. ex L.

Zygophyllidium

A segregate of *Euphorbia*, based on sect. *Zygophyllidium* Boiss. (part of the present subgen.
Agaloma).

Synonyms:
Zygophyllidium Small === **Euphorbia** L.
Zygophyllidium delicatulum Wooton & Standl. === **Euphorbia bifurcata** Engelm.

Zygospermum

Synonyms:
Zygospermum Thwaites ex Baill. === **Margaritaria** L.f.
Zygospermum zeylanicum Thwaites ex Baill. === **Margaritaria cyanosperma** (Gaertn.)
  Airy Shaw

Pandaceae

4 genera, 19 species. Shrubs or small to large trees of tropical W. & C. Africa, Asia and Malesia. As a family first recognised by Pierre in 1896, inclusive of *Panda* and *Microdesmis*, and the following year extended to include *Galearia*. Ordinal rank was proposed by Engler (1912; see **Panda**). Valid publication of the family dates from 1913 with its entry by Engler and Gilg in their *Syllabus der Pflanzenfamilien* (7. Aufl., 1913), but there they limited it to *Panda*; this course was in turn followed by Mildbraed (1931) for *Die natürlichen Pflanzenfamilien*. The remaining genera continued to hold a place in Euphorbiaceae (Pax & Hoffmann 1931). Not until the 1960s did more support became available for extending its circumscription as had been suggested by Pierre. At a later date *Centroplacus* was also transferred to Pandaceae. Particular features include a greater or lesser expression of 'phyllanthoid' branching and drupaceous fruits with a more or less stony endocarp; both are indications of evolutionary advance. The nature of the branches caused some workers in the past to regard them as compound leaves, resulting in placements at different times in Burseraceae, Sapindaceae and Anacardiaceae. The leaves, often more or less crenate, are in addition reminiscent of some Flacourtiaceae. For some authors, however, this has been insufficient to keep them apart from Euphorbiaceae (Webster, Classification 1994); in that worker's system of the family (Webster, Synopsis, 1994) the four genera are partially dispersed. Recent work in molecular systematics, however, supports the distinctiveness of the family (M. Chase, personal communication, July 1998).

General

The major paper on the family is Forman (1966) who also includes a history.

- Boissier, E. and J. Mueller (Argoviensis) (1862-66). Euphorbiaceae. In A. de Candolle (ed.), Prodromus systematis naturalis regni vegetabilis, 15(2): 1-1286. Paris: Masson. La. — For description, see **Euphorbiaceae**.
- Mildbraed, J. (1931). Pandaceae. In A. Engler (ed.), Die natürlichen Pflanzenfamilien, 2. Aufl., 19a: 1-3, illus. Leipzig. Ge. — Family treatment, with list of references, and account of *Panda*. [The family was presented as the sole member of Pandales, and preceded the Geraniales.]
- Pax, F. and K. Hoffmann (1931). Euphorbiaceae. In A. Engler (ed.), Die natürlichen Pflanzenfamilien, 2. Aufl., 19c: 11-233, illus. Leipzig. Ge. — *Centroplacus* at no. 3 in Phyllantheae; *Galearia* and *Microdesmis* at nos. 212 and 214 in Cluytieae subtribe Galeariinae. [For *Panda* see Mildbraed (1931).]
- Airy-Shaw, H. K. (1965). Diagnoses of new families, new names, etc., for the seventh edition of Willis's 'Dictionary'. Kew Bull. 18: 249-273. En. — Includes several taxonomic and nomenclatural revisions relating to Euphorbiaceae s.l.
- Forman, L. (1966). The reinstatement of *Galearia* and *Microdesmis* in the Pandaceae, with appendices by C. Metcalfe and N. Parameswaran. Kew Bull. 20: 309-321, illus. En. — Arguments, with evidence from vegetative and wood anatomy, for transfer to *Galearia* to Pandaceae; survey of the history of *Panda* and comparison of it with *Galearia* and *Microdesmis* (table, p. 312); character review; conclusions; key to genera and, in *Galearia*, subgenera (p. 318).
- Webster, G. L. (1975). Conspectus of a new classification of the Euphorbiaceae. Taxon 24: 593-601. En. — A new system of the family proposed; key to 5 subfamilies and synopsis of 52 tribes (and many additional subtribes) with included genera. The 'Stenolobieae' of Mueller are dismembered. Pollen morphology was seen as especially useful in suprageneric classification. [For historical discussion, see Webster (1987).]
- Jury, S. L., T. Reynolds, D. F. Cutler and F. J. Evans (eds) (1987). The Euphorbiales: chemistry, taxonomy and economic botany. [iv], 326 pp., illus. London: Academic

Press. (Repr. from Bot. J. Linn. Soc. 94(1/2).) En. — The first 2 articles are separately accounted for here.

Radcliffe-Smith, A. (1987). Segregate families from the Euphorbiaceae. In S. L. Jury et al. (eds), The Euphorbiales: 47-66. London. En. — Discussion of 11 older and more recent segregates as well as the tribe Paivaeusinae of Pax and Hoffmann. [All Paivaeusinae now in Oldfieldioideae. Of the segregate families, Buxaceae, Aextoxicaceae, Didymelaceae, Daphniphyllaceae and Pandaceae are definitely excluded, but the last-named is retained within Euphorbiales.]

- Webster, G. L. (1987). The saga of the spurges: a review of classification and relationships in the Euphorbiales. In S. L. Jury et al. (eds), The Euphorbiales: 3-46. London. En. — Historical survey and review, including recent systems of classification; recognition of certain segregate families not advocated. Table 1 gives summary characters of the subfamilies of Euphorbiaceae and the tribes of Pandaceae.

Meeuse, A. D. J. (1990). The Euphorbiaceae auct. plur.: an unnatural taxon. 38 pp. Delft: Eburon. En. — The author accepts the existing 'consensus' for recognition of Pandaceae.

- Webster, G. L. (1994). Classification of the Euphorbiaceae. Ann. Missouri Bot. Gard. 81: 3-32, maps. En. — Pandaceae not accepted as distinct.

- Webster, G. L. (1994). Synopsis of the genera and suprageneric taxa of Euphorbiaceae. Ann. Missouri Bot. Gard. 81: 33-144. En. — Genera of Pandaceae are treated as tribe Galearieae (nos. 100-102) in Euphorbiaceae-Acalyphoideae with *Centroplacus* as *incertae sedis* in Phyllanthoideae (no. 59).

Special

See also Forman (1966) under **General**.

Nowicke, J. (1984). A palynological study of the Pandaceae. Pollen Spores 26: 31-42, illus. En. — Representatives of *Galearia*, *Microdesmis* and *Panda* studied; close connection between first two as earlier established confirmed, but *Panda* was less obviously related. The pollen is easily distinguished from that of Euphorbiaceae.

Africa

The wetter parts of west and west-central tropical Africa are home to *Centroplacus*, *Microdesmis* (in part) and *Panda*. Four recent flora accounts are cited below.

Robyns, W. (1957). Pandaceae. Fl. Congo Belge 7: 1-4, illus. Brussels. Fr. — Flora treatment (*Panda* only) with description, synonymy, references and citations, vernacular names, localities with exsiccatae, indication of distribution and habitat, and notes on special features, uses, properties, taxonomy, etc.

Léonard, J. (1958). Euphorbiaceae, [I]. Fl. Congo Belge 8(1): 1-214, illus. Brussels. Fr. — Flora treatment (*Microdesmis*, pp. 102-115) with descriptions, synonymy, references and citations, vernacular names, localities with exsiccatae, indication of distribution and habitat, and notes on special features, uses, properties, taxonomy, etc. [5 confirmed and 1 insufficiently known species of *Microdesmis*).

Villiers, J.-F. (1973). Pandaceae. Fl. Gabon 22: 14-30, illus. Paris. Fr. — Flora treatment (*Panda* 1, *Microdesmis* 5; *Centroplacus* not included) with keys, descriptions, synonymy, references and citations, vernacular names, localities with exsiccatae, indication of distribution and habitat, and notes on special features, uses, properties, taxonomy, etc.

Villiers, J.-F. (1975). Pandaceae. Fl. Cameroun 19: 42-58, illus. Paris. Fr. — Flora treatment (*Centroplacus* not included) with keys, descriptions, synonymy, references and citations, vernacular names, localities with exsiccatae, indication of distribution and habitat, and notes on special features, uses, properties, taxonomy, etc.

Bennettia

of R. Brown (1852), non S.F. Gray (1821; = *Saussurea* DC., Asteraceae) nec Miq. (1859; = *Bennettiodendron* Merr., Flacourtiaceae).

Synonyms:
Bennettia R.Br. === **Galearia** Zoll. & Moritzi
Bennettia affinis R.Br. === **Galearia fulva** (Tul.) Miq.
Bennettia finlaysoniana R.Br. === **Galearia fulva** (Tul.) Miq.
Bennettia jackiana R.Br. === **Galearia fulva** (Tul.) Miq.
Bennettia javanica R.Br. === **Galearia fulva** (Tul.) Miq.
Bennettia pedicellata R.Br. === **Galearia fulva** (Tul.) Miq.
Bennettia phlebocarpa R.Br. === **Galearia fulva** (Tul.) Miq.
Bennettia subulata Müll.Arg. === **Galearia fulva** (Tul.) Miq.
Bennettia wallichii R.Br. === **Galearia fulva** (Tul.) Miq.

Centroplacus

1 species, WC. Africa. *C. glaucinus* is a forest tree to 20 m, in distribution overlapping with *Panda oleosa*. The genus, in the past referred with doubt to Flacourtiaceae (Gilg 1908), was included in Euphorbiaceae, firstly by Pax and Hoffmann in 1931 within Phyllanthoideae between *Spondianthus* and *Dicoelia*. It was retained therein by Webster (1994, no. 59, as *incertae sedis*). Kew includes it with Pandaceae. The past association with *Dicoelia* is significant as that Malesian genus, too, might be referable to Pandaceae.

    Gilg, E. (1908). Flacourtiaceae africanae. Bot. Jahrb. Syst. 40: 444-518, illus. (Beiträge zur Flora von Afrika XXXII.) Ge. —*Centroplacus*, pp. 516-518 including fig. 3; note on *C. glaucinus* Pierre of which the author had received much additional material. [Current opinion refers it to Pandaceae.]

    Pax, F. (with K. Hoffmann) (1911). *Microdesmis*. In A. Engler (ed.), Das Pflanzenreich, IV 147 III (Euphorbiaceae-Cluytieae): 105-108. Berlin. (Heft 47.) La/Ge. — Includes under 'Species excludendae' (p. 108) *M. paniculata* Pax, synonymised with *Centroplacus glaucinus* Pierre following Gilg (1908; see above) who referred the genus, dubiously, to Flacourtiaceae.

Centroplacus Pierre, Bull. Mens. Soc. Linn. Paris, II, 1: 114 (1899).
    WC. Trop. Africa. 23. Phan.

Centroplacus glaucinus Pierre, Bull. Mens. Soc. Linn. Paris, II, 1: 115 (1899).
    Cameroon, Equatorial Guinea, Gabon. 23 CMN EQG GAB. Phan.
    Microdesmis paniculata Pax, Bot. Jahrb. Syst. 28: 25 (1899).

Cremostachys

Synonyms:
Cremostachys Tul. === **Galearia** Zoll. & Moritzi
Cremostachys fulva Tul. === **Galearia fulva** (Tul.) Miq.
Cremostachys lindlaeana Tul. === **Galearia fulva** (Tul.) Miq.

Galearia

6 species, SE. Asia, Malesia and Solomon Is.; shrubs or small trees with more or less 'phyllanthoid' foliage and drupaceous fruits, the endocarp stony. Flowers appear in spikes, those in *G. aristifera* almost grass-like. The position of this genus in Pandaceae, first argued by Pierre in the 1890s, was reaffirmed following studies by Forman (1966). Its closest relative is *Microdesmis* as Pierre early realised. This is followed in most 'modern' systems of

flowering plants. Webster (1994, as no. 101), however, retains it in Euphorbiaceae, assigned to Acalyphoideae. Pax (1911) covered 16 species (2 illustrated) in 2 sections (these latter being retained by Pax & Hoffmann (1931; see General)). Forman (1971), however, considered there to be but 6 species; he placed them in two sections *Galearia* and *Orthopetalum*. A new revision of Malesian species has been prepared by Metilistina Sasinggala. The generic name has been conserved against *Galearia* C. Presl (Leguminosae).

Pax, F. (with K. Hoffmann) (1911). *Galearia*. In A. Engler (ed.), Das Pflanzenreich, IV 147 III (Euphorbiaceae-Cluytieae): 97-104. Berlin. (Heft 47.) La/Ge. — 14 species in 2 sections; a further two species of uncertain position (now referred to a third taxon, subgenus *Orthopetalum*). [Genus since transferred to Pandaceae; see Forman 1966.]

Forman, L. (1960). Notes on *Galearia* Zoll. & Mor. Kew Bull. 14: 311-317. En. — Notes on two species left by Pax & Hoffmann as *incertae sedis*, *GG. maingayi* and *celebica*; redescriptions included along with exsiccatae, vernacular names and habitats; general notes on genus including proposal of a subgenus *Orthopetalum* for the two species under consideration.

Forman, L. (1966). The reinstatement of *Galearia* and *Microdesmis* in the Pandaceae, with appendices by C. Metcalfe and N. Parameswaran. Kew Bull. 20: 309-321, illus. En. — Arguments, with evidence from vegetative and wood anatomy, for transfer to *Galearia* to Pandaceae; survey of the history of *Panda* and comparison of it with *Galearia* and *Microdesmis* (table, p. 312); character review; conclusions; key to genera and, in *Galearia*, subgenera (p. 318).

• Forman, L. (1971). A synopsis of *Galearia* Zoll. & Mor. Kew Bull. 26: 153-165, illus. En. — Synoptic revision (6 species) with key, descriptions of novelties, synonymy, references, localities with exsiccatae, indication of distribution, and commentary; list of collections seen at end. *G. fulva* has a very long synonymy reflecting imperfect material, overlapping studies, and formerly narrow conceptions of species. [For extension of range to Thailand and Indo-China, see *idem* in Kew Bull. 29: 212 (1974).]

Galearia Zoll. & Moritzi in A.Moritzi, Syst. Verz.: 19 (1846), nom. cons.
Trop. Asia. 41 42.
Bennettia R.Br. in J.J.Bennett, Pl. Jav. Rar.: 249 (1838).
Cremostachys Tul., Ann. Sci. Nat., Bot., III, 15: 259 (1851).

Galearia aristifera Miq., Fl. Ned. Ind., Eerste Bijv.: 471 (1861).
Pen. Malaysia, Sumatera, Borneo (Sarawak, W. Kalimantan). 42 BOR MLY SUM. Phan.
Galearia leptostachya Pax in H.G.A.Engler, Pflanzenr., IV, 147, III: 102 (1911).

Galearia celebica Koord., Meded. Lands Plantentuin 19: 584, 626 (1898).
NE. Sulawesi, New Guinea, Bismarck Archip., Solomon Is. 42 BIS NWG SUL 60 SOL. Phan.

var. **celebica**
NE. Sulawesi, New Guinea, Bismarck Archip., Solomon Is. 42 BIS NWG SUL 60 SOL. Phan.

var. **pubescens** Forman, Kew Bull. 14: 315 (1960).
New Guinea, Bismarck Archip. 42 BIS NWG. Phan.

Galearia filiformis (Blume) Boerl., Handl. Fl. Ned. Ind. 3: 282 (1890).
Pen. Malaysia, Sumatera, Jawa. 42 JAW MLY SUM. Nanophan.
* *Antidesma filiforme* Blume, Bijdr.: 1124 (1827).
Galearia pedicellata Zoll. & Moritzi in A.Moritzi, Syst. Verz.: 19 (1846).
Galearia sessilis Zoll. & Moritzi in A.Moritzi, Syst. Verz.: 19 (1846).

Galearia fulva (Tul.) Miq., Fl. Ned. Ind. 1(2): 430 (1859).
Indo-China, W. & C. Malesia. 41 BMA THA VIE 42 BOR MLY PHI SUM. Nanophan. or phan.
* *Cremostachys fulva* Tul., Ann. Sci. Nat., Bot., III, 15: 261 (1851).

Cremostachys lindlaeana Tul., Ann. Sci. Nat., Bot., III, 15: 262 (1851). *Galearia lindlaeana* (Tul.) Hook.f., Fl. Brit. India 5: 379 (1887).

Bennettia affinis R.Br. in J.J.Bennett, Pl. Jav. Rar.: 249 (1852). *Galearia affinis* (R.Br.) Miq., Fl. Ned. Ind. 1(2): 430 (1859).

Bennettia finlaysoniana R.Br. in J.J.Bennett, Pl. Jav. Rar.: 251 (1852). *Galearia finlaysoniana* (R.Br.) Miq., Fl. Ned. Ind. 1(2): 429 (1859).

Bennettia jackiana R.Br. in J.J.Bennett, Pl. Jav. Rar.: 251 (1852). *Galearia jackiana* (R.Br.) Miq., Fl. Ned. Ind. 1(2): 430 (1859).

Bennettia javanica R.Br. in J.J.Bennett, Pl. Jav. Rar.: 249 (1852).

Bennettia pedicellata R.Br. in J.J.Bennett, Pl. Jav. Rar.: 251 (1852). *Galearia pedicellata* (R.Br.) Miq., Fl. Ned. Ind. 1(2): 430 (1859), nom. illeg. *Galearia caudata* Forman, Kew Bull. 14: 316 (1960).

Bennettia phlebocarpa R.Br. in J.J.Bennett, Pl. Jav. Rar.: 251 (1852). *Galearia phlebocarpa* (R.Br.) Miq., Fl. Ned. Ind. 1(2): 429 (1859).

Bennettia wallichii R.Br. in J.J.Bennett, Pl. Jav. Rar.: 251 (1852). *Galearia wallichii* (R.Br.) Kurz, Prelim. Rep. Forest Pegu, App. A: cxiii (1875).

Galearia angustifolia Miq., Fl. Ned. Ind., Eerste Bijv.: 470 (1861).

Galearia elliptica Miq., Fl. Ned. Ind., Eerste Bijv.: 469 (1861).

Galearia splendens Miq., Fl. Ned. Ind., Eerste Bijv.: 469 (1861).

Galearia sumatrana Miq., Fl. Ned. Ind., Eerste Bijv.: 469 (1861).

Bennettia subulata Müll.Arg. in A.P.de Candolle, Prodr. 15(2): 1039 (1866). *Galearia subulata* (Müll.Arg.) Hook.f., Fl. Brit. India 5: 379 (1887).

Galearia helferi Hook.f., Fl. Brit. India 5: 378 (1887).

Galearia philippinensis Merr., Philipp. J. Sci., C 9: 482 (1914 publ. 1915).

Galearia minor Gage, Rec. Bot. Surv. India 9: 234 (1922).

Galearia ridleyi Gage, Rec. Bot. Surv. India 9: 235 (1922).

Galearia sessiliflora Merr., J. Straits Branch Roy. Asiat. Soc. 86: 320 (1922).

Galearia dongnaiensis Pierre ex Gagnep., Bull. Soc. Bot. France 71: 1025 (1924 publ. 1925).

Galearia fusca Ridl., Fl. Malay Penins. 3: 257 (1924).

Galearia dolichobotrys Merr., Philipp. J. Sci. 29: 384 (1926).

Galearia maingayi Hook.f., Fl. Brit. India 5: 377 (1887).
Pen. Malaysia, S. Sumatera, Borneo. 42 BOR MLY SUM. Phan.

Galearia stenophylla Merr., J. Straits Branch Roy. Asiat. Soc. 86: 320 (1922).
Pen. Malaysia, Borneo (Sarawak, Sabah). 42 BOR MLY. Nanophan. or phan.
   Galearia lancifolia Ridl., Bull. Misc. Inform. Kew 1926: 476 (1926).

Synonyms:
Galearia affinis (R.Br.) Miq. === **Galearia fulva** (Tul.) Miq.
Galearia angustifolia Miq. === **Galearia fulva** (Tul.) Miq.
Galearia caudata Forman === **Galearia fulva** (Tul.) Miq.
Galearia dolichobotrys Merr. === **Galearia fulva** (Tul.) Miq.
Galearia dongnaiensis Pierre ex Gagnep. === **Galearia fulva** (Tul.) Miq.
Galearia elliptica Miq. === **Galearia fulva** (Tul.) Miq.
Galearia finlaysoniana (R.Br.) Miq. === **Galearia fulva** (Tul.) Miq.
Galearia fusca Ridl. === **Galearia fulva** (Tul.) Miq.
Galearia helferi Hook.f. === **Galearia fulva** (Tul.) Miq.
Galearia jackiana (R.Br.) Miq. === **Galearia fulva** (Tul.) Miq.
Galearia lancifolia Ridl. === **Galearia stenophylla** Merr.
Galearia leptostachya Pax === **Galearia aristifera** Miq.
Galearia lindlaeana (Tul.) Hook.f. === **Galearia fulva** (Tul.) Miq.
Galearia minor Gage === **Galearia fulva** (Tul.) Miq.
Galearia pedicellata Zoll. & Moritzi === **Galearia filiformis** (Blume) Boerl.

Galearia pedicellata (R.Br.) Miq. === **Galearia fulva** (Tul.) Miq.
Galearia philippinensis Merr. === **Galearia fulva** (Tul.) Miq.
Galearia phlebocarpa (R.Br.) Miq. === **Galearia fulva** (Tul.) Miq.
Galearia ridleyi Gage === **Galearia fulva** (Tul.) Miq.
Galearia sessiliflora Merr. === **Galearia fulva** (Tul.) Miq.
Galearia sessilis Zoll. & Moritzi === **Galearia filiformis** (Blume) Boerl.
Galearia splendens Miq. === **Galearia fulva** (Tul.) Miq.
Galearia subulata (Müll.Arg.) Hook.f. === **Galearia fulva** (Tul.) Miq.
Galearia sumatrana Miq. === **Galearia fulva** (Tul.) Miq.
Galearia wallichii (R.Br.) Kurz === **Galearia fulva** (Tul.) Miq.

Microdesmis

11 species, Africa, Asia; shrubs or small trees with more or less 'phyllanthoid' foliage, *M. puberula* in Africa showing a preference for moist forest. Flowers develop in small axillary clusters, while the drupaceous fruits feature a stony endocarp. Two sections were recognised by Pax & Hoffmann (1931; see **General**). Opinions vary on the placement of this genus, although it has long been considered close to *Galearia*. Pierre included it in his concept of Pandaceae in the 1890s, but this was not generally taken up. Its position as part of Pandaceae was effectively reaffirmed only in the 1960s following studies by Forman. This is followed in most 'modern' systems of the flowering plants. Webster (1994, as no. 100), however, includes it in Galearieae within Euphorbiaceae-Acalyphoideae. The revision by Léonard (1961) for Africa, recognising 8 species, remains standard there; however, research towards a new revision is at present in progress at the Real Jardín Botánico, Madrid.

Pax, F. (with K. Hoffmann) (1911). *Microdesmis*. In A. Engler (ed.), Das Pflanzenreich, IV 147 III (Euphorbiaceae-Cluytieae): 105-108. Berlin. (Heft 47.) La/Ge. — 2 species, each in its own section (one in Africa, the other in tropical Asia). [Now entirely out of date for Africa; see Léonard 1961.]

• Léonard, J. (1961). Notulae systematicae XXXI. Révision des espèces africaines de *Microdesmis* (Euphorbiacées). Bull. Jard. Bot. État 31: 159-187, illus., maps. Fr. — Synoptic treatment, with descriptions of novelties but no keys, of 8 species; synoptic table and geographical review, pp. 191-194, with 3 maps, followed by conclusions (among them that it was necessary to have a sound idea of species before drawing inferences with respect to ecology, phytosociology or phytogeography; also, that the species recognised were distinguished one from another by a correlation of several small constant features). The synopsis proper is preceded by a historical survey and detailed observations on certain species.

Forman, L. (1966). The reinstatement of *Galearia* and *Microdesmis* in the Pandaceae, with appendices by C. Metcalfe and N. Parameswaran. Kew Bull. 20: 309-321. En. — Arguments, with evidence from vegetative and wood anatomy, for transfer of *Microdesmis* to Pandaceae; survey of the history of *Panda* and comparison of it with *Galearia* and *Microdesmis* (table, p. 312); character review; conclusions; key to genera (p. 318).

Microdesmis Planch., Hooker's Icon. Pl. 8: t. 758 (1848).
　Trop. Africa, Trop. Asia. 22 23 25 26 36 41 42.
　　Tetragyne Miq., Fl. Ned. Ind., Eerste Bijv.: 463 (1861).
　　Worcesterianthus Merr., Philipp. J. Bot., C 9: 288 (1914).

Microdesmis afrodecandra Floret, A.M.Louis & J.M.Reitsma, Bull. Mus. Natl. Hist. Nat., B, Adansonia 11: 104 (1989).
　Gabon. 23 GAB.

Microdesmis camerunensis J.Léonard, Bull. Jard. Bot. État 31: 189 (1961).
　Cameroon. 23 CMN. Nanophan. or phan.

Microdesmis caseariifolia Planch. ex Hook., Hooker's Icon. Pl. 8: 758 (1848). *Microdesmis caseariifolia* f. *genuina* Pax in H.G.A.Engler, Pflanzenr., IV, 147, III: 106 (1911), nom. inval.
 S. China, Hainan, Indo-China, W. & C. Malesia. 36 CHC CHH CHS 41 BMA CBD LAO THA VIE 42 BOR MLY PHI SUM. Nanophan. or phan.
  Tetragyne acuminata Miq., Fl. Ned. Ind., Eerste Bijv.: 464 (1861).
  Microdesmis caseariifolia f. *sinensis* Pax in H.G.A.Engler, Pflanzenr., IV, 147, III: 106 (1911).
  Microdesmis philippinensis Elmer, Leafl. Philipp. Bot. 4: 10 (1911).

Microdesmis haumaniana J.Léonard, Bull. Jard. Bot. État 31: 185 (1961).
 Congo, Zaire, Gabon, Angola. 23 CON GAB ZAI 26 ANG. Nanophan. or phan.

Microdesmis kasaiensis J.Léonard, Bull. Jard. Bot. État 31: 182 (1961).
 Zaire. 23 ZAI. Phan.

Microdesmis keayana J.Léonard, Bull. Jard. Bot. État 31: 180 (1961).
 W. Trop. Africa. 22 BEN GHA GUI IVO LBR NGA SIE TOG. Nanophan.
  Microdesmis puberula var. *chevalieri* Beille, Bull. Soc. Bot. France 55(8): 84 (1908).

Microdesmis klainei J.Léonard, Bull. Jard. Bot. État 31: 189 (1961).
 Gabon. 23 GAB. Nanophan. or phan.

Microdesmis magallanensis (Elmer) Steenis, Acta Bot. Neerl. 4: 480 (1955).
 Philippines. 42 PHI.
  * *Flacourtia magallanensis* Elmer, Leafl. Philipp. Bot. 4: 1519 (1912). *Worcesterianthus magallanensis* (Elmer) Merr., Philipp. J. Bot., C 10: 270 (1915).
  Worcesterianthus caseariifoides Merr., Philipp. J. Bot., C 9: 288 (1914).

Microdesmis pierlotiana J.Léonard, Bull. Jard. Bot. État 31: 171 (1961).
 Zaire. 23 ZAI. Nanophan. or phan.

Microdesmis puberula Hook.f. ex Planch., Hooker's Icon. Pl. 8: t. 758 (1848).
 Trop. Africa. 22 NGA 23 CAB CAF CMN CON EQG GAB GGI ZAI 25 UGA 26 ANG. Nanophan. or phan.
  Microdesmis zenkeri Pax, Bot. Jahrb. Syst. 23: 531 (1897).

Microdesmis yafungana J.Léonard, Bull. Jard. Bot. État 31: 176 (1961).
 Zaire. 23 ZAI. Nanophan. or phan.

Synonyms:
Microdesmis caseariifolia f. *genuina* Pax === **Microdesmis caseariifolia** Planch. ex Hook.
Microdesmis caseariifolia f. *sinensis* Pax === **Microdesmis caseariifolia** Planch. ex Hook.
Microdesmis paniculata Pax === **Centroplacus glaucinus** Pierre
Microdesmis philippinensis Elmer === **Microdesmis caseariifolia** Planch. ex Hook.
Microdesmis puberula var. *chevalieri* Beille === **Microdesmis keayana** J.Léonard
Microdesmis puberula var. *macrocarpa* Pax & K.Hoffm. === ?
Microdesmis zenkeri Pax === **Microdesmis puberula** Hook.f. ex Planch.

Panda

1 species, W. and WC. Africa (Ivory Coast to central Congo). *P. oleosa* becomes a big tree to 35 by 1 m, with a dense, evergreen crown. The flowers appear in large, sometimes clustered axillary or ramiflorous spikes, while the seeds, which yield an edible oil but have a stony endocarp, are dispersed by elephants. Germination is reportedly slow (W. Hawthorne, *Ecological profiles of Ghanian forest trees*: 236 (1995). For illustrations, see Engler (1912; reused in *Pflanzenwelt Afrikas*, 3(1), as fig. 317 (1915)). The genus is the type of Pandaceae, given its own order in 1912 but until the 1960s usually assigned elsewhere to a place preceding

Geraniales and so revised for *Die natürlichen Pflanzenfamilien* by Mildbraed (1931; see **General**). Forman (1966) showed it to be related to *Microdesmis* and *Galearia*, confirming earlier thoughts by Pierre. Within this alliance, it has been shown to have the largest number of advanced features (Forman 1966, 1968). Webster (Synopsis, 1994, as no. 102) referred it to Galearieae within Euphorbiaceae-Acalyphoideae.

Chevalier, A. & A. Guillaumin (1911). Pandacées. Mém. Soc. Bot. France 8: 202-205. (A. Chevalier, Novitates florae africanae.) Fr. — Features a detailed treatment of *Panda* (including historical review), with reduction of *Porphyranthus*.

Engler, A. (1912). Ölsamenbaum Westafrikas. Notizbl. Bot. Gart. Mus. Berlin-Dahlem 5: 274-276, illus. Ge. — Detailed treatment with a historical outline, description and illustration; reduction of *Sorindeia rubriflora*; notes on relationship along with proposal of a new order, Pandales; indication of distribution in Gabon and elsewhere.

Forman, L. (1966). The reinstatement of *Galearia* and *Microdesmis* in the Pandaceae, with appendices by C. Metcalfe and N. Parameswaran. Kew Bull. 20: 309-321, illus. En. — Includes a survey of the history of *Panda* and comparison of it with *Galearia* and *Microdesmis* (table, p. 312).

Forman, L. (1968). The systematic position of *Panda*. Proc. Linn. Soc. London 179: 269-270. En. — Review of evidence with description of anatomical features shared by *Panda* with *Galearia* and *Microdesmis*.

Panda Pierre, Bull. Mens. Soc. Linn. Paris 2: 1255 (1896).
    W. & WC. Trop. Africa. 22 23.
        Porphyranthus Engl., Bot. Jahrb. Syst. 26: 367 (1899).

Panda oleosa Pierre, Bull. Mens. Soc. Linn. Paris 2: 1255 (1896).
    W. & WC. Trop. Africa. 22 GHA IVO LBR NGA 23 CAB CMN EQG GAB ZAI. Phan.
        Porphyranthus zenkeri Engl., Bot. Jahrb. Syst. 26: 367 (1899).
        Sorindeia rubriflora Engl., Bot. Jahrb. Syst. 46: 338 (1911).

Porphyranthus

Synonyms:
Porphyranthus Engl. === **Panda** Pierre

Tetragyne

Synonyms:
Tetragyne Miq. === **Microdesmis** Planch.
Tetragyne acuminata Miq. === **Microdesmis caseariifolia** Planch. ex Hook.

Worcesterianthus

The name honours Dean C. Worcester, Interior Secretary and government patron of science during the early years of United States administration in the Philippines.

Synonyms:
Worcesterianthus Merr. === **Microdesmis** Planch.
Worcesterianthus casearioides Merr. === **Microdesmis magallanensis** (Elmer) Steenis
Worcesterianthus magallanensis (Elmer) Merr. === **Microdesmis magallanensis** (Elmer) Steenis

Summary of unplaced taxa and names

Names included here represent good taxa without at present a satisfactory generic placement, taxa with no modern evaluation, taxa (e.g. those of Rafinesque) without vouchers, or *nomina nuda*. Certain details are given where possible.

Acalypha bracteata Miq., Fl. Ned. Ind. 1(2): 406 (1859).

=== ? – Indonesia (Sumbawa, Bima), *Zollinger*. Not in the regional review of Airy-Shaw (1982; see **Malesia**). Transferred to *Ricinocarpus* by Kuntze (see below)

Acalypha celebica Koord., Meded. Lands Plantentuin 19: 624 (1898).

=== ? – Indonesia (Sulawesi, NE. part), *Koorders 16782* (BO)

Acalypha digyneia Raf., Fl. Ludov.: 112 (1817).

=== ? – USA (Louisiana), based on 'Acalyphe 2' of Robin, Travels 3: 516 (1806). Not disposed of, though transferred to *Ricinocarpus* by Kuntze (see below); remains in, for example, the latest checklist for Kentucky

Acalypha divaricata Raf., New Fl. 1: 44 (1836).

=== ? – USA ('Virginia and Kentucky'), annual. Not disposed of

Acalypha fruticulosa Raf., Fl. Ludov.: 112 (1817).

=== ? – USA (Louisiana); based on 'Acalyphe monoique' of Robin, Travels 3: 516 (1806). Not disposed of, though transferred to *Ricinocarpus* by Kuntze (see below)

Acalypha hirsuta Mart. ex Colla, Herb. Pedem. 5: 114 (1836).

=== ? – Brazil (Rio de Janeiro, Serra d'Estrella), *Martius* (TO?)

Acalypha jamaicensis Raf., New Fl. 1: 46 (1836).

=== ? – 'Jamaica and possibly Florida'; not disposed of

Acalypha jardinii Müll.Arg., Linnaea 34: 36 (1865).

=== ? – 'Kouma', *E. Jardin* in herb. Lenormand (P). Close to *A. cuspidata*. Florence, *Flore de la Polynésie française*, 1 (1997) thinks plant and label are mixed, and does not accept it as part of that flora

Acalypha muralis Zipp. ex Span., Linnaea 15: 350 (1841).

=== ? – Indonesia (Timor), *Zippel*. Miquel in *Florae indiae batavae* (1859) treated it as 'dubia'.

Acalypha ruderalis Mart. ex Colla, Herb. Pedem. 5: 114 (1836), nom. nud.

=== ? – Brazil (Rio de Janeiro, Serra d'Estrella), *Martius* (TO?)

Acalypha scleropumila A.Chev., Rev. Int. Bot. Appl. Agric. Trop. 31: 268 (1951), no latin descr.

=== ? – Congo-Brazzaville, Pays Batéké, *Chevalier s/n*. Near *A. polymorpha*

Acalypha tenera Royle, Ill. Bot. Himal. Mts.: 327 (1836), nom. nud.

=== ? – Himalaya; mentioned in running text as appearing only in rainy season

Acalypha tiliifolia Poir. in J.B.A.M.de Lamarck, Encycl. 6: 203 (1804).

=== Acalypha sp. – Hispaniola, *herb. Jussieu* (fide Poiret); a shrub or small tree. Not disposed of

Acalypha urticifolia Poir. in J.B.A.M.de Lamarck, Encycl. 6: 208 (1804).

=== Acalypha sp. – 'America', *herb. Lamarck* (fide Poiret); an annual. Not disposed of

Actinostemon anisandrus (Griseb.) Pax in H.G.A.Engler, Pflanzenr., IV, 147, V: 79 (1912).

=== Sebastiania sp. (Jablonski 1969; see **Actinostemon**). – Argentina (Andine region, Oran), *Lorentz & Hieronymus 310*. Pax & Hoffmann in 1912 had already expressed doubt about placement in *Actinostemon*

Actinostemon verrucosum Glaz., Bull. Soc. Bot. France 59(3): 634 (1912 publ. 1913), nom. nud.

=== ? – Brazil, *Glaziou 13481* (P)

Adelia myrtifolia Vent. ex Spreng., Syst. Veg. 3: 147 (1826).

=== Adelia sp. – Origin unknown

Adelia patens Baill., Étude Euphorb.: 418 (1858), nom. nud.

=== Adelia sp. – Based on material in herb. Delessert (G)

Adelia scandens Span., Companion Bot. Mag. 1: 350 (1835), nom. nud.

=== ? – Indonesia (Timor and neighboring islands); no supporting data

Adelia timoriana Span., Companion Bot. Mag. 1: 350 (1835), nom. nud.

=== ? – Indonesia (Timor and neighboring islands); no supporting data

Adriana gaudichaudii var. thomasiifolia Baill., Adansonia 6: 312 (1866).

=== ? – Australia, *Gaudichaud* (Voyage de l'Uranie) *20, 21* (P). Overlooked by Pax (1910)

Agyneia lanceolata F.Dietr., Vollst. Lex. Gärtn. 1: 186 (1802).

=== ? – Of horticultural origin

Aleurites erratica O.Deg., I.Deg. & K.Hummel, Phytologia 38: 362 (1978).

=== ? – Canton, Phoenix Is., *Degener & Degener 24627*; drift seeds referable not to *Aleurites moluccana* but to a non-aleuretoid uniovulate Euphorbiacea (Stuppy et al., 1999)

Anisophyllum humboldtii (Donn) Haw., Syn. Pl. Succ.: 160 (1812).

=== ? – Origin unknown

*Euphorbia humboldtii Donn, Hort. Cantab. ed. 6: 132 (1811), nom. nud.

Antidesma ponapense Kaneh., J. Dept. Agric. Kyushu Imp. Univ. 4: 347 (1935), nom. nud.

=== ? – Micronesia (Ponape); not disposed of

Antidesma praegrandifolium S.Moore, J. Bot. 61(Suppl.): 46 (1923).

=== Aporusa sp. – Papua New Guinea (Central: Sogere area), *Forbes 250*. Disposition awaits the outcome of current research on that genus by Anne Schot

Antidesma puberum Zipp. ex Span., Linnaea 15: 350 (1841), nom. nud.

=== ? – Indonesia (Timor), *Zippel*; not disposed of

Antidesma rugosum Wall. ex Voigt, Hort. Suburb. Calcutt.: 295 (1845), nom. nud.

=== ? – India (Indian Botanic Garden, ex-Silhet area, Bangladesh); based on cultivated plants

Aporusa aberrans Gagnep., Bull. Soc. Bot. France 70: 232 (1923).

=== Antidesma sp. – Laos (southern part, Sé Lamphao), *Harmand* (P)

Aporusa australiana F.Muell., Syst. Census Pl. Austral.: 20 (1883), nom. nud.

=== ? – Australia; no specimens known (Bailey, Queensland Flora 5: 1431 (1902))

Aporusa lanceolata Hance, J. Bot. 17: 14 (1879), nom. illeg.

=== ? – China (Guangdong, along West River), *Sampson*. Overlooked in *Flora reipublicae popularis sinicae* (vol. 44). **Not** a new species, but an application by Hance of *Aporusa lanceolata* (Tul.) Thwaites

*Chiropetalum argentinense Skottsb.

Argythamnia pauciflora Mirb. ex Steud., Nomencl. Bot., ed. 2, 1: 128 (1840), nom. nud.

=== ? – 'Amer. austr.'; no supporting data

Banalia muricata Raf., Autik. Bot.: 50 (1840).

=== Croton sp. – USA (Florida); not disposed of

Bertya neglecta Dummer, J. Bot. 52: 151 (1914).

=== Bertya sp. – Australia (Barrens, north of Arbuthnot Range), *Fraser* (CGE). Related to *B. rosmarinifolia* Planch.

Blachia chunii Y.T.Chang & P.T.Li, Guihaia 8: 53 (1988), two types cited.

=== ? – China (Hainan), *Wang 32737* (female flowers), *Lau 27874* (fruits)(both SCBI)

Breynia paniculata Spreng., Pl. Min. Cogn. Pug. 2: 93 (1815).

=== ? – S. Asia; probably a synonym but not yet disposed of

Claoxylon ubanghense A.Chev., Etudes Fl. Afr. Centr. Franc. 1: 277 (1913), nom. nud.

=== ? – Central African Republic (Oubangui), *Chevalier 5122, 10854* (P)

Cleidion populifolium Zipp. ex Span., Linnaea 15: 349 (1841), nom. nud.

=== ? – Indonesia (Timor), *Zippel*

Cluytiandra perrieri Leandri ex Humbert, Notul. Syst. (Paris) 7: 193 (1939).

=== ? – Madagascar, Masaola, *Perrier 9721* (P); shrub 3-4 m

Colliguaja patagonica Speg., Revista Fac. Agron. Univ. Nac. La Plata 3: 572 (1897).

=== ? – Argentina (Chubut, Santa Cruz). A largely leafless, glabrous, stem-succulent shrub. Removed from *Sapium* by Kruijt (1996; see that genus); position problematic, with either a new genus required or inclusion in, and redefinition of, *Stillingia* (Esser 1994: 62-63; see **Euphorbioideae (except Euphorbieae)**)

Croton alegrensis Glaz., Bull. Soc. Bot. France 59(3): 616 (1912 publ. 1913), nom. nud.

=== ? – Brazil (Espirito Santo), *Glaziou 11511*

Croton althaeifolius Noronha, Verh. Batav. Genootsch. Kunsten 5(4): 12 (1790), nom. illeg.

=== ? – Indonesia (Java), *Noronha*; no specimens extant

Croton briquetianus Herter, Estud. Bot. Reg. Uruguay 4: 79 (1931), nom. nud.

=== ? – Uruguay, without supporting data. [Not in the author's later *Flora ilustrada del Uruguay*.]

Croton chevalieri Gagnep., Bull. Soc. Bot. France 68: 550 (1921 publ. 1922), nom. illeg.

=== ? – Vietnam (near Bien Hoa), *Chevalier 32048* (P). Accepted by Nguyen Nghia Thin in *Euphorbiaceae of Vietnam* (1995)

Croton collinus Phil., Fl. Atacam.: 49 (1860), nom. illeg.

=== ? – Chile (Atacama Desert); 'fruticosus erectus'

Croton colobocarpus Airy Shaw, Kew Bull. 23: 76 (1969).

=== Trigonostemon sp. – NE. Thailand, *Kerr 8479* (K). A caudiciform hemicryptophyte.

Croton decarianus Pilg. ex Glaziou, Bull. Soc. Bot. France 59(3): 620 (1912 publ. 1913), nom. nud.

=== ? – Brazil (Goiás), *Glaziou 22106*

Croton digitatus Fisch. ex Steud., Nomencl. Bot. 1: 240 (1821), nom. nud. *Jatropha fischeri* Steud., Nomencl. Bot., 2nd edn., 1: 799 (1841), nom. nud.

=== ?

Jatropha fischeri Steud.; Manihot digitata Sweet (q.v.)

Croton dioicus Roxb., Fl. Ind. ed. 1832, 3: 680 (1832), nom. illeg.

=== ? – 'Malay Archipelago', *Roxburgh*. No disposition known

Croton dracena Larrañaga, Escritos 2: 258 (1923), nom. nud.

=== ? – Uruguay, without supporting data; not disposed of

Croton erythraema Mart. ex Peckolt, Arch. Pharm. (Berlin) 158: 142 (1861).

=== ? – Brazil, without precise locality; extensive commentary but no essential botanical or topographical details

Croton exiliatus Croizat, Darwiniana 6: 454 (1944), nom. nud.

=== ? – Angola; merely mentioned in commentary under *Croton microgyne*

Croton furcellatus Baill., Bull. Mens. Soc. Linn. Paris 2: 967 (1891).

=== ? – Madagascar, Toamasina (Fort-Dauphin), *Scott Elliott 180*; not disposed of

Croton glaber Glaz., Bull. Soc. Bot. France 59(3): 619 (1912 publ. 1913), nom. nud.

=== ? – Brazil (Minas Gerais), *Glaziou 15403*

Croton glandulatus Vell., Fl. Flumin. 10: t. 73 (1831); descr. in Arq. Mus. Nac. Rio de Janeiro 5: 412 (1881).

=== ? – Brazil; not disposed of

Croton insignis Glaz., Bull. Soc. Bot. France 59(3): 617 (1912 publ. 1913), nom. nud.

=== ? – Brazil (Goiás), *Glaziou 15403*

Croton luridus Geiseler, Croton. Monogr.: 26 (1807).

=== ? – Origin unknown, *Commerson* (P)

Croton macrourus Mart. ex Colla, Herb. Pedem. 5: 110 (1836).

=== ? – Brazil (Rio de Janeiro), Cabo Frio, *Martius*

Croton nettoanus Glaz., Bull. Soc. Bot. France 59(3): 615 (1912 publ. 1913), nom. nud.

=== ? – Brazil (Espirito Santo), *Glaziou 11521*

Croton pauciflorus Span., Linnaea 15: 348 (1841), nom. nud.

=== ? – Indonesia (Timor), *Spanoghe*

Croton polygamus Geiseler, Croton. Monogr.: 24 (1807).

=== ? – Origin unknown; described from cultivated plants raised at Kew

Croton reticulatus Thunb., Fl. Jav.: 23 (1825), nom. illeg.

=== ? – Jawa, *Thunberg* (UPS); leaves opposite, by contrast with *C. balsamifer* Jacq. (= *C. flavens* L.)

Croton rotundifolius Glaz., Bull. Soc. Bot. France 59(3): 620 (1912 publ. 1913), nom. nud.

=== ? – Brazil (Minas Gerais), *Glaziou 17759*

Croton sublanceolatus Herter, Revista Sudamer. Bot. 3: 165 (1936), nom. nud.

=== ? – Uruguay, without supporting data; not disposed of

Croton thebaloides Teijsm. & Binn., Cat. Hort. Bot. Bogor.: 392 (1866), nom. nud.

=== ? – Origin unknown; described from cultivated plants in Kebun Raya, Bogor, Indonesia, earlier received from Sydney Botanic Gardens

Croton urophylla Wall. ex Voigt, Hort. Suburb. Calcutt.: 157 (1845), nom. nud.

=== ? – NE. Bengal or Assam; described from cultivated plants in the Indian Botanic Garden, Sibpur, India

Croton vahlii Geiseler, Croton. Monogr.: 41 (1807) .

=== ? – Origin unknown; in herb. Vahl (C) according to author

Cunuria casiquiarensis Croizat, J. Arnold Arbor. 26: 192 (1945).

=== ? – Venezuela (Amazonas), Alto Casiquiare, *Ll. Williams 15690*. Not certainly in *Cunuria* according to Croizat as the inflorescence was aberrant.

Dalechampia bathiana Leandri, Notul. Syst. (Paris) 11: 35 (1943), nom. provis.

=== ? – Madagascar (western part), *Perrier 1686, 9914, 13461*; scandent small liana, the basal leaves in a rosette. No disposition available.

Drypetes glauca Griseb., Mem. Amer. Acad. Arts, n.s., 8: 157 (1860), misapp.

=== Drypetes spp.; based on *Wright 593*, Cuba. [For further details, see Krug & Urban in Bot. Jahrb. Syst. **15**: 286-361 (1892).]

Emblica pisiformis Buch.-Ham., Trans. Linn. Soc. London 13: 506 (1822).

=== Phyllanthus sp. – India, 'Shiray in Carnata, in sylvis durioribus', *Buchanan*; scandent shrub

Eriococcus caudatus Zoll., Tijdschr. Ned.-Indië 14: 173 (1857).

=== Phyllanthus sp. – Indonesia, 'in fruticetis littoralibus inter Buer et Allas ins. Sumbawa', *Zollinger 1116*, 29 September 1847

Erythrochilus multiflorus Zipp. ex Span., Linnaea 15: 349 (1841), nom. nud.

=== Claoxylon sp. – Indonesia (Timor), *Zippel* ; not disposed of

Esula multicorymbosa Haw., Syn. Pl. Succ.: 156 (1812).

=== Euphorbia sp. – Origin unknown

Euphorbia abbreviata Thuill., Fl. Env. Paris, ed. 2: 239 (1799).

=== ? – France (Paris region, Mancoussis). Evidently overlooked in subsequent French floras

Euphorbia abortiva Forssk., Fl. Aegypt.-Arab.: cxii (1775), nom. nud.

=== ? – 'Arabia Felix', Bert el Sakih, *Forskål* (C); in checklist only

Euphorbia acuta Bellardi ex Colla, Herb. Pedem. 5: 132 (1836).

=== ? – Origin unknown (collection in TO?)

Euphorbia adenophylla Steud., Nomencl. Bot., ed. 2, 1: 609 (1840), nom. nud.

=== ? – Of horticultural origin

Euphorbia angustifolia Glaz., Bull. Soc. Bot. France 59(3): 636 (1912 publ. 1913), nom. nud.

=== ? – Brazil (Minas Gerais), *Glaziou 19826*

Euphorbia ciliata Spreng., Mant. Prim. Fl. Hal.: 42 (1807).

=== ? – Origin unknown; 'ex herbario Römeri Turic.'

Euphorbia cotylophora Spreng., J. Bot. (Schrader) 4(2): 197 (1800).

=== ? – Of horticultural origin; received from Schott in Brno, Czech Republic

Euphorbia erythrocarpa Klotzsch in M.R.Schomburgk, Reis. Br.-Guiana: 848 (1848), nom. nud.

=== ? – Guyana; annual in sugar fields, *Schomburgk*

Euphorbia filifolia Glaz., Bull. Soc. Bot. France 59(3): 637 (1912 publ. 1913), nom. nud.

=== ? – Brazil (Goiás), *Glaziou 22081, 22082, 22083*

Euphorbia flerowii Woronow ex Flerov, Uchen. Zap. Rostov. Donu Gosud. Univ. V.M.Molotova, Trudy Biol. Fak. 4: 133 (1940), no latin descr.

=== Euphorbia sp. – Georgian Republic

Euphorbia geminispina Haw. ex Loud., Hort. Brit., ed. 2: 190 (1832).

=== ? – Of horticultural origin.

Euphorbia glauca Steud., Nomencl. Bot. 1: 325 (1821), nom. illeg.

=== Euphorbia moenchiana Steud. (**q.v.**)

*Tithymalus glaucus Moench

Euphorbia gomesii Croizat ex Gomes, Subsid. Estud. Fl. Niassa Portug.: 29 (1935), nom. nud.

=== ? – Mozambique, Massangulo, *Gomes*; named by Croizat but never described. A succulent

Euphorbia haematodes Boiss. in A.P.de Candolle, Prodr. 15(2): 179 (1862), nom. nud.

=== ? – Introduced for *E. sanguinea* hort. Berol., non Hochst. & Steud. (*Alectoroctonium sanguineum* Klotzsch & Garcke)

Euphorbia lanifera Haw. ex Loud., Hort. Brit., ed. 2: 190 (1832).

=== ? – Of horticultural origin.

Euphorbia × lyttoniana Dexter, Euphorbia Rev. 1: 14 (1935).

=== ?

Euphorbia magnimamma Haw. ex Loud., Hort. Brit., ed. 2: 190 (1832).

=== ? – Of horticultural origin

Euphorbia mananarensis Leandri, Notul. Syst. (Paris) 12: 69 (1945), nom. provis.

=== Euphorbia sp. – Madagascar, *Decary 9413*. A shrub in sect. *Denisophorbia*; not in *Famata Malagasy* list by M. Singer (Euphorbia Journal 3)

Euphorbia moenchiana Steud., Nomencl. Bot., ed. 2, 1: 613 (1840).

=== ? – Described from cool-house plants of unknown origin

*Tithymalus glaucus Moench.

Euphorbia montana Raf., Amer. Monthly Mag. & Crit. Rev. 1: 450 (1817).

=== ? – Sicily; not disposed of

Euphorbia multicorymbosa (Haw.) Sweet, Hort. Brit.: 357 (1818).

=== ? – Origin unknown

*Esula multicorymbosa Haw.

Euphorbia octogona Steud., Nomencl. Bot., ed. 2, 1: 613 (1840), nom. nud.

=== ? – Of horticultural origin

Euphorbia pauciflora Hill, Veg. Syst. 10: 57 (1765).

=== ? – 'Native of the East Indies and America'; in subgen. *Chamaesyce*

Euphorbia phagriformis Graessn., Hauptverz. 1937: 16 (1937), nom. nud.

=== ? – Of horticultural origin

Euphorbia pratensis Gromov, Trudy Obshch. Nauk Charkov. 1: 146 (1817), nom. nud.

=== ? – Not disposed of

Euphorbia reflexa Spreng., Syst. Veg. 3: 802 (1826).

=== ? – Origin unknown; not disposed of

Euphorbia schleinitzii Engl. ex Pax in H.G.A.Engler & K.A.E.Prantl, Nat. Pflanzenfam. 3(5): 105 (1890), nom. nud.

=== ? – Australia; not disposed of. Collected on 'Gazelle' voyage in 1875 (commanded by von Schleinitz)

Euphorbia scordioides Deflers ex Blatter, Rec. Bot. Surv. India 8: 431 (1923), nom. nud.

=== ? – Arabian Peninsula, oasis of Lahaj, *Deflers 121*

Euphorbia spathulifolia (Haw.) Steud., Nomencl. Bot. 1: 328 (1821).

=== ? – Of horticultural origin

*Galarhoeus spathulifolius Haw.

Euphorbia theodosia Sennen, Campagn. Bot.: 117 (1930-1935 publ. 1936), nom. nud.

=== ? – Not disposed of, mentioned in an itinerary as 'intermédiaire à *E. medicaginea* et à *E. pithyusa*' by Fr. Mauricio in eastern Morocco

Euphorbia × turneri Druce, Bot. Soc. Exch. Club Brit. Isles 4: 428 (1916 publ. 1917).

=== ? – Britain, Bath, Prior Park; parents given as *E. amygdaloides* and *E. pilosa*

Euphorbia ucrainica Andrz. ex Trautv., Trudy Imp. S.-Peterburgsk. Bot. Sada 9: 163 (1884), nom. nud.

=== ? – Ukraine, Kiev region; not disposed of

Euphorbia uniflora (Haw.) G.Don in J.C.Loudon, Suppl. Hort. Brit.: 588 (1850).

=== ? – Of horticultural origin

*Tithymalus uniflorus Haw.

Euphorbia verticillata Desf., Tabl. École Bot.: 205 (1804).

=== ? – Antilles; cultivated at Jardin des Plantes, Paris, in a warm house

Excoecaria diandra Müll.Arg. in C.F.P.von Martius, Fl. Bras. 11(2): 615 (1874).

=== Sebastiania sp. (fide Kruijt 1996; see **Sapium**)

*Omphalea diandra Vell. (**q.v.**)

Excoecaria glandulosa Millsp., Publ. Field Columb. Mus. 1: 305 (1896), nom. illeg.

=== ? – Mexico (Yucatán, Silam), *Gaumer 615* (F)
  Sebastiania adenophora Pax & K.Hoffm.

Excoecaria herbertiifolia Jacq., Hort. Universel 5: 289 (1844).

=== ? – Of horticultural origin

Excoecaria myrioneura Airy Shaw, Kew Bull. 33: 537 (1979).

=== Euphorbia sp.? – Indonesia (Morotai, Irian Jaya); trees with elegantly parallel-veined leaves. No flowering material is known, but the branching infructescences suggest a place at least in Euphorbieae (D. Frodin). [Illustrated on p. 1621.]

Excoecaria polyandra Griseb., Nachr. Königl. Ges. Wiss. Georg-Augusts-Univ. 1: 180 (1865).

=== ? – Cuba, *Wright 2008*. According to Pax (1912: 173-174) probably not euphorbiaceous (flowers 'monstrose', laticifers absent and vegetative anatomy aberrant). Müller earlier had proposed inclusion in Oleaceae (cf. *Forestiera*); Eggers, in *Drypetes*; Bentham, in *Gymnanthes* (cf. *Gymnanthes polyandra* below)

Galarhoeus spathulifolius Haw., Syn. Pl. Succ.: 146 (1812).

=== ? – Origin unknown; probably Euphorbia sp.

Gelonium scaligerianum A.Massal., Stud. Fl. Foss. Senigall.: 101 (1859), nom. nud., fossil.

––– ? – Described from fossils collected in Italy

Glochidion bifarium Royle, Ill. Bot. Himal. Mts.: 327 (1835 publ. 1836), nom. nud.

=== ? – Mentioned in running text as common in Himalaya; observed at 'Jureepanee'

Glochidion laurifolium Heynh., Nom. Bot. Hort. 2: 266 (1841).

=== ? – Of horticultural origin

Glochidion ovalifolium Zipp. ex Span., Linnaea 15: 346 (1841), nom. nud.

=== ? – Indonesia (Timor), *Zippel*

Glochidion puberulum Hosok., Trans. Nat. Hist. Soc. Taiwan 25: 23 (1935), nom. illeg.

=== Glochidion sp. – Micronesia (Truk or Chuuk group, Ponape). [Accepted by Fosberg, Sachet & Oliver in *A geographical checklist of the Micronesian Dicotyledoneae* (1979, in Micronesica 15: 41-295). Glassman in Flora of Ponape (1952), however, reduced it to *G. ramiflorum* J.R.Forst & G.Forst.]

Glochidion tsusimense Nakai, Proc. Third Pan-Pacif. Sc. Congr. 1: 896 (1926 publ. 1928), nom. nud.

=== Glochidion sp. – Japan (Tsushima); mentioned in a running account of island floristics

Gymnanthes polyandra (Griseb.) Benth. & Hook.f. ex B.D. Jackson, Index Kewensis 1: 1073 (1893).

=== ?

  *Excoecaria polyandra Griseb. (**q.v.**)

Hevea camargoana Pires, Bol. Mus. Paraense Emilio Goeldi, N. S., Bot. [1](52): 4 (1981).

=== Hevea sp. – Brazil (Pará, Ilha Marajó, *Rosa & Rosário 50* (MG 62200) with several paratypes

Hevea gracilis Ducke, Rev. Bot. Appl. Agric. Trop. 11: 28 (1931).

=== Hevea sp. – Brazil (Amazonas, Manaus), *Ducke*

Hevea huberiana Ducke, Rev. Bot. Appl. Agric. Trop. 9: 627 (1929).

=== Hevea sp. – Brazil (Amazonas, lower Rio Negro), Ducke; photograph

Hieronyma boliviana Pax, Repert. Spec. Nov. Regni Veg. 7: 109 (1909).

=== ? – Bolivia. According to Pax & Hoffmann (1922: 40; see **Hieronyma**) not in the genus and, moreover, the flowers are juvenile

Hieronyma reticulata (Planch.) Britton ex Rusby, Mem. Torrey Bot. Club 4: 255 (1895), nom. nud.

=== ? – Peru; allegedly based on *Antidesma reticulata* Planch. but that name merely exists *in sched.* on *Mathews 1562* (K) and was never described. Britton and Rusby added *Bang 383* (Bolivia; by Pax & Hoffmann referred to their *H. andina*, here a synonym of *H. oblonga* (Tul.) Müll.Arg.)

Hippomane cerifera Sessé & Moç., Fl. Mexic., ed. 2: 226 (1894).

=== ? – Cuba; a source of wax

Hippomane fruticosa Sessé & Moç., Fl. Mexic., ed. 2: 226 (1894).

=== ? – 'in montibus Puruanderi'. Might be *Sapium glandulosum* (L.) Morong

Hymenocardia intermedia Dinkl. ex Mildbr., Repert. Spec. Nov. Regni Veg. 41: 255 (1937), nom. nud.

=== ? – Liberia, *Dinklage 2483, 2686*. Imperfectly known, according to *Flora of West Tropical Africa*

Jatropha angustifolia Steud., Nomencl. Bot., ed. 2, 1: 799 (1840), nom. nud.

=== ? – Recommended as a 'nomen delendum' by Pax (1910; see **Jatropha**)

Jatropha arborea Glaz., Bull. Soc. Bot. France 59(3): 629 (1912 publ. 1913), nom. nud.

=== ? – Brazil (Rio de Janeiro: Larangeira), *Glaziou 16328*

Jatropha berteri Spreng., Syst. Veg. 3: 76 (1826).

=== ? – Brazil: 'ad fl. Magdalenae', *Bertero*. Later segregated as *Adenoropium berteri* by Pohl, Pl. Bras. 1: 14 (1827)

Jatropha calyculata Vahl ex Steud., Nomencl. Bot., ed. 2, 1: 799 (1840), nom. nud.

=== ? – Recommended as a 'nomen delendum' by Pax (1910; see **Jatropha**)

Jatropha diversifolia Steud., Nomencl. Bot., ed. 2, 1: 799 (1840), nom. nud.

=== ? – Recommended as a 'nomen delendum' by Pax (1910; see **Jatropha**)

Jatropha fischeri Steud., Nomencl. Bot., ed. 2, 1: 799 (1840), nom. nud.

=== ? – Recommended as a 'nomen delendum' by Pax (1910; see **Jatropha**)

*Croton digitatus Fisch. ex Steud.; Manihot digitata Sweet

Jatropha hastata Griseb., Mem. Amer. Acad. Arts, n.s., 8: 158 (1860).

=== ? – Cuba, *Wright 574, 575*

Jatropha loeflingii Aresch., Kongl. Svenska Vetenskapsakad. Handl. 39(2): 43 (1905), nom. nud.

=== ? – Included in an anatomical paper; based on cultivated material from Bogor (Indonesia)

Jatropha palmata Sessé & Moç. ex Cerv., Supl. Gaz. Lit. Mexico: 1 (2 July 1794).

=== Manihot sp. – Mexico. Incompletely described according to Pax (1910); disposition to a species not possible (Rogers & Appun 1973; see **Manihot**)

Jatropha palustris Sessé & Moç., Fl. Mexic., ed. 2: 225 (1894).

=== ? – 'in paludosis calidarium Regionum N.H. ut Acaponeta'; annual

Jatropha peltata Kunth in F.W.H.von Humboldt, A.J.A.Bonpland & C.S.Kunth, Nov. Gen. Sp. 2: 83 (1817), nom. illeg.

=== ? – Brazil, on Amazon River near Tomependa, *Humboldt*; no. 34 in Pax (1910).

Jatropha peltata C.Wright ex Sauvalle, Anales Acad. Ci. Méd. Habana 7: 155 (1870), nom. illeg.

=== ? – Cuba, San Cristobal, *Wright 3689*

Jatropha viminea Retz. ex Steud., Nomencl. Bot. 1: 426 (1821), nom. nud.

=== ? – No supporting data

Julocroton ostenii Herter, Estud. Bot. Reg. Uruguay 4: 80 (1931), nom. nud.

=== ? – Uruguay (Artigas, Salto); without supporting data. [Not in the author's later *Flora ilustrada del Uruguay*.]

Julocroton rufescens Klotzsch ex Baill., Étude Euphorb.: 376 (1858), nom. nud.

=== ? – Brazil, *Guillemin 337*

Julocroton valenzuellae (Chodat & Hassl.) Croizat, Revista Argent. Agron. 10: 136 (1943).

=== Croton sp.

*Julocroton villosissimus var. valenzuellae Chodat & Hassl. (**q.v.**)

Julocroton villosissimus var. **valenzuellae** Chodat & Hassl., Bull. Herb. Boissier, II, 5: 499 (1905).

=== Croton sp. – Paraguay, Chololo, *Hassler 6771*; Valenzuela, *6771a*

Leptopus orinocensis Klotzsch & Garcke, Monatsber. Königl. Preuss. Akad. Wiss. Berlin 1859: 250 (1859), nom. nud.

=== Euphorbia sp. (sect. (Cyrrarospermum). – Venezuela?

Macaranga anjuanensis Leandri, Notul. Syst. (Paris) 10: 162 (1942), nom. provis.
=== ? – Madagascar

Macaranga coursi Leandri, Notul. Syst. (Paris) 10: 140 (1942), nom. provis.
=== ? – Madagascar

Macaranga danguyana Leandri, Notul. Syst. (Paris) 10: 146 (1942), nom. provis.
=== ? – Madagascar

Macaranga madagascariensis Steud., Nomencl. Bot., ed. 2, 2: 86 (1841), nom. nud.
=== ? – Madagascar

Macaranga perrieri Leandri, Notul. Syst. (Paris) 10: 161 (1942), nom. provis.
=== ? – Madagascar

Macaranga thouarsii Baill., Étude Euphorb.: 432 (1858), nom. nud.
=== ? – Madagascar

Manihot berroana Beauverd, Bull. Soc. Bot. Genève 4: 69 (1912), nom. nud.
=== ? – Uruguay, Berro (G). Not mentioned by Hester, *Florula uruguayensis* (1930)

Manihot digitata Sweet, Hort. Brit., ed. 2: 458 (1830)
=== ? – Of horticultural origin; referred to *Jatropha fischeri* Steud. by Pax (1910).
*Croton digitatus Fisch. ex Steud.

Manihot diversifolia (Steud.) Sweet, Hort. Brit., ed. 2: 458 (1830)
=== ? – Of horticultural origin; referred to *Jatropha diversifolia* Steud. by Pax (1910).
*Jatropha diversifolia Steud.

Manihot japonica Semler, Trop. Agric. (Ceylon) 2: 614 (1887), nom. nud.
=== ?

Manihot neoglaziovii Pax & K.Hoffm. ex Luetzelb., Estud. Bot. Nordéste 3: 146 (1923), nom. nud.
=== ? – Brazil (Rio Grande do Norte), *Luetzelburg*; three localities cited

Manihot spinosissima Mill. ex Steud., Nomencl. Bot., ed. 2, 2: 99 (1841), nom. nud.
=== ? – By Steudel referred to *Cridoscolus vitifolius* (Mill.) Pohl.

Manihot teissonnieri A.Chev., J. Agric. Trop. 7: 57 (1907).
=== ? – West Africa, without specimen citations; treated as a 'species dubia' by Rogers & Appun (1973)

Melanthesa canescens Zipp. ex Span., Linnaea 15: 347 (1841), nom. nud.
=== Breynia sp. – Indonesia (Timor), *Zippel*

Melanthesa glaucescens Miq., J. Bot. Néerl. 1: 97 (1861).
=== Breynia sp. – China (Guangdong), by sea, *B. Krone*

Mercurialis glabra M.E.Jones, Contr. W. Bot. 18: 68 (1933).
=== ? – Mexico (Jalisco), *M.E.Jones s/n*, coll. 1930; annual

Mercurialis tarraconensis Sennen, Ann. Soc. Linn. Lyon, II, 72: 16 (1925 publ. 1926).
=== ? – Spain (Tarragón), *Sennen s/n*, coll. 1917; coastal

Mercurialis × taurica Juz., Bot. Mater. Gerb. Bot. Inst. Komarova Akad. Nauk S.S.S.R. 14: 17 (1951).
=== ? – Ukraine (Crimea), *Juzepezuk & Kuprianova 1546* (LE); perennial

Microdesmis puberula var. **macrocarpa** Pax & K.Hoffm. in H.G.A. Engler, Pflanzenr., IV, 147, VII: 408 (1914).
=== ? – Cameroon, Ebolowa, *Mildbraed 5588*.

Omphalea diandra Vell., Fl. Flumin. 10: 12 (1831), nom. illeg.; descr. in Arch. Mus. Nac. Rio de Janeiro 5: 396 (1881).
=== Sebastiania sp. (fide Kruijt 1996; see **Sapium**). – Brazil (Rio de Janeiro), 'Labitat maritimis', *Vellozo*, icon tantum

Paivaeusa gabonensis A.Chev., Veg. Ut. Afr. Trop. Franc. 9: 298 (1917).
=== ? – Gabon, on Ogowé River, *Chevalier 26598*; tree 25-30 m, sterile

Pedilanthus pectinatus Baker, J. Linn. Soc., Bot. 25: 343 (1890).
=== Euphorbia sp. – Madagascar, northwestern part, *Baron 5461*; not found again (Denis 1921; see **Euphorbia**). Stems angular, fleshy; angles spiny

Pera corcovadensis Glaz., Bull. Soc. Bot. France 59(3): 626 (1912 publ. 1913), nom. nud.
=== ? – Brazil (Rio de Janeiro), *Glaziou 9357, 12156, 15414*

Phyllanthus albus Noronha, Verh. Batav. Genootsch. Kunsten 5(4): 22 (1790), nom. nud.

=== ? – Indonesia (Java), *Noronha*; local name 'Nasi-nasi'. No specimens extant

Phyllanthus alegrensis Glaz., Bull. Soc. Bot. France 59(3): 612 (1912 publ. 1913), nom. nud.

=== ? – Brazil (Espirito Santo), *Glaziou 11505*

Phyllanthus atropurpureus Bojer, Hortus Maurit.: 280 (1837), nom. nud.

=== ? – Comoro Islands. Not accounted for in *Flore des Mascareignes*.

Phyllanthus casticum var. **roxburghianus** Müll.Arg. in A.P.de Candolle, Prodr. 15(2): 348 (1866)

=== Phyllanthus sp. – 'Hortus Botanicus Calcuttensis', *Roxburgh* (G-DC).

Phyllanthus casuarinoides Glaz., Bull. Soc. Bot. France 59(3): 614 (1912 publ. 1913), nom. nud.

=== ? – Brazil (Goiás), *Glaziou 22092*

Phyllanthus crotalaroides Zipp. ex Span., Linnaea 15: 347 (1841), nom. nud.

=== ? – Indonesia (Timor), *Zippel*

Phyllanthus ellipticus Desf. ex Poir. in J.B.A.M.de Lamarck, Encycl., Suppl. 4: 404 (1816).

=== ? – Origin unknown; described from plants cultivated in Jardin des Plantes, Paris

Phyllanthus elongatus Mart. ex Colla, Herb. Pedem. 5: 105 (1836), nom. illeg.

=== ? – Brazil (Rio de Janeiro, Cabo Frio), *Martius* (TO?)

Phyllanthus essequiboensis Klotzsch in R.Schomburgk, Reis. Br.-Guiana: 1187 (1849), nom. nud.

=== ? – Guyana, *Schomburgk*

Phyllanthus finlaysonianus Wall., Numer. List: 7907 (1847), nom. inval.

=== ? – Thailand, 'hort. Finlayson', *Finlayson* (K-WALL)

Phyllanthus humilis Pax, Bot. Jahrb. Syst. 10: 34 (1888), nom. illeg.

=== ? – South Africa (Cape Prov. through Free State to the north); accepted in *Plants of Southern Africa: names and distribution* (Arnold & de Wet 1993). Based on *Marloth 1087*

Phyllanthus jardinii Müll.Arg., Linnaea 32: 21 (1863)

=== ? – Marquesas (Muku Hiva), *Jardin s/n* (G-DC, O). Florence, *Flore de la Polynésie francaise*, 1 (1997), considers this species to be of uncertain provenance, with resolution possible only with monographic treatment (cf. *Acalypha jardinii*)

Phyllanthus macahensis Glaz., Bull. Soc. Bot. France 59(3): 612 (1912 publ. 1913), nom. nud.

=== ? – Brazil (Rio de Janeiro), in restingas, *Glaziou 1157*

Phyllanthus minensis Glaz., Bull. Soc. Bot. France 59(3): 614 (1912 publ. 1913), nom. nud.

=== ? – Brazil (Minas Gerais), *Glaziou 13193*

Phyllanthus minusculus Glaz., Bull. Soc. Bot. France 59(3): 613 (1912 publ. 1913), nom. nud.

=== ? – Brazil (Goiás), *Glaziou 22089*

Phyllanthus parahybensis Glaz., Bull. Soc. Bot. France 59(3): 612 (1912 publ. 1913), minimal descr.

=== ? – Brazil (Rio de Janeiro), *Glaziou 12189*; a rheophyte.

Phyllanthus pedunculatus Warb., Bot. Jahrb. Syst. 13: 357 (1891), nom. illeg.

=== Glochidion sp. – Maluka (Kai Kecil), *Warburg*; in secondary forest

Phyllanthus poilanei Beille in H.Lecomte, Fl. Indo-Chine 5: 593 (1927), nom. illeg.

=== ? – Vietnam (Nha Trang, Han Heo), *Poilane*; accepted by Ngyuen Nghia Thin in *Euphorbiaceae of Vietnam* (1995).

Phyllanthus scoparius Müll.Arg. in C.F.P.von Martius, Fl. Bras. 11(2): 74 (1873), nom. illeg.

=== Phyllanthus sp. – Brazil (Serra da Lapa), *Riedel 1353*

Phyllanthus senaei Glaz., Bull. Soc. Bot. France 59(3): 614 (1912 publ. 1913), minimal descr.

=== ? – Brazil (Minas Gerais), *Glaziou 19842*

Phyllanthus smilacifolius Griseb., Fl. Brit. W. I.: 710 (1864), nom. nud.

=== ? – Trinidad, *Crueger*; not disposed of in *Flora of Trinidad and Tobago*.

Phyllanthus ternauxii Glaz., Bull. Soc. Bot. France 59(3): 612 (1912 publ. 1913), minimal descr.

=== ? – Brazil (Rio de Janeiro), *Glaziou 8928*

Phyllanthus tonkinensis Gentil, Pl. Cult. Serres Jard. Bot. Brux.: 150 (1907), nom. nud.

=== ? – ? Vietnam; no supporting data

Phyllanthus udicola Mart. ex Colla, Herb. Pedem. 5: 105 (1836)

=== ? – Brazil, without locality, *Martius* (TO?)

Phyllanthus valeriae M.Schmid, in Fl. N. Caled. & Depend. 17: 217 (1991), nom. illeg.

=== ? – New Caledonia (northwestern part); accepted by Schmid (1991; see **Phyllanthus**)

Phyllanthus zygophylloides Müll.Arg. in A.P.de Candolle, Prodr. 15(2): 336 (1866).

=== ? – Origin unknown (perhaps from Africa); described from plants cultivated in Hort. Berol. around 1844

Ricinocarpus bracteatus (Miq.) Kuntze, Revis. Gen. Pl. 2: 617 (1891).

=== Acalypha sp.

*Acalypha bracteata Miq. (**q.v.**)

Ricinocarpus digyneius (Raf.) Kuntze, Revis. Gen. Pl. 2: 617 (1891).

=== ?

*Acalypha digyneia Raf. (**q.v.**)

Ricinocarpus fruticulosus (Raf.) Kuntze, Revis. Gen. Pl. 2: 618 (1891).

=== ?

*Acalypha fruticulosa Raf. (**q.v.**)

Ricinus morifolius Noronha, Verh. Batav. Genootsch. Kunsten 5(4): 2 (1790), nom. nud.

=== ? – Indonesia (Java), *Noronha*; Javanese name: 'Kerúpuc'. No specimens extant

Ricinus pulchellus Noronha, Verh. Batav. Genootsch. Kunsten 5(4): 2 (1790), nom. nud.

=== ? – Indonesia (Java), *Noronha*; Javanese name: 'Zaràc-gorita'. No specimens extant

Sapium atrobadiomaculatum F.P.Metcalf, Lingnan Sci. J. 10: 490 (1931).

=== Neoshirakia sp. (aff. N. japonica). – China (Fujian), *Hu 1819; Ching 2224* (A, US)

Sapium balansae Parodi, Anales Soc. Ci. Argent. 11: 53 (1881).

=== Sebastiania ? – Paraguay (Corrientes), *Parodi s/n.* Type not obtained by Kruijt (1996; see **Sapium**)

Sapium brasiliense Spreng. ex Pax in H.G.A.Engler, Pflanzenr., IV, 147, V: 256 (1912), nom. nud., pro syn.

=== Croton sp. (fide Pax)

Sapium diandrum (Müll.Arg.) Pax in H.G.A.Engler, Pflanzenr., IV, 147, V: 256 (1912).

=== Sebastiania sp. (fide Kruijt 1996; see **Sapium**)

*Excoecaria diandra Müll.Arg. (**q.v.**)

Sapium herbertiifolium Jacques, Hort. Universel 5: 289 (1844).

=== ? – Of horticultural origin

Sapium martii var. **peruvianum** J.F.Macbr., Field Mus. Nat. Hist., Bot. Ser. 13(3a): 196 (1951).

=== Sapium sp. – Peru (San Martín & Loreto), *Ll. Williams 6658, 6909, 8032* (F). Type and other materials not studied by Kruijt (1996; see **Sapium**). Accepted in *Catalogue of the flowering plants and gymnosperms of Peru* (Brako & Zarucchi 1993).

Sapium parvifolium Alain, Contr. Ocas. Mus. Hist. Nat. Colegio ''De La Salle'' 11: 10 (1952)

=== Gymnanthes ? – Cuba (eastern part, Moa region), *León & Clemente 23163* (G, HAC (formerly LS), NY). Material not seen by Kruijt (1996; see **Sapium**) but thought by him to be in *Ateramnus* (= *Gymnanthes*).

Sapium patagonicum (Speg.) D.J.Rogers, Ann. Missouri Bot. Gard. 38: 251 (1951)

=== ?

*Colliguaja patagonica Speg. (**q.v.**)

Sapium peruvianum (J.F.Macbr.) Jabl., Phytologia 14: 451 (1967), nom. illeg.

=== ?

*Sapium martii var. peruvianum J.F.Macbr. (**q.v.**)

Sapium rotundifolium Hemsl., J. Linn. Soc., Bot. 26: 445 (1891).

=== Triadica sp. – S. China (Guangdong), Vietnam; tree. Described and illustrated in Vu Van Dung (ed.), *Vietnam Forest Trees*: 246 (1996); still to be evaluated (Esser 1999; see *Triadica*)

Sapium ruizii Hemsl., Hooker's Icon. Pl. 29: t. 2894 (1909).

=== Stillingia sp. (fide Kruijt 1996; see **Sapium**). – 'Central America', *Ruiz & Pavón* (BM)

Sapium simile Hemsl., Hooker's Icon. Pl. 29: t. 2894 (1909).

=== Sebastiania sp. (fide Kruijt 1996; see **Sapium**). – 'Central America', *Ruiz & Pavón* (BM)

Sapium steyermarkii Jabl., Phytologia 14: 450 (1967).

=== Tetrorchidium sp. (fide Kruijt 1996; see **Sapium**). – Ecuador (SE of El Pan), *Steyermark 53508* (NY)

Sapium veracruzense Lundell, Wrightia 5: 79 (1975).

=== Sapium sp. – Mexico (Veracruz), *Matuda 1334* (LL, MEXU)

Scepasma parvifolia Rchb.f. & Zoll. ex Teijsm. & Binn., Cat. Hort. Bot. Bogor.: 229 (1866).

=== ? – Indonesia (Madura); recorded from plants cultivated in Hortus Bogoriensis

Sebastiania adenophora Pax & K.Hoffm. in H.G.A.Engler, Pflanzenr., IV, 147, V: 145 (1912)

=== ?

*Excoecaria glandulosa Millsp. (**q.v.**)

Sebastiania anisodonta Müll.Arg. in C.F.P.von Martius, Fl. Bras. 11(2): 550 (1874).

=== Microstachys sp. – Brazil (Goiás), *Pohl 1555*. Not seen by Pax & Hoffmann

Sebastiania diamantinensis Glaz., Bull. Soc. Bot. France 59(3): 632 (1912 publ. 1913), nom. nud.

=== ? – Brazil (Minas Gerais), *Glaziou 19460*

Sebastiania glaucophylla Müll.Arg. in C.F.P.von Martius, Fl. Bras. 11(2): 571 (1874).

=== Gymnanthes sp. – Brazil (eastern part), *Neuwied*. Treelet

Sebastiania hexaptera Urb., Symb. Antill. 3: 303 (1902).

=== ? – Guadeloupe, *Duss 3239*; Martinique, *Duss 890*. Shrub to 4 m

Sebastiania lanceifolia Steenis, Bull. Jard. Bot. Buitenzorg, III, 17: 410 (1948).

=== Gymnanthes sp. – Indonesia (Lingga Is.), *Bünnemeijer 6925* (BO)

Sebastiania larensis Croizat & Tamayo, Lilloa 17: 1 (1949).

=== Bonania sp. – Venezuela (Lara), *Tamayo 3333*

Sebastiania picardae Urb., Symb. Antill. 3: 304 (1902).

=== ? – Hispaniola (Haiti, Morne de l'Hôpital), *Picarda 842*. Much-branched shrub

Sebastiania salicifolia (Mart.) Pax in H.G.A.Engler, Pflanzenr., IV, 147, V: 103 (1912).

=== Microstachys sp. – Brazil (Rio de Janeiro, Minas Gerais, Goiás), in campos, the varieties mostly in Minas Gerais

*Cnemidostachys salicifolia Mart.

Sebastiania salicifolia var. **genuina** Pax in H.G.A.Engler, Pflanzenr., IV, 147, V: 104 (1912), nom. inval.

=== Microstachys sp. – Brazil (Minas Gerais); based on *Martius s/n*, Ouro Preto (M)

*Cnemidostachys salicifolia Mart.

Sebastiania salicifolia var. **glaziovii** Pax & K.Hoffm. in H.G.A.Engler, Pflanzenr., IV, 147, V: 104 (1912).

=== Microstachys sp. – Brazil, *Glaziou 15406* p.p.; *16340*

Sebastiania salicifolia var. **leptoclada** (Müll.Arg.) Pax in H.G.A.Engler, Pflanzenr., IV, 147, V: 104 (1912).

=== Microstachys sp. – Brazil (Rio de Janeiro), *Riedel 1200, Glaziou 11541*

*Sebastiania corniculata var. leptoclada Müll.Arg.

Sebastiania salicifolia var. **longifolia** (Mart.) Pax in H.G.A.Engler, Pflanzenr., IV, 147, V: 104 (1912).

=== Microstachys sp. – Brazil (Minas Gerais), *Martius s/n* (M)

*Cnemidostachys longifolia Mart.

Sebastiania salicifolia var. **similis** Pax & K.Hoffm. in H.G.A.Engler, Pflanzenr., IV, 147, V: 104 (1912).

=== Microstachys sp. – Brazil (Minas Gerais), *Mendonça 253*

Sebastiania singularis Rizzini, Leandra 3-4(4-5): 7 (1974).

=== ? – Brazil (Bahia, Casa Nova), *Ramalho 190* (RB). Small tree to 8 m

Securinega guaraiuva Kuhlm., Arq. Inst. Biol. Veg. 1: 241 (1935)

=== ? – Brazil (Santa Catarina to Rio de Janeiro and Minas Gerais); tree. Illustrated and described in H. Lorenzi, *Árvores Brasileiras*: 112 (1992). Not properly in *Securinega* but no alternative disposition established.

Stillingia crotonoides (Mart.) Baill., Étude Euphorb.: 516 (1858)

=== Microstachys sp. – Brazil (Minas Gerais, São Paulo), *Martius 906* and other collections. Pax & Hoffmann (1912; see **Sebastiania**) referred this plant to *Sebastiania hispida* var. *crotonoides* (Mart.) Pax. It may form part of *Microstachys corniculata* (Vahl) Griseb. sens. lat. (see that genus).

*Cnemidostachys crotonoides Mart.

Stillingia patagonica (Speg.) Pax & K.Hoffm. in H.G.A.Engler, Pflanzenr., IV, 147, V: 188 (1912).

=== ?

*Colliguaja patagonica Speg. (**q.v.**)

Stillingia patula Baill., Étude Euphorb.: 516 (1858).

=== Microstachys sp. – Brazil (Bahia, Rio de Janeiro), *Martius 427* and other collections. Pax & Hoffmann (1912; see **Sebastiania**) referred this plant to *Sebastiania hispida* var. *patula* (Mart.) Müll.Arg. It may form part of *Microstachys corniculata* (Vahl) Griseb. sens. lat. (see that genus)

Synadenium ballyi Werderm. ex Ball., Repert. Spec. Nov. Regni Veg. Beih. 102: 35 (1938), nom. nud.

=== ? – Tanzania, *Bally*. Not listed in *Flora of Tropical East Africa*, Euphorbiaceae

Tetraplandra grandifolia Glaz., Bull. Soc. Bot. France 59(3): 630 (1912 publ. 1913), nom. nud.

___ ? – Brazil (Rio de Janeiro), *Glaziou 13170*

Tithymalus cinerea Raf., Autik. Bot.: 91 (1840).

=== ? – Germany and Austria; not disposed of

Tithymalus floridana Raf., Autik. Bot.: 91 (1840).

=== ? – USA (Florida and Georgia); not disposed of

Tithymalus glaucus Moench, Suppl. Meth.: 282 (1802).

=== Euphorbia moenchiana Steud. (**q.v.**)

Tithymalus mucronata Raf., Autik. Bot.: 91 (1840).

=== ? – Sicily and Crete; not disposed of

Tithymalus uniflorus Haw., Philos. Mag. Ann. Chem. 10: 418 (1831).

=== ?

Euphorbia uniflora (Haw.) G.Don (**q.v.**).

Tragia pubescens Glaz., Bull. Soc. Bot. France 59(3): 625 (1912 publ. 1913), nom. nud.

=== ? – Brazil (Espirito Santo, Itapemirim), *Glaziou 11531*

Trewia discolor Sm. in A.Rees, Cycl. 36: Trewia no. 4 (1817).

=== Mallotus sp. – 'East Indies', *herb. Linnaeus f.*; possibly in herb. Smith (LINN)

Trewia pubescens Sm. in A.Rees, Cycl. 36: Trewia no. 3 (1817).

=== Macaranga sp. – Indonesia (Moluccas, Ambon), *Roxburgh*; possibly in herb. Smith (LINN)

Trewia rusciflora B.Heyne ex Roth, Nov. Pl. Sp.: 374 (1821).

=== Mallotus sp. –'India orientali', *B. Heyne*. By Hooker in *Flora of British India* 5: 424 (1887) throught to be *Macaranga indica* Wight.

Xylophylla media Lodd. ex G.Don in J.C.Loudon, Hort. Brit.: 391 (1830), minimal descr.

=== Phyllanthus ?

Summary of excluded taxa

Acalypha japonica Houtt. ex Steud., Nomencl. Bot. 1: 4 (1821).
=== Boehmeria platyphylla D. Don (Urticaceae)

Actinostemon verrucosum Glaz., Bull. Soc. Bot. France 59(3): 634 (1912 publ. 1913), nom. nud.
=== ? (Celastraceae). Minas Gerais, Brazil, *Glaziou 13481*

Aerisilvaea serrata Radcl.-Sm., Kew Bull. 47: 677 (1992).
=== Maytenus undata (Thunb.) Blakelock (Celastraceae)

Amanoa schweinfurthii Baker & Hutch., Bull. Misc. Inform. Kew 1910: 56 (1910).
=== Erythroxylum fischeri Engl. (Erythroxylaceae)

Amperea subnuda Nees in J.G.C.Lehmann, Pl. Preiss. 2: 229 (1848).
=== Gyrostemon subnudus (Nees) Baill. (Gyrostemonaceae)

Andrachne frutescens Ehret, Philos. Trans. 57: 114 (1768).
=== Arbutus andrachne L. (Ericaceae)

Anisophyllum trapezoidale Baill., Adansonia 3: 24 (1862).
=== Anisophyllea disticha (Jack) Baill. (Anisophylleaceae)

Antidesma alnifolium Hook., Hooker's Icon. Pl. 10: t. 481 (1842).
=== Trimeria grandifolia (Hochst.) Warb. (Flacourtiaceae)

Antidesma crenatum H.St.John, Pacific Sci. 26: 279 (1972).
=== Xylosma crenatum (H.St. John) H.St. John (Flacourtiaceae)

Antidesma grossularia Raeusch., Nomencl. Bot.: 287 (1789).
=== Embelia ribes Burm.f. (Myrsinaceae)

Antidesma litorale Blume, Bijdr.: 1123 (1827).
=== Polyosma integrifolia Blume (Escalloniaceae)

Antidesma megalocarpum S.Moore, J. Bot. 61(Suppl.): 46 (1923).
=== Rhyticaryum longifolium K.Schum. & Lauterb. (Icacinaceae)

Antidesma nervosum Wall., Numer. List: 7289 (1832).
=== Gironniera nervosa Planch. (Ulmaceae)

Antidesma parasitica Dillwyn, Rev. Hortus Malab.: 33 (1839).
=== Scleropyrum pentandrum (Dennst.) D.J.Mabb. (Santalaceae)

Antidesma ribes (Burm.f.) Raeusch., Nomencl. Bot.: 287 (1797).
=== Embelia ribes (Burm.f.) (Myrsinaceae)

Antidesma scandens Lour., Fl. Cochinch.: 617 (1790).
=== Humulus scandens (Lour.) Merr. (Cannabaceae)

Aporusa incisa Airy Shaw, Kew Bull. 25: 477 (1971).
=== Prunus arborea var. montana (Hook. f.) Kalkman (Rosaceae)

Argyrodendron petersii Klotzsch in W.C.H.Peters, Naturw. Reise Mossambique: 101 (1861).
=== Combretum imberbe Wawra (Combretaceae)

Argythamnia acalyphifolia (Griseb.) Kuntze, Revis. Gen. Pl. 2: 593 (1891).
=== Byttneria sp. (Sterculiaceae)
 *Caperonia acalyphifolia Griseb. (**q.v.**)

Baccaurea capensis Spreng. ex Pax & K.Hoffm. in H.G.A.Engler, Pflanzenr., IV, 147, XV: 71 (1922)
=== ? (Flacourtiaceae). Collection in herb. Zeyher

Baccaurea pyrrhodasya (Miq.) Müll.Arg. in A.P.de Candolle, Prodr. 15(2): 462 (1866).
=== Terminalia sp. (Combretaceae)
 *Pierardia pyrrhodasya Miq. (**q.v.**)

Beyeria loranthoides Baill., Étude Euphorb.: 403 (1858).
=== Drimys sp. (Winteraceae). Australia, *Le Guillou* (P).

Beyeria uncinata Baill., Adansonia 6: 306 (1866).
=== Cryptandra uncinata (Baill.) Grüning (Rhamnaceae)

Bradleia nephroia Lour. ex Steud., Nomencl. Bot. 1: 116 (1821).
=== Cocculus trilobus (Thunb.) DC. (Menispermaceae)

Bridelia horrida (Kostel.) Dillwyn, Rev. Hortus Malab.: 15 (1839).
=== Scleropyrum pentandrum (Dennst.) D.J.Mabb. (Santalaceae)

Bridelia spinosa DC., Mém. Soc. Phys. Genève 6: 563 (1833).

=== Damnacanthus sp. (Rubiaceae) (fide Jablonski, 1915; see **Bridelia**).

Caperonia acalyphifolia Griseb., Abh. Königl. Ges. Wiss. Göttingen 24: 58 (1879).

=== Byttneria sp. (Sterculiaceae). Argentina (former Oran Province), *Grisebach*

Ceratogynum concolor Pritz., Icon. Bot. Index: 245 (1854), orth. var.

=== Ceratolobus concolor Blume (Arecaceae)

Cicca macrostachya (Müll.Arg.) Benth., Bot. Voy. Sulphur: 166 (1846).

=== Picramnia antidesma Sw. (Picramiaceae; Simaroubaceae s.l.)

*Phyllanthus macrostachyus Müll.Arg. (q.v. etiam)

Claoxylon remyi Sherff, Publ. Field Mus. Nat. Hist., Bot. Ser. 17: 557 (1939).

=== Platydesma remyi (Sherff) O.Deg., I.Deg., Sherff & B.C.Stone (Rutaceae)

Croton bentzoe L., Mant. Pl. 2: 297 (1771).

=== Terminalia bentzoë (L.) L.f. (Combretaceae)

Croton benzoe L. in J.A.Murray (ed.), Syst. Veg. ed. 13: 721 (1774), nom. illeg.

=== Terminalia bentzoë (L.) L.f. (Combretaceae)

Croton curviflorus Elmer, Leafl. Philipp. Bot. 1: 310, 363 (1908).

=== Sycopsis dunnii Hemsl. (Hamamelidaceae)

Croton eriospermus Lam., Encycl. 2: 211 (1786).

=== Trigonia crotonoides Cambess. (Trigoniaceae)

Croton incanus Blume ex Baill., Étude Euphorb.: 470 pl. 11, Fig. 14 (1858), in obs., pro syn. 'Praetoria incana Baill.'

=== Pipturus sp. (Urticaceae); in 'herb Leyd. et herb. Ventenat'

Cyclostemon cuspidatus Blume, Bijdr.: 599 (1826).

=== Aphananthe cuspidata (Blume) Planch. (Ulmaceae)

Cyclostemon reticulatus Schltr., Bot. Jahrb. Syst. 39: 148 (1906).

=== Lasiochlamys reticulata (Schltr.) Pax & K.Hoffm. (Flacourtiaceae)

Dalechampia houlletiana Baill., Étude Euphorb.: 486 (1858), nom. nud.

=== Cissus sp. (Vitaceae); 'herb. Houllet' (P)

Drypetes laevigata Griseb. ex Eggers, Fl. St. Croix: 90 (1879).

=== Forestiera rhamnifolia Griseb. (Oleaceae); refers to *Eggers 735*, Virgin Islands

Drypetes talbotii S.Moore in A.B.Rendle & al., Cat. Pl. Oban: 97 (1913).

=== Uvariopsis bakeriana (Hutch. & Dalziel) Robyns & Ghesguière (Annonaceae), leaves; Drypetes sp., inflorescences

Endospermum eglandulosum Pax & K.Hoffm. in H.G.A.Engler, Pflanzenr., IV, 147, VII: 418 (1914).

=== Sterculia macrophylla Vent. (Sterculiaceae)

Euphorbia antunesii Pax, Bot. Jahrb. Syst. 34: 79 (1904).

=== Tavaresia barklyi (Dyer) N.E.Br. (Asclepiadaceae)

Euphorbia geminispina Haw. ex Loud., Hort. Brit., ed. 2: 190 (1832).

=== ? (Cactaceae). Mexico; introduced 1823

Euphorbia gracilis Pav. ex Moq. in A.P.de Candolle, Prodr. 15(2): 269 (1862), pro syn.

=== Amaranthus urceolatus Benth. (Amaranthaceae), pro syn.

Euphorbia lanifera Haw. ex Loud., Hort. Brit., ed. 2: 190 (1832).

=== ? (Cactaceae). Mexico; introduced 1823

Euphorbia lucidissima H.Lév. & Vaniot, Bull. Herb. Boissier, II, 6: 763 (1906).

=== Canscora lucidissima H.Lév. & Vaniot) Hand.-Mazz. (Gentianaceae)

Euphorbia magnimamma Haw. ex Loud., Hort. Brit., ed. 2: 190 (1832)

=== ? (Cactaceae). Mexico; introduced 1823

Euphorbia masafuerae Phil., Bot. Zeitung (Berlin) 14: 647 (1856).

=== Wahlenbergia tuberosa Hook.f. (Campanulaceae)

Euphorbia pendula (Haw.) Sweet, Hort. Suburb. Lond.: 107 (1818).

=== Sarcostemma sp. (Asclepiadaceae)

*Tithymalus pendulus Haw. (q.v.)

Euphorbia radlkoferi Oudejans, Phytologia 67: 48 (1989).

=== Dimocarpus longan var. malesianus Leenh. (Sapindaceae)

*Euphoria gracilis Radlk.

Euphorbia schizoclada Baill. ex Poiss., Rech. Fl. Mérid. Madagascar: 33 (1912).
=== Didierea sp. (Didiereaceae). SW. Madagascar, *Grandidier* (P); illus., p. 34

Euphorbia viminalis L., Sp. Pl.: 452 (1753).
=== Sarcostemma viminale R.Br. (Asclepiadaceae)

Excoecaria ilicifolia Spreng., Neue Entd. 2: 117 (1821).
=== Sahagunia sp. (Moraceae), Brazil, *Sello*

Excoecaria ovatifolia Noronha, Verh. Batav. Genootsch. Kunsten 5(4): 1 (1790).
=== Cerbera odollam Gaertn. (Apocynaceae)

Excoecaria polyandra Griseb., Nachr. Königl. Ges. Wiss. Georg-Augusts-Univ. 1: 180 (1865).
=== Forestiera polyandra (Griseb.) Alain (Oleaceae)

Flueggea javanica Blume, Bijdr.: 580 (1826).
=== Streblus spinosus (Blume) Corner (Moraceae)

Flueggea serrata Miq., Fl. Ned. Ind. 1(2): 356 (1859).
=== Celastrus hindsii Benth. (Celastraceae)

Gelonium arboreum (J.F.Gmel.) Kuntze, Revis. Gen. Pl. 2: 144 (1891).
=== Molinaea arborea J.F.Gmel. (Sapindaceae)
   *Molinaea arborea J.F.Gmel.

Gelonium brevipes (Radlk.) Kuntze, Revis. Gen. Pl. 2: 144 (1891).
=== Molinaea brevipes Radlk. (Sapindaceae)
   *Molinaea brevipes Radlk.

Gelonium glandulosum Elmer, Leafl. Philipp. Bot. 3: 917 (1910).
=== Rinorea bengalensis (Wall.) Kuntze (Violaceae)

Gelonium pinatubense Elmer, Leafl. Philipp. Bot. 9: 3186 (1934).
=== Casearia trivalvis (Blanco) Merr. (Flacourtiaceae)

Gelonium trifidum Elmer, Leafl. Philipp. Bot. 3: 918 (1910).
=== Rinorea bengalensis (Wall.) Kuntze (Violaceae)

Glochidion cinerascens Miq., Fl. Ned. Ind., Eerste Bijv.: 451 (1861).
=== Alphitonia cinerascens (Miq.) Hoogl. (Rhamnaceae)

Glochidion sieboldianum (Miq.) Koidz., Fl. Symb. Orient.-Asiat.: 25 (1930).
=== Actinodaphne sieboldiana Miq. (Lauraceae)

Guarania suberosa Standl., J. Wash. Acad. Sci. 15: 461 (1925), orth. var.
=== Gurania suberosa Standl. (Cucurbitaceae)

Guya integrifolia (Willd.) H.Perrier, Mém. Mus. Natl. Hist. Nat. 13: 269 (1940).
=== Aphloia theiformis Bennett (Flacourtiaceae)
   *Lightfootia integrifolia Willd.

Gymnanthes texana Standl., Proc. Biol. Soc. Wash. 39: 135 (1926).
=== Forestiera reticulata Torr. (Oleaceae)

Hymenocardia grandis Hutch., Bull. Misc. Inform. Kew 1911: 184 (1911).
=== Holoptelea grandis (Hutch.) Mildbr. (Ulmaceae)

Jatropha australis Lodd. ex G.Don in J.C.Loudon, Hort. Brit.: 391 (1820).
=== Brachychiton populneus (Schott & Endl.) R.Br. subsp. populneus (Sterculiaceae)

Jatropha tomentosa Spreng., Syst. Veg. 3: 77 (1826.)
=== Vitex cymosa Bert. ex Spreng. (Lamiaceae)

Macaranga dawei Prain, Bull. Misc. Inform. Kew 1911: 232 (1911).
=== Myrica salicifolia Hochst. ex A. Rich. (Myricaceae)

Macaranga maudsleyi Horne, Year Fiji: 264 (1881), nom. nud.
=== Trichospermum calyculatum (Seem.) Burret (Tiliaceae)

Mallotus derbyensis W.Fitzg., J. Roy. Soc. W. Australia 3: 165 (1918).
=== Grewia breviflora Benth. (Tiliaceae)

Margaritaria oppositifolia L., Pl. Surin.: 16 (1775), nom. inval.
=== ? (Combretaceae, fide Müller Argov.)

Mercurialis afra L., Mant. Pl. 2: 298 (1771).
=== Centella villosa L. (Apiaceae)

Monadenium lunulatum Chiov., Fl. Somala 1: 298 (1929).
=== Kleinia subulifolia (Chiov.) P.Halliday (Asteraceae)

Monadenium subulifolium Chiov., Fl. Somala 1: 298 (1929).

=== Kleinia subulifolia (Chiov.) P.Halliday (Asteraceae)

Neoboutonia musculiformis K.Schum.ex T.Durand & B.D.Jacks., Ind. Kew. suppl. 1: 291 (1903), nom. inval. (error for *Neubergia muscutiformis* (Gaertn.) Miq.)

=== Neuburgia moluccana (Boerl.) Leenh. (Strychnaceae)

Omphalea bracteata var. **pedicellaris** Airy Shaw, Kew Bull., Addit. Ser. 4: 180 (1975).

=== Erycibe sp. (Convolvulaceae). Borneo (Sarawak), *Jacobs 5306* (K, L)

Ostodes appendiculata Hook.f., Fl. Brit. India 5: 401 (1887).

=== Lepisanthes tetraphylla (Vahl) Radlk. (Sapindaceae)

Ostodes helferi Müll.Arg., Linnaea 34: 215 (1865).

=== Popowia pauciflora Maing. ex Hook.f. & Thoms. (Annonaceae)

Phyllanthus ceramica Pers., Syn. Pl. 2: 591 (1807).

=== Exocarpos longifolius (L.) Endl. (Santalaceae)

Phyllanthus cinerascens (Miq.) Müll.Arg., Flora 48: 390 (1865), nom. illeg.

=== Alphitonia cinerascens (Miq.) Hoogl. (Rhamnaceae)

*Glochidion cinerascens Miq. (**q.v. etiam**)

Phyllanthus cinereoviridis Pax, Bot. Jahrb. Syst. 43: 76 (1909).

=== Cissampelos capensis L.f. (Menispermaceae)

Phyllanthus macrostachyus Müll.Arg., Linnaea 32: 16 (1863).

=== Picramnia antidesma Sw. (Picramniaceae; Simaroubaceae s.l.))

Phyllanthus nitidus Müll.Arg., Flora 48: 371 (1865).

=== Eurya japonica Thunb. (Theaceae)

Picrodendron arboreum (Mill.) Planch., London J. Bot. 5: 580 (1846).

=== Allophylus cobbe (L.) Raeusch. (Sapindaceae)

*Toxicodendron arboreum Mill.

Picrodendron calunga Mart. ex Engl. in C.F.P.von Martius, Fl. Bras. 12(2): 214 (1874).

=== Simaba ferruginea A.St.Hil. (Simaroubaceae)

Pierardia pyrrhodasya Miq., Fl. Ned. Ind., Eerste Bijv.: 442 (1861).

=== Terminalia sp. (Combretaceae). Sumatera, *Junghuhn*

Ricinodendron staudtii Pax, Bot. Jahrb. Syst. 23: 532 (1897).

=== leaves, Cola pachycarpa K. Schum. (Sterculiaceae); inflorescence, Lannea welwitschii (Hiern) Engl. (Anacardiaceae)

Ricinus furfuraceus Wall., Numer. List: 7805 in part (1847), nom. inval.

=== Daphniphyllum majus Müll.Arg. (Daphniphyllaceae)

Ricinus odoratus Noronha, Verh. Batav. Genootsch. Kunsten 5(4): 2 (1790), nom. nud.

=== Commersonia bartramia (L.) Merr. (Sterculiaceae)

Sagotia gardenioides Vieill. ex Guillaumin, Ann. Inst. Bot.-Géol. Colon. Marseille, II, 9: 168 (1911), in obs.

=== Morierina montana Viell. (Rubiaceae)

Sapium drummondii Jacques, Hort. Universel 5: 289 (1844).

=== Scolopia zeyheri (Nees) Harv. (Flacourtiaceae)

Savia volubilis Raf., Med. Repos. 5: 352 (1808).

=== Amphicarpaea bracteata (L.) Fern. (Fabaceae)

Suregada glandulosa (Elmer) Croizat, Bull. Jard. Bot. Buitenzorg, III, 17: 216 (1942).

=== Rinorea bengalensis (Wall.) Kuntze (Violaceae)

*Gelonium glandulosum Elmer (**q.v. etiam**)

Suregada pinatubensis (Elmer) Croizat, Bull. Jard. Bot. Buitenzorg, III, 17: 217 (1942).

=== Casearia trivialis (Blanco) Merr.) (Flacourtiaceae)

*Gelonium pinatubense Elmer (**q.v. etiam**)

Suregada trifida (Elmer) Croizat, Bull. Jard. Bot. Buitenzorg, III, 17: 216 (1942).

=== Rinorea bengalensis (Wall.) Kuntze (Violaceae)

*Gelonium trifidum Elmer (**q.v. etiam**)

Tithymalus pendulus Haw., Syn. Pl. Succ.: 138 (1812).

=== Sarcostemma sp. (Asclepiadaceae). Origin not known; introduced 1808.

Toxicodendrum acutifolium Benth., J. Linn. Soc., Bot. 17: 214 (1878).
=== Xymalos monospora (Harv.) Baill. (Monimiaceae)
Tragia erosa Hochst., Flora 28: 86 (1845).
=== Pyrenacantha scandens Planch. ex Harv. (Icacinaceae)
Tragia interrupta Spreng., Syst. Veg. 3: 833 (1826).
=== Laportea interrupta (L.) Chew (Urticaceae)
Tragia luzoniensis Merr. Philipp. J. Sci. 16: 564 (1920).
=== Pyrenacantha sp. (Icacinaceae)
Tragia scandens L., Herb. Amb.: 18 (1754).
=== Tetracera scandens (L.) Merr. (Dilleniaceae)
Xylophylla artensis Montrouz., Mém. Acad. Roy. Sci. Lyon, Sect. Sci. 10: 250 (1860).
=== Exocarpos sp. (Santalaceae). [Not mentioned in Stauffer, *Revisio Anthobolearum* (1959)]
Xylophylla ensifolia Bojer ex Drake in A.Grandidier, Hist. Phys. Madagascar 1: 191 (1897).
=== Phylloxylon ensifolius Baill. (Fabaceae)
Xylophylla ensifolia Bojer ex Baker, J. Linn. Soc., Bot. 20: 249 (1883).
=== Exocarpos xylophylloides Baker (Santalaceae)
Xylophylla longifolia L., Mant. Pl. 2: 221 (1771).
=== Exocarpos longifolius (L.) Endl. (Santalaceae)

Addendum, Euphorbiaceae

Guya

Synonyms:

Guya Frapp. ex Cordem. === **Drypetes** Vahl

Guya caustica Frapp. ex Cordem. === **Drypetes caustica** (Frapp. ex Cordem.) Airy Shaw

Guya integrifolia (Willd.) H. Perrier === **Aphloia theiformis** Bennett (Flabourtiaceae)

A Euphorbiaceous enigma

An enigma among spurge-trees; without flowers its proper place remains indeterminable.

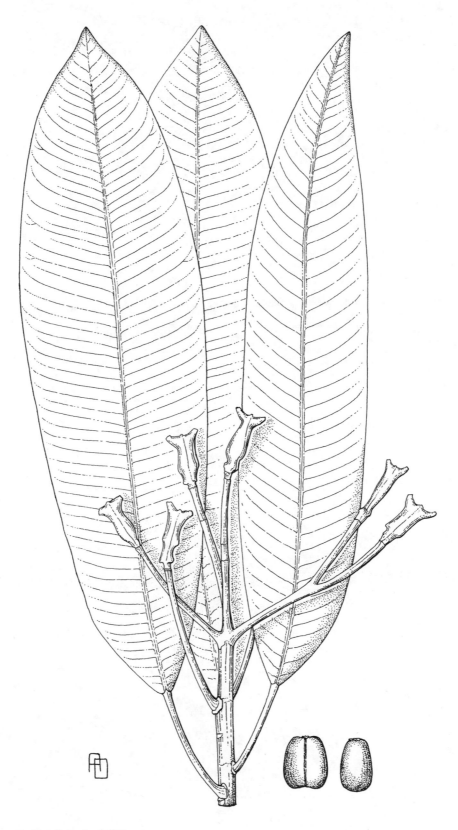

Excoecaria myrioneura Airy Shaw [p. 1608]
Artist: Ann Davies
Airy-Shaw, *Euphorbiaceae of New Guinea* (1980)
KEW ILLUSTRATIONS COLLECTION